Gregory L. Naber
Quantum Mechanics

Also of Interest

Hypersymmetry
Physics of the Isotopic Field-Charge Spin Conservation
György Darvas, 2021
ISBN 978-3-11-071317-6, e-ISBN (PDF) 978-3-11-071318-3,
e-ISBN (EPUB) 978-3-11-071348-0

Data Science
Time Complexity, Inferential Uncertainty, and Spacekime Analytics
Ivo D. Dinov, Milen Velchev Velev, 2021
ISBN 978-3-11-069780-3, e-ISBN (PDF) 978-3-11-069782-7,
e-ISBN (EPUB) 978-3-11-069797-1

Elementary Particle Theory
Volume 1: Quantum Mechanics
Eugene Stefanovich, 2019
ISBN 978-3-11-049088-6, e-ISBN (PDF) 978-3-11-049213-2,
e-ISBN (EPUB) 978-3-11-049103-6

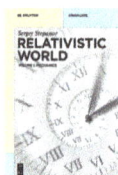

Relativistic World
Volume 1: Mechanics
Sergey Stepanov, 2018
ISBN 978-3-11-051587-9, e-ISBN (PDF) 978-3-11-051588-6,
e-ISBN (EPUB) 978-3-11-051600-5

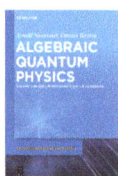

Algebraic Quantum Physics
Volume 1: Quantum Mechanics via Lie Algebras
Arnold Neumaier, Dennis Westra, to be published in 2022
ISBN 978-3-11-040610-8, e-ISBN (PDF) 978-3-11-040620-7,
e-ISBN (EPUB) 978-3-11-040624-5

Gregory L. Naber

Quantum Mechanics

An Introduction to the Physical Background and
Mathematical Structure

DE GRUYTER

Author
Prof. Dr. Gregory L. Naber
111 N 9th Street Unit 701
Philadelphia PA 19107
USA
gregorynaber@gmail.com

ISBN 978-3-11-075161-1
e-ISBN (PDF) 978-3-11-075194-9
e-ISBN (EPUB) 978-3-11-075204-5

Library of Congress Control Number: 2021940402

Bibliographic information published by the Deutsche Nationalbibliothek
The Deutsche Nationalbibliothek lists this publication in the Deutsche Nationalbibliografie;
detailed bibliographic data are available on the Internet at http://dnb.dnb.de.

© 2021 Walter de Gruyter GmbH, Berlin/Boston
Cover image: Egor Suvorov / iStock / Getty Images Plus
Typesetting: VTeX UAB, Lithuania
Printing and binding: CPI books GmbH, Leck

www.degruyter.com

―――――

This is for all of the young ones in my life: Amber, Emily, Garrett, Gracie, Lacey, Maya, Miles, Phoenix and Vinnie

Preface

The goal here is to provide an introduction to the physical and mathematical foundations of quantum mechanics. It is addressed to those who have been trained in modern mathematics and whose background in physics may not extend much beyond $F = mA$, but for whom the following sorts of questions require more than a perfunctory response. What are the physical phenomena that forced physicists into such a radical re-evaluation of the very beautiful and quite successful ideas bequeathed to them by the founding fathers of classical physics? Why did this re-evaluation culminate in a view of the world that is probabilistic and formulated in terms of Hilbert spaces and self-adjoint operators? Where did the Planck constant come from? What are the basic assumptions of quantum mechanics? Are they consistent? What motivated them? What objections might be raised to them? Where did the Heisenberg algebra come from? What motivated Feynman to introduce his path integral? Is it really an "integral"? Does it admit a rigorous mathematical definition? Why does one distinguish two "types" of particles in quantum mechanics (bosons and fermions)? Why and how are they treated differently? In what sense does supersymmetry provide a more unified picture of the two types? One need not know the answers to all of these questions in order to study the mathematical formalism of quantum mechanics, but for those who would like to know we will try to provide some answers or, at least, some food for thought. As to the mathematical formalism itself, we will provide careful, detailed and rigorous treatments of just a few of the simplest and most fundamental systems with which quantum mechanics deals in the hope that this will lay the foundation for a deeper study of the physical applications to be found in the literature.

In a sense, the harmonic oscillator is to physics what the set of natural numbers is to mathematics. It is a simple, almost "trivial" system, but one which conceals much subtlety and beauty and from which a great deal of what is of interest in the subject evolves. We will follow some of this evolution from the simple classical problem through its canonical quantization and path integral to its fermionic and supersymmetric versions and will pause along the way to consider the tools and thought processes employed by physicists to construct their theoretical models. We will make a concerted effort to rephrase, whenever possible, these tools and thought processes in a form more congenial to those trained in modern mathematics, for this is our intended audience. However, we will also make a concerted effort to discourage the view that physics can simply be translated into mathematics. We take seriously Einstein's dictum that "... as far as the propositions of mathematics refer to reality, they are not certain; and, as far as they are certain, they do not refer to reality." Our very modest goal is to alleviate, in some small measure, the stress that generally accompanies the mathematically inclined when they stray into the very foreign world of bosons and fermions, Lagrangians and path integrals, nonexistent measures on infinite-dimensional spaces and supersymmetry. The best we can offer to the mathematician interested in dipping his or her toes into the murky waters of physics is an honest attempt, at each

https://doi.org/10.1515/9783110751949-201

stage, to clearly distinguish those items that are accessible to mathematical definition and proof from those that are not, and an honest admission that the rigor we so
earnestly strive for can do violence to the intentions of the physicists. Elegance and
brevity, while admirable traits, will play very little role here; the goal is communica-
tion and if we feel that this is best accomplished by an ugly argument in coordinates,
or a treatment that is perversely elementary, then so be it.

In broad strokes, here is the plan. We begin by briefly reviewing the part of the
classical story that we all learn as undergraduates (masses oscillating on springs and
simple harmonic motion) and then say a few words about why this rather special and
seemingly uninteresting problem is so important (Chapter 1). We will see how the problem and its solution can be rephrased in both Lagrangian and Hamiltonian form and
argue that each of these has its advantages and that both provide a conceptually more
satisfactory framework for our problem and for physics in general (Chapter 2). This
last point is brought home with particular force when the length scale of the harmonic oscillator is sufficiently small that classical mechanics fails entirely and one
must treat the problem quantum mechanically. To see how this is done we begin with
some motivation for the formalism of quantum mechanics (Chapter 3). This formalism
is quite unlike anything in classical physics and evolved historically over many years
in a highly nonlinear fashion from the brilliant insights and inspired guesswork of its
creators. In the end we are forced to concede that the conceptual apparatus that has
evolved in our species over eons in response to the macroscopic world in which we live
is simply not adequate for the description of the microscopic, quantum world, which
operates according to entirely different rules. For example, the intuitively all too familiar distinction between a particle and a wave disappears and we are required to
regard these as simply dual aspects of the same underlying physical object. But if familiar, classical concepts fail us, there is still mathematics, which does not require that
the objects with which it deals correspond to any ready-made, familiar concepts. The
formalism of quantum mechanics provides a mathematical, not a conceptual model
of what goes on in the world, but the model has proved to be remarkably successful.
In this world the states of a physical system are represented by elements of a Hilbert
space and the things we observe (measure) are represented by self-adjoint operators
on this Hilbert space. We will not pretend that this formalism can be "deduced" log-
ically from a few simple physical principles, for it cannot. However, by taking a very
general view of what constitutes a mathematical model for a physical system, it is
possible to argue that, in hindsight at least, the formalism has a certain element of
"naturalness" to it and we will attempt to make this argument.

Even so, to do justice to the formalism, both physically and mathematically, requires considerable preparation. Section 3.2 presents a very general view of what constitutes a mathematical model of a physical system in the hope that the model we will
eventually propose for quantum mechanics might appear somewhat less outrageous.
Section 3.3 continues this theme by briefly describing a mathematical model of classical statistical mechanics due to Koopman [Koop] in which Hilbert spaces and self-

adjoint operators arise naturally. We then examine some of the experimental facts of life which suggest that mechanics at the atomic and subatomic levels is more akin to classical statistical mechanics than to classical particle mechanics. Specifically, a brief tutorial on electromagnetic radiation (Section 4.2) is followed by discussions of black-body radiation and the photoelectric effect (Section 4.3) and two-slit experiments (Section 4.4); in particular, we go to some lengths in Section 4.3 to track down the origin of the ubiquitous "Planck constant." Following this there is a rather lengthy synopsis (Chapter 5) of the required functional analysis (unbounded self-adjoint operators, spectral theory and Stone's theorem); here the definitions and theorems are all stated precisely, but in lieu of proofs we generally offer only a few detailed examples relevant to quantum mechanics and plentiful references. With this we are in a position to describe the mathematical skeleton of quantum mechanics. In Chapter 6 we follow the usual procedure of describing this skeleton in the form of a set of postulates, but we devote considerably more time than is customary to discussions of what these postulates are supposed to mean, where they came from and what one might find questionable about them. Chapter 6 includes also a discussion of various uncertainty relations (Section 6.3) and the so-called Heisenberg picture of quantum mechanics (Section 6.4). The path that led Heisenberg to his formulation of quantum mechanics is elaborated more fully in Section 7.1, not only because it is a fascinating story, but also because it is here that one sees most clearly the emergence of the algebraic underpinnings of "canonical quantization." Section 7.2 describes these algebraic structures in more detail as well as the problem of representing them as self-adjoint operators on a Hilbert space. The famous Groenewold–Van Hove theorem, which restricts the extent to which this can be done, is also discussed.

With the formalism of quantum mechanics in hand one can consider the problem of "quantizing" a classical mechanical system such as the harmonic oscillator, that is, constructing a quantum mechanical model that reflects the essential features of the classical system. For instance, a diatomic molecule is very much like a mass on a spring, but, because of its size, it behaves very differently and requires a quantum mechanical treatment. Just what these essential features are, how they are to be described classically and how they are to be reflected in the quantum model are issues that we will have to discuss. Many schemes for arriving at such a quantum model for a classical system have been proposed. We will consider only two and will apply them only to the free particle and the harmonic oscillator. *Canonical quantization* (Chapter 7) is based on the Hamiltonian picture of the classical system, while the *Feynman path integral* (Chapter 8) evolves from the Lagrangian picture. We will work out both of these in detail for the free particle (Sections 7.3 and 8.2) and the harmonic oscillator (Sections 7.4 and 8.3). Both of these approaches raise serious mathematical issues, and in Chapter 9 we will survey a few of the rigorous theorems that have been proved in order to address some of these. Included are some self-adjointness theorems for quantum Hamiltonians (Section 9.2), Brownian motion and the Wiener measure (Section 9.3) and a rigorous approach to the Feynman integral via analytic continuation (Section 9.4).

Canonical quantization leads to a rather algebraic view of the quantization process and suggests certain variants of the quantum harmonic oscillator which, although they are legitimate and meaningful quantum systems, cannot be regarded as the quantization of any classical system. These are the so-called *fermionic harmonic oscillator* and *supersymmetric harmonic oscillator*, and we will take a look at each in Chapter 10, beginning with a discussion of the Stern–Gerlach experiment and the quantum mechanical notion of *spin* in Section 10.1. In Section 10.2 we briefly discuss the *spin-statistics theorem* and the *Pauli exclusion principle* in the hope of motivating Pascual Jordan's extraordinary idea of replacing commutation relations by anticommutation relations for the description of fermionic systems. It is this idea that gives rise to a fermionic analogue of the quantum harmonic oscillator and to the anticommuting or Grassmann variables with which one can build a "quasi-classical" system corresponding to it. In Section 10.3 we make a few (very few) general remarks on the notion of *supersymmetry* and then describe the simplest possible system in which this symmetry is exhibited. This is the so-called supersymmetric harmonic oscillator. We will then abstract the essential features of this example to define what is called $N = 2$ *supersymmetry*. We will see how the notion of a *Lie superalgebra* arises naturally in this context and, in Section 10.4, how old and venerable parts of mathematics (such as Hodge theory) offer additional examples.

At this point it would be best to make some simple declarative statement about the prerequisites required to read and understand this material. That would be best, but it is not going to happen. The reason is simply that these prerequisites vary widely from section to section, ranging from almost nothing at all in Chapter 1 to various aspects of analysis and functional analysis that one could reasonably expect only a specialist to know in Chapters 8, 9 and 10. Supplying all of this background here would not only result in a manuscript with essentially unbounded pagination, but would also be pointless since there are available many excellent sources for the material that we could not hope to improve upon. We will try to handle this problem in the following way. Beyond $\mathbf{F} = m\mathbf{A}$ any basic physics that needs to be explained *will* be explained; for the not-so-basic physics we will do our best to provide some intuition and some readable references. On the mathematical side, we will make judgment calls concerning what it is reasonable to assume that a graduate student in mathematics will know. Basic measure theory certainly qualifies, but not Wiener measure; functional analysis through the spectral theorem for compact, self-adjoint operators, but perhaps not the unbounded case of the spectral theorem; Banach and Hilbert spaces, but not Fréchet spaces; differentiable manifolds, basic Lie groups and Lie algebras, but not symplectic manifolds; basic partial differential equations, but not heat kernels or elliptic regularity. For those items we need that have been judged not to be in everyone's cache we will spend some time introducing them precisely and illustrating them with *relevant* examples. Then we will go to some lengths to provide detailed, explicit and accessible references in which the specific material we require is treated in a similar spirit and at a comparable level. The hope is that rigorous statements and carefully worked out examples will clarify the concepts, but then pursuing the subject in more depth can be left to the reader's discretion. All of this is entirely subjective, of course, and may fail to meet the needs of anyone except the author, but something had to be done. A number of appendices are included either to establish notation and terminology, or because we felt a particular need to have the material readily available. One can consult these as the need arises. The 238 exercises interspersed throughout the text are generally fairly routine opportunities for the reader to get involved and solidify the new ideas; these are not collected at the end of each section, but are placed at precisely the point at which they can be solved with optimal benefit.

It goes without saying that, beyond a few minor issues of expository style, there is nothing original in anything that follows. The manuscript arose simply from an attempt on the part of the author to organize some things for himself in a language he could understand and the hope that what emerged might be of use to someone else as well. Except for those items that are, by now, completely standard, we have tried to be clear on the sources from which the material was drawn. One of these sources, however, requires special attention at the outset. At many points along the way I found myself needing to understand mathematics that I either should have understood decades ago or, perhaps, did understand decades ago, but forgot. At all of these points I was patiently instructed by my son, Aaron, who always knew the mathematics and very often grasped its significance for physics long before I did.

Contents

1 The classical harmonic oscillator

The "trivial" side of the harmonic oscillator is known to every calculus student. One considers a mass m attached to one end of a spring with the other end held fixed. When set in motion the mass is free to oscillate along a straight line about its equilibrium position (where it sat, at rest, when the spring was unstretched). Making our first concession to the conventions of the physicists, we call the line along which the oscillations take place the q-axis and fix the origin at the equilibrium position of m. At this point one borrows a few "laws" (that is, assumptions) from classical physics. The first is called *Hooke's law* and asserts that the spring exerts a force $\mathbf{F}$ on m that tends to restore it to its equilibrium position and has a magnitude that is (under a certain range of conditions) proportional to the distance from the mass to the equilibrium position. Thus, for some positive constant k (called the *spring constant* and determined by the material the spring is made of and how tightly it is wound),

$$\mathbf{F} = -kq\mathbf{i},$$

where we use $\mathbf{i}$ for the unit vector in the positive q-direction. Next, *Newton's second law* asserts that the total force $\mathbf{F}$ acting on a mass m is proportional to the acceleration $\mathbf{A}$ it will experience as a result of the force and that the constant of proportionality is just m:

$$\mathbf{F} = m\mathbf{A}. \tag{1.1}$$

Assuming that no force other than that exerted by the spring is acting on m (no friction along the q-axis, no gravity, no one blowing on it), we conclude that at each instant t of time, $m\ddot{q}(t) = -kq(t)$ ($\ddot{q}$ is the second t-derivative of q). We will write this as

$$\ddot{q}(t) + \omega^2 q(t) = 0, \tag{1.2}$$

where $\omega = \sqrt{k/m}$.

Remark 1.0.1. We should say at the outset that t is to be thought of as Newton thought of it and as you have been thinking of it all of your life, as a universal time coordinate with the property that everyone agrees on the time lapse $t_2 - t_1$ between two events. There is no operational definition, for it does not exist, but this will not matter as long as we choose not to take relativistic effects into account.

Equation (1.2) is called the *harmonic oscillator equation*. It is a simple homogeneous, second order, linear equation with constant coefficients, the general solution to which can be written

$$q(t) = A\,\cos\,(\omega t + \varphi), \tag{1.3}$$

https://doi.org/10.1515/9783110751949-001

where A is the *amplitude*, ω is the *natural frequency* and φ is the *phase* of the motion. One can spice the problem up a bit by including the effects of additional forces (damping or driving forces) and some interesting phenomena emerge, but basically this is the whole story. That being the case it would seem incumbent upon us to offer just a few words on why we intend to make such a fuss about such a simple problem. In truth, this is really the issue we would like to address in the remainder of the manuscript, but a brief prologue would not be amiss. We will begin by thinking about just a few more simple problems and for this it is best to forget where the harmonic oscillator equation (1.2) came from and remember only that q is, in some sense, a position coordinate and ω is a positive constant.

Let us consider first a pendulum consisting of a string of length l and negligible mass with one end attached to the ceiling and a mass m attached to the other end. Suppose the mass is displaced from its equilibrium position (hanging straight down) and released, or perhaps given some initial velocity that lies in the plane of the string and its original, vertical position. Then the pendulum will move in this vertical plane. Let $\phi(t)$ denote the angle between the string and the vertical at time t. The forces acting on the mass are the vertical gravitational force with a magnitude of mg (g is the acceleration due to gravity near the surface of the earth, which is about $9.8\,\text{m/sec}^2$) and the tension in the string. The component of the gravitational force parallel to the string cancels the tension, while the component perpendicular to the string provides the tangential restoring force which causes the pendulum to oscillate. The magnitude of the tangential force at time t is $mg \sin \phi(t)$ and the magnitude of the velocity is $l\dot{\phi}(t)$, so Newton's second law gives

$$\ddot{\phi}(t) + \omega^2 \sin \phi(t) = 0, \tag{1.4}$$

where $\omega = \sqrt{g/l}$. Now, the *pendulum equation* (1.4) is, of course, not the harmonic oscillator equation. However, if we assume the oscillations (that is, the values of ϕ) are small, the Taylor series expansion for $\sin \phi$ at $\phi = 0$ gives the approximation $\sin \phi \approx \phi$ and we may consider instead

$$\ddot{\phi}(t) + \omega^2 \phi(t) = 0, \tag{1.5}$$

which is precisely equation (1.2). Of course, the solutions to (1.5) only approximate the motion of the pendulum for small displacements.

There is something quite general going on in this last example that we would like to discuss before moving on to a few more examples. This is most efficiently done if we recall a bit of vector calculus. We consider a single particle of mass m moving in $\mathbb{R}^n$ under the influence of some time-independent force $\mathbf{F}(x) = \mathbf{F}(x^1, \dots, x^n)$; generally, n will be 1, 2 or 3. We assume that $\mathbf{F}$ is *conservative* in the sense that it is the gradient of some smooth real-valued function $-V$ on $\mathbb{R}^n$ (the minus sign is conventional). Note that V is determined only up to the addition of a real constant a since $\nabla(-V + a) =$

$V(-V)$. Then Newton's second law asserts that the motion of the particle, represented by the curve $x(t) = (x^1(t), \ldots, x^n(t))$ in $\mathbb{R}^n$, is determined by

$$m \frac{d^2x(t)}{dt^2} = -\nabla V(x(t)).$$

Here $V(x(t))$ (without the minus sign) is called the *potential energy* at time t. One also defines the *kinetic energy* at time t to be

$$\frac{1}{2}m \left\| \frac{dx(t)}{dt} \right\|^2 = \frac{1}{2}m \frac{dx(t)}{dt} \cdot \frac{dx(t)}{dt},$$

where the · indicates the usual inner product on $\mathbb{R}^n$. The *total energy* at time t is then

$$E(x(t)) = \frac{1}{2}m \left\| \frac{dx(t)}{dt} \right\|^2 + V(x(t)).$$

The rationale behind the word "conservative" is that, although the kinetic and potential energies change along the trajectory $x(t)$, E does not because

$$\begin{aligned}
\frac{dE(x(t))}{dt} &= \frac{d}{dt}\left(\frac{1}{2}m \frac{dx(t)}{dt} \cdot \frac{dx(t)}{dt} + V(x(t)) \right) \\
&= m \frac{dx(t)}{dt} \cdot \frac{d^2x(t)}{dt^2} + \nabla V(x(t)) \cdot \frac{dx(t)}{dt} \\
&= \frac{dx(t)}{dt} \cdot \left(m \frac{d^2x(t)}{dt^2} + \nabla V(x(t)) \right) \\
&= \frac{dx(t)}{dt} \cdot \mathbf{0} = 0.
\end{aligned}$$

For *conservative forces* the total energy E is *conserved* during the motion. The fact that physical systems that evolve in time can nevertheless leave certain "observable quantities" unchanged is of profound significance to physics and will recur again and again in the course of our discussions here.

When $n = 1$ it is customary to assume everything is written in terms of the standard basis for $\mathbb{R}$, drop all of the vectorial notation and write, for example, $F(x) = -\frac{dV}{dx}$ for a conservative force, $m\ddot{x}(t) = -\frac{dV}{dx}(x(t))$ for Newton's second law, and so on. For the mass on a spring example, $x = q$ and the potential can be taken to be $V(q) = \frac{1}{2}kq^2$ since $F(q) = -kq = -\frac{d}{dq}(\frac{1}{2}kq^2)$. The total energy of the mass at time t is therefore $\frac{1}{2}m\dot{q}(t)^2 + \frac{1}{2}kq(t)^2$. Note that, using primes to denote derivatives with respect to q,

$$V(0) = V'(0) = 0 \quad \text{and} \quad V''(0) > 0.$$

The potential has a relative minimum value of 0 at $q = 0$ and so the restoring force $F(q) = -V'(q)$ vanishes at $q = 0$ with a negative derivative there and this accounts for the *stable equilibrium point* of the mass–spring system at $q = 0$. For the pendulum, $x = \phi$ and one can take $V(\phi) = -\frac{mg}{l}(\cos\phi - 1)$ since $F(\phi) = -\frac{mg}{l}\sin\phi = -\frac{d}{d\phi}(-\frac{mg}{l}(\cos\phi - 1))$; the -1 is not necessary, but ensures that the zero potential level occurs at the equilibrium position. Once again using a prime for differentiation with respect to ϕ we obtain

$$V(0) = V'(0) = 0 \quad \text{and} \quad V''(0) > 0$$

and, again, this is a reflection of the fact that the pendulum has a stable equilibrium point at $\phi = 0$. Moreover, this is precisely what gives rise to the fact that small oscillations of the pendulum are modeled by the harmonic oscillator equation. We began by claiming that there is something quite general going on and this is it, as we now show.

Suppose we have a one-dimensional system on which a force $F(x) = -\frac{dV}{dx}$ acts and that $V(x)$ has a relative minimum value of 0 at $x = 0$ so that

$$V(0) = V'(0) = 0 \quad \text{and} \quad V''(0) > 0.$$

The Taylor series for $V(x)$ at $x = 0$ then has the form

$$V(x) = \frac{1}{2}V''(0)x^2 + \cdots.$$

Thus, for small x, $V(x) \approx \frac{1}{2}kx^2$, where $k = V''(0) > 0$ and so the potential is approximately that of a harmonic oscillator and the system behaves, for small x, like a harmonic oscillator. Note that there is no reason for the equilibrium point to be at $x = 0$ since the same argument using the Taylor series at $x = x_0$ yields the same result. Moreover, since the potential function is determined only up to an additive constant, it can always be chosen to vanish at any given point. Thus, any x_0 at which $V'(x_0) = 0$ and $V''(x_0) > 0$ is a stable equilibrium point for the system and, near x_0, the potential is approximately that of a harmonic oscillator.

> The essential reason for the significance of the harmonic oscillator is that any conservative one-dimensional system with a stable point of equilibrium behaves like a harmonic oscillator near the equilibrium point.

Remark 1.0.2. We have been rather cavalier in our use of the term *energy* and should offer something in the way of an apology, or, at least, an explanation. There is nothing at all ambiguous in our notions of kinetic energy or potential energy; these are *defined* by the formulas we recorded above. One might ask, however, *why* they are defined by these formulas. Why, for example, is kinetic energy $\frac{1}{2}m\|\frac{dx(t)}{dt}\|^2$ and not, say, $m\|\frac{dx(t)}{dt}\|^2$? The answer is quite simple; without the $\frac{1}{2}$ in the definition of kinetic energy, the total energy (kinetic plus potential) would not be conserved. Energy is arguably the most fundamental concept in physics and it appears in many guises, but it is a subtle one

and the unmodified word itself is *never* defined (except in high school where it is intuitively identified with "the ability to do work"). Various types of energy are defined in
various contexts, but always with the sole objective of isolating a number that does not
change as the (isolated) physical system evolves. The existence of such numbers is an
extraordinarily powerful thing in physics. With them one can make statements about
how a system evolves without any real understanding of the detailed interactions that
occur during the evolution; we do not need to know what *really* goes on inside two
billiard balls when they collide in order to compute how they will move.

"It is important to realize that in physics today, we have no knowledge of what energy is."

Richard Feynman (Section 4-1, Volume I, [FLS])

The last two examples we would like to look at are intended to give a brief suggestion of a few of the things to come. For the first we will consider what are called
diatomic molecules. As the name suggests, these are molecules in which precisely two
atoms are bound together. The atoms may be the same, for example, oxygen in O_2, or
nitrogen in N_2, or they may be different, for example, carbon and oxygen in carbon
monoxide (CO). Such molecules are extremely common in nature. Indeed, O_2 and N_2
together comprise 99 % of the earth's atmosphere (fortunately, CO is not so prevalent).
The bond between the atoms in such a molecule is not rigid. Rather, the distance between the nuclei of the atoms (the *internuclear distance*) varies periodically around
some equilibrium value and the potential energy of the molecule is proportional to
the *square* of the displacement from equilibrium (at least for small displacements).
Because of this quadratic dependence, if we view the molecule from the perspective
of one of the atoms, the other appears to be very much like a mass on a spring (again,
for small displacements). It may seem then that we have not really described a new
example at all. The reason that the example is, in fact, new is that *diatomic molecules
are small*; indeed, they are so small that one cannot expect them to behave according
to the rules of classical Newtonian physics and they do not. The behavior of such a
system falls within the purview of quantum mechanics.

Eventually, we will describe procedures for quantizing the classical harmonic oscillator and this quantum system does, in fact, accurately describe the small vibra-
tions of diatomic molecules. For the present we would simply like to point out that the
classical and quantum descriptions are very different. Here is one particularly striking instance. Note that the potential energy for the classical harmonic oscillator can
be written

$$V(q) = \frac{1}{2}kq^2 = \frac{1}{2}m(\sqrt{k/m})^2 q^2 = \frac{1}{2}m\omega^2 q^2, \tag{1.6}$$

where ω is the natural frequency of vibration of the oscillator. Observe that, as q varies
over the interval $[-A, A]$, where A is the amplitude, $V(q)$ takes *every* value between 0
and $\frac{1}{2}m\omega^2 A^2$ (twice, in fact). In particular, the energy can take on continuously many

values. By contrast, we will find that, in the corresponding quantum system, the energy of the oscillator can assume only the values

$$\frac{1}{2}\hbar\omega, \frac{3}{2}\hbar\omega, \frac{5}{2}\hbar\omega, \dots, \left(n + \frac{1}{2}\right)\hbar\omega, \dots, \tag{1.7}$$

where $\hbar$ is a certain positive constant that we will discuss later. In the quantum system the energy is *quantized* so any transition from one energy level to another cannot occur continuously and must be the result of a *quantum jump*. Perhaps even more interesting is the fact that 0 is not in this list. Unlike its classical counterpart, the quantum harmonic oscillator cannot have zero energy. The smallest possible value of the energy is $\frac{1}{2}\hbar\omega$, which is called the *ground state energy*.

Remark 1.0.3. We will eventually *derive* the energy spectrum (1.7) from the basic postulates of quantum mechanics, but it is worth pointing out that, historically, the order was reversed. The fundamental idea that underlies quantum mechanics was discovered by Max Planck in his study of what is called *blackbody radiation*. Here the predictions of classical physics do not correspond at all to what is actually observed in the laboratory and Planck found that he could construct a model that yielded very accurate predictions under the *hypothesis* that harmonic oscillators exist only at discrete energy levels. This hypothesis was totally inconsistent with classical physics, but it worked and, as we shall see, led to an entirely new way of thinking about the world. In Section 4.3 we will describe all of this in much more detail.

Now we present our final example. Thinking of a complex system as being built out of simple, well-understood systems can be quite useful and we would like to conclude this section with a rather extreme example. We intend to reinterpret the classical *vibrating string problem* in terms of a countably infinite family of harmonic oscillators. This interpretation provided the physicists with a means of quantizing the vibrating string and the result was essentially the first example of a quantum field. The same ideas will be put to use in Section 4.3, where we study blackbody radiation.

In the vibrating string problem one is asked to describe the small transverse displacements $u(t, x)$ of an elastic string, tightly stretched along the x-axis between $x = 0$ and $x = l$ assuming that no external forces act on it. A bit of physics, which one sees in any elementary course on partial differential equations (for example, Section 25 of [BC], or Chapter 8 of [Sp3]), shows that $u(t, x)$ satisfies the *one-dimensional wave equation*

$$a^2\frac{\partial^2 u}{\partial x^2} = \frac{\partial^2 u}{\partial t^2}, \quad 0 < x < l, \, t > 0, \tag{1.8}$$

where a is a positive constant determined by the material the string is made of and how tightly it is stretched (more precisely, $a = \sqrt{\tau/\rho}$, where ρ is the mass per unit length and τ is the tension, both of which will be assumed constant). For the moment let us focus on solutions $u(t, x)$ that are continuous on the boundary of $[0, l] \times [0, \infty)$. Then, since the string is fixed at $x = 0$ and $x = l$, $u(t, x)$ must satisfy the boundary conditions

$$u(t, 0) = u(t, l) = 0, \quad t \geq 0. \tag{1.9}$$

Since we do not need them at the moment we will not be explicit about the initial displacement $u(0, x)$ and initial velocity $\frac{\partial u}{\partial t}(0, x)$ that would be required to produce a well-posed problem. The usual procedure (Section 32 of [BC]) is to separate variables $u(t, x) = T(t)X(x)$ and obtain a Sturm–Liouville problem

$$X''(x) + \lambda X(x) = 0, \tag{1.10}$$
$$X(0) = X(l) = 0 \tag{1.11}$$

for $X(x)$ and an ordinary differential equation

$$\ddot{T}(t) + \lambda a^2 T(t) = 0 \tag{1.12}$$

for $T(t)$. The Sturm–Liouville problem has eigenvalues $\lambda_n = \frac{n^2 \pi^2}{l^2}$, $n = 1, 2, 3, \ldots$, and corresponding orthonormal eigenfunctions $X_n(x) = \sqrt{\frac{2}{l}} \sin \frac{n \pi x}{l}$. We recall what this means. Think of (1.10) as the eigenvalue equation $(\frac{d^2}{dx^2} + \lambda) X(x) = 0$ for the one-dimensional Laplacian $\frac{d^2}{dx^2}$ on $[0, l]$ subject to the boundary conditions $X(0) = X(l) = 0$. Then λ_n are the only values of λ for which nontrivial solutions exist and $X_n(x)$ are corresponding solutions, that is, $(\frac{d^2}{dx^2} + \lambda_n) X_n(x) = 0$, $n = 1, 2, 3, \ldots$. "Orthonormal" means in the L^2-sense, that is,

$$\int_0^l X_n(x) X_m(x) dx = \begin{cases} 1, & \text{if } m = n, \\ 0, & \text{if } m \neq n. \end{cases} \tag{1.13}$$

Exercise 1.0.1. If you have never verified all of this before, do so now.

For each $n = 1, 2, 3, \ldots$, we let

$$\omega_n = \frac{n \pi a}{l},$$

so that (1.12) becomes

$$\ddot{T}(t) + \omega_n^2 T(t) = 0, \tag{1.14}$$

which is, for each $n = 1, 2, 3, \ldots$, an instance of the harmonic oscillator equation. Standard operating procedure would now have us write down the general solution to (1.14), multiply by $X_n(x) = \sqrt{\frac{2}{l}} \sin \frac{n \pi x}{l}$ and then superimpose (sum over $n = 1, 2, 3, \ldots$) to obtain a general expression for $u(t, x)$. Instead, let us denote by $q_n(t)$ any nontrivial solution to (1.14) and write

$$u(t, x) = \sum_{n=1}^{\infty} q_n(t) \sqrt{\frac{2}{l}} \sin \frac{n\pi x}{l} \qquad (1.15)$$

(since our purpose here is purely motivational we will skirt all of the obvious convergence issues). Now look at some fixed term $u_n(t, x) = q_n(t) \sqrt{\frac{2}{l}} \sin \frac{n\pi x}{l}$ in (1.15). Here $q_n(t)$, being a solution to (1.14), represents a simple harmonic motion with natural frequency ω_n and some amplitude. But the harmonic oscillator equation is linear so, if we fix some $x_0 \in (0, l)$ with $\sin \frac{n\pi x_0}{l} \neq 0$, $u_n(t, x_0)$ also represents a simple harmonic motion with natural frequency ω_n, but with a different amplitude. Consequently, $u_n(t, x)$ represents a motion of the string in which each nonstationary point along the length of the string is executing simple harmonic motion with the same frequency. If $a_1, a_2, \ldots$ are constants, then the solution $u(t, x) = \sum_{n=1}^{\infty} a_n u_n(t, x)$ is a superposition of these *harmonics*. What is essential from our point of view is that, since the eigenfunctions $X_n(x) = \sqrt{\frac{2}{l}} \sin \frac{n\pi x}{l}$ are fixed, all of the information about a solution $u(t, x)$ to the vibrating string problem is contained in a *countable sequence* of classical harmonic oscillators $q_n(t)$ as opposed to a continuum of oscillators, one for each point along the length of the string. Here, briefly, is a simple, concrete illustration of this.

Exercise 1.0.2. Classical physics provides the following expression for the total energy (kinetic plus potential) of our vibrating string at time t:

$$E(t) = \int_0^l \frac{1}{2} \left[\rho \left(\frac{\partial u}{\partial t} \right)^2 + \tau \left(\frac{\partial u}{\partial x} \right)^2 \right] dx$$

(recall that ρ is the mass density and τ is the tension). Take this for granted. Then substitute (1.15) and use the orthonormality relations (1.13) for the functions $X_n(x) = \sqrt{\frac{2}{l}} \sin \frac{n\pi x}{l}$ to show that

$$E(t) = \sum_{n=1}^{\infty} \frac{\rho}{2} [\dot{q}_n(t)^2 + \omega_n^2 q_n(t)^2].$$

Note that this is just the sum of the energies of the harmonic oscillators $\{q_n(t)\}_{n=1}^{\infty}$ if they are all taken to have mass ρ.

A vibrating string is therefore essentially a sequence of classical harmonic oscillators. These same ideas can be applied in much more significant contexts and we will see some of them later. In particular, we will see in Section 4.3 how they can be used to study *blackbody radiation* and trace the ideas that led Max Planck to the *quantum hypothesis* that eventually evolved into our topic here.

2 Lagrangian and Hamiltonian mechanics

2.1 Introduction

Our entire discussion of the classical harmonic oscillator in the previous section was "Newtonian" (basically, just $\mathbf{F} = m\mathbf{A}$). While this Newtonian picture is perfectly adequate for a great many purposes, there are certain aspects of the picture that limit its usefulness to us. Fields such as the electromagnetic field do not fit into the picture at all, for example. More significantly, there are no natural techniques for quantizing a classical mechanical system described in Newtonian terms. In this section we will introduce two alternative pictures, each of which encompasses both mechanics and field theory and for each of which there are procedures for producing quantum analogues of classical systems. Our objectives here are, as always, quite modest; we hope only to introduce the fundamental ideas required to understand what is to come later on. For a thorough grounding in Lagrangian and Hamiltonian mechanics we direct the reader to [Arn2], [Sp3], [CM] [GS1], or the standard physics text [Gold].

2.2 State space and Lagrangians

Let us consider again a particle of mass m moving in $\mathbb{R}^n$. We denote its position at time t by $\alpha(t) = (q^1(t), \ldots, q^n(t))$. We assume that the particle is moving under the influence of a time-independent force $\mathbf{F}(q) = \mathbf{F}(q^1, \ldots, q^n)$ and that the force is conservative, that is, $\mathbf{F}(q) = -\nabla V(q)$ for some smooth, real-valued function $V(q)$ on $\mathbb{R}^n$. The basic assumption of Newtonian mechanics is that, along the trajectory $\alpha(t)$ of the particle, Newton's second law $m\frac{d^2\alpha(t)}{dt^2} = -\nabla V(\alpha(t))$ is satisfied. For the moment, all we want to do is find an equivalent way of saying "Newton's second law is satisfied." We will do this by introducing a few definitions and performing a little calculation. When this is done we will describe the much more general context in which these ideas live and that will eventually allow us to leave the confines of Newtonian mechanics.

Fix two points $a, b \in \mathbb{R}^n$ and an interval $[t_0, t_1]$ in $\mathbb{R}$. We consider the set $C^\infty_{a,b}([t_0, t_1], \mathbb{R}^n)$ of all smooth (infinitely differentiable) curves $\alpha(t)$ in $\mathbb{R}^n$ from $\alpha(t_0) = a$ to $\alpha(t_1) = b$; this has the structure of an infinite-dimensional affine space. For each element α of $C^\infty_{a,b}([t_0, t_1], \mathbb{R}^n)$ we define the *kinetic energy function* $K_\alpha(t)$ and the *potential energy function* $V_\alpha(t)$ by

$$K_\alpha(t) = \frac{1}{2} m \left\| \dot{\alpha}(t) \right\|^2,$$

where $\dot{\alpha}(t)$ denotes the velocity (tangent) vector to α at t and $\| \ \|$ is the usual norm on $\mathbb{R}^n$, and

$$V_\alpha(t) = V(\alpha(t)).$$

https://doi.org/10.1515/9783110751949-002

The kinetic energy plus the potential energy is the *total energy* $E_\alpha(t)$ and we have seen that Newton's second law implies that this is constant *along the actual trajectory of the particle*. Instead of the sum, we would now like to consider the difference of the kinetic and potential energies (we will have a few words to say about the physical interpretation of this a bit later). Specifically, we define

$$L_\alpha(t) = K_\alpha(t) - V_\alpha(t) = \frac{1}{2}m\,\|\dot\alpha(t)\|^2 - V(\alpha(t)).$$

Next we define the *action functional* to be the real-valued function

$$S: C^\infty_{a,b}([t_0, t_1], \mathbb{R}^n) \to \mathbb{R} \tag{2.1}$$

given by

$$S(\alpha) = \int_{t_0}^{t_1} L_\alpha(t)dt = \int_{t_0}^{t_1} \frac{1}{2}m\,\|\dot\alpha(t)\|^2 - V(\alpha(t))\,dt. \tag{2.2}$$

We propose to characterize the trajectory α of a particle moving under the influence of $\mathbf{F}(q) = -\nabla V(q)$ from $\alpha(t_0) = a$ to $\alpha(t_1) = b$ as a "critical point" of this action functional S, that is, a point where the "derivative" of S vanishes. First, of course, we must isolate the appropriate notion of "derivative." For any $\alpha \in C^\infty_{a,b}([t_0, t_1], \mathbb{R}^n)$ we define a *(fixed endpoint) variation* of α to be a smooth map

$$\Gamma: [t_0, t_1] \times (-\epsilon, \epsilon) \to \mathbb{R}^n$$

for some $\epsilon > 0$ such that

$$\Gamma(t, 0) = \alpha(t), \quad t_0 \le t \le t_1,$$
$$\Gamma(t_0, s) = \alpha(t_0) = a, \quad -\epsilon < s < \epsilon,$$
$$\Gamma(t_1, s) = \alpha(t_1) = b, \quad -\epsilon < s < \epsilon.$$

For any fixed $s \in (-\epsilon, \epsilon)$ the map

$$\gamma_s : [t_0, t_1] \to \mathbb{R}^n$$

defined by

$$\gamma_s(t) = \Gamma(t, s)$$

is an element of $C^\infty_{a,b}([t_0, t_1], \mathbb{R}^n)$ so the action $S(\gamma_s)$ is defined. Moreover, $\gamma_0 = \alpha$ so $S(\gamma_0) = S(\alpha)$. Intuitively, we think of Γ as giving rise to a one-parameter family of curves γ_s in $C^\infty_{a,b}([t_0, t_1], \mathbb{R}^n)$ near α and, for this family of curves, the action functional becomes a real-valued function of the real variable s and so is something whose rate of change we can compute. Specifically, we will say that a curve $\alpha \in C_{a,b}^{\infty}([t_0, t_1], \mathbb{R}^n)$ is a *stationary, or critical point* of the action functional S if

$$\frac{d}{ds} S(\gamma_s)\Big|_{s=0} = 0 \tag{2.3}$$

for every variation Γ of α. In particular, this must be true for variations of the form

$$\Gamma(t, s) = \alpha(t) + s h(t), \tag{2.4}$$

where $h : [t_0, t_1] \to \mathbb{R}^n$ is an arbitrary smooth function satisfying $h(t_0) = h(t_1) = 0 \in \mathbb{R}^n$ (physicists would be inclined to write h as $\delta\alpha$). What we propose to prove now is that Newton's second law is satisfied along $\alpha(t)$ if and only if (2.3) is satisfied for all variations of α of the form (2.4). The procedure will be to prove

$$\frac{d}{ds} S(\gamma_s)\Big|_{s=0} = \int_{t_0}^{t_1} \left[-m\ddot{\alpha}(t) - \nabla V(\alpha(t))\right] \cdot h(t)\, dt \tag{2.5}$$

and then appeal to the following lemma.

Lemma 2.2.1 (Basic lemma of the calculus of variations). *Let $t_0 < t_1$ be real numbers and let $f : [t_0, t_1] \to \mathbb{R}^n$ be a continuous function that satisfies*

$$\int_{t_0}^{t_1} f(t) \cdot h(t)\, dt = 0$$

for every smooth function $h : [t_0, t_1] \to \mathbb{R}^n$ with $h(t_0) = h(t_1) = 0 \in \mathbb{R}^n$. Then $f(t) = 0 \in \mathbb{R}^n$ for every $t \in [t_0, t_1]$.

Proof. The general result follows easily from the $n = 1$ case so we will prove only this. Thus, we assume $f : [t_0, t_1] \to \mathbb{R}$ is continuous and satisfies

$$\int_{t_0}^{t_1} f(t) h(t)\, dt = 0$$

for all smooth functions $h : [t_0, t_1] \to \mathbb{R}$ with $h(t_0) = h(t_1) = 0$. Assume that f is nonzero at some point in $[t_0, t_1]$ and, without loss of generality, that it is positive there. Then, by continuity, f is positive on some relatively open interval in $[t_0, t_1]$ and therefore on some open interval (α, β) in $\mathbb{R}$ contained in $[t_0, t_1]$. Now one can select a smooth real-valued function h on $\mathbb{R}$ that is positive on (α, β) and zero elsewhere (see, for example, Exercise 2-26 of [Sp1]). But then $\int_{t_0}^{t_1} f(t) h(t)\, dt = \int_{\alpha}^{\beta} f(t) h(t)\, dt > 0$ and this is a contradiction. Consequently, $f(t) = 0\ \forall t \in [t_0, t_1]$. $\square$

Now we turn to the proof of (2.5). For this we compute

$$\frac{d}{ds}S(y_s)\Big|_{s=0} = \frac{d}{ds}\int_{t_0}^{t_1} \frac{1}{2}m\dot{y}_s(t)\cdot\dot{y}_s(t) - V(y_s(t))\,dt\,\Big|_{s=0}$$

$$= \int_{t_0}^{t_1} \frac{d}{ds}\left[\frac{1}{2}m\dot{y}_s(t)\cdot\dot{y}_s(t) - V(y_s(t))\right]\Big|_{s=0}\,dt$$

$$= \int_{t_0}^{t_1}\left[m\dot{y}_s(t)\cdot\frac{d}{ds}\dot{y}_s(t) - \nabla V(y_s(t))\cdot\frac{d}{ds}y_s(t)\right]\Big|_{s=0}\,dt$$

$$= \int_{t_0}^{t_1}\left[m\dot{y}_s(t)\cdot\frac{d}{ds}\dot{y}_s(t) - \nabla V(y_s(t))\cdot h(t)\right]\Big|_{s=0}\,dt. \qquad (2.6)$$

Now, note that

$$\frac{d}{ds}\dot{y}_s(t) = \frac{d}{ds}\frac{d}{dt}y_s(t) = \frac{d}{dt}\frac{d}{ds}y_s(t) = \frac{d}{dt}h(t),$$

so (2.6) becomes

$$\frac{d}{ds}S(y_s)\Big|_{s=0} = \int_{t_0}^{t_1}\left[m\dot{y}_s(t)\cdot\frac{d}{dt}h(t) - \nabla V(y_s(t))\cdot h(t)\right]\Big|_{s=0}\,dt$$

$$= \int_{t_0}^{t_1}\left[m\dot{\alpha}(t)\cdot\frac{d}{dt}h(t) - \nabla V(\alpha(t))\cdot h(t)\right]dt$$

$$= \int_{t_0}^{t_1}\left[-m\ddot{\alpha}(t)\cdot h(t) - \nabla V(\alpha(t))\cdot h(t)\right]dt + m\dot{\alpha}(t)\cdot h(t)\Big|_{t_0}^{t_1}$$

$$= \int_{t_0}^{t_1}\left[-m\ddot{\alpha}(t) - \nabla V(\alpha(t))\right]\cdot h(t)\,dt \quad (\text{since } h(t_0) = h(t_1) = 0)$$

and this is (2.5). Now we apply Lemma 2.2.1 to conclude that $m\ddot{\alpha}(t) = -\nabla V(\alpha(t))$ exactly when α is a stationary point for the action S.

Generally, although not always (see Example 2.2.4), a stationary point α for an action functional S will correspond to a relative minimum value of S. In this case one can think of our result intuitively as saying that Newton's second law dictates that the trajectory of our particle is that particular curve that minimizes the average kinetic minus the average potential energy, that is, the energy of motion minus the energy available for motion; in some sense, nature wants to see as little energy expended on motion as possible. For a much more illuminating discussion of this interpretation of our result, a few instances of how such ideas appear in other parts of physics, and just a fun read, see Chapter 19, Volume II, of [FLS]. This, incidentally, is our first exposure to what is often called in physics the *principle of least action*, although it would more properly be called the *principle of stationary action*. We will see more before we are through.

The point to this calculation is that Newton's second law has been rephrased as a *variational principle* and this suggests the possibility that other basic laws of physics might be similarly rephrased (*the actual evolution of the system is that which "minimizes" something*). If this is the case, then the calculus of variations might provide a general perspective for viewing a wider swath of physics than does Newtonian mechanics. This is, in fact, true and we would now like to describe this perspective.

To ease the transition to our new abstract setting, let us first rephrase a bit of what we have just done in order to see how the main player in the drama, the Lagrangian, enters the picture. The particle we have been discussing moves in

$$M = \mathbb{R}^n$$

and we will refer to this as the *configuration space* of the particle (space of possible positions). We will let $q^1, \ldots, q^n$ denote standard coordinate functions on the manifold $M = \mathbb{R}^n$. The potential V can then be thought of as a function of $q^1, \ldots, q^n$. The coordinate velocity vector fields $\partial_{q^1}, \ldots, \partial_{q^n}$ corresponding to $q^1, \ldots, q^n$ (also often written $\frac{\partial}{\partial q^1}, \ldots, \frac{\partial}{\partial q^n}$, or simply $\partial_1, \ldots, \partial_n$) provide a basis $\partial_{q^1}|_p, \ldots, \partial_{q^n}|_p$ for the tangent space $T_p(M)$ at each $p \in M$. Bowing once again to the conventions of the physicists we will denote the corresponding component functions on $T_p(M)$ by $\dot{q}^1, \ldots, \dot{q}^n$ so that $v_p \in T_p(M)$ is written $v_p = \sum_{i=1}^{n} \dot{q}^i(v_p)\partial_{q^i}|_p$ or, better yet, with the Einstein summation convention, $v_p = \dot{q}^i(v_p)\partial_{q^i}|_p$. Thus, $q^1, \ldots, q^n, \dot{q}^1, \ldots, \dot{q}^n$ are coordinate functions for the tangent bundle

$$TM = \mathbb{R}^n \times \mathbb{R}^n$$

of $\mathbb{R}^n$. We refer to TM as the *state space* of the particle; it is the space of pairs consisting of a possible position and a possible velocity.

Remark 2.2.1. The structure of the tangent and cotangent bundles of a manifold is reviewed in Appendix D.

One must take care not to interpret the dot in $\dot{q}^i$ as signifying a derivative with respect to t. There is no t here; $\dot{q}^i$ is simply a name for a coordinate in the tangent bundle. The reason for this rather odd notational convention will become clear in a moment when we lift curves in the configuration space to curves in the state space. In terms of these coordinates we define a function,

$$L : TM \to \mathbb{R},$$

called the *Lagrangian*, by

$$L(q, \dot{q}) = L(q^1, \ldots, q^n, \dot{q}^1, \ldots, \dot{q}^n) = \sum_{i=1}^{n} \frac{1}{2} m(\dot{q}^i)^2 - V(q^1, \ldots, q^n). \tag{2.7}$$

For $t_0 < t_1$ in $\mathbb{R}$ and $a, b \in M = \mathbb{R}^n$ the *path space* $C_{a,b}^{\infty}([t_0, t_1], M)$ is the space of all smooth curves $\alpha : [t_0, t_1] \to M$ with $\alpha(t_0) = a$ and $\alpha(t_1) = b$. Every α in $C_{a,b}^{\infty}([t_0, t_1], M)$ has a unique lift to a smooth curve

$$\tilde{\alpha} : [t_0, t_1] \to TM$$

in the tangent bundle defined by

$$\tilde{\alpha}(t) = (\alpha(t), \dot{\alpha}(t)),$$

where $\dot{\alpha}(t)$ denotes the velocity (tangent) vector to α at t. In coordinates we will simplify the notation a bit and write $q^i(\tilde{\alpha}(t)) = q^i(\alpha(t))$ as $q^i(t)$ and $\dot{q}^i(\tilde{\alpha}(t)) = \dot{q}^i(\dot{\alpha}(t))$ as $\dot{q}^i(t)$. Thus,

$$\tilde{\alpha}(t) = (q^1(t), \ldots, q^n(t), \dot{q}^1(t), \ldots, \dot{q}^n(t)).$$

Remark 2.2.2. *Now* you can think of the dot in $\dot{q}^i(t)$ as signifying the derivative with respect to t of $q^i(\alpha(t))$.

Consequently, the functions $L_\alpha(t)$ representing the kinetic minus potential energy along $\alpha(t)$ and whose t-integral is the action $S(\alpha)$ can be described as

$$L_\alpha = L \circ \tilde{\alpha},$$

where L is the Lagrangian defined on $TM = \mathbb{R}^n \times \mathbb{R}^n$ by (2.7). The essential information is contained in the function L defined on the state space (tangent bundle) and, as we move now to the general setting, our focus will shift to it.

We begin with a smooth (C^∞) manifold M of dimension n which we will refer to as the *configuration space* (space of positions). This might be, for example, $\mathbb{R}^n$, $n = 1, 2, 3$, for a single particle moving along a line, in a plane, or in 3-space. For k particles moving in 3-space one would take $M = \mathbb{R}^{3k}$ (three position coordinates for each of the k particles). For a particle constrained to move on the surface of the earth one might take $M = S^2$. One can imagine many more exotic possibilities.

Exercise 2.2.1. Describe a physical system whose configuration space is the torus $S^1 \times S^1$.

The tangent bundle of M is denoted TM and called the *state space*. This is the space of pairs consisting of a possible configuration and a possible rate of change. The topology of the state space, however, need not be that of a product. This is the case, for example, when $M = S^2$ (see Appendix D). Any smooth real-valued function $L : TM \to \mathbb{R}$ on the state space TM is called a *Lagrangian* on M. Such a function can be described *locally* in natural coordinates. We adopt the usual custom of writing these local coordinate representations as

$$L(q^1, \ldots, q^n, \dot{q}^1, \ldots, \dot{q}^n) = L(q, \dot{q}).$$

For $t_0 < t_1$ in $\mathbb{R}$ and $a, b \in M$ the *path space* $C_{a,b}^{\infty}([t_0, t_1], M)$ is the space of all smooth curves $\alpha : [t_0, t_1] \to M$ with $\alpha(t_0) = a$ and $\alpha(t_1) = b$. Every α in $C_{a,b}^{\infty}([t_0, t_1], M)$ has a unique lift to a smooth curve

$$\tilde{\alpha} : [t_0, t_1] \to TM$$

in the tangent bundle defined by

$$\tilde{\alpha}(t) = (\alpha(t), \dot{\alpha}(t)),$$

where $\dot{\alpha}(t)$ denotes the velocity (tangent) vector to α at t. The *action functional* associated with the Lagrangian L is the real-valued function

$$S_L : C_{a,b}^{\infty}([t_0, t_1], M) \to \mathbb{R} \tag{2.8}$$

defined by

$$S_L(\alpha) = \int_{t_0}^{t_1} L(\tilde{\alpha}(t)) dt = \int_{t_0}^{t_1} L(\alpha(t), \dot{\alpha}(t)) dt. \tag{2.9}$$

Since M is now an arbitrary smooth manifold rather than $\mathbb{R}^n$, $C_{a,b}^{\infty}([t_0, t_1], M)$ no longer has the algebraic structure of an affine space. It does, however, have the structure of an infinite-dimensional *Fréchet manifold*. This sort of structure is thoroughly discussed in the first four sections of [Ham], but we will make no use of it just yet, except to intuitively relate some of the following definitions to familiar objects in the finite-dimensional situation. For example, thinking of the curves in $C_{a,b}^{\infty}([t_0, t_1], M)$ as *points* in some sort of manifold, one can imagine smooth curves in this manifold (that is, "curves of curves"), tangent vectors to such curves of curves, and so on. For instance, a smooth curve in $C_{a,b}^{\infty}([t_0, t_1], M)$ through some point $\alpha \in C_{a,b}^{\infty}([t_0, t_1], M)$ is what we will now call a "variation" of α.

For any $\alpha \in C_{a,b}^{\infty}([t_0, t_1], M)$ we define a *(fixed endpoint) variation of* α to be a smooth map

$$\Gamma : [t_0, t_1] \times (-\epsilon, \epsilon) \to M$$

for some $\epsilon > 0$ such that

$$\Gamma(t, 0) = \alpha(t), \quad t_0 \le t \le t_1,$$

$$\Gamma(t_0, s) = \alpha(t_0) = a, \quad -\epsilon < s < \epsilon,$$
$$\Gamma(t_1, s) = \alpha(t_1) = b, \quad -\epsilon < s < \epsilon.$$

For any fixed $s \in (-\epsilon, \epsilon)$ the map

$$\gamma_s : [t_0, t_1] \to M$$

defined by

$$\gamma_s(t) = \Gamma(t, s)$$

is an element of $C_{a,b}^{\infty}([t_0, t_1], M)$; γ_s is a "point" along the "curve of curves" represented by the variation Γ. Then $S_L(\gamma_s)$ is a real-valued function of the real variable s whose value at $s = 0$ is $S_L(\alpha)$. We say that $\alpha \in C_{a,b}^{\infty}([t_0, t_1], M)$ is a *stationary, or critical point* of the action functional S_L if

$$\frac{d}{ds} S_L(\gamma_s) \Big|_{s=0} = 0$$

for every variation Γ of α. Intuitively, the rate of change of S_L along every curve in $C_{a,b}^{\infty}([t_0, t_1], M)$ through the point α is zero at α or, better yet, if S_L is thought of as a real-valued function on the manifold $C_{a,b}^{\infty}([t_0, t_1], M)$, its derivative is zero in every direction at α.

For curves α that lie in some coordinate neighborhood in M one can write down explicit equations that are *necessary* conditions for α to be a stationary point of S_L. We will derive these now and then look at some examples. Thus, we suppose (U, ϕ) is a chart on M and denote its coordinate functions $q^1 \ldots, q^n$. The corresponding natural coordinates on $\tilde{U} \subseteq TM$ are denoted $q^1 \ldots, q^n, \dot{q}^1, \ldots, \dot{q}^n$. We consider a smooth curve $\alpha : [t_0, t_1] \to U \subseteq M$ in $C_{a,b}^{\infty}([t_0, t_1], M)$ whose image lies in U. The lift of α to $TU \subseteq TM$ is written in these natural coordinates as $\tilde{\alpha}(t) = (q^1(t) \ldots, q^n(t), \dot{q}^1(t), \ldots, \dot{q}^n(t))$, where we recall that $q^i(t)$ is a notational shorthand for $q^i(\alpha(t))$ and similarly $\dot{q}^i(t)$ means $\dot{q}^i(\dot{\alpha}(t))$. Now we construct some specific variations of α.

Remark 2.2.3. We are looking for *necessary* conditions for stationary points so we are free to select any particular variations we choose.

Let $h : [t_0, t_1] \to \mathbb{R}^n$ be any smooth map which satisfies $h(t_0) = h(t_1) = 0$ and write the coordinate functions of h as $h(t) = (h^1(t), \ldots, h^n(t))$. For $\epsilon > 0$ sufficiently small and $-\epsilon < s < \epsilon$, $(q^1(t) + sh^1(t), \ldots, q^n(t) + sh^n(t))$ will be in the open set $\phi(U)$ and this gives a variation of α whose lift is given in natural coordinates by $(q^1(t) + sh^1(t), \ldots, q^n(t) + sh^n(t), \dot{q}^1(t) + s\dot{h}^1(t), \ldots, \dot{q}^n(t) + s\dot{h}^n(t))$. To ease the typography a bit we will write this as $(\alpha(t) + sh(t), \dot{\alpha}(t) + s\dot{h}(t))$. Thus,

$$\frac{d}{ds} S_L(\gamma_s) \Big|_{s=0} = \frac{d}{ds} \int_{t_0}^{t_1} L(\alpha(t) + sh(t), \dot{\alpha}(t) + s\dot{h}(t)) \, dt \Big|_{s=0}$$

$$= \int_{t_0}^{t_1} \frac{d}{ds} L(\alpha(t) + sh(t), \dot{\alpha}(t) + s\dot{h}(t)) \Big|_{s=0} dt$$

$$= \int_{t_0}^{t_1} \left(\frac{\partial L}{\partial q^k}(\alpha(t), \dot{\alpha}(t)) h^k(t) + \frac{\partial L}{\partial \dot{q}^k}(\alpha(t), \dot{\alpha}(t)) \dot{h}^k(t) \right) dt$$

$$= \int_{t_0}^{t_1} \left[\frac{\partial L}{\partial q^k}(\alpha(t), \dot{\alpha}(t)) - \frac{d}{dt}\left(\frac{\partial L}{\partial \dot{q}^k}(\alpha(t), \dot{\alpha}(t)) \right) \right] h^k(t)\, dt,$$

so, appealing to Lemma 2.2.1, we conclude that if α is a stationary point of S_L, then

$$\frac{\partial L}{\partial q^k}(\alpha(t), \dot{\alpha}(t)) - \frac{d}{dt}\left(\frac{\partial L}{\partial \dot{q}^k}(\alpha(t), \dot{\alpha}(t)) \right) = 0, \quad 1 \le k \le n. \qquad (2.10)$$

These are the famous *Euler–Lagrange equations*, which one often sees written simply as

$$\frac{\partial L}{\partial q^k} - \frac{d}{dt}\left(\frac{\partial L}{\partial \dot{q}^k} \right) = 0, \quad 1 \le k \le n. \qquad (2.11)$$

These equations are necessarily satisfied along any stationary curve for S_L whose image lies in *any* local coordinate neighborhood U. By compactness, we can cover the image of any stationary curve $\alpha(t)$, $t_0 \le t \le t_1$, by finitely many coordinate neighborhoods and the Euler–Lagrange equations are satisfied on each so it is customary to say simply that they are satisfied "on α."

Note that the derivation of the Euler–Lagrange equations was carried out for an *arbitrary* local coordinate system $(q^1, \ldots, q^n)$ on M so, unlike Newton's second law, these equations take exactly the same form in *every* coordinate system. This coordinate independence is one of their great advantages. It is instructive to check this with a direct computation.

Exercise 2.2.2. Let $(Q^1, \ldots, Q^n)$ be another local coordinate system on M defined on an open set that intersects the domain of $(q^1, \ldots, q^n)$. On this intersection transform the Euler–Lagrange equations (2.11) to the new local coordinates $(Q^1, \ldots, Q^n)$ and show that the resulting equations are equivalent to

$$\frac{\partial L}{\partial Q^k} - \frac{d}{dt}\left(\frac{\partial L}{\partial \dot{Q}^k} \right) = 0, \quad 1 \le k \le n.$$

Note that we have *not* asserted that stationary curves joining any two points in M must exist, nor that they are unique even when they do exist and, indeed, neither of these is true in general, as we will see in Example 2.2.4.

Remark 2.2.4. We have defined a Lagrangian to be a function on the tangent bundle TM, but it is sometimes convenient to allow it to depend explicitly on t as well, that is, to define a Lagrangian to be a smooth map $L : \mathbb{R} \times TM \to \mathbb{R}$. Then, for any path in the domain of a coordinate neighborhood on M, one would write $L = L(t, \alpha(t), \dot{\alpha}(t))$ in natural coordinates. The action associated with this path is defined in the same way as the integral of $L(t, \alpha(t), \dot{\alpha}(t))$ over $[t_0, t_1]$. Stationary points for the action are also defined in precisely the same way and a glance back at the calculations leading to the Euler–Lagrange equations shows that the additional t-dependence has no effect at all on the end result, that is, stationary curves satisfy

$$\frac{\partial L}{\partial q^k}(t, \alpha(t), \dot{\alpha}(t)) - \frac{d}{dt}\left(\frac{\partial L}{\partial \dot{q}^k}(t, \alpha(t), \dot{\alpha}(t))\right) = 0, \quad 1 \leq k \leq n.$$

Note that one can draw a rather remarkable conclusion just from the form in which the Euler–Lagrange equations (2.11) are written, namely, that if the Lagrangian L happens not to depend on one of the coordinates, say q^{k_0}, then $\frac{\partial L}{\partial q^{k_0}} = 0$ everywhere and (2.11) implies that, along any stationary curve, $\frac{\partial L}{\partial \dot{q}^{k_0}}$ is constant. In more colloquial terms, $\frac{\partial L}{\partial \dot{q}^{k_0}}$ is *conserved* as the system evolves.

$$\frac{\partial L}{\partial q^{k_0}} = 0 \quad \Longrightarrow \quad \frac{\partial L}{\partial \dot{q}^{k_0}} \text{ is conserved along any stationary path.}$$

This is the simplest instance of one of the most important features of the Lagrangian formalism, that is, the deep connection between the symmetries of a Lagrangian (in this case, its invariance under translations of q^{k_0}) and the existence of quantities that are conserved during the evolution of the system. This feature is entirely absent from the Newtonian picture and we will have more to say about it shortly when we discuss what is called "Noether's theorem." For the moment, however, we would just like to look at a few examples.

Example 2.2.1. Let us have another look at the example that motivated all of this in the first place. For our configuration space we take $M = \mathbb{R}^n$ and choose global standard coordinates on $\mathbb{R}^n$; to emphasize this special choice we will revert to $x^1, \dots, x^n$ for the coordinate functions. The state space is then $T\mathbb{R}^n = \mathbb{R}^n \times \mathbb{R}^n$ and the corresponding natural coordinates are $x^1, \dots, x^n, \dot{x}^1, \dots, \dot{x}^n$. Letting $V(x^1, \dots, x^n)$ denote an arbitrary smooth, real-valued function on $\mathbb{R}^n$ and m a positive constant, we take our Lagrangian to be

$$L(x^1, \dots, x^n, \dot{x}^1, \dots, \dot{x}^n) = \frac{1}{2}m\sum_{i=1}^{n}(\dot{x}^i)^2 - V(x^1, \dots, x^n).$$

To write down the Euler–Lagrange equations we note that

$$\frac{\partial L}{\partial x^i} = -\frac{\partial V}{\partial x^i}, \quad i = 1, \dots, n,$$

and

$$\frac{\partial L}{\partial \dot{x}^i} = m\dot{x}^i, \quad i = 1, \ldots, n.$$

Thus, (2.11) becomes

$$-\frac{\partial V}{\partial x^i} - \frac{d}{dt}(m\dot{x}^i) = 0, \quad i = 1, \ldots, n,$$

that is,

$$m\frac{d^2 x^i}{dt^2} = -\frac{\partial V}{\partial x^i}, \quad i = 1, \ldots, n,$$

and these are, as expected, Newton's second law.

Remark 2.2.5. Although it is merely a very special case of what we have just done, we will record, for future reference, what this looks like for the harmonic oscillator. For this we take the configuration space to be $M = \mathbb{R}$ with standard coordinate q. The state space is therefore $TM = \mathbb{R} \times \mathbb{R}$ with natural coordinates $(q, \dot{q})$. The potential is $V(q) = \frac{1}{2}kq^2$, where $k > 0$ is a constant and the Lagrangian is

$$L(q, \dot{q}) = \frac{1}{2}m\dot{q}^2 - \frac{1}{2}kq^2.$$

The Euler–Lagrange equation is therefore

$$\frac{\partial L}{\partial q} - \frac{d}{dt}\left(\frac{\partial L}{\partial \dot{q}}\right) = 0,$$
$$-kq - m\ddot{q} = 0,$$
$$\ddot{q} + \omega^2 q = 0 \quad (\omega = \sqrt{k/m}),$$

and we are right back where we started in Chapter 1.

For future reference (Sections 7.3 and 7.4) we would like to write out the action for the free particle and the harmonic oscillator along a solution curve.

Exercise 2.2.3. Write the solution to the one-dimensional free particle equation $m\ddot{q} = 0$ as $\alpha(t) = at + b$.

1. Suppose $t_0 < t_1$. Show that the solution $\alpha(t)$ to $m\ddot{q} = 0$ satisfying the boundary conditions $\alpha(t_0) = q_0$ and $\alpha(t_1) = q_1$ is

$$\alpha(t) = \frac{q_1 - q_0}{t_1 - t_0}(t - t_0) + q_0.$$

2. Show that the action $S(\alpha)$ is given by

$$S(\alpha) = \frac{m}{2(t_1 - t_0)}(q_1 - q_0)^2.$$

3. Show that, with $p = m\dot{\alpha}(t)$, the action can be written

$$S(\alpha) = p(q_1 - q_0) - \frac{t_1 - t_0}{2m} p^2.$$

Remark 2.2.6. The function $p(q_1 - q_0) - \frac{t-t_0}{2m} p^2$ will put in another appearance in Section 7.3 when we compute the propagator for a free quantum particle (see (7.44)).

Exercise 2.2.4. Write the solution to the harmonic oscillator equation $\ddot{q} + \omega^2 q = 0$ as $\alpha(t) = A \cos \omega t + B \sin \omega t$.

1. Suppose $T > 0$ and assume ωT is not an integer multiple of π. Show that for the solution $\alpha(t)$ to $\ddot{q} + \omega^2 q = 0$ satisfying the boundary conditions $\alpha(0) = q_0$ and $\alpha(T) = q_T$,

$$A = q_0$$

and

$$B = \frac{q_T - q_0 \cos \omega T}{\sin \omega T}.$$

2. Show that the action $S(\alpha)$ can be written as

$$S(\alpha) = \int_0^T \left[\frac{1}{2} m\dot{\alpha}(t)^2 - \frac{1}{2} m\omega^2 \alpha(t)^2 \right] dt$$

$$= \frac{m\omega}{2} [(B^2 - A^2) \sin \omega T \cos \omega T - 2AB \sin^2 \omega T]$$

$$= \frac{m\omega}{2 \sin \omega T} [(q_0^2 + q_T^2) \cos \omega T - 2q_0 q_T].$$

Remark 2.2.7. We will see the function $\frac{m\omega}{2 \sin \omega T} [(q_0^2 + q_T^2) \cos \omega T - 2q_0 q_T]$ again in Section 7.4 when we compute the propagator for the quantum harmonic oscillator (see (7.61)).

Before leaving Example 2.2.1, let us suppose, for example, that the potential $V(x^1, \ldots, x^n)$ happens not to depend on, say, the ith coordinate x^i so that $\frac{\partial L}{\partial x^i} = 0$ everywhere. Since $\frac{\partial L}{\partial \dot{x}^i} = m\dot{x}^i$, we conclude that $m\dot{x}^i$ is constant along the trajectory of the particle. Now, $m\dot{x}^i$ is what physicists call the ith component of the particle's *(linear) momentum*. Thus, if the potential V is independent of the ith coordinate, then the ith component of momentum is conserved during the motion.

Spatial translation symmetry implies conservation of (linear) momentum

In particular, for a particle that is not subject to any forces (a *free particle*), all of $m\dot{x}^1, \ldots, m\dot{x}^n$ remain unchanged during the motion. This, naturally enough, is called the *conservation of (linear) momentum*.

Motivated by the last example we introduce some terminology that will turn out to be more significant than it might appear at first. Let M be any manifold and let $L : TM \rightarrow \mathbb{R}$ be a Lagrangian on it. If q^i, $i = 1, \ldots, n$, is a local coordinate system on M and if we write L in natural coordinates as $L(q^1, \ldots, q^n, \dot{q}^1, \ldots, \dot{q}^n)$, then

$$p_i = \frac{\partial L}{\partial \dot{q}^i}$$

is called the *momentum conjugate to* q^i, even though it need not correspond to "momentum" in the usual sense at all. We have shown that p_i is conserved along stationary paths if L is independent of q^i and we will soon see that such pairs (q^i, p_i) of so-called *conjugate coordinates* play an essential role in the Hamiltonian formulation of classical mechanics as well as in quantum mechanics; we will also see why the superscript was turned into a subscript (Remark 2.3.1).

Next we would like to write out at least one physically interesting example for which the configuration space is a nontrivial manifold so that local coordinates are actually required.

Example 2.2.2. We will describe what is called the *spherical pendulum*. This is basically the same as the pendulum we discussed in Section 1 except that there is no ceiling and the motion is not restricted to a plane. Specifically, we consider a pendulum with a bob of mass m suspended from a fixed point by a massless string of length l that is set in motion and free to move on a sphere of radius l about this fixed point under the influence of the earth's gravitational field (acceleration g). We will arrange Cartesian coordinate axes x^1, x^2, x^3 with x^3 vertical and the pendulum bob moving on the sphere $(x^1)^2 + (x^2)^2 + (x^3)^2 = l^2$, which is therefore the configuration space M; M is topologically the 2-sphere S^2. The state space TM can be identified with the set of pairs (x, v) in $\mathbb{R}^3 \times \mathbb{R}^3$ with $x \in M$ and $x^1 v^1 + x^2 v^2 + x^3 v^3 = 0$ (the velocity vector of the mass is tangent to the sphere). On M we introduce spherical coordinates denoted

$$q^1 = \phi,$$
$$q^2 = \theta$$

and defined by

$$x^1 = l \sin \phi \cos \theta,$$
$$x^2 = l \sin \phi \sin \theta, \qquad\qquad (2.12)$$
$$x^3 = -l \cos \phi$$

(so ϕ is measured *up* from the negative x^3-axis and θ is measured in the $x^1 x^2$-plane from the positive x^1-axis). By restricting (ϕ, θ) to $(0, \pi) \times (0, 2\pi)$ and then to $(0, \pi) \times (-\pi, \pi)$ one obtains two charts that cover all of M except the north and south poles and these points can be covered by defining analogous coordinates measured from some other coordinate axis. The associated natural coordinates on TM will be denoted $(\phi, \theta, \dot{\phi}, \dot{\theta})$. The Lagrangian L is taken to be the kinetic minus the potential energy associated with any state. In rectangular coordinates, the kinetic energy is just $\frac{1}{2}m\left((\dot{x}^1)^2 + (\dot{x}^2)^2 + (\dot{x}^3)^2\right)$. The potential is taken to be mgx^3 (keep in mind that the potential is defined only up to an additive constant). Using (2.12) to convert the Lagrangian to spherical coordinates gives

$$L = \frac{1}{2}m\left((\dot{x}^1)^2 + (\dot{x}^2)^2 + (\dot{x}^3)^2\right) - mgx^3$$
$$= \frac{1}{2}ml^2\left(\dot{\phi}^2 + \dot{\theta}^2 \sin^2 \phi\right) + mgl \cos \phi.$$

Note that the Cartesian coordinate version is independent of x^1 and x^2 so we already know that the x^1- and x^2-components of the linear momentum are conserved. In spherical coordinates we have

$$\frac{\partial L}{\partial q^1} = \frac{\partial L}{\partial \phi} = ml^2\dot{\theta}^2 \sin \phi \cos \phi - mgl \sin \phi,$$

$$\frac{\partial L}{\partial q^2} = \frac{\partial L}{\partial \theta} = 0,$$

$$\frac{\partial L}{\partial \dot{q}^1} = \frac{\partial L}{\partial \dot{\phi}} = ml^2\dot{\phi},$$

$$\frac{\partial L}{\partial \dot{q}^2} = \frac{\partial L}{\partial \dot{\theta}} = ml^2\dot{\theta} \sin^2 \phi.$$

The $k = 1$ Euler–Lagrange equation (2.11) therefore becomes

$$\ddot{\phi} + \omega^2 \sin \phi = \dot{\theta}^2 \sin \phi \cos \phi, \tag{2.13}$$

where, as we did for the simple pendulum in Chapter 1, we have written ω^2 for g/l. Note, incidentally, that when $\dot{\theta} = 0$, this reduces to the simple pendulum equation (1.4), as it should.

Since L is independent of θ and $\frac{\partial L}{\partial \dot{\theta}} = ml^2\dot{\theta} \sin^2 \phi$ we conclude that

$$ml^2\dot{\theta} \sin^2 \phi$$

is conserved during the motion and therefore so is $\dot{\theta} \sin^2 \phi$. One might (indeed, should) wonder about the physical interpretation of any quantity that is conserved during the evolution of a physical system. What exactly is $ml^2\dot{\theta} \sin^2 \phi$ and why should it remain constant during the motion? We will answer this question soon (see (2.21)), but for the moment we would like to simply record two remarks. First note that a bit of playing around with (2.12) shows that

$$\dot{\theta} \sin^2 \phi = x^1\dot{x}^2 - x^2\dot{x}^1.$$

Next observe that, since L is independent of θ, the Lagrangian is invariant under rotations in the x^1x^2-plane. We will see soon that this "symmetry" of the Lagrangian and the fact that $x^1\dot{x}^2 - x^2\dot{x}^1$ is conserved during the motion are intimately related by what is called "Noether's theorem."

Particles moving in space that are constrained to remain on some surface (for example, the spherical pendulum bob) are so constrained by various forces acting on them (for example, string tension). Such constraints can be imposed in a variety of ways and their precise physical nature can be quite complicated. One of the beauties of the Lagrangian (and Hamiltonian) formalism is that, whatever their physical nature, such constraints can often be incorporated directly by simply decreeing that the configuration space is the surface to which the particles are constrained. This assertion is generally known as *d'Alembert's principle* and is assumed by physicists to hold whenever the constraint forces are *holonomic*, which means that they do no work (for example, when they are normal to the constraint surface, as in the case of string tension). As a reality check, one should perhaps compute the equations of motion for the spherical pendulum the "old fashioned" way ($\mathbf{F} = m\mathbf{A}$) to see that two of the three components reduce to our Euler–Lagrange equations and the third simply says what the constraint force must be to keep the bob on the sphere (however this is accomplished physically). This is actually a pretty routine (albeit messy) exercise so we will simply sketch the procedure in the next example and leave the calculus and algebra to those who feel morally obligated to supply it.

Exercise 2.2.5. Fill in the details of the following example.

Example 2.2.3. We will not specify what the constraint force is, but only that it is normal to the sphere. We would like to keep this as close to a calculus experience as possible so, for this example, we will write (x, y, z) for the Cartesian coordinates in $\mathbb{R}^3$, while the usual spherical coordinates in space will be denoted (ρ, ϕ, θ); they are related by

$$x = \rho \sin \phi \cos \theta,$$
$$y = \rho \sin \phi \sin \theta,$$
$$z = \rho \cos \phi.$$

We will use $\hat{e}_x, \hat{e}_y, \hat{e}_z$ and $\hat{e}_\rho, \hat{e}_\phi, \hat{e}_\theta$ for the unit vector fields in the x-, y-, z- and ρ-, ϕ- and θ directions, respectively, at each point. These are related at each point by

$$\hat{e}_\rho = (\sin \phi \cos \theta)\,\hat{e}_x + (\sin \phi \sin \theta)\,\hat{e}_y + (\cos \phi)\,\hat{e}_z,$$
$$\hat{e}_\phi = (\cos \phi \cos \theta)\,\hat{e}_x + (\cos \phi \sin \theta)\,\hat{e}_y - (\sin \phi)\,\hat{e}_z,$$
$$\hat{e}_\theta = (-\sin \theta)\,\hat{e}_x + (\cos \theta)\,\hat{e}_y$$

and

$$\hat{e}_x = (\sin \phi \cos \theta)\,\hat{e}_\rho + (\cos \phi \cos \theta)\,\hat{e}_\phi - (\sin \theta)\,\hat{e}_\theta,$$

$$\hat{e}_y = (\sin \phi \, \sin \theta) \, \hat{e}_\rho + (\cos \phi \, \sin \theta) \, \hat{e}_\phi + (\cos \theta) \, \hat{e}_\theta,$$
$$\hat{e}_z = (\cos \phi) \, \hat{e}_\rho - (\sin \phi) \, \hat{e}_\phi.$$

Write the Cartesian components of acceleration as $\mathbf{A} = \ddot{x}\hat{e}_x + \ddot{y}\hat{e}_y + \ddot{z}\hat{e}_z$. With $\rho = l$ so that $\dot{\rho} = \ddot{\rho} = 0$ these components are given in spherical coordinates by

$$\ddot{x} = -2l \cos \phi \, \sin \theta \, \dot{\phi}\dot{\theta} - l \sin \phi \, \sin \theta \, \ddot{\theta} + l \cos \phi \, \cos \theta \, \ddot{\phi} - l \sin \phi \, \cos \theta \, (\dot{\phi}^2 + \dot{\theta}^2),$$
$$\ddot{y} = 2l \cos \phi \, \cos \theta \, \dot{\phi}\dot{\theta} + l \sin \phi \, \cos \theta \, \ddot{\theta} + l \cos \phi \, \sin \theta \, \ddot{\phi} - l \sin \phi \, \sin \theta \, (\dot{\phi}^2 + \dot{\theta}^2),$$
$$\ddot{z} = -l \sin \phi \, \ddot{\phi} - l \cos \phi \, \dot{\phi}^2.$$

Now, for any force $\mathbf{F}$ acting on m, the ρ-, ϕ- and θ-components of $\mathbf{F} = m\mathbf{A}$ are

$$\mathbf{F} \cdot \hat{e}_\rho = m\mathbf{A} \cdot \hat{e}_\rho = (m \sin \phi \, \cos \theta)\ddot{x} + (m \sin \phi \, \sin \theta)\ddot{y} + (m \cos \phi)\ddot{z},$$
$$\mathbf{F} \cdot \hat{e}_\phi = m\mathbf{A} \cdot \hat{e}_\phi = (m \cos \phi \, \cos \theta)\ddot{x} + (m \cos \phi \, \sin \theta)\ddot{y} - (m \sin \phi)\ddot{z},$$
$$\mathbf{F} \cdot \hat{e}_\theta = m\mathbf{A} \cdot \hat{e}_\theta = -(m \sin \theta)\ddot{x} + (m \cos \theta)\ddot{y}.$$

Now assume that $\mathbf{F} = F_\rho \hat{e}_\rho - mg\hat{e}_z$, where $F_\rho \hat{e}_\rho$ is the radial constraint force holding m on the sphere. Then

$$\mathbf{F} = (F_\rho - mg \, \cos \phi)\hat{e}_\rho - (mg \, \sin \phi)\hat{e}_\phi.$$

Writing out the ϕ-component of $\mathbf{F} = m\mathbf{A}$ and simplifying gives

$$\ddot{\phi} + \omega^2 \sin \phi = \dot{\theta}^2 \sin \phi \, \cos \phi,$$

where $\omega^2 = g/l$ and this is (2.13). Similarly, the θ-component of $\mathbf{F} = m\mathbf{A}$ gives

$$(\sin \phi)\ddot{\theta} + (2 \cos \phi)\dot{\phi}\dot{\theta} = 0,$$

which is equivalent to our conservation law

$$\frac{d}{dt}(ml^2 \dot{\theta} \sin^2 \phi) = 0.$$

Finally, the ρ-component of $\mathbf{F} = m\mathbf{A}$ is the only one that involves F_ρ and can simply be solved for F_ρ and therefore regarded as a specification of what $F_\rho \hat{e}_\rho$ must be in order for m to remain on the sphere.

Since (holonomic) constraints are built into Lagrangian mechanics through the choice of the configuration space, the notion of a "constraining force" (like string tension) essentially disappears from the picture. As a result, it makes sense to discuss particle motion that is "free" *except* for whatever is constraining the particle to remain in the configuration space. The Lagrangian is simply the kinetic energy, that is, $\frac{1}{2}m$ times the squared magnitude of the velocity vector. Note that the velocity vectors are now tangent vectors to the configuration manifold so one can generalize this scenario from constraint surfaces in space to any manifold in which each tangent space is provided with an inner product with which to compute these squared magnitudes, that is, to *Riemannian manifolds* (see Remark 2.2.8 below). Although it is not our practice here to strive for optimal generality, this particular example is worth doing generally since it brings us face-to-face with a reinterpretation of a very fundamental notion in differential geometry and provides insight into the nature of stationary curves.

Remark 2.2.8. A *Riemannian metric* on a manifold M is just an assignment to each tangent space $T_p(M)$ of a positive definite inner product $\langle\,,\,\rangle_p$ that varies smoothly with p in the sense that if X is a smooth vector field on M, then $\langle X(p), X(p)\rangle_p$ is a smooth real-valued function on M. This then gives rise to a smooth, real-valued function on the tangent bundle which assigns to every (p, v_p) the squared magnitude $\langle v_p, v_p\rangle_p$ of v_p. A manifold together with a fixed Riemannian metric is called a *Riemannian manifold*. Riemannian metrics are introduced in Section 5.11 of [Nab3]. A much more detailed introduction to Riemannian geometry can be found in [Lee1] or Chapter 9 of [Sp2]. In the following example we will view Riemannian metrics simply as particular types of Lagrangians and will require no information about them except that they exist on every smooth manifold (Theorem 4, Chapter 9, of [Sp2]).

Example 2.2.4. Here we will discuss *free motion on a Riemannian manifold*, that is, *free motion with constraints*. Specifically, our configuration space is an arbitrary smooth, n-dimensional manifold M equipped with a Riemannian metric $\langle\,,\,\rangle_p$. The Lagrangian $L : TM \to \mathbb{R}$ is defined by $L(p, v_p) = \frac{1}{2}m\langle v_p, v_p\rangle_p$ for each $(p, v_p) \in TM$, where m is some positive constant; this is often called the *kinetic energy metric* on M. If $q^1,\ldots,q^n$ are local coordinates on M, then, in the corresponding natural coordinates on TM,

$$L(q, \dot q) = \frac{1}{2}mg_{ij}(q)\dot q^i \dot q^j \quad \text{(summation convention)} \tag{2.14}$$

for some positive definite, symmetric matrix $(g_{ij}(q))$ of smooth functions on the open subset of M on which $q^1,\ldots,q^n$ are defined. From this we compute

$$\frac{\partial L}{\partial \dot q^i} = mg_{ij}(q)\dot q^j,$$

$$\frac{\partial L}{\partial q^i} = \frac{1}{2}m\frac{\partial g_{kj}}{\partial q^i}\dot q^k \dot q^j,$$

and, along any smooth curve in M,

$$\frac{d}{dt}\left(\frac{\partial L}{\partial \dot q^i}\right) = mg_{ij}\ddot q^j + m\frac{\partial g_{ij}}{\partial q^k}\dot q^k \dot q^j.$$

Thus, the Euler–Lagrange equations become

$$g_{ij}\ddot q^j + \frac{\partial g_{ij}}{\partial q^k}\dot q^k \dot q^j - \frac{1}{2}\frac{\partial g_{kj}}{\partial q^i}\dot q^k \dot q^j = 0, \quad i = 1,\ldots,n.$$

We interchange k and j in this last equation to get

$$g_{ij}\ddot{q}^j + \frac{\partial g_{ik}}{\partial q^j}\dot{q}^k\dot{q}^j - \frac{1}{2}\frac{\partial g_{kj}}{\partial q^i}\dot{q}^k\dot{q}^j = 0, \quad i = 1,\ldots,n.$$

Now we add the last two equations and divide by 2 to get

$$g_{ij}\ddot{q}^j + \frac{1}{2}\left(\frac{\partial g_{ik}}{\partial q^j} + \frac{\partial g_{ij}}{\partial q^k} - \frac{\partial g_{kj}}{\partial q^i}\right)\dot{q}^k\dot{q}^j = 0, \quad i = 1,\ldots,n.$$

The matrix (g_{ij}) is invertible and we will denote its inverse by (g^{ij}). Thus, if we multiply the last equation by g^{li} and sum over $i = 1,\ldots,n$ the result is

$$\ddot{q}^l + \frac{1}{2}g^{li}\left(\frac{\partial g_{ik}}{\partial q^j} + \frac{\partial g_{ij}}{\partial q^k} - \frac{\partial g_{kj}}{\partial q^i}\right)\dot{q}^k\dot{q}^j = 0, \quad l = 1,\ldots,n.$$

The coefficient of $\dot{q}^k\dot{q}^j$ is generally denoted Γ^l_{kj} and called a *Christoffel symbol* for the given Riemannian metric. With this the Euler–Lagrange equations assume the form

$$\ddot{q}^l + \Gamma^l_{kj}\dot{q}^k\dot{q}^j = 0, \quad l = 1,\ldots,n. \tag{2.15}$$

These are the familiar *geodesic equations* of Riemannian geometry. Their solutions, that is, the stationary curves for the Lagrangian L, are called the *geodesics* of the Riemannian manifold M. For certain Riemannian manifolds the equations can be solved explicitly. The geodesics of $\mathbb{R}^n$ with its standard Riemannian metric, for example, are the arc length parametrizations of straight lines so, in particular, there is a unique stationary curve joining any two points and, moreover, this curve has minimal length among all curves joining these two points. For the sphere S^n with the metric it inherits from $\mathbb{R}^{n+1}$ the geodesics are the arc length parametrizations of the great circles so any two points are joined by *two* stationary curves and, unless the points are diametrically opposite, only one of them minimizes length. On the other hand, the geodesics of the punctured plane $\mathbb{R}^2 - \{(0,0)\}$ with the metric inherited from $\mathbb{R}^2$ are still unit speed straight lines so, for example, $(-1,0)$ and $(1,0)$ cannot be joined by any stationary curve.

All of the calculations in the preceding example are equally valid if $\langle\,,\,\rangle_p$ is only assumed to be a nondegenerate, symmetric, bilinear form on $T_p(M)$, but not necessarily positive definite. A manifold equipped with such a $\langle\,,\,\rangle_p$ at each $p \in M$, varying smoothly with p, is called a *semi-Riemannian manifold* and these are of fundamental importance in many aspects of mathematical physics, particularly general relativity. They also have geodesics, but the analysis and interpretation of these is much more subtle (see [Nab1] for a brief encounter with this and [O'N] for a more thorough treatment).

Exercise 2.2.6. Let M be an arbitrary manifold. Another type of geometrical object on M that can be thought of as a Lagrangian is a 1-form θ. These are simply smooth real-valued functions on TM that are linear on each fiber $\pi^{-1}(p) \cong \{p\} \times T_p(M)$ so that, in any local coordinate system, $\theta(q, \dot{q}) = \theta_i(q)\dot{q}^i$ for some smooth real-valued functions $\theta_i(q) = \theta_i(q^1, \ldots, q^n)$, $i = 1, \ldots, n$.

1. Show that the Euler–Lagrange equations for the Lagrangian θ can be written in the form

$$\iota_{\dot{\alpha}(t)}\, d\theta = 0,$$

where $d\theta$ is the exterior derivative of θ and $\iota_{\dot{\alpha}(t)}\, d\theta$ is the interior product (contraction) of the 2-form $d\theta$ with $\dot{\alpha}(t)$, that is, $(\iota_{\dot{\alpha}(t)}\, d\theta)(v(t)) = d\theta\,(\dot{\alpha}(t), v(t))$ for any tangent vector $v(t)$ to M at $\alpha(t)$.

2. Let $L : TM \to \mathbb{R}$ be an arbitrary Lagrangian on M and let θ be a 1-form. Define a new Lagrangian $L' : TM \to \mathbb{R}$ on M by $L' = L + \theta + c$, where c is a real constant. Show that L and L' have the same Euler–Lagrange equations if and only if θ is a closed 1-form (that is, if and only if $d\theta = 0$). In particular, if $g : M \to \mathbb{R}$ is any smooth, real-valued function on M, then L and $L + dg$ have the same Euler–Lagrange equations.

Next we would like to look at an example of a slightly different sort.

Example 2.2.5. We mentioned in Remark 2.2.4 that the Euler–Lagrange equations are still satisfied even if one allows the Lagrangian to depend explicitly on time t. We have chosen not to do this; our Lagrangians are all functions on TM. Thought of somewhat differently, we are really considering only "time-dependent" Lagrangians $L(t, q^1, \ldots, q^n, \dot{q}^1, \ldots, \dot{q}^n)$ possessing the time translation symmetry $\frac{\partial L}{\partial t} = 0$. We show now that this symmetry also gives rise to a quantity that is conserved along stationary paths and, indeed, to one that we have already seen in a special case.

Remark 2.2.9. Note that there is no t floating around so we cannot arrive at this conserved quantity as we did in the previous examples by computing "$\partial L / \partial t$."

Write the local coordinates of the stationary curve as $q^1(t), \ldots, q^n(t)$ and the Lagrangian evaluated on (the lift of) this curve as $L(t, q^1(t), \ldots, q^n(t), \dot{q}^1(t), \ldots, \dot{q}^n(t))$. Now compute the rate of change of L along the curve. We have

$$
\begin{aligned}
\frac{dL}{dt} &= \frac{\partial L}{\partial q^i}\frac{dq^i}{dt} + \frac{\partial L}{\partial \dot{q}^i}\frac{d\dot{q}^i}{dt} + \frac{\partial L}{\partial t} \\
&= \frac{d}{dt}\left(\frac{\partial L}{\partial \dot{q}^i}\right)\frac{dq^i}{dt} + \frac{\partial L}{\partial \dot{q}^i}\frac{d}{dt}\left(\frac{dq^i}{dt}\right) + 0 \\
&= \frac{d}{dt}\left(\frac{\partial L}{\partial \dot{q}^i}\frac{dq^i}{dt}\right)
\end{aligned}
$$

$$= \frac{d}{dt}(p_i \dot{q}^i).$$

In other words,

$$\frac{d}{dt}(p_i \dot{q}^i - L) = 0,$$

so $p_i \dot{q}^i - L$ is conserved along a stationary path for *any* time-independent Lagrangian. To see how one should interpret this conserved quantity, let us write it out in the case of the Lagrangian (2.7). For $L(q^1, \ldots, q^n, \dot{q}^1, \ldots, \dot{q}^n) = \sum_{i=1}^n \frac{1}{2} m(\dot{q}^i)^2 - V(q^1, \ldots, q^n)$ we have

$$p_i \dot{q}^i - L = \frac{\partial L}{\partial \dot{q}^i} \dot{q}^i - L = \sum_{i=1}^n m(\dot{q}^i)^2 - L = \sum_{i=1}^n \frac{1}{2} m(\dot{q}^i)^2 + V(q^1, \ldots, q^n)$$

and this is just the total energy (kinetic plus potential). For any time-independent Lagrangian L we define the *total energy* E_L by

$$E_L = p_i \dot{q}^i - L = \frac{\partial L}{\partial \dot{q}^i} \dot{q}^i - L \tag{2.16}$$

and we summarize what we have just shown by saying the following.

Time translation symmetry implies conservation of energy.

The energy function E_L is defined in terms of natural coordinates on TM, but one can check that the definitions agree on the intersection of any two coordinate neighborhoods so E_L is a well-defined real-valued function on TM. Another way of seeing the same thing is to check that the following invariant definition agrees with the coordinate definition. Let R be the vector field on TM that is "radial" on each $T_p(M)$. More explicitly, for each $(p, v_p) \in T(M)$, let $R(p, v_p) = \frac{d}{dt}(p, tv_p)|_{t=0}$; in local coordinates, $R = \dot{q}^i \partial_{\dot{q}^i}$. Then $E_L = dL(R) - L$.

One could continue this list of examples indefinitely, but this is not really our business here (many more are available in, for example, [Arn2], [Sp3] and [Gold]). Next we would like to look into the relation to which we alluded earlier between "symmetries" and conserved quantities. We will define first a "symmetry" of a Lagrangian L and then an "infinitesimal symmetry" of L; it is the latter notion that is related to conservation laws by Noether's theorem. To motivate the definitions we first consider a few simple examples.

Example 2.2.6. We consider a free particle moving in $\mathbb{R}^n$. Thus, the configuration space is $M = \mathbb{R}^n$, on which we choose standard coordinates $x^1, \ldots, x^n$. The state space is $T\mathbb{R}^n = \mathbb{R}^n \times \mathbb{R}^n$ with natural coordinates $x^1, \ldots, x^n, \dot{x}^1, \ldots, \dot{x}^n$, and the Lagrangian is just $L(x, \dot{x}) = \frac{1}{2} m((\dot{x}^1)^2 + \cdots + (\dot{x}^n)^2) = \frac{1}{2} m \|\dot{x}\|^2$ because the particle is free. Now fix an $a = (a^1, \ldots, a^n)$ in $\mathbb{R}^n$ and define a map

$$F_a : \mathbb{R}^n \to \mathbb{R}^n$$

by

$$F_a(x) = x + a$$

for every $x \in \mathbb{R}^n$ (translation by a). Then F_a is a diffeomorphism and its derivative $(F_a)_{*x} : T_x(\mathbb{R}^n) \to T_{x+a}(\mathbb{R}^n)$ at any x is just the identity map when both tangent spaces are canonically identified with $\mathbb{R}^n$. Thus, we have an induced map on the state space

$$TF_a : TM = \mathbb{R}^n \times \mathbb{R}^n \to TM = \mathbb{R}^n \times \mathbb{R}^n$$

given by

$$(TF_a)(x, \dot{x}) = (F_a(x), (F_a)_{*x}(\dot{x})) = (x + a, \dot{x}).$$

This is a diffeomorphism of TM onto TM that clearly satisfies

$$L \circ TF_a = L$$

and it is this property that will qualify F_a as a symmetry of L once we have formulated the precise definition. One says simply that L is *invariant under spatial translation*.

Example 2.2.7. Next we will consider a particle moving in a spherically symmetric potential in $\mathbb{R}^3$. More precisely, our configuration space is $M = \mathbb{R}^3$ on which we again choose standard coordinates x^1, x^2, x^3, the state space is $TM = \mathbb{R}^3 \times \mathbb{R}^3$ with natural coordinates $x^1, x^2, x^3, \dot{x}^1, \dot{x}^2, \dot{x}^3$ and we take as our Lagrangian

$$L(x, \dot{x}) = \frac{1}{2}m\|\dot{x}\|^2 - V(\|x\|), \tag{2.17}$$

where V is a smooth function on $\mathbb{R}^3$ that depends only on $\|x\|$. Now fix an element g of the rotation group SO(3), that is, a 3×3 matrix that is orthogonal ($g^T g = \mathrm{id}_{3\times 3}$) and has $\det(g) = 1$. Define a map

$$F_g : \mathbb{R}^3 \to \mathbb{R}^3$$

by

$$F_g(x) = g \cdot x,$$

where $g \cdot x$ means matrix multiplication with x thought of as a column vector. Since g is invertible, F_g is a diffeomorphism of M onto M. Moreover, since F_g is linear, its derivative at each point is the same linear map (multiplication by g) once the tangent spaces are canonically identified with $\mathbb{R}^3$. Thus, the induced map on state space

$$TF_g : TM = \mathbb{R}^3 \times \mathbb{R}^3 \to TM = \mathbb{R}^3 \times \mathbb{R}^3$$

is given by

$$(TF_g)(x, \dot{x}) = \big(F_g(x), (F_g)_{*x}(\dot{x})\big) = (g \cdot x, g \cdot \dot{x}).$$

This is again a diffeomorphism of TM onto TM. Moreover, since g is orthogonal, $\|g \cdot x\| = \|x\|$ and $\|g \cdot \dot{x}\| = \|\dot{x}\|$ so, once again,

$$L \circ TF_g = L$$

and we will say that L is *invariant under rotation* and that F_g is a symmetry of L for every $g \in SO(3)$.

The general definition is as follows. If $L : TM \rightarrow \mathbb{R}$ is a Lagrangian on a smooth manifold M, then a *symmetry* of L is a diffeomorphism

$$F : M \rightarrow M$$

of M onto itself for which the induced map

$$TF : TM \rightarrow TM$$

given by

$$TF(p, v_p) = \big(F(p), F_{*p}(v_p)\big)$$

satisfies

$$L \circ TF = L.$$

Symmetries often arise from the action of a Lie group G on the configuration space M. Recall that a *(left) action* of G on M is a smooth map $\sigma : G \times M \rightarrow M$, usually written $\sigma(g, p) = g \cdot p$, that satisfies $e \cdot p = p \,\forall p \in M$, where e is the identity element of G, and $g_1 \cdot (g_2 \cdot p) = (g_1 g_2) \cdot p$ for all $g_1, g_2 \in G$ and all $p \in M$. Given such an action one can define, for each $g \in G$, a diffeomorphism $\sigma_g : M \rightarrow M$ by $\sigma_g(p) = g \cdot p$. If a Lagrangian is given on M, then it *may* be possible to find a Lie group G and an action σ of G on M for which these diffeomorphisms are symmetries. This was the case for both of the previous examples; in the first, G was the additive group $\mathbb{R}^3$, thought of as the translation group of $\mathbb{R}^3$, while in the second it was $SO(3)$. Topological groups, group actions and Lie groups are discussed in Sections 1.6 and 5.8 of [Nab3].

The theorem of Noether to which we have alluded several times refers not to symmetries of the Lagrangian, but rather to "infinitesimal" symmetries. Again, we precede the precise definition with a simple, but very important example.

Example 2.2.8. We will continue the discussion of a particle moving in a spherically symmetric potential in $\mathbb{R}^3$ and will use the notation established in Example 2.2.7. We will also need a more explicit description of the elements of the rotation group SO(3). The following result, which essentially says that the exponential map on the Lie algebra of SO(3) is surjective, is proved on pages 393–395 of [Nab3]. We will denote by $\mathfrak{so}(3)$ the Lie algebra of SO(3), that is, the set of all 3×3, skew-symmetric, real matrices with entrywise linear operations and matrix commutator as bracket (see Section 5.8 of [Nab3] for Lie algebras).

Theorem 2.2.2. *Let A be an element of $\mathfrak{so}(3)$. Then the matrix exponential e^A is in SO(3). Conversely, if g is any element of SO(3), then there is a unique $t \in [0, \pi]$ and a unit vector $\hat{n} = (n^1, n^2, n^3)$ in $\mathbb{R}^3$ for which*

$$g = e^{tN} = \mathrm{id}_{3 \times 3} + (\sin t)N + (1 - \cos t)N^2,$$

where N is the element of $\mathfrak{so}(3)$ given by

$$N = \begin{pmatrix} 0 & -n^3 & n^2 \\ n^3 & 0 & -n^1 \\ -n^2 & n^1 & 0 \end{pmatrix}.$$

Geometrically, one thinks of $g = e^{tN}$ as the rotation of $\mathbb{R}^3$ through t radians about an axis along $\hat{n}$ in a sense determined by the right-hand rule from the direction of $\hat{n}$. Now fix an $\hat{n}$ and the corresponding N in $\mathfrak{so}(3)$. For any $x \in \mathbb{R}^3$,

$$t \rightarrow e^{tN} \cdot x$$

is a curve in $\mathbb{R}^3$ passing through x at $t = 0$ with velocity vector

$$\frac{d}{dt}(e^{tN} \cdot x)\Big|_{t=0} = N \cdot x.$$

Doing this for each $x \in \mathbb{R}^3$ gives a smooth vector field X_N on $\mathbb{R}^3$ defined by

$$X_N(x) = \frac{d}{dt}(e^{tN} \cdot x)\Big|_{t=0} = N \cdot x.$$

Like any (complete) vector field on $\mathbb{R}^3$, X_N determines a *one-parameter group of diffeomorphisms*

$$\varphi_t : \mathbb{R}^3 \rightarrow \mathbb{R}^3, \quad -\infty < t < \infty,$$

where φ_t pushes each point of $\mathbb{R}^3$ t units along the integral curve of X_N that starts there (this one-parameter group of diffeomorphisms is also called the *flow* of the vector field). In this case,

$$\varphi_t(x) = e^{tN} \cdot x.$$

Remark 2.2.10. Integral curves and one-parameter groups of diffeomorphisms are discussed on pages 270–275 of [Nab3].

Note that each φ_t, being multiplication by some element of SO(3), is a symmetry of L by Example 2.2.7. This, according to the definition we will formulate in a moment, makes the vector field X_N an "infinitesimal symmetry" of L. One can think of it intuitively as an object that determines not a single symmetry of L, but rather a one-parameter family of symmetries. We will conclude this example by writing a few of these vector fields out explicitly. Choose, for example, $\hat{n} = (0,0,1) \in \mathbb{R}^3$. Then

$$N = \begin{pmatrix} 0 & -1 & 0 \\ 1 & 0 & 0 \\ 0 & 0 & 0 \end{pmatrix},$$

so

$$X_N(x) = N \cdot x = \begin{pmatrix} -x^2 \\ x^1 \\ 0 \end{pmatrix}.$$

We conclude then that the vector field X_N is just

$$X_{12} = x^1 \partial_{x^2} - x^2 \partial_{x^1},$$

which is generally referred to as the *infinitesimal generator for rotations in the $x^1 x^2$-plane*. Taking $\hat{n}$ to be $(1,0,0)$ and $(0,1,0)$ one obtains, in the same way, vector fields

$$X_{ij} = x^i \partial_{x^j} - x^j \partial_{x^i}, \quad i,j = 1,2,3, \ i \ne j,$$

which is called the *infinitesimal generator for rotations in the $x^i x^j$-plane*.

The general definition of an infinitesimal symmetry is complicated just a bit by the fact that, unlike the examples we have discussed thus far, not every vector field on a smooth manifold is complete, that is, has integral curves defined for all $t \in \mathbb{R}$. For such vector fields one has only a *local* one-parameter group of diffeomorphisms (see pages 272–273 of [Nab3]). In order not to cloud the essential issues we will give the definition twice, once for vector fields that are complete and once for those that need not be complete (naturally, the first definition is a special case of the second). Let L be a Lagrangian on a smooth manifold M. A complete vector field X on M is said to be an *infinitesimal symmetry* of L if each φ_t in its one-parameter group of diffeomorphisms is a symmetry of L. Now we drop the assumption that X is complete. For each $p \in M$, let α_p be the maximal integral curve of X through p (see Theorem 5.7.2 of [Nab3]). For each $t \in \mathbb{R}$, let $\mathcal{D}_t$ be the set of all $p \in M$ for which α_p is defined at t and define $\varphi_t : \mathcal{D}_t \to M$ by $\varphi_t(p) = \alpha_p(t)$. By Theorem 5.7.4 of [Nab3], each $\mathcal{D}_t$ is an open set (perhaps empty) and φ_t is a diffeomorphism of $\mathcal{D}_t$ onto $\mathcal{D}_{-t}$ with inverse φ_{-t}.

Now we will say that X is an *infinitesimal symmetry* of L if, for each t with $\mathcal{D}_t \neq \emptyset$, the induced map $T\varphi_t : T\mathcal{D}_t \to T\mathcal{D}_{-t}$, defined by $(T\varphi_t)(p, v_p) = (\varphi_t(p), (\varphi_t)_{*p}(v_p))$, satisfies $L \circ T\varphi_t = L$ on $T\mathcal{D}_t$.

Our objective is to show that every infinitesimal symmetry X of L on M gives rise to a "conserved quantity," that is, a function that is constant along every stationary curve. For this we need to write a bit more explicitly what it means for X to be an infinitesimal symmetry. Let $q^1, \ldots, q^n$ be local coordinates on the open set $U \subseteq M$ and $q^1, \ldots, q^n, \dot{q}^1, \ldots, \dot{q}^n$ the corresponding natural coordinates on $TU \subseteq TM$. In these coordinates we write the vector field X as

$$X = X^i \partial_{q^i},$$

where $X^i = X^i(q^1, \ldots, q^n)$, $i = 1, \ldots, n$, are the local component functions of X. Each local diffeomorphism φ_t lifts to the tangent bundle by $(T\varphi_t)(p, v_p) = (\varphi_t(p), (\varphi_t)_{*p}(v_p))$. For each fixed (p, v_p), the curve $t \to (T\varphi_t)(p, v_p)$ lifts the integral curves of X through p. These curves determine a vector field $\tilde{X}$ on the tangent bundle whose value at any point (p, v_p) is the tangent vector to $t \to (T\varphi_t)(p, v_p)$ at (p, v_p).

Exercise 2.2.7. Show that, in natural coordinates $q^1, \ldots, q^n, \dot{q}^1, \ldots, \dot{q}^n$, this vector field is

$$\tilde{X} = X^i \partial_{q^i} + \left(\frac{\partial X^i}{\partial q^j} \dot{q}^j \right) \partial_{\dot{q}^i}.$$

Now, by definition, X is an infinitesimal symmetry of L if and only if the rate of change of L along each integral curve of this lifted vector field is zero, that is,

$$\tilde{X}L = \left(X^i \partial_{q^i} + \left(\frac{\partial X^i}{\partial q^j} \dot{q}^j \right) \partial_{\dot{q}^i} \right) L = 0,$$

or

$$X^i \frac{\partial L}{\partial q^i} + \frac{\partial X^i}{\partial q^j} \dot{q}^j \frac{\partial L}{\partial \dot{q}^i} = 0.$$

Now note that, along the lift of a solution $t \to (q^1(t), \ldots, q^n(t))$ to the Euler–Lagrange equations ($\frac{\partial L}{\partial q^i} = \frac{d}{dt}(\frac{\partial L}{\partial \dot{q}^i})$), this can be written

$$X^i \frac{d}{dt} \left(\frac{\partial L}{\partial \dot{q}^i} \right) + \frac{\partial L}{\partial \dot{q}^i} \left(\frac{\partial X^i}{\partial q^j} \dot{q}^j \right) = 0$$

or, better yet,

$$\frac{d}{dt} \left(X^i \frac{\partial L}{\partial \dot{q}^i} \right) = 0,$$

so $X^i \frac{\partial L}{\partial \dot{q}^i} = X^i p_i$ is conserved. We summarize all of this as follows.

Theorem 2.2.3. *Let $L : TM \to \mathbb{R}$ be a Lagrangian on a smooth manifold M and suppose X is an infinitesimal symmetry of L. Let $q^1, \ldots, q^n$ be any local coordinate system for M with corresponding natural coordinates $q^1, \ldots, q^n, \dot{q}^1, \ldots, \dot{q}^n$. Write $X = X^i \partial_{q^i}$. Then*

$$X^i \frac{\partial L}{\partial \dot{q}^i} = X^i p_i \tag{2.18}$$

is constant along every stationary curve in the coordinate neighborhood on which $q^1, \ldots, q^n$ are defined.

This is (the simplest version of) *Noether's theorem*. In a nutshell, it says that every infinitesimal symmetry gives rise to a conserved quantity. Let us see now how all of this works out for a few examples.

Example 2.2.9. Suppose we have a Lagrangian L that is independent of one of the coordinates in M, say, q^i. Then certainly $X = \partial_{q^i}$ is an infinitesimal symmetry. Since the only component of X relative to $\partial_{q^1}, \ldots, \partial_{q^n}$ is the ith and this is 1 we find that the corresponding Noether conserved quantity is the same as the one we found earlier, namely, the conjugate momentum $p_i = \frac{\partial L}{\partial \dot{q}^i}$.

Example 2.2.10. Here we will continue the discussion in Example 2.2.8 and we will use the notation established there. Specifically, we will consider, for each $i, j = 1, 2, 3$, $i \neq j$, the vector field on $\mathbb{R}^3$ given by

$$X_{ij} = x^i \partial_{x^j} - x^j \partial_{x^i}.$$

Each of these is an infinitesimal symmetry for the Lagrangian $L(x, \dot{x}) = \frac{1}{2} m \|\dot{x}\|^2 - V(\|x\|)$ on $\mathbb{R}^3$. To find the corresponding Noether conserved quantity in standard coordinates we compute

$$X_{ij}^k \frac{\partial L}{\partial \dot{x}^k} = X_{ij}^1 \frac{\partial L}{\partial \dot{x}^1} + X_{ij}^2 \frac{\partial L}{\partial \dot{x}^2} + X_{ij}^3 \frac{\partial L}{\partial \dot{x}^3} = -x^j (m\dot{x}^i) + x^i (m\dot{x}^j) = m(x^i \dot{x}^j - x^j \dot{x}^i).$$

Thus, on any stationary curve,

$$\begin{aligned} m[x^1(t)\dot{x}^2(t) - x^2(t)\dot{x}^1(t)], \\ m[x^3(t)\dot{x}^1(t) - x^1(t)\dot{x}^3(t)], \\ m[x^2(t)\dot{x}^3(t) - x^3(t)\dot{x}^2(t)] \end{aligned} \tag{2.19}$$

are all constant. Note that these are precisely the components of the cross-product

$$\mathbf{L}(t) = \mathbf{r}(t) \times (m\mathbf{v}(t)) = \mathbf{r}(t) \times \mathbf{p}(t) \tag{2.20}$$

of the position and momentum vectors of the particle and this is what physicists call its *angular momentum* (with respect to the origin). Note also that the constancy of this vector along the trajectory of the particle implies that the motion takes place entirely in a two-dimensional plane in $\mathbb{R}^3$, namely, the plane with this normal vector. The existence of these three conserved quantities arose from the infinitesimal symmetries of the spherically symmetric Lagrangian corresponding to rotations about the various coordinate axes so we may summarize all of this as follows.

Rotational symmetry implies conservation of angular momentum

It is important to understand what is really going on in this example. We have a Lagrangian $L : TM \rightarrow \mathbb{R}$ on a smooth manifold M and a (matrix) Lie group G which acts on M in such a way that each diffeomorphism $\sigma_g : M \rightarrow M, g \in G$, is a symmetry of the Lagrangian; under such circumstances we refer to G as a *symmetry group* of L; G has a Lie algebra $\mathfrak{g}$ and each generator (basis element) N_i of $\mathfrak{g}$ gives rise to an infinitesimal symmetry X_{N_i} defined by

$$ X_{N_i}(p) = \frac{d}{dt}(e^{tN_i} \cdot p)\Big|_{t=0} $$

and each of these in turn gives rise, via Noether's theorem, to a conserved quantity. The Lie algebra of the symmetry group is where the conservation laws come from.

Note, incidentally, that it is entirely possible for a Lagrangian to have an infinitesimal symmetry that requires the conservation of one of the components of angular momentum, but not the others. Indeed, we can now see that this is precisely what occurred in our discussion of the spherical pendulum in Example 2.2.2 where we found that the θ-independence of the Lagrangian gave rise to the conserved quantity

$$ m\dot{\theta} \sin^2 \phi = m[x^1 \dot{x}^2 - x^2 \dot{x}^1], \tag{2.21} $$

which we now recognize as the x^3-component of the angular momentum. The moral of this example is that the proper choice of coordinates can uncover symmetries, and therefore conservation laws, that are not otherwise apparent.

Linear and angular momentum conservation in $\mathbb{R}^3$ both arise from a certain symmetry group; in the first case this is the spatial translation group $\mathbb{R}^3$ and in the second it is the rotation group $SO(3)$. We would now like to show these two can be combined into a single group.

Example 2.2.11. Fix an element R of $SO(3)$ and an a in $\mathbb{R}^3$. Define a mapping $(a, R) : \mathbb{R}^3 \rightarrow \mathbb{R}^3$ by

$$ x \in \mathbb{R}^3 \rightarrow (a, R)(x) = R \cdot x + a \in \mathbb{R}^3. $$

Thus, (a, R) rotates by R and then translates by a so it is an isometry of $\mathbb{R}^3$. The composition of two such mappings is given by

$$ x \rightarrow R_1 \cdot x + a_1 \rightarrow R_2 \cdot (R_1 \cdot x + a_1) + a_2 = (R_2 R_1) \cdot x + (R_2 \cdot a_1 + a_2). $$

Since $R_2 R_1 \in SO(3)$ and $R_2 \cdot a_1 + a_2 \in \mathbb{R}^3$, this composition is just

$$(a_2, R_2) \circ (a_1, R_1) = (R_2 \cdot a_1 + a_2, R_2 R_1)$$

so this set of mappings is closed under composition. Moreover, $(0, \mathrm{id}_{3\times3})$ is clearly an identity element and every (a, R) has an inverse given by

$$(a, R)^{-1} = (-R^{-1} a, R^{-1})$$

so this collection of maps forms a group under composition. This group is the *semi-direct product* of $\mathbb{R}^3$ and $SO(3)$. We will denote it $ISO(3)$ and refer to it as the *inhomogeneous rotation group*. Its elements are diffeomorphisms of $\mathbb{R}^3$ onto itself and we can think of it as defining a group action on $\mathbb{R}^3$. We have

$$(a, R) \cdot x = R \cdot x + a.$$

Note that the maps $a \rightarrow (a, \mathrm{id}_{3\times3})$ and $R \rightarrow (0, R)$ identify $\mathbb{R}^3$ and $SO(3)$ with subgroups of $ISO(3)$ and that $\mathbb{R}^3$ is a normal subgroup since it is the kernel of the projection $(a, R) \rightarrow (0, R)$ and this is a homomorphism (the projection onto $\mathbb{R}^3$ is not a homomorphism).

We would like to find an explicit matrix model for $ISO(3)$. For this we identify $\mathbb{R}^3$ with the subset of $\mathbb{R}^4$ consisting of (column) vectors of the form

$$\begin{pmatrix} x^1 \\ x^2 \\ x^3 \\ 1 \end{pmatrix} = \begin{pmatrix} x \\ 1 \end{pmatrix},$$

where $x = (x^1 \ x^2 \ x^3)^T \in \mathbb{R}^3$. Now consider the set G of 4×4 matrices of the form

$$\begin{pmatrix} R & a \\ 0 & 1 \end{pmatrix} = \begin{pmatrix} R^1_{\ 1} & R^1_{\ 2} & R^1_{\ 3} & a^1 \\ R^2_{\ 1} & R^2_{\ 2} & R^2_{\ 3} & a^2 \\ R^3_{\ 1} & R^3_{\ 2} & R^3_{\ 3} & a^3 \\ 0 & 0 & 0 & 1 \end{pmatrix},$$

where $R \in SO(3)$ and $a \in \mathbb{R}^3$. Note that

$$\begin{pmatrix} R & a \\ 0 & 1 \end{pmatrix} \begin{pmatrix} x \\ 1 \end{pmatrix} = \begin{pmatrix} Rx + a \\ 1 \end{pmatrix}$$

and

$$\begin{pmatrix} R_2 & a_2 \\ 0 & 1 \end{pmatrix} \begin{pmatrix} R_1 & a_1 \\ 0 & 1 \end{pmatrix} = \begin{pmatrix} R_2 R_1 & R_2 a_1 + a_2 \\ 0 & 1 \end{pmatrix}$$

so we can identify $ISO(3)$ with G and its action on $\mathbb{R}^3$ with matrix multiplication.

Here G is a matrix Lie group of dimension 6. Its Lie algebra can be identified with the set of 4×4 real matrices that arise as velocity vectors to curves in G through the identity with matrix commutator as bracket (see Section 5.8 of [Nab3]). We find a basis for this Lie algebra (otherwise called a set of *generators*) by noting that if

$$\alpha_a(t) = \begin{pmatrix} \mathrm{id}_{3\times3} & ta \\ 0 & 1 \end{pmatrix},$$

then

$$\alpha_a'(0) = \begin{pmatrix} 0 & a \\ 0 & 0 \end{pmatrix}$$

and if

$$\alpha_N(t) = \begin{pmatrix} e^{tN} & 0 \\ 0 & 1 \end{pmatrix},$$

then

$$\alpha_N'(0) = \begin{pmatrix} N & 0 \\ 0 & 0 \end{pmatrix}$$

(see Theorem 2.2.2 for N). Taking $a = (1, 0, 0), (0, 1, 0), (0, 0, 1)$ and $\hat{n} = (1, 0, 0), (0, 1, 0),$ $(0, 0, 1)$ (again, see Theorem 2.2.2 for $\hat{n}$) we obtain a set of six generators for the Lie algebra $\mathfrak{iso}(3)$ of $\mathrm{ISO}(3)$ that we will write as follows:

$$N_1 = \begin{pmatrix} 0 & 0 & 0 & 0 \\ 0 & 0 & -1 & 0 \\ 0 & 1 & 0 & 0 \\ 0 & 0 & 0 & 0 \end{pmatrix},$$

$$N_2 = \begin{pmatrix} 0 & 0 & 1 & 0 \\ 0 & 0 & 0 & 0 \\ -1 & 0 & 0 & 0 \\ 0 & 0 & 0 & 0 \end{pmatrix},$$

$$N_3 = \begin{pmatrix} 0 & -1 & 0 & 0 \\ 1 & 0 & 0 & 0 \\ 0 & 0 & 0 & 0 \\ 0 & 0 & 0 & 0 \end{pmatrix},$$

$$P_1 = \begin{pmatrix} 0 & 0 & 0 & 1 \\ 0 & 0 & 0 & 0 \\ 0 & 0 & 0 & 0 \\ 0 & 0 & 0 & 0 \end{pmatrix},$$

$$P_2 = \begin{pmatrix} 0 & 0 & 0 & 0 \\ 0 & 0 & 0 & 1 \\ 0 & 0 & 0 & 0 \\ 0 & 0 & 0 & 0 \end{pmatrix},$$

$$P_3 = \begin{pmatrix} 0 & 0 & 0 & 0 \\ 0 & 0 & 0 & 0 \\ 0 & 0 & 0 & 1 \\ 0 & 0 & 0 & 0 \end{pmatrix}.$$

N_1, N_2 and N_3 are called the *generators of rotations*, while P_1, P_2 and P_3 are called the *generators of translations*. Their significance for us is that if M is a configuration space on which a Lagrangian $L : TM \to \mathbb{R}$ is defined and on which ISO(3) acts, then each N_i and each P_i determines a vector field on M and, depending on the Lagrangian, these may (or may not) be infinitesimal symmetries of L.

We will write $[A, B]_- = AB - BA$ for the matrix commutator (the reason for the apparently unnecessary subscript is that later we will need the anticommutator $[A, B]_+ = AB + BA$ as well). Using ϵ_{ijk} for the Levi-Civita symbol (1 if ijk is an even permutation of 123, -1 if ijk is an odd permutation of 123 and 0 otherwise) we record, for future reference, the following *commutation relations* for these generators, all of which can be verified by simply computing the matrix products:

$$[N_i, N_j]_- = \sum_{k=1}^{3} \epsilon_{ijk} N_k, \quad i, j = 1, 2, 3, \tag{2.22}$$

$$[P_i, P_j]_- = 0, \quad i, j = 1, 2, 3, \tag{2.23}$$

$$[N_i, P_j]_- = \sum_{k=1}^{3} \epsilon_{ijk} P_k, \quad i, j = 1, 2, 3. \tag{2.24}$$

Every element of ISO(3) can be written as $e^{\sum_{i=1}^{3}(\alpha_i N_i + \beta_i P_i)}$ for some real number α_i and β_i, but one should keep in mind that for matrices A and B, $e^{A+B} = e^A e^B$ if and only if A and B commute. When $[A, B]_- \neq 0$ the situation is not so simple. Indeed, the *Baker–Campbell–Hausdorff formula* gives a series expansion of the form

$$e^A e^B = e^{A+B+\frac{1}{2}[A,B]+\frac{1}{12}[A,[A,B]]-\frac{1}{12}[B,[A,B]]+\cdots} \tag{2.25}$$

for $e^A e^B$ (see Chapter 3 of [Hall1]). An analogous result for certain self-adjoint operators on a Hilbert space, called the *Lie–Trotter–Kato product formula*, will play an essential role in Chapter 8 when we introduce the Feynman path integral.

Next we will consider an important example that has ISO(3) as a symmetry group.

Example 2.2.12. We consider the classical *two-body problem*. Thus, we have two masses m_1 and m_2 moving in space under a conservative force that depends only on the distance between them and is directed along the line joining them (for example, two planets, each of which exerts a gravitational force on the other). The configuration space is taken to be $M = \mathbb{R}^3 \times \mathbb{R}^3 = \mathbb{R}^6$ so the state space is $TM = (\mathbb{R}^3 \times \mathbb{R}^3) \times (\mathbb{R}^3 \times \mathbb{R}^3) = \mathbb{R}^{12}$. Let x_1 and x_2 denote the position vectors of m_1 and m_2, respectively, and write their Cartesian components $x_j = (x_j^1, x_j^2, x_j^3), j = 1, 2$. We take the Lagrangian L to be

$$
\begin{aligned}
L &= \frac{1}{2} m_1 \sum_{i=1}^{3} (\dot{x}_1^i)^2 + \frac{1}{2} m_2 \sum_{i=1}^{3} (\dot{x}_2^i)^2 - V\left(\left(\sum_{i=1}^{3} (x_1^i - x_2^i)^2 \right)^{1/2} \right) \\
&= \frac{1}{2} m_1 \| \dot{x}_1 \|^2 + \frac{1}{2} m_2 \| \dot{x}_2 \|^2 - V(\| x_1 - x_2 \|),
\end{aligned}
\tag{2.26}
$$

where V is any smooth function of the distance $\| x_1 - x_2 \|$ between m_1 and m_2. ISO(3) acts on $\mathbb{R}^3 \times \mathbb{R}^3$ by acting on each factor separately and it is clear from (2.26) that L is invariant under this action. In particular, this is true for the translation and rotation subgroups so we expect some sort of "momentum" and "angular momentum" conservation (the quotation marks are due to the fact that we are no longer talking about a single particle moving in space so these terms are being used in some new sense that we have not yet made explicit). We will begin with invariance under the SO(3) subgroup. Specifically, for any $g \in$ SO(3), $L(g \cdot x_1, g \cdot x_2, g \cdot \dot{x}_1, g \cdot \dot{x}_2) = L(x_1, x_2, \dot{x}_1, \dot{x}_2)$ so, in particular, for any $N \in \mathfrak{so}(3)$,

$$
X_N(x_1, x_2) = \frac{d}{dt}\left(e^{tN} \cdot (x_1, x_2) \right)\Big|_{t=0} = (N x_1, N x_2)
$$

is an infinitesimal symmetry of L. Now, any vector field on $\mathbb{R}^3 \times \mathbb{R}^3$ is a linear combination of

$$
\partial_{x_1^1}, \partial_{x_1^2}, \partial_{x_1^3}, \partial_{x_2^1}, \partial_{x_2^2}, \partial_{x_2^3}
$$

and if we take N to be one of the infinitesimal generators of rotations ($N = N_1, N_2, N_3$), then, just as in Example 2.2.8 , we obtain the vector fields

$$
(x_1^i \partial_{x_1^j} - x_1^j \partial_{x_1^i}) + (x_2^i \partial_{x_2^j} - x_2^j \partial_{x_2^i}),
$$

from which we can read off the components of the infinitesimal symmetries. To write down the corresponding Noether conserved quantities (2.18) we note that $\partial L / \partial \dot{x}_1^i = m_1 \dot{x}_1^i$ and $\partial L / \partial \dot{x}_2^i = m_2 \dot{x}_2^i$. Thus, (2.18) gives

$$
m_1(x_1^i \dot{x}_1^j - x_1^j \dot{x}_1^i) + m_2(x_2^i \dot{x}_2^j - x_2^j \dot{x}_2^i)
$$

for the conserved quantities. Consequently, rotational invariance for the two-body problem implies conservation of the *total angular momentum* of the system.

One can deal with translational invariance in the same way, but for the sake of variety and because we will need the ideas when we discuss the quantum two-body problem, that is, the hydrogen atom (Example 9.1.1), we will describe instead the traditional approach in physics. The idea is to replace the Cartesian coordinates with the so-called *center of mass coordinates*. For this we replace x_1 and x_2 with the displacement vector

$$x = x_1 - x_2$$

and the *center of mass vector*

$$R = \frac{m_1 x_1 + m_2 x_2}{m_1 + m_2}.$$

Then

$$x_1 = R + \frac{m_2 x}{m_1 + m_2}$$

and

$$x_2 = R - \frac{m_1 x}{m_1 + m_2}.$$

Computing derivatives and substituting into (2.26) gives

$$L = \frac{1}{2}(m_1 + m_2)\|\dot{R}\|^2 + \frac{1}{2}\frac{m_1 m_2}{m_1 + m_2}\|\dot{x}\|^2 - V(\|x\|). \tag{2.27}$$

Now note that this expression for L depends on $x^1, x^2, x^3, \dot{x}^1, \dot{x}^2, \dot{x}^3, \dot{R}^1, \dot{R}^2, \dot{R}^3$, but not on R^1, R^2, R^3. Since

$$\frac{\partial L}{\partial \dot{R}^i} = (m_1 + m_2)\dot{R}^i = m_1\dot{x}_1^i + m_2\dot{x}_2^i, \quad i = 1, 2, 3,$$

and these must be constant along a stationary curve we conclude that the *total momentum*

$$P = (m_1 + m_2)\dot{R} = m_1\dot{x}_1 + m_2\dot{x}_2$$

of the system is conserved.

Finally, we come to an example that is quite important to physicists and also to us, but for rather different reasons. Rigid body dynamics is an old and venerable part of physics. We will begin to set up the formalism for the subject, but will find ourselves immediately distracted by a rather curious possibility that this formalism suggests. Curious or not, we will find before we are through that this possibility is directly relevant to the behavior of certain elementary particles called "fermions" (see Section 10.1). The example (or, rather, the distraction) requires a bit of standard topology, specifically, fundamental groups. Everything we need is available in a great many sources (for example, Sections 2.1–2.4 of [Nab3], or Part I of [Gre]).

Example 2.2.13. Intuitively, a rigid body is a physical system consisting of a finite number of point masses each one of which maintains a constant distance from all of the rest as the system evolves in time, that is, as it moves in space. Such things can exist only as idealizations, of course, and, indeed, special relativity prohibits their existence even as idealizations, but we are now ignoring relativistic effects and so we will not worry about it. We will be interested in the motion of such a rigid body that is constrained to pivot about some fixed point. Thus, we begin with $n + 1$ positive real numbers $m_0, m_1, \ldots, m_n$ (the masses) and $n + 1$ distinct points $x_0^0, x_1^0, \ldots, x_n^0$ in $\mathbb{R}^3$ (the positions of the masses $m_0, m_1, \ldots, m_n$, respectively, at time $t = 0$). We will assume that $n > 3$ and that not all of the masses lie in a single plane in $\mathbb{R}^3$. Furthermore, we will take the fixed point about which the rigid body is to pivot to be x_0^0 and choose coordinate axes so that $x_0^0 = (0,0,0) \in \mathbb{R}^3$. For $i, j = 0, 1, \ldots, n$, $i \neq j$, we let $\|x_i^0 - x_j^0\| = c_{ij} > 0$.

We refer to $\{x_0^0 = (0,0,0), x_1^0, \ldots, x_n^0\}$ as the *initial configuration* of the rigid body. By assumption, any other configuration $\{x_0 = (0,0,0), x_1, \ldots, x_n\}$ of the masses $m_0, m_1, \ldots, m_n$ must satisfy $\|x_i - x_j\| = c_{ij}$, $i, j = 0, 1, \ldots, n$, $i \neq j$. Although one could define the configuration space of our rigid body in these terms, a much clearer picture emerges in the following way. Since there are at least four masses and not all of them lie in a single plane, some three of $x_1^0, \ldots, x_n^0$ must be linearly independent. We can assume the masses are labeled so that x_1^0, x_2^0, x_3^0 are linearly independent and therefore form a basis for $\mathbb{R}^3$. Now let A be the unique linear transformation of $\mathbb{R}^3$ that carries x_i^0 onto x_i for $i = 1, 2, 3$.

Exercise 2.2.8. Show that A is an orthogonal transformation of $\mathbb{R}^3$ and that $Ax_i^0 = x_i$ for $i = 4, \ldots, n$.

Consequently, we can identify any configuration of the rigid body with an orthogonal transformation A of $\mathbb{R}^3$ that carries the initial configuration onto it. Every such A is in the orthogonal group $O(3)$ and therefore has determinant ± 1. We would like to conclude that $\det A = 1$ so that, in fact, A is in the rotation group $SO(3)$. This *does not* follow mathematically from the assumptions we have made, but rests on additional physical input. The reasoning by which we arrive at this additional physical input lies at the heart of the point we wish to make in this example so we will try to explain carefully. Pick up a rigid body (a Rubik's cube, for example), hold it in some initial configuration and then, keeping one of its points fixed, move it (in any way you like) to a new configuration. This is, of course, a physical process. The rigid body proceeds through a "continuous" sequence of physical configurations (parametrized by some interval of time values, for example) beginning with the initial configuration and ending with the final one. At each instant t the configuration of the rigid body corresponds to an element $A(t)$ of $O(3)$ that carries the initial configuration onto the configuration at time t with, of course, $A(0) = \text{id}_{3\times3} \in O(3)$. Thus, we have a curve $t \to A(t)$ in $O(3)$ that models our physical process. Now, $O(3)$ is a Lie group (see Section 5.8 of [Nab3]) and therefore a topological space and the curve $t \to A(t)$ is continuous with respect to this topology (and the usual topology on the real line) if and only if the entries in $A(t)$ relative to any basis are continuous real-valued functions of the real variable t. Let us now *assume that this is the appropriate interpretation of a "continuous" sequence of physical configurations.*

Remark 2.2.11. If you are saying to yourself that this is patently obvious and unworthy of being singled out as a "physical assumption" we would ask that you reserve judgment until you have seen some of the consequences.

Then $t \rightarrow \det A(t)$ is a continuous real-valued function of a real variable, defined on an interval and taking only the values 1 and -1. The intermediate value theorem implies that $\det A(t)$ must take exactly one of these values for all t. Since $\det A(0) = 1$, we find that $\det A(t) = 1$ for every t, that is, $A(t) \in SO(3)$ for all t. Every configuration of the rigid body can therefore be identified with an element of $SO(3)$. Let us summarize the conclusions to which this discussion has led us.

The configuration space of a rigid body constrained to pivot about some fixed point is $SO(3)$ *and any such motion of the rigid body is represented by a continuous curve* $t \in \mathbb{R} \rightarrow A(t) \in SO(3)$ *in* $SO(3)$.

Exercise 2.2.9. Describe the configuration space of a rigid body that is *not* constrained to pivot about some fixed point.

The state space for a rigid body with one point fixed is therefore the tangent bundle $T(SO(3))$ which happens to be the product $SO(3) \times \mathfrak{so}(3)$ of $SO(3)$ and its Lie algebra $\mathfrak{so}(3)$ because any Lie group is parallelizable (Exercise 5.8.17 of [Nab3]). If one now wishes to understand the dynamics of rigid body motion one must specify a Lagrangian $L : SO(3) \times \mathfrak{so}(3) \rightarrow \mathbb{R}$ and study the Euler–Lagrange equations. For free motion the Lagrangian is simply the total kinetic energy of the point masses and it can be shown that one can reduce the problem of calculating trajectories and conserved quantities to that of computing geodesics and symmetries of a *kinetic energy metric* on $SO(3)$ (compare Example 2.2.4). This is quite a beautiful story, but not the one we wish to tell here (see Chapter 6 of [Arn2] or Chapter 13 of [CM]). Instead we would like to focus our attention on what might appear to be a rather odd question that arises in the following way.

We have seen that the physical process of rotating a rigid body from one configuration to another is modeled mathematically by a continuous curve in $SO(3)$. Let us consider the special case of a motion that begins and ends at the initial configuration. The corresponding curve in $SO(3)$ is then a *loop* in $SO(3)$ at the identity element $\mathrm{id}_{3\times3}$, that is, a continuous curve $A : [0, \beta] \rightarrow SO(3)$ with $A(0) = A(\beta) = \mathrm{id}_{3\times3}$. Here are two examples. The loop $A_1 : [0, 2\pi] \rightarrow SO(3)$ defined by

$$A_1(t) = \begin{pmatrix} \cos t & -\sin t & 0 \\ \sin t & \cos t & 0 \\ 0 & 0 & 1 \end{pmatrix}$$

corresponds to rotating our rigid body $360°$ about the z-axis, whereas $A_2 : [0, 4\pi] \rightarrow$ SO(3), defined by the same formula

$$A_2(t) = \begin{pmatrix} \cos t & -\sin t & 0 \\ \sin t & \cos t & 0 \\ 0 & 0 & 1 \end{pmatrix},$$

corresponds to a $720°$ rotation about the z-axis ("A_1 traversed twice"). Both of these motions leave the rigid body in precisely the same state since $(A_1(2\pi), \dot{A}_1(2\pi)) = (A_2(4\pi), \dot{A}_2(4\pi))$.

Mathematically, however, the two loops A_1 and A_2 in SO(3) are not at all equivalent. The fundamental group $\pi_1(\text{SO}(3))$ of the topological space SO(3) classifies the continuous loops at $\text{id}_{3\times3}$ according to (path) homotopy type. As it happens, $\pi_1(\text{SO}(3))$ is isomorphic to the additive group $\mathbb{Z}_2 = \{[0], [1]\}$ of integers mod 2 (there are two proofs of this in Appendix A of [Nab3]). There are precisely two homotopy classes of loops, [0] and [1], the first consisting of those that are homotopically trivial (they can be "continuously deformed to a point" in SO(3)) and the second consisting of those that are not. A_1 is not homotopically trivial (Exercise 2.4.14 of [Nab3]) so $A_1 \in [1]$, but, since $[1] + [1] = [0]$ in $\mathbb{Z}_2$, $A_2 \in [0]$. In particular, A_1 and A_2 are not homotopically equivalent; neither can be continuously deformed into the other in SO(3). From a topological point of view, A_1 and A_2 are certainly not equivalent.

Now for the "rather odd" question we would like to pose. Is it possible that, when thought of as physical motions of a rigid body, two nonhomotopic loops such as A_1 and A_2 exhibit different physical effects? Stated otherwise, is it possible that the rigid body is somehow physically different at the end of the two motions despite the fact that the "states" are the same? This may seem silly. Really, just look at it! At the end of either motion the Rubik's cube is in exactly the same configuration in space with the same angular velocity so its state is the same; what could possibly be different?

Nevertheless, there are physical systems in nature that behave in just this sort of bizarre way. Indeed, this is the case for any of the elementary particles classified by the physicists as fermions. We will have more to say about this in Section 10.1 (for informative, nontechnical previews by very authoritative sources one might have a look at [AS] and/or ['t Ho1]). For the present we cannot resist the temptation to mention an ingenious demonstration, devised by Paul Dirac, of a perfectly mundane macroscopic physical system in which "something" appears to be altered by a rotation through $360°$, but returned to its original status by a $720°$ rotation. It is called the *Dirac scissors*. The demonstration involves a pair of scissors, a piece of elastic string and a chair. Pass the string through one finger hole of the scissors, then around one leg of the chair, then through the other finger hole and around the other leg of the chair and then tie the two ends of the string together (see Figure 2.1[1]).

---

1 Reproduced from [Nab5] with the permission of Springer.

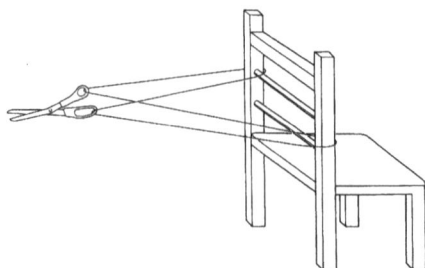

Figure 2.1: Dirac scissors 1. [Nab5].

Now rotate the scissors about its axis of symmetry through one complete revolution (360°). The string becomes entangled and the problem is to disentangle it by moving only the string, holding the scissors and chair fixed (the string needs to be elastic so it can be moved around the objects if desired). Try it! No amount of maneuvering, simple or intricate, will return the string to its original, disentangled state. This, in itself, is not particularly surprising perhaps, but now repeat the exercise, this time rotating the scissors through two complete revolutions (720°). The string now appears even more hopelessly tangled, but looping the string just once over the pointed end of the scissors will return it to its original condition. There is clearly "something different" about the state of the system when it has undergone a rotation of 360° and when it has been rotated by 720°. Note, however, that now the "system" is somehow more than just an isolated rigid body, but includes, in some sense, the way in which the object is "entangled" with its environment. We will return to this in Section 10.1, where we will also see what it has to do with fermions.

Before leaving this behind for a while, we should reap one more dividend from Dirac's remarkable little game. As with any good magic trick, some of the paraphernalia is present only to divert the attention of the audience. Note that none of the essential features are altered if we imagine the string glued (in an arbitrary manner) to the surface of an elastic belt so that we may discard the string altogether in favor of a belt connecting the scissors and the chair (see Figure 2.2[1]). Rotate the scissors through 360° and the belt acquires one twist that cannot be untwisted by moving the belt alone. Rotate through 720° and the belt has two twists that can be removed by looping the belt once around the scissors. In either case imagine the scissors free to slide along

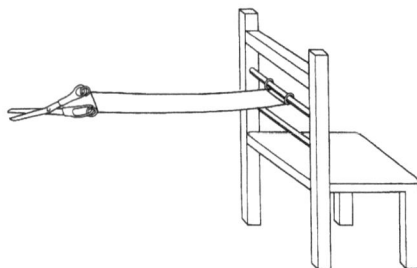

Figure 2.2: Dirac scissors 2. [Nab5].

the belt toward the chair and, at the end of the trip, translate the scissors (without rotation) back to its original position. In the first case the scissors will have accomplished a 360° rotation, and in the second a 720° rotation. In both cases, the belt itself is a physical model of the loop in SO(3) representing the rotation (with no twists the belt represents the trivial loop). Now imagine yourself manipulating the belt (without moving the scissors or chair), trying to undo the twists. At each instant the position of the belt represents a loop in SO(3) so your manipulations represent a continuous sequence of loops in SO(3) parametrized by time. In the second case you will succeed in creating a continuous sequence of loops (that is, a homotopy) from the loop representing a 720° rotation to the trivial loop (no rotation, that is, no twists, at all). In the first case you will not succeed in doing this because no such homotopy exists. Dirac has given us a physical picture of the two homotopy classes of loops in SO(3). Needless to say, tireless manipulation of a belt does not constitute a proof. For those who would like to see this discussion carried out rigorously we recommend [Bolker] and then [Fadell].

We will leave the reader with an example to work out from scratch. It is still a relatively simple physical system, but one that exhibits an extraordinary range of motions, from the expected to the chaotic. Here we will ask you simply to write out the Lagrangian and the Euler–Lagrange equations. In the next section you will continue the analysis by writing out the Hamiltonian formulation of the same system (Exercise 2.3.5) and then we will say a few words about the sort of motions one can observe by supplying various initial conditions. Before doing any of this, however, you should either visit an undergraduate physics lab and watch the system in action or, better yet, you should build one yourself (instructions are available at http://makezine.com/projects/double-pendulum/). At the very least, have a look at http://gizmodo.com/5869648/spend-the-next-22-minutes-mesmerized-by-this-near+perpetual-motion-double-pendulum.

Exercise 2.2.10. The *double pendulum* consists of two pendulums (pendula, if you prefer) with the first suspended from a point in space and the second suspended from the end of the first. We will assume that the initial conditions are such that the motion remains in a plane (see Figure 2.3).

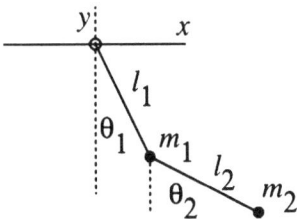

Figure 2.3: Double pendulum.

1. Show that the Lagrangian is given by

$$L(\theta_1, \theta_2, \dot\theta_1, \dot\theta_2) = \frac{1}{2}(m_1 + m_2)l_1^2\dot\theta_1^2 + \frac{1}{2}m_2l_2^2\dot\theta_2^2 + m_2l_1l_2\dot\theta_1\dot\theta_2\cos(\theta_1 - \theta_2)$$
$$+ (m_1 + m_2)gl_1\cos\theta_1 + m_2gl_2\cos\theta_2. \qquad (2.28)$$

2. Show that the Euler–Lagrange equations are

$$(m_1 + m_2)l_1^2\ddot\theta_1 + m_2l_1l_2\ddot\theta_2\cos(\theta_1 - \theta_2)$$
$$+ m_2l_1l_2\dot\theta_2^2\sin(\theta_1 - \theta_2) + l_1(m_1 + m_2)g\sin\theta_1 = 0 \qquad (2.29)$$

and

$$m_2l_2^2\ddot\theta_2 + m_2l_1l_2\ddot\theta_1\cos(\theta_1 - \theta_2) - m_2l_1l_2\dot\theta_1^2\sin(\theta_1 - \theta_2) + l_2m_2g\sin\theta_2 = 0. \qquad (2.30)$$

2.3 Phase space and Hamiltonians

The Lagrangian approach to mechanics clearly affords insights that are not readily uncovered in the Newtonian picture. Moreover, this approach generalizes to include classical field theory and forms the basis for Feynman's path integral approach to quantization (Chapter 8). Nevertheless, there is another approach that affords at least as much insight into mechanics, also generalizes to classical field theory and is the basis for canonical quantization (Chapter 7). It is to this Hamiltonian picture of mechanics that we now turn (a more concise and no doubt more elegant synopsis of Hamiltonian mechanics in geometrical terms is available in the classic paper [MacL]).

The view of classical mechanics that we would now like to describe evolves from what would appear to be an innocuous attempt to change coordinates in the state space. Suppose that we are given a Lagrangian $L : TM \to \mathbb{R}$ and a local coordinate system $q^1, \ldots, q^n$ on M with associated natural coordinates $q^1, \ldots, q^n, \dot q^1, \ldots, \dot q^n$ on TM. We have defined the corresponding conjugate momenta $p_1 = \frac{\partial L}{\partial \dot q^1}, \ldots, p_n = \frac{\partial L}{\partial \dot q^n}$ and now ask if one can use $q^1, \ldots, q^n, p_1, \ldots, p_n$ as local coordinates on TM. One might take as motivation here the simple fact that some things would look nicer in such coordinates, for example, the Noether conserved quantities would be $X^i p_i$ rather than $X^i \frac{\partial L}{\partial \dot q^i}$. As it happens, the dividends are substantially greater than just this.

At least one condition must clearly be imposed on the Lagrangian if

$$(q^1, \ldots, q^n, \dot q^1, \ldots, \dot q^n) \to (q^1, \ldots, q^n, p_1, \ldots, p_n)$$

is to be a legitimate change of coordinates on TM. Specifically, we must be able to solve (at least locally) the system of equations

$$p_1 = \frac{\partial L}{\partial \dot q^1}(q^1, \ldots, q^n, \dot q^1, \ldots, \dot q^n)$$

$$\vdots \qquad\qquad\qquad (2.31)$$

$$p_n = \frac{\partial L}{\partial \dot{q}^n}(q^1,\ldots,q^n,\dot{q}^1,\ldots,\dot{q}^n)$$

for $\dot{q}^i$ in terms of $q^1,\ldots,q^n,p_1,\ldots,p_n$. By the implicit function theorem, this will be possible if the matrix

$$\mathcal{H}_L(q,\dot{q}) = \left(\frac{\partial^2 L}{\partial \dot{q}^i \partial \dot{q}^j}(q,\dot{q})\right)_{i,j=1,\ldots,n}$$

is nonsingular at each point in the domain of the natural coordinates. We will say that a Lagrangian L is *nondegenerate* if, for every chart on an open set $U \subseteq M$ with coordinate functions $q^1,\ldots,q^n$, the matrix $\mathcal{H}_L(q,\dot{q})$ is nonsingular at each point of $TU \subseteq TM$.

Example 2.3.1. For $M = \mathbb{R}^n$ we have a global coordinate system $q^1,\ldots,q^n = x^1,\ldots,x^n$ so it clearly suffices to check the nondegeneracy condition for these. For example, if $L = \frac{1}{2}m((\dot{x}^1)^2 + \cdots + (\dot{x}^n)^2) - V(x^1,\ldots,x^n)$ with $m > 0$, then

$$\left(\frac{\partial^2 L}{\partial \dot{x}^i \partial \dot{x}^j}\right)_{i,j=1,\ldots,n} = m\,(\mathrm{id}_{3\times 3}),$$

so L is certainly nondegenerate. Note that, in this case, $p_i = \frac{\partial L}{\partial \dot{x}^i} = m\dot{x}^i$, so inverting the equations (2.31) is not so hard. We have

$$\dot{x}^i = \frac{1}{m}p_i, \quad i = 1,\ldots,n. \qquad\qquad (2.32)$$

Consequently, for nondegenerate Lagrangians, we have the state space TM covered by coordinate systems $(q^1,\ldots,q^n,p_1,\ldots,p_n)$ of the desired type. These are *not* natural coordinates on TM, of course. In fact, we would like to argue that it is somehow wrongheaded to think of them as coordinates on TM at all. The key to understanding what is behind this rather obscure comment is to look at two such coordinate systems and see how they are related.

Suppose then that we have two charts on M with coordinate functions $q^1,\ldots,q^n$ and $Q^1,\ldots,Q^n$, respectively, and suppose that their coordinate neighborhoods intersect. On this intersection we will write the coordinate transformation functions as

$$Q^i = Q^i(q^1,\ldots,q^n), \quad i = 1,\ldots,n,$$

and

$$q^i = q^i(Q^1,\ldots,Q^n), \quad i = 1,\ldots,n.$$

Each of these coordinate systems gives rise to natural coordinates

$$q^1,\ldots,q^n,\dot{q}^1,\ldots,\dot{q}^n$$

and

$$Q^1, \ldots, Q^n, \dot{Q}^1, \ldots, \dot{Q}^n$$

on TM. The chain rule implies that

$$\dot{q}^i = \frac{\partial q^i}{\partial Q^j} \dot{Q}^j, \quad i = 1, \ldots, n, \tag{2.33}$$

and

$$\dot{Q}^i = \frac{\partial Q^i}{\partial q^j} \dot{q}^j, \quad i = 1, \ldots, n. \tag{2.34}$$

These are, of course, just the standard transformation laws for the components of tangent vectors (*contravariant vectors*, as the physicists would say). Now suppose we have a nondegenerate Lagrangian $L : TM \to \mathbb{R}$. The relationship between the coordinate expressions for L in our two natural coordinate systems can be written

$$L(Q^1, \ldots, Q^n, \dot{Q}^1, \ldots, \dot{Q}^n)$$
$$= L\left(q^1(Q^1, \ldots, Q^n), \ldots, q^n(Q^1, \ldots, Q^n), \frac{\partial q^1}{\partial Q^j} \dot{Q}^j, \ldots, \frac{\partial q^n}{\partial Q^j} \dot{Q}^j \right).$$

From this we compute the relationship between the conjugate momenta p_i and P_i:

$$\begin{aligned}
P_i &= \frac{\partial L}{\partial \dot{Q}^i} = \frac{\partial L}{\partial q^j} \frac{\partial q^j}{\partial \dot{Q}^i} + \frac{\partial L}{\partial \dot{q}^j} \frac{\partial \dot{q}^j}{\partial \dot{Q}^i} \\
&= \frac{\partial L}{\partial q^j} \cdot 0 + \frac{\partial L}{\partial \dot{q}^j} \frac{\partial \dot{q}^j}{\partial \dot{Q}^i} \\
&= \frac{\partial q^j}{\partial Q^i} \frac{\partial L}{\partial \dot{q}^j} \quad \text{(by (2.33))} \\
&= \frac{\partial q^j}{\partial Q^i} p_j.
\end{aligned}$$

Consequently, the conjugate momenta transform under the transposed inverse of the (Jacobian) matrix that transforms $\dot{Q}^i$ (see (2.34)). As a result, it is more natural to regard them as components of not an element of the tangent space $T_p(M)$, but rather an element of its dual $T_p^*(M)$, the so-called *cotangent space*. The elements of $T_p^*(M)$ are called *covectors* or *covariant vectors*. We intend to view $(q^1, \ldots, q^n, p_1, \ldots, p_n)$ as coordinate functions on the cotangent bundle T^*M (see Appendix D).

Since conjugate momenta transform under a change of coordinates in M in the same way as cotangent vectors, it seems more natural to regard them as coordinates on T^*M and we now describe how to do this. We assume that we are given a nondegenerate Lagrangian $L : TM \to \mathbb{R}$ on M and will use it to construct a map

$$\mathcal{L} = \mathcal{L}_L : TM \to T^*M.$$

For each $p \in M$, the fiber $\pi_{TM}^{-1}(p)$ of TM above p is a submanifold of TM that is naturally identified with $T_p(M)$. Thus, we can restrict L to this submanifold and compute the differential of this restriction at any point $v_p \in T_p(M)$. Furthermore, since $T_p(M)$ is a vector space, the tangent space to $T_p(M)$ at any $v_p \in T_p(M)$ is also naturally identified with $T_p(M)$. Then $d(L|_{T_p(M)})_{v_p}$ can be regarded as a real-valued, linear function on $T_p(M)$. In other words, it is an element of $T_p^*(M)$, whose value at any $w_p \in T_p(M)$ is given by

$$d(L|_{T_p(M)})_{v_p}(w_p) = \frac{d}{dt} L(p, v_p + tw_p)\Big|_{t=0}.$$

If $q^1, \ldots, q^n$ are local coordinates on M and L is expressed in terms of the corresponding natural coordinates on TM, then this can be written

$$d(L|_{T_p(M)})_{v_p}(w_p) = \frac{d}{dt} L(p, v_p + tw_p)\Big|_{t=0} = \frac{\partial L}{\partial \dot{q}^1}(p, v_p)\, dq^1(w_p) + \cdots$$

$$+ \frac{\partial L}{\partial \dot{q}^n}(p, v_p)\, dq^n(w_p). \tag{2.35}$$

We define the *Legendre transformation* $\mathcal{L} = \mathcal{L}_L : TM \to T^*M$ associated with L by

$$\mathcal{L}(p, v_p) = \mathcal{L}_L(p, v_p) = (p, d(L|_{T_p(M)})_{v_p}). \tag{2.36}$$

We should note that the Legendre transformation can be introduced in a much more general context. This is thoroughly discussed in Chapter 16 of [Sp3].

If $q^1, \ldots, q^n$ are local coordinates on M then, in terms of natural coordinates on TM and T^*M, (2.35) implies that

$$\mathcal{L}_L(q, \dot{q}) = \mathcal{L}_L(q^1, \ldots, q^n, \dot{q}^1, \ldots, \dot{q}^n) = \left(q^1, \ldots, q^n, \frac{\partial L}{\partial \dot{q}^1}, \ldots, \frac{\partial L}{\partial \dot{q}^n} \right)$$

$$= (q^1, \ldots, q^n, p_1, \ldots, p_n). \tag{2.37}$$

Thus, $\mathcal{L}_L$ is defined by

$$\pi_{T^*M} \circ \mathcal{L}_L = \pi_{TM} \tag{2.38}$$

and

$$\xi_i \circ \mathcal{L}_L = p_i = \frac{\partial L}{\partial \dot{q}^i}, \quad i = 1, \ldots, n, \tag{2.39}$$

where $(q^1, \ldots, q^n, \xi_1, \ldots, \xi_n)$ are the natural coordinates on T^*M (see (D.1)). Our non-degeneracy assumption therefore implies that the derivative $(\mathcal{L}_L)_*$ of $\mathcal{L}_L$ is nonsingular at each point so $\mathcal{L}_L$ is a local diffeomorphism. Consequently, $(q, p) = (q^1, \ldots, q^n, p_1, \ldots, p_n)$ are local coordinates on a neighborhood of each point in the image of $\mathcal{L}_L$. We will refer to these as *canonical coordinates* on T^*M determined by the Lagrangian L.

Remark 2.3.1. Note, incidentally, that thinking of the conjugate momenta as components of a cotangent vector explains why we chose to label them with subscripts rather
than superscripts when we first introduced them.

The Legendre transformation $\mathcal{L}_L : TM \rightarrow T^*M$ is generally not surjective. For a
nondegenerate Lagrangian it is a local diffeomorphism and so the image is an open
set $\mathcal{O}$ in T^*M. If one assumes a bit more of the Lagrangian one can say a bit more. If,
for example, the matrices

$$\left(\frac{\partial^2 L}{\partial \dot{q}^i \partial \dot{q}^j} (q, \dot{q}) \right)_{i,j=1,\dots,n} \tag{2.40}$$

are not only nonsingular, but are positive definite at each point, then one can prove
that $\mathcal{L}_L$ is a diffeomorphism of TM onto $\mathcal{O}$. This is, in fact, the common state of affairs
so, in order to prune away some inessential technical issues in the development, we
will restrict our attention to Lagrangians for which the corresponding Legendre transformation is a diffeomorphism. More precisely, let us say that a Lagrangian $L : TM \rightarrow$
$\mathbb{R}$ is *regular* if the Legendre transformation $\mathcal{L}_L : TM \rightarrow T^*M$ is a diffeomorphism onto
its image $\mathcal{O} \subseteq T^*M$.

Henceforth we assume that all Lagrangians are regular.

Of course, we will need to verify regularity for any particular example.

Example 2.3.2. We will compute the Legendre transformation corresponding to the
kinetic energy Lagrangian on a Riemannian manifold M (Example 2.2.4), taking $m = 1$
for convenience. Thus,

$$L(p, v_p) = \frac{1}{2} \langle v_p, v_p \rangle_p \quad \forall (p, v_p) \in TM.$$

For any $(p, v_p) \in TM$,

$$\mathcal{L}(p, v_p) = \mathcal{L}_L(p, v_p) = (p, d(L|_{T_p(M)})_{v_p})$$

and, for any $w_p \in T_p(M)$,

$$d(L|_{T_p(M)})_{v_p}(w_p) = \frac{d}{dt} L(p, v_p + t w_p) \Big|_{t=0}$$

$$= \frac{d}{dt} \left(\frac{1}{2} \langle v_p + t w_p, v_p + t w_p \rangle_p \right) \Big|_{t=0}$$

$$= \frac{d}{dt} \left(\frac{1}{2} \langle v_p, v_p \rangle_p + t \langle v_p, w_p \rangle_p + \frac{1}{2} t^2 \langle w_p, w_p \rangle_p \right) \Big|_{t=0}$$

$$= \langle v_p, w_p \rangle_p.$$

We conclude that $d(L|_{T_p(M)})_{v_p} = \langle v_p, \cdot \rangle_p$ so

$$\mathcal{L}_L(p, v_p) = (p, \langle v_p, \cdot \rangle_p).$$

Thus, for each $p \in M$ the Legendre transformation is just the canonical isomorphism of $T_p(M)$ onto $T_p^*(M)$ determined by the inner product $\langle , \rangle_p$. In particular, $\mathcal{L}_L$ is a bijection from TM onto T^*M. Since it is also a local diffeomorphism, $\mathcal{L}_L$ is a global diffeomorphism of TM onto T^*M.

Exercise 2.3.1. Show that any Lagrangian $L : T\mathbb{R}^n \to \mathbb{R}$ of the form

$$L(q^1, \ldots, q^n, \dot{q}^1, \ldots, \dot{q}^n) = \sum_{i=1}^{n} \frac{1}{2} m(\dot{q}^i)^2 - V(q^1, \ldots, q^n),$$

where V is smooth, is regular.

Lagrangian mechanics takes place on the state space TM. The cotangent bundle T^*M is called the *phase space* and this is where *Hamiltonian mechanics* takes place. Our objective now is to move what we know about mechanics from state space to phase space. "Why?" you may ask. The most compelling reason we can offer is that, somewhat miraculously, T^*M has a more natural mathematical structure than TM and this additional structure greatly clarifies the geometrical picture of mechanics in particular and physics in general. Let us see how this additional structure comes about.

If M is an n-dimensional smooth manifold, then its cotangent bundle T^*M is a $2n$-dimensional smooth manifold and the projection $\pi : T^*M \to M$ is a smooth surjection (in fact, a submersion). Consequently, one may speak of 1-forms on T^*M. We now define what is called the *natural 1-form* θ on T^*M. The idea is very simple. A 1-form θ on T^*M should assign to each point $(p, \eta_p) \in T^*M$ a real-valued linear function on $T_{(p,\eta_p)}(T^*M)$. But the derivative $\pi_{*(p,\eta_p)}$ carries $T_{(p,\eta_p)}(T^*M)$ linearly onto $T_p(M)$ and η_p is a real-valued, linear function on $T_p(M)$ so we will just compose them, that is, we define

$$\theta_{(p,\eta_p)} = \eta_p \circ \pi_{*(p,\eta_p)}.$$

This assigns to each (p, η_p) in T^*M a real-valued, linear function on $T_{(p,\eta_p)}(T^*M)$ so one need only check that it is smooth. The best way to do this is to write it in natural coordinates $(q^1, \ldots, q^n, \xi_1, \ldots, \xi_n)$ on T^*M. We claim that, on an open set on which these coordinates are defined,

$$\theta = \xi_1 dq^1 + \cdots + \xi_n dq^n = \xi_i dq^i \qquad (2.41)$$

and, from this, smoothness is clear. To prove (2.41) we note that, since $\pi(q^1, \ldots, q^n, \xi_1, \ldots, \xi_n) = (q^1, \ldots, q^n)$, the matrix of $\pi_{*(p,\eta_p)}$ in these coordinates is

$$
\begin{pmatrix}
1 & 0 & \cdots & 0 & 0 & \cdots & 0 \\
0 & 1 & \cdots & 0 & 0 & \cdots & 0 \\
\vdots & & & \vdots & \vdots & & \vdots \\
0 & & \cdots & 1 & 0 & \cdots & 0
\end{pmatrix}.
$$

Thus, $\pi_{*(p,\eta_p)}(\partial_{q^i}|_{(p,\eta_p)}) = \partial_{q^i}|_p$ and $\pi_{*(p,\eta_p)}(\partial_{\xi_i}|_{(p,\eta_p)}) = 0$. Consequently,

$$
\begin{aligned}
\theta_{(p,\eta_p)}(\partial_{q^i}|_{(p,\eta_p)}) &= \eta_p \circ \pi_{*(p,\eta_p)}(\partial_{q^i}|_{(p,\eta_p)}) \\
&= \eta_p(\partial_{q^i}|_p) \\
&= \xi_i(p, \eta_p),
\end{aligned}
$$

so $\theta(\partial_{q^i}) = \xi_i$. Similarly, $\theta_{(p,\eta_p)}(\partial_{\xi_i}|_{(p,\eta_p)}) = 0$, so $\theta(\partial_{\xi_i}) = 0$. This proves (2.41), so we have a well-defined, smooth 1-form θ on T^*M which, in natural coordinates, is given by (2.41).

The exterior derivative $d\theta$ of θ is a 2-form on T^*M given locally in natural coordinates by

$$
d\theta = d(\xi_i dq^i) = d\xi_i \wedge dq^i.
$$

The 2-form $\omega = -d\theta$ is called the *natural symplectic form* on T^*M (see Remark 2.3.2 below)). In coordinates,

$$
\omega = -d\theta = dq^i \wedge d\xi_i.
$$

Since θ is invariantly defined, so is ω. In natural coordinates on T^*M, ω takes on the particularly simple form $dq^i \wedge d\xi_i$. In an arbitrary coordinate system (Q^i, P_i) on T^*M, ω will generally be some linear combination of $dQ^i \wedge dQ^j$, $dQ^i \wedge dP_j$ and $dP_i \wedge dP_j$. However, it may happen that for some particular coordinate systems, $\omega = dQ^i \wedge dP_i$. In this case, (Q^i, P_i) are called *canonical coordinates* on T^*M. In particular, natural coordinates are canonical coordinates.

Exercise 2.3.2. Show that if L is a regular Lagrangian, then the canonical coordinates (q^i, p_i) defined from the corresponding Legendre transformation $\mathcal{L}_L$ by $p_i = \xi_i \circ \mathcal{L}_L = \frac{\partial L}{\partial \dot{q}^i}$, deserve the name, that is,

$$
\omega = dq^i \wedge dp_i.
$$

Remark 2.3.2. A *symplectic form* on a smooth manifold X is a 2-form ω on X that is closed ($d\omega = 0$) and nondegenerate ($\iota_V \omega = 0 \Rightarrow V = 0$); the pair (X, ω) is then called a *symplectic manifold*. Both of these conditions are clearly satisfied by the ω we have defined on T^*M so this is, indeed, a symplectic form. The formalism we will describe for Hamiltonian mechanics is based almost entirely on the existence of this natural symplectic form on T^*M together with the Legendre transform $H : T^*M \to \mathbb{R}$ of the energy function $E_L : TM \to \mathbb{R}$ on TM (which we will introduce shortly). It turns out that the entire formalism extends to an arbitrary symplectic manifold (X, ω) together with a choice of some distinguished real-valued function $H : X \to \mathbb{R}$ on X. We will not require this much generality, but we will nevertheless endeavor to phrase our discussions in terms that will ease the transition to symplectic geometry and Hamiltonian mechanics (some standard references for this material are [Abra], [Arn2], [CM], [Gold], [GS1] and [Sp3]).

Let us be explicit about the context in which we will make our move from state space to phase space. We are given a smooth n-dimensional manifold M, called the configuration space. The tangent bundle TM is called the state space and we assume that we are given a *regular* Lagrangian $L : TM \to \mathbb{R}$ on M. The cotangent bundle T^*M is called the phase space and the corresponding Legendre transformation $\mathcal{L}_L : TM \to T^*M$ is a diffeomorphism of TM onto an open subset $\mathcal{O}$ of T^*M. The natural 1-form on T^*M is denoted θ and the natural symplectic form on T^*M is $\omega = -d\theta$. Local coordinates on M are written $q^1,\ldots,q^n$ and these give rise to natural coordinates $q^1,\ldots,q^n,\dot{q}^1,\ldots,\dot{q}^n$ on TM and $q^1,\ldots,q^n,\xi_1,\ldots,\xi_n$ on T^*M. In such natural coordinates, the energy function on TM is given by

$$E_L(q,\dot{q}) = \frac{\partial L}{\partial \dot{q}^i}\dot{q}^i - L(q,\dot{q})$$

(see (2.16)), the natural 1-form on T^*M is

$$\theta = \xi_i\, dq^i$$

and the natural symplectic form on T^*M is

$$\omega = dq^i \wedge d\xi_i.$$

We begin by moving the energy function $E_L : TM \to \mathbb{R}$ to T^*M. Note that, since $\mathcal{L}_L$ is a diffeomorphism of TM onto its image $\mathcal{O} \subseteq T^*M$, there is clearly a function $H : \mathcal{O} \to \mathbb{R}$ whose pullback to TM under $\mathcal{L}_L$ is E_L, that is, which satisfies

$$\mathcal{L}_L^*(H) = H \circ \mathcal{L}_L = E_L = \frac{\partial L}{\partial \dot{q}^i}\dot{q}^i - L(q,\dot{q}).$$

We will describe H in canonical coordinates $(q^1,\ldots,q^n,p_1,\ldots,p_n)$ on T^*M. Since $\mathcal{L}_L$ is a diffeomorphism,

$$H = E_L \circ \mathcal{L}_L^{-1} = (\mathcal{L}_L^{-1})^* E_L = (\mathcal{L}_L^{-1})^* \left(\frac{\partial L}{\partial \dot{q}^i}\dot{q}^i - L(q,\dot{q}) \right) = p_i\,(\dot{q}^i \circ \mathcal{L}_L^{-1}) - L \circ \mathcal{L}_L^{-1}. \quad (2.42)$$

One often sees $H = p_i (\dot{q}^i \circ \mathcal{L}_L^{-1}) - L \circ \mathcal{L}_L^{-1}$ written as

$$H(q^1,\ldots,q^n,p_1,\ldots,p_n) = p_i\dot{q}^i - L(q^1,\ldots,q^n,\dot{q}^1,\ldots,\dot{q}^n), \qquad (2.43)$$

where it is understood that, on the right-hand side, each $\dot{q}^i$ is expressed as a function of $q^1,\ldots,q^n,p_1,\ldots,p_n$ (by solving the system (2.31)). The energy function H on T^*M is called the *Hamiltonian*.

Every curve in M has a natural lift to the state space TM. If the local coordinate functions of the curve are $q^1(t),\ldots,q^n(t)$, then the natural coordinates of the lift are $q^1(t),\ldots,q^n(t),\dot{q}^1(t),\ldots,\dot{q}^n(t)$ and, for a stationary curve, these satisfy the Euler–Lagrange equations

$$\frac{\partial L}{\partial q^k} - \frac{d}{dt}\left(\frac{\partial L}{\partial \dot{q}^k}\right) = 0, \quad 1 \le k \le n.$$

These stationary curves describe the evolution *in state space* of the physical system whose Lagrangian is L. The Legendre transformation, being a diffeomorphism, moves lifted curves in TM to curves in phase space T^*M by simply composing with $\mathcal{L}_L$. The image of a stationary curve can be thought of as describing the evolution of the physical system *in phase space* rather than in state space. We would like to show now that, in terms of canonical coordinates on T^*M, the differential equations that describe such curves assume a particularly simple and symmetrical form. Begin by computing the differential dH. By definition,

$$dH = \frac{\partial H}{\partial q^i} dq^i + \frac{\partial H}{\partial p_i} dp_i. \qquad (2.44)$$

But also we compute from $H = p_i (\dot{q}^i \circ \mathcal{L}_L^{-1}) - L \circ \mathcal{L}_L^{-1}$ that

$$
\begin{aligned}
dH &= d\left(p_i\left(\dot{q}^i \circ \mathcal{L}_L^{-1}\right) - L \circ \mathcal{L}_L^{-1}\right) \\
&= \left(\dot{q}^i \circ \mathcal{L}_L^{-1}\right) dp_i + p_i\, d(\dot{q}^i \circ \mathcal{L}_L^{-1}) - d(L \circ \mathcal{L}_L^{-1}) \\
&= \left(\dot{q}^i \circ \mathcal{L}_L^{-1}\right) dp_i + p_i\, d\left((\mathcal{L}_L^{-1})^*(\dot{q}^i)\right) - d\left((\mathcal{L}_L^{-1})^*(L)\right) \\
&= \left(\dot{q}^i \circ \mathcal{L}_L^{-1}\right) dp_i + p_i (\mathcal{L}_L^{-1})^*(d\dot{q}^i) - (\mathcal{L}_L^{-1})^*(dL) \\
&= \left(\dot{q}^i \circ \mathcal{L}_L^{-1}\right) dp_i + p_i (\mathcal{L}_L^{-1})^*(d\dot{q}^i) - (\mathcal{L}_L^{-1})^*\left(\frac{\partial L}{\partial q^i} dq^i + \frac{\partial L}{\partial \dot{q}^i} d\dot{q}^i\right) \\
&= \left(\dot{q}^i \circ \mathcal{L}_L^{-1}\right) dp_i + p_i (\mathcal{L}_L^{-1})^*(d\dot{q}^i) - \left(\frac{\partial L}{\partial q^i} \circ \mathcal{L}_L^{-1}\right) dq^i - p_i (\mathcal{L}_L^{-1})^*(d\dot{q}^i) \\
&= \left(\dot{q}^i \circ \mathcal{L}_L^{-1}\right) dp_i - \left(\frac{\partial L}{\partial q^i} \circ \mathcal{L}_L^{-1}\right) dq^i. \qquad (2.45)
\end{aligned}
$$

Comparing (2.44) and (2.45) we find that

$$\frac{\partial H}{\partial q^i} = -\frac{\partial L}{\partial q^i} \circ \mathcal{L}_L^{-1} \quad \text{and} \quad \frac{\partial H}{\partial p_i} = \dot{q}^i \circ \mathcal{L}_L^{-1}, \quad i = 1,\ldots,n. \qquad (2.46)$$

Now, suppose $\alpha(t) = (q^1(t), \ldots, q^n(t))$ is a stationary curve in M. Then the lift $\tilde\alpha(t) = (q^1(t), \ldots, q^n(t), \dot q^1(t), \ldots, \dot q^n(t))$ of α to TM satisfies the Euler–Lagrange equations. Its image $\mathcal{L}_L \circ \tilde\alpha$ in T^*M under the Legendre transformation is written

$$\mathcal{L}_L \circ \tilde\alpha(t) = (q^1(t), \ldots, q^n(t), p_1(t), \ldots, p_n(t)).$$

From (2.46) we find that $\frac{\partial H}{\partial p_i}(\mathcal{L}_L \circ \tilde\alpha(t)) = \dot q^i(t)$ and

$$
\begin{aligned}
-\frac{\partial H}{\partial q^i}(\mathcal{L}_L \circ \tilde\alpha(t)) &= \frac{\partial L}{\partial q^i}(\tilde\alpha(t)) \\
&= \frac{d}{dt}\left(\frac{\partial L}{\partial \dot q^i}(\tilde\alpha(t))\right) \quad (Euler\text{–}Lagrange) \\
&= \frac{d}{dt}(p_i(t)) \\
&= \dot p_i(t).
\end{aligned}
\tag{2.47}
$$

We conclude that the differential equations that determine the evolution of the system in phase space are

$$\frac{dq^i}{dt} = \frac{\partial H}{\partial p_i} \quad \text{and} \quad \frac{dp_i}{dt} = -\frac{\partial H}{\partial q^i}, \quad i = 1, \ldots, n. \tag{2.48}$$

These are called *Hamilton's equations.* We will write out a few examples momentarily, but first we would like to observe that, even though Hamilton's equations do the same job as the Euler–Lagrange equations, they are geometrically much nicer, even aside from their obvious symmetry. Euler–Lagrange is a system of second order ordinary differential equations with no really apparent geometrical interpretation. On the other hand, Hamilton's equations form a system of first order ordinary differential equations in the coordinates on phase space and so the solutions can be viewed as integral curves of a vector field on T^*M, specifically, the so-called *Hamiltonian vector field*

$$X_H = \frac{\partial H}{\partial p_i}\,\partial_{q^i} - \frac{\partial H}{\partial q^i}\,\partial_{p_i}. \tag{2.49}$$

The time evolution of the system can be viewed as the flow of this vector field on phase space; this is very pretty and, as we will see, very useful.

Example 2.3.3. As a consistency check, let us look once more at a single particle moving in $\mathbb{R}^n$. As usual, we choose standard coordinates $x^1, \ldots, x^n$ on $\mathbb{R}^n$ and take the Lagrangian to be

$$L(x, \dot x) = L(x^1, \ldots, x^n, \dot x^1, \ldots, \dot x^n) = \frac{1}{2}m\sum_{i=1}^{n}(\dot x^i)^2 - V(x^1, \ldots, x^n),$$

where V is an arbitrary smooth function on $\mathbb{R}^n$. Then $p_i = \frac{\partial L}{\partial \dot{x}^i} = m\dot{x}^i$ so $\dot{x}^i = \frac{p_i}{m}$. The energy function on the tangent bundle is therefore

$$E_L(x, \dot{x}) = \frac{1}{2}m\|\dot{x}\|^2 + V(x)$$
$$= \frac{1}{2}m\left((\dot{x}^1)^2 + \cdots (\dot{x}^n)^2\right) + V(x^1, \ldots, x^n)$$

and so the Hamiltonian on phase space is

$$H(x, p) = \frac{1}{2}m\left(\left(\frac{p_1}{m}\right)^2 + \cdots + \left(\frac{p_n}{m}\right)^2\right) + V(x^1, \ldots, x^n)$$
$$= \frac{1}{2m}\|p\|^2 + V(x).$$

Hamilton's equations therefore give $\dot{x}^i = \frac{\partial H}{\partial p_i} = \frac{p_i}{m}$ and $\dot{p}_i = -\frac{\partial H}{\partial x^i} = -\frac{\partial V}{\partial x^i}$. Combining these two gives

$$m\ddot{x}^i = -\frac{\partial V}{\partial x^i}, \quad i = 1, \ldots, n,$$

and, as expected, this is just Newton's second law in components.

Let us also write out a special case of interest to us and have a look at its Hamiltonian vector field.

Example 2.3.4. For the harmonic oscillator the Hamiltonian on $T^*\mathbb{R}$ is given by

$$H(q, p) = \frac{1}{2m}p^2 + \frac{m\omega^2}{2}q^2, \tag{2.50}$$

where we have used q as the standard coordinate on $\mathbb{R}$ and $\omega = \sqrt{k/m}$. Thus, $\frac{\partial H}{\partial p} = \frac{p}{m}$ and $\frac{\partial H}{\partial q} = m\omega^2 q$ so the Hamiltonian vector field is

$$X_H = \frac{p}{m}\partial_q - m\omega^2 q\,\partial_p. \tag{2.51}$$

In this case, of course, Hamilton's equations are easy to solve explicitly (they are equivalent to Newton's second law $\ddot{q}(t) + \omega^2 q(t) = 0$ and we solved this long ago). However, this is generally not the case and one would like to retrieve at least some of the qualitative behavior of the system from X_H itself, that is, from qualitative information about the integral curves of X_H in the qp-phase plane. This is an old and venerable part of mathematics, but one that we will not make use of here (a nice introduction is available in [Arn1]). Suffice it to say that one can read off directly from $\frac{p}{m}\partial_q - m\omega^2 q\,\partial_p$ the existence of one stable point of equilibrium (at $(q, p) = (0, 0)$) and the fact that the integral curves are ellipses about the equilibrium point so that, in particular, they do not approach equilibrium and the periodicity of the system's behavior is manifest.

Exercise 2.3.3. The Lagrangian for the spherical pendulum (Example 2.2.2) is

$$L(\phi, \theta, \dot{\phi}, \dot{\theta}) = \frac{1}{2}ml^2(\dot{\phi}^2 + \dot{\theta}^2 \sin^2 \phi) + mgl \cos \phi.$$

Denote by p_ϕ and p_θ the conjugate momenta of ϕ and θ, respectively. Show that the Hamiltonian is given by

$$H(\phi, \theta, p_\phi, p_\theta) = \frac{1}{2ml^2}(p_\phi^2 + p_\theta^2 \csc^2 \phi) - mgl \cos \phi$$

and that Hamilton's equations are

$$\dot{\phi} = \frac{1}{ml^2} p_\phi,$$

$$\dot{\theta} = \frac{1}{ml^2} (\csc^2 \phi) p_\theta,$$

$$\dot{p}_\phi = \frac{1}{ml^2} (\cos \phi)(\csc^2 \phi) p_\theta^2 - mgl \sin \phi \quad \text{and}$$

$$\dot{p}_\theta = 0.$$

Our objective in the remainder of this section is to describe what has proved to be the essential mathematical structure of classical Hamiltonian mechanics. This structure is much more general than the context in which we currently find ourselves (see Remark 2.3.2) and it is the structure that is generalized to produce the standard formalism of quantum mechanics in Chapters 3 and 7. We will, whenever feasible, provide arguments that exhibit this generality, even if they are not the most elementary possible.

An inner product $\langle , \rangle$ on a vector space V determines a natural isomorphism of V onto its dual V^* given by $v \in V \to \langle v, \cdot \rangle \in V^*$ and this depends only on the bilinearity and nondegeneracy of $\langle , \rangle$. At each point (p, η_p) of T^*M the canonical symplectic form ω on T^*M is a nondegenerate, bilinear form $\omega_{(p,\eta_p)} : T_{(p,\eta_p)}(T^*M) \times T_{(p,\eta_p)}(T^*M) \to \mathbb{R}$ on $T_{(p,\eta_p)}(T^*M)$ so it induces an isomorphism

$$x \in T_{(p,\eta_p)}(T^*M) \to \omega_{(p,\eta_p)}(x, \cdot) \in T^*_{(p,\eta_p)}(T^*M)$$

of $T_{(p,\eta_p)}(T^*M)$ onto $T^*_{(p,\eta_p)}(T^*M)$. This then extends to an isomorphism of the infinite-dimensional vector space of smooth vector fields on T^*M onto the space of 1-forms on T^*M given by

$$X \to \iota_X \omega,$$

where $\iota_X \omega$ is the contraction of ω with X, that is, $(\iota_X \omega)(V) = \omega(X, V)$ for any smooth vector field V.

Example 2.3.5. Let us compute the image of the Hamiltonian vector field X_H under this isomorphism. In canonical coordinates, $\omega = dq^i \wedge dp_i$ and $X_H = \frac{\partial H}{\partial p_i}\partial_{q^i} - \frac{\partial H}{\partial q^i}\partial_{p_i}$, so

$$\iota_{X_H}\omega = \iota_{X_H}(dq^i \wedge dp_i)$$
$$= \iota_{X_H}(dq^i \otimes dp_i - dp_i \otimes dq^i)$$
$$= dq^i(X_H)dp_i - dp_i(X_H)dq^i$$
$$= -X_H[p_i]dq^i + X_H[q^i]dp_i$$
$$= \frac{\partial H}{\partial q^i}dq^i + \frac{\partial H}{\partial p_i}dp_i$$

and therefore

$$\iota_{X_H}\omega = dH.$$

We generalize this last example in the following way. The Hamiltonian H is, in particular, a smooth real-valued function on T^*M. For any $f \in C^\infty(T^*M)$, df is a 1-form on T^*M so there exists a smooth vector field X_f on T^*M satisfying

$$\iota_{X_f}\omega = df.$$

Exercise 2.3.4. Show that

$$X_f = \frac{\partial f}{\partial p_i}\partial_{q^i} - \frac{\partial f}{\partial q^i}\partial_{p_i}. \tag{2.52}$$

We will refer to the vector field X_f in (2.52) as the *symplectic gradient* of f; X_f is often called the *Hamiltonian vector field* of f, but we will reserve this terminology for the symplectic gradient X_H of the Hamiltonian. Now note that if g is another smooth, real-valued function on T^*M, then

$$X_f[g] = \frac{\partial f}{\partial p_i}\frac{\partial g}{\partial q^i} - \frac{\partial f}{\partial q^i}\frac{\partial g}{\partial p_i} = -X_g[f].$$

Moreover,

$$\omega(X_f, X_g) = \iota_{X_g} \circ \iota_{X_f}\omega$$
$$= -\iota_{X_f} \circ \iota_{X_g}\omega$$
$$= -\iota_{X_f}(dg) = -dg(X_f) = -X_f[g]$$
$$= \frac{\partial f}{\partial q^i}\frac{\partial g}{\partial p_i} - \frac{\partial f}{\partial p_i}\frac{\partial g}{\partial q^i}.$$

Now we will consolidate this information by defining, for all $f, g \in C^\infty(T^*M)$, the *Poisson bracket* $\{f, g\}$ of f and g by

$$\{f, g\} = \omega(X_f, X_g) = -X_f[g] = X_g[f] = \frac{\partial f}{\partial q^i}\frac{\partial g}{\partial p_i} - \frac{\partial f}{\partial p_i}\frac{\partial g}{\partial q^i}. \tag{2.53}$$

The Poisson bracket $\{\,,\,\} : C^\infty(T^*M) \times C^\infty(T^*M) \to C^\infty(T^*M)$ provides $C^\infty(T^*M)$ with a certain mathematical structure that is central to everything we will say about Hamiltonian mechanics as well as to the formulation of quantum mechanics that we will describe in Chapters 3 and 7 so we would now like to spell out this structure in detail. We will refer to the elements of $C^\infty(T^*M)$ as *classical observables*.

Here is the idea behind this last definition. Intuitively, an observable is something we can measure and the value we actually do measure depends on the state of the system at the time we do the measuring. The result of the measurement is a real number such as an energy, a component of linear or angular momentum, a coordinate of the position of some particle, etc. If we represent the states of the system as points in phase space, each of these observables is a real-valued function on T^*M. One can assume, at least provisionally, that these functions vary smoothly with the state and are therefore elements of $C^\infty(T^*M)$. Now, it is certainly not the case that every element of $C^\infty(T^*M)$ can be naturally identified with something we might actually set up an experiment to measure. Nevertheless, every such function in $C^\infty(T^*M)$ is, locally at least, expressible in terms of position coordinates q^i and momentum coordinates p_i, which certainly should qualify as physical observables; measuring these, in effect, measures everything in $C^\infty(T^*M)$.

Note that $C^\infty(T^*M)$ has the structure of a unital algebra over $\mathbb{R}$ with pointwise addition $((f+g)(x) = f(x) + g(x))$, scalar multiplication $((af)(x) = af(x))$ and multiplication $((fg)(x) = f(x)g(x))$ and for which the multiplicative unit element is the constant function on T^*M whose value is $1 \in \mathbb{R}$. The following three properties of the Poisson bracket show that it provides $C^\infty(T^*M)$ with the structure of a Lie algebra:

$$\{\,,\,\} : C^\infty(T^*M) \times C^\infty(T^*M) \to C^\infty(T^*M) \quad \text{is } \mathbb{R}\text{-bilinear,} \tag{2.54}$$

$$\{g,f\} = -\{f,g\} \quad \forall f,g \in C^\infty(T^*M), \tag{2.55}$$

$$\{f,\{g,h\}\} + \{h,\{f,g\}\} + \{g,\{h,f\}\} = 0 \quad \forall f,g,h \in C^\infty(T^*M). \tag{2.56}$$

The first two of these are obvious from the fact that $\{f,g\} = \omega(X_f, X_g)$ and ω is a 2-form and therefore bilinear and skew-symmetric. The proof of the *Jacobi identity* (2.56) takes a bit more work. One could, of course, compute everything in local coordinates to produce a great mass of partial derivatives and then watch everything cancel, but this is not particularly enlightening. We prefer to give an argument that involves a bit more machinery, but indicates clearly that the result depends only on the fact that ω is a nondegenerate, closed 2-form and the definition of the symplectic gradient.

Since ω is closed, the 3-form $d\omega$ is identically zero so, in particular, $d\omega(X_f, X_g, X_h) = 0$. We begin by writing this out in coordinate free fashion in terms of the Lie bracket (see Exercise 4.4.1 of [Nab4]):

$$0 = X_f[\omega(X_g, X_h)] - X_g[\omega(X_f, X_h)] + X_h[\omega(X_f, X_g)]$$
$$- \omega([X_f, X_g], X_h) + \omega([X_f, X_h], X_g) - \omega([X_g, X_h], X_f). \tag{2.57}$$

Next note that

$$X_f[\omega(X_g, X_h)] = X_f(\{g, h\}) = -\{f, \{g, h\}\}$$

and similarly for the second and third terms in (2.57). From this we conclude that

$$
\begin{aligned}
X_f[\omega(X_g, X_h)] &- X_g[\omega(X_f, X_h)] + X_h[\omega(X_f, X_g)] \\
&= -\{f, \{g, h\}\} + \{g, \{f, h\}\} - \{h, \{f, g\}\} \\
&= -\{f, \{g, h\}\} - \{h, \{f, g\}\} - \{g, \{h, f\}\}.
\end{aligned}
\tag{2.58}
$$

In order to handle the remaining three terms in (2.57) we require a preliminary result that we will see is of independent interest. Specifically, we will show that

$$[X_f, X_g] = X_{\{g, f\}}.\tag{2.59}$$

Remark 2.3.3. The next few arguments will require some basic facts about the Lie derivative. Recall that if X is a vector field, then the Lie derivative $\mathcal{L}_X$ computes rates of change along the integral curves of X. If f is a smooth real-valued function, then $\mathcal{L}_X f = X[f]$; if Y is a vector field, then $\mathcal{L}_X Y = [X, Y]$, the Lie bracket of X and Y; if η is a differential form, then $\mathcal{L}_X \eta$ can be computed from the Cartan formula $\mathcal{L}_X \eta = (\iota_X \circ d)\,\eta + (d \circ \iota_X)\,\eta$. All of this can be found in Volume I of [Sp2]; most of it is in Chapter 5, but the Cartan formula is Exercise 18(e).

We begin by noting that

$$\mathcal{L}_{X_f} \omega = 0$$

because $\mathcal{L}_{X_f} \omega = \iota_{X_f}(d\omega) + d(\iota_{X_f}\omega) = \iota_{X_f}(0) + d(df) = 0 + d^2f = 0$. This, together with the identity

$$\mathcal{L}_X(\iota_Y \eta) - \iota_Y(\mathcal{L}_X \eta) = \iota_{[X,Y]}\,\eta,$$

which is satisfied by all vector fields X and Y and all differential forms η, gives

$$
\begin{aligned}
\iota_{[X_f, X_g]}\omega &= \mathcal{L}_{X_f}(\iota_{X_g}\omega) = (d \circ \iota_{X_f} + \iota_{X_f} \circ d)(\iota_{X_g}\omega) \\
&= d(\iota_{X_f} \circ \iota_{X_g}\omega) + \iota_{X_f}(d(\iota_{X_g}\omega)) \\
&= d(\omega(X_g, X_f)) + \iota_{X_f}(d^2 g) \\
&= d(\omega(X_g, X_f)) + 0,
\end{aligned}
$$

so that

$$\iota_{[X_f, X_g]}\omega = d(\{g, f\}),$$

and this is precisely the statement (2.59). With this we can finish the proof of the Jacobi identity (2.56). Note that we can now write

$$-\omega([X_f, X_g], X_h) = -\omega(X_{\{g,f\}}, X_h) = -\{\{g,f\}, h\} = \{h, \{g,f\}\} = -\{h, \{f,g\}\} \qquad (2.60)$$

and similarly for the fifth and sixth terms in (2.57). Thus, we find that

$$-\omega([X_f, X_g], X_h) + \omega([X_f, X_h], X_g) - \omega([X_g, X_h], X_f)$$
$$= -\{f, \{g, h\}\} - \{h, \{f, g\}\} - \{g, \{h, f\}\}. \qquad (2.61)$$

Inserting (2.58) and (2.61) into (2.57) gives (2.56).

We have proved (2.54), (2.55) and (2.56) and therefore that the Poisson bracket provides the algebra $C^\infty(T^*M)$ of classical observables with the structure of a Lie algebra. It does even more, however. In general, a Lie algebra is a vector space, but it has no multiplicative structure other than that provided by the bracket. However, $C^\infty(T^*M)$ is itself an algebra under pointwise multiplication and we now show that this algebra structure is consistent with the Lie algebra structure in that they are related by the following *Leibniz rule*:

$$\{f, gh\} = \{f, g\}h + g\{f, h\} \quad \forall f, g, h \in C^\infty(T^*M). \qquad (2.62)$$

The proof of this is easy:

$$\{f, gh\} = \iota_{X_{gh}} \circ \iota_{X_f}\omega = -\iota_{X_f}(\iota_{X_{gh}}\omega) = -\iota_{X_f}(d(gh))$$
$$= -\iota_{X_f}((dg)h + gdh) = -\iota_{X_f}((dg)h) - \iota_{X_f}(gdh)$$
$$= -(\iota_{X_f}(dg))h - g(\iota_{X_f}(dh)) = -(dg(X_f))h - g(dh(X_f))$$
$$= -X_f[g]h - gX_f[h]$$
$$= \{f, g\}h + g\{f, h\}.$$

This provides $C^\infty(T^*M)$ with the structure of what is called a *Poisson algebra* and this makes T^*M itself a *Poisson manifold*.

Before returning to mechanics to see what all of this is good for, we pause to make a few observations and introduce a little terminology. We have shown that for any classical observable $f \in C^\infty(T^*M)$, the vector field X_f satisfies $\mathcal{L}_{X_f}\omega = 0$. In general, a vector field X on T^*M is said to be *symplectic* if the Lie derivative of the canonical symplectic form ω with respect to X is zero ($\mathcal{L}_X\omega = 0$) and this is equivalent to $\varphi_t^*\omega = \omega$ for every φ_t in the (local) one-parameter group of diffeomorphisms of X. A diffeomorphism φ of T^*M onto itself satisfying $\varphi^*\omega = \omega$ is called a *symplectomorphism* (by mathematicians) or a *canonical transformation* (by physicists). Consequently, any symplectic gradient is a symplectic vector field. In particular, the Hamiltonian vector field X_H is symplectic and in this case we think of $\varphi_t^* \omega = \omega$ as saying that ω is constant along the integral curves of the Hamiltonian, that is, along the trajectories of the system. Finally, we note that, since

$$\mathcal{L}_X \omega = d(\iota_X \omega) + \iota_X(d\omega) = d(\iota_X \omega),$$

the vector field X is symplectic if and only if the 1-form $\iota_X \omega$ is closed.

To get back to the physics, let us note first that, even if they served no other useful purpose, Poisson brackets make Hamilton's equations (2.48) very pretty. We have

$$\frac{dq^i}{dt} = \{q^i, H\} \quad \text{and} \quad \frac{dp_i}{dt} = \{p_i, H\}, \quad i = 1, \dots, n. \tag{2.63}$$

There is more, however. In the Hamiltonian picture the state of a physical system is described by a point in phase space T^*M and the system evolves along the integral curves of the Hamiltonian vector field X_H. We will say that an observable $f \in C^\infty(T^*M)$ is *conserved* if it is constant along each of these integral curves, that is, if $X_H[f] = 0$, which we now know is equivalent to

$$\{f, H\} = 0. \tag{2.64}$$

Consequently, *conserved quantities are precisely those observables that (Poisson) commute with the Hamiltonian*. In particular, the Hamiltonian itself (that is, the total energy) is clearly conserved since $\{H, H\} = 0$ follows from the skew-symmetry (2.55) of the Poisson bracket. Moreover, it follows from the Jacobi identity (2.56) that *the Poisson bracket of two conserved quantities is also conserved*. Indeed,

$$\{f, H\} = \{g, H\} = 0 \quad \Rightarrow \quad \{\{f, g\}, H\} = \{g, \{H, f\}\} + \{f, \{g, H\}\} = \{g, 0\} + \{f, 0\} = 0.$$

More generally, even if f is not conserved, the Poisson bracket keeps track of how it evolves with the system in the sense that, along an integral curve of X_H,

$$\frac{df}{dt} = \{f, H\} \tag{2.65}$$

(because $X_H[f] = -\{H, f\} = \{f, H\}$). Still more generally, if f and g are two observables, then the rates of change of f along the integral curves of X_g and of g along the integral curves of X_f are, according to (2.53), both encoded in the bracket $\{f, g\}$. The point here is that the Poisson structure of the algebra of classical observables contains a great deal of information about the dynamics of the system. As we will see in Chapter 7, it also provides the key to a process known as "canonical quantization."

If you will grant the importance of the Poisson bracket $\{f, g\}$ and recall that, locally at least, any observable can be written as a function of canonical coordinates, then it may come as no great surprise that a particular significance attaches to the brackets of the observables $q^1, \ldots, q^n, p_1, \ldots, p_n : T^*M \to \mathbb{R}$. These are called the *canonical commutation relations (for classical mechanics)* and are simply

$$\{q^i, q^j\} = \{p_i, p_j\} = 0 \text{ and } \{q^i, p_j\} = \delta^i{}_j, \quad i, j = 1, \ldots, n, \tag{2.66}$$

where $\delta^i{}_j$ is the Kronecker delta. We will have much more to say about these later.

Exercise 2.3.5. In Exercise 2.2.10 you derived the Lagrangian and Euler–Lagrange equations for the double pendulum. The objective now is to write out the Hamiltonian formulation. We will denote the conjugate momenta by $p_1 = \partial L/\partial \dot{\theta}_1$ and $p_2 = \partial L/\partial \dot{\theta}_2$.
1. Show that

$$p_1 = (m_1 + m_2) l_1^2 \dot{\theta}_1 + m_2 l_1 l_2 \dot{\theta}_2 \cos(\theta_1 - \theta_2)$$

and

$$p_2 = m_2 l_2^2 \dot{\theta}_2 + m_2 l_1 l_2 \dot{\theta}_1 \cos(\theta_1 - \theta_2).$$

2. Compute the Hamiltonian and show that

$$H(\theta_1, \theta_2, p_1, p_2) = \frac{l_2^2 m_2 p_1^2 + l_1^2 (m_1 + m_2) p_2^2 - 2 m_2 l_1 l_2 p_1 p_2 \cos(\theta_1 - \theta_2)}{2 l_1^2 l_2^2 m_2 [m_1 + m_2 \sin^2(\theta_1 - \theta_2)]}$$
$$- (m_1 + m_2) g l_1 \cos\theta_1 - m_2 g l_2 \cos\theta_2.$$

3. Show that Hamilton's equations are as follows:

$$\dot{\theta}_1 = \frac{l_2 p_1 - l_1 p_2 \cos(\theta_1 - \theta_2)}{l_1^2 l_2 [m_1 + m_2 \sin^2(\theta_1 - \theta_2)]},$$

$$\dot{\theta}_2 = \frac{l_1 (m_1 + m_2) p_2 - l_2 m_2 p_1 \cos(\theta_1 - \theta_2)}{l_1 l_2^2 m_2 [m_1 + m_2 \sin^2(\theta_1 - \theta_2)]},$$

$$\dot{p}_1 = -(m_1 + m_2) g l_1 \sin\theta_1 - \frac{p_1 p_2 \sin(\theta_1 - \theta_2)}{l_1 l_2 [m_1 + m_2 \sin^2(\theta_1 - \theta_2)]}$$
$$+ \frac{l_2^2 m_2 p_1^2 + l_1^2 (m_1 + m_2) p_2^2 - l_1 l_2 m_2 p_1 p_2 \cos(\theta_1 - \theta_2)}{2 l_1^2 l_2^2 [m_1 + m_2 \sin^2(\theta_1 - \theta_2)]^2} \sin 2(\theta_1 - \theta_2),$$

$$\dot{p}_2 = -m_2 g l_2 \sin\theta_2 + \frac{p_1 p_2 \sin(\theta_1 - \theta_2)}{l_1 l_2 [m_1 + m_2 \sin^2(\theta_1 - \theta_2)]}$$
$$- \frac{l_2^2 m_2 p_1^2 + l_1^2 (m_1 + m_2) p_2^2 - l_1 l_2 m_2 p_1 p_2 \cos(\theta_1 - \theta_2)}{2 l_1^2 l_2^2 [m_1 + m_2 \sin^2(\theta_1 - \theta_2)]^2} \sin 2(\theta_1 - \theta_2).$$

Needless to say, Hamilton's equations are generally solvable only numerically. When this is done one discovers a great variety of possible motions depending on the initial conditions. This motion might be periodic, nearly periodic or even chaotic.

Remark 2.3.4. There is an important aspect of Hamiltonian mechanics, called *Hamilton–Jacobi theory*, that we will not discuss here. There is a brief introduction to the idea in Section 11 of [MacL] and an elementary, but more detailed treatment in Chapter 18 of [Sp3]; the latter also contains a few remarks on the relevance of Hamilton–Jacobi theory to Schrödinger's derivation of his wave equation for quantum mechanics. For a more sophisticated, mathematically rigorous treatment one might consult, for example, [Abra].

2.4 Segue

Before leaving classical particle mechanics behind we would like describe one more example that may ease the transition into our next topic in Chapter 3.

Example 2.4.1. One liter of the earth's atmosphere at standard temperature (0°C) and standard pressure (approximately the average atmospheric pressure at sea level) contains about $k = 2.68 \times 10^{22}$ molecules. Suppose the system that interests us consists of these molecules and that they are free to roam without interference wherever they please in space. Then we have k particles with masses $m_1, \ldots, m_k$ and positions specified by $q_1 = (q^1, q^2, q^3), q_2 = (q^4, q^5, q^6), \ldots, q_k = (q^{3k-2}, q^{3k-1}, q^{3k})$. The configuration space is $M = \mathbb{R}^{3k}$ and the phase space is $T^*M = \mathbb{R}^{3k} \times \mathbb{R}^{3k}$. Canonical coordinates on T^*M are $q^1, \ldots, q^{3k}, p_1, \ldots, p_{3k}$ and the interaction among the molecules is specified by some Hamiltonian $H(q^1, \ldots, q^{3k}, p_1, \ldots, p_{3k})$. The evolution of the system in phase space is then determined by Hamilton's equations (2.48). All we need to do is supply this system of first order equations with some initial conditions and it will tell us how the system evolves. That is all! We simply need to determine the positions and momenta for 2.68×10^{22} molecules at some instant.

This is absurd, of course. While everything we said in the last example is quite true, it simply provides no effective means of actually studying such large systems. Here we get a glimpse into the subject of *statistical mechanics* which, as the name suggests, concedes that we may have access to only statistical information about the state of such a system and therefore can reasonably expect to make only statistical statements about the observables we measure. Such a scenario must be modeled mathematically in a rather different way. The "state" of such a system is no longer identified with an experimentally meaningless point in phase space, but rather with a Borel probability measure ν on T^*M.

Remark 2.4.1. Recall that, in any topological space X, a *Borel set* is an element of the σ-algebra generated by the open sets in X. A *Borel probability measure* on X is a measure ν on the σ-algebra of Borel sets for which $\nu(X) = 1$.

The ν-measure of a subset of phase space is thought of as the probability that the "actual state" is in that subset. Note, however, that the "actual state" of such a system is something of a Platonic ideal. One imagines, for example, that each molecule in our sample of the atmosphere really does have a well-defined position and momentum at each instant so that "in principle" the state of the system really is described by a point in $\mathbb{R}^{5.36\times10^{22}}$, even though we have no chance at all of determining what it actually is. It is easy to be persuaded by such "in principle" arguments, but it is wise to exercise just a bit of caution. If two firecrackers explode nearby one might reasonably ask whether or not the explosions were simultaneous. If two supernovae occur at distant points in the galaxy, special relativity asserts that it makes no sense at all, even "in principle," to ask if the explosions were or were not simultaneous (see Section 1.3 of [Nab5]). More to the point, we shall see that, when the rules of quantum mechanics take effect, the assertion that even a single particle has a well-defined position and momentum at each instant has no meaning and so its classical phase space has no meaning. The moral here is that extrapolation beyond our immediate range of experience can be dangerous; one should be aware of the assumptions one is making and open to the possibility that they may eventually need to be abandoned.

Given such a state/measure v and an observable f (like the total energy, for example) one obtains a Borel probability measure μ_{vf} on $\mathbb{R}$ by defining

$$\mu_{vf}(S) = v(f^{-1}(S))$$

for any Borel set $S \subseteq \mathbb{R}$. One then interprets $\mu_{vf}(S)$ as the *probability* that we will measure the value of f to lie in the set S if the system is known to be in the state v. Such probabilistic statements are generally the best one can hope for in statistical mechanics. For example, one can compute the expectation value of a given observable in a given state, but one has no information about the precise value of any given measurement of the observable.

At first glance this scheme may strike one as excessively abstract and rather exotic, but we will argue in Chapter 3 that, in fact, it presents a rather natural way of viewing the general problem of describing physical systems mathematically. Even the classical particle mechanics that we have been discussing can be phrased in these terms, although in this case there is no particular reason to do so. Such probabilistic models *are* appropriate for statistical mechanics and, more significantly for us, for quantum mechanics as well, since, in the quantum world, we will see that phenomena are *inherently* statistical and probabilistic and not simply because of our technological inability to deal with 2.68×10^{22} particles. We will see, for example, that the more you know about an electron's position, the less it is possible to know about its momentum so it is, *in principle*, impossible to represent the state of an electron by a point in T^*M.

In anticipation of our move toward probabilistic models in Chapter 3 we will conclude by noting that the structure of phase space determines a naturally associated volume form and therefore a measure. Indeed, the natural symplectic form ω is a nondegenerate 2-form on the $2n$-dimensional manifold T^*M, so $\frac{1}{n!}\omega^n = \frac{1}{n!}\omega \wedge \overset{n}{\cdots} \wedge \omega$ is a $2n$-form on T^*M. Nondegeneracy of ω implies that $\frac{1}{n!}\omega^n$ is nonvanishing and therefore is a volume form on T^*M. In particular, T^*M is orientable (Theorem 4.3.1 of [Nab4]). Note that, since pullback commutes with the wedge product and $\varphi_t^*\omega = \omega$ for every φ_t in the (local) one-parameter group of diffeomorphisms of the Hamiltonian vector field X_H, $\varphi_t^*(\frac{1}{n!}\omega^n) = \frac{1}{n!}\omega^n$ as well, so $\mathcal{L}_{X_H}(\frac{1}{n!}\omega^n) = 0$, and this volume form is, like ω, invariant under the flow of X_H. On the space of continuous functions with compact support on T^*M the integral

$$ f \to \int_{T^*M} f\left(\frac{1}{n!}\omega^n\right) $$

defines a positive linear functional and so there is a Borel measure μ on T^*M such that for every such function f,

$$ \int_{T^*M} f\, d\mu = \int_{T^*M} f\left(\frac{1}{n!}\omega^n\right) \tag{2.67} $$

(see Theorem D, Section 56, of [Hal1]); μ is called the *Liouville measure on T^*M*. In canonical coordinates on T^*M, $\frac{1}{n!}\omega^n = dq^1 \wedge dp_1 \wedge dq^2 \wedge dp_2 \wedge \cdots \wedge dq^n \wedge dp_n$ and

$$ d\mu = dq^1 dp_1 \cdots dq^n dp_n, $$

by which we mean simply Lebesgue measure on the image in $\mathbb{R}^{2n}$ of the canonical coordinate neighborhood in T^*M. *Liouville's theorem* asserts that this measure is also invariant under the Hamiltonian flow in the sense that for any Borel set B in T^*M, $\mu(\varphi_t(B)) = \mu(B)$ for every φ_t in the one-parameter group of diffeomorphisms of X_H (for a proof of Liouville's theorem for $M = \mathbb{R}^{2n}$ see pages 69–70 of [Arn2]).

We will have more to say about this measure shortly, but we should point out that the objects of particular interest in statistical mechanics are certain measures obtained from it. We will simply sketch the idea. The Hamiltonian is a smooth real-valued function $H : T^*M \to \mathbb{R}$ on phase space thought of as the total energy function. Let $E \in \mathbb{R}$ be a regular value of H (see Section 5.6 of [Nab3]). Then $\Omega_E - H^{-1}(E)$ is (either empty or) a smooth submanifold of T^*M of dimension $2n-1$ (Corollary 5.6.7 of [Nab3]). Since H is conserved, the evolution of the system always takes place in such a *constant energy hypersurface* Ω_E in phase space. Since E is a regular value of H, dH is a nonzero 1-form on a neighborhood of Ω_E and it can be shown that locally one can write

$$ \frac{1}{n!}\omega^n = \eta \wedge dH $$

for some (nonunique) $(2n-1)$-form η on T^*M. Moreover, the restriction of η to Ω_E is independent of the choice of η and is a volume form on Ω_E. This volume form on Ω_E defines a measure on Ω_E called the *Liouville measure on Ω_E*. Since the Hamiltonian flow preserves ω, dH and Ω_E, it preserves the Liouville measure on Ω_E as well.

3 The formalism of quantum mechanics: motivation

3.1 Introduction

Classical mechanics is remarkably successful in describing a certain limited range of phenomena, but there is no doubt that this range *is* limited. We have spoken often about "particles," but a precise, or even imprecise definition of just what the word means was conspicuously absent. Depending on the circumstances, a particle might be a planet, or a baseball, or a grain of sand. One cannot, however, continue to diminish the size of these objects indefinitely for it has been found (experimentally) that one reaches a point where the rules change and the predictions of classical mechanics do not even remotely resemble what actually happens. In the late years of the nineteenth and early years of the twentieth centuries the technology available to experimental physicists led to the discovery that matter is composed of atoms and that these atoms, in turn, are composed of still smaller, electrically charged objects that came to be known as protons and electrons. The experiments strongly suggested a picture of the atom that resembled the solar system with a positively charged nucleus playing the role of the sun and the negatively charged electrons orbiting like planets. Unfortunately, this picture is completely incompatible with classical physics, which predicts that the charged, orbiting electrons must radiate energy and, as a result, spiral into the nucleus in short order. According to the rules of classical physics, such atoms could not be stable and the world as we perceive it could not exist. Clinging to the presumption that the world as we perceive it does exist, some adjustments to the classical picture would appear to be in order. It is to these adjustments that we now turn our attention.

Remark 3.1.1. Still other adjustments are necessitated by quite different experimental discoveries of the late nineteenth century. These two independent adjustments to classical physics are known as *quantum mechanics* and the *special theory of relativity*. Attempts to reconcile these to produce a more unified picture of the physical world led to what is called *quantum field theory*. Incidentally, we say "more unified" rather than simply "unified" because all of this fails to take any account of gravitational fields, which are described by the *general theory of relativity* (for a brief glimpse into what general relativity looks like we might suggest [Nab1] or Chapter 4 of [Nab5], but for a serious study of the subject one should proceed to [Wald]). Attempts to include general relativity in the picture, that is, to build a quantum theory of gravity, are ongoing, but concerning these we will maintain a discreet and humble silence.

Quantum mechanics is not a modest tweaking of classical mechanics, but a subtle and profoundly new view of how the physical universe works. It did not spring full blown into the world, but evolved over many years in fits and starts. Eventually, however, the underlying structure emerged and crystallized into a set of "postulates" that capture at least the formal aspects of the subject. It is entirely possible to simply write

https://doi.org/10.1515/9783110751949-003

down the axioms and start proving theorems, but here we will operate under the assumption that manipulating axioms for their own sake without any appreciation of where they came from or what they mean is rather sterile. We will therefore spend a bit of time discussing some of the experimental background in the hope of motivating certain aspects of the model which might otherwise give the impression of being something of a *deus ex machina*. We will not discuss these experiments in anything like historical order, nor will we try to provide the sort of detailed analysis one would expect to find in a physics text. Most importantly, it must be understood that one cannot *derive* the axioms of quantum mechanics any more than one can derive the axioms of group theory. One can only hope to provide something in the way of motivation.

Finally, we would like to have some mathematical context in which to carry out these discussions. Ideally, this would be a mathematical structure general enough to encompass anything that might reasonably be viewed as a model of some physical system, or at least those physical systems of interest to us at the moment. The choice we have made, which was briefly suggested at the end of Section 2.3, was introduced and studied by Mackey in [Mack2], which, together with [ChM2], or Lecture 7 in [Mar2], contains everything we will say here and much more. A more concise outline of Mackey's view of quantum mechanics is available in [Mack1].

3.2 States, observables and dynamics

We take the point of view that a mathematical model of a physical system consists of (at least) the following items:

1. a set $\mathcal{S}$, the elements of which are called *states*;
2. a set $\mathcal{O}$, the elements of which are called *observables*;
3. a mapping from $\mathcal{S} \times \mathcal{O}$ to the set of Borel probability measures on $\mathbb{R}$

$$(\psi, A) \rightarrow \mu_{\psi, A}$$

with the following physical interpretation: for any Borel set S in $\mathbb{R}$, $\mu_{\psi, A}(S)$ is the probability that we will measure the value of A to lie in the set S if the system is known to be in the state ψ;

4. a one-parameter family $\{U_t\}_{t \in \mathbb{R}}$ of mappings

$$U_t : \mathcal{S} \rightarrow \mathcal{S},$$

called *evolution operators*, satisfying

$$U_0 = \mathrm{id}_{\mathcal{S}},$$
$$U_{t+s} = U_t \circ U_s$$

and having the following physical interpretation: for any state ψ and any $t \in \mathbb{R}$, $U_t(\psi)$ is the state of the system at time t if its state at time 0 is ψ. Note that each U_t

is necessarily a bijection of S with inverse U_{-t}; $\{U_t\}_{t\in\mathbb{R}}$ is called a *one-parameter group* of transformations of S and it describes the *dynamics* of the system (how the states change with time). The physical meaning of $U_0 = \mathrm{id}_S$ is clear, whereas $U_{t+s} = U_t \circ U_s$ is a *causality* statement (the state of the system at any time s uniquely determines its state at any other time t since $U_t(\psi) = U_{t-s}(U_s(\psi))$).

Remark 3.2.1. The term "causality" often carries with it a great deal of metaphysical baggage, none of which is intended here. We have assumed only that the *states* of the system evolve deterministically and this need not imply anything at all about *events* occurring within the system. Soon we will see that the state of an atom of uranium ^{235}U is described in quantum mechanics at any time by a unit vector in a Hilbert space and that this state evolves deterministically according to the Schrödinger equation. Nevertheless, the emission of an alpha particle by the atom in radioactive decay is entirely random and spontaneous and one can attribute to it no "cause."

One can also formulate local versions of 4. in which t is restricted to some interval about 0 in $\mathbb{R}$, but for the sake of clarity we will, at least for the time being, restrict our attention to systems for which an evolution operator is defined for every $t \in \mathbb{R}$.

This model is extremely general, of course, and one cannot expect that a random specification of items 1.–4. will represent a physically reasonable system. It is possible to formulate a rigorous set of axioms that the states, observables and measures should satisfy in order to be deemed "reasonable" (see [Mack2]) and this is a useful thing to do, but may give the wrong impression of our objective here, which is to motivate, not axiomatize. For this reason we will be a bit more lighthearted about the additional assumptions we make, introducing them as needed. Here, for example, is one. We will assume that *two states that have the same probability distributions for **all** observables are, in fact, the same state*. More precisely, if $\psi_1, \psi_2 \in S$ have the property that $\mu_{\psi_1,A}(S) = \mu_{\psi_2,A}(S)$ for all $A \in \mathcal{O}$ and all Borel sets $S \subseteq \mathbb{R}$, then $\psi_1 = \psi_2$. Similarly, *two observables that have the same probability distributions for **all** states are the same observable*, that is, if $A_1, A_2 \in \mathcal{O}$ have the property that $\mu_{\psi,A_1}(S) = \mu_{\psi,A_2}(S)$ for all $\psi \in S$ and all Borel sets $S \subseteq \mathbb{R}$, then $A_1 = A_2$. Having made these assumptions we can now think of a state as an injective mapping from the set $\mathcal{O}$ of observables to the probability measures on $\mathbb{R}$. From this point of view one could even suppress the ψ altogether and think of a state as a family of probability measures

$$\mu_A, \quad A \in \mathcal{O},$$

on $\mathbb{R}$ parametrized by the set $\mathcal{O}$ of observables. Alternatively, one can think of an observable as an injective mapping from states to probability measures, or as a family $\mu_\psi, \psi \in S$, of probability measures on $\mathbb{R}$ parametrized by the set S of states.

Example 3.2.1. Regardless of the number of classical particles in the system of interest one begins the construction of a statistical model with a configuration space M, which is a smooth manifold of (potentially huge) dimension n, its phase space T^*M, a Hamiltonian $H : T^*M \to \mathbb{R}$ and the corresponding flow $\{\varphi_t\}_{t\in\mathbb{R}}$ of the Hamiltonian vector field X_H. We persist in the notion that the "actual state" of the system at any instant has some meaning and can be identified with a point in T^*M, despite the fact that all we can say about it is that it has a certain probability of lying in any given Borel subset of T^*M. Thus, we take $\mathcal{S}$ to be the set of all Borel probability measures on T^*M.

The set $\mathcal{O}$ of observables consists of all real-valued, Borel measurable functions on T^*M. If $v \in \mathcal{S}$ and $f \in \mathcal{O}$, then we define a Borel probability measure $\mu_{v,f}$ on $\mathbb{R}$ by

$$\mu_{v,f}(S) = v(f^{-1}(S))$$

for every Borel set $S \subseteq \mathbb{R}$. The interpretation is that $\mu_{v,f}(S)$ is the probability that a measurement of the observable f when the system is in state v will yield a value in S.

Remark 3.2.2. If $B \subseteq T^*M$ is a Borel set, then its characteristic function $\chi_B : T^*M \to \mathbb{R}$ is Borel measurable and is therefore an observable; Mackey [Mack2] refers to these observables as *questions*. Since any observable $f : T^*M \to \mathbb{R}$ is uniquely determined by its level sets $f^{-1}(a), a \in \mathbb{R}$, and these, in turn, are uniquely determined by their characteristic functions, there is a sense in which questions are the most fundamental observables. Questions will re-emerge in quantum mechanics as projection operators and describing arbitrary observables in terms of them is what the spectral theorem does.

Next we must specify, for each $t \in \mathbb{R}$, an evolution operator $U_t : \mathcal{S} \to \mathcal{S}$ that carries an initial state $v_0 \in \mathcal{S}$ to the corresponding state $v_t = U_t(v_0)$ at time t. The reasoning is as follows. For any Borel set B in T^*M, the probability that the actual state of the system at time t lies in B is exactly the same as the probability that the actual state of the system at time 0 lies in $\varphi_t^{-1}(B) = \varphi_{-t}(B)$ and this is $v_0(\varphi_{-t}(B))$. Thus, we define

$$U_t(v_0) = (\phi_t)_* v_0,$$

where $(\phi_t)_* v_0$ is the pushforward measure $((\phi_t)_* v_0)(B) = v_0(\phi_{-t}(B))$. This clearly defines a one-parameter group $\{U_t\}_{t\in\mathbb{R}}$ of transformations of $\mathcal{S}$ and therefore completes Example 3.2.1.

Exercise 3.2.1. Show that Hamiltonian mechanics, as we described it in Section 2.3, can be viewed in these same terms, although perhaps a bit artificially. Begin with a smooth, n-dimensional manifold M (configuration space), its cotangent bundle T^*M (phase space) and a Hamiltonian $H : T^*M \to \mathbb{R}$. Previously, we identified a state of the corresponding classical mechanical system with a point $x \in T^*M$, but now observe that the points of T^*M are in one-to-one correspondence with the point measures v_x

on T^*M defined, for every Borel set $B \subseteq T^*M$, by

$$v_x(B) = \begin{cases} 1 & \text{if } x \in B, \\ 0 & \text{if } x \notin B. \end{cases}$$

Take S to be the set of all such point measures and transfer the structure of T^*M to it. Take the set $\mathcal{O}$ of observables to consist of all real-valued, smooth (or simply Borel measurable) functions A on S. Define $U_t : S \to S$, $t \in \mathbb{R}$, by

$$U_t(v_x) = v_{\varphi_t(x)}, \quad t \in \mathbb{R}, x \in T^*M,$$

where $\{\varphi_t\}_{t \in \mathbb{R}}$ is the one-parameter family of diffeomorphisms for the Hamiltonian vector field X_H and convince yourself that the properties required of U_t in 4. are just restatements of known properties of the flow.

Except for one more brief encounter, we do not intend to pursue statistical mechanics beyond the description of the model in Example 3.2.1. Nevertheless, there are things that remain to be said about the model itself (Section 1-5 of [Mack2] contains a bit more information on how statistical mechanics works). First, however, we recall a few basic notions from probability theory (see Chapter 9 of [Hal1] for the details).

Let X be a topological space and v a Borel probability measure on X (see Remark 2.4.1). A real-valued Borel measurable function $f : X \to \mathbb{R}$ on X is called a *random variable* (so the observables in Example 3.2.1 are random variables). If f is integrable, then the *expectation value* (or *expected value*, or *mean value*) of f is defined by

$$E(f) = \int_X f \, dv.$$

The *variance* (or *dispersion*) of f is defined by

$$\sigma^2(f) = E\big([f - E(f)]^2 \big) = E(f^2) - E(f)^2$$

and is regarded as a measure of the extent to which the values of f cluster around its expected value (the smaller it is, the more clustered they are). The same interpretation is ascribed to the nonnegative square root of the dispersion, denoted $\sigma(f)$ and called the *standard deviation* of f. The *distribution function* of f is the real-valued function F of a real variable λ defined by

$$F(\lambda) = v\left(f^{-1}(-\infty, \lambda] \right).$$

Note that F is of bounded variation and both $E(f)$ and $\sigma^2(f)$ can be computed as Riemann–Stieltjes integrals with respect to F (see Appendix H.2). Specifically,

$$E(f) = \int_{\mathbb{R}} \lambda \, dF(\lambda) \tag{3.1}$$

and

$$\sigma^2(f) = \int_{\mathbb{R}} (\lambda - E(f))^2 \, dF(\lambda). \tag{3.2}$$

Finally, if μ is an arbitrary Borel measure on X, then a *probability density function* for μ is an integrable, real-valued function ρ that is nonnegative and satisfies

$$\int_X \rho \, d\mu = 1.$$

Given a probability density function ρ for μ one defines, for any Borel set $B \subseteq X$,

$$\nu_\rho(B) = \int_B \rho \, d\mu.$$

Then ν_ρ is a probability measure on X and

$$\int_X g \, d\nu_\rho = \int_X g\rho \, d\mu$$

for every nonnegative measurable function g.

Now let us return to Example 3.2.1. Suppose ν_0 is the initial state of our statistical mechanical system. Then ν_0 is intended to be a probability measure representing the extent of our knowledge of the "actual initial state" in T^*M. Physically, this knowledge would generally be expressed in terms of some sort of density function. Here, and on occasion in the future, we will rely on various analogies with fluid flow. Fluids consist of a huge (but finite) number of particles of various masses. Nevertheless, one generally idealizes and describes the density of a fluid with a *continuous* real-valued function ρ whose integral over any region B contained in the fluid is taken to be the mass of fluid in B. This leads us to assume that the initial states of physical interest are of a particular form. Specifically, we let μ denote the Liouville measure on T^*M (see (2.67)). Then, for any probability density function ρ_0 for μ we obtain a probability measure ν_{ρ_0} on T^*M, that is,

$$\nu_{\rho_0}(B) = \int_B \rho_0 \, d\mu,$$

for every Borel set B in T^*M. Aside from the fact that it is the most natural measure on T^*M, the reason for choosing the Liouville measure μ is that if the initial state has a probability density function for μ, then the same is true of every state throughout the evolution of the system. Specifically, if

$$\nu_0 = \nu_{\rho_0},$$

then we claim that

$$U_t(v_0) = U_t(v_{\rho_0}) = v_{\rho_t}, \tag{3.3}$$

where

$$\rho_t = \rho_0 \circ \varphi_{-t}. \tag{3.4}$$

To see this we recall that for any Borel set $B \subseteq T^*M$,

$$U_t(v_{\rho_0})(B) = ((\phi_t)_* v_0)(B) = v_{\rho_0}(\varphi_{-t}(B)) = \int_{\varphi_{-t}(B)} \rho_0 \, d\mu = \int_B (\rho_0 \circ \varphi_{-t}) \, d\mu,$$

where the last equality follows from the change of variables formula and the fact that φ_{-t} preserves the Liouville measure. Since

$$\int_B (\rho_0 \circ \varphi_{-t}) \, d\mu = \int_B \rho_t \, d\mu = v_{\rho_t}(B),$$

the proof of (3.3) is complete. We conclude from this that probability measures of the form

$$v_\rho(B) = \int_B \rho \, d\mu, \tag{3.5}$$

where ρ is a probability density function for the Liouville measure μ, are preserved by the Hamiltonian flow. According to the Radon–Nikodym theorem (Theorem B, Section 31, page 128, of [Hal1]), these are just the probability measures that are absolutely continuous with respect to the Liouville measure. Of course, there are probability measures, such as the point measures associated with Dirac deltas and convex combinations of them, that are not included among these. Nevertheless, we will, from this point on, focus our attention on measures of the form (3.5). We have several reasons for this, most of which are essentially physical. Point measures on T^*M represent absolute certainty regarding the actual state of the system and this is unattainable for systems that are "genuinely statistical," that is, very large. More significantly, we will soon describe some of the experimental evidence that led physicists to the conclusion that, in the quantum world, *all* systems, even those containing a single particle, are "genuinely statistical." Furthermore, from a practical point of view restricting attention to measures of the form (3.5) will permit us to shift the focus from probability measures (v_ρ) to probability densities (ρ) and these are much simpler objects, that is, just functions.

The model of statistical mechanics we have described has the advantage of being quite intuitive, but also the disadvantages of being technically rather difficult to deal with and still seemingly far removed from the generally accepted formalism of quantum mechanics that we are trying to motivate (Hilbert spaces and self-adjoint operators). We propose now to overcome both of these disadvantages by rephrasing the model in the more familiar and much more powerful language of functional analysis.

3.3 Mechanics and Hilbert spaces

The theory of Hilbert spaces has its roots deep in classical analysis and mathematical physics, but its birth as an independent discipline is to be found in the work of Hilbert on integral equations and von Neumann on quantum mechanics. By 1930, the theory was highly evolved and making its presence felt throughout analysis and theoretical physics. We will have a great deal more to say about its impact on quantum mechanics as we proceed, but for the moment we would like to focus our attention on the rather brief and not altogether well-known paper [Koop] of Koopman in 1931, where our probabilistic model of statistical mechanics is rephrased in functional analytic terms. This done we will try to show in the next chapter that at the quantum level, even the mechanics of a single particle is more akin to classical statistical mechanics than to classical particle mechanics and so, we hope, minimize the shock of introducing massive amounts of functional analysis to describe it.

We must assume a basic familiarity with Banach and Hilbert spaces and bounded operators on them. This information is readily available in essentially any functional analysis text such as, for example, [Fried], [RiSz.N], [TaylA] or [Rud1]; other good sources are [Prug], [RS1] or [vonNeu], which are specifically focused on the needs of quantum theory. For this material we will provide only a synopsis of the notation and terminology we intend to employ. We will, however, also require some rather detailed information about unbounded operators on Hilbert spaces and, on occasion, certain results that one cannot expect to find just anywhere. For this material we will provide some background, a few illustrative examples, precise statements of the results and either an accessible reference to a proof of what we need or, if such a reference does not seem to be readily available, a proof. To avoid sensory overload, we will try to introduce all of this material only as needed, although our initial foray into unbounded operators in Chapter 5 is necessarily rather lengthy.

Remember that $\mathcal{H}$ will always denote a *separable* Hilbert space and we will write its inner product as $\langle\,,\,\rangle_\mathcal{H}$ or simply $\langle\,,\,\rangle$ if there is no likelihood of confusion. Generally, $\mathcal{H}$ will be *complex*. Lest there be any confusion, we point out that we intend to adopt the physicist's convention of taking $\langle\,,\,\rangle$ to be complex linear in the *second* slot and *conjugate linear* in the first rather than the other way around. Specifically, if $\phi, \psi \in \mathcal{H}$ and $a, b \in \mathbb{C}$, then

$$\langle a\phi, \psi \rangle = \bar{a}\langle \phi, \psi \rangle$$

and

$$\langle \phi, b\psi \rangle = b\langle \phi, \psi \rangle.$$

The norm of $\psi \in \mathcal{H}$ is denoted $\|\psi\|_\mathcal{H}$ or simply $\|\psi\|$ and defined by $\|\psi\| = \sqrt{\langle \psi, \psi \rangle}$. The algebra of bounded linear operators on $\mathcal{H}$ will be denoted $\mathcal{B}(\mathcal{H})$. This is a Banach space (in fact, a Banach algebra) if the norm of any $T \in \mathcal{B}(\mathcal{H})$ is defined by

$$\|T\| = \sup_{\psi \neq 0} \frac{\|T\psi\|}{\|\psi\|} = \sup_{\|\psi\|=1} \|T\psi\|.$$

The topology induced on $\mathcal{B}(\mathcal{H})$ by this norm is called the *uniform operator topology*, or the *norm topology*, and convergence in this topology is called *uniform convergence*, or *norm convergence*. We will have occasion to consider various other notions of operator convergence as we proceed.

If $\mathcal{H}_1$ and $\mathcal{H}_2$ are two complex Hilbert spaces with inner products $\langle\,,\,\rangle_1$ and $\langle\,,\,\rangle_2$, respectively, then a linear map $T : \mathcal{H}_1 \to \mathcal{H}_2$ that satisfies $\langle T\phi, T\psi\rangle_2 = \langle\phi, \psi\rangle_1$ for all $\phi, \psi \in \mathcal{H}_1$ is said to be an *isometry*. If T maps *into* $\mathcal{H}_2$, then it is called an *isometric embedding* of $\mathcal{H}_1$ into $\mathcal{H}_2$; if T maps *onto* $\mathcal{H}_2$, then it is called a *unitary equivalence*, or an *isometric isomorphism* between $\mathcal{H}_1$ and $\mathcal{H}_2$. In particular, an operator $U \in \mathcal{B}(\mathcal{H})$ that is a unitary equivalence of $\mathcal{H}$ onto itself is called a *unitary operator*. Unitary operators have norm 1 and are characterized by the fact that $UU^* = U^*U = \mathrm{id}_{\mathcal{H}}$, that is, their inverses are the same as their adjoints so $\langle U\phi, \psi\rangle = \langle\phi, U^{-1}\psi\rangle$ for all $\phi, \psi \in \mathcal{H}$ (recall that for any bounded operator $T : \mathcal{H} \to \mathcal{H}$, there exists a unique bounded operator $T^* : \mathcal{H} \to \mathcal{H}$, called the *adjoint* of T, satisfying $\langle T\phi, \psi\rangle = \langle\phi, T^*\psi\rangle$ for all $\phi, \psi \in \mathcal{H}$).

Now let us begin Koopman's translation. Here are the essential features of the model that we wish to translate. We have a manifold that we choose to denote Ω since it might be either T^*M or one of the constant energy hypersurfaces Ω_E introduced earlier (see Section 2.4). On Ω we have a Liouville measure that we will denote by μ and a one-parameter group $\{\varphi_t\}_{t\in\mathbb{R}}$ of diffeomorphisms of Ω that leave μ invariant. The states of the system are probability measures on Ω of the form (3.5) and the observables are the real-valued, Borel measurable functions on Ω.

Let $L^2(\Omega, \mu)$ denote the Hilbert space of (equivalence classes of) complex-valued, Borel measurable functions on Ω that are square integrable with respect to μ. The inner product on $L^2(\Omega, \mu)$ is given by

$$\langle\phi, \psi\rangle = \int_{\Omega} \bar{\phi}\psi \, d\mu,$$

so

$$\|\psi\|^2 = \int_{\Omega} |\psi|^2 \, d\mu.$$

Note that any element $\psi \in L^2(\Omega, \mu)$ with $\|\psi\| = 1$ gives rise to a probability measure ν_ψ on Ω by taking $\rho = |\psi|^2$ as the probability density in (3.5). Conversely, for any measure of the form (3.5), one can find a $\psi \in L^2(\Omega, \mu)$ for which $\rho = |\psi|^2$. Note, however, that the ψ corresponding to a given ρ is not unique since, for any $\theta \in \mathbb{R}$, $|e^{i\theta}\psi|^2 = |\psi|^2$. Thus, we can identify a state with what is called a "unit ray" in $L^2(\Omega, \mu)$, that is, a set of unit vectors of the form $\{e^{i\theta}\psi : \theta \in \mathbb{R} \text{ and } \|\psi\| = 1\}$.

Remark 3.3.1. This identification of probability density functions with unit vectors in $L^2(\Omega, \mu)$ clearly does not depend on having chosen complex, as opposed to real, square integrable functions on Ω. The choice of the complex Hilbert space $L^2(\Omega, \mu)$ is motivated primarily by things that are yet to come. On the mathematical side, we will make heavy use of certain results such as the spectral theorem and Stone's theorem that live much more naturally in the complex world. Physically, we will find that, in quantum mechanics, interference effects such as one encounters in wave motion in fluids are fundamental and that such effects are much more conveniently described in terms of complex numbers. One should also consult Lecture 2 of [Mar2] for additional motivation.

Next we observe that the dynamics $\varphi_t : \Omega \to \Omega$, $t \in \mathbb{R}$, on Ω induces a dynamics $U_t : L^2(\Omega, \mu) \to L^2(\Omega, \mu)$, $t \in \mathbb{R}$, on $L^2(\Omega, \mu)$. Indeed, according to (3.4) we should define

$$U_t(\psi) = \psi \circ \varphi_{-t}.$$

We will discuss the observables shortly, but would like to pause for a moment to point out that our new perspective already promises to yield some dividends. Each U_t in the one-parameter group $\{U_t\}_{t \in \mathbb{R}}$ is, of course, a bijection of $L^2(\Omega, \mu)$ onto itself, but it is, in fact, much more. Clearly, U_t is a *linear* map on $L^2(\Omega, \mu)$ since

$$U_t(a_1\psi_1 + a_2\psi_2) = (a_1\psi_1 + a_2\psi_2) \circ \varphi_{-t} = a_1(\psi_1 \circ \varphi_{-t}) + a_2(\psi_2 \circ \varphi_{-t})$$
$$= a_1 U_t(\psi_1) + a_2 U_t(\psi_2).$$

But, in fact, each U_t is actually a *unitary* operator on $L^2(\Omega, \mu)$. To see this we let $\phi, \psi \in L^2(\Omega, \mu)$ and compute

$$\langle U_t\phi, U_t\psi \rangle = \langle \phi \circ \varphi_{-t}, \psi \circ \varphi_{-t} \rangle$$
$$= \int_\Omega \overline{\phi(\varphi_{-t}(x))}\, \psi(\varphi_{-t}(x))\, d\mu(x)$$
$$= \int_{\varphi_t(\Omega)} \overline{\phi(y)}\, \psi(y)\, d\mu(\varphi_t(y))$$
$$= \int_\Omega \overline{\phi(y)}\, \psi(y)\, d\mu(y)$$
$$= \langle \phi, \psi \rangle,$$

where we have used the change of variables formula and the invariance of Ω and μ under φ_t. We have then what is called a one-parameter group of unitary operators on $L^2(\Omega, \mu)$ (see Remark 3.3.2 below).

Remark 3.3.2. Let us recall a few more notions from functional analysis. Let $\mathcal{H}$ be a complex, separable Hilbert space and suppose that $\{U_t\}_{t \in \mathbb{R}}$ is a family of unitary operators $U_t : \mathcal{H} \to \mathcal{H}$ on $\mathcal{H}$ satisfying $U_{t+s} = U_t U_s$ for all $t, s \in \mathbb{R}$ and $U_0 = \mathrm{id}_{\mathcal{H}}$. Then

$\{U_t\}_{t\in\mathbb{R}}$ is called a *one-parameter group of unitary operators on* $\mathcal{H}$. $\{U_t\}_{t\in\mathbb{R}}$ is said to be *strongly continuous* if, for every $\psi \in \mathcal{H}$, $t \to t_0$ in $\mathbb{R} \Rightarrow U_t\psi \to U_{t_0}\psi$ in $\mathcal{H}$. By the group property of $\{U_t\}_{t\in\mathbb{R}}$, this is equivalent to $t \to 0$ in $\mathbb{R} \Rightarrow U_t\psi \to \psi$ in $\mathcal{H}$. $\{U_t\}_{t\in\mathbb{R}}$ is said to be *weakly continuous* if, for all $\phi, \psi \in \mathcal{H}$, $t \to t_0$ in $\mathbb{R} \Rightarrow \langle\phi, U_t\psi\rangle \to \langle\phi, U_{t_0}\psi\rangle$ in $\mathbb{C}$. Again, this is equivalent to $t \to 0$ in $\mathbb{R} \Rightarrow \langle\phi, U_t\psi\rangle \to \langle\phi, \psi\rangle$ in $\mathbb{C}$. Certainly, strong continuity implies weak continuity, but, despite the terminology, the converse is also true (this depends heavily on the fact that each U_t is unitary). To see this, suppose $\{U_t\}_{t\in\mathbb{R}}$ is weakly continuous and $t \to 0$ in $\mathbb{R}$. Then

$$\|U_t\psi - \psi\|^2 = \langle U_t\psi - \psi, U_t\psi - \psi\rangle = \|U_t\psi\|^2 - \langle U_t\psi, \psi\rangle - \langle\psi, U_t\psi\rangle + \|\psi\|^2$$
$$= \|\psi\|^2 - 2\operatorname{Re}\langle\psi, U_t\psi\rangle + \|\psi\|^2$$
$$\to 2\|\psi\|^2 - 2\|\psi\|^2 = 0,$$

as $t \to 0$. There is also a much stronger result due to von Neumann. Let us say that $\{U_t\}_{t\in\mathbb{R}}$ is *weakly measurable* if, for all $\phi, \psi \in \mathcal{H}$, the complex-valued function of the real variable t given by $t \to \langle\phi, U_t\psi\rangle$ is Lebesgue measurable. Von Neumann showed that a weakly measurable one-parameter group of unitary operators on a separable, complex Hilbert space is strongly continuous. We will not prove this here, but will simply refer to Theorem VIII.9 of [RS1]. Finally, we should also point out that, because of the *polarization identity*

$$\langle\alpha, \beta\rangle = \frac{1}{4}[\langle\alpha + \beta, \alpha + \beta\rangle - \langle\alpha - \beta, \alpha - \beta\rangle$$
$$- i\langle\alpha + i\beta, \alpha + i\beta\rangle + i\langle\alpha - i\beta, \alpha - i\beta\rangle], \tag{3.6}$$

it is enough to prove weak continuity and weak measurability in the case $\phi = \psi$. As it turns out, one can explicitly describe *all* of the strongly continuous one-parameter groups of unitary operators on $\mathcal{H}$. This is Stone's theorem, which we will discuss in Section 5.5.

Example 3.3.1. We will show that the one-parameter group we have defined on $L^2(\Omega, \mu)$ by $U_t(\psi) = \psi \circ \varphi_{-t}$ is strongly continuous. To see this we fix a $\psi \in L^2(\Omega, \mu)$ and suppose $t \to 0$ in $\mathbb{R}$. We must show that $U_t\psi \to \psi$ in $L^2(\Omega, \mu)$. The proof is based on two observations. First, $L^2(\Omega, \mu)$ contains a dense set of continuous functions (those with compact support, for example). Thus, given an $\epsilon > 0$ we can select a continuous $\psi_\epsilon \in L^2(\Omega, \mu)$ with $\|\psi_\epsilon - \psi\| < \epsilon/3$. Next, we appeal to a standard result from the theory of ordinary differential equations on continuous dependence on initial conditions (Theorem 4.26 of [CM]) which implies that for any *continuous* $\phi \in L^2(\Omega, \mu)$, $t \to U_{-t}\phi = \phi \circ \varphi_{-t}$ is a continuous map of $\mathbb{R}$ into $L^2(\Omega, \mu)$. Thus, we can select $\delta > 0$ such that $|t| < \delta$ implies $\|U_t\psi_\epsilon - \psi_\epsilon\| < \epsilon/3$. Now write

$$U_t\psi - \psi = U_t(\psi - \psi_\epsilon) + (U_t\psi_\epsilon - \psi_\epsilon) + (\psi_\epsilon - \psi).$$

Since each U_t is unitary, $\|U_t(\psi - \psi_\epsilon)\| = \|\psi - \psi_\epsilon\|$, so the triangle inequality gives $\|U_t\psi - \psi\| < \epsilon$, as required.

Finally, let us see how the observables fit into our new picture. Let A be a real-valued, Borel measurable function on Ω, that is, an observable in the old picture. For any state, thought of now as a unit vector ψ in $L^2(\Omega, \mu)$, A is just a random variable for the probability measure ν_ψ and, assuming it is integrable, its expected value is

$$E(A) = \int_\Omega A \, d\nu_\psi = \int_\Omega A \, |\psi|^2 d\mu = \int_\Omega \overline{A\psi} \, \psi d\mu = \int_\Omega \overline{\psi} \, A\psi d\mu$$

$$= \langle A\psi, \psi \rangle \quad = \langle \psi, A\psi \rangle.$$

This suggests thinking of A as a *multiplication operator* on $L^2(\Omega, \mu)$. Thought of in this way the operator would appear to be self-adjoint ($\langle A\psi, \psi \rangle = \langle \psi, A\psi \rangle$). There is an issue, however. As a multiplication operator, A will be defined only for those $\psi \in L^2(\Omega, \mu)$ for which $A\psi$ is also in $L^2(\Omega, \mu)$, that is, for which $A\psi$ is square integrable on Ω. As a result, A is not defined everywhere and therefore is *not* a bounded operator on $L^2(\Omega, \mu)$, even though it is clearly linear on its domain. It is, in fact, what is known as an "unbounded, self-adjoint operator on $L^2(\Omega, \mu)$." We will provide a synopsis of what we need to know about such unbounded operators in Chapter 5. The bottom line of this section, however, is that Koopman has rephrased classical statistical mechanics in such a way that the states are represented by unit vectors in a complex Hilbert space and the observables are unbounded, self-adjoint operators on that Hilbert space. As it happens, this is precisely how quantum mechanics is generally formulated. We will spend quite a bit of time describing this formalism, but first we should try to come to some understanding of why quantum mechanics is like this. Why, in other words, is the quantum mechanics of even a single particle more akin to classical statistical mechanics, where we can follow Koopman's lead to construct a mathematical model, than to classical particle mechanics? This is the subject of the next chapter.

4 Physical background

4.1 Introduction

By the end of the nineteenth century classical particle mechanics, statistical mechanics and electromagnetic theory were very finely tuned instruments capable of treating an enormous variety of physical problems with remarkable success. Some even believed that there was little left to do.

> "The more important fundamental laws and facts of physical science have all been discovered, and these are now so firmly established that the possibility of their ever being supplanted in consequence of new discoveries is exceedingly remote ... Our future discoveries must be looked for in the sixth place of decimals."
>
> Albert. A. Michelson, speech at the dedication of Ryerson Physics Lab, University of Chicago, 1894

> "When I began my physical studies [in Munich in 1874] and sought advice from my venerable teacher Philipp von Jolly ... he portrayed to me physics as a highly developed, almost fully matured science ... Possibly in one or another nook there would perhaps be a dust particle or a small bubble to be examined and classified, but the system as a whole stood there fairly secured, and theoretical physics approached visibly that degree of perfection which, for example, geometry has had already for centuries."
>
> from a 1924 lecture by Max Planck

Needless to say, this optimism regarding the then current state of physics was somewhat premature. Michelson himself had, in 1887, unearthed something of a conundrum for classical physics that was only resolved 18 years later by Einstein, who pointed out that classical physics had the concepts of space and time entirely wrong. Planck struggled for many years with the problem of the equilibrium distribution of electromagnetic radiation for which classical physics provided a perfectly explicit, and quite incorrect, solution. In the end he obtained a solution that was in complete accord with the experimental data, but only at the expense of what he himself called an "act of desperation." He postulated, in flat contradiction to the requirements of classical physics, that harmonic oscillators can exist only at certain discrete energy levels determined by their frequency (see Remark 1.0.3). Planck did not use the term, but today we would credit this as the birth of the *quantum hypothesis*, although one can argue that there are precursors in the work of Boltzmann on statistical mechanics (for some references, see [Flamm]). It was left to Einstein, however, in his analysis of what is called the "photoelectric effect," to transform this provisional hypothesis into a revolutionary new view of physics.

One can find this story, both its history and the physics behind it, told concisely and elegantly in Chapters II and VI of [Pais] and we will not offer a pale imitation

https://doi.org/10.1515/9783110751949-004

here. Nevertheless, it seems disingenuous to introduce the ubiquitous Planck constant without providing some sense of what it is and where it came from, or to simply insist that atoms are so unlike baseballs that one must abandon long cherished notions of causality just to say something reliable about how they behave. We will therefore devote this chapter to an attempt to come to a rudimentary understanding of some of the physical facts of life that necessitate this profound revision of the physicist's *Weltanschauung*. We will begin at the historical beginning with Planck and Einstein, but will then abandon chronology to describe a number of experimental facts that may not have been available to the founding fathers of the subject, but seem to express most clearly the essential nature of the quantum world. Our discussions will necessarily be rather brief and certainly not at the level one would expect to find in a physics text, but we will try to provide sufficient references for those who wish to pursue these matters more seriously.

The phenomena we would like to discuss first are those of blackbody radiation and the photoelectric effect. Both of these deal with the interaction of electromagnetic radiation with matter (what really goes on when the rays of the summer sun burn your skin). A necessary prerequisite then is to come to terms with electromagnetic radiation.

4.2 Electromagnetic radiation

We will accept the view that matter, at least all matter within the current range of our experience, is composed of *atoms* and that these atoms can be visualized, somewhat naively perhaps, as something of a mini-solar system with a *nucleus* composed of objects called *protons* and *neutrons* playing the role of the sun and a collection of *electrons* orbiting the nucleus like planets. The essential difference is that, whereas we think we know what holds the solar system together (gravity does that), we will not pretend, at this stage, to have any idea of what holds an atom together. This has something to do with the fact that protons and electrons possess a physical characteristic called *electric charge*, which comes in two flavors, *positive* and *negative*. Two positive charges, or two negative charges, brought near each other will exert a force, each on the other, that pushes the charges apart, whereas a positive and a negative charge will attract each other. Physics offers no explanation for this behavior in terms of some more fundamental phenomenon. Some things are charged and some things are not (like the neutron); just deal with it. What physics does offer is a very detailed understanding of how these *electric forces* act and how they can be used. In the course of acquiring this understanding it was discovered that electric force is very closely related to the analogous, but seemingly distinct phenomenon of *magnetic force*. Every child knows that if you bring a magnet near a compass, the arrow on the compass will spin, but may not know that the same thing happens if the compass is brought near a stream of electrons flowing through a wire. On the other hand, a charged particle placed *at rest* between the poles of a horseshoe magnet will just sit there unaware of the magnet's presence, but, if it is *thrown* between the poles, its path will be deflected. Indeed, electric and magnetic forces are more than just analogous; there is a very real sense in which they are the same thing, but viewed from different perspectives. The appropriate context within which to understand this is the special theory of relativity (see, for example, [Nab5]).

The part of classical physics that deals with all of this is called *electrodynamics* and the best place to go to understand it is Volume II of [FLS]. It is fortunate for us that physicists understand this subject so well that they were able, or rather one of them, named James Clerk Maxwell, was able, to encode essentially all of the relevant information in just a few equations and that for our purposes, only a very special case of these equations will be required. One should keep the following picture in mind. We have a collection of charges, some being stationary, some moving willy-nilly through space. We are interested in the cumulative effect these will have on some other "test charge" moving around in a "charge-free" region of space.

Remark 4.2.1. A few remarks are in order. Physicists will choose some favored system of units in which to describe all of the relevant quantities, but which system is favored depends heavily on the context. On rare occasions we will be forced to be explicit about the choice of units (for example, when trying to make sense of statements like "Planck's constant is small," or "the speed of light is large"). Generally, we will be inclined to use what are called *SI units* (*Le Système international d'unités*), in which length is in meters (m), time is in seconds (s), mass is in kilograms (kg), force is in Newtons (N $=$ (kg)ms^{-2}), frequency is in hertz (Hz $=$ s^{-1}), energy is in Joules (J $=$ Nm $=$ m^2(kg)s^{-2}), current is in amperes (A), charge is in coulombs (C $=$ sA), *und so weiter und so fort*. Whatever system of units is chosen, a "test charge" is, by definition, one that is sufficiently small that it has a negligible effect on the other charges. The seemingly contradictory assertion that the test charge moves in a "charge-free" region simply means that it is, at each instant, in the complement of the region occupied by the original distribution of charges at that instant; such a region is generally assumed to have nothing at all in it except the test charge and is then referred to as a *charge-free vacuum*.

Physicists describe the effect we are after with two (generally time-dependent) vector fields on some region in $\mathbb{R}^3$. The *electric field* **E** and the *magnetic field* **B** are both functions of (t, x, y, z), where t is time and (x, y, z) are Cartesian coordinates in space. Denoting the spatial gradient operator by ∇, the spatial divergence operator by $\nabla\cdot$ and the spatial curl by $\nabla\times$ and writing $\frac{\partial \mathbf{B}}{\partial t}$ and $\frac{\partial \mathbf{E}}{\partial t}$ for the componentwise derivatives of **B** and **E** with respect to t, *Maxwell's equations* in a charge-free vacuum are

$$\nabla \cdot \mathbf{E} = 0, \tag{4.1}$$

$$\nabla \times \mathbf{E} = -\frac{\partial \mathbf{B}}{\partial t}, \tag{4.2}$$

$$\nabla \cdot \mathbf{B} = 0, \tag{4.3}$$

$$\nabla \times \mathbf{B} = \mu_0 \epsilon_0 \frac{\partial \mathbf{E}}{\partial t}, \tag{4.4}$$

where μ_0 and ϵ_0 are two universal constants called, respectively, the *vacuum permeability* and *vacuum permittivity*. We will not go into the rather convoluted story of how these two constants are defined (see Volume II of [FLS]), but will point out only that their product in SI units is approximately

$$\mu_0 \epsilon_0 \approx 1.1126499 \times 10^{-17} \ \mathrm{s}^2/\mathrm{m}^2. \tag{4.5}$$

One more equation, called the *Lorentz force law*, gives the force $\mathbf{F}$ experienced by a test charge q moving with velocity $\mathbf{V}$ in the presence of the electric field $\mathbf{E}$ and magnetic field $\mathbf{B}$. Using $\times$ to denote the vector (cross) product in $\mathbb{R}^3$, this can be written

$$\mathbf{F} = q \, [\, \mathbf{E} + (\mathbf{V} \times \mathbf{B}) \,]. \tag{4.6}$$

There are a great many things to be said about this set of equations, but we will mention only the few items we specifically need later on. We begin with a few simple observations. Equation (4.1) says that our region contains no *sources* for the electric field and this simply reflects our decision to work in a charge-free region. In the general form of Maxwell's equations the zero on the right-hand side of (4.1) is replaced by a function describing the charge density of the region. Equation (4.3) says the same thing about the magnetic field, but this equation remains the same in the general form of Maxwell's equations; there are no *magnetic charges* in the electromagnetic theory of Maxwell. That is not to say that magnetic charges (more commonly called *magnetic monopoles*) cannot exist. Dirac considered the possibility that they might and drew some rather remarkable conclusions from the assumption that they do (see Chapter 0 of [Nab3]).

Equations (4.2) and (4.4) imply, among other things, that a time-varying magnetic field is always accompanied by a nonzero electric field and a time-varying electric field is always accompanied by a nonzero magnetic field. One can therefore envision the following scenario. An electric charge setting at rest in space gives rise to a static (that is, time-independent) electric field in its vicinity (this is described by *Coulomb's law*, which is no doubt familiar from calculus). But suppose we wiggle the charge. Now the electric field nearby is varying with time and so must give rise to a magnetic field. The magnetic field also varies with time so it, in turn, must give rise to an electric field, which varies in time giving rise to a magnetic field, and so on and so on. Intuitively, at least, one sees the effects of wiggling (that is, accelerating) the charge as propagating away from the charge through space in some sort of wave disturbance. We will make this more precise momentarily by showing that any solutions $\mathbf{E}$ and $\mathbf{B}$ to (4.1)–(4.4) have components that satisfy a wave equation.

Next we mention that the Lorentz force law (4.6) has, as the equations of physics often do, a dual character. If you know what $\mathbf{E}$ and $\mathbf{B}$ are, you can calculate the force $\mathbf{F}$

on a charge. On the other hand, if you measure forces you can determine **E** and **B**. This latter point of view provides an operational definition of the electric and magnetic fields. For example, $\mathbf{E}(t, x, y, z)$ is the force experienced by a unit charge setting at rest at (x, y, z) at the instant t. If this dual character seems circular to you, that is because it is; definitions in physics are generally not at all like definitions in mathematics.

Now we will show that the components of any solutions **E** and **B** to Maxwell's equations (4.1)–(4.4) in a charge-free vacuum satisfy the same wave equation. Since all of the components are treated in exactly the same way we will lump them all together and write $\frac{\partial \mathbf{E}}{\partial t}$ and $\frac{\partial \mathbf{B}}{\partial t}$ for the componentwise partial derivatives of **E** and **B** with respect to t, $\nabla \mathbf{E}$ and $\nabla \mathbf{B}$ for the componentwise spatial gradients of **E** and **B** and $\nabla^2 \mathbf{E} = \nabla \cdot \nabla \mathbf{E}$ and $\nabla^2 \mathbf{B} = \nabla \cdot \nabla \mathbf{B}$ for the componentwise spatial Laplacians. For example, if $\mathbf{E} = (E_x, E_y, E_z)$, then

$$\nabla^2 \mathbf{E} = \nabla \cdot \nabla \mathbf{E} = \left(\nabla^2 E_x, \nabla^2 E_y, \nabla^2 E_z \right) = (\nabla \cdot \nabla E_x, \nabla \cdot \nabla E_y, \nabla \cdot \nabla E_z).$$

Begin by taking the curl of $\nabla \times \mathbf{E} = -\frac{\partial \mathbf{B}}{\partial t}$ to obtain

$$\nabla \times (\nabla \times \mathbf{E}) = -\nabla \times \left(\frac{\partial \mathbf{B}}{\partial t} \right) = -\frac{\partial}{\partial t} (\nabla \times \mathbf{B}) = -\frac{\partial}{\partial t} \left(\mu_0 \epsilon_0 \frac{\partial \mathbf{E}}{\partial t} \right) = -\mu_0 \epsilon_0 \frac{\partial^2 \mathbf{E}}{\partial t^2}.$$

Now use the vector identity $\nabla \times (\nabla \times \mathbf{A}) = \nabla(\nabla \cdot \mathbf{A}) - \nabla^2 \mathbf{A}$ and the fact that $\nabla \cdot \mathbf{E} = 0$ to write this as

$$\nabla^2 \mathbf{E} = \mu_0 \epsilon_0 \frac{\partial^2 \mathbf{E}}{\partial t^2}. \tag{4.7}$$

The same argument, starting with (4.4) rather than (4.2) shows that

$$\nabla^2 \mathbf{B} = \mu_0 \epsilon_0 \frac{\partial^2 \mathbf{B}}{\partial t^2}. \tag{4.8}$$

What we have then are six copies of the same wave equation, one for each component of **E** and **B**, all of which describe a wave propagating with speed

$$\frac{1}{\sqrt{\mu_0 \epsilon_0}}.$$

Note that, from (4.5),

$$\frac{1}{\sqrt{\mu_0 \epsilon_0}} \approx 2.99792 \times 10^8 \text{ m/s}.$$

Now for the really good part. Maxwell published his famous paper *A Dynamical Theory of the Electromagnetic Field* in 1865. At that time, physicists had no reason to suspect that electromagnetic effects might propagate as waves, but, as we have just seen, Maxwell's equations seemed to suggest that they can. It was not until 1886 that Hertz verified this prediction of Maxwell by detecting what we would today call *radio waves*. But there is much more. Three years prior to the appearance of Maxwell's paper, in 1862, Foucault had made the most accurate measurement to date of the speed of light *in vacuo*. His value was 2.99796×10^8 m/s, and it did not escape Maxwell's attention that, modulo experimental errors, this is precisely the predicted propagation speed of his electromagnetic waves. This raised the possibility, never before imagined, that light itself is an electromagnetic phenomenon. Because of the way light behaves, it was generally accepted at the time that light represents some sort of wave propagation (we will have a bit more to say about this in Section 4.4), but it was certainly not viewed as the sort of electromagnetic wave that we have just seen emerge from Maxwell's equations.

But, of course, not all light is the same. It comes in different colors that can be separated out of the "white light" we generally encounter by sending it through a prism; this, of course, was known long before Maxwell. But if all of these colors are really electromagnetic waves, then they can differ from each other only in various wave characteristics, such as wavelength λ (or, equivalently, frequency ν). Carrying this speculation just a bit further, it is not difficult to imagine that the ocular sense that has evolved in our species is sensitive only to those wavelengths that we must be sensitive to in order to survive (and so Hertz could not "see" his radio waves). One would then imagine electromagnetic waves of every possible frequency, some of which our eyes can see and some of which we can perceive only by doing more than just looking, despite the fact that they are all really the same phenomenon. All of this is, in fact, true and physicists now display the range of possibilities for electromagnetic waves in a continuous *electromagnetic spectrum* labeled by wavelength and/or frequency (see Figure 4.1). The visible (to humans) part of this spectrum is quite small, but it had been the object of study by physicists ever since Newton.

Remark 4.2.2. Newton, however, believed that light was composed of *particles* of different colors and that different colored particles moved at different speeds through the glass of the prism, resulting in different angles of refraction, thus creating the spectrum of colors. Classically, the wave and particle pictures of light are quite inconsistent, but in 1905 Einstein [Ein1] proposed that for a proper understanding of the properties of light, both were necessary (see Section 4.3). This was the birth of the *wave/particle duality* that became the hallmark of modern quantum theory.

In 1814, Fraunhofer made a particularly interesting discovery when he noted that, in the spectrum of light coming from the sun, certain frequencies were missing (there were dark lines where one would expect to see a color); see Figure 4.2. Somewhat later it was observed that, in a sense, the opposite can occur. For example, when hydrogen gas is heated it gives off light which, when sent through a prism, exhibits just a few bright lines on an otherwise dark background (see Figure 4.3). Other elements behave in the same way, but the visible lines are different; each has a characteristic *emission*

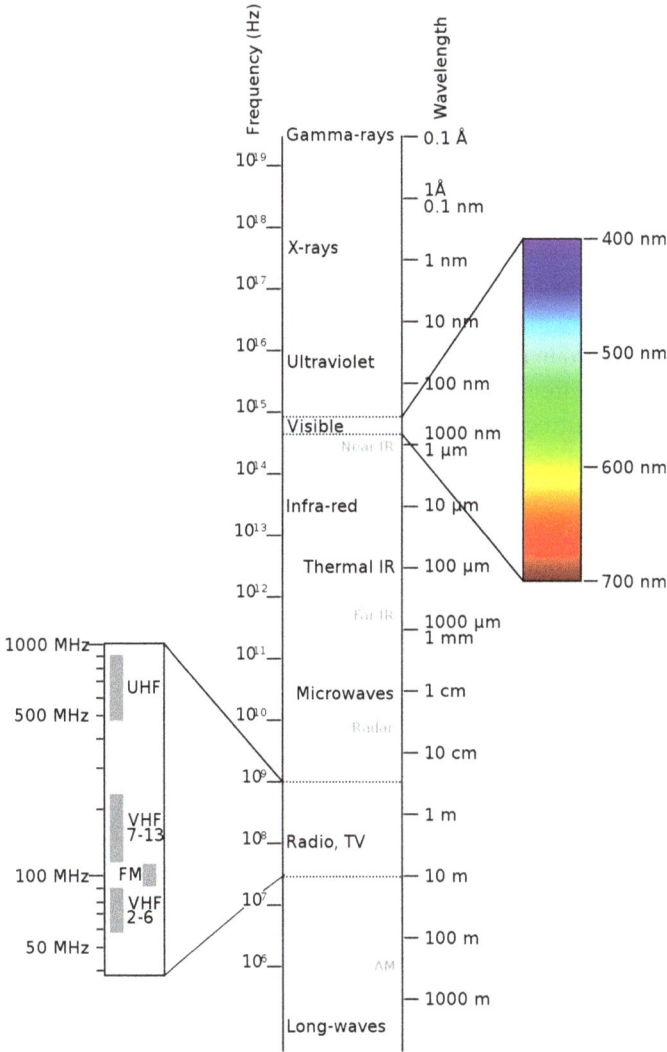

Figure 4.1: Electromagnetic spectrum. https://commons.wikimedia.org/wiki/File:Electromagnetic-Spectrum.svg (last access date: 21.05.2021).

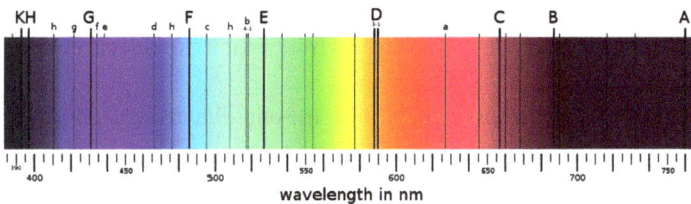

Figure 4.2: Fraunhofer lines. https://commons.wikimedia.org/wiki/File:Fraunhofer_lines.svg (last access date: 21.05.2021).

Figure 4.3: Hydrogen emission spectrum. https://commons.wikimedia.org/wiki/File:Emission_spectrum-H.svg (last access date: 21.05.2021).

spectrum. Furthermore, certain elements (such as sodium) were found to have emission spectra that exactly matched certain of the dark Fraunhofer lines in the solar spectrum. The conclusion drawn was that every element both emits and absorbs light (electromagnetic waves) of certain characteristic frequencies and that, for example, sodium is either entirely absent from the sun (unlikely), or whatever sodium is present in the hot interior regions is emitting its frequencies only to have them reabsorbed (and probably emitted again, but in different directions, that is, scattered) by sodium in the cooler exterior regions so that they never reach us and appear as Fraunhofer lines. The question that was left unanswered and had to await the advent of quantum theory was why atoms can emit and absorb only a *discrete* set of frequencies.

We will take all of these experimental facts for granted without further comment, but there are still issues we need to address. We begin by introducing the usual symbol

$$c \approx 2.99792 \times 10^8 \text{ m/s} \tag{4.9}$$

for the speed of light *in vacuo* and rewriting our wave equations as

$$\nabla^2 \mathbf{E} = \frac{1}{c^2} \frac{\partial^2 \mathbf{E}}{\partial t^2} \tag{4.10}$$

and

$$\nabla^2 \mathbf{B} = \frac{1}{c^2} \frac{\partial^2 \mathbf{B}}{\partial t^2}. \tag{4.11}$$

Now, (4.10) and (4.11) are really six independent copies of the wave equation, so producing solutions to them is easy; just select your six favorite solutions to the wave equation and make them the components in anyway you like. It is unlikely that you will produce solutions to Maxwell's equations this way, however. We need to see what additional constraints are imposed by the full set of Maxwell's equations and the best way to do this is to look at some very simple solutions from which the rest can be obtained by superposition.

Just to establish some notation, let us review a bit of the one-dimensional situation from calculus. The one-dimensional wave equation for $u(t, x)$ is $\frac{1}{a^2} \frac{\partial^2 u}{\partial t^2} = \frac{\partial^2 u}{\partial x^2}$, where a is a positive constant. Looking for some simple solutions one considers the family of sinusoidal waves of the form

$$u(t, x) = A_0 \cos(kx - \omega t + \varphi) = \text{Re}\left(A_0 e^{i(kx - \omega t + \varphi)}\right),$$

where A_0, k, ω are positive real constants and φ is an arbitrary real constant: A_0 is the *amplitude* of the wave and φ is the *phase*; ω and k are related to the period T, wavelength λ and frequency $\nu = \frac{1}{T}$ of the wave by $\omega = 2\pi\nu = \frac{2\pi}{T}$ and $k = \frac{2\pi}{\lambda}$; ω is called the *(angular) frequency* and k is the *(angular) wavenumber* (the adjective *angular* simply means that we are counting the number of cycles/wavelengths per 2π units of time/distance and it is very often dropped). The speed of propagation of the wave is $\frac{\omega}{k} = \frac{\lambda}{T}$. Substituting into the wave equation, one finds that $u(t, x)$ is a solution if and only if $\frac{\omega}{k} = a$, so a is the speed of propagation of the wave. One could, of course, replace cos by sin, but this simply amounts to shifting the phase by $\frac{\pi}{2}$. Since it is easier, algebraically and analytically, to deal with exponentials, one generally focuses on the complex solution $A_0 e^{i(kx-\omega t+\varphi)}$, keeping in mind that it is the real (or imaginary) part that is of interest. Going a step further, one can split off the phase $A_0 e^{i(kx-\omega t+\varphi)} = A_0 e^{i\varphi} e^{i(kx-\omega t)}$, absorb it into the coefficient and deal with the complex solution

$$U(t, x) = U_0 \, e^{i\,(kx-\omega t)},$$

where the constant $U_0 = A_0 e^{i\varphi}$ is also complex. To keep track of the direction of propagation and the direction in which the oscillations take place one can introduce the standard Euclidean basis vectors in the plane and define vectors $\mathbf{k} = (k, 0)$, $\mathbf{x} = (x, 0)$, $\mathbf{U}_0 = (0, U_0)$ and $\mathbf{U}(t, \mathbf{x}) = (0, U(t, x))$ and write

$$\mathbf{U}(t, \mathbf{x}) = \mathbf{U}_0 \, e^{i\,(\mathbf{k}\cdot\mathbf{x}-\omega t)},$$

where $\mathbf{k} \cdot \mathbf{x}$ is the usual Euclidean inner product of $\mathbf{k}$ and $\mathbf{x}$.

The appropriate generalization to the three-dimensional wave equation is now clear. We fix an arbitrary nonzero vector $\mathbf{k} = (k_1, k_2, k_3)$ in $\mathbb{R}^3$ and a positive real number ω and write $k = \|\mathbf{k}\|$. Let $\mathbf{x} = (x, y, z)$ denote an arbitrary vector in $\mathbb{R}^3$. For any $\mathbf{U}_0 = (A_1 e^{i\varphi_1}, A_2 e^{i\varphi_2}, A_3 e^{i\varphi_3})$, where A_1, A_2, A_3 are fixed positive real numbers and φ_1, φ_2, φ_3 are fixed real numbers, define

$$\mathbf{U}(t, \mathbf{x}) = \mathbf{U}_0 \, e^{i\,(\mathbf{k}\cdot\mathbf{x}-\omega t)}.$$

A little calculus proves the divergence and curl formulas

$$\nabla \cdot \mathbf{U}(t, \mathbf{x}) = i\,\mathbf{k} \cdot \mathbf{U}(t, \mathbf{x}) \tag{4.12}$$

and

$$\nabla \times \mathbf{U}(t, \mathbf{x}) = i\,\mathbf{k} \times \mathbf{U}(t, \mathbf{x}). \tag{4.13}$$

Although we have taken $\mathbf{U}$ to be complex it will do no harm, and will aid the intuition, if we treat it formally as if it were a vector in $\mathbb{R}^3$. Thus, the $\cdot$ and $\times$ in $\mathbf{k} \cdot \mathbf{U}(t, \mathbf{x})$ and $\mathbf{k} \times$

$U(t, \mathbf{x})$ here are the usual $\mathbb{R}^3$-dot and cross-products, but with complex components. Substituting into the three-dimensional wave equation

$$\nabla^2 \mathbf{U} = \frac{1}{a^2} \frac{\partial^2 \mathbf{U}}{\partial t^2}$$

one finds that $\mathbf{U}$ is a solution if and only if

$$\frac{\omega}{k} = a.$$

For such a solution, $\mathbf{k}$ is called the *wavevector* and k is the *wavenumber*. Choosing $\frac{\omega}{k} = c$ we can therefore write down lots of solutions

$$\mathbf{E}(t, \mathbf{x}) = \mathbf{E}_0 \, e^{i(\mathbf{k} \cdot \mathbf{x} - \omega t)}$$

and

$$\mathbf{B}(t, \mathbf{x}) = \mathbf{B}_0 \, e^{i(\mathbf{k} \cdot \mathbf{x} - \omega t)}$$

to (4.10) and (4.11). We will now see what additional constraints are imposed by the full set of Maxwell equations (4.1)–(4.4) which, for these particular functions $\mathbf{E}$ and $\mathbf{B}$, become

$$\mathbf{k} \cdot \mathbf{E} = 0, \tag{4.14}$$

$$\mathbf{k} \times \mathbf{E} = \omega \mathbf{B}, \tag{4.15}$$

$$\mathbf{k} \cdot \mathbf{B} = 0, \tag{4.16}$$

$$\mathbf{k} \times \mathbf{B} = -\frac{\omega}{c^2} \mathbf{E}. \tag{4.17}$$

Exercise 4.2.1. Note that *a priori* the phase factors in $\mathbf{E}_0 = (E_1 e^{i\varphi_1^E}, E_2 e^{i\varphi_2^E}, E_3 e^{i\varphi_3^E})$ and $\mathbf{B}_0 = (B_1 e^{i\varphi_1^B}, B_2 e^{i\varphi_2^B}, B_3 e^{i\varphi_3^B})$ could be different. Show that (4.17) and the fact that $\mathbf{k}$ is real imply $\varphi_j^E = \varphi_j^B \bmod 2\pi$, $j = 1, 2, 3$.

By (4.14) and (4.16), both $\mathbf{E}$ and $\mathbf{B}$ are orthogonal to $\mathbf{k}$ for each $(t, \mathbf{x})$ and, by (4.15), they are orthogonal to each other. Looking just at the real part of (4.15) we conclude from this that $\|\mathbf{k}\| \, \|\mathbf{E}\| = \omega \|\mathbf{B}\|$, or

$$\|\mathbf{E}\| = c \, \|\mathbf{B}\|.$$

Note also that, by (4.15), $\mathbf{B} \cdot (\mathbf{k} \times \mathbf{E}) = \omega \|\mathbf{B}\|^2 > 0$, so $\mathbf{k} \times \mathbf{E}$ is in the direction of $\mathbf{B}$ and we can visualize $\{\mathbf{k}, \mathbf{E}, \mathbf{B}\}$ as a right-handed orthogonal basis at each point. Next let ϕ be some real constant and consider all of the $(t, \mathbf{x})$ for which the phase $\mathbf{k} \cdot \mathbf{x} - \omega t$ takes this constant value, that is,

$$k_1 x + k_2 y + k_3 z = \omega t + \phi.$$

For each fixed t this is a plane orthogonal to $\mathbf{k}$ on which $\mathbf{E}$ and $\mathbf{B}$ are both constant. As t varies over $-\infty < t < \infty$ these planes move in the direction of $\mathbf{k}$ with speed $\frac{\omega}{k} = c$. The planes of constant phase are called *wavefronts* and the electromagnetic wave itself is called a *plane electromagnetic wave* or a *linearly polarized electromagnetic wave*. The term *linearly polarized* refers to the fact that the electric field vector oscillates in a single direction (that is, along a single *line*). If one imagines oneself situated at some fixed point along a line on which the wave is propagating and if one could see the tip of the electric field vector, then, as the wavefronts pass through this point, the tip would look just like a mass on a spring. This is true of the magnetic field vector as well, but, since $\mathbf{E}$ and $\mathbf{B}$ are always orthogonal to $\mathbf{k}$ and to each other, it is conventional to mention only the electric component. The direction of the electric field vector $\mathbf{E}$ is called the *direction of polarization*.

These are, of course, very special electromagnetic waves and much more complicated behavior results when the wave is a superposition (sum) of two or more plane waves. For example, the superposition of two orthogonal plane waves propagating in the same direction, of equal magnitude, but differing in phase by $\pi/2$ is *circularly polarized*. For these the tip of the electric vector approaching you along the direction of propagation would appear to rotate (either clockwise or counterclockwise) around a circle. Similarly, the sum of two orthogonal plane waves propagating in the same direction of different magnitude and which differ in phase by $\pi/2$ is *elliptically polarized*. More complicated superpositions of plane waves need not have any of these characteristics. Indeed, the electric vectors approaching you along the direction of propagation can be randomly distributed and, in this case, the light is said to be *unpolarized*. This is true, for example, of the light coming from the sun or from a light bulb. However, Fourier analysis guarantees that any electromagnetic wave can be viewed as a superposition of (perhaps infinitely many) plane waves, each with its own polarization direction. We will write out a concrete example in Section 4.3 when we attempt to track down what was behind Max Planck's quantum hypothesis.

Next we introduce a notion that simplifies many computations and, moreover, provides the prototypical example of what is called a *gauge field*. We begin by returning to Maxwell's equations

$$\nabla \cdot \mathbf{E} = 0, \tag{4.18}$$

$$\nabla \times \mathbf{E} = -\frac{\partial \mathbf{B}}{\partial t}, \tag{4.19}$$

$$\nabla \cdot \mathbf{B} = 0, \tag{4.20}$$

$$\nabla \times \mathbf{B} = \frac{1}{c^2}\frac{\partial \mathbf{E}}{\partial t} \tag{4.21}$$

and considering solutions defined and smooth *for all* $(t, x, y, z) \in \mathbb{R} \times \mathbb{R}^3$. Suppose that there exists a smooth, time-dependent vector field $\mathbf{A}(t, x, y, z)$ on $\mathbb{R}^3$ and a smooth real-valued function $\phi(t, x, y, z)$ for which

$$\mathbf{B} = \nabla \times \mathbf{A} \tag{4.22}$$

and

$$\mathbf{E} = -\nabla\phi - \frac{\partial\mathbf{A}}{\partial t} \tag{4.23}$$

(keep in mind that ∇ denotes the *spatial* gradient operator). Then, since the curl of a gradient is zero and the divergence of a curl is zero, (4.19) and (4.20) are satisfied automatically. In this case, $\mathbf{A}$ and ϕ are called, respectively, *vector* and *scalar potentials* for $\mathbf{B}$ and $\mathbf{E}$. Furthermore, (4.18) and (4.21) become

$$\nabla^2\phi + \frac{\partial}{\partial t}(\nabla \cdot \mathbf{A}) = 0 \tag{4.24}$$

and

$$\nabla\left(\nabla \cdot \mathbf{A} + \frac{1}{c^2}\frac{\partial\phi}{\partial t}\right) - \left(\nabla^2\mathbf{A} - \frac{1}{c^2}\frac{\partial^2\mathbf{A}}{\partial t^2}\right) = \mathbf{0}, \tag{4.25}$$

respectively (for (4.25) we have used $\nabla \times (\nabla \times \mathbf{A}) = \nabla(\nabla \cdot \mathbf{A}) - \nabla^2\mathbf{A}$).

The existence of $\mathbf{A}$ and ϕ is not at all obvious and depends crucially on the topology of the region on which the solutions are assumed to exist (which we are here taking to be all of $\mathbb{R} \times \mathbb{R}^3$). This is best viewed from the relativistic point of view, where one can prove the existence of $\mathbf{A}$ and ϕ at the same time. Here we will offer a less elegant argument based on the relationship between the usual vector calculus on $\mathbb{R}^3$ and the exterior calculus of differential forms on $\mathbb{R}^3$; this relationship is spelled out in detail in Exercise 4.4.8 of [Nab4]. Thus, we fix a $t \in \mathbb{R}$ and let β denote the 1-form on $\mathbb{R}^3$ corresponding to the vector field $\mathbf{B}$ at time t. Then $\nabla \cdot \mathbf{B} = 0$ implies $^*d^*\beta = 0$, where * is the Hodge star operator on $\mathbb{R}^3$ determined by the standard metric and orientation of $\mathbb{R}^3$. Thus, $d^*\beta = 0$, so $^*\beta$ is a closed 2-form on $\mathbb{R}^3$. By the Poincaré lemma, $^*\beta$ is exact on $\mathbb{R}^3$, that is, there exists a smooth 1-form α on $\mathbb{R}^3$ with $^*\beta = d\alpha$. Consequently, $\beta = {}^{**}\beta = {}^*d\alpha$ and, if $\mathbf{A}$ is the vector field on $\mathbb{R}^3$ corresponding to the 1-form α, we have $\mathbf{B} = \nabla \times \mathbf{A}$, as required. Now, to obtain ϕ we note that

$$\nabla \times \left(-\mathbf{E} - \frac{\partial\mathbf{A}}{\partial t}\right) = -\nabla \times \mathbf{E} - \frac{\partial}{\partial t}(\nabla \times \mathbf{A}) = \frac{\partial\mathbf{B}}{\partial t} - \frac{\partial\mathbf{B}}{\partial t} = \mathbf{0}.$$

Thus, if ϵ is the 1-form on $\mathbb{R}^3$ corresponding to the vector field $-\mathbf{E} - \frac{\partial\mathbf{A}}{\partial t}$, then $^*d\epsilon = 0$ and so $d\epsilon = 0$. The Poincaré lemma then implies that $\epsilon = d\phi$ for some 0-form (real-valued function) on $\mathbb{R}^3$ and this translates into $-\mathbf{E} - \frac{\partial\mathbf{A}}{\partial t} = \nabla\phi$, which is what we wanted. We should also point out that there are physically interesting magnetic fields $\mathbf{B}$ on open regions $U \subseteq \mathbb{R}^3$ to which the Poincaré lemma does not apply and for which there is no vector potential **A** defined on all of U. As it turns out, this leads to interesting things (see pages 2–3 of [Nab3]).

We now know that **A** and ϕ exist, but they are certainly not unique, since if $\lambda(t, x, y, z)$ is any smooth function on $\mathbb{R} \times \mathbb{R}^3$, then, for each t,

$$\mathbf{A}' = \mathbf{A} - \nabla\lambda \quad \text{and} \quad \phi' = \phi + \frac{\partial\lambda}{\partial t} \tag{4.26}$$

also satisfy

$$\nabla \times \mathbf{A}' = \nabla \times \mathbf{A} - \nabla \times (\nabla\lambda) = \nabla \times \mathbf{A} = \mathbf{B}$$

because the curl of a gradient is zero, and

$$-\nabla\phi' - \frac{\partial\mathbf{A}'}{\partial t} = -\nabla\phi + \frac{\partial}{\partial t}(\nabla\lambda) - \frac{\partial\mathbf{A}}{\partial t} - \frac{\partial}{\partial t}(\nabla\lambda) = -\nabla\phi - \frac{\partial\mathbf{A}}{\partial t} = \mathbf{E}.$$

A transformation $(\mathbf{A}, \phi) \rightarrow (\mathbf{A}', \phi')$ of the form (4.26) is called a *gauge transformation* and the freedom to make such a transformation of potentials is called *gauge freedom*. Note, in particular, that one can add an arbitrary constant vector to any vector potential and an arbitrary real constant to any scalar potential and the results will still be potentials for **E** and **B**. We would now like to show that one can use this freedom to make some particularly convenient choices for the potentials. In the process we will need to be sure that certain partial differential equations have smooth solutions, but we will save the discussion of the theorems that ensure the existence of these solutions for Appendix E.

We will begin by selecting arbitrary potentials **A** and ϕ. For any smooth function λ, the gauge transformation (4.26) yields new potentials $(\mathbf{A}', \phi')$ that satisfy

$$\nabla \cdot \mathbf{A}' + \frac{1}{c^2}\frac{\partial\phi'}{\partial t} = \left(\nabla \cdot \mathbf{A} + \frac{1}{c^2}\frac{\partial\phi}{\partial t}\right) - \left(\nabla^2\lambda - \frac{1}{c^2}\frac{\partial^2\lambda}{\partial t^2}\right)$$

which will be zero if λ satisfies

$$\nabla^2\lambda - \frac{1}{c^2}\frac{\partial^2\lambda}{\partial t^2} = \nabla \cdot \mathbf{A} + \frac{1}{c^2}\frac{\partial\phi}{\partial t}. \tag{4.27}$$

The right-hand side of (4.27) is a known, smooth function so (4.27) is just the inhomogeneous wave equation and the existence of a smooth solution λ is ensured (see Appendix E). With such a choice of λ we have potentials that satisfy the so-called *Lorenz condition*

$$\nabla \cdot \mathbf{A}' + \frac{1}{c^2}\frac{\partial\phi'}{\partial t} = 0. \tag{4.28}$$

Physicists refer to a set of potentials $(\mathbf{A}', \phi')$ satisfying (4.28) as a *Lorenz gauge*.

Remark 4.2.3. One might also see this called a *Lorentz gauge*, but these are two different guys. The gauge condition is named for Ludwig Lorenz, who introduced it, but its
most important property is that it happens to be "Lorentz invariant" and this is named
for Hendrik Lorentz. Take your pick.

Note that if we add on to λ any solution λ' to the homogeneous wave equation

$$\Delta\lambda' - \frac{1}{c^2}\frac{\partial^2\lambda'}{\partial t^2} = 0,$$

of which there are many (see Appendix E), then the resulting potentials clearly still
satisfy the Lorenz condition. Consequently, there is a great deal of freedom in choosing
a Lorenz gauge. Also note that, in a Lorenz gauge, both the vector and scalar potentials
satisfy homogeneous wave equations. Indeed, the coupled equations (4.24) and (4.25)
decouple in a Lorenz gauge and become

$$\nabla^2\phi' - \frac{1}{c^2}\frac{\partial^2\phi'}{\partial t^2} = 0$$

and

$$\nabla^2\mathbf{A}' - \frac{1}{c^2}\frac{\partial^2\mathbf{A}'}{\partial t^2} = \mathbf{0}.$$

Now, let us begin again with some arbitrary potentials $\mathbf{A}$ and ϕ. For any gauge
transformation (4.26) we have, from $\mathbf{A}' = \mathbf{A} - \nabla\lambda$,

$$\nabla\cdot\mathbf{A}' = \nabla\cdot\mathbf{A} - \nabla^2\lambda.$$

Since $\nabla\cdot\mathbf{A}$ is known, we can ensure that

$$\nabla\cdot\mathbf{A}' = 0 \tag{4.29}$$

by taking λ to be any smooth solution to the *Poisson equation*

$$\nabla^2\lambda = \nabla\cdot\mathbf{A},$$

and, again, there are many of these (see Appendix E). Potentials satisfying (4.29) are
said to be a *Coulomb gauge* and, for these, (4.24) and (4.25) become

$$\nabla^2\phi' = 0$$

and

$$\nabla^2\mathbf{A}' - \frac{1}{c^2}\frac{\partial^2\mathbf{A}'}{\partial t^2} = \frac{1}{c^2}\nabla\left(\frac{\partial\phi'}{\partial t}\right).$$

In particular, in a Coulomb gauge, the scalar potential must be a solution to the *Laplace equation*, that is, it must be *harmonic* on $\mathbb{R}^3$ for each t.

In the physics literature one might find the Coulomb gauge defined by $\nabla \cdot \mathbf{A}' = 0$ and $\phi' = 0$. Now, it does not follow from what we have said that the scalar potential ϕ' must be zero. It is, however, harmonic on $\mathbb{R}^3$ for each t and if one imposes the additional *physical assumption* that it should be bounded on $\mathbb{R}^3$ for each t, then, in fact, it must be constant. This follows from Liouville's theorem, which says that any bounded, harmonic function on any $\mathbb{R}^n$ is constant (if this version of Liouville's theorem is unfamiliar to you, consult [Nel2] for the shortest paper you are ever likely to see). Since potentials are determined only up to additive constants, one can take it to be zero. This physical assumption is satisfied in the case of particular interest to us in Section 4.3 (electromagnetic radiation in a black box) so we will say that a pair of potentials $\mathbf{A}$ and ϕ satisfying

$$\nabla \cdot \mathbf{A} = 0 \quad \text{and} \quad \phi = 0 \tag{4.30}$$

is a *radiation gauge*. In such a gauge,

$$\mathbf{B} = \nabla \times \mathbf{A} \quad \text{and} \quad \mathbf{E} = -\frac{\partial \mathbf{A}}{\partial t}, \tag{4.31}$$

and $\mathbf{A}$ is determined by

$$\nabla^2 \mathbf{A} - \frac{1}{c^2}\frac{\partial^2 \mathbf{A}}{\partial t^2} = \mathbf{0} \quad \text{and} \quad \nabla \cdot \mathbf{A} = 0. \tag{4.32}$$

Such an $\mathbf{A}$ is simply a divergence-free solution to the wave equation.

Remark 4.2.4. The electric and magnetic fields can be computed directly from *any* pair of potentials and, as we have seen, a clever choice of potentials can significantly simplify Maxwell's equations. Classical electrodynamics makes considerable use of these potentials as computational tools, but no physical significance was ascribed to $\mathbf{A}$ and ϕ themselves (essentially because they are highly nonunique). The situation is dramatically different in quantum mechanics. We will see some of the reasons for this as we proceed, but a more complete picture, described in elementary terms, is available in Chapter 0 of [Nab3].

The final topic we need to address in this section is, physically at least, rather subtle because it deals with the rather elusive notion of energy (see Remark 1.0.2). Intuitively, it seems clear that electromagnetic radiation must, in some sense, contain energy since it can warm you on a sunny day, give life to plants through photosynthesis and even air condition your home. How is the energy associated with an electromagnetic field to be defined? The objective of any definition of energy is a conservation law (see Remark 1.0.2). In classical mechanics this conservation law takes the form of an assertion that a certain number (the sum of the kinetic and potential energies) remains constant along the trajectory of the particle. In other contexts, conservation laws take the form of what are called *continuity equations*. A familiar example from calculus concerns the flow of a fluid. If the mass density of the fluid is ρ and its velocity vector field is $\mathbf{V}$, then the continuity equation is

$$\frac{\partial \rho}{\partial t} + \nabla \cdot (\rho \mathbf{V}) = 0. \tag{4.33}$$

To understand why this qualifies as a conservation law, suppose U is any bounded, open region in $\mathbb{R}^3$ with smooth boundary ∂U. Integrating (4.33) over the closure $\mathrm{cl}_{\mathbb{R}^3} U$ of U in $\mathbb{R}^3$ and using the divergence theorem gives

$$\frac{\partial}{\partial t} \iiint_{\mathrm{cl}_{\mathbb{R}^3} U} \rho \, dV = - \iint_{\partial U} \rho \mathbf{V} \cdot d\mathbf{S},$$

which says that the rate at which mass enters or leaves $\mathrm{cl}_{\mathbb{R}^3} U$ is equal to the flux of mass through the boundary of U and so, since U is arbitrary, mass is conserved (neither created nor destroyed anywhere).

To find an analogue of (4.33) for electromagnetic radiation we return to the vacuum Maxwell equations (4.18)–(4.21), but now we will write $\frac{1}{c^2} = \mu_0 \epsilon_0$, where μ_0 is the vacuum permeability and ϵ_0 is the vacuum permittivity (see (4.4)). Dot both sides of (4.19) with $\mathbf{B}$ and both sides of (4.21) with $\mathbf{E}$ and add to obtain

$$\mathbf{E} \cdot (\nabla \times \mathbf{B}) + \mathbf{B} \cdot (\nabla \times \mathbf{E}) = - \left(\mu_0 \epsilon_0 \, \mathbf{E} \cdot \frac{\partial \mathbf{E}}{\partial t} + \mathbf{B} \cdot \frac{\partial \mathbf{B}}{\partial t} \right).$$

The identities $\mathbf{V} \cdot (\nabla \times \mathbf{W}) + \mathbf{W} \cdot (\nabla \times \mathbf{V}) = \nabla \cdot (\mathbf{V} \times \mathbf{W})$ and $\frac{\partial}{\partial t} \|\mathbf{V}\|^2 = 2\mathbf{V} \cdot \frac{\partial \mathbf{V}}{\partial t}$ and a little algebra reduce this to

$$\nabla \cdot \left(\frac{1}{\mu_0} \mathbf{E} \times \mathbf{B} \right) = - \frac{\partial}{\partial t} \left(\frac{\epsilon_0}{2} \|\mathbf{E}\|^2 + \frac{1}{2\mu_0} \|\mathbf{B}\|^2 \right).$$

Now, defining

$$\mathbf{S} = \frac{1}{\mu_0} \mathbf{E} \times \mathbf{B}$$

and

$$\mathcal{E} = \frac{\epsilon_0}{2} \|\mathbf{E}\|^2 + \frac{1}{2\mu_0} \|\mathbf{B}\|^2, \tag{4.34}$$

this becomes

$$\frac{\partial \mathcal{E}}{\partial t} + \nabla \cdot \mathbf{S} = 0. \tag{4.35}$$

We find then that Maxwell's equations determine a very natural continuity equation and therefore a conservation law. By analogy with (4.33), one would be inclined to identify $\mathcal{E}$ with the thing being conserved and $\mathbf{S}$ with a vector describing the direction and rate at which this thing is being transported by the field. In physics, $\mathcal{E}$ is called the *energy density* of the electromagnetic field and $\mathbf{S}$ is the *Poynting vector*; (4.35) is a special case of what is called *Poynting's theorem*.

Remark 4.2.5. Not every "thing" that is conserved can reasonably be interpreted as an energy (one has, for example, momentum, angular momentum, etc.) and simply calling $\mathcal{E}$ the energy density of the field does not justify the use of the term. The intuition we were asked to accept in high school is at least morally correct; energy should somehow be associated with the ability to do work. That the terminology we have introduced really is appropriate should be checked by relating $\mathcal{E}$ and $\mathbf{S}$ to the work the field is capable of doing. The full Maxwell equations contain the electric charge ρ and current $\mathbf{J}$ densities responsible for creating the field and with these and the Lorentz force law (4.6) one can compute the work done by the field on the charges and in this way motivate our interpretations of $\mathcal{E}$ and $\mathbf{S}$. Since this is all done carefully and clearly in Sections 27-1 through 27-3, Volume II, of [FLS] we will simply refer those interested in the details to the exposition by one of the great physicists of the twentieth century.

In addition to carrying energy, electromagnetic radiation exerts pressure on any surface it falls upon and therefore should also carry momentum. This is rather convincingly demonstrated by a device called a *Nichols radiometer*, which you can now buy in almost any toy store. It is simply a very delicate pinwheel in a vacuum that will spin if you shine light on it.

The *momentum density* of an electromagnetic field is, of course, a vector at each point in space and at each instant of time and is identified by the physicists with a multiple of the Poynting vector

$$\frac{1}{c^2}\mathbf{S} = \epsilon_0 \mathbf{E} \times \mathbf{B}$$

(the rationale for this is discussed in Section 27-6, Volume II, of [FLS]). By analogy with classical particle mechanics (see (2.20)) one then defines the *angular momentum density* (with respect to the origin) of the electromagnetic field by

$$\mathbf{r} \times (\epsilon_0 \mathbf{E} \times \mathbf{B}),$$

where $\mathbf{r}$ is the position vector in $\mathbb{R}^3$.

Exercise 4.2.2. Write out the energy density, Poynting vector, momentum density and angular momentum density for a plane electromagnetic wave.

We should conclude this section by saying that quite soon we will be forced by circumstances (in Section 4.3) to adopt quite a different view of electromagnetic radiation and the energy and momentum it contains and that this different view of electromagnetic radiation will lead us inexorably to a different view of everything.

4.3 Blackbody radiation and the photoelectric effect

Place a bar of iron in the summer sun for a few hours. When you return to retrieve your iron bar and reach to pick it up you find to your chagrin that it is emitting thermal energy (it is hot). The electromagnetic radiation coming from the sun, which contains energy, has communicated some of this energy to the iron and "heated" it. But what is this "heat" that we perceive? Here is a hint. Suppose we could move the iron bar off the surface of the earth and closer and closer to the sun. Of course, it would get hotter and hotter, but it would do something else as well; it would change color. Close enough to the sun it would glow red hot, closer still, orange, then yellow and finally blue. But what our eyes perceive as color is simply a particular frequency of light so it would seem that the heat we sense coming from the iron bar is again just electromagnetic radiation. The rays from the sun supply energy to the molecules and atoms of the iron which vibrate in response, thus causing the electrons in the atoms near the surface to vibrate and these, as all accelerating charges do, generate electromagnetic radiation. This, in turn, supplies energy to the molecules and atoms of our skin which we sense as thermal energy, that is, heat. It is important to note that we sense this heat long before the iron has started to glow red hot. For these more moderate temperatures the frequency of the electromagnetic radiation being emitted by the bar is not in the (to us) visible range, but rather in the infrared (see Figure 4.1). It is a fact of nature that *every body at a temperature above absolute zero emits electromagnetic radiation.* Mercifully, the human eye perceives only a minute portion of this radiation.

Remark 4.3.1. A precise physical definition of absolute zero or, indeed, even of temperature, would involve a rather lengthy digression into thermodynamics and we have neither the time nor the competence to do this properly here (there are many introductory texts available if this interests you, or you may prefer the concise exposition [Fermi] by a Nobel Laureate). Fortunately, the physicists have relieved some of this burden by agreeing to *define absolute zero* to be $-273.15°C$ (or, equivalently, $-459.67°F$) and we will take this as our definition as well (of course, this presumes that you know what temperature means when measured on the Celsius or Fahrenheit scales and it completely evades the issue of the physical significance of this particular value). This value is also taken to be zero on the *Kelvin scale* so that absolute zero is $0\,K$ (physicists have apparently also agreed that writing $0°\,K$ is not to be tolerated (see http://en.wikipedia.org/wiki/Kelvin).

We will often be confronted in this section with physical statements the theoretical justification of which requires sophisticated ideas and techniques from not only thermodynamics, but statistical mechanics and electromagnetic theory as well. In these cases we will not presume to offer sound bites that pretend to be explanations, but will try to provide ample references for those who would like to *really* understand. A good place to begin is Chapter 1 of [Bohm], which contains a detailed and very readable account of everything we will have to say in this section and much more together with a number of (admittedly rather old) references to discussions of the thermodynamics and statistical mechanics.

All objects absorb and emit electromagnetic radiation, but they do not all do it in the same way or at the same rate. A red fire truck is red because the paint on its surface absorbs every frequency of light except those that we perceive as red (around 4.3×10^{14} Hz), which it reflects back to our eyes. A substance that is very black, like graphite or soot, absorbs essentially all of the electromagnetic radiation falling on it. On the other hand, it is the case that at a given temperature, a body always emits radiation of a given frequency exactly as well as it absorbs radiation of that frequency (this is an experimental fact, but also a consequence of the second law of thermodynamics). Consequently, graphite is not only a nearly perfect absorber of electromagnetic radiation, but a nearly perfect emitter as well. A *blackbody* is an (idealized) physical object that absorbs all incident electromagnetic radiation. In this section we are interested in the spectrum of radiation emitted by such a blackbody (we will define more precisely what this means in just a moment).

An object with very special and interesting thermodynamic properties that has been investigated since the nineteenth century is what we will call a *black box*. This is essentially an oven with black walls and with a tiny hole drilled in one of the walls. Turn the oven on. The temperature of the walls increases and so they emit radiation of every possible frequency at a rate that depends on the temperature. These same walls, in turn, absorb this radiation at a rate that depends on the intensity of the radiation in the interior of the oven. Eventually the emission and absorption balance and a state of thermal equilibrium is achieved in which the temperature T is constant (in this section T will always be measured in Kelvin). The object we are interested in is the function $\rho_T(\lambda)$ or, equivalently, $\rho_T(\nu)$, that gives the energy density of the equilibrium radiation of wavelength λ, or of frequency $\nu = c/\lambda$.

Physicists often consider instead the *intensity* I_T of the radiation as a function of λ or ν. The intensity is defined to be the energy which the radiation carries per second across a $1\,\text{m}^2$ area normal to the direction of propagation. As it happens, the intensity and energy density are proportional with a constant of proportionality that does not depend on the wavelength/frequency or T. The function $I_T(\lambda)$ can be measured experimentally. The radiation escaping from the small hole is passed through a diffraction grating (high tech prism) sending the different wavelengths in different directions, all toward a screen. A detector is moved along the screen to determine the intensity emitted at each wavelength. It has been found that for a given wavelength, the intensity depends only on T and not on the details of the oven's construction (size, shape,

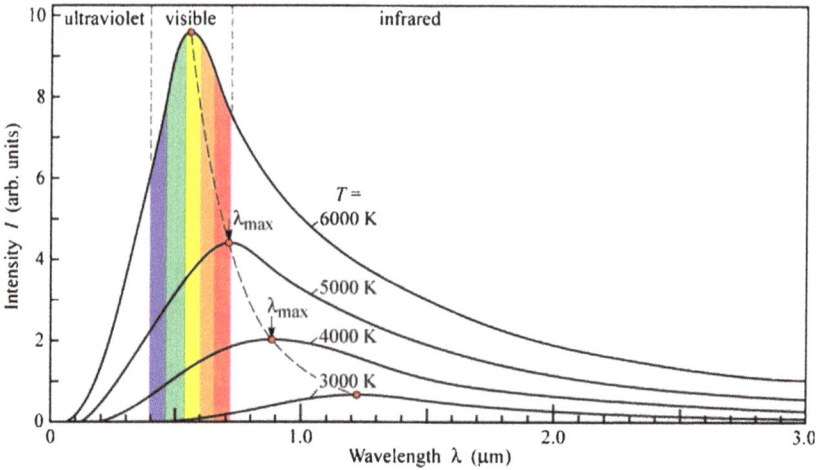

Figure 4.4: Blackbody radiation curves. http://hyperphysics.phy-astr.gsu.edu/hbase/mod6.html (last access date: 21.05.2021).

material, etc.). Figure 4.4[1] shows the graphs of $I_T(\lambda)$ for T = 3000 K, 4000 K, 5000 K and 6000 K. At any given temperature T the intensity of the radiation increases rather rapidly with the wavelength λ until it reaches a maximum value at some wavelength λ_{max} that depends on T and at this point it begins to decrease with λ. The integral

$$I_T = \int_0^\infty I_T(\lambda)\,d\lambda$$

represents the total intensity of the radiation emitted over all wavelengths at temperature T. In 1879, Jožef Stefan deduced from the empirical data that I_T is proportional to the fourth power of T:

$$I_T = \sigma T^4,$$

where σ = 5.670400×10^{-8} Jm^{-2}s^{-1}K^{-4} is the so-called *Stefan–Boltzmann constant*; this was later derived on theoretical grounds by Ludwig Boltzmann.

In 1893, Wilhelm Wien showed using thermodynamic arguments that there is a universal function f for which

$$\rho_T(\lambda) = \frac{f(\lambda T)}{\lambda^5}. \tag{4.36}$$

Thermodynamics alone, however, cannot determine the function f. This is essentially because thermodynamic arguments are based on very general principles that apply

---

1 Reproduced by permission of Professor Carl R. Nave of the HyperPhysics Project, Georgia State University. See http://hyperphysics.phy-astr.gsu.edu/hbase/index.html

to all physical systems and often do not take into account the specific details of any particular system. Nevertheless, plotting $\lambda^5 \rho_T(\lambda)$ versus λT for the empirical data one finds that the points lie on the same curve for any T so there is solid experimental evidence to support *Wien's law* (4.36).

Although thermodynamics cannot identify the function f, electrodynamics and classical statistical mechanics combine to give the completely explicit prediction

$$f(\lambda T) = 8\pi\kappa_B\,(\lambda T), \quad \text{(Rayleigh–Jeans)} \tag{4.37}$$

where $\kappa_B = 1.3806488 \times 10^{-23}\,\text{JK}^{-1}$ is the *Boltzmann constant*. Thus,

$$\rho_T(\lambda) = \frac{8\pi\kappa_B T}{\lambda^4}. \quad \text{(Rayleigh–Jeans)} \tag{4.38}$$

This result was derived by Lord Rayleigh and Sir James Jeans in 1905 (except for the precise value of the constant of proportionality $[8\pi\kappa_B]$ the result was actually established by Rayleigh in 1900). The argument involved quite nontrivial aspects of classical physics and we will briefly describe how it was done later in this section (see (4.55)).

The only issue one might want to take with the Rayleigh–Jeans argument is that its conclusion is *totally incorrect*. One can see this by simply comparing its predictions with the empirical data (see Figure 4.5[2]). Even without any delicate experimental data, however, one can see that the result cannot be correct since it implies that the total energy contained in the black box is, by Wien's law (4.36),

$$\int_0^\infty \rho_T(\lambda)\,d\lambda = 8\pi\kappa_B T \int_0^\infty \frac{d\lambda}{\lambda^4} = \frac{8\pi}{3}\kappa_B T \lim_{\lambda \to 0^+} \frac{1}{\lambda^3},$$

and this is infinite unless $T = 0$ (Paul Ehrenfest referred to this as the *ultraviolet catastrophe*). Since the logic of the Rayleigh–Jeans argument was considered unassailable, one is forced to question the premises on which the argument is based. But these premises were believed to be among the most firmly established principles of classical physics. One can see a storm on the horizon.

Max Planck set himself the task of deriving a formula for the energy spectrum that agreed with the experimental data. Eventually, he succeeded, but only by straying outside the confines of classical physics with an *ad hoc* assumption that he himself regarded as an "act of desperation." We would like to have a look, admittedly a rather cursory and informal one, at a path one can follow that leads to Planck's formula since it is along such a path that one finds for the first time the ubiquitous "Planck constant" and we really should have some idea of where this comes from. First, however, let us simply record the formula to see where we are headed. Planck determined that the

---

2 Reproduced by permission of Professor Carl R. Nave of the HyperPhysics Project, Georgia State University. See http://hyperphysics.phy-astr.gsu.edu/hbase/index.html

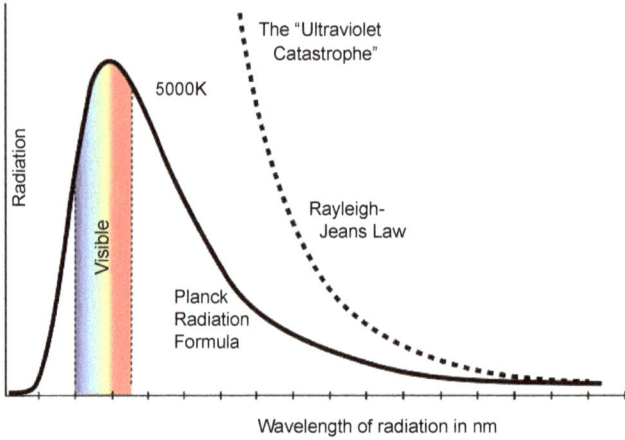

Figure 4.5: Rayleigh–Jeans. http://hyperphysics.phy-astr.gsu.edu/hbase/mod6.html (last access date: 21.05.2021).

function $f(\lambda T)$ in Wien's law (4.36) is given not by the Rayleigh–Jeans law (4.37), but rather by

$$f(\lambda T) = \frac{8\pi hc}{e^{hc/\kappa_B \lambda T} - 1},\tag{4.39}$$

where c is the speed of light and h is a positive constant that must be determined to fit the data. Thus,

$$\rho_T(\lambda) = \frac{8\pi hc}{\lambda^5}\frac{1}{e^{hc/\kappa_B \lambda T} - 1},\tag{4.40}$$

or, in terms of the frequency ν,

$$\rho_T(\nu) = \frac{8\pi h\nu^3}{c^3}\frac{1}{e^{h\nu/\kappa_B T} - 1}.\tag{4.41}$$

This is known as *Planck's law*, which we will derive later in this section (see (4.56)). We will see that these are to be regarded as density functions. For example, the amount of electromagnetic energy per unit volume accounted for by radiation with wavelengths in $[\lambda_0, \lambda_1]$ is

$$\int_{\lambda_0}^{\lambda_1} \rho_T(\lambda)d\lambda = \int_{\lambda_0}^{\lambda_1} \frac{8\pi hc}{\lambda^5}\frac{1}{e^{hc/\kappa_B \lambda T} - 1} d\lambda.\tag{4.42}$$

Exercise 4.3.1. Show that, when λT is large, Planck's formula (4.40) is approximately given by the Rayleigh–Jeans formula (4.37).

Note that h is, of course, the famous *Planck constant*. Its currently accepted value is $6.62606957 \times 10^{-34}$ m^2(kg)/s. Because it occurs so frequently, one also defines the *normalized Planck constant* $\hbar = h/2\pi$. Note that h has the same units as the action in classical mechanics (see (2.1)) so one often sees Planck's constant referred to as the *quantum of action*. What we would like to do now is try to understand where h comes from.

The physics behind Planck's formula is deep and we will not pretend to offer more than a rather pedestrian synopsis. We begin by setting up the problem we need to solve. The basic object of interest is a black box (oven) filled with electromagnetic radiation in thermal equilibrium (that is, at some constant temperature T). We have already mentioned that it has been shown both experimentally and theoretically that the energy spectrum is independent of the shape and material construction of the box so we are free to choose this as we please. For the black box we will choose a cube $R = [0,L]^3 = [0,L] \times [0,L] \times [0,L] \subseteq \mathbb{R}^3$ of side length $L > 0$ in $\mathbb{R}^3$. Here is what we must do.

1. Prescribe appropriate boundary conditions and solve Maxwell's equations for fields **E** and **B** that represent electromagnetic waves in thermal equilibrium with the boundary at temperature T. We will be assuming that the black box contains nothing but electromagnetic radiation so by "Maxwell's equations" we mean the empty space version (4.1)–(4.4).
2. Compute the total electromagnetic energy (see (4.34))

$$E = \int_R \mathcal{E}\, dV = \int_R \left(\frac{\epsilon_0}{2} \|\mathbf{E}\|^2 + \frac{1}{2\mu_0} \|\mathbf{B}\|^2 \right) dV \qquad (4.43)$$

contained in R. Assuming, as we shall, that the system is isolated, this total energy is constant.
3. Determine how this total electromagnetic energy is distributed among the various frequencies of radiation present in R.

The question of appropriate boundary conditions can be a subtle one, depending on the specific physical circumstances of the problem. However, if we once again appeal the fact that the energy spectrum at thermal equilibrium is generally independent of these specifics we are able to choose boundary conditions that will confine the radiation within the box and simplify the calculations. To this end physicists generally adopt "periodic boundary conditions" according to which the unknown fields take the same values at corresponding points on opposite faces of the cube (we will describe this a bit more precisely in a moment).

By translation in the directions of the coordinate axes we can partition all of $\mathbb{R}^3$ into a countable family of copies of R intersecting only along their common boundaries. Because the boundary conditions are periodic, the sought-after fields can then be thought of as defined on all of $\mathbb{R}^3$, and on $\mathbb{R}^3$ we have shown that we can work in a Coulomb gauge (see (4.29)) with potentials $\mathbf{A}$ and ϕ. Moreover, ϕ is harmonic and, in particular, continuous and therefore bounded on R. As a result, it is bounded everywhere on $\mathbb{R}^3$, so we can actually work in a radiation gauge (see (4.30)) with $\phi = 0$. Thus, the only field we need to find is the vector potential $\mathbf{A}$ and this is determined by (4.32). The electric and magnetic fields are then given by (4.31) and the total energy by (4.43). However, to carry out Step 3. in the program described above we will need all of this expressed in terms of the radiation frequencies, and this means Fourier analysis. Because of the periodic boundary conditions, however, the proper context for this Fourier analysis is not really $\mathbb{R}^3$, but rather a certain three-dimensional "flat torus." We will therefore need to digress and sketch some of the background.

Remark 4.3.2. Let $\{\mathbf{v}_1, \ldots, \mathbf{v}_N\}$ be a basis for $\mathbb{R}^N$. Associated with this basis is a lattice Γ consisting of all of the integer linear combinations of the basis vectors:

$$\Gamma = \left\{ \mathbf{v} = \sum_{i=1}^{N} n^i \mathbf{v}_i : n^i \in \mathbb{Z}, i = 1, \ldots, N \right\}.$$

Identify Γ with a (discrete, Abelian) subgroup of $\mathbb{R}^N$. One can show that the quotient space $\mathbb{R}^N/\Gamma$ admits a unique differentiable manifold structure for which the canonical projection $\pi : \mathbb{R}^N \to \mathbb{R}^N/\Gamma$ is a smooth submersion; this follows, for example, from Theorem 7.10 of [Lee2]. Indeed, if we let $\mathbb{T}^N = S^1 \times \overset{N}{\cdots} \times S^1$ denote the N-dimensional torus, then the map $\varphi : \mathbb{R}^N \to \mathbb{T}^N$ defined by

$$\varphi(\mathbf{x}) = \varphi \left(\sum_{i=1}^{N} x^i \mathbf{v}_i \right) = \left(e^{2\pi i x^1}, \ldots, e^{2\pi i x^N} \right)$$

is constant on each fiber $\pi^{-1}([\mathbf{x}]) = \mathbf{x} + \Gamma$, $[\mathbf{x}] \in \mathbb{R}^N/\Gamma$, so it descends to a map $\tilde{\varphi} : \mathbb{R}^N/\Gamma \to \mathbb{T}^N$ and one can show that this is a diffeomorphism.

$$\mathbb{R}^N/\Gamma \cong \mathbb{T}^N.$$

Furthermore, since Γ is discrete, $\pi : \mathbb{R}^N \to \mathbb{R}^N/\Gamma$ is a smooth covering map (see Theorem 7.13 of [Lee2]). In particular, its derivative $\pi_* : T\mathbb{R}^N \to T\mathbb{T}^N$ is an isomorphism on each fiber and therefore induces, from the standard Riemannian metric g of $\mathbb{R}^N$, a Riemannian metric g_Γ on $\mathbb{T}^N \cong \mathbb{R}^N/\Gamma$. These are locally isometric so that, since g is flat, that is, has zero Riemannian curvature, the same is true of g_Γ. With this Riemannian metric $\mathbb{R}^N/\Gamma$ is called the *flat torus* determined by Γ.

The convex hull $D(\Gamma)$ of $\{\mathbf{v}_1, \ldots, \mathbf{v}_N\}$ in $\mathbb{R}^N$ is the set of all convex linear combinations of $\mathbf{v}_1, \ldots, \mathbf{v}_N$ and is called the *fundamental domain* of $\mathbb{R}^N/\Gamma$. This is a closed interval when $N = 1$, a closed parallelogram when $N = 2$, a closed parallelepiped when $N = 3$, and so on. Translates by elements of Γ of the interior of $D(\Gamma)$ are pairwise disjoint in $\mathbb{R}^N$, but the translates of $D(\Gamma)$ itself cover $\mathbb{R}^N$. The torus $\mathbb{R}^N/\Gamma$ can be viewed as the fundamental domain $D(\Gamma)$ with points on its boundary identified if they differ by an element of Γ. The standard Lebesgue measure μ on $D(\Gamma)$ induces a pushforward measure $\tilde{\mu} = \pi_*(\mu)$ on the torus $\mathbb{R}^N/\Gamma$.

Remark 4.3.3. Recall that if $(X, \mathcal{A}, \mu)$ is a measure space, $(Y, \mathcal{C})$ is a measurable space and $f : (X, \mathcal{A}) \to (Y, \mathcal{C})$ is a measurable function, then the *pushforward measure* $f_*(\mu)$ on $(Y, \mathcal{C})$ is defined by

$$(f_*(\mu))(C) = \mu(f^{-1}(C)) \quad \forall C \in \mathcal{C}.$$

There is an abstract change of variables formula (Theorem C, Section 39, of [Hal1]) that asserts the following. If F is any extended real- or complex-valued measurable function on $(Y, \mathcal{C})$, then F is integrable with respect to $f_*(\mu)$ if and only if $F \circ f$ is μ-integrable and, in this case,

$$\int_Y F(y)\, d(f_*(\mu))(y) = \int_X (F \circ f)(x)\, d\mu(x). \tag{4.44}$$

Furthermore, (4.44) holds in the stronger sense that if either side is defined (even if it is not finite), then the other side is defined and they agree.

The measure $\tilde{\mu}$ has the following property. Any real- or complex-valued integrable function ϕ on $\mathbb{R}^N$ that is Γ-periodic ($\phi(\mathbf{x} + \mathbf{v}) = \phi(\mathbf{x})\ \forall \mathbf{v} \in \Gamma$) descends to a unique integrable function $\tilde{\phi}$ on $\mathbb{R}^N/\Gamma$ and

$$\int_{D(\Gamma)} \phi\, d\mu = \int_{\mathbb{R}^N/\Gamma} \tilde{\phi}\, d\tilde{\mu}. \tag{4.45}$$

Conversely, any real- or complex-valued function $\tilde{\phi}$ on $\mathbb{R}^N/\Gamma$ lifts uniquely to a Γ-periodic function ϕ on $\mathbb{R}^N$ and (4.45) is satisfied. As long as the lattice Γ is fixed it does no real harm to adopt the usual custom and blur the distinction between the Γ-periodic functions on $\mathbb{R}^N$ and the functions to which they descend on $\mathbb{R}^N/\Gamma$ and even to identify $\tilde{\mu}$ with μ and write simply $\mathbb{T}^N$ for $\mathbb{R}^N/\Gamma$. We will therefore tend to drop the tildes and write such things as $\int_{\mathbb{R}^N/\Gamma} \phi\, d\mu$, $\int_{\mathbb{T}^N} \phi\, d\mu$ and $\int_{D(\Gamma)} \phi\, d\mu$ interchangeably. Finally, note that, since the faces of $D(\Gamma)$ have measure zero in $\mathbb{R}^N$, we can identify $L^p(\mathbb{T}^N)$ with $L^p(D(\Gamma))$ for any $1 \leq p \leq \infty$.

Now let us adapt this last remark to our black box $[0, L]^3$. Let $\{\mathbf{e}_1, \mathbf{e}_2, \mathbf{e}_3\}$ be the standard basis for $\mathbb{R}^3$ and define a new (orthogonal, but not orthonormal) basis $\{\mathbf{v}_1, \mathbf{v}_2, \mathbf{v}_3\}$ by $\mathbf{v}_i = L\mathbf{e}_i$ for $i = 1, 2, 3$. The corresponding lattice

$$\Gamma = \{\mathbf{v} = n^1\mathbf{v}_1 + n^2\mathbf{v}_2 + n^3\mathbf{v}_3 = (n^1 L, n^2 L, n^3 L) : n^1, n^2, n^3 \in \mathbb{Z}\}$$

determines a flat torus $\mathbb{T}^3 = \mathbb{R}^3/\Gamma$ and the fundamental domain $D(\Gamma)$ is just the cube $[0, L]^3$. A function ϕ on $\mathbb{T}^3$ is identified with a function on $\mathbb{R}^3$ that is Γ-periodic, that is, one that satisfies

$$\phi(\mathbf{x} + \mathbf{v}) = \phi(\mathbf{x})$$

for all $\mathbf{v} \in \Gamma$. In more detail,

$$\phi(x^1 + n^1 L, x^2 + n^2 L, x^3 + n^3 L) = \phi(x^1, x^2, x^3)$$

for all $(x^1, x^2, x^3) \in \mathbb{R}^3$ and all $(n^1, n^2, n^3) \in \mathbb{Z}^3$. We will write out some particularly important examples. For this we consider another lattice $[\frac{2\pi}{L}\mathbb{Z}]^3$ in $\mathbb{R}^3$ that we will use to index the functions. Each $\mathbf{k} \in [\frac{2\pi}{L}\mathbb{Z}]^3$ is of the form

$$\mathbf{k} = (k^1, k^2, k^3) = \frac{2\pi}{L}(m^1, m^2, m^3),$$

where $(m^1, m^2, m^3) \in \mathbb{Z}^3$. For each such $\mathbf{k}$ we define

$$\phi_{\mathbf{k}}(\mathbf{x}) = L^{-3/2} e^{i\mathbf{k}\cdot\mathbf{x}} = L^{-3/2} e^{i(k^1 x^1 + k^2 x^2 + k^3 x^3)}.$$

Exercise 4.3.2. Show that for each $\mathbf{k} \in [\frac{2\pi}{L}\mathbb{Z}]^3$,
1. $\phi_{\mathbf{k}}$ is Γ-periodic,
2. $\phi_{\mathbf{k}}$ is in $L^2(\mathbb{T}^3)$ and, in fact, $\|\phi_{\mathbf{k}}\|_{L^2}^2 = 1$,
3. $\phi_{\mathbf{k}_1}$ and $\phi_{\mathbf{k}_2}$ are orthogonal in $L^2(\mathbb{T}^3)$ if $\mathbf{k}_1 \neq \mathbf{k}_2$, and
4. $-\Delta\phi_{\mathbf{k}}(\mathbf{x}) = \|\mathbf{k}\|^2 \phi_{\mathbf{k}}(\mathbf{x})$.
   Note: In 4., Δ is the Laplacian on $\mathbb{T}^3$ which, locally, is the same as the Laplacian on $\mathbb{R}^3$ because $\mathbb{T}^3$ is locally isometric to $\mathbb{R}^3$.

Thus, each $\phi_{\mathbf{k}}$ is an eigenfunction for $-\Delta$ on $L^2(\mathbb{T}^3)$ with eigenvalue $\|\mathbf{k}\|^2$. According to 2. and 3., $\{\phi_{\mathbf{k}} : \mathbf{k} \in [\frac{2\pi}{L}\mathbb{Z}]^3\}$ is an orthonormal set in $L^2(\mathbb{T}^3)$, but one can show that it is, in fact, an *orthonormal basis* for $L^2(\mathbb{T}^3)$; this follows from the Stone–Weierstrass theorem (see Proposition 3.1.16 of [Graf]). It follows from this that $\{\|\mathbf{k}\|^2 : \mathbf{k} \in [\frac{2\pi}{L}\mathbb{Z}]^3\}$ contains *all* of the eigenvalues of $-\Delta$ on $L^2(\mathbb{T}^3)$. We will explain this in more detail in Section 5.2, but briefly the reason is that $-\Delta$ defines a self-adjoint operator on $L^2(\mathbb{T}^3)$ so that eigenfunctions corresponding to distinct eigenvalues are orthogonal. Thus, any eigenfunction corresponding to some other eigenvalue would have to be orthogonal to everything in an orthonormal basis for $L^2(\mathbb{T}^3)$ and this would force it to be zero. We find then that any $f \in L^2(\mathbb{T}^3)$ can be written as

$$f(\mathbf{x}) = \sum_{\mathbf{k}\in[\frac{2\pi}{L}\mathbb{Z}]^3} a_{\mathbf{k}} e^{i\mathbf{k}\cdot\mathbf{x}}, \tag{4.46}$$

where

$$a_{\mathbf{k}} = L^{-3} \int_{\mathbb{T}^3} f(\mathbf{x}) e^{-i\mathbf{k}\cdot\mathbf{x}} d\mu(\mathbf{x}) = L^{-3} \int_{[0,L]^3} f(\mathbf{x}) e^{-i\mathbf{k}\cdot\mathbf{x}} d^3\mathbf{x} \tag{4.47}$$

and the convergence of the series is in $L^2(\mathbb{T}^3)$. The series (4.46) is called the *Fourier series* for $f(\mathbf{x})$ in $L^2(\mathbb{T}^3)$ and $a_{\mathbf{k}}$ are the *Fourier coefficients* of $f(\mathbf{x})$ (note that $a_{\mathbf{k}} =$

$L^{-3/2}\langle\phi_{\mathbf{k}},f\rangle_{L^2}$). One might wonder about the definition of the partial sums of the Fourier series since there is no unique natural ordering of the elements of $[\frac{2\pi}{L}\mathbb{Z}]^3$. In the general theory of Fourier series this is, indeed, an issue with which one must deal (see Definition 3.1.12 of [Graf] for some of the options). However, for an L^2 function f, Parseval's theorem (see, for example, Theorem 6.4.5 of [Fried]) asserts that the sum of the squares of the Fourier coefficients converges to (L^{-3} times) the square of the L^2 norm of f. Absolute convergence then implies that the same is true of any rearrangement, so the order in which one defines the partial sums is irrelevant.

The rate at which the Fourier coefficients of f converge to zero with $\|\mathbf{k}\|$ is directly related to the degree of regularity of f (see Section 3.2 of [Graf]). The only result of this sort that we will appeal to states that if f is smooth (C^∞), then the Fourier coefficients $a_{\mathbf{k}}$ decay at a rate sufficient to ensure that the convergence of the Fourier series is uniform and that the same is true after differentiating any number of times with respect to x^1, x^2 and x^3. In our discussion of blackbody radiation (to which we now return) we will restrict our attention to smooth functions so that we can perform all of the calculations one generally sees in the physics literature (for example, Chapter 1 of [Bohm]) with impunity.

Example 4.3.1. Let us try to get some orientation for what is to come next by using these ideas to search for smooth solutions $A(t,\mathbf{x})$ to the wave equation

$$\Delta A - \frac{1}{c^2}\frac{\partial^2 A}{\partial t^2} = 0$$

on $\mathbb{T}^3$. Separating variables $A(t,\mathbf{x}) = T(t)X(\mathbf{x})$ in the usual way leads to two eigenvalue problems $\Delta X = \lambda X$ and $\ddot{T} = \lambda c^2 T$. For the first of these we now know the eigenvalues. For each

$$\mathbf{k} = \frac{2\pi}{L}(m^1, m^2, m^3) \in \left[\frac{2\pi}{L}\mathbb{Z}\right]^3$$

we have the eigenvalue

$$\lambda_{\mathbf{k}} = -\|\mathbf{k}\|^2 = -\frac{4\pi^2}{L^2}\left((m^1)^2 + (m^2)^2 + (m^3)^2\right)$$

and a corresponding eigenfunction

$$X_{\mathbf{k}}(\mathbf{x}) = e^{i\mathbf{k}\cdot\mathbf{x}}.$$

Letting

$$\omega_{\mathbf{k}}^2 = \|\mathbf{k}\|^2 c^2$$

the T-equation is therefore $\ddot{T} = -\omega_{\mathbf{k}}^2 T$, for which we have the solution

$$T_{\mathbf{k}}(t) = e^{-i\omega_{\mathbf{k}}t}.$$

The corresponding (complex) solution to the wave equation is therefore

$$A_{\mathbf{k}}(t, \mathbf{x}) = T_{\mathbf{k}}(t)X_{\mathbf{k}}(\mathbf{x}) = e^{i(\mathbf{k}\cdot\mathbf{x}-\omega_{\mathbf{k}}t)}.$$

Superimposing these gives

$$A(t, \mathbf{x}) = \sum_{\mathbf{k}\in[\frac{2\pi}{L}\mathbb{Z}]^3} a_{\mathbf{k}} e^{i(\mathbf{k}\cdot\mathbf{x}-\omega_{\mathbf{k}}t)},$$

where $a_{\mathbf{k}}$ must be the $\mathbf{k}$th Fourier coefficient of the initial wave $A(0, \mathbf{x})$. We find then that our solution to the wave equation is a superposition of *plane waves*; it might be useful at this point to review our earlier discussion of plane electromagnetic waves (see Section 4.2).

Now we would like to apply these same ideas to the vector potential $\mathbf{A}$ for the electromagnetic radiation in our black box $[0, L]^3$, which we recall is determined by

$$\Delta \mathbf{A} - \frac{1}{c^2}\frac{\partial^2 \mathbf{A}}{\partial t^2} = \mathbf{0} \quad \text{and} \quad \nabla \cdot \mathbf{A} = 0.$$

There are two minor complications. The wave equation is, in this case, a *vector* equation and the physical interpretation requires that the solutions be *real*. We handle these issues in the following way. Motivated by our experience in the previous example, we will begin by looking for complex solutions that are superpositions of plane waves, that is, of the form

$$\mathbf{A}_c(t, \mathbf{x}) = \sum_{\mathbf{k}\in[\frac{2\pi}{L}\mathbb{Z}]^3} \mathbf{A}_{\mathbf{k}} e^{i(\mathbf{k}\cdot\mathbf{x}-\omega_{\mathbf{k}}t)},$$

where $\mathbf{A}_{\mathbf{k}}$ is some constant vector with three complex components for each $\mathbf{k}$ and $\omega_{\mathbf{k}} = \|\mathbf{k}\| c$. Generally it will be more convenient to write this as

$$\mathbf{A}_c(t, \mathbf{x}) = \sum_{\mathbf{k}\in[\frac{2\pi}{L}\mathbb{Z}]^3} \mathbf{A}_{\mathbf{k}}(t) e^{i\mathbf{k}\cdot\mathbf{x}},$$

where

$$\mathbf{A}_{\mathbf{k}}(t) = \mathbf{A}_{\mathbf{k}} e^{-i\omega_{\mathbf{k}}t}.$$

We are not aiming for rigorous theorems in this section, but only for some appreciation of what led Planck to his "quantum hypothesis." As a result we will be somewhat cavalier in the following computations, basically doing everything we need to do term-by-term in the series. Computing the divergence term-by-term we find that the condition $\nabla \cdot \mathbf{A}_c = 0$ becomes

$$0 = \nabla \cdot \mathbf{A}_c(t, \mathbf{x}) = \sum_{\mathbf{k}\in[\frac{2\pi}{L}\mathbb{Z}]^3} i\mathbf{k}\cdot\mathbf{A}_{\mathbf{k}} e^{i(\mathbf{k}\cdot\mathbf{x}-\omega_{\mathbf{k}}t)} = \sum_{\mathbf{k}\in[\frac{2\pi}{L}\mathbb{Z}]^3} i\mathbf{k}\cdot\mathbf{A}_{\mathbf{k}}(t) e^{i\mathbf{k}\cdot\mathbf{x}},$$

so the uniqueness of Fourier expansions implies that

$$\mathbf{k} \cdot \mathbf{A_k}(t) = 0$$

for every $\mathbf{k} \in [\frac{2\pi}{L}\mathbb{Z}]^3$ and every $t \in \mathbb{R}$. Thus, each of the coefficients $\mathbf{A_k}(t)$ is orthogonal to the corresponding wavevector $\mathbf{k}$, that is, to the direction of propagation of the corresponding plane wave, for every t. We therefore choose, for each $\mathbf{k}$, two orthogonal unit vectors $\epsilon_\mathbf{k}^1$ and $\epsilon_\mathbf{k}^2$ in the plane perpendicular to $\mathbf{k}$ in $\mathbb{R}^3$ such that $\{\epsilon_\mathbf{k}^1, \epsilon_\mathbf{k}^2, \mathbf{k}\}$ is an orthogonal basis for $\mathbb{R}^3$ consistent with the usual orientation for $\mathbb{R}^3$. Now we can write

$$\mathbf{A_k}(t) = a_{\mathbf{k}1}(t)\epsilon_\mathbf{k}^1 + a_{\mathbf{k}2}(t)\epsilon_\mathbf{k}^2 = \sum_{\alpha=1}^{2} a_{\mathbf{k}\alpha}(t)\epsilon_\mathbf{k}^\alpha,$$

where $a_{\mathbf{k}\alpha}(t)$ are generally complex. Recall from Section 4.2 that the electric and magnetic field vectors of an electromagnetic plane wave are also orthogonal to each other and to the wavevector and that the direction of the electric field is called the direction of polarization. For this reason $\epsilon_\mathbf{k}^1$ and $\epsilon_\mathbf{k}^2$ are also called *polarization directions*. Note that

$$\dot{\mathbf{A}}_\mathbf{k}(t) = -i\omega_\mathbf{k}\mathbf{A_k}(t),$$

so

$$\ddot{\mathbf{A}}_\mathbf{k}(t) = -\omega_\mathbf{k}^2 \mathbf{A_k}(t).$$

Consequently,

$$\ddot{a}_{\mathbf{k}\alpha}(t) + \omega_\mathbf{k}^2 a_{\mathbf{k}\alpha}(t) = 0$$

for every $\mathbf{k}$ and each $\alpha = 1, 2$. Thus, each $a_{\mathbf{k}\alpha}(t)$ satisfies the harmonic oscillator equation with angular frequency $\omega_\mathbf{k} = \|\mathbf{k}\|c$.

With this the complex solution becomes

$$\mathbf{A}_c(t, \mathbf{x}) = \sum_{\mathbf{k}\in[\frac{2\pi}{L}\mathbb{Z}]^3} \sum_{\alpha=1}^{2} \epsilon_\mathbf{k}^\alpha a_{\mathbf{k}\alpha}(t)e^{i\mathbf{k}\cdot\mathbf{x}}.$$

Finally, to get a real solution we take the real parts:

$$\mathbf{A}(t, \mathbf{x}) = \frac{1}{2} \sum_{\mathbf{k}\in[\frac{2\pi}{L}\mathbb{Z}]^3} [\mathbf{A_k}(t)e^{i\mathbf{k}\cdot\mathbf{x}} + \overline{\mathbf{A}}_\mathbf{k}(t)e^{-i\mathbf{k}\cdot\mathbf{x}}]$$

$$= \frac{1}{2} \sum_{\mathbf{k}\in[\frac{2\pi}{L}\mathbb{Z}]^3} \sum_{\alpha=1}^{2} \epsilon_\mathbf{k}^\alpha [a_{\mathbf{k}\alpha}(t)e^{i\mathbf{k}\cdot\mathbf{x}} + \overline{a}_{\mathbf{k}\alpha}(t)e^{-i\mathbf{k}\cdot\mathbf{x}}].$$

From (4.31) we obtain the electric field

$$
\begin{aligned}
\mathbf{E}(t,\mathbf{x}) = -\frac{\partial \mathbf{A}}{\partial t} &= -\frac{1}{2} \sum_{\mathbf{k} \in [\frac{2\pi}{L}\mathbb{Z}]^3} [\dot{\mathbf{A}}_{\mathbf{k}}(t)e^{i\mathbf{k}\cdot\mathbf{x}} + \dot{\overline{\mathbf{A}}}_{\mathbf{k}}(t)e^{-i\mathbf{k}\cdot\mathbf{x}}] \\
&= \frac{i}{2} \sum_{\mathbf{k} \in [\frac{2\pi}{L}\mathbb{Z}]^3} \omega_{\mathbf{k}} [\mathbf{A}_{\mathbf{k}}(t)e^{i\mathbf{k}\cdot\mathbf{x}} - \overline{\mathbf{A}}_{\mathbf{k}}(t)e^{-i\mathbf{k}\cdot\mathbf{x}}] \\
&= \frac{i}{2} \sum_{\mathbf{k} \in [\frac{2\pi}{L}\mathbb{Z}]^3} \sum_{\alpha=1}^{2} \omega_{\mathbf{k}} \boldsymbol{\epsilon}_{\mathbf{k}}^{\alpha} [a_{\mathbf{k}\alpha}(t)e^{i\mathbf{k}\cdot\mathbf{x}} - \overline{a}_{\mathbf{k}\alpha}(t)e^{-i\mathbf{k}\cdot\mathbf{x}}].
\end{aligned}
$$

For the total electromagnetic energy contained in the black box (see (4.43)) we need to compute

$$
\begin{aligned}
\int_R &\frac{\epsilon_0}{2} \|\mathbf{E}(t,\mathbf{x})\|^2 \, d^3\mathbf{x} \\
&= \frac{\epsilon_0}{2} \int_R \mathbf{E}(t,\mathbf{x}) \cdot \mathbf{E}(t,\mathbf{x}) \, d^3\mathbf{x} \\
&- \frac{\epsilon_0}{8} \int_R \left(\sum_{\mathbf{k}} \omega_{\mathbf{k}} [\mathbf{A}_{\mathbf{k}}(t)e^{i\mathbf{k}\cdot\mathbf{x}} - \overline{\mathbf{A}}_{\mathbf{k}}(t)e^{-i\mathbf{k}\cdot\mathbf{x}}] \right) \\
&\quad \cdot \left(\sum_{\mathbf{k}'} \omega_{\mathbf{k}'} [\mathbf{A}_{\mathbf{k}'}(t)e^{i\mathbf{k}'\cdot\mathbf{x}} - \overline{\mathbf{A}}_{\mathbf{k}'}(t)e^{-i\mathbf{k}'\cdot\mathbf{x}}] \right) d^3\mathbf{x} \\
&= -\frac{\epsilon_0}{8} \sum_{\mathbf{k}} \sum_{\mathbf{k}'} \omega_{\mathbf{k}}\omega_{\mathbf{k}'} \int_R [\mathbf{A}_{\mathbf{k}}(t)e^{i\mathbf{k}\cdot\mathbf{x}} - \overline{\mathbf{A}}_{\mathbf{k}}(t)e^{-i\mathbf{k}\cdot\mathbf{x}}] \cdot [\mathbf{A}_{\mathbf{k}'}(t)e^{i\mathbf{k}'\cdot\mathbf{x}} - \overline{\mathbf{A}}_{\mathbf{k}'}(t)e^{-i\mathbf{k}'\cdot\mathbf{x}}] \, d^3\mathbf{x}.
\end{aligned}
$$

Now, fix a $\mathbf{k} \in [\frac{2\pi}{L}\mathbb{Z}]^3$. By the L^2 orthogonality of the exponentials $e^{i\mathbf{k}\cdot\mathbf{x}}$, all of these integrals will be zero except when either $\mathbf{k}' = \mathbf{k}$ or $\mathbf{k}' = -\mathbf{k}$. Consider $\mathbf{k}' = \mathbf{k}$. Then

$$
\begin{aligned}
-\frac{\epsilon_0}{8} \omega_{\mathbf{k}}^2 &\int_R [\mathbf{A}_{\mathbf{k}}(t)e^{i\mathbf{k}\cdot\mathbf{x}} - \overline{\mathbf{A}}_{\mathbf{k}}(t)e^{-i\mathbf{k}\cdot\mathbf{x}}] \cdot [\mathbf{A}_{\mathbf{k}}(t)e^{i\mathbf{k}\cdot\mathbf{x}} - \overline{\mathbf{A}}_{\mathbf{k}}(t)e^{-i\mathbf{k}\cdot\mathbf{x}}] \, d^3\mathbf{x} \\
&= -\frac{\epsilon_0}{8} \omega_{\mathbf{k}}^2 \Big[\mathbf{A}_{\mathbf{k}}(t) \cdot \mathbf{A}_{\mathbf{k}}(t) \int_R e^{2i\mathbf{k}\cdot\mathbf{x}} d^3\mathbf{x} \\
&\quad - 2\mathbf{A}_{\mathbf{k}}(t) \cdot \overline{\mathbf{A}}_{\mathbf{k}}(t) \int_R 1 \, d^3\mathbf{x} + \overline{\mathbf{A}}_{\mathbf{k}}(t) \cdot \overline{\mathbf{A}}_{\mathbf{k}}(t) \int_R e^{-2i\mathbf{k}\cdot\mathbf{x}} d^3\mathbf{x} \Big].
\end{aligned}
$$

The first integral on the right-hand side is zero since it is the L^2 inner product of $e^{i(3\mathbf{k})\cdot\mathbf{x}}$ and $e^{i\mathbf{k}\cdot\mathbf{x}}$ and similarly for the third integral. The second integral is just the volume of R, that is, L^3. Furthermore, $\mathbf{A}_{\mathbf{k}}(t) \cdot \overline{\mathbf{A}}_{\mathbf{k}}(t) = \|\mathbf{A}_{\mathbf{k}}(t)\|^2 = |a_{\mathbf{k}1}(t)|^2 + |a_{\mathbf{k}2}(t)|^2$, so we obtain

$$
\begin{aligned}
-\frac{\epsilon_0}{8} \omega_{\mathbf{k}}^2 &\int_R [\mathbf{A}_{\mathbf{k}}(t)e^{i\mathbf{k}\cdot\mathbf{x}} - \overline{\mathbf{A}}_{\mathbf{k}}(t)e^{-i\mathbf{k}\cdot\mathbf{x}}] \cdot [\mathbf{A}_{\mathbf{k}}(t)e^{i\mathbf{k}\cdot\mathbf{x}} - \overline{\mathbf{A}}_{\mathbf{k}}(t)e^{-i\mathbf{k}\cdot\mathbf{x}}] \, d^3\mathbf{x} \\
&= \frac{\epsilon_0 L^3 \omega_{\mathbf{k}}^2}{4} \|\mathbf{A}_{\mathbf{k}}(t)\|^2
\end{aligned}
$$

$$= \frac{\epsilon_0 L^3 \omega_k^2}{4} \left(\left| a_{k1}(t) \right|^2 + \left| a_{k2}(t) \right|^2 \right).$$

The same term arises from $\mathbf{k'} = -\mathbf{k}$, so adding them and summing over $\mathbf{k}$, we obtain

$$\int_R \frac{\epsilon_0}{2} \left\| \mathbf{E}(t, \mathbf{x}) \right\|^2 d^3x = \frac{L^3}{2} \sum_{\mathbf{k} \in [\frac{2\pi}{L} \mathbb{Z}]^3} \epsilon_0 \omega_k^2 \left\| \mathbf{A}_k(t) \right\|^2.$$

Similar, but algebraically a bit more labor intensive computations give

$$\int_R \frac{1}{2\mu_0} \left\| \mathbf{B}(t, \mathbf{x}) \right\|^2 d^3x = \frac{L^3}{2} \sum_{\mathbf{k} \in [\frac{2\pi}{L} \mathbb{Z}]^3} \frac{\|\mathbf{k}\|^2}{\mu_0} \left\| \mathbf{A}_k(t) \right\|^2.$$

However,

$$\frac{\|\mathbf{k}\|^2}{\mu_0} = \frac{\omega_k^2}{c^2 \mu_0}$$

$$= \frac{\epsilon_0 \mu_0 \omega_k^2}{\mu_0} = \epsilon_0 \omega_k^2,$$

so

$$\int_R \frac{1}{2\mu_0} \left\| \mathbf{B}(t, \mathbf{x}) \right\|^2 d^3x = \frac{L^3}{2} \sum_{\mathbf{k} \in [\frac{2\pi}{L} \mathbb{Z}]^3} \epsilon_0 \omega_k^2 \left\| \mathbf{A}_k(t) \right\|^2.$$

The energies contributed by the electric and magnetic fields are therefore the same. From (4.43) we then obtain the total electromagnetic energy within the black box:

$$E = \int_R \left(\frac{\epsilon_0}{2} \left\| \mathbf{E}(t, \mathbf{x}) \right\|^2 + \frac{1}{2\mu_0} \left\| \mathbf{B}(t, \mathbf{x}) \right\|^2 \right) d^3x = \sum_{\mathbf{k} \in [\frac{2\pi}{L} \mathbb{Z}]^3} L^3 \epsilon_0 \omega_k^2 \left\| \mathbf{A}_k(t) \right\|^2$$

$$= \sum_{\mathbf{k} \in [\frac{2\pi}{L} \mathbb{Z}]^3} \sum_{\alpha=1}^{2} L^3 \epsilon_0 \omega_k^2 \left| a_{k\alpha}(t) \right|^2.$$

Now recall that each $a_{k\alpha}$ satisfies the harmonic oscillator equation $\ddot{a}_{k\alpha} + \omega_k^2 a_{k\alpha} = 0$ and therefore the same is true of their real parts. We will now make a change of variable to exhibit these real parts more explicitly. Specifically, for each $\mathbf{k} \in [\frac{2\pi}{L} \mathbb{Z}]^3$ and each $\alpha = 1, 2$, we let

$$Q_{k\alpha} = a_{k\alpha} + \bar{a}_{k\alpha} \quad \text{and} \quad P_{k\alpha} = \frac{L^3 \epsilon_0 \omega_k}{2i} (a_{k\alpha} - \bar{a}_{k\alpha}).$$

Then

$$a_{k\alpha} = \frac{1}{2} Q_{k\alpha} + \frac{i}{L^3 \epsilon_0 \omega_k} P_{k\alpha} \quad \text{and} \quad \bar{a}_{k\alpha} = \frac{1}{2} Q_{k\alpha} - \frac{i}{L^3 \epsilon_0 \omega_k} P_{k\alpha},$$

so

$$L^3 \epsilon_0 \, \omega_{\mathbf{k}}^2 \, | a_{\mathbf{k}\alpha} |^2 = L^3 \epsilon_0 \, \omega_{\mathbf{k}}^2 \, a_{\mathbf{k}\alpha} \, \bar{a}_{\mathbf{k}\alpha} = \frac{L^3 \epsilon_0 \omega_{\mathbf{k}}^2}{4} \, Q_{\mathbf{k}\alpha}^2 + \frac{1}{L^3 \epsilon_0} \, P_{\mathbf{k}\alpha}^2.$$

Now let

$$M = \frac{L^3 \epsilon_0}{2}.$$

Then

$$L^3 \epsilon_0 \, \omega_{\mathbf{k}}^2 \, | a_{\mathbf{k}\alpha} |^2 = \frac{1}{2M} P_{\mathbf{k}\alpha}^2 + \frac{M \omega_{\mathbf{k}}^2}{2} \, Q_{\mathbf{k}\alpha}^2$$

and we can write the total electromagnetic energy in the box as

$$E = \sum_{\mathbf{k} \in [\frac{2\pi}{L} \mathbb{Z}]^3} \sum_{\alpha=1}^{2} \left(\frac{1}{2M} P_{\mathbf{k}\alpha}^2 + \frac{M \omega_{\mathbf{k}}^2}{2} \, Q_{\mathbf{k}\alpha}^2 \right). \tag{4.48}$$

From Example 2.3.4 we recall that the total energy of the classical harmonic oscillator of mass m and angular frequency ω is given in canonical coordinates (q, p) by $\frac{1}{2m} p^2 + \frac{m\omega^2}{2} q^2$. This leads us to the following interpretation of (4.48). For each $\mathbf{k} \in [\frac{2\pi}{L} \mathbb{Z}]^3$ and each $\alpha = 1, 2$ we have a real-valued function $Q_{\mathbf{k}\alpha}$ of t satisfying the harmonic oscillator equation. If we interpret this as a classical harmonic oscillator with mass $M = L^3 \epsilon_0 / 2$ and angular frequency $\omega_{\mathbf{k}}$ and interpret $P_{\mathbf{k}\alpha}$ as the momentum conjugate to $Q_{\mathbf{k}\alpha}$, then the energy of the oscillator is $\frac{1}{2M} P_{\mathbf{k}\alpha}^2 + \frac{M\omega_{\mathbf{k}}^2}{2} Q_{\mathbf{k}\alpha}^2$. The sum of the energies of this countable family of harmonic oscillators is precisely the total electromagnetic energy contained in the box. Since $a_{\mathbf{k}\alpha}$ completely determine the electromagnetic potential $\mathbf{A}$ and therefore also the electric and magnetic fields $\mathbf{E}$ and $\mathbf{B}$, one can, at least for our purposes at the moment, identify the electromagnetic field with a countable family of harmonic oscillators; this is what Fourier analysis does for you. These harmonic oscillators are generally called *radiation oscillators* and it is important to observe that they are *independent* in the sense that the potential energy $\frac{M\omega_{\mathbf{k}}^2}{2} Q_{\mathbf{k}\alpha}^2$ of each contains no "interaction term" coupling it to any of the others. Soon we will see that these oscillators can profitably be viewed as analogous to the molecules of an ideal gas in thermal equilibrium.

Aside from the physical input provided by Maxwell's equations the development to this point has been entirely mathematical. We have resolved the relevant fields into superpositions of plane waves $e^{i(\mathbf{k}\cdot\mathbf{x} - \omega_{\mathbf{k}} t)}$, calculated the total electromagnetic energy in R and expressed it as the sum of the energies of a countable family of radiation oscillators. What remains is to determine how the total energy is distributed among the various frequencies of radiation present in R, and for this the physics becomes rather more serious and dominates the discussion.

We will begin with a few remarks on the sort of radiation one would typically expect to find in a black box. When you turn on your oven and set it at the desired temperature it will heat up and eventually reach thermal equilibrium and maintain that temperature. What it will generally not do, however, is visibly glow. At the relatively moderate temperatures the oven can produce, the electromagnetic radiation within the box is in the infrared and not in the visible range of the spectrum (see Figure 4.1). Infrared radiation has wavelengths on the order of 10^{-6} to 10^{-4} m. For an oven of typical size, L is a great deal larger than this so the number of waves one can "fit in the box" is correspondingly very large. This suggests adopting a procedure analogous to the usual one in the study of fluids by regarding the number of oscillators as "virtually continuous" (as a function of the wavelength or frequency) and representing it in terms of a density function. The next step in our program is to determine this density function.

Each plane wave $e^{i(\mathbf{k}\cdot\mathbf{x}-\omega_{\mathbf{k}}t)}$ is uniquely determined by its wavevector $\mathbf{k} \in [\frac{2\pi}{L}\mathbb{Z}]^3$. Its angular frequency $\omega_{\mathbf{k}}$ is uniquely determined by $k = \|\mathbf{k}\|$ ($\omega_{\mathbf{k}} = kc$). The frequency, however, does not uniquely determine the wavevector since distinct elements of $[\frac{2\pi}{L}\mathbb{Z}]^3$ can have the same k. Consequently, there are, in general, many plane waves with the same frequency (but different directions of propagation). We construct a geometrical picture of this in the following way. View $[\frac{2\pi}{L}\mathbb{Z}]^3$ as a lattice in a copy of $\mathbb{R}^3$ (generally referred to as k $space$ in the physics literature). Any point in the lattice then corresponds to a plane wave whose angular frequency is just (c times) its distance to the origin. Consequently, the problem of counting all of the plane waves of a given frequency amounts to counting the number of lattice points on a sphere of some radius about the origin. Generically, the answer is zero since any radius k for which the sphere contains a lattice point must have a square k^2 for which $L^2k^2/4\pi^2$ is an integer that is expressible as a sum of three squares. Even in this case one must then know the number of ways in which $L^2k^2/4\pi^2$ can be represented as a sum of three squares in order to count lattice points. For the purpose of finding our density function we are more interested in the counting the number $S(k)$ of lattice points in a solid ball of radius k about the origin. Needless to say, one cannot simply write down an explicit formula, but there are asymptotic results for large k of the form

$$S(k) = \frac{4}{3}\pi k^3 + O(k^\theta),$$

where θ is a positive real number. For example, this is known when $\theta = \frac{29}{22}$ and it is conjectured to be true when $\theta = 1$. These are deep number theoretic results and we will have to content ourselves with a reference to [CI] for those who are interested in learning more about them. For our purposes we will need only a very crude estimate. Note that one can establish a one-to-one correspondence between the plane waves and the cubes in the partition of $\mathbb{R}^3$ determined by the lattice points (for example, each such cube is uniquely determined by the vertex $(2\pi/L)(m^1, m^2, m^3)$ with smallest m^1, m^2 and m^3). Consequently, we can count cubes instead of lattice points and the number of cubes can be measured by the volume they take up. Now, let k be the radius of some ball centered at the origin. Since the side length of each cube is $2\pi/L$ we will define

$$N(k) = \frac{\frac{4}{3}\pi k^3}{(\frac{2\pi}{L})^3} = \frac{L^3}{6\pi^2} k^3$$

and regard this as an approximate measure of the number of lattice cubes that fit inside the ball (it is generally not an integer, of course).

Exercise 4.3.3. Show that $N(k) = \frac{4}{3}\pi k^3 + O(k^2)$. *Hint*: Note that

$$N(k - \epsilon) < \frac{4}{3}\pi k^3 < N(k + \epsilon)$$

for any $\epsilon > 0$ and consider $N(k + \epsilon) - N(k - \epsilon)$.

Regarding $N(k)$ as the integral of a continuous density function we find that this density function is given by

$$\frac{V}{2\pi^2} k^2,$$

where we now write V for the volume L^3 of the box. One often sees this density expressed as a measure

$$\frac{V}{2\pi^2} k^2 dk$$

and, still more often, as a measure expressed in terms of $v = \frac{\omega}{2\pi} = \frac{kc}{2\pi}$, that is,

$$\frac{4\pi V}{c^3} v^2 dv.$$

Finally, we note that each of the plane waves, determined by $\mathbf{k}$, actually determines *two* radiation oscillators corresponding to the two independent polarization states ($\alpha = 1, 2$) so the number of oscillators in a given frequency range is determined by

$$\frac{8\pi V}{c^3} v^2 dv. \tag{4.49}$$

If you are keeping track, this is (29) in Chapter 1 of [Bohm], where it is described, in the fashion of the physicists, as "the total number of oscillators [in the box] between v and $v + dv$." A given choice of $\mathbf{k} \in [\frac{2\pi}{L}\mathbb{Z}]^3$ and of $\alpha \in \{1, 2\}$ corresponds to what is called a *mode* of oscillation so the process we have just gone through is called *counting modes*.

It will not have escaped your attention that, although we have had quite a bit to say about electromagnetic radiation confined to a box, we have yet to mention the characteristic feature of the radiation we are particularly interested in, that is, the fact that it is in thermal equilibrium at some constant temperature T. This is what we must contend with now as we try to understand how the total energy of the electromagnetic radiation contained in the box is distributed among the various radiation oscillators (that is, among the various modes). It would be disingenuous to pretend that the tools we require for this are as elementary as those we have needed so far. The main player in the remainder of the story is a result from classical statistical mechanics due to Ludwig Boltzmann. This is not a mathematical theorem so we can offer no proof, nor is it a consequence of anything we have said to this point. It is the result of a deep analysis of the statistical behavior of certain very special types of physical systems. We will try to explain where this result (called the *Boltzmann distribution*) came from, what it is intended to describe, how it can be used to get the wrong answer (4.37) and how it, together with Planck's "act of desperation," can be used to get the right answer (4.41).

To get some idea of where the Boltzmann distribution comes from and what it means, it is probably best to put aside electromagnetic radiation for a moment and consider instead the somewhat more familiar system that Boltzmann himself studied. Let us suppose then that our box $R = [0, L]^3$ contains not radiation, but a gas consisting of N particles (molecules of oxygen, for example); N will generally be huge (on the order of 10^{23}). The so-called *macrostate* of this system is specified by such quantities as the number N of molecules, the volume V, pressure P, temperature T and total energy E of the gas. These are not all independent, of course (you may remember, for example, the *ideal gas law* from chemistry). This macrostate is determined by the states of the individual molecules (their positions and momenta), but the size of N makes these inaccessible to us. Moreover, many different configurations of the particle states (many different *microstates*) can give rise to the same macrostate. The basic operating principle of statistical mechanics is that, even though one cannot know the microstates, one can sometimes know their statistical distribution (their probabilities) and this is often enough to compute useful information. We will try to illustrate how this comes about.

Example 4.3.2. We begin with an oversimplified, but nevertheless instructive example. Consider a container of volume V in which there is a gas consisting of N independent, noninteracting point particles. We assume that the system is isolated so that the total energy E is constant and that it has settled down into thermal equilibrium at temperature T. Although the amount of energy associated to a given particle cannot be determined, we will investigate how the energy is distributed among the particles *on average*. Now let us oversimplify. We will, for the moment, assume that each particle can assume only one of finitely many, evenly spaced energy levels

$$\varepsilon_0 = 0, \quad \varepsilon_1 = \varepsilon_0 + \Delta\varepsilon, \quad \varepsilon_2 = \varepsilon_1 + \Delta\varepsilon, \quad \ldots, \quad \varepsilon_K = \varepsilon_{K-1} + \Delta\varepsilon.$$

A typical microstate will have n_0 particles of energy ε_0, n_1 particles of energy ε_1, ..., and n_K particles of energy ε_K, where

$$\sum_{i=0}^{K} n_i = N \tag{4.50}$$

and

$$\sum_{i=0}^{K} n_i \varepsilon_i = E. \tag{4.51}$$

Collisions between the particles will generally change the so-called *occupation numbers* $n_0, n_1, \ldots, n_K$ (or they could remain the same, but with different particles occupying the energy levels $\varepsilon_0, \varepsilon_2 \ldots, \varepsilon_K$). We assume that all possible divisions of the total energy among the particles occur with the same probability and we seek the configuration of particle energies that is most likely to occur, that is, the configuration that can be achieved in the largest number of ways. Here is an example that you can work out by hand.

Exercise 4.3.4. Suppose $N = 3$ and $K = 4$ so that there are three particles P_1, P_2 and P_3 and five possible energy states

$$\varepsilon_0 = 0, \quad \varepsilon_1 = \varepsilon_0 + \Delta\varepsilon, \quad \varepsilon_2 = \varepsilon_1 + \Delta\varepsilon, \quad \varepsilon_3 = \varepsilon_2 + \Delta\varepsilon, \quad \varepsilon_4 = \varepsilon_3 + \Delta\varepsilon$$

for the particles. Assume, however, that the total energy E of the system is $3\Delta\varepsilon$. Find all possible configurations of the particle energies and the number of ways in which each can occur. What configuration is most likely and what is the probability that it will occur? *Hint*: The columns below represent three possible configurations of the particle energies. Find the other possible energy configurations.

ε_4 :	∅	∅	∅
ε_3 :	∅	∅	P_2
ε_2 :	P_1	P_2	∅
ε_1 :	P_2	P_3	∅
ε_0 :	P_3	P_1	$\{P_1, P_3\}$

Answer: The most likely configuration has one particle of energy ε_0, one particle of energy ε_1 and one particle of energy ε_2; its probability is 0.6.

The number of ways to take N particles and choose n_0 of them to assign to the energy level ε_0 (without regard to the order in which they are chosen) is given by the binomial coefficient $\binom{N}{n_0} = \frac{N!}{n_0!(N-n_0)!}$. Then the number of ways to take the remaining $N - n_0$ particles and choose n_1 of them to assign to the energy level ε_1 is $\binom{N-n_0}{n_1} = \frac{(N-n_0)!}{n_1!(N-(n_0+n_1))!}$. Thus, the number of ways to do both of these is

$$\binom{N}{n_0}\binom{N-n_0}{n_1} = \frac{N!}{n_0! n_1! (N - (n_0 + n_1))!}.$$

Continuing inductively one finds that the number of configurations of the particle energies that give rise to a microstate with n_0 particles of energy ε_0, n_1 particles of energy $\varepsilon_1, \ldots$, and n_K particles of energy ε_K is

$$\frac{N!}{n_0! n_1! \cdots n_K!} \tag{4.52}$$

and this is called the *weight* of the configuration and denoted $W = W(n_0, n_1, \ldots, n_K)$. The most probable configuration is the one with maximal weight, so we want to determine the values of $n_0, n_1, \ldots, n_K$ for which (4.52) is as large as possible, subject to the constraints (4.50) and (4.51). Needless to say, with integers on the order of 10^{23} the explicit expression (4.52) for W is hopeless so we will need to approximate. The usual procedure is to take logarithms and apply *Sterling's formula*

$$\ln(n!) = n \ln n - n + O(\ln n) \quad \text{as } n \to \infty.$$

Crudely put, $\ln(n!) \approx n \ln n - n$. From this we obtain

$$\ln W = \ln(N!) - \sum_{i=0}^{K} \ln(n_i!)$$

$$\approx N \ln N - N - \sum_{i=0}^{K} (n_i \ln n_i - n_i) = N \ln N - \sum_{i=0}^{K} n_i \ln n_i.$$

The bottom line then is that the most probable configuration of particle energies is the one for which the occupation numbers $n_0, n_1, \ldots, n_K$ maximize

$$\ln W \approx N \ln N - \sum_{i=0}^{K} n_i \ln n_i$$

subject to the constraints

$$\sum_{i=0}^{K} n_i = N$$

and

$$\sum_{i=0}^{K} n_i \varepsilon_i = E.$$

This has all the earmarks of a problem in Lagrange multipliers except, of course, for the fact that the variables take only integer values. One possibility is simply to proceed formally, regarding $n_0, n_1, \ldots, n_K$ as *real* variables, and applying the usual Lagrange multiplier procedure; this is fairly straightforward and is done in considerable detail on pages 582–583 of [AdeP] (although certainly not written for mathematicians, Chapter 16 of this book contains a lot of interesting and easy reading on the topics that we are breezing through rather quickly). In this way one obtains the following (approximate) formulas for the occupation numbers of the most probable configuration:

$$n_i = N \frac{e^{-\beta \varepsilon_i}}{\sum_{j=0}^{K} e^{-\beta \varepsilon_j}}, \quad i = 1, 2, \ldots, K,$$

where β is a constant. For systems in thermal equilibrium at temperature T (measured in Kelvin), thermodynamic considerations (Section 16.3(b) of [AdeP]) identify β as

$$\beta = \frac{1}{\kappa_B T},$$

where κ_B is the Boltzmann constant. The probability that a particle has energy ε_i is therefore

$$p_i = \frac{n_i}{N} = \frac{e^{-\varepsilon_i/\kappa_B T}}{\sum_{j=0}^{K} e^{-\varepsilon_j/\kappa_B T}} = \frac{e^{-\varepsilon_i/\kappa_B T}}{Z}, \quad i = 1, 2, \ldots, K, \tag{4.53}$$

where Z is the so-called *partition function*

$$Z = \sum_{j=0}^{K} e^{-\varepsilon_j/\kappa_B T}.$$

The utility of these computations resides in the fact that for the extremely large values of N that typically occur, this most probable configuration is much more than most probable; statistically, it is essentially inevitable so that one can study an equilibrium gas by studying this configuration.

The objection that classical physics would make to our last example is that the particle energies are not restricted to finitely many evenly spaced values $\varepsilon_0, \varepsilon_1, \ldots, \varepsilon_K$, but rather can take on *continuously many values* so ε should be regarded as a real variable in $[0, \infty)$. In this view the probabilities $p_i, i = 1, \ldots, K$, would be replaced by a probability distribution $p = p(\varepsilon)$ given by the continuous analogue of (4.53), that is,

$$p = p(\varepsilon) = \frac{e^{-\varepsilon/\kappa_B T}}{\int_0^\infty e^{-\varepsilon/\kappa_B T} d\varepsilon} = \frac{1}{\kappa_B T} e^{-\varepsilon/\kappa_B T}. \tag{4.54}$$

This is the so-called *Boltzmann distribution*. From it one can compute the *mean energy* $\bar{E}$ of the particles by weighting each energy ε with its probability $p(\varepsilon)$ and integrating over $[0, \infty)$. The resulting integral is completely elementary and one obtains

$$\bar{E} = \frac{1}{\kappa_B T} \int_0^\infty \varepsilon e^{-\varepsilon/\kappa_B T} d\varepsilon = \frac{(\kappa_B T)^2}{\kappa_B T} = \kappa_B T.$$

Now it is time to get back to the task at hand. We have been discussing gases in thermal equilibrium, but only because they are rather familiar and intuitively accessible. The arguments we have sketched are quite general and apply to any isolated physical system in thermal equilibrium which can be thought of as consisting of a very large number of independent subsystems. The system we have in mind is the electromagnetic radiation in a black box in thermal equilibrium at temperature T. We have seen that this can be regarded as a family of independent harmonic oscillators with total energy given by (4.48). If we assume that these radiation oscillators behave in the way we would expect from our experience with masses on springs and pendulums, that is, that their energies can take on continuously many values, then the mean energy of the radiation is determined entirely by the temperature according to $\overline{E} = \kappa_B T$. Now recall the density function (4.49) for the number of radiation oscillators in the box in a given frequency range. Dividing out the volume V we obtain the density of oscillators per unit volume

$$\frac{8\pi}{c^3} \nu^2.$$

Approximating the energy of each oscillator by the mean energy $\overline{E} = \kappa_B T$ we arrive at the density function

$$\frac{8\pi \kappa_B T}{c^3} \nu^2$$

for the energy per unit volume contained in a given frequency range. In terms of the wavelength this becomes

$$\frac{8\pi \kappa_B T}{\lambda^4}, \tag{4.55}$$

which is precisely the Rayleigh–Jeans law (4.38).

Well, this is lovely. We now know exactly how to get the wrong answer (see Figure 4.5). This is basically the conundrum that faced Max Planck in the last years of the nineteenth century. The argument leading to the Rayleigh–Jeans law seemed watertight and yet the formula to which it led was quite wrong. What to do?

Remark 4.3.4. We should preface the remainder of our discussion by saying that the path we will follow is not precisely the same as the path followed by Planck, who focused his attention not on radiation oscillators, but rather on material oscillators in the walls of the box and how they interact with the radiation. Those who would prefer to see Planck's arguments are referred to his paper [Planck] of 1900 (there is an English translation available at http://web.ihep.su/dbserv/compas/src/planck00b/eng.pdf). Also highly recommended are Chapters 18 and 19 of the wonderful book [Pais] by Abraham Pais.

The argument we have given leads directly to the Rayleigh–Jeans formula, which works quite nicely for large wavelengths/low frequencies, but fails miserably for small wavelengths/high frequencies, where it seriously overestimates the energy contribution of the modes. One needs to modify the argument in such a way so as to reduce this contribution at high frequencies. Planck's idea (or rather, his "act of desperation") was to assume that the oscillator energies did not vary continuously as classical physics would demand, but were restricted to be integral multiples of some basic unit of energy that is proportional to their frequencies. The constant of proportionality is denoted h and its value would need to be determined by comparison with the experimental results (Figure 4.4). Thus, we formulate *Planck's hypothesis* in the following way.

The energy of a radiation oscillator of frequency v can assume only one of the following values:

$$\varepsilon_n = nhv, \quad n = 0, 1, 2, \ldots$$

Planck's hypothesis places us in a situation not unlike that of Example 4.3.2. We have a discrete, albeit infinite, set of equally spaced ($\Delta\varepsilon = hv$) allowed energy levels. A countable version of the Boltzmann distribution gives the probability that an oscillator is in the energy level ε_n as

$$p_n = \frac{e^{-\varepsilon_n/\kappa_B T}}{\sum_{j=0}^{\infty} e^{-\varepsilon_j/\kappa_B T}} = \frac{e^{-nhv/\kappa_B T}}{\sum_{j=0}^{\infty} e^{-jhv/\kappa_B T}} = e^{-nhv/\kappa_B T}(1 - e^{-hv/\kappa_B T}),$$

where the last equality follows from the fact that the sum in the denominator is geometric. As before, the mean energy $\bar{E}$ is obtained by weighting each energy level ε_n with its probability p_n and summing over all of the allowed energy levels. We have

$$\bar{E} = \sum_{n=0}^{\infty} \varepsilon_n p_n = hv(1 - e^{-hv/\kappa_B T}) \sum_{n=0}^{\infty} n e^{-nhv/\kappa_B T}.$$

We sum this series as follows. The function

$$\sum_{n=0}^{\infty} e^{-nx} = \sum_{n=0}^{\infty} (e^{-x})^n = \frac{1}{1 - e^{-x}}$$

is real analytic for $x > 0$, so

$$\frac{d}{dx} \sum_{n=0}^{\infty} e^{-nx} = -\sum_{n=0}^{\infty} n e^{-nx} = -\frac{e^{-x}}{(1 - e^{-x})^2}$$

for $x > 0$. In particular,

$$\sum_{n=0}^{\infty} n e^{-nhv/\kappa_B T} = \frac{e^{-hv/\kappa_B T}}{(1 - e^{-hv/\kappa_B T})^2}.$$

From this we obtain

$$\overline{E} = \frac{hv e^{-hv/\kappa_B T}}{1 - e^{-hv/\kappa_B T}} = \frac{hv}{e^{hv/\kappa_B T} - 1}.$$

Approximating, as we did before, the energy of each radiation oscillator by the mean energy $\overline{E}$ we obtain the density function

$$\rho_T(v) = \left(\frac{8\pi}{c^3} v^2 \right) \left(\frac{hv}{e^{hv/\kappa_B T} - 1} \right) = \frac{8\pi hv^3}{c^3} \frac{1}{e^{hv/\kappa_B T} - 1} \qquad (4.56)$$

for the energy per unit volume at frequency v, and this is precisely Planck's law (4.41). Note that the mean energy is no longer constant, but decreases with increasing frequency, thus reducing the contribution at high frequencies as we hoped to do. Comparing this $\rho_T(v)$ with the experimental results one finds that, by taking the value of h to be $6.62606957 \times 10^{-34}$ m^2(kg)/s, the fit is, within the limits of experimental error, extremely precise for all T. So, now we understand where Planck's constant came from.

When Planck presented his formula in 1900 no one doubted that it was correct. His derivation of the formula, however, was regarded by the community of physicists (and by Planck himself) as nothing more than a mathematical artifice for arriving at the correct relation and, so everyone thought, would inevitably be superseded by an argument consistent with the cherished principles of classical physics.

> "The general attitude toward Planck's theory was to state that 'everything behaves as if' the energy exchanges between radiation and the black body occur by quanta, and to try to reconcile this ad hoc hypothesis with the wave theory [of light]."

> Albert Messiah [Mess1]

With one exception, no one took Planck's hypothesis to be the harbinger of a new perspective on physics. The exception, of course, was Einstein.

In 1905, Albert Einstein published four extraordinary papers in the *Annalen der Physik*. All of these papers were revolutionary, but one of them [Ein1], entitled *On a Heuristic Point of View about the Creation and Conversion of Light*, bordered on the heretical (there is an English translation available in [ter H]). At a time when Maxwell's theory of electromagnetic radiation (and therefore of light) was virtually sacrosanct one reads the following proposal that is in flat contradiction to Maxwell's equations and the myriad phenomena (diffraction, reflection, dispersion, ...) that they so beautifully describe.

> "According to the assumption considered here, when a light ray starting from a point is propagated, the energy is not continuously distributed over an ever increasing volume, but it consists of a finite number of energy quanta, localized in space, which move without being divided and which can be absorbed or emitted only as a whole."

> Albert Einstein [Ein1]

Einstein postulated that a beam of monochromatic light could be treated as a stream
of particles (today we call them *photons*) moving with speed *c in vacuo* and with an energy that depended only on the frequency and was given by the Planck relation $E = h\nu$.
In hindsight, it is perhaps a little surprising that Einstein who, in that same year, also
introduced the special theory of relativity, did not at that time also associate a momentum to the photon since relativity specifies an essentially unique way to define this.
Nevertheless, Einstein did eventually do this (in 1916). Specifically, the *momentum* of
a photon is a vector **p** in the direction of its motion with magnitude

$$p = \|\mathbf{p}\| = h\nu/c.$$

The reaction to Einstein's proposal was immediate, universal and decidedly negative and the reason for this is not at all difficult to discern. Then, as now, the concepts
of "particle" and "wave" seemed irreconcilably distinct. One "thing" cannot be both
and, considering the success of Maxwell's theory, light is a wave; end of story. But, of
course, that is not the end of the story. If there is a single, underlying, philosophical
lesson that quantum theory requires us to learn it is that the conceptual apparatus
that has evolved in our species over eons of experience with the macroscopic world
around us is simply not up to the task of describing what goes on at the atomic and
subatomic levels. Among the concepts that we come to the game equipped with are
"particle" and "wave" and, as we shall see, they just will not do.

None of this was clear yet in 1905. It was clear, however, that certain electromagnetic phenomena seemed to resist inclusion into Maxwell's theory. Einstein applied his "heuristic point of view" to a number of these and we will briefly describe the
one that is best known. This is the so-called *photoelectric effect*. This phenomenon was
first observed by Hertz in 1887 and was subsequently studied by Hallwachs, Thomson
and Lenard (brief descriptions of these experiments can be found on pages 379–380
of [Pais]). The experiments themselves are delicate, but the conclusions to which they
led are easy to describe. A metal surface is illuminated by visible light or ultraviolet
radiation. The radiation communicates energy to electrons in the atoms of the metal
and, if the amount of energy is sufficient, the electrons are ejected and can be detected
outside the metal. This, in itself, is easy to understand on the basis of the wave theory of electromagnetic radiation, but there is more. It was found that the speed of the
ejected electrons (that is, their kinetic energy) did not depend at all on the intensity
of the radiation, but only on its frequency. Moving the light source far from the metal
(that is, decreasing the intensity of the radiation) decreased only the number of electrons detected per second, but not their energies. This would suggest that an electron
must absorb a certain critical amount of energy in order to be freed from the metal surface. Viewing the radiation as a wave, which conveys energy *continuously*, one would
be forced to conclude that for a very low intensity beam, the electrons would not be
observed immediately since it would take some time to store the necessary energy.
However, this simply did not occur. An extremely low intensity beam of the proper

frequency produced the so-called *photoelectrons* instantaneously. Despite many ingenious attempts, no one has ever succeeded in reconciling this behavior with a wave
theory of light. From Einstein's heuristic point of view, however, the explanation is
simple. When a photon of energy hv collides with an electron in the metal it is entirely
absorbed and the electron acquires the energy hv regardless of how far the photon
traveled. If the frequency v is sufficiently high (how high depends on the metal) the
collision will eject the electron.

Einstein, of course, did much more than offer this sort of qualitative explanation of
the photoelectric effect. His assumptions made very strong, testable predictions about
the behavior of the photoelectrons. Robert A. Millikan, one of the greatest experimental physicists of the twentieth century, devoted a decade of his life to testing these predictions and, in the end and despite his firm conviction that Einstein must be wrong,
validated them all.

What emerges from all of this is rather disconcerting. Light would appear to resist
classification as either "wave" or "particle." Rather, it seems to have a dual character,
behaving in some circumstances as though it fit very nicely into our classical concep-
tion of a wave and yet, in other circumstances, looking for all the world like a particle.

4.4 Two-slit experiments

In the previous section we saw how, in 1900, Max Planck, confronted with the experimental data on the spectrum of electromagnetic energy in thermal equilibrium, was
forced into what he regarded as the *ad hoc* assumption that harmonic oscillators can
emit and absorb energy only in discrete amounts nhv determined by their frequency
of oscillation. We saw also how Einstein, in 1905, elevated this hypothesis to a general
principle to account for the dual nature of electromagnetic radiation, which behaves
sometimes like a wave and sometimes like a particle. Then, in his 1924 PhD thesis,
Louis de Broglie suggested that this duality may be a characteristic feature of nature
at the quantum level so that even a presumed "particle" like an electron would, under
certain circumstances, exhibit "wave-like" behavior. He postulated that the energy E
of such a particle would be related to the frequency v of an "associated" wave (now
called its *de Broglie wave*) by the Planck–Einstein relation $E = hv$ and that the wavelength λ of the wave was related to the magnitude p of the linear momentum of the
particle by $p = h/\lambda$. Shortly thereafter this wave-like behavior of electrons was con-
firmed in experiments performed by Davisson and Germer, who found that a beam of
electrons fired at a crystalline target experienced the same sort of diffraction as would
a beam of X-rays scattered from a crystal. It may not, indeed, should not be clear at
the moment what sort of "thing" this de Broglie wave is, that is, what exactly is "waving." We hope that this will appear somewhat less obscure soon and then quite clear
in Section 6.2, but for the present we would like to examine another very beautiful
experiment that not only exhibits this wave-like behavior, but

"... has in it the heart of quantum mechanics. In reality, it contains the only mystery."

Richard Feynman (Volume III, page 1-1, of [FLS])

Let us begin with something familiar. Imagine plane water waves approaching a barrier in which there is a small gap. Emerging on the other side of the gap one sees not plane, but rather circular waves (this is the phenomenon known as *diffraction*). Now imagine that the barrier has two small gaps. Circular waves of the same amplitude emerge from each and, when they meet, they do what waves do, that is, they interfere with each other, constructively when they meet in phase and destructively when they meet out of phase. In particular, one sees radial lines where the two circular waves meet with the same amplitude, but with a phase difference of π and therefore cancel (a cork placed anywhere else would bob up and down, but not here).

Remark 4.4.1. There are some nice pictures available at https://wiki.anton-paar.com/en/double-slit-experiment/.

Now let us station a team of graduate students along a straight line parallel to the barrier and instruct them to measure the intensity of the wave that arrives at their location. The intensity of a wave can be defined as the energy per unit volume times the speed of the wave. Now, in general, a water wave is a very complicated thing, in which the combination of both longitudinal and transverse waves can produce quite complicated motions of the water molecules (circles, for example). Things are a bit simpler if the waves occur in a *shallow* body of water where, as a first approximation, one can assume that the molecules exhibit the same simple harmonic vertical motion as a cork placed nearby. In this case the energy is proportional to the square of the maximum displacement, that is, to the square of the height of the wave. The intensity is therefore also proportional to the square of the wave's amplitude and, for simplicity, we will make this assumption.

What sort of data would we expect our graduate students to collect? For a single gap one would probably expect to get essentially a bell curve with a maximum directly opposite the gap. Now, this is not entirely accurate. Even in the case of a single slit (gap) there may be some interference effects present. These depend on the width of the slit. One can understand this in the following way. As the wave enters the slit we can regard each point within the slit as the source of a circular wave leaving from there (this is essentially a consequence of what is called *Huygens' principle*). Such waves leaving from different points within the slit travel different distances to arrive at the same point and so may meet in phase or out of phase and this is just what interference means. One can compute the first points at which the waves cancel to give zero intensity and push these off to infinity by taking the width of the slit equal to the wavelength of the impinging plane waves and then one really does see something like a bell curve. To keep our discussion as simple as possible we will assume this has been done. Figure 4.6 shows such a (light) bell-shaped intensity curve for each of the

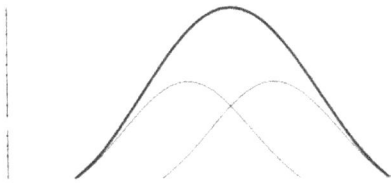

Figure 4.6: Sum of one-slit intensities.

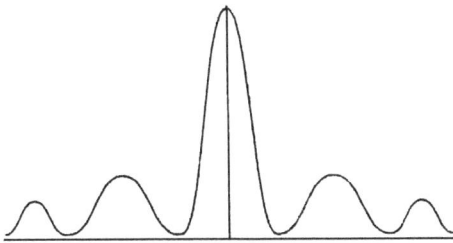

Figure 4.7: Two-slit intensity.

two gaps individually, assuming that the other is covered up, as well as their (dark) pointwise sum.

With both gaps uncovered the circular waves leaving them interfere and as a result the intensities certainly do not add. That is to say, the dark curve in Figure 4.6 does *not* represent the data our students would record in this case. What we would actually get is a curve of the sort shown in Figure 4.7 with a maximum opposite the midpoint between the two gaps and local maxima decreasing as we recede from this point and alternating with points of zero intensity corresponding to the radial lines where the circular waves cancel. Since the intensities for the two gaps individually do not simply add to give the intensity corresponding to the situation in which both gaps are open, one wonders if there is some other mathematical device that takes the interference into account and "predicts" Figure 4.7. Indeed, there is, but we will save this until we have looked at a few variants of this experiment with water waves.

Now let us replace the water waves rushing toward a barrier with a light source aimed at a wall with two small slits. The best choice for the light source is a laser since it can produce light that is monochromatic (of essentially one frequency) and coherent (constant phase difference over long intervals of space and time). We can also replace our conscripted students measuring intensities with a screen that registers the arrival of the light (a high resolution camera, for example). We will gauge the intensity of the light at the screen by its brightness.

Remark 4.4.2. We have already mentioned that the intensity of a wave is a well-defined, physically measurable quantity. Brightness, however, is not since it is really a function of how the light is *perceived* (by you, or by me or by an owl). Nevertheless, for the sort of qualitative considerations you will find in this section we only need to "gauge" the intensity, not measure it, and brightness is a reasonable and very intuitive way of doing this. In particular, we do not need to graph a brightness function; we only need to look at the screen.

What you actually see on the screen is precisely the same as for water waves: a region of maximal intensity/brightness in the center with regions of decreasing brightness as one recedes from the center separated by dark regions where the intensity/brightness is zero. This should not come as a surprise if you are thinking of the light as electromagnetic waves; waves are waves, after all, whether water or light. However, in Section 4.3 we saw that the light source can equally well be regarded as emitting a stream of particles (photons) and this would seem to present us with a problem. To our classically conditioned minds, a particle approaching the wall will either hit the wall or go through one of the two slits so that one would expect to see quite a different image on the screen. How does one reconcile this classical picture with what actually happens without giving up photons altogether (an unacceptable option in view of the success of Einstein's analysis of the photoelectric effect)? One can speculate and philosophize or one can take the view that the only honest way through this impasse (if there is one) is more careful observation, that is, more refined experiments. Fortunately, experimental physicists are very clever and they have given us a few remarkable experiments to think about.

We would now like to describe the results of two experiments, both of which were envisioned by Feynman, but neither of which was performed exactly as Feynman saw it until much later. We describe them at the same time because, although the specifics of the experimental apparatus in the two experiments are quite different, the results are precisely the same. In one of the experiments the "particles" are photons while in the other they are electrons. In particular, electrons will be seen to behave in exactly the same way as photons, thus confirming de Broglie's hypothesis concerning the wave/particle duality of even massive particles at the quantum level. We will not be concerned with the details of the experimental apparatus or procedure, but only with the results. For those who would like to learn more about the hardware and the techniques involved, the photon experiment is described at http://www.sps.ch/en/articles/progresses/wave particle duality of light for the classroom 13/ and, for the electron experiment, one can consult [BPLB].

In each of the experiments the source (of photons or electrons) is attenuated (weakened) to such an extent that the particles leave the source, and therefore arrive at the wall, essentially one at a time. Such a particle might well run into the wall and disappear, of course, but if it does not then what is observed in the experiment is the appearance of a dot on the screen (just the sort of thing one would expect of a classical particle). However, the dot is not necessarily where we would expect it to be classically (near a point directly opposite some point in a slit). Indeed, allowing another, and then another, and then another particle to emerge from the source one obtains dots on the screen that, at first, appear to be almost randomly distributed on the screen. But if we continue, allowing more and more particles to emerge over longer

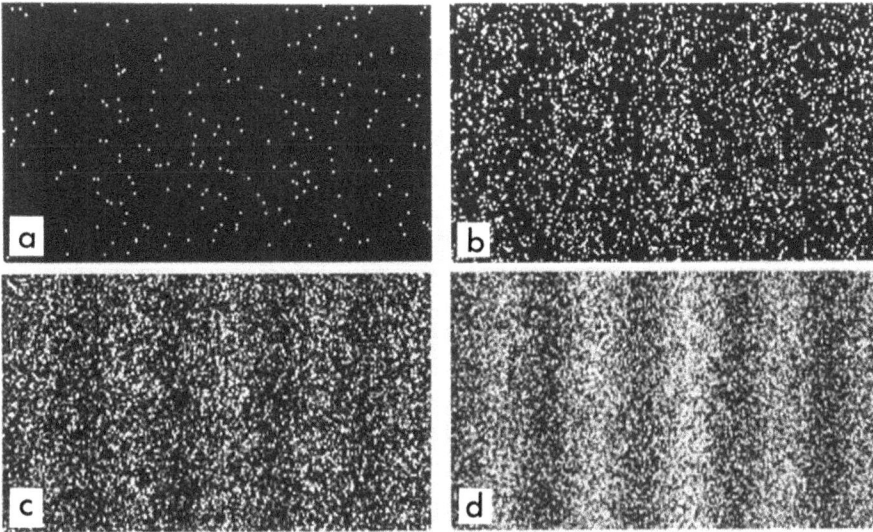

Figure 4.8: Single-particle two-slit experiment. https://commons.wikimedia.org/wiki/File:Double-slit_experiment_results_Tanamura_2.jpg (last access date: 21.05.2021).

and longer periods of time, a remarkable pattern begins to appear on the screen (see Figure 4.8).

What we witness is the gradual build-up, point by point, of the same pattern of light and dark strips that we would have obtained had we not bothered to send the particles one at a time. Certainly, some sort of interference is taking place, but how does one account for it? This is quite mysterious, and it gets worse. Suppose we place some sort of detector near each slit that will flash, or beep, or something when a particle passes through that slit. Then we can keep track of where the corresponding dot appears on the screen and maybe figure out how this strange interference pattern is coming about; we will just watch the particles. This sounds like a great idea, but if you actually do it you may walk away from your experiment with the feeling that nature is thumbing her nose at you because *the interference pattern disappears altogether and the image on the screen is exactly what classical mechanics would predict* (this is described by Feynman in Section 1-6, Volume III, of [FLS], which is so much fun that it should be on everyone's "must read" list).

So, how do you explain all of this rather bizarre behavior? If by "explain" you mean string together some elegant and imaginative combination of classical concepts like "particle" and "wave," then the answer is quite simple: You do not! Everyone has given up trying.

"It is all quite mysterious. And the more you look at it the more mysterious it seems."

Richard Feynman (Volume III, page 1-6, of [FLS])

An electron is not a particle and it is not a wave. To say that it behaves sometimes like a particle (for example, when you detect it) and sometimes like a wave (when you do not) is a little better, but rather difficult to take seriously as an "explanation."

> "It's the way nature works. If you wanna know the way nature works, we looked at it carefully, that is the way it looks. If you don't like it, go somewhere else."

> Richard Feynman, Lectures on Quantum Electrodynamics, University of Auckland (available at http://vega.org.uk/video/subseries/8)

Like it or not, the rules that govern the quantum world are entirely different than those that govern the world in which our conceptual apparatus and our languages evolved. We have no built-in concepts ("wavicle"?) that will come to our aid since we are encountering these behaviors for the first time in our long, and not altogether illustrious, evolutionary history. In lieu of a conceptual model that "explains" these behaviors in terms of some more fundamental physical principles we will have to be content with a mathematical model that accounts for the behaviors we have seen and predicts an enormous number that we have not yet seen. This mathematical model is really the subject of the rest of this manuscript, but we will take the first baby steps toward its construction now.

Since it carries along with it so much classical baggage we would like to avoid using the term "particle" as much as possible. The following discussion is based on our rather limited experience with photons and electrons and everything we say is equally true of either, but for the sake of having a single word to use we will focus on electrons. A word of caution is in order, however. A proper discussion of electrons must take into account another very important quantum mechanical property they possess and that we have not seen and will not see until Section 10.1. This is called "spin" and one cannot fully understand the behavior of electrons without it. Even so, a detour at this point to introduce the notion of spin would only serve to muddy the already rather murky waters so *until further notice we will ignore the effects of spin, and in Section* 10.1 *we will make any adjustments required to include these effects.*

What have we learned from the two-slit experiment with electrons? Sending a single electron from the source and assuming it makes it past the wall we obtain a dot at the point where it is detected on the screen. Ideally, we would like to be able to predict where this dot will appear. As we have seen, however, this does not seem to be in the cards. Sending a few more electrons from the source we obtain dots that seem almost randomly distributed on the screen. But if we let the experiment run full course we find that the distribution of dots is not entirely random (Figure 4.8). The electrons seem most likely to land in the vertical strip in the center, not at all likely to land in the dark vertical strips and progressively less likely to land in the bright strips receding from the center. This suggests that we might want to lower our expectations somewhat and try to predict only the *probability* that an electron will be detected at some given point on the screen. The usual procedure in classical probability theory would go something like this. Fix some point X on the screen. An electron leaving the source can arrive at X either by passing through the first slit S_1 or by passing through the second slit S_2. If P_1 is the probability that it arrives at X via S_1 and P_2 is the probability that it arrives at X via S_2, then the probability P_{12} that it arrives at X at all is $P_{12} = P_1 + P_2$. But we know already that this cannot be the case (identify the intensity graphs in Figure 4.6 and Figure 4.7 with probability distributions). The problem, of course, is that these classical probabilities take no account of the very interference effects that we are trying to describe. How can we do this? For a hint let us return for a moment to water waves where the interference is simple and clear. This interference comes about because of the *phase difference* of the two waves meeting at a point. If the phase difference is zero (that is, if they are *in phase*), then the resulting intensity is a maximum. If the phase difference is π, then the resulting intensity is a minimum. For any other phase difference there will be some, but not complete canceling and the resultant intensity will be something between the minimum and the maximum intensities. Now, complex numbers $z = |z| e^{i\varphi}$ have an amplitude/modulus $|z|$ and a phase φ and have the property that

$$|z_1 + z_2|^2 = |z_1|^2 + |z_2|^2 + 2|z_1|\,|z_2|\,\cos(\varphi_1 - \varphi_2).$$

Consequently, $|z_1 + z_2|^2$ achieves a maximum when $\varphi_1 - \varphi_2 = 0$, a minimum when $\varphi_1 - \varphi_2 = \pi$ and is otherwise something between the minimum and the maximum. The idea then is to represent the probabilities P_1 and P_2 as the squared moduli of complex numbers $\psi_1 = |\psi_1| e^{i\varphi_1}$ and $\psi_2 = |\psi_2| e^{i\varphi_2}$, that is,

$$P_1 = |\psi_1|^2 \quad \text{and} \quad P_2 = |\psi_2|^2.$$

Here ψ_i, $i = 1, 2$, is called the *probability amplitude* for the electron to reach P via S_i. One then assumes that, unlike the probabilities themselves, the probability amplitudes add so that the probability amplitude that the electron arrives at P at all is $\psi_1 + \psi_2$ and so the probability is

$$|\psi_1 + \psi_2|^2 = |\psi_1|^2 + |\psi_2|^2 + 2|\psi_1|\,|\psi_2|\,\cos(\varphi_1 - \varphi_2).$$

How one assigns probability amplitudes to events is, of course, what quantum mechanics is all about, so this will have to wait for a while, but we are at least in a position to record what Feynman calls the first principles of quantum mechanics.

First principles of quantum mechanics

Richard Feynman (Volume III, page 1–10, of [FLS])

1. Because of the wave-like attributes of particles in quantum mechanics (de Broglie waves) and the resultant interference effects, the probability P that a particular event (such as the arrival of an electron at some location on a screen) will occur

is represented as the squared modulus $P = |\psi|^2$ of a complex number ψ called the *probability amplitude* of the event.

2. When there are several classical alternatives for the way in which the event can occur and no measurements are made on the system, the probability amplitude of the event is the sum of the probability amplitudes for each alternative considered separately. In particular, if there are two possibilities with amplitudes ψ_1 and ψ_2 and probabilities $P_1 = |\psi_1|^2$ and $P_2 = |\psi_2|^2$, then the probability amplitude of the event is $\psi_1 + \psi_2$ and the probability of the event is

$$P_{12} = |\psi_1 + \psi_2|^2 = |\psi_1|^2 + |\psi_2|^2 + 2\,\mathrm{Re}(\psi_1\overline{\psi_2}).$$

The last term represents the effect of interference.

3. When a measurement is made to determine whether one or another of the classical alternatives in 2. is actually taken, the interference is lost and the probability P of the event is the sum of the probabilities for each alternative taken separately. In particular, when there are just two alternatives, $P_{12} = P_1 + P_2$.

We begin to see then that what is "waving" in the case of an electron is a probability amplitude. This will be made precise in Chapter 6 with Schrödinger's notion of a "wave function." We will see in Section 8.1 how Feynman envisioned a generalization of the two-slit experiment in which there are many walls, each with many slits, and applied these first principles to arrive at his notion of a *path integral*.

5 Synopsis of self-adjoint operators, spectral theory and Stone's theorem

5.1 Introduction

Now that we have some modest appreciation of the sort of physical phenomena with which quantum mechanics deals and why these necessitate a shift away from the paradigm of classical particle mechanics, it is time to confront the rather substantial mathematical background required for this paradigm shift. In this admittedly rather lengthy section we will begin the process by providing a synopsis of the three pillars of this mathematical structure: unbounded self-adjoint operators, the spectral theorem and Stone's theorem. Despite its length, however, it is only a synopsis. The definitions are given precisely and the theorems are stated precisely, but in lieu of proofs we generally offer only a few detailed examples relevant to quantum mechanics and plentiful references. *Those who are comfortable with these topics might do well to proceed directly to Chapter 6 and refer back as the need arises.*

Throughout Chapter 5 all Hilbert spaces are complex and separable.

Everyone is familiar with the spectral theorem for self-adjoint operators on a *finite-dimensional*, complex inner product space $\mathcal{H}$ (a particularly nice treatment, with an eye on the infinite-dimensional generalizations, is available in Chapter Eleven of [Simm1]). In a nutshell, the theorem says simply that any such operator A is a finite linear combination

$$A = \sum_{i=1}^{n} \lambda_i P_i$$

of orthogonal projections $P_1, P_2, \ldots, P_n$, where the coefficients $\lambda_1, \lambda_2, \ldots, \lambda_n$ are just the (necessarily real) eigenvalues of A. There is a relatively straightforward generalization of the finite-dimensional theorem to operators A on a complex, separable, infinite-dimensional Hilbert space $\mathcal{H}$ that are self-adjoint and *compact* (see Remark 5.5.2 for the definitions). Here one simply replaces the finite sum with an infinite series

$$A = \sum_{i=1}^{\infty} \lambda_i P_i, \tag{5.1}$$

where $P_1, P_2, \ldots$ are still orthogonal projections, $\lambda_1, \lambda_2, \ldots$ are still the (necessarily real) eigenvalues of A and the convergence is understood to be *strong convergence* in $\mathcal{H}$, that is,

$$\left\| \sum_{i=1}^{n} \lambda_i P_i \psi - A\psi \right\| \to 0 \quad \text{as } n \to \infty \ \forall \psi \in \mathcal{H}.$$

https://doi.org/10.1515/9783110751949-005

We will go through this version of the theorem a bit more carefully soon (a direct proof is available in Section 93 of [RiSz.N]). There is also a generalization to arbitrary bounded, self-adjoint operators on $\mathcal{H}$, but this is less straightforward. In this case there need not be any eigenvalues at all and the sum is replaced by an integral over the spectrum of the operator, but the integration is with respect to a "measure" taking values in the set of projection operators on $\mathcal{H}$. One can get some rough sense of how these funny integrals arise by rewriting (5.1). Introduce projections

$$E_{\lambda_0} = 0,$$
$$E_{\lambda_1} = P_1,$$
$$E_{\lambda_2} = P_1 + P_2,$$
$$\vdots$$
$$E_{\lambda_n} = P_1 + P_2 + \cdots + P_n.$$
$$\vdots$$

Then (5.1) becomes

$$A = \sum_{i=1}^{\infty} \lambda_i (E_{\lambda_i} - E_{\lambda_{i-1}}) = \lim_{n\to\infty} \sum_{i=1}^{n} \lambda_i \Delta E_{\lambda_i}$$

and one obtains something that at least looks like a Riemann–Stieltjes integral (see Appendix H). One might be tempted to write this as

$$A = \int \lambda \, dE_\lambda.$$

As it happens, one can make rigorous sense of such Stieltjes integrals with respect to "projection-valued measures" and thereby justify this sort of integral representation for bounded, self-adjoint operators. The details are available in Section 6.7 of [Fried], Section 107 of [RiSz.N], Chapters 5–6 of [TaylA], Chapter VII of [RS1], Chapter 12 of [Rud1], Section 3, Chapter XI, of [Yosida] and a great many other places as well.

Regrettably, even this result is not adequate for our purposes since, as we indicated at the close of Section 3.3, identifying observables in statistical mechanics with operators on $L^2(\Omega, \mu)$ requires us to consider maps that are linear only on certain subspaces of $L^2(\Omega, \mu)$ and none of the results described above apply to these. Von Neumann faced the same problem in his study of the mathematical foundations of quantum mechanics where essentially all of the relevant operators are of this type. He solved the problem by developing yet another generalization of the spectral theorem that very closely resembled the result for bounded, self-adjoint operators, but applied also to just the sort of operator we now have in mind. Von Neumann's "Spectral Theorem for Unbounded Self-Adjoint Operators" is a bit technical, but it also lies at the heart of quantum mechanics and we must try to provide some sense of what it says and what it means. In the following synopsis we will try to provide enough narrative to instill a reasonable level of comfort with the ideas, a number of examples relevant to quantum mechanics to illustrate these ideas and enough references to facilitate access to the proofs we do not supply. Here are the sources from which we will draw these references. Chapter III of [Prug] is very detailed and readable and contains everything we will need. The work [RiSz.N] is a classic, also very readable, and contains a great deal more than we will need. Chapter 13 of [Rud1] contains an elegant and very clear discussion, but one that will require a bit of backtracking that ultimately leads to the Gelfand–Naimark theorem from the theory of Banach algebras (a detour well worth taking, by the way). Chapter VIII, Sections 1–3, of [RS1] contains a number of different formulations of the spectral theorem and many interesting examples and is specifically geared to the needs of quantum field theory (which is discussed in [RS2]), but is also a bit more condensed. Also rather condensed is the treatment in Sections 5–6, Chapter XI, of [Yosida], but the book itself is very authoritative and contains a wealth of important material.

5.2 Unbounded self-adjoint operators

As always, $\mathcal{H}$ will denote a separable, complex Hilbert space and we will write its inner product as $\langle \, , \, \rangle$, assumed to be complex linear in the second slot and conjugate linear in the first. The corresponding norm is defined by $\|\psi\| = \sqrt{\langle \psi, \psi \rangle}$ for every $\psi \in \mathcal{H}$. Recall that a *bounded operator* on $\mathcal{H}$ is a linear map $T : \mathcal{H} \to \mathcal{H}$ for which there is a constant K such that $\|T\psi\| \leq K\|\psi\|\ \forall \psi \in \mathcal{H}$ and that these are precisely the linear maps from $\mathcal{H}$ to $\mathcal{H}$ that are continuous in the norm topology of $\mathcal{H}$. We will use the unmodified term *operator* to stand for a map

$$A : \mathcal{D}(A) \to \mathcal{H},$$

where $\mathcal{D}(A)$ is a *dense, linear subspace* of $\mathcal{H}$ and A is linear on $\mathcal{D}(A)$. Two such operators A and B are considered *equal* only if $\mathcal{D}(B) = \mathcal{D}(A)$ and $B\psi = A\psi$ for all ψ in this common domain.

Example 5.2.1. For any bounded operator $T : \mathcal{H} \to \mathcal{H}$ and any dense, linear subspace $\mathcal{D}$ of $\mathcal{H}$, the restriction $A = T|_{\mathcal{D}}$ of T to $\mathcal{D}$ is an operator in this sense. These, of course, have bounded extensions to all of $\mathcal{H}$, but this is generally not the case, as we will now see.

Example 5.2.2. Let $\mathcal{H} = L^2(\mathbb{R})$ be the space of (equivalence classes of) complex-valued functions on $\mathbb{R}$ that are square integrable with respect to Lebesgue measure. Let $\mathcal{D} = C_0^\infty(\mathbb{R})$ be the dense linear subspace of $L^2(\mathbb{R})$ consisting of smooth functions with compact support (if density is not clear to you, this is proved in Theorem 5.6, Chapter II, of [Prug]). Define $X : C_0^\infty(\mathbb{R}) \to L^2(\mathbb{R})$ by

$$(X\psi)(x) = x\psi(x) \quad \forall x \in \mathbb{R}.$$

Then X is clearly linear and therefore an operator on $C_0^\infty(\mathbb{R})$. To see that it does not have a bounded extension to $L^2(\mathbb{R})$ we will exhibit a sequence ψ_n, $n = 0, 1, \ldots$, in $C_0^\infty(\mathbb{R})$ for which $\|\psi_n\| = 1$, but $\|X\psi_n\| \to \infty$ as $n \to \infty$. Let ψ_0 be any smooth, real-valued, nonnegative function on $\mathbb{R}$ that has support contained in $[0, 1]$ and $\|\psi_0\| = 1$.

Exercise 5.2.1. If you have not constructed such a function before, do so now. *Hint:* Use Exercise 2-26 of [Sp1].

For any $n = 1, 2, \ldots$, define ψ_n by $\psi_n(x) = \psi_0(x - n)$. Then ψ_n is in $C_0^\infty(\mathbb{R})$ and has its support in $[n, n + 1]$. Moreover, in $L^2(\mathbb{R})$, $\|\psi_n\| = \|\psi_0\| = 1 \ \forall n \geq 1$ and

$$\|X\psi_n\|^2 = \int_\mathbb{R} x^2 \psi_0^2(x - n)dx = \int_{[n,n+1]} x^2 \psi_0^2(x - n)dx$$

$$\geq n^2 \int_\mathbb{R} \psi_n^2(x)dx = n^2 \|\psi_n\|^2 = n^2 \to \infty$$

as $n \to \infty$.

Example 5.2.3. Note that there was no particularly compelling reason to choose $C_0^\infty(\mathbb{R})$ as the domain in Example 5.2.2. Indeed, it might have been more natural to take the domain to be the set of all ψ in $L^2(\mathbb{R})$ for which the function $x \to x\psi(x)$ is also in $L^2(\mathbb{R})$. This is certainly a linear subspace and, since it contains $C_0^\infty(\mathbb{R})$, it is also dense. With this domain the operator will play a particularly important role in our discussion of quantum mechanics so we will introduce a special name for it and will revert to the custom in physics of using q rather than x for the independent variable in $\mathbb{R}$. Thus, we define an operator Q on

$$\mathcal{D}(Q) = \left\{ \psi \in L^2(\mathbb{R}) : \int_\mathbb{R} q^2 |\psi(q)|^2 dq < \infty \right\}$$

by $(Q\psi)(q) = q\psi(q)$. The argument given in Example 5.2.2 again shows that Q has no bounded extension to $L^2(\mathbb{R})$.

Although the reason may not be clear just yet we will refer to the operator in Example 5.2.3 as the *position operator* on $\mathbb{R}$. We will try to clarify the origin of the name as we gain some understanding of quantum mechanics.

It turns out that many issues that need to be resolved concerning an operator A are very sensitive to the choice of domain and that the proper choice is very often dictated by the problem (or the physics) with which one is coping at the moment. We will write out some simple examples, all of which will figure heavily in our subsequent discussion of quantum mechanics.

Example 5.2.4. Let $\mathcal{H} = L^2(\mathbb{R})$. The operator P that we wish to consider is given by

$$(P\psi)(q) = -i\hbar\frac{d\psi}{dq} \quad \forall q \in \mathbb{R}$$

(the reason for the factor $-i$ will become clear shortly [see Remark 5.2.2] and the reason for introducing the normalized Planck constant $\hbar$ will emerge in Chapter 7). Of course, the derivative is not defined for all ψ in $L^2(\mathbb{R})$. Before specifying the domain we have in mind for P we will review a few facts from real analysis.

Remark 5.2.1. A good reference for the material we are about to review is [RiSz.N], Sections 25 and 26. Let $a < b$ be real numbers and consider a complex-valued function $f : [a, b] \to \mathbb{C}$ on $[a, b]$. Then f is said to be *absolutely continuous* on $[a, b]$ if $\forall \epsilon > 0 \,\exists \delta > 0$ such that for any finite family $\{[a_1, b_1], \dots, [a_n, b_n]\}$ of nonoverlapping intervals in $[a, b]$, $\sum_{i=1}^{n}(b_i - a_i) < \delta$ implies $\sum_{i=1}^{n}|f(b_i) - f(a_i)| < \epsilon$ (*nonoverlapping* means that the open intervals (a_i, b_i) are pairwise disjoint). Any such function is uniformly continuous and of bounded variation on $[a, b]$. The set of all absolutely continuous functions on $[a, b]$ is denoted $AC[a, b]$ and it is a complex vector space. For our purposes the two most important facts about $AC[a, b]$ are the following:
1. $f \in AC[a, b] \Leftrightarrow f$ is differentiable almost everywhere, $f' \in L^1[a, b]$ and $f(q) - f(a) = \int_a^q f'(t)dt$ for all $q \in [a, b]$;
2. for all $f, g \in AC[a, b]$,

$$\int_a^b f(q)\frac{dg}{dq}dq = f(b)g(b) - f(a)g(a) - \int_a^b \frac{df}{dq}g(q)dq \quad \text{(integration by parts)}.$$

Now, as our domain for P we take

$$\mathcal{D}(P) = \left\{\psi \in L^2(\mathbb{R}) : \psi \in AC[a, b] \,\forall a < b \text{ in } \mathbb{R} \text{ and } \frac{d\psi}{dq} \in L^2(\mathbb{R})\right\}$$

(motivation for this choice is to be found in Example 5.2.7). Clearly, $\mathcal{D}(P)$ is a linear subspace of $L^2(\mathbb{R})$ and it is dense because it contains $C_0^\infty(\mathbb{R})$. To see that P does not have a bounded extension to $L^2(\mathbb{R})$ we will exhibit a sequence ψ_n, $n = 1, 2, \dots$, in $\mathcal{D}(P)$ for which $\|\psi_n\|^2 = \sqrt{\pi}$ and $\|P\psi_n\|^2 \to \infty$ as $n \to \infty$. One possible choice is as follows. Let

$$\psi_n(q) = e^{inq}e^{-q^2/2}, \quad n = 1, 2, \dots.$$

Then, from the Gaussian integral (A.1) in Appendix A,

$$\|\psi_n\|^2 = \int_\mathbb{R} \overline{\psi_n(q)}\psi_n(q)dq = \int_\mathbb{R} e^{-q^2}dq = \sqrt{\pi},$$

so each ψ_n is a smooth element of $L^2(\mathbb{R})$.

Exercise 5.2.2. Show that $\psi_n \in \mathcal{D}(P)$ for each $n = 1, 2, \ldots$.

Moreover,

$$(P\psi_n)(q) = \hbar\, n\, \psi_n(q) + i\hbar\, q\psi_n(q) = \hbar(n + qi)\psi_n(q).$$

By Exercise A.0.3.2 in Appendix A,

$$\int_{\mathbb{R}} q^2 e^{-q^2}\, dq = \frac{\sqrt{\pi}}{2},$$

so

$$\|P\psi_n\|^2 = \hbar^2 \sqrt{\pi}\left(n^2 + \frac{1}{2}\right)$$

and therefore $P\psi_n$ is in $L^2(\mathbb{R})$, but $\|P\psi_n\|^2 \to \infty$ as $n \to \infty$.

The operator in Example 5.2.4 is called the *momentum operator* on $\mathbb{R}$. As for the position operator Q we will try to explain the reason for the name as we develop some understanding of quantum mechanics.

An operator $A : \mathcal{D}(A) \to \mathcal{H}$, such as X, Q or P in the last three examples, defined on a dense linear subspace of $\mathcal{H}$ that does *not* have a bounded extension to all of $\mathcal{H}$ is called an *unbounded operator*. Next we must try to make sense of "self-adjoint" for such operators.

Recall that for any bounded operator $T : \mathcal{H} \to \mathcal{H}$, there exists a unique bounded operator $T^* : \mathcal{H} \to \mathcal{H}$, called the *adjoint* of T, satisfying $\langle T\phi, \psi \rangle = \langle \phi, T^*\psi \rangle$ for all $\phi, \psi \in \mathcal{H}$. For unbounded operators the situation is a bit more delicate. We will define the adjoint for an unbounded operator $A : \mathcal{D}(A) \to \mathcal{H}$ by first specifying its domain. Specifically, we let $\mathcal{D}(A^*)$ be the set of all $\phi \in \mathcal{H}$ for which there exists an $\eta \in \mathcal{H}$ such that $\langle A\psi, \phi \rangle = \langle \psi, \eta \rangle$ for every $\psi \in \mathcal{D}(A)$. Note that $\mathcal{D}(A^*)$ contains at least 0 and is a linear subspace. For each $\phi \in \mathcal{D}(A^*)$, η is uniquely determined because $\mathcal{D}(A)$ is dense. Thus, we can define the *adjoint* A^* of A by

$$A^*\phi = \eta \quad \forall \phi \in \mathcal{D}(A^*)$$

so that

$$\langle A\psi, \phi \rangle = \langle \psi, A^*\phi \rangle \quad \forall \psi \in \mathcal{D}(A)\ \forall \phi \in \mathcal{D}(A^*),$$

where A^* is linear on $\mathcal{D}(A^*)$, but $\mathcal{D}(A^*)$ need not be dense in $\mathcal{H}$ so A^* need not be densely defined. Next we introduce a condition on A that ensures the density of $\mathcal{D}(A^*)$ and therefore ensures that A^* is an "operator" in our sense of the term.

An operator $A : \mathcal{D}(A) \to \mathcal{H}$ is said to be *closed* if, whenever $\psi_1, \psi_2, \ldots \in \mathcal{D}(A)$ converges to a vector ψ in $\mathcal{H}$ and $A\psi_1, A\psi_2, \ldots$ converges to some vector ϕ in $\mathcal{H}$, then

ψ is in $\mathcal{D}(A)$ and $\phi = A\psi$. This can be phrased more picturesquely in the following way. Recall that the algebraic direct sum $\mathcal{H} \oplus \mathcal{H}$ of the vector space $\mathcal{H}$ with itself has a natural Hilbert space structure with the inner product defined by

$$\langle (\phi_1, \psi_1), (\phi_2, \psi_2) \rangle = \langle \phi_1, \phi_2 \rangle + \langle \psi_1, \psi_2 \rangle.$$

The *graph* $\mathrm{Gr}(A)$ of the operator $A : \mathcal{D}(A) \to \mathcal{H}$ is the linear subspace of $\mathcal{H} \oplus \mathcal{H}$ defined by

$$\mathrm{Gr}(A) = \{(\phi, A\phi) : \phi \in \mathcal{D}(A)\}.$$

Then A is a closed operator if and only if $\mathrm{Gr}(A)$ is a closed linear subspace of $\mathcal{H} \oplus \mathcal{H}$. The *closed graph theorem* from functional analysis (see, for example, Theorem 4.64 of [Fried]) implies the following.

Theorem 5.2.1. *A closed operator* $A : \mathcal{H} \to \mathcal{H}$ *defined on the entire Hilbert space* $\mathcal{H}$ *is bounded.*

An operator $A : \mathcal{D}(A) \to \mathcal{H}$ that is not closed may nevertheless have a closed extension, that is, there may be a closed operator $\tilde{A} : \mathcal{D}(\tilde{A}) \to \mathcal{H}$ with $\mathcal{D}(A) \subseteq \mathcal{D}(\tilde{A})$ and $\tilde{A}|_{\mathcal{D}(A)} = A$. In this case we say that A is *closable*. Any closable operator has a minimal closed extension $\overline{A} : \mathcal{D}(\overline{A}) \to \mathcal{H}$, that is, one for which any other closed extension $\tilde{A} : \mathcal{D}(\tilde{A}) \to \mathcal{H}$ of A is also an extension of $\overline{A}$; $\overline{A}$ is called the *closure* of A and, in fact, the graph of $\overline{A}$ is just the closure in $\mathcal{H} \oplus \mathcal{H}$ of the graph of A

$$\mathrm{Gr}(\overline{A}) = \overline{\mathrm{Gr}(A)}$$

(this is the proposition on page 250 of [RS1]). The following is Theorem VIII.1 of [RS1].

Theorem 5.2.2. *Let* $A : \mathcal{D}(A) \to \mathcal{H}$ *be an operator on a Hilbert space* $\mathcal{H}$. *Then:*
1. A^* *is closed;*
2. A *is closable if and only if* $\mathcal{D}(A^*)$ *is dense in* $\mathcal{H}$ *and, in this case,* $\overline{A} = A^{**}$;
3. *if* A *is closable, then* $(\overline{A})^* = A^*$.

In particular, it is precisely for closable operators A that the adjoint A^* qualifies as an "operator" according to our definition. An operator $A : \mathcal{D}(A) \to \mathcal{H}$ is said to be *self-adjoint* if it equals its adjoint in the sense that $\mathcal{D}(A^*) = \mathcal{D}(A)$ and $A^*\psi = A\psi \ \forall\psi \in \mathcal{D}(A)$. In the formalism of quantum mechanics (Chapter 6) self-adjoint operators will play the role of the observables. We will write out some important examples shortly.

A self-adjoint operator A is a closed operator and it satisfies

$$\langle A\psi, \phi \rangle = \langle \psi, A\phi \rangle \quad \forall\psi, \phi \in \mathcal{D}(A). \tag{5.2}$$

Note, however, that (5.2) is *not* equivalent to self-adjointness. An operator A that satisfies (5.2) is said to be *symmetric* (or *formally self-adjoint*). For any symmetric operator A

it is clear that $\mathcal{D}(A) \subseteq \mathcal{D}(A^*)$. However, it is possible for the containment to be proper, in which case A is not self-adjoint. We will sketch one example; the details can be found on pages 257–259 of [RS1].

Example 5.2.5. We let $\mathcal{H} = L^2[0,1]$ and consider the operator $A = -i\hbar \frac{d}{dx}$ on the following subspace of the space $AC[0,1]$ of complex-valued, absolutely continuous functions on $[0,1]$:

$$\mathcal{D}(A) = \left\{ \psi \in AC[0,1] : \frac{d\psi}{dx} \in L^2[0,1], \ \psi(0) = \psi(1) = 0 \right\}.$$

Now, for $\phi, \psi \in \mathcal{D}(A)$ we compute, from the integration by parts formula (Remark 5.2.1),

$$\langle A\psi, \phi \rangle = \int_0^1 \overline{\left(-i\hbar \frac{d\psi}{dx} \right)} \phi(x) dx = i\hbar \int_0^1 \phi(x) \frac{d\overline{\psi}}{dx} dx$$

$$= i\hbar \left[\phi(1)\overline{\psi}(1) - \phi(0)\overline{\psi}(0) \right] - i\hbar \int_0^1 \frac{d\phi}{dx} \overline{\psi}(x) dx$$

$$= \int_0^1 \overline{\psi}(x) \left(-i\hbar \frac{d\phi}{dx} \right) dx$$

$$= \langle \psi, A\phi \rangle$$

and conclude that $A : \mathcal{D}(A) \to L^2[0,1]$ is symmetric.

Remark 5.2.2. Note that we needed the factor of i in A for this to work; the minus sign is conventional.

In [RS1] it is shown that the adjoint A^* of A is the operator $A^* = -i\hbar \frac{d}{dx}$ on $\mathcal{D}(A^*) = \{\psi \in AC[0,1] : \frac{d\psi}{dx} \in L^2[0,1]\}$. Since $\mathcal{D}(A^*)$ properly contains $\mathcal{D}(A)$, A is not self-adjoint despite the fact that A and A^* agree where they are both defined. It is also shown in [RS1] that A is closed and has (lots of) self-adjoint extensions. By way of contrast, Problem 5, Chapter VIII, of [RS1] gives a densely defined, symmetric operator with *no* self-adjoint extensions.

There are some useful criteria for determining when a symmetric operator is self-adjoint. The following is Theorem VIII.3 of [RS1]; here, and henceforth, we will use Kernel (A) to denote the kernel of an operator A, we will use Image (A) to denote its image (range) and will write simply "k" for the operator that is multiplication by the constant $k \in \mathbb{C}$.

Theorem 5.2.3. *Let $A : \mathcal{D}(A) \to \mathcal{H}$ be a symmetric operator. Then the following are equivalent:*
1. *A is self-adjoint;*
2. *A is closed and* Kernel $(A^* \pm i) = \{0\}$;
3. Image $(A \pm i) = \mathcal{H}$.

Since the proof uses only ideas that we have already described, you may like to try it yourself with a few hints.

Exercise 5.2.3. Prove Theorem 5.2.3. *Hints:*

1. For 1. $\Rightarrow$ 2., assume A is self-adjoint and show that $A^*\psi = i\psi$ implies $-i\langle\psi, \psi\rangle = i\langle\psi, \psi\rangle$. Similarly for $A^*\psi = -i\psi$.
2. For 2. $\Rightarrow$ 3., show, for example, that if $A^*\psi = -i\psi$ has no nonzero solutions, then Image $(A - i)^\perp = 0$, so Image $(A - i)$ is dense. Then show that since A is closed and $\|(A - i)\psi\|^2 = \|A\psi\|^2 + \|\psi\|^2$, Image $(A - i)$ is closed so that Image $(A - i) = \mathcal{H}$.
3. For 3. $\Rightarrow$ 1., let $\phi \in \mathcal{D}(A^*)$. Select $\eta \in \mathcal{D}(A)$ such that $(A - i)\eta = (A^* - i)\phi$ and show that $(A^* - i)(\phi - \eta) = 0$, so $\phi = \eta$ and therefore $\mathcal{D}(A^*) = \mathcal{D}(A)$.

For a symmetric operator $A : \mathcal{D}(A) \to \mathcal{H}$, $\mathcal{D}(A) \subseteq \mathcal{D}(A^*)$ and $\mathcal{D}(A)$ is dense so $\mathcal{D}(A^*)$ is also dense. Consequently, a symmetric operator is always closable since A^* is always closed. If the closure $\overline{A} : \mathcal{D}(\overline{A}) \to \mathcal{H}$ of A is self-adjoint, then A is said to be *essentially self-adjoint*. In this case, $\overline{A}$ is the *unique* self-adjoint extension of A and, in fact, the converse is also true, that is, *a symmetric operator is essentially self-adjoint if and only if it has a unique self-adjoint extension* (see page 256 of [RS1]). From the previous theorem one obtains the following criteria for essential self-adjointness.

Corollary 5.2.4. *Let* $A : \mathcal{D}(A) \to \mathcal{H}$ *be a symmetric operator. Then the following are equivalent:*

1. *A is essentially self-adjoint;*
2. *Kernel* $(A^* \pm i) = \{0\}$;
3. *Image* $(A \pm i)$ *are both dense in* $\mathcal{H}$.

Exercise 5.2.4. Prove Corollary 5.2.4 from Theorem 5.2.3.

Exercise 5.2.5. Show that a self-adjoint operator is *maximally symmetric*, that is, has no proper symmetric extensions.

Example 5.2.6. We consider the operator $Q : \mathcal{D}(Q) \to L^2(\mathbb{R})$ of Example 5.2.3. Note that Q is symmetric since $\psi, \phi \in \mathcal{D}(Q)$ implies

$$\langle Q\psi, \phi\rangle = \int_{\mathbb{R}} \overline{Q\psi(q)}\phi(q)dq = \int_{\mathbb{R}} q\,\overline{\psi}(q)\phi(q)dq$$
$$= \int_{\mathbb{R}} \overline{\psi}(q)(q\phi(q))dq = \langle\psi, Q\phi\rangle.$$

We will give two proofs that Q is self-adjoint. For the first we will appeal directly to the definition and for the second we will show that Image $(Q \pm i) = L^2(\mathbb{R})$ and use Theorem 5.2.3.3. For the first proof we note that, by symmetry, $\mathcal{D}(Q) \subseteq \mathcal{D}(Q^*)$, so it will suffice to show that $\mathcal{D}(Q^*) \subseteq \mathcal{D}(Q)$. By definition, $\mathcal{D}(Q^*)$ consists of all $\phi \in L^2(\mathbb{R})$ for which there is an $\eta \in L^2(\mathbb{R})$ with

$$\int_{\mathbb{R}} (\, q\bar{\psi}(q)\,)\,\phi(q)dq = \int_{\mathbb{R}} \bar{\psi}(q)\eta(q)dq \quad \forall \psi \in \mathcal{D}(Q),$$

that is,

$$\int_{\mathbb{R}} \bar{\psi}(q)\,[q\phi(q) - \eta(q)]\,dq = 0 \quad \forall \psi \in \mathcal{D}(Q).$$

Since this last equation must be satisfied *for all* ψ in the dense set $\mathcal{D}(Q)$, we must have $\eta(q) = q\,\phi(q)$ almost everywhere. In particular, $q\,\phi(q)$ is in $L^2(\mathbb{R})$ and this puts ϕ in $\mathcal{D}(Q)$, so $\mathcal{D}(Q^*) \subseteq \mathcal{D}(Q)$, as required.

Now for the second proof we note that

$$(Q \pm i)\psi(q) = (q \pm i)\psi(q)$$

and that $q \pm i \neq 0$ since $q \in \mathbb{R}$. Now, for any $\varphi \in L^2(\mathbb{R})$, $\frac{1}{q \pm i}\varphi(q)$ is in $\mathcal{D}(Q)$ since

$$q^2 \left| \frac{1}{q \pm i}\varphi(q) \right|^2 = \frac{q^2}{q^2 + 1}|\varphi(q)|^2 \leq |\varphi(q)|^2.$$

Since $(Q \pm i)(\frac{1}{q \pm i}\varphi(q)) = \varphi(q)$, Image $(Q \pm i) = L^2(\mathbb{R})$, as required.

Exercise 5.2.6. Show similarly that if $g : \mathbb{R} \to \mathbb{R}$ is a *real-valued* measurable function, then the multiplication operator $Q_g : \mathcal{D}(Q_g) \to L^2(\mathbb{R})$ defined by $(Q_g\psi)(q) = g(q)\psi(q)$ on $\mathcal{D}(Q_g) = \{\psi \in L^2(\mathbb{R}) : \int_{\mathbb{R}} g(q)^2|\psi(q)|^2dq < \infty\}$ is self-adjoint.

Exercise 5.2.7. Show that Q is essentially self-adjoint on the Schwartz space $\mathcal{S}(\mathbb{R})$ (see Section G.2).

Somewhat later (Theorem 5.3.1) we will also need the following analogous result.

Exercise 5.2.8. Let $\mathbb{N} = \{1, 2, \ldots\}$ denote the set of natural numbers and consider the Hilbert space $\ell^2(\mathbb{N})$ of square summable sequences of complex numbers, that is,

$$\ell^2(\mathbb{N}) = \left\{ x = (x_1, x_2, \ldots) : x_i \in \mathbb{C}, i = 1, 2, \ldots, \text{ and } \sum_{i=1}^{\infty} |x_i|^2 < \infty \right\},$$

with coordinatewise algebraic operations and $\langle x, y \rangle = \sum_{i=1}^{\infty} \bar{x}_i y_i$. Note that this is just $L^2(\mathbb{N}, \mu)$, where μ is the point measure (also called the counting measure) on $\mathbb{N}$. Let $\lambda = (\lambda_1, \lambda_2, \ldots)$ be a sequence of real numbers and define a multiplication operator Q_λ on $\mathcal{D}(Q_\lambda) = \{x = (x_1, x_2, \ldots) \in \ell^2(\mathbb{N}) : \sum_{i=1}^{\infty} \lambda_i^2|x_i|^2 < \infty\}$ by $Q_\lambda(x) = Q_{(\lambda_1, \lambda_2, \ldots)}(x_1, x_2, \ldots) = (\lambda_1 x_1, \lambda_2 x_2, \ldots)$. Show that Q_λ is self-adjoint.

Example 5.2.7. Now we wish to consider the operator P in Example 5.2.4. One can find a direct proof that P is self-adjoint on $\mathcal{D}(P)$ in Example 3, page 198, of [Yosida], but we would like to follow a different route. Essentially, we would like to turn Example 5.2.4 into Example 5.2.3 by applying the Fourier transform (see Section G.2). Before long we will see that this is standard operating procedure in quantum mechanics for switching back and forth between "position space" and "momentum space" so it will be nice to get a taste of it early on. A review of the Schwartz space, Fourier transform and tempered distributions on $\mathbb{R}$ is available in Appendix G.

Now we will consider again the position Q and momentum P operators on $\mathbb{R}$. Fix some $\phi \in \mathcal{S}(\mathbb{R})$. Then $\mathcal{F}\phi = \hat{\phi}$ is in $\mathcal{S}(\mathbb{R})$ and so it is in $\mathcal{D}(Q)$. Moreover, $(Q\hat{\phi})(p) = p\hat{\phi}(p) = -i\mathcal{F}(d\phi/dq)$. Applying $\mathcal{F}^{-1}$ we find that $[(\mathcal{F}^{-1}Q\mathcal{F})\phi](q) = -id\phi/dq$. We conclude that

$$\phi \in \mathcal{S}(\mathbb{R}) \quad \Rightarrow \quad (\mathcal{F}^{-1}(\hbar Q)\mathcal{F})\phi = P\phi,$$

so P agrees with $\mathcal{F}^{-1}(\hbar Q)\mathcal{F}$ on $\mathcal{S}(\mathbb{R})$. Now, $\mathcal{F}$ and $\mathcal{F}^{-1}$ are unitary operators on $L^2(\mathbb{R})$ and $\hbar Q$ is a self-adjoint operator on $\mathcal{D}(Q)$ so the following lemma implies that $\mathcal{F}^{-1}(\hbar Q)\mathcal{F}$ is self-adjoint on $\mathcal{F}^{-1}(\mathcal{D}(Q))$. Consequently, $\mathcal{F}^{-1}(\hbar Q)\mathcal{F}$ is a self-adjoint extension of $P\,|_{\mathcal{S}(\mathbb{R})}$.

Lemma 5.2.5. *Let $\mathcal{H}$ be a complex, separable Hilbert space, $A : \mathcal{D}(A) \rightarrow \mathcal{H}$ a self-adjoint operator and $U : \mathcal{H} \rightarrow \mathcal{H}$ a unitary operator on $\mathcal{H}$. Then $B = UAU^{-1}$ is a self-adjoint operator on $\mathcal{D}(B) = U(\mathcal{D}(A))$.*

Proof. Since $\mathcal{D}(A)$ is dense in $\mathcal{H}$ by assumption and U is unitary (and therefore a homeomorphism of $\mathcal{H}$ onto itself), $\mathcal{D}(B)$ is also dense. Since A is self-adjoint on $\mathcal{D}(A)$, $B = UAU^{-1} = UA^*U^{-1}$ on $\mathcal{D}(B)$. Now, let $\phi, \psi \in \mathcal{D}(B)$. Then

$$\langle \phi, B\psi \rangle = \langle \phi, UA^*U^{-1}\psi \rangle = \langle U^{-1}\phi, A^*U^{-1}\psi \rangle$$
$$= \langle AU^{-1}\phi, U^{-1}\psi \rangle = \langle UAU^{-1}\phi, \psi \rangle = \langle B\phi, \psi \rangle,$$

so B is symmetric. Consequently, $\mathcal{D}(B) \subseteq \mathcal{D}(B^*)$ and $B^*\,|_{\mathcal{D}(B)} = B$. Thus, it will suffice to show that $\mathcal{D}(B^*) \subseteq \mathcal{D}(B)$.

Note that $\psi \in \mathcal{D}(A) \Rightarrow U\psi \in \mathcal{D}(B) \Rightarrow BU\psi = UAU^{-1}(U\psi) = UA\psi$ so, on $\mathcal{D}(A)$, $BU = UA$. Now, let $\phi, \psi \in \mathcal{D}(A)$. Then

$$\langle \phi, U^{-1}B^*U\psi \rangle = \langle \phi, U^{-1}BU\psi \rangle = \langle U\phi, BU\psi \rangle = \langle U\phi, UA\psi \rangle = \langle \phi, A\psi \rangle = \langle A\phi, \psi \rangle$$

and therefore, A^* is an extension of $U^{-1}B^*U$. Consequently, UA^*U^{-1} is an extension of B^*. But, on $\mathcal{D}(B)$, $UA^*U^{-1} = UAU^{-1} = B$, so B is an extension of B^* and, in particular, $\mathcal{D}(B^*) \subseteq \mathcal{D}(B)$, as required. □

Exercise 5.2.9. Show that the same argument establishes the following seemingly more general result. Let $U : \mathcal{H}_1 \rightarrow \mathcal{H}_2$ be a unitary equivalence and $A : \mathcal{D}(A) \rightarrow \mathcal{H}_1$ a self-adjoint operator on the dense linear subspace $\mathcal{D}(A)$ of $\mathcal{H}_1$. Then $\mathcal{D}(B) = U(\mathcal{D}(A))$ is dense in $\mathcal{H}_2$ and $B = UAU^{-1}$ is self-adjoint on $\mathcal{D}(B)$. Two operators A and B related in this way by a unitary equivalence U are said to be *unitarily equivalent*. Thus, self-adjointness is preserved by unitary equivalence.

As mentioned above, we now have that $\mathcal{F}^{-1}(\hbar Q)\mathcal{F}$ is a self-adjoint extension of $P|_{S(\mathbb{R})}$. However, we claim that $P|_{S(\mathbb{R})}$ is essentially self-adjoint and therefore has a *unique* self-adjoint extension, namely, its closure. Having already noted the direct proof in [Yosida] that $P : \mathcal{D}(P) \to \mathcal{H}$ is self-adjoint, we find ourselves with two self-adjoint extensions of $P|_{S(\mathbb{R})}$ and conclude that these must be the same, so $\mathcal{F}^{-1}(\hbar Q)\mathcal{F} = P$.

To prove essential self-adjointness of $P|_{S(\mathbb{R})}$ we will apply Corollary 5.2.4.3, that is, we show that $P|_{S(\mathbb{R})}$ is symmetric and that the images of $P|_{S(\mathbb{R})} + i$ and $P|_{S(\mathbb{R})} - i$ are both dense in $L^2(\mathbb{R})$.

Exercise 5.2.10. Prove that $P|_{S(\mathbb{R})}$ is symmetric.

For the rest it is enough to show that the images of $P|_{S(\mathbb{R})}+i$ and $P|_{S(\mathbb{R})}-i$ contain $S(\mathbb{R})$ since this is dense. But this simply amounts to the statement that for any $\psi_0 \in S(\mathbb{R})$, the first order linear differential equations

$$-i\hbar \frac{d\psi}{dq} \pm i\psi = \psi_0$$

have solutions in $S(\mathbb{R})$.

Exercise 5.2.11. Prove this.

One often sees the momentum operator P *defined* by $P = \mathcal{F}^{-1}(\hbar Q)\mathcal{F}$, in which case its self-adjointness follows immediately from that of Q and Lemma 5.2.5.

We will work through another example that is quite analogous to what we have just done, but is worth doing carefully not only because it plays an important role in quantum mechanics, but also because it presents us with the opportunity to make a few observations that are important in the general scheme of things.

Example 5.2.8. Introduce another positive constant m and define an operator H_0 on $L^2(\mathbb{R})$ by specifying that, on $S(\mathbb{R})$, it is given by

$$H_0 = -\frac{\hbar^2}{2m}\frac{d^2}{dq^2} = -\frac{\hbar^2}{2m}\Delta,$$

where we use Δ for the one-dimensional Laplacian $\frac{d^2}{dq^2}$. We begin with a few general remarks on H_0. First note that two integrations by parts together with the fact that elements of $S(\mathbb{R})$ approach zero as $q \to \pm\infty$ show that H_0 is symmetric on $S(\mathbb{R})$. Specifically, since it clearly suffices to prove that Δ is symmetric on $S(\mathbb{R})$, we let $\psi, \phi \in$

$S(\mathbb{R})$ and compute

$$\langle \Delta \psi, \phi \rangle = \langle \psi'', \phi \rangle = \int_{\mathbb{R}} \overline{\psi}''(q)\phi(q)dq$$

$$= - \int_{\mathbb{R}} \overline{\psi}'(q)\phi'(q)dq = \int_{\mathbb{R}} \overline{\psi}(q)\phi''(q)dq = \langle \psi, \Delta\phi \rangle.$$

On the other hand, a single integration by parts shows that H_0 is a positive operator on $S(\mathbb{R})$, that is, $\langle H_0\psi, \psi \rangle \geq 0$ for any $\psi \in S(\mathbb{R})$. Indeed, it is enough to prove this for $-\Delta$ and

$$\langle -\Delta\psi, \psi \rangle = \langle -\psi'', \psi \rangle = \int_{\mathbb{R}} -\overline{\psi}''(q)\psi(q)dq = - \int_{\mathbb{R}} -\overline{\psi}'(q)\psi'(q)dq = \|\psi'\|^2 \geq 0.$$

We point out these two properties of H_0 because they provide an excuse to mention a theorem of Friedrichs according to which any densely defined, positive, symmetric operator on a Hilbert space has a positive, self-adjoint extension, called the *Friedrichs extension* (this is Theorem X.23 of [RS2]). This is not quite enough for our purposes since we need to know that H_0 has a *unique* self-adjoint extension and Friedrichs' theorem does not guarantee uniqueness. However, once we have proved that H_0 has a unique self-adjoint extension we will be assured that this extension is a positive operator and this is important for the role H_0 will play in quantum mechanics. We begin with the fact that the Fourier transform $\mathcal{F}$ is a unitary operator on $L^2(\mathbb{R})$ (see Section G.2) and compute, for any $\psi \in S(\mathbb{R})$,

$$(\mathcal{F}H_0\mathcal{F}^{-1})(\psi)(p) = (\mathcal{F}H_0)(\check{\psi})(p) = \mathcal{F}\left(-\frac{\hbar^2}{2m}\check{\psi}''\right)(p)$$

$$= \frac{\hbar^2}{2m}\mathcal{F}(-\check{\psi}'')(p) = \frac{\hbar^2}{2m}p^2\psi(p). \qquad (5.3)$$

We conclude that, on $S(\mathbb{R})$, $\mathcal{F}H_0\mathcal{F}^{-1}$ agrees with the multiplication operator Q_g, where $g(p) = \frac{\hbar^2}{2m}p^2$. We know, by Exercise 5.2.6, that this multiplication operator is self-adjoint on

$$\mathcal{D}(Q_g) = \{\psi \in L^2(\mathbb{R}) : g\psi \in L^2(\mathbb{R})\} = \left\{\psi \in L^2(\mathbb{R}) : \int_{\mathbb{R}} p^4|\psi(p)|^2 dp < \infty\right\}.$$

Consequently, $\mathcal{F}^{-1}Q_g\mathcal{F}$ is a self-adjoint extension of H_0 defined on

$$\mathcal{F}^{-1}(\mathcal{D}(Q_g)) = \{\psi \in L^2(\mathbb{R}) : \Delta\psi \in L^2(\mathbb{R})\},$$

where $\Delta\psi$ means the second derivative of ψ thought of as a tempered distribution in $L^2(\mathbb{R})$ (see Section G.2). Thus, H_0 does, indeed, have a self-adjoint extension and we will now prove that it has only one (which must therefore be the Friedrichs extension).

For this it will suffice to show that Δ is essentially self-adjoint on $\mathcal{S}(\mathbb{R})$ and we will do this by showing that Image $(\Delta|_{\mathcal{S}(\mathbb{R})} \pm i)$ are both dense in $L^2(\mathbb{R})$ (Corollary 5.2.4.3), that is, that the orthogonal complements Image $(\Delta|_{\mathcal{S}(\mathbb{R})} \pm i)^\perp$ are both zero. Let z denote either i or $-i$ and suppose $\phi \in$ Image $(\Delta|_{\mathcal{S}(\mathbb{R})} + z)^\perp$. Then, for every $\psi \in \mathcal{S}(\mathbb{R})$,

$$0 = \langle \phi, (\Delta + z)\psi \rangle = \langle \mathcal{F}(\phi), \mathcal{F}((\Delta + z)\psi) \rangle = \langle \check{\phi}, (-p^2 + z)\check{\psi} \rangle$$
$$= (-p^2 + z)\langle \check{\phi}, \check{\psi} \rangle = (-p^2 + z)\langle \phi, \psi \rangle.$$

But p^2 is real-valued so $-p^2 + z$ is never zero and we conclude that $\langle \phi, \psi \rangle = 0$ for every $\psi \in \mathcal{S}(\mathbb{R})$. Since $\mathcal{S}(\mathbb{R})$ is dense in $L^2(\mathbb{R})$, $\langle \phi, \psi \rangle = 0$ for every ψ in $L^2(\mathbb{R})$ and therefore $\phi = 0 \in L^2(\mathbb{R})$, as required.

We conclude that H_0 is essentially self-adjoint on $\mathcal{S}(\mathbb{R})$ and so has a unique self-adjoint extension to $\{\psi \in L^2(\mathbb{R}) : \Delta\psi \in L^2(\mathbb{R})\}$. This self-adjoint extension is also denoted H_0 and, for reasons that are yet to come, is called the *free particle Hamiltonian*.

5.3 $H_B = \frac{1}{2m}P^2 + \frac{m\omega^2}{2}Q^2$

The operators we have considered thus far all play a particularly prominent role in quantum mechanics, as we will soon see. We would like to include another such example that will eventually bring us back to the harmonic oscillator. In this case the idea we employ to prove self-adjointness is different, but it is an idea that will lead us naturally into our next topic (spectral theory). We begin with a few definitions. An *eigenvalue* of an operator $A : \mathcal{D}(A) \to \mathcal{H}$ is a complex number λ for which there exists a nonzero vector $\psi \in \mathcal{D}(A)$ which satisfies $A\psi = \lambda\psi$; such a nonzero vector ψ is called an *eigenvector* corresponding to the eigenvalue λ. Operators need not have eigenvalues at all, of course. Consider, for example, the operator $Q : \mathcal{D}(Q) \to L^2(\mathbb{R})$ of Example 5.2.3. For any fixed $\lambda \in \mathbb{C}$, the equation $Q\psi = \lambda\psi$ would imply that $(q - \lambda)\psi(q) = 0$ almost everywhere and therefore $\psi(q) = 0$ almost everywhere, so $\psi = 0$ in $L^2(\mathbb{R})$. For the operator $P : \mathcal{D}(P) \to L^2(\mathbb{R})$ of Example 5.2.5 the equation $-i\hbar\frac{d\psi}{dq} = \lambda\psi$ does have solutions, even nice smooth solutions such as $\psi(q) = e^{i\lambda q/\hbar}$, but, alas, they are not in $L^2(\mathbb{R})$ and therefore certainly not in $\mathcal{D}(P)$, so they do not count as eigenvectors. If it should happen that $A : \mathcal{D}(A) \to \mathcal{H}$ does have an eigenvalue λ, then the corresponding *eigenspace* is Kernel $(\lambda - A) = \{\psi \in \mathcal{D}(A) : A\psi = \lambda\psi\}$, which is clearly a linear subspace of $\mathcal{D}(A)$. If A is a symmetric (in particular, self-adjoint) operator, then any eigenvalue is necessarily real and eigenvectors corresponding to distinct eigenvalues are orthogonal in $\mathcal{H}$. The proofs are exactly as in the finite-dimensional case. To wit, if $A\psi = \lambda\psi$ with $\psi \neq 0$, then $\langle \psi, A\psi \rangle = \langle \psi, \lambda\psi \rangle = \lambda\langle \psi, \psi \rangle$, but also $\langle \psi, A\psi \rangle = \langle A\psi, \psi \rangle = \langle \lambda\psi, \psi \rangle = \bar{\lambda}\langle \psi, \psi \rangle$, so $\bar{\lambda} = \lambda$. Next, if $A\psi_1 = \lambda_1\psi_1$ and $A\psi_2 = \lambda_2\psi_2$ with $\lambda_2 \neq \lambda_1$ and neither ψ_1 nor ψ_2 is zero, then, since λ_1 and λ_2 are real, $\langle A\psi_1, \psi_2 \rangle = \langle \psi_1, A\psi_2 \rangle \Rightarrow \lambda_1\langle \psi_1, \psi_2 \rangle = \lambda_2\langle \psi_1, \psi_2 \rangle \Rightarrow (\lambda_2 - \lambda_1)\langle \psi_1, \psi_2 \rangle = 0 \Rightarrow \langle \psi_1, \psi_2 \rangle = 0$. Now, here is the point. A symmetric operator with a large enough supply of eigenvectors must be essentially self-adjoint. More precisely, we have the following theorem.

Theorem 5.3.1. *Let $\mathcal{H}$ be a separable, complex Hilbert space and $A : \mathcal{D}(A) \to \mathcal{H}$ a symmetric, unbounded operator on $\mathcal{H}$. Assume that there exists an orthonormal basis $\{e_1, e_2, \ldots\}$ for $\mathcal{H}$ consisting of eigenvectors for A, that is, such that each e_i is in $\mathcal{D}(A)$ and $Ae_i = \lambda_i e_i$ for some $\lambda_i \in \mathbb{R}$ and for each $i = 1, 2, \ldots$. Then A is essentially self-adjoint on $\mathcal{D}(A)$.*

Proof. Let $\mathbb{N} = \{1, 2, \ldots\}$ denote the set of natural numbers and consider the Hilbert space $\ell^2(\mathbb{N})$ of square summable sequences of complex numbers, that is,

$$\ell^2(\mathbb{N}) = \left\{ x = (x_1, x_2, \ldots) : x_i \in \mathbb{C}, i = 1, 2, \ldots, \text{ and } \sum_{i=1}^{\infty} |x_i|^2 < \infty \right\},$$

with coordinatewise algebraic operations and $\langle x, y \rangle = \sum_{i=1}^{\infty} \bar{x}_i y_i$. The orthonormal basis $\{e_1, e_2, \ldots\}$ for $\mathcal{H}$ consisting of eigenvectors for A determines a unitary equivalence $T : \ell^2(\mathbb{N}) \to \mathcal{H}$ given by

$$T(x) = T((x_1, x_2, \ldots)) = \sum_{i=1}^{\infty} x_i e_i,$$

where convergence of the series is in $\mathcal{H}$. The eigenvalues of A determine a (real-valued) function $\lambda : \mathbb{N} \to \mathbb{C}$ defined by

$$\lambda(i) = \lambda_i, \quad i = 1, 2, \ldots,$$

that is,

$$\lambda = (\lambda_1, \lambda_2, \ldots).$$

Thus, the multiplication operator Q_λ defined by $Q_\lambda(x) = Q_\lambda(x_1, x_2, \ldots) = (\lambda_1 x_1, \lambda_2 x_2, \ldots)$ on $\mathcal{D}(Q_\lambda) = \{x = (x_1, x_2, \ldots) \in \ell^2(\mathbb{N}) : \sum_{i=1}^{\infty} \lambda_i^2 |x_i|^2 < \infty\}$ is self-adjoint (see Exercise 5.2.8). According to Exercise 5.2.9, $TQ_\lambda T^{-1}$ is self-adjoint on $T(\mathcal{D}(Q_\lambda)) \subseteq \mathcal{H}$. To prove that A is essentially self-adjoint it will suffice to show that its closure $\bar{A}$ is $TQ_\lambda T^{-1}$ (see Exercise 5.2.5).

Let $(v, w) = (v, Av)$ be a point in the graph $\mathrm{Gr}(A)$ of A. Since A is symmetric (in particular, $\mathcal{D}(A) \subseteq \mathcal{D}(A^*)$) and each e_i is in $\mathcal{D}(A)$,

$$\langle e_i, w \rangle = \langle e_i, Av \rangle = \langle Ae_i, v \rangle = \lambda_i \langle e_i, v \rangle, \quad i = 1, 2, \ldots.$$

Since $\{e_i\}$ is an orthonormal basis for $\mathcal{H}$,

$$w = \sum_{i=1}^{\infty} \langle e_i, w \rangle e_i = \sum_{i=1}^{\infty} \lambda_i \langle e_i, v \rangle e_i,$$

and, in particular,

$$\sum_{i=1}^{\infty} \lambda_i^2 \, |\langle e_i, v \rangle|^2 = \|w\|^2 < \infty.$$

Thus, for any $v = \sum_{i=1}^{\infty} \langle e_i, v \rangle e_i$ in $\mathcal{D}(A)$, $T^{-1}v = (\langle e_1, v \rangle, \langle e_2, v \rangle, \ldots)$ is in $\mathcal{D}(Q_\lambda)$ and $TQ_\lambda T^{-1}v = w = Av$. In other words, $TQ_\lambda T^{-1}$ is a self-adjoint extension of A.

All that remains is to show that $\mathrm{Gr}(A)$ is dense in $\mathrm{Gr}(TQ_\lambda T^{-1})$. For this we let (V, W) be a point in $\mathrm{Gr}(TQ_\lambda T^{-1})$. Then

$$V = \sum_{i=1}^{\infty} \langle e_i, V \rangle e_i$$

and

$$W = \sum_{i=1}^{\infty} \lambda_i \, \langle e_i, V \rangle e_i.$$

Now, define, for each $n = 1, 2, \ldots$,

$$v_n = \sum_{i=1}^{n} \langle e_i, V \rangle e_i.$$

Then each v_n is in $\mathcal{D}(A)$ and, since the sum is finite and A is linear,

$$Av_n = \sum_{i=1}^{n} \langle e_i, V \rangle A e_i = \sum_{i=1}^{n} \lambda_i \langle e_i, V \rangle e_i.$$

Now, in the norm on $\mathcal{H} \times \mathcal{H}$,

$$\left\| (V, W) - (v_n, Av_n) \right\|^2 = \|V - v_n\|^2 + \|W - Av_n\|^2$$

and, as $n \to \infty$, both terms approach zero so $(v_n, Av_n) \to (V, W)$ in $\mathcal{H} \times \mathcal{H}$, as required. $\quad \square$

Now we will begin our application of Theorem 5.3.1 to an operator that will soon emerge as one of the central characters in our drama. Its significance may not yet be apparent, but it is real and for this reason we intend to carry out this discussion in excruciating detail.

Example 5.3.1. The operator of interest contains three positive constants m, ω and $\hbar$. Each of these has some physical significance, but we will worry about that later. We define an unbounded operator $H_B : \mathcal{D}(H_B) \to L^2(\mathbb{R})$ by first specifying that, on the Schwartz space $\mathcal{S}(\mathbb{R})$, it is given by

$$H_B = \frac{1}{2m} P^2 + \frac{m\omega^2}{2} Q^2 = -\frac{\hbar^2}{2m} \frac{d^2}{dq^2} + \frac{m\omega^2}{2} q^2. \tag{5.4}$$

The subscript B is there for a reason, but this reason will not be apparent for some time (if you must know, it stands for "bosonic"). Before moving on you may also want to glance back at Example 2.3.4. The objective now is to show that H_B is essentially self-adjoint on $\mathcal{S}(\mathbb{R})$ (then $\mathcal{D}(H_B)$ will be the domain of its unique self-adjoint extension). To apply Theorem 5.3.1 we must show that H_B is symmetric on $\mathcal{S}(\mathbb{R})$ and then exhibit an orthonormal basis for $L^2(\mathbb{R})$ consisting of eigenvectors for H_B living in $\mathcal{S}(\mathbb{R})$. For symmetry it is clear that we can consider each of the summands individually since $\langle H_B \psi, \phi \rangle = \langle -\frac{\hbar^2}{2m} \frac{d^2 \psi}{dq^2}, \phi \rangle + \langle \frac{m\omega^2}{2} q^2 \psi, \phi \rangle$. The first operator is just H_0, which we already know is symmetric, and the second is clearly symmetric since $\frac{m\omega^2}{2} q^2$ is just a real-valued multiplication operator (see Exercise 5.2.6).

Theorem 5.3.1 now requires that we find an orthonormal basis $\{\psi_0, \psi_1, \ldots\}$ for $L^2(\mathbb{R})$ with each ψ_i an eigenvector for H_B in $\mathcal{S}(\mathbb{R})$. A priori there is no reason to suppose that such an orthonormal basis exists, but it does and we now set about finding it. There are two ways to do this. We are looking for nonzero elements ψ of $\mathcal{S}(\mathbb{R})$ that satisfy

$$H_B \psi = -\frac{\hbar^2}{2m} \frac{d^2 \psi}{dq^2} + \frac{m\omega^2}{2} q^2 \psi = \mathcal{E} \psi \tag{5.5}$$

for some real constant $\mathcal{E}$ (I know, you were expecting a λ, but the eigenvalues will eventually turn out to be energy levels so $\mathcal{E}$ seemed a better choice). The obvious thing to do is simply try to solve the differential equation. As it happens, this is relatively straightforward, albeit rather tedious. It is, however, not particularly informative in that it sheds no real light on the general structure of this sort of eigenvalue problem. We will see that every application of quantum mechanics begins with a problem of this sort ($H\psi = \mathcal{E}\psi$) so we would like to do better than this. There is a much more enlightening "algebraic" approach, due to Dirac, that we will work out in considerable generality and detail. First, however, we will sketch the straightforward solution to (5.5) in enough detail that anyone interested in doing so should be able to fill in the gaps; if you would prefer to read it rather than do it, Appendix B, Chapter 5, of [Simm2] contains most of the details.

Rather than carry around all of the constants in (5.5) we begin by making the change of variable

$$x = \sqrt{\frac{m\omega}{\hbar}}\, q,$$

which converts (5.5) into

$$\frac{d^2 \psi}{dx^2} + \left(\frac{2\mathcal{E}}{\hbar\omega} - x^2 \right) \psi = 0. \tag{5.6}$$

To simplify just a bit more we will let $p + 1 = \frac{2\mathcal{E}}{\hbar\omega}$ and write (5.6) as

$$\frac{d^2 \psi}{dx^2} + (p + 1 - x^2) \psi = 0. \tag{5.7}$$

Motivated by the fact that we need solutions in the Schwartz space $\mathcal{S}(\mathbb{R})$ we will seek $\psi(x)$ of the form

$$\psi(x) = \phi(x)e^{-x^2/2}, \tag{5.8}$$

where $\phi(x)$ is to be determined. Substituting this into (5.7) gives

$$\frac{d^2\phi}{dx^2} - 2x\frac{d\phi}{dx} + p\phi = 0, \tag{5.9}$$

which is the famous *Hermite equation*. One notices that $x = 0$ is an ordinary point for the equation and so we seek analytic solutions of the form $\phi(x) = \sum_{n=0}^{\infty} a_n x^n$. Substituting this into (5.9) and equating the coefficients of the resulting power series to zero gives the recurrence relation

$$a_{n+2} = \frac{2n - p}{(n + 2)(n + 1)} a_n, \quad n \geq 0.$$

One solves for the coefficients in the usual way and arrives at the following formal solution:

$$\phi(x) = a_0\left[1 - \frac{p}{2!}x^2 - \frac{(4 - p)p}{4!}x^4 - \frac{(8 - p)(4 - p)p}{6!}x^6 - \cdots\right]$$

$$+ a_1\left[x + \frac{2 - p}{3!}x^3 + \frac{(6 - p)(2 - p)}{5!}x^5 + \frac{(10 - p)(6 - p)(2 - p)}{7!}x^7 + \cdots\right]$$

$$= a_0\,\phi_0(x) + a_1\,\phi_1(x).$$

These series, in fact, are easily seen to converge for all x (ratio test) and, since the Wronskian of ϕ_0 and ϕ_1 is clearly nonzero (evaluate it at $x = 0$), we have found the general solution to the Hermite equation on $(-\infty, \infty)$. By comparing the series expansions of $\phi_0(x)$ and $\phi_1(x)$ with $e^{x^2/2} = \sum_{n=0}^{\infty} \frac{1}{2^n n!}x^{2n}$ one finds that $\phi_0(x)e^{-x^2/2}$ and $\phi_1(x)e^{-x^2/2}$ do *not* approach zero as $|x| \to \pm\infty$ and so cannot be in $\mathcal{S}(\mathbb{R})$ *unless* the series for $\phi_0(x)$ and $\phi_1(x)$ terminate and are therefore polynomials. Since we seek two independent solutions in $\mathcal{S}(\mathbb{R})$ we must take either $a_1 = 0$ and $p = 0, 4, 8, \ldots$, or $a_0 = 0$ and $p = 2, 6, 10 \ldots$. In each case, one chooses the nonzero coefficient a_i in such a way so as to ensure that the resulting polynomials $H_0(x), H_1(x), H_2(x), \ldots, H_n(x), \ldots$ satisfy

$$\int_{\mathbb{R}} [H_m(x)e^{-x^2/2}][H_n(x)e^{-x^2/2}]\,dx = \int_{\mathbb{R}} H_m(x)H_n(x)e^{-x^2}\,dx = \sqrt{\pi}\,2^n\,n!\,\delta_{mn}$$

(here the subscript indicates the degree of the polynomial). The reason for this rather strange normalization will emerge soon. These $H_n(x)$, $n = 0, 1, 2, \ldots$, are called the *Hermite polynomials*. These have all sorts of wonderful properties of which we will need just a few (these are all proved in, for example, [AAR]). One can show that *every* polynomial is a linear combination of Hermite polynomials, that each H_n is given by

$$H_n(x) = (-1)^n e^{x^2} \frac{d^n}{dx^n}(e^{-x^2})$$

and that $e^{2xz-z^2/2}$ is a generating function for the sequence $H_0(x), H_1(x), H_2(x), \ldots$, that is,

$$e^{2xz-z^2/2} = \sum_{n=0}^{\infty} H_n(x)\frac{z^n}{n!}. \tag{5.10}$$

More to the point for us is that we can now build solutions to (5.5) from $H_n(x)e^{-x^2/2}$ that are in $S(\mathbb{R})$. We revert to the original variable q, related to x by $x = \sqrt{\frac{m\omega}{\hbar}}\,q$, and again renormalize to obtain the solutions $\psi_n(q)$ to (5.5) we were after. Specifically, we have one solution for each $p_n = 2n$, $n = 0, 1, 2, \ldots$, that is, for each eigenvalue

$$\mathcal{E}_n = \left(n + \frac{1}{2}\right)\hbar\omega, \quad n = 0, 1, 2, \ldots$$

(recall that $p + 1 = \frac{2\mathcal{E}}{\hbar\omega}$ and have another look at (1.7)). These solutions are given by

$$\psi_n(q) = \frac{1}{\sqrt{2^n n!}}\left(\frac{m\omega}{\hbar\pi}\right)^{1/4}e^{-m\omega q^2/2\hbar}H_n\left(\sqrt{\frac{m\omega}{\hbar}}\,q\right) \tag{5.11}$$

and they satisfy

$$H_B\psi_n = \mathcal{E}_n\psi_n = \left(n + \frac{1}{2}\right)\hbar\omega\,\psi_n, \quad n = 0, 1, 2, \ldots. \tag{5.12}$$

Being eigenvectors of a symmetric operator corresponding to distinct eigenvalues, ψ_m and ψ_n are orthogonal in $L^2(\mathbb{R})$ whenever $n \neq m$. The odd looking normalizations are intended to ensure that each ψ_n is a unit vector in $L^2(\mathbb{R})$ so that

$$\langle \psi_m, \psi_n \rangle = \int_{\mathbb{R}} \overline{\psi_m(q)}\psi_n(q)\,dq = \delta_{mn}.$$

One must still show that the orthonormal set $\{\psi_n\}_{n=0}^{\infty}$ in $L^2(\mathbb{R})$ is a basis and for this it is enough to show that the orthogonal complement of its closed span in $L^2(\mathbb{R})$ consists of the zero element alone. The argument is not so difficult, but we will save it until we have given a complete and very different derivation of the eigenfunctions ψ_n.

This concludes our sketch of the traditional solution to the eigenvalue problem (5.5) in $S(\mathbb{R})$. It produces an orthonormal basis for $L^2(\mathbb{R})$ consisting of eigenvectors for H_B in $S(\mathbb{R})$ and so allows one to conclude from Theorem 5.3.1 that H_B is essentially self-adjoint on $S(\mathbb{R})$. It does not, however, uncover the underlying algebraic structure that lies hidden here and it is precisely this algebraic structure that we need to understand since it plays a fundamental role in quantum mechanics. Before getting under way we must sound a cautionary note, not only for this example, but for the remainder of our work. The algebraic structure we have in mind is determined by *commutation relations* for the operators of interest and these, as the name suggests, involve the *commutator* and, later on, the *anticommutator* of a pair of operators. These are defined by $[A, B]_- = AB - BA$ and $[A, B]_+ = AB + BA$, respectively; here AB and BA stand for the *compositions* $A \circ B$ and $B \circ A$, respectively. For bounded operators, which are defined on all of $\mathcal{H}$, these definitions present no problem, but for unbounded operators defined only on dense subspaces of $\mathcal{H}$ the sum or difference of two operators is defined only on the intersection of their domains and the product (composition) AB is defined only on the subset of $\mathcal{D}(B)$ that B maps into $\mathcal{D}(A)$. All of these might very well consist only of the zero element in $\mathcal{H}$. These and many other such *domain issues* cannot be regarded as simply a minor technical inconvenience, but are often crucial to the mathematics and even to the physics (see Section X.1 of [RS2]). The moral is that we must be careful about domains.

> For the remainder of this example, all of our calculations will be carried out in the Schwartz space $\mathcal{S}(\mathbb{R})$.

Now we begin our new derivation of the eigenvalues and eigenvectors for the operator

$$H_B = \frac{1}{2m}P^2 + \frac{m\omega^2}{2}Q^2 = -\frac{\hbar^2}{2m}\frac{d^2}{dq^2} + \frac{m\omega^2}{2}q^2$$

of (5.4). First note that $Q : \mathcal{S}(\mathbb{R}) \to \mathcal{S}(\mathbb{R})$ and $P : \mathcal{S}(\mathbb{R}) \to \mathcal{S}(\mathbb{R})$, so their commutators are well-defined on $\mathcal{S}(\mathbb{R})$ and, moreover,

$$[Q, Q]_- = 0,$$
$$[P, P]_- = 0, \tag{5.13}$$
$$[P, Q]_- = -i\hbar.$$

The first two are obvious and the third follows by computing, for any $\psi \in \mathcal{S}(\mathbb{R})$,

$$([P, Q]_-\psi)(q) = (P(Q\psi) - Q(P\psi))(q) = -i\hbar\frac{d}{dq}(q\psi(q)) - q\left(-i\hbar\frac{d\psi}{dq}\right)$$
$$= -i\hbar q\frac{d\psi}{dq} - i\hbar\psi(q) + i\hbar q\frac{d\psi}{dq}$$
$$= -i\hbar\psi(q),$$

so

$$[P, Q]_-\psi = -i\hbar\psi.$$

It might be worthwhile at this point to compare (5.13) and (2.66).

For the remainder of this discussion it will be very important to notice that, with one exception that we will point out explicitly, the analysis depends *only* on the fact

that the operators Q and P are symmetric on $S(\mathbb{R})$ and satisfy the commutation relations (5.13) there and *not* on their definitions as multiplication by q and $-i\hbar\frac{d}{dq}$, respectively. In Chapter 7 we will describe the abstract algebraic setting in which all of this
lives most naturally.

Now we define two new operators b and $b^\dagger$ on $S(\mathbb{R})$ by

$$b = \frac{1}{\sqrt{2m\omega\hbar}}(m\omega Q + iP) \tag{5.14}$$

and

$$b^\dagger = \frac{1}{\sqrt{2m\omega\hbar}}(m\omega Q - iP). \tag{5.15}$$

Many sources use a and $a^\dagger$, or a^- and a^+, respectively, rather than our b and $b^\dagger$, but
our choice is intended to distinguish this "bosonic" example from the "fermionic" and
"supersymmetric" generalizations that are yet to come. The dagger $\dagger$ is used to indicate
that b and $b^\dagger$ are *formal adjoints* of each other in the sense that, on $S(\mathbb{R})$, they satisfy

$$\langle b^\dagger\psi, \phi\rangle = \frac{1}{\sqrt{2m\omega\hbar}}\langle m\omega Q\psi - iP\psi, \phi\rangle$$

$$= \frac{1}{\sqrt{2m\omega\hbar}}[\langle m\omega Q\psi, \phi\rangle - \langle iP\psi, \phi\rangle]$$

$$= \frac{1}{\sqrt{2m\omega\hbar}}[m\omega\langle\psi, Q\phi\rangle + i\langle\psi, P\phi\rangle]$$

$$= \langle\psi, b\phi\rangle$$

and $\langle b\phi, \psi\rangle = \langle\phi, b^\dagger\psi\rangle$, which follows by taking conjugates. Now, let us compute $bb^\dagger$
on $S(\mathbb{R})$. We have

$$bb^\dagger = \frac{1}{2m\omega\hbar}[m^2\omega^2 Q^2 + im\omega[PQ - QP] + P^2]$$

$$= \frac{m\omega}{2\hbar}Q^2 + \frac{1}{2} + \frac{1}{2m\omega\hbar}P^2$$

since $[P, Q]_- = -i\hbar$. Similarly,

$$b^\dagger b = \frac{m\omega}{2\hbar}Q^2 - \frac{1}{2} + \frac{1}{2m\omega\hbar}P^2$$

so that $[b, b^\dagger]_- = bb^\dagger - b^\dagger b = 1$. Adding to this a few more relations that are obvious
we have

$$[b, b]_- = 0,$$
$$[b^\dagger, b^\dagger]_- = 0, \tag{5.16}$$
$$[b, b^\dagger]_- = 1.$$

Now, note that if we define an operator N_B on $S(\mathbb{R})$ by

$$N_B = b^\dagger b,$$

then

$$H_B = \frac{1}{2}\hbar\omega[b^\dagger, b]_+ = \hbar\omega\left(N_B + \frac{1}{2}\right).$$

For reasons that will emerge quite soon N_B is called the *(bosonic) number operator*. Furthermore, $[N_B, b^\dagger]_- = N_B b^\dagger - b^\dagger N_B = b^\dagger b b^\dagger - b^\dagger b^\dagger b = b^\dagger [b, b^\dagger]_- = b^\dagger$, so

$$[N_B, b^\dagger]_- = b^\dagger$$

and, similarly,

$$[N_B, b]_- = -b.$$

What, you may ask, is the point of all this arithmetic? Recall that the objective here is to derive (again) all of the eigenvalues and eigenvectors for H_B on $S(\mathbb{R})$. From $H_B = \hbar\omega(N_B + \frac{1}{2})$ it is clear that, on $S(\mathbb{R})$,

$$N_B\psi = \lambda\psi \quad \Leftrightarrow \quad H_B\psi = \left(\lambda + \frac{1}{2}\right)\hbar\omega\psi,$$

so this is equivalent to finding the eigenvalues and eigenvectors of N_B on $S(\mathbb{R})$ and this we can now do with ease. First note that any nonzero $\psi \in S(\mathbb{R})$ satisfying $b\psi = 0$ also satisfies $N_B\psi = 0$ and so is an eigenvector of N_B with eigenvalue $\lambda = 0$ and consequently an eigenvector of H_B with eigenvalue $\frac{1}{2}\hbar\omega$. But how do we know that there is such a ψ in $S(\mathbb{R})$? This is the only point at which we require some information about the operators Q and P beyond their commutation relations which, by themselves, cannot answer the question. However, if we write $b\psi = 0$ as

$$\frac{1}{\sqrt{2m\omega\hbar}}(m\omega Q\psi + iP\psi) = 0$$

and then recall that $(Q\psi)(q) = q\psi(q)$ and $(P\psi)(q) = -i\hbar\frac{d\psi}{dq}$, we obtain

$$\frac{d\psi}{dq} + \left(\frac{m\omega}{\hbar}\right)q\psi(q) = 0.$$

This is a simple first order, linear equation whose general solution is $\psi(q) = Ce^{-m\omega q^2/2\hbar}$, where C is an arbitrary constant. Computing the $L^2(\mathbb{R})$ norm of $\psi(q) = Ce^{-m\omega q^2/2\hbar}$ we can choose C in such a way so as to ensure that the solution has norm 1. Specifically, one obtains

$$\psi_0(q) = \left(\frac{m\omega}{\hbar\pi}\right)^{1/4} e^{-m\omega q^2/2\hbar}, \tag{5.17}$$

which is clearly in $S(\mathbb{R})$ and is the unique L^2-normalized eigenvector of H_B with eigenvalue $\frac{1}{2}\hbar\omega$:

$$H_B\psi_0 = \frac{1}{2}\hbar\omega\psi_0.$$

We will see now that all the rest follows from the commutation relations alone.

First note that $\frac{1}{2}\hbar\omega$ is the *smallest eigenvalue of H_B on $S(\mathbb{R})$*. Indeed, suppose $H_B\psi = \lambda\psi$ with $\psi \neq 0$. Then

$$\lambda\|\psi\|^2 = \lambda\langle\psi,\psi\rangle = \langle\psi,\lambda\psi\rangle = \langle\psi,H_B\psi\rangle = \left\langle\psi,\hbar\omega N_B\psi + \frac{1}{2}\hbar\omega\psi\right\rangle$$

$$= \hbar\omega\langle\psi,N_B\psi\rangle + \frac{1}{2}\hbar\omega\langle\psi,\psi\rangle = \hbar\omega\langle\psi,b^\dagger b\psi\rangle + \frac{1}{2}\hbar\omega\|\psi\|^2$$

$$= \hbar\omega\langle b\psi,b\psi\rangle + \frac{1}{2}\hbar\omega\|\psi\|^2$$

$$= \hbar\omega\|b\psi\|^2 + \frac{1}{2}\hbar\omega\|\psi\|^2.$$

Consequently,

$$\lambda = \frac{1}{2}\hbar\omega + \hbar\omega(\|b\psi\|^2/\|\psi\|^2) \geq \frac{1}{2}\hbar\omega,$$

with equality holding if and only if $b\psi = 0$.

Next, let us consider the vector

$$\psi_1 = b^\dagger\psi_0.$$

Then

$$N_B\psi_1 = N_B(b^\dagger\psi_0) = (N_B b^\dagger\psi_0 - b^\dagger N_B\psi_0) + b^\dagger N_B\psi_0$$

$$= ([N_B,b^\dagger]_- + b^\dagger N_B)\psi_0 = (b^\dagger + b^\dagger N_B)\psi_0 = b^\dagger\psi_0 + 0$$

$$= \psi_1,$$

so ψ_1 is an eigenvector of N_B in $S(\mathbb{R})$ with eigenvalue $\lambda = 1$. As a result, ψ_1 is an eigenvector of H_B in $S(\mathbb{R})$ with eigenvalue $(1 + \frac{1}{2})\hbar\omega$:

$$H_B\psi_1 = \frac{3}{2}\hbar\omega\psi_1.$$

The operator $b^\dagger$ carries a $\lambda = 0$ eigenvector of N_B to a $\lambda = 1$ eigenvector of N_B. Also note that ψ_1 has norm 1 since

$$\|\psi_1\|^2 = \langle\psi_1,\psi_1\rangle = \langle b^\dagger\psi_0,b^\dagger\psi_0\rangle = \langle bb^\dagger\psi_0,\psi_0\rangle$$

$$= \langle[b,b^\dagger]_-\psi_0 + b^\dagger b\psi_0,\psi_0\rangle = \langle\psi_0 + 0,\psi_0\rangle = \|\psi_0\|^2$$

$$= 1.$$

Let us try this again. Let

$$\phi_2 = b^\dagger \psi_1 = (b^\dagger)^2 \psi_0.$$

Then

$$N_B \phi_2 = N_B(b^\dagger \psi_1) = ([N_B, b^\dagger]_- + b^\dagger N_B)\psi_1 = (b^\dagger + b^\dagger N_B)\psi_1$$
$$= b^\dagger \psi_1 + b^\dagger N_B \psi_1 = b^\dagger \psi_1 + b^\dagger \psi_1$$
$$= 2\phi_2,$$

so ϕ_2 is an eigenvector of N_B with eigenvalue $\lambda = 2$ and therefore it is an eigenvector of H_B with eigenvalue $(2+\frac{1}{2})\hbar\omega$. Unfortunately, we lose the normalization this time since

$$\|\phi_2\|^2 = \langle \phi_2, \phi_2 \rangle = \langle b^\dagger \psi_1, b^\dagger \psi_1 \rangle = \langle bb^\dagger \psi_1, \psi_1 \rangle = \langle [b, b^\dagger]_- \psi_1 + b^\dagger b \psi_1, \psi_1 \rangle$$
$$= \langle \psi_1 + \psi_1, \psi_1 \rangle = 2\|\psi_1\|^2 = 2.$$

Consequently,

$$\psi_2 = \frac{1}{\sqrt{2}} b^\dagger \psi_1 = \frac{1}{\sqrt{2}}(b^\dagger)^2 \psi_0$$

is a normalized eigenvector of H_B with eigenvalue $(2 + \frac{1}{2})\hbar\omega$:

$$H_B \psi_2 = \frac{5}{2}\hbar\omega\psi_2.$$

Exercise 5.3.1. Show in the same way that if $N_B\phi = \lambda\phi$, then $N_B(b^\dagger\phi) = (\lambda + 1)b^\dagger\phi$ for any λ. Thus, $b^\dagger$ carries eigenvectors of N_B to eigenvectors of N_B, increasing the eigenvalue by one. Equivalently, $b^\dagger$ carries eigenvectors of H_B to eigenvectors of H_B, increasing the eigenvalue by $\hbar\omega$. Replace $b^\dagger$ by b and show that if $N_B\phi = \lambda\phi$, then $N_B(b\phi) = (\lambda - 1)b\phi$.

For this reason, $b^\dagger$ and b are called *raising* and *lowering operators*, respectively; together they are called *ladder operators*. We will find that, in quantum mechanics, they are viewed as raising and lowering the energy level of a harmonic oscillator. Analogous operators exist in quantum field theory, where they are called *creation* and *annihilation operators* because there they are viewed as creating and annihilating particles (more precisely, *quanta*) of a specific energy. The eigenvalues of the number operator N_B count the number of such quanta; hence the name.

Continuing inductively we find that for each $n = 0, 1, 2, \ldots$,

$$\psi_n = \frac{1}{\sqrt{n!}}(b^\dagger)^n \psi_0 \tag{5.18}$$

is a normalized eigenvector of N_B with eigenvalue $\lambda = n$ and so is also a normalized eigenvector of H_B with eigenvalues $(n + \frac{1}{2})\hbar\omega$:

$$H_B\psi_n = \left(n + \frac{1}{2}\right)\hbar\omega\psi_n, \quad n = 0, 1, 2, \ldots. \tag{5.19}$$

Note that $(n + \frac{1}{2})\hbar\omega$, $n = 0, 1, 2, \ldots$, are the *only* eigenvalues of H_B on $\mathcal{S}(\mathbb{R})$. Indeed, we have already seen that $\frac{1}{2}\hbar\omega$ is the smallest eigenvalue so, in particular, all eigenvalues are positive. But if λ were some positive eigenvalue that was not of the form $(n + \frac{1}{2})\hbar\omega$, $n = 0, 1, 2, \ldots$, then repeated application of b would eventually produce a negative eigenvalue (see Exercise 5.3.1) so such a λ cannot exist.

The rather remarkable conclusion is that if we happen to know just one eigenvector for H_B in $\mathcal{S}(\mathbb{R})$, then we can produce all of the rest simply by successively applying the ladder operators. For instance, if we identify Q and P with multiplication by q and $-i\hbar\frac{d}{dq}$, then ψ_0 is given by (5.17) and we can proceed to grind out all of the remaining eigenvectors by repeatedly applying $\frac{1}{\sqrt{2m\omega\hbar}}(m\omega q - \hbar\frac{d}{dq})$. The result, of course, will be (5.11), which can be proved using induction and the properties of the Hermite polynomials.

It is useful to observe that, since $\psi_n = \frac{1}{\sqrt{n!}}(b^\dagger)^n\psi_0$,

$$b^\dagger\psi_n = \sqrt{n+1}\,\psi_{n+1}, \quad n = 0, 1, 2, \ldots,$$

and similarly,

$$b\psi_n = \sqrt{n}\,\psi_{n-1}, \quad n = 1, 2, \ldots.$$

Exercise 5.3.2. Prove these.

With this we can show that *all of the eigenspaces of H_B are one-dimensional*, that is, all of the eigenvalues are *simple*. Suppose this is not the case. Then there is a *least* nonnegative integer n with two independent normalized eigenvectors ψ_n and ψ_n'. Then $n \geq 1$, since we have seen that ψ_0 is the *unique* normalized eigenvector of H_B with eigenvalue $\frac{1}{2}\hbar\omega$. Now,

$$n\psi_n' = N_B\psi_n' = b^\dagger b\psi_n',$$

so

$$n(b\psi_n') = bb^\dagger(b\psi_n') = (1 + N_B)(b\psi_n')$$

and $b\psi_n'$ is an eigenvector of N_B with eigenvalue $n - 1$. Since n is minimal, $b\psi_n'$ must be a nonzero multiple of ψ_{n-1}, say, $b\psi_n' = k\psi_{n-1}$. Then

$$n\psi_n' = b^\dagger(b\psi_n') = kb^\dagger\psi_{n-1} = k\sqrt{n}\,\psi_n,$$

so

$$\psi_n' = \frac{k}{\sqrt{n}} \psi_n$$

and this is a contradiction so the proof is complete. These one-dimensional eigenspaces are, of course, orthogonal.

There is just one loose end that remains to be tied up. We must show that our orthonormal sequence of eigenfunctions for H_B is actually complete, that is, forms a basis for $L^2(\mathbb{R})$. To ease the typography a bit we will temporarily revert to our original variable $x = \sqrt{\frac{m\omega}{\hbar}} q$ and consider the so-called *Hermite functions*

$$\phi_n(x) = \frac{1}{\sqrt{2^n n! \sqrt{\pi}}} e^{-x^2/2} H_n(x), \ n = 0, 1, 2, \dots.$$ Note that these are all in $S(\mathbb{R})$. We will show that they form a basis for $L^2(\mathbb{R})$. To prove this it is clearly sufficient to show that the orthogonal complement of the closed linear span of $\{\phi_n\}_{n=0}^{\infty}$ consists of the zero vector in $L^2(\mathbb{R})$ alone. This, in turn, will follow if we show that any $f \in L^2(\mathbb{R})$ satisfying $\langle \phi_n, f \rangle = 0 \ \forall n = 0, 1, 2, \dots$ must be the zero element of $L^2(\mathbb{R})$. Suppose then that $\langle \phi_n, f \rangle = 0 \ \forall n = 0, 1, 2, \dots.$ Since every polynomial is a (finite) linear combination of Hermite polynomials, it follows that f is orthogonal to every function of the form $P(x)e^{-x^2/2}$, where $P(x)$ is a polynomial. Consequently,

$$\int_{\mathbb{R}} [f(x)e^{-x^2/2}] e^{-i\xi x} dx = \sum_{k=0}^{\infty} \int_{\mathbb{R}} f(x) \left[\frac{(-i\xi x)^k}{k!} e^{-x^2/2} \right] dx = 0,$$

where the interchange of summation and integration is justified because the product of two L^2 functions is an L^1 function (Theorem 3.2.1 of [Fried]). But this shows that the Fourier transform of $f(x)e^{-x^2/2}$ is zero. Since the Fourier transform $\mathcal{F} : L^2(\mathbb{R}) \to L^2(\mathbb{R})$ is an isometry, $f(x)e^{-x^2/2} = 0$ almost everywhere and it follows that $f(x) = 0$ almost everywhere, that is, $f = 0$ in $L^2(\mathbb{R})$, as required.

Exercise 5.3.3. Show that the Hermite function $\phi_n(x) = \frac{1}{\sqrt{2^n n! \sqrt{\pi}}} e^{-x^2/2} H_n(x)$ is an eigenvector of the Fourier transform operator $\mathcal{F} : L^2(\mathbb{R}) \to L^2(\mathbb{R})$ with eigenvalue $(-i)^n$, that is,

$$\mathcal{F}\left(e^{-x^2/2} H_n(x)\right) = (-i)^n e^{-\xi^2/2} H_n(\xi).$$

Hint: Begin with the generating function (5.10) for the Hermite polynomials, multiply through by $e^{-x^2/2}$, compute the Fourier transform of both sides, use the generating function again and equate the coefficients of z^n in the two series.

Finally, it is instructive to use the orthonormal basis $\{\psi_0, \psi_1, \dots\}$ for $L^2(\mathbb{R})$ to write out matrix representations for the operators b, $b^\dagger$, N_B and H_B. Somewhat more precisely, we use the basis to establish an isometric isomorphism $\psi = \sum_{n=0}^{\infty} c_n \psi_n \in L^2(\mathbb{R}) \to (c_n)_{n=0}^{\infty} \in \ell^2(\mathbb{N})$ of $L^2(\mathbb{R})$ onto $\ell^2(\mathbb{N})$ and regard them as operators on $\ell^2(\mathbb{N})$.

One need only read off the coefficients in the expressions we have derived above to obtain

$$b^\dagger = \begin{pmatrix} 0 & 0 & 0 & 0 & \cdots \\ \sqrt{1} & 0 & 0 & 0 & \cdots \\ 0 & \sqrt{2} & 0 & 0 & \cdots \\ 0 & 0 & \sqrt{3} & 0 & \cdots \\ \vdots & \vdots & \vdots & \vdots & \end{pmatrix},$$

$$b = \begin{pmatrix} 0 & \sqrt{1} & 0 & 0 & \cdots \\ 0 & 0 & \sqrt{2} & 0 & \cdots \\ 0 & 0 & 0 & \sqrt{3} & \cdots \\ \vdots & \vdots & \vdots & \vdots & \end{pmatrix},$$

$$N_B = \begin{pmatrix} 0 & 0 & 0 & 0 & \cdots \\ 0 & 1 & 0 & 0 & \cdots \\ 0 & 0 & 2 & 0 & \cdots \\ 0 & 0 & 0 & 3 & \cdots \\ \vdots & \vdots & \vdots & \vdots & \end{pmatrix},$$

$$H_B = \begin{pmatrix} \frac{1}{2}\hbar\omega & 0 & 0 & 0 & \cdots \\ 0 & \frac{3}{2}\hbar\omega & 0 & 0 & \cdots \\ 0 & 0 & \frac{5}{2}\hbar\omega & 0 & \cdots \\ 0 & 0 & 0 & \frac{7}{2}\hbar\omega & \cdots \\ \vdots & \vdots & \vdots & \vdots & \end{pmatrix}.$$

This brings to a close our rather extended introduction of the operator $H_B = \frac{1}{2m}P^2 + \frac{m\omega^2}{2}Q^2$, but do not despair; we will have a great deal more to say about H_B in Examples 5.4.5 and 5.5.4, and then again in Chapters 7 and 8, where its unique self-adjoint extension, also denoted $H_B : \mathcal{D}(H_B) \to L^2(\mathbb{R})$, will be known as the *Hamiltonian* for the *(bosonic) quantum harmonic oscillator*. For a description of $\mathcal{D}(H_B)$, see Example 5.4.5.

5.4 The spectrum

Although our interest at the moment is in operators on a Hilbert space we should point out that all of the basic definitions in this section apply equally well to operators on a Banach space, or even a linear topological space (see Section VIII.1 of [Yosida]). Now note that if λ is an eigenvalue of the operator $A : \mathcal{D}(A) \to \mathcal{H}$, then $\lambda - A$ has a nontrivial kernel and so $\lambda - A$ is not injective and therefore not invertible. As it happens, $\lambda - A$ can fail to have a *bounded* inverse for a variety of reasons even if λ is not an eigenvalue of A and this is what we need to discuss now. For this we need a few definitions. We will say that a $\lambda \in \mathbb{C}$ is in the *resolvent set* $\rho(A)$ of A if $\lambda - A$ is injective on $\mathcal{D}(A)$, its range

Image $(\lambda - A)$ is dense in $\mathcal{H}$ and $R_\lambda(A) = (\lambda - A)^{-1}$ is continuous on Image $(\lambda - A)$. The operator $R_\lambda(A) = (\lambda - A)^{-1}$ is called the *resolvent* of A at λ. If A happens to be closed (in particular, self-adjoint), then, for any $\lambda \in \rho(A)$, $R_\lambda(A) = (\lambda - A)^{-1}$ is actually defined on all of $\mathcal{H}$ (Theorem, Section VIII.1, of [Yosida]) and so is necessarily bounded by the closed graph theorem (Theorem 4.6.4 of [Fried]). Conversely, if $(\lambda - A)^{-1}$ is bounded and defined on all of $\mathcal{H}$, then $(\lambda - A)^{-1}$ is closed and it follows that $\lambda - A$ is closed (Proposition 3, Section II.6, of [Yosida]) and therefore A is closed.

The complement $\sigma(A) = \mathbb{C} - \rho(A)$ of the resolvent set of A is called the *spectrum* of A. The spectrum certainly contains any eigenvalues that A might have, but, in general, it contains more. Indeed, we will decompose $\sigma(A)$ into three disjoint sets

$$\sigma(A) = P_\sigma(A) \sqcup C_\sigma(A) \sqcup R_\sigma(A)$$

defined as follows. First, $P_\sigma(A)$ is called the *point spectrum* of A and consists of all $\lambda \in \mathbb{C}$ for which $\lambda - A$ is not injective on $\mathcal{D}(A)$ and therefore has no inverse at all; thus, $P_\sigma(A)$ consists precisely of the eigenvalues of A. Second, $C_\sigma(A)$ is called the *continuous spectrum* of A and consists of all $\lambda \in \mathbb{C}$ for which $(\lambda-A)^{-1}$ exists and has a dense domain Image $(\lambda-A)$, but is not continuous on Image $(\lambda-A)$. Finally, $R_\sigma(A)$ is called the *residual spectrum* of A and consists of all $\lambda \in \mathbb{C}$ for which $(\lambda - A)^{-1}$ exists, but whose domain Image $(\lambda - A)$ is not dense in $\mathcal{H}$. For self-adjoint operators, $R_\sigma(A) = \emptyset$ (Theorem 1, part (iv), Section XI.8, of [Yosida]). According to Theorem 1, Section VIII.2, of [Yosida], the resolvent set $\rho(A)$ of a *closed* operator A is an *open* subset of the complex plane $\mathbb{C}$ and, consequently, the spectrum $\sigma(A)$ is closed. For self-adjoint operators one can say even more. In this case, the spectrum is a (closed) subset of the real line $\mathbb{R}$; this is Example 4, Chapter VIII, Section 1 of [Yosida], but this fact is of such fundamental importance to quantum mechanics that we will include a proof.

Remark 5.4.1. In quantum mechanics, a self-adjoint operator A represents an observable and the possible measured values of the observable are the points in the operator's spectrum. Since the result of a measurement is always a real number it is essential that $\sigma(A) \subseteq \mathbb{R}$.

Theorem 5.4.1. *Let* $A : \mathcal{D}(A) \rightarrow \mathcal{H}$ *be a self-adjoint operator on a separable, complex Hilbert space* $\mathcal{H}$. *Then the resolvent set* $\rho(A)$ *contains all complex numbers* λ *with nonzero imaginary part so the spectrum* $\sigma(A)$ *is a subset of the real line* $\mathbb{R}$.

Proof. Note first that for $\psi \in \mathcal{D}(A)$, $\langle A\psi, \psi \rangle$ is real since $\langle A\psi, \psi \rangle = \langle \psi, A\psi \rangle = \overline{\langle A\psi, \psi \rangle}$; indeed, this is clearly true for any symmetric operator. Consequently,

$$\text{Im} \langle (\lambda - A)\psi, \psi \rangle = \text{Im} \langle \lambda\psi, \psi \rangle - \text{Im} \langle A\psi, \psi \rangle = \text{Im} (\bar{\lambda}) \|\psi\|^2 = -\text{Im} (\lambda) \|\psi\|^2.$$

From this and the Schwarz inequality we obtain

$$\|(\lambda - A)\psi\| \|\psi\| \geq |\langle (\lambda - A)\psi, \psi \rangle| \geq |\text{Im}(\lambda)| \|\psi\|^2,$$

so

$$\|(\lambda - A)\psi\| \geq |\mathrm{Im}(\lambda)| \, \|\psi\|. \tag{5.20}$$

From this we conclude that $\lambda - A$ is injective and therefore invertible if $\mathrm{Im}(\lambda) \neq 0$. Moreover, we claim that, in this case, $\mathcal{D}((\lambda - A)^{-1}) = \mathrm{Image}\,(\lambda - A)$ must be dense in $\mathcal{H}$. To see this we suppose not. Then $\mathrm{Image}\,(\lambda - A)$ would have a nontrivial orthogonal complement so we could select a $\phi \neq 0$ in $\mathcal{H}$ with $\langle (\lambda - A)\psi, \phi \rangle = 0$ for all $\psi \in \mathcal{D}(A) = \mathcal{D}(\lambda - A)$. This, by definition, forces ϕ into the domain of the adjoint $(\lambda - A)^*$ of $\lambda - A$. Since the adjoint of $\lambda - A$ is $\bar{\lambda} - A$ on $\mathcal{D}(A^*) = \mathcal{D}(A)$ we find that $\langle \psi, (\bar{\lambda} - A)\phi \rangle = 0 \; \forall \psi \in \mathcal{D}(A)$. But $\mathcal{D}(A)$ is assumed dense in $\mathcal{H}$, so $\langle \psi, (\bar{\lambda} - A)\phi \rangle = 0 \; \forall \psi \in \mathcal{H}$ and the nondegeneracy of $\langle\,,\,\rangle$ implies that $(\bar{\lambda} - A)\phi = 0$, that is, $A\phi = \bar{\lambda}\phi$. But then $\langle A\phi, \phi \rangle = \langle \bar{\lambda}\phi, \phi \rangle = \lambda\|\phi\|^2$, which is not real since $\mathrm{Im}(\lambda) \neq 0$ and $\phi \neq 0$, and this is a contradiction. Thus, we have shown that if $\mathrm{Im}(\lambda) \neq 0$, then $(\lambda - A)^{-1}$ exists and is densely defined on $\mathrm{Image}\,(\lambda - A)$. To conclude the proof that $\lambda \in \rho(A)$ we need only show that $(\lambda - A)^{-1}$ is bounded on $\mathrm{Image}\,(\lambda - A)$. But, for any $\xi \in \mathrm{Image}\,(\lambda - A)$, (5.20) gives

$$\|\xi\| = \|(\lambda - A)(\lambda - A)^{-1}\xi\| \geq |\mathrm{Im}(\lambda)| \, \|(\lambda - A)^{-1}\xi\|,$$

so

$$\|(\lambda - A)^{-1}\xi\| \leq \frac{1}{|\mathrm{Im}(\lambda)|}\|\xi\|,$$

as required. □

Theorem X.1 in [RS2] is a much more general result of von Neumann who showed that if A is a closed, symmetric operator, then the spectrum $\sigma(A)$ must be one of the following:
1. the closed upper half-plane $\{z \in \mathbb{C} : \mathrm{Im}\, z \geq 0\}$,
2. the closed lower half-plane $\{z \in \mathbb{C} : \mathrm{Im}\, z \leq 0\}$,
3. the entire complex plane $\mathbb{C}$ or
4. a closed subset of the real line $\mathbb{R}$ (in which case A is self-adjoint).

Having just noted at the beginning of the proof of Theorem 5.4.1 that $\langle \psi, A\psi \rangle$ is real for any symmetric operator, this would seem an appropriate moment to introduce a little notation that will play a role in our discussion of the spectral theorem. We define, for any symmetric operator A,

$$m(A) = \inf\langle \psi, A\psi \rangle \tag{5.21}$$

and

$$M(A) = \sup\langle \psi, A\psi \rangle, \tag{5.22}$$

where the infimum and supremum are taken over all $\psi \in \mathcal{D}(A)$ with $\|\psi\| = 1$. If A is defined on all of $\mathcal{H}$, then these are both finite if and only if A is bounded and, in this case, $\|A\| = \max(\,|m(A)|, |M(A)|\,)$; this is Theorem 6.11-C of [TaylA].

We will see in just a moment that the spectrum of an unbounded operator might well be empty, but this cannot happen for a bounded operator; this is the first corollary in Section VIII.2 of [Yosida] and also the corollary in Section VI.3 of [RS1], but one should also look at Theorem A, Section 67, of [Simm1], which proves the same result for the spectrum of an element in any Banach algebra. The essential ingredient in all of the proofs is Liouville's theorem from complex analysis and a byproduct of the proof is that the spectrum of a bounded operator is a bounded subset of $\mathbb{C}$. Finally, we remark that unitarily equivalent operators (Remark 5.2.9) have the same point, continuous and residual spectra.

We will now look at a few examples. The first is Example 5, Section VIII.1, of [RS1]. Its moral is that the spectrum (and almost everything else) is very sensitive to the choice of domain.

Example 5.4.1. We define two operators A_1 and A_2 on $L^2[0, 1]$. Their domains are

$$\mathcal{D}(A_1) = \left\{\psi \in L^2[0,1] : \psi \in AC[0,1] \text{ and } \frac{d\psi}{dq} \in L^2[0,1]\right\}$$

and

$$\mathcal{D}(A_2) = \{\psi \in \mathcal{D}(A_1) : \psi(0) = 0\}.$$

Each of these contains the smooth functions with compact support contained in $(0, 1]$ and so each is dense in $L^2[0, 1]$; A_1 and A_2 are both defined on their respective domains by $A_j = -i\hbar \frac{d}{dq}, j = 1, 2$. We claim that

$$\sigma(A_1) = \mathbb{C}, \tag{5.23}$$

but

$$\sigma(A_2) = \emptyset. \tag{5.24}$$

To prove (5.23) we need only observe that for any $\lambda \in \mathbb{C}$, $e^{i\lambda q/\hbar} \in \mathcal{D}(A_1)$ and $(\lambda - A_1)e^{i\lambda q/\hbar} = 0$, so $\lambda - A_1$ fails to be invertible and $\lambda \in \sigma(A_1)$. For the proof of (5.24), we must show that for every $\lambda \in \mathbb{C}$, $\lambda - A_2$ is invertible with a bounded inverse. We will do this by exhibiting the inverse explicitly and then showing that it is bounded. For this we define $S_\lambda : L^2[0, 1] \to \mathcal{D}(A_2)$ by

$$(S_\lambda g)(q) = \frac{-i}{\hbar} \int_0^q e^{i\lambda(q-s)/\hbar} g(s)\,ds$$

for every $g \in L^2[0,1]$ and every $q \in [0,1]$. Now we compute $(\lambda - A_2)S_\lambda$ and $S_\lambda(\lambda - A_2)$. For the record, we will use the following formula from calculus for differentiating integrals. If $f(q,s)$ and its q-derivative are continuous and $u(q)$ and $v(q)$ are continuously differentiable, then

$$\frac{d}{dq}\int_{u(q)}^{v(q)} f(q,s)ds = f(q,v(q))\frac{dv}{dq} - f(q,u(q))\frac{du}{dq} + \int_{u(q)}^{v(q)} \frac{\partial f}{\partial q}(q,s)ds.$$

Since both $L^2[0,1]$ and $\mathcal{D}(A_2)$ contain a dense set of smooth functions, it will suffice to show $(\lambda - A_2)S_\lambda g = g$ and $S_\lambda(\lambda - A_2)g = g$ for any smooth function g in $L^2[0,1]$ and $\mathcal{D}(A_2)$, respectively. For $(\lambda - A_2)S_\lambda$ we note that

$$(\lambda S_\lambda g)(q) = -\frac{i\lambda}{\hbar}\int_0^q e^{i\lambda(q-s)/\hbar}g(s)ds$$

and

$$(A_2 S_\lambda g)(q) = (-i\hbar)\left(\frac{-i}{\hbar}\right)\frac{d}{dq}\left(\int_0^q e^{i\lambda(q-s)/\hbar}g(s)ds\right)$$

$$= -\left[g(q) - 0 + \frac{i\lambda}{\hbar}\int_0^q e^{i\lambda(q-s)/\hbar}g(s)ds\right]$$

$$= -g(q) - \frac{i\lambda}{\hbar}\int_0^q e^{i\lambda(q-s)/\hbar}g(s)ds,$$

so $(\lambda - A_2)S_\lambda$ is the identity on $L^2[0,1]$, as required. To show that $S_\lambda(\lambda - A_2)$ is the identity on $\mathcal{D}(A_2)$ we will need the fact that $g \in \mathcal{D}(A_2)$ satisfies $g(0) = 0$. Of course, $(S_\lambda \lambda g)(q)$ is the same as $(\lambda S_\lambda g)(q)$, but integrating by parts gives

$$(S_\lambda A_2 g)(q) = \left(\frac{-i}{\hbar}\right)(-i\hbar)\int_0^q e^{i\lambda(q-s)/\hbar}\frac{dg}{ds}ds$$

$$= -\left[e^{i\lambda(q-s)/\hbar}g(s)\Big|_0^q - \frac{-i\lambda}{\hbar}\int_0^q e^{i\lambda(q-s)/\hbar}g(s)ds\right]$$

$$= -g(q) + 0 - \frac{i\lambda}{\hbar}\int_0^q e^{i\lambda(q-s)/\hbar}g(s)ds.$$

Thus, we find that $S_\lambda(\lambda - A_2)$ is the identity on $\mathcal{D}(A_2)$, as required.

Having shown that S_λ is the inverse of $\lambda - A_2$ we can finish the proof by showing that S_λ is bounded for any fixed $\lambda \in \mathbb{C}$. But, for any $g \in L^2[0,1]$,

$$\|S_\lambda g\|_{L^2}^2 = \int_0^1 |(S_\lambda g)(q)|^2 dq$$

$$\leq \left(\sup_{0 \leq q \leq 1} |(S_\lambda g)(q)| \right)^2$$

$$\leq \left(\sup_{0 \leq q \leq 1} \int_0^q |e^{i\lambda(q-s)/\hbar} g(s)| \, ds \right)^2$$

$$\leq \left(\sup_{0 \leq q \leq 1} \int_0^q |e^{i\lambda(q-s)/\hbar}|^2 ds \right) \left(\sup_{0 \leq q \leq 1} \int_0^q |g(s)|^2 ds \right)$$

$$\leq C(\lambda) \|g\|_{L^2}^2,$$

where $C(\lambda)$ is a constant that depends only on λ. Consequently, S_λ is bounded.

Example 5.4.2. Next we will consider the operator Q defined on $\mathcal{D}(Q) = \{\psi \in L^2(\mathbb{R}) : \int_\mathbb{R} q^2|\psi(q)|^2 dq < \infty\}$ by $(Q\psi)(q) = q\psi(q)$. Here is what we know so far (see Example 5.2.6). The operator Q is unbounded and self-adjoint (and therefore closed). Moreover, for any $\lambda \in \mathbb{C}$, $(\lambda - Q)\psi = 0 \Rightarrow \psi = 0 \in L^2(\mathbb{R})$, so $\lambda - Q$ is invertible. In particular, the point spectrum $P_\sigma(Q)$ is empty. Moreover, since Q is self-adjoint, the residual spectrum $R_\sigma(Q)$ is also empty (we will prove this directly in a moment). All that remains is the continuous spectrum $C_\sigma(Q)$ and, since Q is self-adjoint, this must be a closed (possibly empty) subset of $\mathbb{R}$ (Theorem 5.4.1). What we will now show is that, in fact, $C_\sigma(Q)$ is all of $\mathbb{R}$, that is, for every $\lambda \in \mathbb{R}$, $(\lambda - Q)^{-1}$, which we know exists ($P_\sigma(Q) = \emptyset$) and will show is densely defined, is an unbounded operator on Image $(\lambda - Q)$.

Note that, with $\lambda \in \mathbb{R}$ fixed, any $\psi \in L^2(\mathbb{R})$ that vanishes on some interval $U_\psi(\lambda)$ about λ is in Image $(\lambda - Q)$ because we can solve the equation $(\lambda - q)\phi(q) = \psi(q)$ by taking $\phi(q) = 0$ on $U_\psi(\lambda)$ and $\phi(q) = \psi(q)/(\lambda - q)$ outside of $U_\psi(\lambda)$. The resulting function ϕ is in $\mathcal{D}(Q)$ since ψ is in $L^2(\mathbb{R})$. In particular, for each $n = 1, 2, \dots$ and each $\psi \in L^2(\mathbb{R})$, Image $(\lambda - Q)$ contains $\chi(J_n)\psi$, where $\chi(J_n)$ is the characteristic function of the set $J_n = (-\infty, -\frac{1}{n}] \cup [\frac{1}{n}, \infty)$. But $\chi(J_n)\psi$ converge pointwise almost everywhere to ψ as $n \to \infty$ and so, by the dominated convergence theorem (Theorem 2.9.1 of [Fried]), $\chi(J_n)\psi \to \psi$ in $L^2(\mathbb{R})$. Consequently, Image $(\lambda - Q)$ is dense in $L^2(\mathbb{R})$. In particular, the residual spectrum $R_\sigma(A)$ is empty.

Finally, we show that $(\lambda - Q)^{-1}$ is unbounded on Image $(\lambda - Q)$. Suppose to the contrary that $(\lambda - Q)^{-1}$ is bounded. Then there exists a positive constant M such that $\|(\lambda - Q)^{-1}\psi\| \leq M\|\psi\|$ for all $\psi \in$ Image $(\lambda - Q)$. In particular, for any ϕ in $\mathcal{D}(\lambda - Q) = \mathcal{D}(Q)$ we have

$$\|\phi\| = \|(\lambda - Q)^{-1}(\lambda - Q)\phi\| \leq M\|(\lambda - Q)\phi\|. \tag{5.25}$$

We arrive at a contradiction by constructing, for each $n = 1, 2, \ldots$, an element ϕ_n of $\mathcal{D}(\lambda - Q) = \mathcal{D}(Q)$ satisfying $\|\phi_n\| = 1$ and $\|(\lambda - Q)\phi_n\| < \frac{1}{n}$ so that for sufficiently large n, (5.25) cannot be satisfied. For this we let $I_n = [\lambda - \frac{1}{2n}, \lambda + \frac{1}{2n}]$ and take $\phi_n = \sqrt{n}\chi(I_n)$ so that ϕ_n takes the constant value $\sqrt{n}$ on $[\lambda - \frac{1}{2n}, \lambda + \frac{1}{2n}]$ and is zero outside $[\lambda - \frac{1}{2n}, \lambda + \frac{1}{2n}]$. Then ϕ_n is clearly in $\mathcal{D}(Q)$ and satisfies $\|\phi_n\| = 1$. Furthermore, $[(\lambda - Q)\phi_n](q)$ is the linear function $(\lambda - q)\sqrt{n}$ on $[\lambda - \frac{1}{2n}, \lambda + \frac{1}{2n}]$ and is zero outside $[\lambda - \frac{1}{2n}, \lambda + \frac{1}{2n}]$. The maximum value of $|(\lambda - q)\sqrt{n}|$ on $[\lambda - \frac{1}{2n}, \lambda + \frac{1}{2n}]$ is $\frac{\sqrt{n}}{2n} = \frac{1}{2\sqrt{n}}$, so

$$\|(\lambda - Q)\phi_n\|^2 = \int_{\mathbb{R}} |[(\lambda - q)\phi_n](q)|^2 dq \leq \int_{\lambda - \frac{1}{2n}}^{\lambda + \frac{1}{2n}} \left(\frac{1}{2\sqrt{n}}\right)^2 dq = \frac{1}{4n}\left(\frac{1}{n}\right) < \frac{1}{n^2},$$

as required. The conclusion is that

$$\sigma(Q) = C_\sigma(Q) = \mathbb{R}.$$

Remark 5.4.2. The ideas in this last example can be employed to yield a more general result. Let $g : \mathbb{R} \to \mathbb{R}$ be a real-valued, measurable function on $\mathbb{R}$ that is finite almost everywhere with respect to Lebesgue measure μ and consider the self-adjoint multiplication operator $Q_g : \mathcal{D}(Q_g) \to L^2(\mathbb{R})$ defined on $\mathcal{D}(Q_g) = \{\psi \in L^2(\mathbb{R}) : g\psi \in L^2(\mathbb{R})\}$ by $(Q_g\psi)(q) = g(q)\psi(q) \, \forall \psi \in \mathcal{D}(Q_g) \, \forall q \in \mathbb{R}$. By self-adjointness (Exercise 5.2.6), the residual spectrum $R_\sigma(Q_g)$ is always empty. The point spectrum $P_\sigma(Q_g)$ is nonempty if and only if g differs from a constant function only on a set of measure zero, in which case $\sigma(Q_g)$ consists of this constant value alone. In general, the spectrum $\sigma(Q_g)$ is just the essential range of g (recall that a real number λ is in the *essential range* of g if and only if $\mu\{q \in \mathbb{R} : \lambda - \epsilon < g(q) < \lambda + \epsilon\}$ is positive for every $\epsilon > 0$; in particular, if g is continuous, this is just the range of g).

Example 5.4.3. We consider the momentum operator P defined on

$$\mathcal{D}(P) = \left\{\psi \in L^2(\mathbb{R}) : \psi \in AC[a, b] \, \forall a < b \text{ in } \mathbb{R} \text{ and } \frac{d\psi}{dq} \in L^2(\mathbb{R})\right\}$$

by

$$P = -i\hbar\frac{d}{dq}.$$

We have seen (Example 5.2.7) that P is unitarily equivalent, via the Fourier transform $\mathcal{F}$, to the operator Q in the previous example so its spectrum is precisely the same. Specifically, the point spectrum $P_\sigma(P)$ and the residual spectrum $R_\sigma(P)$ are both empty and the continuous spectrum $C_\sigma(P)$ is all of $\mathbb{R}$. We have

$$\sigma(P) = C_\sigma(P) = \mathbb{R}.$$

Example 5.4.4. The free particle Hamiltonian H_0 (Example 5.2.8) defined on $\mathcal{D}(H_0) = \{\psi \in L^2(\mathbb{R}) : \Delta\psi \in L^2(\mathbb{R})\}$ is unitarily equivalent to the self-adjoint multiplication operator Q_g, where $g(p) = \frac{\hbar^2}{2m} p^2$, and so they both have the same spectrum. From Remark 5.4.2 we conclude that the point and residual spectra of H_0 are both empty and the continuous spectrum is the range of g, that is, $C_\sigma(H_0) = [0, \infty)$, so

$$\sigma(H_0) = C_\sigma(H_0) = [0, \infty).$$

For the record we note that this implies, in particular, that

$$\sigma(-\Delta) = C_\sigma(-\Delta) = [0, \infty).$$

We showed earlier (Example 5.2.8) that $\langle \psi, -\Delta\psi \rangle \geq 0$ for all $\psi \in S(\mathbb{R})$. In fact, this is true for all $\psi \in \mathcal{D}(\Delta)$ and this is related to the result we have just proved. A symmetric operator A on a Hilbert space $\mathcal{H}$ is said to be *positive* if $\langle \psi, A\psi \rangle \geq 0$ for all $\psi \in \mathcal{D}(A)$. The following is Lemma 2, Section XII.7.2, of [DSII].

Theorem 5.4.2. *Let $A : \mathcal{D}(A) \to \mathcal{H}$ be a self-adjoint operator on a complex, separable Hilbert space $\mathcal{H}$. Then A is positive if and only if $\sigma(A) \subseteq [0, \infty)$.*

Example 5.4.5. Finally, we return to the operator H_B defined on $S(\mathbb{R})$ by

$$H_B = \frac{1}{2m} P^2 + \frac{m\omega^2}{2} Q^2 = -\frac{\hbar^2}{2m} \frac{d^2}{dq^2} + \frac{m\omega^2}{2} q^2.$$

We have seen (Example 5.3.1) that H_B is essentially self-adjoint on the Schwartz space $S(\mathbb{R})$, so it has a unique self-adjoint extension which we will also denote H_B. This was proved by finding an orthonormal basis $\{\psi_0, \psi_1, \ldots\}$ for $L^2(\mathbb{R})$ consisting of eigenvectors for H_B in $S(\mathbb{R})$ and appealing to Theorem 5.3.1. In particular, all of the corresponding eigenvalues

$$\mathcal{E}_n = \left(n + \frac{1}{2}\right)\hbar\omega, \quad n = 0, 1, 2, \ldots,$$

are elements of the point spectrum $P_\sigma(H_B)$. We claim that, because the eigenvectors of H_B are *complete* in $L^2(\mathbb{R})$, this is the entire spectrum of H_B, that is, for any λ not equal to one of these $\mathcal{E}_n$, the operator $(\lambda - H_B)^{-1}$ exists and is densely defined and bounded on its domain. By Theorem 5.4.1 we know this already for any λ with nonzero imaginary part so we can restrict our attention to real λ.

We begin with a few remarks on H_B itself. Since $\{\psi_n\}_{n=0}^\infty$ is an orthonormal basis for $L^2(\mathbb{R})$ we can write any ψ in $\mathcal{D}(H_B)$ as $\psi = \sum_{n=0}^\infty \langle \psi_n, \psi \rangle \psi_n$ and, since H_B is self-adjoint,

$$H_B\psi = \sum_{n=0}^\infty \langle \psi_n, H_B\psi \rangle \psi_n = \sum_{n=0}^\infty \langle H_B\psi_n, \psi \rangle \psi_n = \sum_{n=0}^\infty \langle \mathcal{E}_n\psi_n, \psi \rangle \psi_n = \sum_{n=0}^\infty \mathcal{E}_n \langle \psi_n, \psi \rangle \psi_n.$$

Thus, *the domain $\mathcal{D}(H_B)$ of H_B consists precisely of those $\psi \in L^2(\mathbb{R})$ for which the series* $\sum_{n=0}^{\infty} \mathcal{E}_n \langle \psi_n, \psi \rangle \psi_n$ *converges in $L^2(\mathbb{R})$.* Moreover, by Parseval's theorem (Theorem 6.4.5 of [Fried]),

$$\|H_B \psi\|^2 = \sum_{n=0}^{\infty} \mathcal{E}_n^2 |\langle \psi_n, \psi \rangle|^2$$

for any such ψ.

Note that H_B is injective and therefore invertible on its domain because 0 is not an eigenvalue. We will find an explicit series representation for H_B^{-1}. Let $\phi \in$ Image (H_B) and write $\phi = \sum_{n=0}^{\infty} \langle \psi_n, \phi \rangle \psi_n$. Then

$$H_B^{-1} \phi = \psi \quad \Leftrightarrow \quad \phi = H_B \psi \quad \Leftrightarrow \quad \langle \psi_n, \phi \rangle = \mathcal{E}_n \langle \psi_n, \psi \rangle \quad \Leftrightarrow \quad \langle \psi_n, \psi \rangle = \frac{1}{\mathcal{E}_n} \langle \psi_n, \phi \rangle$$

for every $n = 0, 1, 2, \ldots$. Thus,

$$H_B^{-1} \phi = \sum_{n=0}^{\infty} \frac{1}{\mathcal{E}_n} \langle \psi_n, \phi \rangle \psi_n. \tag{5.26}$$

This series also converges in $L^2(\mathbb{R})$ since $\phi = \sum_{n=0}^{\infty} \langle \psi_n, \phi \rangle \psi_n$ converges and $0 < \mathcal{E}_0 < \mathcal{E}_1 < \mathcal{E}_2 \cdots \rightarrow \infty$. Indeed, this series converges for *any* ϕ in $L^2(\mathbb{R})$ and the element of $L^2(\mathbb{R})$ it represents satisfies

$$H_B \left(\sum_{n=0}^{\infty} \frac{1}{\mathcal{E}_n} \langle \psi_n, \phi \rangle \psi_n \right) = \sum_{n=0}^{\infty} \mathcal{E}_n \frac{1}{\mathcal{E}_n} \langle \psi_n, \phi \rangle \psi_n = \sum_{n=0}^{\infty} \langle \psi_n, \phi \rangle \psi_n = \phi.$$

We conclude that the image of H_B is not only dense in $L^2(\mathbb{R})$, but is, in fact, all of $L^2(\mathbb{R})$. Moreover, H_B^{-1} is bounded on $L^2(\mathbb{R})$ since, for any $\phi \in L^2(\mathbb{R})$,

$$\|H_B^{-1} \phi\|^2 = \sum_{n=0}^{\infty} \frac{1}{\mathcal{E}_n^2} |\langle \psi_n, \phi \rangle|^2 \leq \frac{1}{\mathcal{E}_0^2} \|\phi\|^2.$$

Let us summarize all of this. We have shown that H_B is an unbounded, self-adjoint, invertible operator whose inverse H_B^{-1} is a bounded operator defined everywhere on $L^2(\mathbb{R})$; we will show in Example 5.5.4 that it is also a compact operator (see Remark 5.5.2 for the definition). Since all of this is equally true of $-H_B$, what we have just shown is that $(0 - H_B)^{-1}$ is a bounded operator defined on all of $L^2(\mathbb{R})$. In other words, $\lambda = 0$ is in the resolvent set $\rho(H_B)$ of H_B. Now, let us deal with all of the remaining $\lambda \neq \mathcal{E}_n$, $n = 0, 1, 2, \ldots$ in $\mathbb{R}$.

Thus, we assume $\lambda \in \mathbb{R}$, $\lambda \neq 0$ and $\lambda \neq \mathcal{E}_n$, $n = 0, 1, 2, \ldots$, and we must show that $(\lambda - H_B)^{-1}$ exists, is densely defined and is bounded on its domain. We will begin by just computing $(\lambda - H_B)^{-1}$ *formally* to get a putative formula to work with and then deal with convergence issues and proving what needs to be proved. Thus, we observe that

$$(\lambda - H_B)^{-1} \phi = \psi \quad \Leftrightarrow \quad \phi = (\lambda - H_B)\psi$$

$$\Leftrightarrow \quad \lambda\psi - H_B\psi = \phi$$
$$\Leftrightarrow \quad \langle\psi_n, \lambda\psi\rangle - \mathcal{E}_n\langle\psi_n, \psi\rangle = \langle\psi_n, \phi\rangle \quad \forall n = 0, 1, 2, \dots$$
$$\Leftrightarrow \quad \langle\psi_n, \psi\rangle = \frac{\langle\psi_n, \phi\rangle}{\lambda - \mathcal{E}_n} \quad \forall n = 0, 1, 2, \dots .$$

Thus, assuming for the moment that the series converges, we have

$$H_B\psi = \sum_{n=0}^{\infty} \mathcal{E}_n \frac{\langle\psi_n, \phi\rangle}{\lambda - \mathcal{E}_n} \psi_n.$$

From this and $\lambda\psi = \phi + H_B\psi$ we arrive at the following potential formula for $(\lambda - H_B)^{-1}$:

$$\psi = (\lambda - H_B)^{-1}\phi = \frac{1}{\lambda}\phi + \frac{1}{\lambda}\sum_{n=0}^{\infty} \mathcal{E}_n \frac{\langle\psi_n, \phi\rangle}{\lambda - \mathcal{E}_n} \psi_n. \tag{5.27}$$

To clean this business up we first show that the series in (5.27) does, indeed, converge in $L^2(\mathbb{R})$ for *any* ϕ. For this we note first that, since $\mathcal{E}_n \to \infty$ as $n \to \infty$, $|\frac{\mathcal{E}_n}{\lambda - \mathcal{E}_n}|$ is bounded and we can let

$$\alpha = \sup_{n \geq 0}\left|\frac{\mathcal{E}_n}{\lambda - \mathcal{E}_n}\right|.$$

Also let

$$\varphi_k = \sum_{n=0}^{k} \mathcal{E}_n \frac{\langle\psi_n, \phi\rangle}{\lambda - \mathcal{E}_n} \psi_n$$

for each $k \geq 0$. Then for $k_1 < k_2$,

$$\|\varphi_{k_2} - \varphi_{k_1}\|^2 = \sum_{n=k_1+1}^{k_2}\left|\frac{\mathcal{E}_n}{\lambda - \mathcal{E}_n}\right|^2 |\langle\psi_n, \phi\rangle|^2 \leq \alpha^2 \sum_{n=k_1+1}^{k_2} |\langle\psi_n, \phi\rangle|^2.$$

Since $\sum_{n=0}^{\infty} |\langle\psi_n, \phi\rangle|^2 = \|\phi\|^2$, the sequence $\{\varphi_k\}_{k=0}^{\infty}$ of partial sums is Cauchy and therefore convergent in $L^2(\mathbb{R})$. Thus, the series in (5.27) converges in $L^2(\mathbb{R})$. Consequently,

$$\psi = \frac{1}{\lambda}\phi + \frac{1}{\lambda}\sum_{n=0}^{\infty} \mathcal{E}_n \frac{\langle\psi_n, \phi\rangle}{\lambda - \mathcal{E}_n} \psi_n$$

is a well-defined element of $L^2(\mathbb{R})$ and it certainly satisfies $(\lambda - H_B)\psi = \phi$. We have therefore shown that $(\lambda - H_B)^{-1}$ is defined for every ϕ in $L^2(\mathbb{R})$ and is given by (5.27). All that remains is to show that $(\lambda - H_B)^{-1}$ is bounded on $L^2(\mathbb{R})$. But, from (5.27) we find that

$$\|(\lambda - H_B)^{-1}\phi\| \leq \frac{1}{|\lambda|}\|\phi\| + \frac{1}{|\lambda|}\alpha\|\phi\| = \frac{1}{|\lambda|}(1 + \alpha)\|\phi\|,$$

so the result follows. Thus,

$$\sigma(H_B) = P_\sigma(H_B) = \left\{\mathcal{E}_n = \left(n + \frac{1}{2}\right)\hbar\omega, \ n = 0, 1, 2, \dots\right\}.$$

In Example 5.5.1 we will show that $(\lambda - H_B)^{-1}$ is, in fact, a compact operator.

5.5 The spectral theorem

Now we turn to the spectral theorem. We have shown that multiplication by q defines a self-adjoint operator $Q : \mathcal{D}(Q) \to L^2(\mathbb{R})$ on $\mathcal{D}(Q) = \{\psi \in L^2(\mathbb{R}) : \int_{\mathbb{R}} q^2 |\psi(q)|^2 dq < \infty\}$ and we noted that an analogous statement is true if q is replaced by any measurable, real-valued function $g(q)$ (Exercise 5.2.6). We have also shown that the operator $P : \mathcal{D}(P) = \{\psi \in L^2(\mathbb{R}) : \psi \in AC[a, b] \,\forall a < b \text{ in } \mathbb{R} \text{ and } \frac{d\psi}{dq} \in L^2(\mathbb{R})\} \to L^2(\mathbb{R})\}$ defined by $P = -i\hbar \frac{d}{dq}$ is self-adjoint because it is unitarily equivalent to Q. There is a sense in which these examples are generic. One version of the *spectral theorem* says roughly that real-valued multiplication operators on an L^2 space are always self-adjoint and that, conversely, any self-adjoint operator is unitarily equivalent to a real-valued multiplication operator on some L^2 space. The following more precise statements are, respectively, Proposition 1, Section VIII.3, and Theorem VIII.4 of [RS1].

Remark 5.5.1. Recall that if g is a real-valued function on a measure space (M, μ), then g is *essentially bounded* if there is a positive constant C for which $|g(m)| \leq C$ almost everywhere and a real number λ is in the *essential range* of g if and only if $\mu \{m \in M : \lambda - \epsilon < g(m) < \lambda + \epsilon\}$ is positive for every $\epsilon > 0$.

Theorem 5.5.1. *Let (M, μ) be a σ-finite measure space and $g : M \to \mathbb{R}$ a measurable, real-valued function on M that is finite almost everywhere. Define the multiplication operator $Q_g : \mathcal{D}(Q_g) \to L^2(M, \mu)$ on $\mathcal{D}(Q_g) = \{\psi \in L^2(M, \mu) : g\psi \in L^2(M, \mu)\}$ by $(Q_g \psi)(m) = g(m)\psi(m) \,\forall m \in M$. Then Q_g is self-adjoint and its spectrum $\sigma(Q_g)$ is the essential range of g; Q_g is a bounded operator if and only if g is essentially bounded.*

Theorem 5.5.2. *Let $A : \mathcal{D}(A) \to \mathcal{H}$ be a self-adjoint operator on a separable, complex Hilbert space $\mathcal{H}$. Then there exist a σ-finite measure space (M, μ) and a real-valued measurable function $g : M \to \mathbb{R}$ on M that is finite almost everywhere such that A is unitarily equivalent to the multiplication operator $Q_g : \mathcal{D}(Q_g) \to L^2(M, \mu)$; that is, there exists a unitary equivalence $U : \mathcal{H} \to L^2(M, \mu)$ for which:*

1. *$\varphi \in \mathcal{D}(A) \Leftrightarrow U\varphi \in \mathcal{D}(Q_g)$ and*
2. *$\psi \in U[\mathcal{D}(A)] \Rightarrow [(UAU^{-1})\psi](m) = g(m)\psi(m) \,\forall m \in M.$*

Dropping the requirement that g be real-valued one obtains the spectral theorem for *normal operators*, that is, operators that commute with their adjoints. Since unitary operators are certainly normal ($U^*U = UU^* = \text{id}$) one obtains such a representation for any unitary operator, but in this case $UU^* = \text{id}$ implies that $|g(m)| = 1$ almost everywhere. Consequently, the essential range of g is contained in the unit circle $\{z \in \mathbb{C} : |z| = 1\}$ and therefore so is the spectrum.

   Theorem 5.5.2 is an elegant way of viewing the essential content of the spectral theorem (see [Hal3] for more on this point of view), but it is not particularly well suited to the physical interpretation of observables in quantum mechanics and this is really our principal objective here. What we need is something more akin to the version of the finite-dimensional spectral theorem that we described in Section 5.1, that is, an explicit representation of any self-adjoint operator in terms of projection operators. What this representation looks like depends on the type of self-adjoint operator at hand. The simplest such result deals with self-adjoint operators that are also compact.

Remark 5.5.2. Recall that a bounded operator $T : \mathcal{H} \to \mathcal{H}$ on a separable, complex Hilbert space $\mathcal{H}$ is said to be *compact* (or *completely continuous*) if, for each bounded sequence $\{\varphi_n\}_{n=1}^{\infty}$ in $\mathcal{H}$, the sequence $\{T\varphi_n\}_{n=1}^{\infty}$ has a subsequence that converges in $\mathcal{H}$ (in other words, T maps bounded sets in $\mathcal{H}$ onto sets with compact closure in $\mathcal{H}$). Finite linear combinations of compact operators are compact, as are products and adjoints of compact operators. Recall also that a bounded operator $P : \mathcal{H} \to \mathcal{H}$ on a separable, complex Hilbert space $\mathcal{H}$ is called a *projection* (more accurately, an *orthogonal projection*) if P is self-adjoint and satisfies $P^2 = P$. Then $M = $ Image (P) is a closed subspace of $\mathcal{H}$, so $\mathcal{H}$ is the orthogonal direct sum $M \oplus M^{\perp}$ of M and its orthogonal complement $M^{\perp}$; P is the orthogonal projection of $\mathcal{H}$ onto M in the sense that every $\psi \in \mathcal{H}$ can be written uniquely as $\psi = \phi + \phi^{\perp}$, with $\phi = P\psi \in M$ and $\phi^{\perp} \in M^{\perp}$. One often writes $P = P_M$ for emphasis. If $P \neq 0$, then $\|P\| = 1$. All of this is discussed in any functional analysis text but, in particular, in Sections 5.1–5.3 and 6.2–6.3 of [Fried]; also see Chapter VI of [RiSz.N] for some nice applications.

The spectral theorem for compact self-adjoint operators is particularly simple and easy to relate to, so, although it is probably already familiar, we intend to linger over it a bit longer than is absolutely necessary for our purposes because it also provides some nice motivation for the more general result that is, perhaps, not so easy to relate to. Probably the best places to find the details that we omit here are Chapter 6 of [TaylA], or Chapter VI of [RiSz.N]; see (5.21) and (5.22) for the notation used in part 2. of the following theorem.

Theorem 5.5.3. *Let T be a nonzero, compact, self-adjoint operator on the separable, complex Hilbert space $\mathcal{H}$. Then:*

1. *T has at least one nonzero (necessarily real) eigenvalue λ with $|\lambda| = \|T\|$.*
2. *The spectrum $\sigma(T)$ is contained in the interval $[m(T), M(T)] \subseteq \mathbb{R}$ and is at most countably infinite.*
3. *Every nonzero element λ of the spectrum $\sigma(T)$ is an eigenvalue of T with a finite-dimensional eigenspace M_λ.*
4. *The eigenvalues of T can accumulate only at 0, and if $\mathcal{H}$ is infinite-dimensional, then 0 must be in $\sigma(T)$, but it need not be an eigenvalue of T.*
5. *If $\lambda_1, \lambda_2, \ldots$ is the (possibly finite) sequence of distinct nonzero eigenvalues of T, then*

$$\mathcal{H} \cong \text{Kernel}\,(T) \oplus M_{\lambda_1} \oplus M_{\lambda_2} \oplus \cdots.$$

6. If $P_{\lambda_n} : \mathcal{H} \to M_{\lambda_n}$ is the orthogonal projection of $\mathcal{H}$ onto M_{λ_n}, then

$$T\psi = \sum_{n \geq 1} \lambda_n P_{\lambda_n} \psi$$

for all $\psi \in \mathcal{H}$; if the sum is infinite, then the convergence is in the norm topology of $\mathcal{H}$.

It is sometimes convenient to arrange the distinct *nonzero* eigenvalues $\lambda_1, \lambda_2, \ldots$ in a sequence in which their absolute values are nonincreasing and each eigenvalue is repeated a number of times equal to its multiplicity. If this is the way we want them listed we will use μ rather than λ to label the eigenvalues. Thus,

$$|\mu_1| \geq |\mu_2| \geq \cdots,$$

where each μ_k is equal to some $\lambda_{n(k)}$ and appears in the sequence $\dim(M_{\lambda_{n(k)}})$ times. Once this is done one can clearly choose an orthonormal basis for each M_{λ_n} and arrange the elements of these bases in a sequence $\psi_1, \psi_2, \ldots$. Then

$$T\psi = \sum_{k \geq 1} \langle \psi_k, T\psi \rangle \psi_k = \sum_{k \geq 1} \langle T\psi_k, \psi \rangle \psi_k = \sum_{k \geq 1} \mu_k \langle \psi_k, \psi \rangle \psi_k$$

for any $\psi \in \mathcal{H}$ (see the theorem on page 233 of [RiSz.N]). Note, however, that, unless Kernel $(T) = \{0\}$, $\psi_1, \psi_2, \ldots$ will only be an orthonormal sequence in $\mathcal{H}$ and not an orthonormal basis. Even so, when phrased in these terms one can formulate a useful converse of Theorem 5.5.3. The following result is proved on pages 234–235 of [RiSz.N].

Theorem 5.5.4. *Let* $T : \mathcal{H} \to \mathcal{H}$ *be a linear map on a separable, complex Hilbert space* $\mathcal{H}$. *Suppose there exists an orthonormal sequence* $\{\psi_k\}_{k=1}^{\infty}$ *in* $\mathcal{H}$ *and a sequence* $\{\mu_k\}_{k=1}^{\infty}$ *of real numbers converging to 0 for which*

$$T\psi = \sum_{k \geq 1} \mu_k \langle \psi_k, \psi \rangle \psi_k$$

for every $\psi \in \mathcal{H}$. *Then* T *is a compact, self-adjoint operator on* $\mathcal{H}$.

Example 5.5.1. Let us return once again to the operator H_B of Example 5.4.5. Here H_B is, of course, not compact (in fact, not even bounded), but it *is* invertible with a globally defined, bounded inverse H_B^{-1} and we claim that this is compact. Indeed, this follows directly from the previous theorem if we recall the expression (5.26) for H_B^{-1} and the fact that the eigenvalues $\mathcal{E}_n$ of H_B are nonzero and satisfy $\mathcal{E}_n \to \infty$ as $n \to \infty$. In fact, if λ is any point in the resolvent set $\rho(H_B)$, then the resolvent operator $(\lambda - H_B)^{-1}$ is compact. This follows in exactly the same way from the fact that (5.27) can be written

$$(\lambda - H_B)^{-1}\phi = \frac{1}{\lambda} \sum_{n=0}^{\infty} \left(1 + \frac{\mathcal{E}_n}{\lambda - \mathcal{E}_n}\right) \langle \psi_n, \phi \rangle \, \psi_n.$$

Compact, self-adjoint operators have important applications, for example, to the Hilbert–Schmidt theory of linear integral equations (where they first arose), but they do not play a major role in the foundations of quantum mechanics. Nevertheless, we would like to spend a little more time with the compact case in order to see how it can be rephrased in a way that admits a generalization to arbitrary bounded and even unbounded self-adjoint operators. Specifically, we would like to turn the sum in Theorem 5.5.3.6 into an integral.

Thus, we consider a nonzero, compact, self-adjoint operator T on a separable, complex Hilbert space $\mathcal{H}$, and we denote by $\lambda_1, \lambda_2, \ldots$ the distinct nonzero (necessarily real) eigenvalues of T and by $\mu_1, \mu_2, \ldots$ these same nonzero eigenvalues arranged in such a way that each λ_n appears $\dim(M_{\lambda_n})$ times and $|\mu_1| \geq |\mu_2| \geq \cdots$. As above, we denote by $\{\psi_1, \psi_2, \ldots\}$ the orthonormal sequence of eigenvectors of T satisfying $T\psi = \sum_{k=1}^{\infty} \mu_k \langle \psi_k, \psi \rangle \psi_k$ for every $\psi \in \mathcal{H}$. Now we will define a one-parameter family of operators E_λ, $\lambda \in \mathbb{R}$, as follows.

1. For $\lambda < 0$ and any $\psi \in \mathcal{H}$,

$$E_\lambda \psi = \sum_{\mu_k \leq \lambda} \langle \psi_k, \psi \rangle \psi_k,$$

where the sum is taken to be 0 if there are no $\mu_k \leq \lambda$. Note that this is a finite sum since $\lambda_k \to 0$ as $k \to \infty$.

2. For $\lambda = 0$ and any $\psi \in \mathcal{H}$,

$$E_0 \psi = \psi - \sum_{\mu_k > 0} \langle \psi_k, \psi \rangle \psi_k,$$

where the sum is taken to be 0 if there are no $\mu_k > 0$. Here the sum need not be finite, but converges in $\mathcal{H}$ by Bessel's inequality.

3. For $\lambda > 0$ and any $\psi \in \mathcal{H}$,

$$E_\lambda \psi = \psi - \sum_{\mu_k > \lambda} \langle \psi_k, \psi \rangle \psi_k,$$

where the sum is taken to be 0 if there are no $\mu_k > \lambda$. Again, the sum is finite since $\lambda_k \to 0$ as $k \to \infty$.

Now one must check a few things. Specifically, one can show that each $E_\lambda, \lambda \in \mathbb{R}$, is a projection and, moreover, the family $\{E_\lambda\}_{\lambda \in \mathbb{R}}$ of projections satisfies:

1. $E_\lambda E_\kappa = E_{\min(\lambda, \kappa)}$,
2. $E_\lambda = 0$ if $\lambda < m(T)$ and $E_\lambda = \mathrm{id}_{\mathcal{H}}$ if $\lambda \geq M(T)$ and
3. $\lim_{\kappa \to \lambda^+} E_\kappa \psi = E_\lambda \psi \, \forall \, \psi \in \mathcal{H}$.

Part 2. is clear from the definition since $\sigma(T) \subseteq [m(T), M(T)]$. Since, as a function of λ, E_λ is constant between any two consecutive eigenvalues of T, Part 3. is clear for $\lambda < 0$

and for $\lambda > 0$, while, for $\lambda = 0$, it is simply a restatement of the definition of E_0. Part 1. is proved by considering the various possibilities for the relative ordering of λ, κ and 0 in $\mathbb{R}$. If you would like to see how this goes you might write out at least the following case.

Exercise 5.5.1. Suppose $\lambda \leq \kappa < 0$ and show that $E_\lambda E_\kappa = E_\lambda$.

It is customary to write $\lim_{\kappa \to \lambda^+} E_\kappa \psi$ as $E_{\lambda+0}\psi$ so that 3. simply says $E_{\lambda+0}\psi = E_\lambda\psi$; this is often abbreviated as $E_{\lambda+0} = E_\lambda$. Similarly, one writes $\lim_{\kappa \to \lambda^-} E_\kappa \psi = E_{\lambda-0}\psi$, but this, in general, is not equal to $E_\lambda\psi$. The family $\{E_\lambda\}_{\lambda \in \mathbb{R}}$ of projections is called a *resolution of the identity* for the operator T. With it one can produce the *integral representation* of T to which we have alluded several times.

Now let us return to our compact, self-adjoint operator T. We begin by noting that T is completely determined by the values of $\langle \psi, T\phi \rangle = \langle T\psi, \phi \rangle$ for $\psi, \phi \in \mathcal{H}$. Next we appeal to Proposition 1, Section XI.5, of [Yosida], according to which the properties of the resolution of the identity $\{E_\lambda\}_{\lambda \in \mathbb{R}}$ we listed above as 1., 2. and 3. imply that for any fixed $\psi, \phi \in \mathcal{H}$,

$$\langle \psi, E_\lambda \phi \rangle$$

is, as a complex-valued function of λ, of bounded variation. Now choose real numbers $a < m(T)$ and $b \geq M(T)$. Then, for any complex-valued, continuous function $f(\lambda)$ on $[a, b]$, the Riemann–Stieltjes integral (see Appendix H.2)

$$\int_a^b f(\lambda) d\langle \psi, E_\lambda \phi \rangle$$

exists and assigns to the pair (ψ, ϕ) of elements of $\mathcal{H}$ a complex number. This is, in particular, true for the identity function $f(\lambda) = \lambda$ on $[a, b]$ and in this case one finds that

$$\int_a^b \lambda \, d\langle \psi, E_\lambda \phi \rangle = \langle \psi, T\phi \rangle. \tag{5.28}$$

Consequently, T is completely determined by the values of these integrals for $\psi, \phi \in \mathcal{H}$. In fact, because of the following version of the polarization identity, satisfied by any operator A on its domain $\mathcal{D}(A)$,

$$\langle \psi, A\phi \rangle = \frac{1}{2i} \langle \psi + i\phi, A(\psi + i\phi) \rangle + \frac{1}{2} \langle \psi + \phi, A(\psi + \phi) \rangle$$
$$- \frac{1-i}{2} [\langle \psi, A\psi \rangle + \langle \phi, A\phi \rangle], \tag{5.29}$$

T is actually determined by the integrals

$$\int_a^b \lambda \, d\langle \psi, E_\lambda \psi \rangle = \langle \psi, T\psi \rangle, \quad \psi \in \mathcal{H}. \tag{5.30}$$

The integral in (5.28) is a limit of Riemann–Stieltjes sums of the form

$$\sum_{k=1}^{n} \tau_k^* [\langle \psi, E_{\tau_k} \phi \rangle - \langle \psi, E_{\tau_{k-1}} \phi \rangle] = \left\langle \psi, \sum_{k=1}^{n} \tau_k^* (E_{\tau_k} - E_{\tau_{k-1}}) \phi \right\rangle$$

so the proof of (5.28) amounts to showing that the operators $\sum_{k=1}^{n} \tau_k^* (E_{\tau_k} - E_{\tau_{k-1}}) = \sum_{k=1}^{n} \tau_k^* \Delta E_{\lambda_k}$ converge weakly to T as $\max(\tau_k - \tau_{k-1}) \to 0$ and independently of the choice of τ_k^*.

Remark 5.5.3. For the record we recall that:
1. $\sum_{k=1}^{n} \tau_k^* \Delta E_{\lambda_k}$ *converges weakly* to T if $\langle \psi, \sum_{k=1}^{n} \tau_k^* \Delta E_{\lambda_k} \phi \rangle \to \langle \psi, T\phi \rangle$ in $\mathbb{C}$ for all $\psi, \phi \in \mathcal{H}$;
2. $\sum_{k=1}^{n} \tau_k^* \Delta E_{\lambda_k}$ *converges strongly* to T if $\sum_{k=1}^{n} \tau_k^* \Delta E_{\lambda_k} \psi \to T\psi$ in $\mathcal{H}$ for all $\psi \in \mathcal{H}$;
3. $\sum_{k=1}^{n} \tau_k^* \Delta E_{\lambda_k}$ *converges uniformly* to T if $\sum_{k=1}^{n} \tau_k^* \Delta E_{\lambda_k} \to T$ in the operator norm topology of $\mathcal{B}(\mathcal{H})$.

Each of these implies the preceding one in the list.

This is true and not so hard to show because E_λ are relatively simple for a compact, self-adjoint operator. However, much more is true. It can be shown that these operators actually converge *uniformly* to T and therefore also converge strongly to T. For this reason one often writes

$$T = \int_a^b \lambda \, dE_\lambda, \tag{5.31}$$

where the integral is the limit in the operator norm of the Riemann–Stieltjes-like sums $\sum_{k=1}^{n} \tau_k^* \Delta E_{\lambda_k}$ (Riemann–Stieltjes-*like* because ΔE_{λ_k} are now projection operators rather than real or complex numbers).

The spectral theorem for compact, self-adjoint operators, when written in the integral forms (5.28) or (5.30), is virtually identical in appearance to the spectral theorem for arbitrary bounded, self-adjoint operators; one need only determine how to associate with such an operator something like $\{E_\lambda\}_{\lambda \in \mathbb{R}}$ with which to define the integrals (however, in the bounded case the convergence in (5.30) will generally not be uniform, but only strong convergence). The same is true of the spectral theorem for unbounded, self-adjoint operators except that the integral is over all of $\mathbb{R}$ rather than a compact interval containing the spectrum since the spectrum of an unbounded operator is not a bounded set.

We trust that this digression on the compact case will make the general form of the spectral theorem that we now record somewhat more palatable. A detailed treatment of the following material is available in Chapter XI of [Yosida] and Chapter III, Section 5, of [Prug].

Let $(\mathcal{M}, \mathcal{A})$ denote a measurable space, that is, a pair consisting of a set $\mathcal{M}$ together with a σ-algebra $\mathcal{A}$ of subsets of $\mathcal{M}$. A *spectral measure*, or *projection-valued measure* on $(\mathcal{M}, \mathcal{A})$ is a function $E : \mathcal{A} \rightarrow \mathcal{B}(\mathcal{H})$ assigning to each measurable set $S \in \mathcal{A}$ a bounded operator $E(S) \in \mathcal{B}(\mathcal{H})$ and satisfying each of the following:

1. for each measurable set $S \in \mathcal{A}$, $E(S)$ is an orthogonal projection, that is, it is idempotent $E(S)^2 = E(S)$ and self-adjoint $E(S)^* = E(S)$;
2. $E(\emptyset) = 0$ and $E(\mathcal{M}) = \mathrm{id}_{\mathcal{H}}$;
3. if $S_1, S_2, \ldots$ is a countable family of pairwise disjoint measurable sets in $\mathcal{A}$ and $S = \bigsqcup_{n=1}^{\infty} S_n$ is their union, then

$$E(S) = \sum_{n=1}^{\infty} E(S_n),$$

where the infinite series converges in the strong sense, that is,

$$E(S)\psi = \lim_{N \to \infty} \sum_{n=1}^{N} E(S_n)\psi,$$

for each $\psi \in \mathcal{H}$.

Exercise 5.5.2. Show that if $S_1 \cap S_2 = \emptyset$, then $E(S_1)$ and $E(S_2)$ project onto orthogonal subspaces of $\mathcal{H}$ and that it follows that

$$S_1, S_2 \in \mathcal{A} \quad \Rightarrow \quad E(S_1 \cap S_2) = E(S_1)E(S_2).$$

We will be interested primarily in the special case in which the measurable space consists of $\mathcal{M} = \mathbb{R}$ and its σ-algebra $\mathcal{A}$ of Borel sets; spectral measures on this measurable space will be called simply *spectral measures on* $\mathbb{R}$.

Now suppose E is a spectral measure on the measurable space $(\mathcal{M}, \mathcal{A})$ and ψ is an element in $\mathcal{H}$. Since each $E(S)$ is idempotent and self-adjoint,

$$\langle \psi, E(S)\psi \rangle = \langle \psi, E(S)^2 \psi \rangle = \langle E(S)\psi, E(S)\psi \rangle = \|E(S)\psi\|^2.$$

As a function of $S \in \mathcal{A}$, $\langle \psi, E(S)\psi \rangle$ is therefore a finite measure on $(\mathcal{M}, \mathcal{A})$ for each fixed $\psi \in \mathcal{H}$ which we will denote

$$\langle \psi, E\psi \rangle,$$

that is,

$$\langle \psi, E\psi \rangle(S) = \langle \psi, E(S)\psi \rangle = \|E(S)\psi\|^2.$$

Note that if ψ is a unit vector in $\mathcal{H}$, then $\langle \psi, E\psi \rangle$ is a *probability measure* on $\mathcal{M}$ since $\langle \psi, E(\mathcal{M})\psi \rangle = \langle \psi, \psi \rangle = 1$. It follows from the polarization identity (5.29) that for any two elements ψ and ϕ in $\mathcal{H}$, $S \rightarrow \langle \psi, E(S)\phi \rangle$ is a complex measure on $\mathcal{M}$ which we will denote

$$\langle \psi, E\phi \rangle.$$

If E is a spectral measure on $\mathbb{R}$ and λ is any real number, then $(-\infty, \lambda]$ is a Borel set, so $E(-\infty, \lambda]$ is a projection on $\mathcal{H}$. We can therefore define a one-parameter family $\{E_\lambda\}_{\lambda \in \mathbb{R}}$ of projections by

$$E_\lambda = E(-\infty, \lambda]$$

for every $\lambda \in \mathbb{R}$. The defining properties of a spectral measure translate into the following properties of $\{E_\lambda\}_{\lambda \in \mathbb{R}}$, where all of the limits are in the strong sense:
1. $E_\lambda E_\kappa = E_{\min(\lambda, \kappa)} \ \forall \lambda, \kappa \in \mathbb{R}$,
2. $\lim_{\lambda \to -\infty} E_\lambda = 0$ and $\lim_{\lambda \to \infty} E_\lambda = \mathrm{id}_\mathcal{H}$,
3. $\lim_{\kappa \to \lambda^+} E_\kappa = E_\lambda \ \forall \lambda \in \mathbb{R}$.

Any one-parameter family $\{E_\lambda\}_{\lambda \in \mathbb{R}}$ of projections on $\mathcal{H}$ satisfying these three properties is called a *resolution of the identity*, or *spectral family* for $\mathcal{H}$. Thus, any spectral measure on $\mathbb{R}$ gives rise to a resolution of the identity. Conversely, since intervals of the form $(-\infty, \lambda]$ generate the σ-algebra of Borel sets in $\mathbb{R}$, any resolution of the identity extends to a unique spectral measure on $\mathbb{R}$. Properties 1.–3. in the definition of a resolution of the identity imply (by Proposition 1, Section XI.5, of [Yosida]) that for any fixed $\psi, \phi \in \mathcal{H}$, $\langle \psi, E_\lambda \phi \rangle$ is, as a function of λ, of bounded variation, so it can be used to define a Stieltjes integral (see Appendix H). One can also check that

$$\lambda_i < \lambda_j \quad \Rightarrow \quad E_{\lambda_j} - E_{\lambda_i} = E(\lambda_i, \lambda_j],$$

where $E(\lambda_i, \lambda_j]$ is the projection associated to $(\lambda_i, \lambda_j]$ by the corresponding spectral measure on $\mathbb{R}$; in particular, $E_{\lambda_j} - E_{\lambda_i}$ is a projection.

Now, let $[a, b]$ be a nondegenerate, closed, bounded interval in $\mathbb{R}$ and suppose $f : \mathbb{R} \to \mathbb{C}$ is a continuous, complex-valued function on $\mathbb{R}$. For any partition $a = \lambda_0 < \lambda_1 < \cdots < \lambda_n = b$ of $[a, b]$ and any choice of $\lambda_k^* \in (\lambda_{k-1}, \lambda_k]$, $k = 1, 2, \ldots, n$, we consider the operator

$$\sum_{k=1}^n f(\lambda_k^*)(E_{\lambda_k} - E_{\lambda_{k-1}}) = \sum_{k-1}^n f(\lambda_k^*) \Delta E_{\lambda_k}.$$

By Proposition 2, Section XI.5, of [Yosida], these Riemann–Stieltjes sums have a strong limit as $\max(\lambda_k - \lambda_{k-1}) \to 0$, independent of the choice of λ_k^*. Thus, we can define an operator, denoted

$$\int_a^b f(\lambda) dE_\lambda,$$

whose value at any $\psi \in \mathcal{H}$ is

$$\left(\int_a^b f(\lambda) dE_\lambda \right) \psi = \lim \sum_{k=1}^n f(\lambda_k^*)(\Delta E_{\lambda_k} \psi),$$

where the limit is taken over finer and finer partitions of $[a, b]$ just as for the Riemann integral. We would also like to define the corresponding improper integral over all of $\mathbb{R}$ in the usual way as

$$
\left(\int_{\mathbb{R}} f(\lambda) dE_\lambda \right) \psi = \left(\int_{-\infty}^{\infty} f(\lambda) dE_\lambda \right) \psi
$$

$$
= \left(\int_{-\infty}^{0} f(\lambda) dE_\lambda \right) \psi + \left(\int_{0}^{\infty} f(\lambda) dE_\lambda \right) \psi
$$

$$
= \lim_{a \to -\infty} \left(\int_{a}^{0} f(\lambda) dE_\lambda \right) \psi + \lim_{b \to \infty} \left(\int_{0}^{b} f(\lambda) dE_\lambda \right) \psi,
$$

but, of course, this will only make sense if both of these limits in $\mathcal{H}$ exist. According to Theorem 1, Section XI.5, of [Yosida], this is the case if and only if

$$
\int_{\mathbb{R}} |f(\lambda)|^2 d\langle \psi, E_\lambda \psi \rangle = \int_{-\infty}^{\infty} |f(\lambda)|^2 d\langle \psi, E_\lambda \psi \rangle = \int_{-\infty}^{\infty} |f(\lambda)|^2 d \|E_\lambda \psi\|^2 < \infty. \tag{5.32}
$$

Consequently, $\int_{\mathbb{R}} f(\lambda) dE_\lambda$ defines an operator on the set of all $\psi \in \mathcal{H}$ satisfying (5.32). We will put all of the information, and a bit more, together in the following theorem.

Theorem 5.5.5. *Let $\mathcal{H}$ be a separable, complex Hilbert space and $E : \mathcal{A} \to \mathcal{B}(\mathcal{H})$ a spectral measure on $\mathbb{R}$ with associated resolution of the identity $\{E_\lambda\}_{\lambda \in \mathbb{R}}$. Let $f : \mathbb{R} \to \mathbb{C}$ be a continuous, complex-valued function on $\mathbb{R}$. Then $\int_{\mathbb{R}} f(\lambda) dE_\lambda$ defines an operator A_f on*

$$
\mathcal{D}(A_f) = \left\{ \psi \in \mathcal{H} : \int_{\mathbb{R}} |f(\lambda)|^2 d\langle \psi, E_\lambda \psi \rangle < \infty \right\}
$$

whose value at any $\psi \in \mathcal{D}(A_f)$ is

$$
A_f \psi = \left(\int_{\mathbb{R}} f(\lambda) dE_\lambda \right) \psi.
$$

The operator A_f is uniquely determined by the condition that

$$
\langle \phi, A_f \psi \rangle = \int_{\mathbb{R}} f(\lambda) \, d\langle \phi, E_\lambda \psi \rangle
$$

for all $\psi \in \mathcal{D}(A_f)$ and all $\phi \in \mathcal{H}$. Furthermore, if $\bar{f}(\lambda) = \overline{f(\lambda)}$ is the conjugate function of f, then

$$
A_{\bar{f}} = A_f^*,
$$

so, in particular, if f is real-valued, then A_f is self-adjoint.

What we conclude from Theorem 5.5.5 is that if we are given a spectral measure on $\mathbb{R}$ with values in $\mathcal{B}(\mathcal{H})$ and a real-valued function on $\mathbb{R}$, then we can manufacture self-adjoint operators on $\mathcal{H}$. Remarkably, every self-adjoint operator on a separable, complex Hilbert space $\mathcal{H}$ can be manufactured this way. This is the content of the *spectral theorem* for self-adjoint operators which we will now record.

Theorem 5.5.6. *Let $\mathcal{H}$ be a separable, complex Hilbert space and A a self-adjoint operator on $\mathcal{H}$ with domain $\mathcal{D}(\mathcal{H})$. Then there exists a unique spectral measure $E^A : \mathcal{A} \to \mathcal{B}(\mathcal{H})$ on $\mathbb{R}$ with associated resolution of the identity $\{E_\lambda^A\}_{\lambda \in \mathbb{R}}$ satisfying each of the following:*

$$\mathcal{D}(A) = \left\{ \psi \in \mathcal{H} : \int_{\mathbb{R}} \lambda^2 d\langle \psi, E_\lambda^A \psi \rangle < \infty \right\},$$

$$A\psi = \left(\int_{\mathbb{R}} \lambda \, dE_\lambda^A \right) \psi.$$

Furthermore, the support of E^A coincides with the spectrum $\sigma(A)$ of A, that is,

$$\lambda \in \sigma(A) \quad \Leftrightarrow \quad E^A(\lambda - \epsilon, \lambda + \epsilon) \neq 0 \quad \forall \epsilon > 0.$$

Remark 5.5.4. There are various approaches to the proof of Theorem 5.5.6. One proof proceeds along the following lines. One first proves an analogous spectral decomposition for *unitary* operators and then appeals to a correspondence between unitary and self-adjoint operators called the Cayley transform. This is carried out in great detail in Chapter III, Section 6, of [Prug]. One can also consult Chapter XI, Sections 4–6, of [Yosida], Chapter VIII of [RiSz.N], Chapter 13 of [Rud1] or Chapter VIII of [RS1].

Note that, given the spectral measure $E^A : \mathcal{A} \to \mathcal{B}(\mathcal{H})$ on $\mathbb{R}$ with associated resolution of the identity $\{E_\lambda^A\}_{\lambda \in \mathbb{R}}$, the operator A is A_f, where $f : \mathbb{R} \to \mathbb{R}$ is the identity map $f(\lambda) = \lambda$ (Theorem 5.5.5). For an arbitrary continuous $f : \mathbb{R} \to \mathbb{C}$ and the specific spectral measure E^A corresponding to the self-adjoint operator A we will write the operator A_f of Theorem 5.5.5 as

$$f(A).$$

We will return to such *functions of A* shortly, but first we will write out concrete examples for the quantum mechanical position operator Q, the momentum operator P and the harmonic oscillator Hamiltonian H_B.

Example 5.5.2. We will define a resolution of the identity $\{E_\lambda\}_{\lambda \in \mathbb{R}}$ on the Hilbert space $\mathcal{H} = L^2(\mathbb{R})$ and then determine the operator $\int_{\mathbb{R}} \lambda \, dE_\lambda$ to which it gives rise. For each $\lambda \in \mathbb{R}$ we define $E_\lambda : L^2(\mathbb{R}) \to L^2(\mathbb{R})$ to be the map that carries $\psi \in L^2(\mathbb{R})$ to $E_\lambda \psi \in L^2(\mathbb{R})$ given by

$$(E_\lambda \psi)(q) = \begin{cases} \psi(q), & \text{if } q \leq \lambda, \\ 0, & \text{if } q > \lambda. \end{cases} \tag{5.33}$$

Each E_λ is a projection since boundedness and $E_\lambda^2 = E_\lambda$ are clear and self-adjointness follows from

$$\langle E_\lambda \psi, \phi \rangle = \int_{-\infty}^{\infty} \overline{E_\lambda \psi(q)} \phi(q) dq = \int_{-\infty}^{\lambda} \overline{\psi(q)} \phi(q) dq = \int_{-\infty}^{\infty} \overline{\psi(q)} E_\lambda \phi(q) dq = \langle \psi, E_\lambda \phi \rangle.$$

Exercise 5.5.3. Verify properties 1., 2. and 3. in the definition of a resolution of the identity for $\{E_\lambda\}_{\lambda \in \mathbb{R}}$.

To compute the relevant integrals, we fix $\psi \in \mathcal{H}$ and consider the bounded variation function

$$\alpha(\lambda) = \langle \psi, E_\lambda \psi \rangle = \int_{-\infty}^{\lambda} \overline{\psi(q)} \psi(q) dq = \int_{-\infty}^{\lambda} |\psi(q)|^2 dq.$$

Then, using (H.2),

$$\int_{\mathbb{R}} \lambda^2 \, d\langle \psi, E_\lambda \psi \rangle = \int_{-\infty}^{\infty} \lambda^2 \, d\alpha(\lambda) = \int_{-\infty}^{\infty} \lambda^2 \alpha'(\lambda) d\lambda$$

$$= \int_{-\infty}^{\infty} \lambda^2 |\psi(\lambda)|^2 d\lambda = \int_{-\infty}^{\infty} |\lambda \psi(\lambda)|^2 d\lambda,$$

which is finite precisely when ψ is in the domain of the position operator $(Q\psi)(q) = q\psi(q)$ of Example 5.2.3. Moreover, for these ψ we have

$$\int_{\mathbb{R}} \lambda \, d\langle \psi, E_\lambda \psi \rangle = \int_{-\infty}^{\infty} \lambda \, d\alpha(\lambda) = \int_{-\infty}^{\infty} \lambda \alpha'(\lambda) d\lambda$$

$$= \int_{-\infty}^{\infty} \lambda |\psi(\lambda)|^2 d\lambda = \int_{-\infty}^{\infty} \overline{\psi(\lambda)} [\lambda \psi(\lambda)] d\lambda$$

$$= \langle \psi, Q\psi \rangle.$$

Polarization in $\mathcal{D}(Q)$ and the density of $\mathcal{D}(Q)$ in $\mathcal{H}$ then imply that for every ψ in its domain, the operator $\int_{\mathbb{R}} \lambda \, dE_\lambda$ agrees with the position operator. But their domains are the same as well so

$$Q = \int_{\mathbb{R}} \lambda \, dE_\lambda.$$

We have therefore found a *spectral decomposition* of the position operator Q on $L^2(\mathbb{R})$.

Exercise 5.5.4. Suppose $U : \mathcal{H}_1 \to \mathcal{H}_2$ is a unitary equivalence. If $A : \mathcal{D}(A) \to \mathcal{H}_1$ is a self-adjoint operator on $\mathcal{H}_1$, then $UAU^{-1} : U(\mathcal{D}(A)) \to \mathcal{H}_2$ is a self-adjoint operator on $\mathcal{H}_2$ (Exercise 5.2.9). Show that if $\{E_\lambda\}_{\lambda \in \mathbb{R}}$ is a resolution of the identity on $\mathcal{H}_1$, then $\{UE_\lambda U^{-1}\}_{\lambda \in \mathbb{R}}$ is a resolution of the identity on $\mathcal{H}_2$. It can be shown that if $\{E_\lambda\}_{\lambda \in \mathbb{R}}$ gives rise to the operator $A = \int_\mathbb{R} \lambda \, dE_\lambda$, then $\{UE_\lambda U^{-1}\}_{\lambda \in \mathbb{R}}$ gives rise to UAU^{-1} (see pages 315–316 of [Yosida]).

Example 5.5.3. The Fourier transform $\mathcal{F} : L^2(\mathbb{R}) \to L^2(\mathbb{R})$ is a unitary equivalence (see Appendix G.2), so if we let $\{E_\lambda^Q\}_{\lambda \in \mathbb{R}}$ denote the resolution of the identity constructed in Example 5.5.2, then $\{\mathcal{F} E_\lambda^Q \mathcal{F}^{-1}\}_{\lambda \in \mathbb{R}}$ is also a resolution of the identity on $L^2(\mathbb{R})$. The corresponding operator is $\mathcal{F} Q \mathcal{F}^{-1}$ and this we know to be the operator $\frac{1}{\hbar}P$ from Example 5.2.7.

Example 5.5.4. For this example we will begin with a self-adjoint operator and find the resolution of the identity that gives rise to it as $\int_\mathbb{R} \lambda dE_\lambda$. Specifically, we will consider the operator H_B of Example 5.4.5 for which we know there exists an orthonormal basis $\{\psi_n\}_{n=0}^\infty$ for $L^2(\mathbb{R})$ consisting of eigenfunctions of H_B with eigenvalues $\mathcal{E}_n = (n + \frac{1}{2})\hbar\omega$, $n = 0, 1, 2, \ldots$:

$$H_B\psi_n = \mathcal{E}_n\psi_n = \left(n + \frac{1}{2}\right)\hbar\omega\,\psi_n.$$

Note that for $\psi \in \mathcal{D}(H_B)$,

$$H_B\psi = \sum_{n=0}^\infty \langle\psi_n, H_B\psi\rangle\psi_n = \sum_{n=0}^\infty \langle H_B\psi_n, \psi\rangle\psi_n = \sum_{n=0}^\infty \langle\mathcal{E}_n\psi_n, \psi\rangle\psi_n = \sum_{n=0}^\infty \mathcal{E}_n\langle\psi_n, \psi\rangle\psi_n,$$

and so

$$\langle\psi, H_B\psi\rangle = \sum_{n=0}^\infty \mathcal{E}_n|\langle\psi_n, \psi\rangle|^2.$$

What we need then is a resolution of the identity $\{E_\lambda\}_{\lambda \in \mathbb{R}}$ for which the value of $\int_\mathbb{R} \lambda \, d\langle\psi, E_\lambda\psi\rangle$ is $\sum_{n=0}^\infty \mathcal{E}_n|\langle\psi_n, \psi\rangle|^2$. For each λ in $\mathbb{R}$ we define $E_\lambda : L^2(\mathbb{R}) \to L^2(\mathbb{R})$ by

$$E_\lambda\psi = \sum_{\mathcal{E}_n \leq \lambda} \langle\psi_n, \psi\rangle\psi_n,$$

where the sum is taken to be zero if there are no $\mathcal{E}_n \leq \lambda$. Thus, $E_\lambda = 0$ for $\lambda < \mathcal{E}_0$ and, for $\lambda \geq \mathcal{E}_0$, E_λ is the projection onto the subspace spanned by the finite number of ψ_n for which $\mathcal{E}_n \leq \lambda$.

Exercise 5.5.5. Show that the projections $E_\lambda, \lambda \in \mathbb{R}$, satisfy the conditions required of a resolution of the identity.

For any fixed $\psi \in L^2(\mathbb{R})$,

$$\langle\psi, E_\lambda\psi\rangle = \sum_{\mathcal{E}_n \leq \lambda} |\langle\psi_n, \psi\rangle|^2,$$

which is zero for $\lambda < \mathcal{E}_0$ and a nondecreasing step function for $\lambda \geq \mathcal{E}_0$, stepping up at each eigenvalue $\mathcal{E}_n$; these occur at intervals of length $\hbar\omega$. Fix $\psi \in \mathcal{D}(H_B)$. Since $\langle \psi, E_\lambda \psi \rangle = 0$ for $\lambda < \mathcal{E}_0 = \frac{1}{2}\hbar\omega$,

$$\int\limits_{\mathbb{R}} \lambda \, d\langle \psi, E_\lambda \psi \rangle = \int\limits_0^\infty \lambda \, d\langle \psi, E_\lambda \psi \rangle = \lim_{b\to\infty} \int\limits_0^b \lambda \, d\langle \psi, E_\lambda \psi \rangle.$$

For the Riemann–Stieltjes sums defining $\int_0^b \lambda \, d\langle \psi, E_\lambda \psi \rangle$ we may consider only partitions $0 = \lambda_0 < \lambda_1 < \cdots < \lambda_k = b$ with $\max |\lambda_i - \lambda_{i-1}| < \hbar\omega$ so that each interval $(\lambda_{i-1}, \lambda_i]$ contains at most one eigenvalue $\mathcal{E}_n$. For those that contain no eigenvalue, $\Delta E_\lambda = 0$, so $\langle \psi, \Delta E_\lambda \psi \rangle = 0$ and there is no contribution to the integral. If $(\lambda_{i-1}, \lambda_i]$ contains an eigenvalue, say, $\mathcal{E}_n$, then ΔE_λ is the projection onto the subspace spanned by ψ_n and we may select $\lambda_i^* = \mathcal{E}_n$ for the corresponding term in the Riemann–Stieltjes sum (the integral is independent of this choice). This term in the sum is therefore $\mathcal{E}_n |\langle \psi_n, \psi \rangle|^2$ and the Riemann–Stieltjes approximation to $\int_0^b \lambda \, d\langle \psi, E_\lambda \psi \rangle$ is

$$\sum_{\mathcal{E}_n \leq b} \mathcal{E}_n |\langle \psi_n, \psi \rangle|^2.$$

Since this is true for *any* sufficiently fine partition,

$$\int\limits_0^b \lambda \, d\langle \psi, E_\lambda \psi \rangle = \sum_{\mathcal{E}_n \leq b} \mathcal{E}_n |\langle \psi_n, \psi \rangle|^2.$$

Taking the limit as $b \to \infty$ gives

$$\int\limits_{\mathbb{R}} \lambda \, d\langle \psi, E_\lambda \psi \rangle = \sum_{n=0}^\infty \mathcal{E}_n |\langle \psi_n, \psi \rangle|^2,$$

as required.

Recall from Theorem 5.5.6 that the support of E^A coincides with the spectrum $\sigma(A)$ of A, that is,

$$\lambda \in \sigma(A) \quad \Leftrightarrow \quad E^A(\lambda - \epsilon, \lambda + \epsilon) \neq 0 \quad \forall \epsilon > 0$$

(see Section 132 of [RiSz.N]). Consequently, if $S \subseteq \mathbb{R}$ is a Borel set, then

$$S \cap \sigma(A) = \emptyset \quad \Rightarrow \quad E^A(S) = 0. \tag{5.34}$$

It follows that the integrals over $\mathbb{R}$ in Theorem 5.5.6 can be replaced by the corresponding integrals over $\sigma(A)$. Consequently, one can regard $f \mapsto f(A)$ as an assignment of an operator $f(A)$ to each element f of the algebra $C^0(\sigma(A))$ of continuous, complex-valued functions on the spectrum $\sigma(A)$ of A. We include in the statement of the next result some basic properties of the assignment $f \mapsto f(A)$ which collectively are referred to as the *(continuous) functional calculus* (see Sections XI.5 and XI.12 of [Yosida] or Chapter IX of [RiSz.N]).

Theorem 5.5.7. *Let $\mathcal{H}$ be a separable, complex Hilbert space and A a self-adjoint operator on $\mathcal{H}$ with dense domain $\mathcal{D}(A)$, perhaps all of $\mathcal{H}$. Let E^A be the unique spectral measure on $\mathbb{R}$ associated with A and $\{E_\lambda^A\}_{\lambda\in\mathbb{R}}$ the corresponding resolution of the identity. For every continuous, complex-valued function f on $\mathbb{R}$, an operator $f(A)$ is densely defined on*

$$\mathcal{D}(f(A)) = \left\{\psi \in \mathcal{H} : \int_\mathbb{R} |f(\lambda)|^2 d\langle\psi, E_\lambda^A\psi\rangle < \infty\right\}$$

by

$$f(A)\psi = \left(\int_\mathbb{R} f(\lambda)dE_\lambda^A\right)\psi$$

for every $\psi \in \mathcal{D}(f(A))$. Moreover, $f(A)$ is characterized by

$$\langle\phi, f(A)\psi\rangle = \int_\mathbb{R} f(\lambda) d\langle\phi, E_\lambda^A\psi\rangle \quad \forall\psi \in \mathcal{D}(f(A)) \, \forall\phi \in \mathcal{H}.$$

If F_A is the map from continuous, complex-valued functions on $\sigma(A)$ to operators on $\mathcal{H}$ given by $F_A(f) = f(A)$, then F_A has the following properties:
1. *F_A is an algebraic *-homomorphism, that is, for $\alpha \in \mathbb{C}$ and f and g continuous, complex-valued functions on $\sigma(A)$,*
   a. *$F_A(\alpha f) = \alpha F_A(f)$, that is, $(\alpha f)(A)\psi = \alpha f(A)\psi$ for $\psi \in \mathcal{D}(f(A)) = \mathcal{D}((\alpha f)(A))$,*
   b. *$F_A(f+g) = F_A(f)+F_A(g)$, that is, $(f+g)(A)\psi = f(A)\psi + g(A)\psi$ for $\psi \in \mathcal{D}(f(A))\cap\mathcal{D}(g(A))$,*
   c. *$F_A(fg) = F_A(f)F_A(g)$, that is, for $\psi \in \mathcal{D}(g(A))$, $g(A)\psi \in \mathcal{D}(f(A))$ is equivalent to $\psi \in \mathcal{D}((fg)(A))$ and, in this case, $(fg)(A)\psi = f(A)g(A)\psi$ (note that the product of two operators is their composition),*
   d. *$F_A(1) = \text{id}_\mathcal{H}$, that is, $1(A)\psi = \psi \, \forall\psi \in \mathcal{H}$, where 1 is the constant function on $\sigma(A)$ whose value is $1 \in \mathbb{R}$,*
   e. *$F_A(\bar{f}) = F_A(f)^*$, that is, for $\psi, \phi \in \mathcal{D}(\bar{f}(A)) = \mathcal{D}(f(A))$, $\langle f(A)\psi, \phi\rangle = \langle\psi, \bar{f}(A)\phi\rangle$; in particular, $f(A)$ is self-adjoint if and only if f is real-valued;*
2. *$f(A)$ is a bounded operator on $\mathcal{H}$ if and only if f is a bounded function on $\sigma(A)$;*
3. *$\sigma(f(A)) = \overline{f(\sigma(A))}$ and $A\psi = \lambda\psi \Rightarrow f(A)\psi = f(\lambda)\psi$; here $\overline{f(\sigma(A))}$ means the closure of $f(\sigma(A))$ in $\mathbb{C}$;*
4. *$f \geq 0$ on $\sigma(A) \Rightarrow \langle\psi, f(A)\psi\rangle \geq 0$ for all $\psi \in \mathcal{D}(f(A))$;*

5. if we let $\{f_n\}_{n=1}^{\infty}$ be a sequence of continuous, complex-valued functions on $\sigma(A)$ that converges pointwise on $\sigma(A)$ to the continuous function f and suppose that $\{\|f_n\|_{\infty}\}_{n=1}^{\infty}$ is bounded, where $\|f_n\|_{\infty} = \sup_{q\in\sigma(A)} |f_n(q)|$, then $\{f_n(A)\}_{n=1}^{\infty}$ converges strongly to $f(A)$, that is, for every $\psi \in \mathcal{H}, f_n(A)\psi \to f(A)\psi$ in $\mathcal{H}$ as $n \to \infty$;

6. if we let B be a bounded operator on $\mathcal{H}$, then the following are equivalent:

    a. B commutes with A in the sense that $B(\mathcal{D}(A)) \subseteq \mathcal{D}(A)$ and $AB\psi = BA\psi \; \forall \psi \in \mathcal{D}(A)$,

    b. B commutes with every projection E_λ^A in the resolution of the identity associated with A,

    c. B commutes with $f(A)$ for every continuous, complex-valued function f on $\sigma(A)$;

7. for every $\psi \in \mathcal{D}(f(A))$,

$$\left\| f(A)\psi \right\|^2 = \int_{\mathbb{R}} |f(\lambda)|^2 d\langle \psi, E_\lambda^A \psi \rangle.$$

For a *bounded* self-adjoint operator A there is a more direct approach to the definition of $f(A)$ for continuous functions f. This is discussed in detail on pages 222–224 of [RS1], but the basic idea is simple. For bounded A, the spectrum $\sigma(A)$ is a compact subset of $\mathbb{R}$ and, since there are no domain issues, one can define $p(A)$ for any polynomial $p(\lambda) = \sum_{n=0}^{N} a_n \lambda^n$ to be simply $p(A) = \sum_{n=0}^{N} a_n A^n$. One shows then that $\sigma(p(A)) = p(\sigma(A))$ and $\|p(A)\| = \sup_{\lambda\in\sigma(A)} |p(\lambda)|$ (Lemmas 1 and 2, Section VII.1, of [RS1]). From this it follows that the function that carries the polynomial p to $p(A) \in \mathcal{B}(\mathcal{H})$ has a unique linear extension to the closure in $C^0(\sigma(A))$ (with the sup norm) of the polynomials which, by the Stone–Weierstrass theorem (Theorem B, Section 36, of [Simm1]) is all of $C^0(\sigma(A))$. In particular, if f is real analytic and given by $f(\lambda) = \sum_{n=0}^{\infty} a_n \lambda^n$ for $|\lambda| < R$, then $f(A) = \sum_{n=0}^{\infty} a_n A^n$ for those A with $\|A\| < R$ and the convergence is uniform. As long as A is bounded, this definition of $f(A)$ for continuous functions f agrees with the definition in Theorem 5.5.7. This approach fails for unbounded A not only because of the usual domain issues, but also because $\sigma(A)$ is not compact.

We point out also that much of the functional calculus described in Theorem 5.5.7 can be extended from continuous to bounded Borel measurable functions on $\mathbb{R}$ (see Section XI.12 of [Yosida] or Theorem VIII.5 of [RS1]). This is a useful and instructive thing to do. For example, it provides a direct link between the functional calculus and the representation of A in terms of spectral measures. Specifically, for any Borel set S in $\mathbb{R}$, one finds that the projection $E^A(S)$ is just the operator $\chi_S(A)$, where χ_S is the characteristic function of S. With a bit more work the functional calculus for A can be extended to Borel measurable functions on $\mathbb{R}$ that are finite and defined almost everywhere with respect to $\{E_\lambda^A\}_{\lambda\in\mathbb{R}}$, that is, with respect to all of the measures $\langle E_\lambda^A \psi, \psi \rangle$ for $\psi \in \mathcal{H}$ (see Sections 127–128 of [RiSz.N]). If f is such a function, then the operator $f(A)$ is bounded if and only if f is bounded almost everywhere with respect to $\{E_\lambda^A\}_{\lambda\in\mathbb{R}}$ (see page 349 of [RiSz.N]). We will make use of this extension to Borel functions only for the statement of a theorem of von Neumann on commuting families of self-adjoint operators (Theorem 5.6.4) so we will say no more about it here. The continuous case will provide our most important application of the functional calculus; in Example 5.5.5 we will define the operator exponentials that will describe the time evolution of quantum states.

Example 5.5.5. Fix an arbitrary real number t and consider the complex-valued, continuous function $f_t(\lambda) = e^{it\lambda}$. For any self-adjoint operator $A : D(A) \to \mathcal{H}$ on $\mathcal{H}$, the functional calculus in Theorem 5.5.7 provides an operator $U_t^A = e^{itA}$ on $\mathcal{H}$. Since $e^{it\lambda}$ is bounded (by 1), e^{itA} is a bounded operator by Theorem 5.5.7.2 and its domain is all of $\mathcal{H}$ because, for any $\psi \in \mathcal{H}$,

$$\int_{\mathbb{R}} |e^{it\lambda}|^2 d\langle \psi, E_\lambda^A \psi \rangle = \int_{-\infty}^{\infty} d\langle \psi, E_\lambda^A \psi \rangle = \langle \psi, \psi \rangle.$$

Exercise 5.5.6. Show that the last equality follows from the fact that the Riemann–Stieltjes sums for $\int_{-\infty}^{\infty} d\langle \psi, E_\lambda^A \psi \rangle$ telescope.

5.6 Stone's theorem

Example 5.5.5 is just the beginning of a very important story. We claim the $\{U_t^A\}_{t \in \mathbb{R}}$ defined there is a strongly continuous one-parameter group of *unitary* operators on $\mathcal{H}$ (see Remark 3.3.2). Specifically, we record the following four properties:
1. $U_0^A = \mathrm{id}_{\mathcal{H}}$,
2. U_t^A is a unitary operator on $\mathcal{H}$ for every $t \in \mathbb{R}$,
3. $U_t^A U_s^A = U_{t+s}^A$ for all $t, s \in \mathbb{R}$,
4. if $t \to 0$ in $\mathbb{R}$, then, for each $\psi \in \mathcal{H}$, $U_t^A \psi \to \psi$ in $\mathcal{H}$.

Although these are all proved in each of our basic references (for example, pages 288–289 of [Prug]), they are so fundamental to the mathematical model of quantum mechanics toward which we are headed that we will pause to give the simple arguments. Note first that U_t^A is uniquely determined by the condition that

$$\langle \phi, U_t^A \psi \rangle = \int_{\mathbb{R}} e^{it\lambda} d\langle \phi, E_\lambda^A \psi \rangle \quad \forall \phi, \psi \in \mathcal{H}.$$

Setting $t = 0$ gives

$$\langle \phi, U_0^A \psi \rangle = \int_{\mathbb{R}} d\langle \phi, E_\lambda^A \psi \rangle = \langle \phi, \psi \rangle$$

for all $\phi, \psi \in \mathcal{H}$, which proves $U_0^A = \mathrm{id}_{\mathcal{H}}$.

To prove that each U_t^A is unitary we must show that $U_t^A(U_t^A)^* = (U_t^A)^*U_t^A = \mathrm{id}_\mathcal{H}$. But this follows immediately from $e^{it\lambda}\overline{e^{it\lambda}} = \overline{e^{it\lambda}}e^{it\lambda} = 1$ and Theorem 5.5.7, parts 1.c–e. To prove $U_t^A U_s^A = U_{t+s}^A$ we use Theorem 5.5.7, part 1.c, to compute

$$\langle \phi, U_t^A U_s^A \psi \rangle = \int_\mathbb{R} e^{it\lambda} e^{is\lambda} d\langle \phi, E_\lambda^A \psi \rangle = \int_\mathbb{R} e^{i(t+s)\lambda} d\langle \phi, E_\lambda^A \psi \rangle = \langle \phi, U_{t+s}^A \psi \rangle$$

for all $\phi, \psi \in \mathcal{H}$, which proves $U_t^A U_s^A = U_{t+s}^A$.

We claim next that the one-parameter group $\{U_t^A\}_{t\in\mathbb{R}} = \{e^{itA}\}_{t\in\mathbb{R}}$ of unitary operators on $\mathcal{H}$ is strongly continuous (Remark 3.3.2). For this we must show that if $t \to 0$ in $\mathbb{R}$, then, for each $\psi \in \mathcal{H}$, $U_t^A \psi \to \psi$ in $\mathcal{H}$. Thus, we consider

$$\left\| e^{itA}\psi - \psi \right\|^2 = \left\| (e^{itA} - \mathrm{id}_\mathcal{H})\psi \right\|^2 = \left\| (e^{itA} - 1)(A)\psi \right\|^2.$$

Now, let $g(t) = e^{it\lambda} - 1$ and note that

$$\|g(A)\psi\|^2 = \langle g(A)\psi, g(A)\psi \rangle = \langle \psi, g(A)^* g(A)\psi \rangle = \langle \psi, \overline{g}(A)g(A)\psi \rangle = \langle \psi, |g|^2(A)\psi \rangle$$
$$= \int_\mathbb{R} |g|^2(\lambda) d\langle \psi, E_\lambda^A \psi \rangle,$$

so

$$\left\| e^{itA}\psi - \psi \right\|^2 = \int_\mathbb{R} \left| e^{it\lambda} - 1 \right|^2 d\langle \psi, E_\lambda^A \psi \rangle.$$

Now, $|e^{it\lambda} - 1|^2 \le 4 \,\forall t, \lambda \in \mathbb{R}$, and the constant function 4 is integrable since

$$\int_\mathbb{R} 4 \, d\langle \psi, E_\lambda^A \psi \rangle = 4\langle \psi, \psi \rangle.$$

Moreover,

$$\left| e^{it\lambda} - 1 \right|^2 \to 0$$

pointwise for each $\lambda \in \mathbb{R}$ as $t \to 0$, so the Lebesgue dominated convergence theorem (Theorem 2.9.1 of [Fried]) implies that $\|e^{itA}\psi - \psi\|^2 \to 0$ as $t \to 0$, as required.

We will record one more important property of $\{e^{itA}\}_{t\in\mathbb{R}}$ that can be proved in much the same way (details are available on pages 289–290 of [Prug]).

Lemma 5.6.1. *Let $A : \mathcal{D}(A) \to \mathcal{H}$ be a self-adjoint operator on the separable, complex Hilbert space $\mathcal{H}$ and $U_t^A = e^{itA}$, $t \in \mathbb{R}$, the corresponding strongly continuous one-parameter group of unitary operators on $\mathcal{H}$. Then ψ is in $\mathcal{D}(A)$ if and only if the limit*

$$\lim_{t\to 0} \frac{U_t^A \psi - \psi}{t}$$

exists and, in this case,

$$\lim_{t\to 0} \frac{U_t^A \psi - \psi}{t} = iA\psi.$$

One can think of this geometrically in the following way. The one-parameter group defines a curve $t \to e^{itA}$ in $\mathcal{B}(\mathcal{H})$ starting at $\mathrm{id}_{\mathcal{H}}$ at $t = 0$. For any $\psi \in \mathcal{D}(A), t \to e^{itA}\psi$ is a curve in $\mathcal{H}$ starting at ψ at $t = 0$; $\mathcal{H}$ is a Hilbert space (in particular, a vector space) and hence can be regarded as an (infinite-dimensional) manifold in which every tangent space can be identified with $\mathcal{H}$. The limit $\lim_{t\to 0} \frac{U_t^A \psi - \psi}{t}$ is the tangent vector to the curve $t \to e^{itA}\psi$ at $t = 0$ which, as an element of $\mathcal{H}$, is $iA\psi$. One might write this more succinctly as

$$\frac{d}{dt}(e^{itA}\psi)\bigg|_{t=0} = iA\psi. \tag{5.35}$$

Note that if it were permitted to expand e^{itA} in a power series (which, when A is unbounded, it is *not*) this is just what we would get by differentiating term-by-term with respect to t. One can check that $e^{itA}(\mathcal{D}(A)) \subseteq \mathcal{D}(A)$ and then use the group property of $\{e^{itA}\}_{t\in\mathbb{R}}$ to show more generally that for any $\psi \in \mathcal{D}(A)$,

$$\frac{d}{dt}(e^{itA}\psi) = iA(e^{itA}\psi) \tag{5.36}$$

for any t, so iA gives the tangent vector at each point along the curve $t \to e^{itA}\psi$. In this way one can think of iA as a vector field on $\mathcal{D}(A) \subseteq \mathcal{H}$ whose integral curves are $t \to e^{itA}\psi$.

We have shown that any self-adjoint operator A gives rise to a strongly continuous one-parameter group $\{e^{itA}\}_{t\in\mathbb{R}}$ of unitary operators. The much deeper converse is due to Marshall Stone.

Theorem 5.6.2 (Stone's theorem). *Let $\{U_t\}_{t\in\mathbb{R}}$ be a strongly continuous one-parameter group of unitary operators on the separable, complex Hilbert space $\mathcal{H}$. Then there exists a unique self-adjoint operator $A : \mathcal{D}(A) \to \mathcal{H}$ on $\mathcal{H}$ such that $U_t = e^{itA}$ for every $t \in \mathbb{R}$. Moreover, $U_t(\mathcal{D}(A)) \subseteq \mathcal{D}(A)$ and $AU_t = U_t A$ on $\mathcal{D}(A)$ for all $t \in \mathbb{R}$.*

Recall (Remark 3.3.2) that for one-parameter groups of unitary operators, strong continuity is equivalent to weak continuity and even to weak measurability. For a proof of Theorem 5.6.2 one can consult Theorem 6.1, Chapter IV, of [Prug], Theorem 1, Section XI.13, of [Yosida], Sections 137–138 of [RiSz.N] or Theorem VIII.8 of [RS1]. We will write out some concrete examples shortly.

The self-adjoint operator A whose existence is asserted by Stone's theorem will play a prominent role in the mathematical formalism of quantum mechanics. The (skew-adjoint) operator iA is called the *infinitesimal generator* of $\{U_t\}_{t\in\mathbb{R}}$ (some sources refer to A itself as the infinitesimal generator). Next we will try to illustrate these ideas by finding the infinitesimal generator for a particularly important strongly continuous one-parameter group of unitary operators.

Example 5.6.1. For this example our Hilbert space is $\mathcal{H} = L^2(\mathbb{R})$ and we define, for every $t \in \mathbb{R}$, the *translation operator*

$$U_t : L^2(\mathbb{R}) \to L^2(\mathbb{R})$$

by

$$(U_t \psi)(q) = \psi(q + t) \quad \forall \psi \in L^2(\mathbb{R}) \, \forall q \in \mathbb{R}.$$

Each U_t is clearly linear and invertible ($U_t^{-1} = U_{-t}$) and is an isometry because the Lebesgue measure on $\mathbb{R}$ is translation invariant. Also clear is the fact that $U_t U_s = U_{t+s}$, so $\{U_t\}_{t \in \mathbb{R}}$ is a one-parameter group of unitary operators on $L^2(\mathbb{R})$. Strong continuity may not be quite so clear, so we will prove it. We show that if $t \to 0$ in $\mathbb{R}$, then $U_t \psi \to \psi$ in $L^2(\mathbb{R})$ for any $\psi \in L^2(\mathbb{R})$. Note that, since each U_t is unitary,

$$\|U_t \psi - \psi\| \leq \|U_t(\psi - \phi)\| + \|U_t \phi - \phi\| + \|\phi - \psi\| = \|U_t \phi - \phi\| + 2\|\psi - \phi\| \quad (5.37)$$

for any $\psi, \phi \in L^2(\mathbb{R})$. From this it follows that it will suffice to prove strong continuity for ϕ in any dense subset of $L^2(\mathbb{R})$. Let us suppose then that $\phi \in C_0^\infty(\mathbb{R})$. Let $\epsilon > 0$ be given. We can find a compact set K in $\mathbb{R}$ for which the support of each $U_t \phi$ with $|t| \leq 1$ is contained in K. Let $d(K)$ denote the diameter of K in $\mathbb{R}$. Since ϕ is uniformly continuous, there is a δ satisfying $0 < \delta \leq 1$ such that $|t| < \delta$ implies $\|U_t \phi - \phi\|_\infty < \epsilon/d(K)^{1/2}$ (here $\|\ \|_\infty$ is just the sup norm). Since the support of $U_t \phi - \phi$ is contained in K,

$$\|U_t \phi - \phi\| \leq d(K)^{1/2} \|U_t \phi - \phi\|_\infty < \epsilon.$$

Thus, $\{U_t\}_{t \in \mathbb{R}}$ is strongly continuous at any $\phi \in C_0^\infty(\mathbb{R})$ and therefore strongly continuous at any $\psi \in L^2(\mathbb{R})$ by (5.37) and the fact that $C_0^\infty(\mathbb{R})$ is dense in $L^2(\mathbb{R})$.

To find the infinitesimal generator iA of $\{U_t\}_{t \in \mathbb{R}}$ we will use Lemma 5.6.1, according to which the domain of A is precisely the set of $\psi \in L^2(\mathbb{R})$ for which the limit $\lim_{t \to 0} \frac{U_t^A \psi - \psi}{t}$ exists and, for these, iA is given by

$$iA\psi = \lim_{t \to 0} \frac{U_t \psi - \psi}{t},$$

where the limit is in $L^2(\mathbb{R})$. Note that the domain of A certainly includes all ψ in the Schwartz space $\mathcal{S}(\mathbb{R})$ (see Section G.2) and for these the right-hand side converges *pointwise* to

$$\lim_{t \to 0} \frac{\psi(q + t) - \psi(q)}{t} = \frac{d\psi}{dq}.$$

Now, recall that if a sequence converges in $L^2(\mathbb{R})$, then a subsequence converges pointwise almost everywhere. Thus, on $\mathcal{S}(\mathbb{R})$, A agrees with $-i\frac{d}{dq}$. But we have already shown that $-i\hbar\frac{d}{dq}$ is essentially self-adjoint on $\mathcal{S}(\mathbb{R})$ and so it has a *unique* self-adjoint extension which we have denoted P (Example 5.5.3). Thus, $-i\frac{d}{dq}$ is essentially self-adjoint on $\mathcal{S}(\mathbb{R})$ and its unique self-adjoint extension is $\frac{1}{\hbar}P$. But A is also self-adjoint by Theorem 5.6.2 and extends $-i\frac{d}{dq}$, so we conclude that $A = \frac{1}{\hbar}P$, that is,

$$(U_t\psi)(q) = \psi(q+t) \quad \forall\psi \in L^2(\mathbb{R})\,\forall q \in \mathbb{R} \quad \Rightarrow \quad U_t = e^{itP/\hbar},$$

where P is the momentum operator on $\mathbb{R}$. We will say simply that *the momentum operator generates spatial translations*.

Example 5.6.2. For every $t \in \mathbb{R}$ define $V_t : L^2(\mathbb{R}) \to L^2(\mathbb{R})$ by

$$(V_t\psi)(q) = e^{itq}\psi(q) \quad \forall\psi \in L^2(\mathbb{R})\,\forall q \in \mathbb{R}.$$

Arguments similar to those in the previous example show that $\{V_t\}_{t\in\mathbb{R}}$ is a strongly continuous one-parameter group of unitary operators on $L^2(\mathbb{R})$ and that

$$V_t = e^{itQ},$$

where Q is the position operator on $\mathbb{R}$.

Proceeding in the other direction, that is, finding $U_t = e^{itA}$ for a given self-adjoint operator A, generally requires information about the spectrum of A. In at least one case it is easy to do.

Example 5.6.3. For an operator A, such as H_B of Example 5.5.4, with the property that $\mathcal{H}$ has an orthonormal basis $\psi_0, \psi_1,\ldots$ of eigenvectors for A ($A\psi_n = \lambda_n\psi_n$, $n = 0,1,\ldots$) one can proceed as follows. For any $\psi \in \mathcal{D}(A)$, write $\psi = \sum_{n=0}^{\infty}\langle\psi_n,\psi\rangle\psi_n$, where the convergence is in $\mathcal{H}$ ($L^2(\mathbb{R})$ for H_B). By the second part of Theorem 5.5.7.3, $U_t\psi_n = e^{itA}\psi_n = e^{it\lambda_n}\psi_n$ and, since $U_t = e^{itA}$ is unitary and therefore bounded (continuous),

$$U_t\psi = e^{itA}\psi = \sum_{n=0}^{\infty}\langle\psi_n,\psi\rangle e^{it\lambda_n}\psi_n. \tag{5.38}$$

This handles e^{itH_B}:

$$e^{itH_B}\psi = \sum_{n=0}^{\infty}\langle\psi_n,\psi\rangle e^{it\mathcal{E}_n}\psi_n. \tag{5.39}$$

We will also need e^{itH_0}, where H_0 is the free particle Hamiltonian in Example 5.2.8. This is most easily done by recalling that H_0 is unitarily equivalent, via the Fourier transform $\mathcal{F}$, to the multiplication operator $Q_{(\hbar^2/2m)p^2}$ (see (5.3)).

Exercise 5.6.1. Use the functional calculus to show that

$$e^{itH_0} = \mathcal{F}^{-1}Q_{g(p)}\mathcal{F}, \tag{5.40}$$

where $g(p) = e^{i(\hbar^2/2m)tp^2}$.

As we have mentioned before, in the formalism of quantum mechanics (Chapter 6) a self-adjoint operator A will represent an observable, a unit vector ψ will represent a state and the probability measure $\mu_{\psi,A} = \langle \psi, E^A \psi \rangle$ on $\mathbb{R}$ is interpreted as assigning to each Borel set S in $\mathbb{R}$ the probability that a measurement of A in the state ψ will yield a value in S. The nonnegative real number $\langle \psi, A\psi \rangle$ will be interpreted as the expected value of the observable A in the state ψ. One more of the basic postulates of quantum mechanics is that the time evolution of the state $\psi(t)$ of an isolated quantum system from some initial state $\psi(0)$ is described by a one-parameter group $\{U_t\}_{t \in \mathbb{R}}$ of unitary operators ($\psi(t) = U_t \psi(0)$). Stone's theorem then provides us with a self-adjoint operator (that is, observable) H that generates this time evolution in the sense that

$$\psi(t) = e^{itH/\hbar} \psi(0).$$

With $A = \frac{1}{\hbar} H$, (5.36) becomes

$$i\hbar \frac{d\psi(t)}{dt} = H(\psi(t)), \tag{5.41}$$

which will be the fundamental equation of nonrelativistic quantum mechanics (the *abstract Schrödinger equation*). We will explain the reason for the $1/\hbar$ in Section 6.2.

We consider next another application of the spectral theorem that will play a prominent role in our discussion of "compatible observables" in quantum mechanics (Section 6.2). Recall that if A and B are two *bounded* operators on $\mathcal{H}$, then both are defined on all of $\mathcal{H}$ and therefore so are the products (compositions) AB and BA. One then says that A and B *commute* if $AB = BA$, that is, if $A(B\psi) = B(A\psi)$ for all $\psi \in \mathcal{H}$. The same definition for unbounded operators does not lead to a useful notion since these are only densely defined and therefore $\mathcal{D}(AB) = \{\psi \in \mathcal{D}(B) : B\psi \in \mathcal{D}(A)\}$ and $\mathcal{D}(BA) = \{\psi \in \mathcal{D}(A) : A\psi \in \mathcal{D}(B)\}$ may intersect in nothing more than the zero vector in $\mathcal{H}$. For self-adjoint operators, however, one can formulate a useful notion in the following way. Note that if A and B are bounded and self-adjoint, then it follows from Theorem 5.5.7.5 that they commute if and only if their spectral resolutions pairwise commute, that is, if and only if $E_\lambda^A E_\mu^B = E_\mu^B E_\lambda^A$ for all $\lambda, \mu \in \mathbb{R}$. Since this latter condition makes sense even for unbounded, self-adjoint operators (because the spectral projections are bounded) we shall adopt the following definition. Let A and B be self-adjoint operators (either bounded or unbounded) on the separable, complex Hilbert space $\mathcal{H}$ and denote by E_λ^A and E_μ^B, $\lambda, \mu \in \mathbb{R}$, their corresponding resolutions of the identity. We will say that A and B *commute* if $E_\lambda^A E_\mu^B = E_\mu^B E_\lambda^A$ for all $\lambda, \mu \in \mathbb{R}$. One reason for the usefulness of this definition is to be found in the fact that two self-adjoint operators commute in this sense if and only if their corresponding one-parameter groups of unitary operators commute (in the usual sense of bounded operators); the following is Theorem VIII.13 of [RS1] and Theorem 6.2, Chapter IV, of [Prug].

Theorem 5.6.3. *Let A and B be self-adjoint operators on the separable, complex Hilbert space $\mathcal{H}$. Then A and B commute if and only if $e^{itA} e^{isB} = e^{isB} e^{itA}$ for all $s, t \in \mathbb{R}$.*

This definition is natural enough and is, as we shall see, the appropriate one for our applications to quantum mechanics, but its intuitive meaning is more elusive than it might seem. For example, one might suppose that if A and B are self-adjoint, $\mathcal{D}$ is a dense subspace in the intersection of their domains, $A : \mathcal{D} \to \mathcal{D}$, $B : \mathcal{D} \to \mathcal{D}$ and $A(B\psi) = B(A\psi)$ for all $\psi \in \mathcal{D}$, then A and B should commute in the sense we have defined. However, this is *not* the case (see Example 1, Section VIII.5, of [RS1]). The converse, however, is true, that is, if A and B commute, then $A(B\psi) = B(A\psi)$ for all ψ in any dense subspace contained in the intersection of their domains that is invariant under both A and B. This follows from the spectral theorem and can provide a useful means of showing that two unbounded, self-adjoint operators *do not* commute.

Example 5.6.4. We will consider the position $Q : \mathcal{D}(Q) \to L^2(\mathbb{R})$ and momentum $P : \mathcal{D}(P) \to L^2(\mathbb{R})$ operators on $\mathbb{R}$. We have seen that both are unbounded, self-adjoint operators on $L^2(\mathbb{R})$ and that the Schwartz space $\mathcal{S}(\mathbb{R})$ is a dense, invariant, linear subspace of both $\mathcal{D}(Q)$ and $\mathcal{D}(P)$. Let $\psi \in \mathcal{S}(\mathbb{R})$. Then $(P(Q\psi))(q) = -i\hbar[q\psi'(q) + \psi(q)]$ and $(Q(P\psi))(q) = -i\hbar q\psi'(q)$, and these are generally not the same, so Q and P do not commute.

Example 5.6.5. It follows from the functional calculus that if B is a self-adjoint operator on $\mathcal{H}$ and f and g are two real-valued continuous, or merely Borel functions on $\sigma(B)$, then $f(B)$ and $g(B)$ are self-adjoint and commute. Consequently, one can build arbitrarily large families of self-adjoint operators that commute in pairs by selecting a family $\{f_\alpha\}_{\alpha \in \mathcal{A}}$ of such real-valued functions on $\sigma(B)$ and building the operators $f_\alpha(B)$.

The principal application of these ideas to quantum mechanics rests on a theorem of von Neumann to the effect that *every* commuting family of self-adjoint operators is of the type described in the previous example. The following result is proved in Sections 130–131 of [RiSz.N].

Theorem 5.6.4. *Let $\mathcal{H}$ be a separable, complex Hilbert space and $\{A_\alpha\}_{\alpha \in \mathcal{A}}$ a family of pairwise commuting self-adjoint operators on $\mathcal{H}$ (A_{α_1} and A_{α_2} commute for all $\alpha_1, \alpha_2 \in \mathcal{A}$). Then there exists a self-adjoint operator B on $\mathcal{H}$ and real-valued Borel functions f_α, $\alpha \in \mathcal{A}$, such that $A_\alpha = f_\alpha(B)$ for every $\alpha \in \mathcal{A}$.*

We will conclude with a result that will be essential when we discuss canonical quantization in Chapter 7. Recall that if A and B are *bounded* operators on $\mathcal{H}$, then one defines their commutator by $[A, B]_- = AB - BA$ and this is a bounded operator defined on all of $\mathcal{H}$ that is the zero operator if and only if A and B commute. If A and B are *unbounded* operators, we have already noted that one can still define $[A, B]_- = AB - BA$, but only on $\mathcal{D}([A, B]_-) = \mathcal{D}(AB) \cap \mathcal{D}(BA)$, and this might well consist of the zero vector alone. However, this generally does not occur for the operators of interest in quantum mechanics. Here is an example.

Example 5.6.6. For the position Q and momentum P operators on $\mathbb{R}$, the domain of $[Q,P]_-$ contains at least $S(\mathbb{R})$. The computations in Example 5.6.4 show that, on $S(\mathbb{R})$,

$$[Q,P]_- = i\hbar,$$

where, as usual, $i\hbar$ means $(i\hbar)\,\mathrm{id}_{S(\mathbb{R})}$.

If A and B are unbounded operators that commute, then their commutator is zero wherever it is defined, but the converse is certainly *not* true, that is, $[A,B]_- = 0$ on $\mathcal{D}([A,B])$ does not imply that A and B commute. The following result (which is Theorem 6.3, Chapter IV, of [Prug]) will allow us to circumvent many such domain issues associated with commutators when discussing the canonical commutation relations in quantum mechanics (Chapter 7).

Theorem 5.6.5. *Let A and B be self-adjoint operators on the separable, complex Hilbert space $\mathcal{H}$ that satisfy*

$$e^{itA}e^{isB} = e^{-ist}e^{isB}e^{itA}$$

for all $s,t \in \mathbb{R}$. Then

$$[A,B]_-\psi = i\psi$$

for all $\psi \in \mathcal{D}([A,B]_-)$.

6 The postulates of quantum mechanics

6.1 Introduction

Physical theories are expressed in the language of mathematics, but these mathematical models are not the same as the physical theories and they are not unique. Here we will build on the physical and mathematical background assembled in the previous chapters to construct one possible mathematical framework in which to do quantum mechanics. The particular model we formulate goes back to von Neumann [vonNeu], but there are others and we will have a look at quite a different idea due to Feynman [Brown] in Chapter 8. We will describe the basic ingredients of the model in a sequence of postulates, each of which will be accompanied by a brief commentary on where it came from, what it means and what issues it raises. You will notice that the term *quantum system* is used repeatedly, but never defined; the same is true of the term *measurement*. We will attempt some clarification of these terms as we proceed, but it is not possible, nor would it be profitable, to try to define them precisely; they are defined by the assumptions we make about them in the postulates. With the postulates in place we will derive some consequences with important physical implications. Concrete examples of such quantum theories are obtained by making specific choices for the items described in the postulates. Some of these are obtained by "quantizing" classical physical systems; our principal examples are the canonical quantizations of the free particle and the harmonic oscillator, both discussed in Chapter 7. Some, on the other hand, have no classical counterpart; the fermionic and supersymmetric harmonic oscillators are of this type and will be described in Chapter 10.

6.2 The postulates

Postulate QM1. *To every quantum system is associated a separable, complex Hilbert space $\mathcal{H}$. The (pure) states of the system are represented by vectors $\psi \in \mathcal{H}$ with $\|\psi\| = 1$ and, for any $c \in \mathbb{C}$ with $|c| = 1$, ψ and $c\psi$ represent the same state.*

We have already spent a fair amount of time trying to prepare the way for this first postulate. We have seen in Chapter 4 that the physical systems for which classical mechanics fails to provide an accurate model are fundamentally probabilistic in nature and so should be regarded as more analogous to the systems treated in classical statistical mechanics. Here states of the system are identified with probability distributions and Koopman showed how to identify these with unit vectors in a Hilbert space; moreover, two such unit vectors that differ only in phase (that is, by a complex multiple of modulus one) give rise to the same probability distribution and therefore the same state. There is, however, a bit more to say about Postulate QM1.

All n-dimensional complex Hilbert spaces are isometrically isomorphic, as are all separable, complex, infinite-dimensional Hilbert spaces (Theorem 6.4.8 of [Fried]), so

https://doi.org/10.1515/9783110751949-006

the choice of $\mathcal{H}$ is very often a matter of convenience. Often $\mathcal{H}$ will be L^2 of some appropriate measure space (such as a classical configuration space), in which case a unit vector ψ representing a state is called a *wave function* for that state. In Koopman's translation of statistical mechanics, the wave functions ψ were auxiliary devices for producing probability distributions $|\psi|^2$, but in quantum theory they play the much more fundamental role of what are called *probability amplitudes*. Being complex-valued, one can incorporate the wave-like interference effects of quantum systems pointed out in Section 4.4 directly into the algebraic operations on the wave functions via

$$|\psi_1 + \psi_2|^2 = |\psi_1|^2 + |\psi_2|^2 + 2\,\mathrm{Re}\langle\psi_1, \psi_2\rangle. \tag{6.1}$$

Here the final term $2\,\mathrm{Re}\langle\psi_1, \psi_2\rangle$ can be viewed as an *interference term* that appears when ψ_1 and ψ_2 interact (like the waves emerging from two slits in Section 4.4).

We should also say something about the parenthetical adjective *pure* in Postulate QM1. In certain contexts, notably quantum statistical mechanics, there is an additional probabilistic element that we have not yet encountered; one might know only that the state of the system is ψ with a certain probability $0 \le t \le 1$. If $\psi_1, \psi_2, \ldots$ are unit vectors in the Hilbert space $\mathcal{H}$ of some quantum system, then a convex combination

$$t_1\psi_1 + t_2\psi_2 + \cdots,$$

with

$$t_n \ge 0 \quad \forall n \ge 1 \quad \text{and} \quad t_1 + t_2 \cdots = 1,$$

is called a *mixed state* for the system and is interpreted as describing a situation in which it is known only that the state of the system is ψ_n with probability t_n for $n = 1, 2, \ldots$. The unmodified term *state* will, for us, always mean *pure state*.

Note that Postulate QM1 does *not* assert that every unit vector in $\mathcal{H}$ represents a possible state of the system, but only that every state is represented by some unit vector. This has to do with what are known in physics as *superselection rules*. Section 1-1 of [SW] contains a brief discussion of this, but roughly the idea is as follows. There are certain conservation laws in physics (of charge, for example) that appear to prohibit the mingling of states for which the value of the conserved quantity is different. For instance, a superposition (sum) of states with different charges is believed not to be physically realizable. This has the consequence of splitting $\mathcal{H}$ up into a direct sum of so-called *coherent subspaces* with the property that only unit vectors in these subspaces represent physically realizable states. However, these superselection rules are generally significant only when the number and type of particles in the system can vary. This is a fundamental feature of quantum field theory (particle creation and annihilation), but will play no role in our lives here. As a result, we will allow ourselves to assume that any unit vector in $\mathcal{H}$ is a possible state of our system.

Finally, we point out that, because unit vectors in $\mathcal{H}$ that differ only by a phase factor describe the same state, one can identify the *state space* $\mathcal{S}$ of a quantum system with the *complex projective space* $\mathbb{CP}(\mathcal{H})$ of $\mathcal{H}$, that is, the quotient of the unit sphere in $\mathcal{H}$ by the equivalence relation that identifies two points if they differ by a complex factor of modulus one (equivalently, the quotient of the set of nonzero elements of $\mathcal{H}$ by the equivalence relation that identifies two points if they differ by a nonzero complex factor). When we have occasion to do this we will write $\mathbf{\Psi}$ for the equivalence class containing ψ and refer to it as a *unit ray* in $\mathcal{H}$. It will sometimes also be convenient to identify the state represented by the unit vector ψ with the operator P_ψ that projects $\mathcal{H}$ onto the one-dimensional subspace of $\mathcal{H}$ spanned by ψ (which clearly depends only on the state and not on the unit vector representing it). In all candor, however, it is customary to be somewhat loose with the terminology and speak of "the state ψ" when one really means "the state $\mathbf{\Psi}$," or "the state P_ψ." Since this is almost always harmless, we will generally adhere to the custom.

Although we will make no use of it, we should mention that there is one other way to view the states (both pure and mixed). Each P_ψ is a projection operator and so, in particular, it is a positive, self-adjoint, trace class operator with trace one. A mixed state, as defined above, can be identified with $\sum_{n=1}^{\infty} t_n P_{\psi_n}$, where $t_n \geq 0 \; \forall n = 1, 2, \ldots$ and $\sum_{n=1}^{\infty} t_n = 1$, and this is also a positive, self-adjoint, trace class operator with trace one; the spectrum consists precisely of the eigenvalues $t_1, t_2, \ldots$. Conversely, every positive, self-adjoint, trace class operator with trace one is a pure or mixed state because it is compact, so the spectral theorem gives a decomposition of this form. One can therefore identify the collection of all pure and mixed states with the positive, self-adjoint, trace class operators with trace one (this is the approach taken in [Takh], for example).

We have seen that in Koopman's model of classical statistical mechanics the observables are identified with unbounded, real multiplication operators on some L^2 (Section 3.3). We take this, together with the fact that self-adjoint operators on a Hilbert space can all be identified with real multiplication operators on some L^2 (Theorems 5.5.1 and 5.5.2), as motivation for the first half of our next postulate.

Postulate QM2. *For a quantum system with Hilbert space $\mathcal{H}$, every observable is identified with a self-adjoint operator $A : \mathcal{D}(A) \to \mathcal{H}$ on $\mathcal{H}$ and any possible outcome of a measurement of the observable is a real number that lies in the spectrum $\sigma(A)$ of A.*

We will use the same symbol for the self-adjoint operator and the physical observable it represents. If A is such an observable and f is a *real-valued* Borel function on $\sigma(A)$, then $f(A)$ is self-adjoint and we will identify it with the physical observable defined in the following way. A measurement of $f(A)$ is accomplished by first measuring A. If the measured value of A is $a \in \sigma(A)$, then the measured value of $f(A)$ is defined to be $f(a)$ (note that $f(a) \in \sigma(f(A))$ by Theorem 5.5.7.3).

The problem of measurement in quantum mechanics is very subtle and, after nearly a century, still controversial. One generally knows what should qualify, physically, as an "observable." For a single particle, for example, the total energy, a coordinate of the position or a component of the momentum should be things that one can measure. The result of a measurement is a real number. Whatever the specifics of the measurement process, it must involve allowing the quantum system to interact with some external and generally macroscopic system (the "measuring device"). We have seen (Section 4.4) that such an interaction has inevitable and unpredictable effects on the quantum system so that repeated measurements on identical systems generally do not give the same results. One can only hope to know the probability that some specific measurement on a system in some specific state will yield a value in some given interval of the real line.

The most significant aspect of Postulate QM2 is its assertion that these measured values, although they appear random in the measurement process, are generally *not* arbitrary, but are constrained to lie in some fixed subset of the real line. This subset was described as the spectrum of some self-adjoint operator A on $\mathcal{H}$. One might think of A as a convenient bookkeeping device. It is the possible measured values of an observable and their probability distributions that are physically significant and not the particular way we choose to keep track of them. Heisenberg's *matrix mechanics* and Schrödinger's *wave mechanics*, which were the original formulations of quantum mechanics, accomplished the same purpose with infinite matrices and differential operators, respectively, and von Neumann's choice of self-adjoint operators on $\mathcal{H}$ was essentially a general, mathematically rigorous way of doing both at once (the first when $\mathcal{H} = \ell^2$ and the second when $\mathcal{H} = L^2$). We should also point out that the assertion about the spectrum of A in Postulate QM2 is, in some statistical sense, redundant. It follows from our next Postulate QM3 and the fact that the spectral measure of A is concentrated on $\sigma(A)$ that the probability of measuring a value outside the spectrum is zero.

We should also emphasize that Postulate QM2 asserts that every physical observable corresponds to some self-adjoint operator, but *not* that every self-adjoint operator corresponds to some physical observable. This is again related to the existence of superselection rules and so, for the reasons we have already discussed, we will, for the time being at least, allow ourselves to regard any self-adjoint operator as corresponding to some physical observable. Also note that the procedures by which one chooses operators to represent specific observables are not addressed by Postulate QM2. These procedures are collectively referred to as *quantization* and physicists have laid down certain rules of the game. We will discuss some of the generally agreed upon principles of the quantization procedure in Chapter 7, but any real understanding of what is involved and how it is done must come from the physicists (see [SN]).

Example 6.2.1. In order to ground this discussion in something a bit more concrete, let us get ahead of ourselves just a bit and anticipate something we will discover in Chapter 7. We will consider a quantum system, such as a diatomic molecule (Chapter 1), whose behavior would classically be approximated by a harmonic oscillator. Assuming that any translational or rotational energy is negligible, the total energy of the system is just the vibrational part that we described classically by the Hamiltonian (2.50). Applying the "generally agreed upon principles" of quantization to the total energy observable of this harmonic oscillator leads one to the operator H_B on $L^2(\mathbb{R})$ that we have discussed in some considerable detail in Chapter 5 (see Examples 5.3.1, 5.4.5, 5.5.1 and 5.5.4). We know, in particular, that the spectrum of H_B consists precisely of the eigenvalues $\mathcal{E}_n = (n + \frac{1}{2})\hbar\omega$, $n = 0, 1, 2, \ldots$, where ω is the natural frequency of the classical oscillator. These then should be (approximately) the possible measured energy levels of the molecule. Whether or not this is true, of course, is something that must be determined in the laboratory. The fact that it actually is true is one of the many circumstances that encourage confidence in this seemingly exotic structure we are building.

Postulate QM3. *Let $\mathcal{H}$ be the Hilbert space of a quantum system, $\psi \in \mathcal{H}$ a unit vector representing a state of the system and $A : D(A) \to \mathcal{H}$ a self-adjoint operator on $\mathcal{H}$ representing an observable. Let E^A be the unique spectral measure on $\mathbb{R}$ associated with A and $\{E_\lambda^A\}_{\lambda \in \mathbb{R}}$ the corresponding resolution of the identity. Denote by $\mu_{\psi,A}$ the probability measure on $\mathbb{R}$ that assigns to every Borel set $S \subseteq \mathbb{R}$ the probability that a measurement of A when the state is ψ will yield a value in S. Then, for every Borel set S in $\mathbb{R}$, we assume that $\mu_{\psi,A}(S)$ is given by the* Born–von Neumann formula

$$\mu_{\psi,A}(S) = \langle \psi, E^A(S)\psi \rangle = \|E^A(S)\psi\|^2. \tag{6.2}$$

If the state ψ is in the domain of A, then the expected value of A in state ψ is

$$\langle A \rangle_\psi = \int_{\mathbb{R}} \lambda \, d\langle \psi, E_\lambda^A \psi \rangle = \langle \psi, A\psi \rangle \tag{6.3}$$

and its dispersion (variance) is

$$\sigma_\psi^2(A) = \int_{\mathbb{R}} \left(\lambda - \langle A \rangle_\psi\right)^2 d\langle \psi, E_\lambda^A \psi \rangle = \left\| \left(A - \langle A \rangle_\psi\right)\psi \right\|^2 = \|A\psi\|^2 - \langle A \rangle_\psi^2. \tag{6.4}$$

It is not an exaggeration to say that Postulate QM3 is the heart and soul of quantum mechanics, so we will linger over it for a while. We should first be clear on how we will interpret the sort of probabilistic statement made in Postulate QM3. Given a quantum system in state ψ and an observable A, *quantum mechanics generally makes no predictions about the result of a single measurement of A.* For example, one cannot predict the location of the dot on the screen when a single electron is sent toward it in the two-slit experiment (Section 4.4). Rather, it is assumed that the state ψ can be replicated over and over again and the measurement performed many times on these identical systems. The probability of a given outcome might then be thought of intuitively as the relative frequency of the outcome for a "large" number of repetitions. More precisely, *the probability of a given outcome for a measurement of some observable in some state is the limit of the relative frequencies of this outcome as the number of repetitions of the measurement in this state approaches infinity.*

Remark 6.2.1. Precisely how one replicates (*prepares*) the state ψ over and over again is a delicate issue that depends a great deal on the particular system under consideration. The preparation of a state is really a particular type of measurement; these are called *preparatory measurements* in [Prug], where one can find a more detailed discussion in Chapter IV, Section 1.4.

Next we point out that if one accepts the interpretation of the probability measure $\langle \psi, E^A \psi \rangle$ in Postulate QM3, then $\langle \psi, E^A_\lambda \psi \rangle$ is the distribution function for the random variable represented by the measurement process. Thus, (6.3) is just the Stieltjes integral formula (3.1) for the expected value of a random variable and the first equality in (6.4) is the analogous formula (3.2) for the variance.

Remark 6.2.2. The second equality is proved below and the third follows from

$$
\begin{aligned}
\left\| (A - \langle A \rangle_\psi) \, \psi \right\|^2 &= \|A\psi\|^2 - \langle A\psi, \langle A \rangle_\psi \psi \rangle - \langle \langle A \rangle_\psi \psi, A\psi \rangle + \|\langle A \rangle_\psi \psi\|^2 \\
&= \|A\psi\|^2 - 2\langle A \rangle_\psi \langle \psi, A\psi \rangle + \langle A \rangle_\psi^2 \|\psi\|^2 \\
&= \|A\psi\|^2 - 2\langle A \rangle_\psi^2 + \langle A \rangle_\psi^2 \\
&= \|A\psi\|^2 - \langle A \rangle_\psi^2.
\end{aligned}
$$

As usual, $\sigma_\psi^2(A)$ measures the dispersion (spread) of the measured values around $\langle A \rangle_\psi$. The dispersion is often identified with a measure of the limitations on the accuracy of the measurements. Note, however, that Postulate QM3 makes no reference to any specific procedure for making the measurements, so $\sigma_\psi^2(A)$ does not in any sense describe the frailties of our instrumentation. Rather, it represents an *intrinsic* limitation on our knowledge of A even when the state ψ of the system is known. It is zero precisely when the probability that the measurement will result in the expected value is 1 (meaning that the relative frequency of the outcome $\langle A \rangle_\psi$ can be made arbitrarily close to 1 with a sufficient number of repetitions of the measurement of A in state ψ). We claim that $\sigma_\psi^2(A) = 0$ *if and only if ψ is an eigenvector of A with eigenvalue* $\langle A \rangle_\psi$. To see this one simply applies part 7. of Theorem 5.5.7 to the continuous function $f(\lambda) = \lambda - \langle A \rangle_\psi$, thereby obtaining

$$
\left\| (A - \langle A \rangle_\psi) \psi \right\|^2 = \int_{\mathbb{R}} (\lambda - \langle A \rangle_\psi)^2 d\langle \psi, E^A_\lambda \psi \rangle = \sigma_\psi^2(A)
$$

(which, incidentally, gives the second equality in (6.4)), so that

$$
\sigma_\psi^2(A) = 0 \quad \Leftrightarrow \quad A\psi = \langle A \rangle_\psi \psi.
$$

For a fixed observable A we think of the measurement process itself as defining a random variable on the state space; in each state ψ a measurement of A results in a real number. The probability measure $\mu_{\psi,A}$ that assigns to a Borel set S in $\mathbb{R}$ the probability that a measurement of A in the state ψ will be in S depends, of course, on the specifics of the quantum system, the observable and the state. The assertion of Postulate QM3 that $\mu_{\psi,A}(S) = \langle \psi, E^A(S)\psi \rangle$ is the presumed link between the physics and the mathematical formalism.

We have made a point of stressing that conceptually quantum mechanics should be more akin to classical statistical mechanics than to classical particle mechanics so that one is not surprised to see the measurement process described in terms of probability measures. Postulate QM3, however, is quite explicit about how these probability measures are determined by A and ψ. The idea is due to Max Born, and von Neumann phrased the idea in the functional analytic terms we have used. To properly understand what led Born to Postulate QM3 one must really follow the struggles of the physicists in 1925, 1926 and 1927 who sought some conceptual basis for the new and very successful, but rather mysterious mechanics of Heisenberg and Schrödinger. In these early years physics was in a rather odd position. Heisenberg formulated his matrix mechanics without knowing what a matrix is and without having a precise idea of what the entries in his rectangular arrays of numbers should mean physically (see Section 7.1 for a bit more on this). Nevertheless, the rules of the game as he laid them down predicted precisely the spectrum of the hydrogen atom. Schrödinger formulated a differential equation for his wave function that yielded the same predictions, but no one had any real idea what the wave function was supposed to represent (Schrödinger himself initially viewed it as a sort of "charge distribution"). It was left to Born, and then Niels Bohr and his school in Copenhagen, to supply the missing conceptual basis for quantum mechanics. It is our good fortune that Born himself, in his Nobel Prize Lecture in 1954, has provided us with a brief and very lucid account of the evolution of his idea and we will simply refer those interested in pursuing this to [Born1]. Interestingly, Born attributes to Einstein the inspiration for the idea, although Einstein never acquiesced to its implications.

Remark 6.2.3. We have already mentioned that there are various approaches to the foundations of quantum mechanics other than the one we are describing here. One such approach, which has a much more algebraic flavor, is formulated in the language of *Banach algebras* or, more specifically, C^*-*algebras* and *von Neumann algebras* (we will not provide the definitions, but if these ideas are unfamiliar we can refer to a very nice introduction in Part Three of [Simm1]). Here each quantum system has an associated C^*-algebra $\mathcal{A}$ and the self-adjoint elements of $\mathcal{A}$ (those satisfying $a^* = a$) represent the observables. Now, there is a very famous result of Gelfand and Naimark [GN] which asserts that any C^*-algebra is isomorphic to a norm closed, self-adjoint subalgebra of the algebra $\mathcal{B}(\mathcal{H})$ of bounded operators on some Hilbert space $\mathcal{H}$. Consequently, one of the advantages to this approach is that unbounded, self-adjoint operators do not arise. Physically, this is possible because all of the relevant physical information is contained in the probability measures $\mu_{\psi,A}(S)$ which, according to Postulate QM3, are given by $\langle \psi, E^A(S)\psi \rangle$ and these are determined by the (bounded) projections $E^A(S)$ corresponding to our unbounded, self-adjoint operators A. However, the bounded operators associated with the elements of the C^*-algebra $\mathcal{A}$ by the Gelfand–Naimark theorem are not canonically determined and hence one loses the direct relations to the physics that we will be exploring.

We have offered nothing in the way of motivation for the Born–von Neumann formula except a reference to [Born1]. By way of compensation, we will write out a number of concrete examples.

Example 6.2.2. Let us once again anticipate one simple, but particularly important special case that we will discuss in more detail in Chapter 7 and that may already be familiar to those who have read anything at all about quantum mechanics elsewhere. We consider the Hilbert space $\mathcal{H} = L^2(\mathbb{R})$ and the operator $Q : \mathcal{D}(Q) \rightarrow L^2(\mathbb{R})$ defined by $(Q\psi)(q) = q\psi(q)$ for all $\psi \in \mathcal{D}(Q)$ (see Example 5.2.3). We have referred to Q as the position operator on $\mathbb{R}$. We found the spectral resolution of Q in Example 5.5.2 and conclude from this that $E^Q(S)\psi = \chi_S \cdot \psi$ for any Borel set $S \subseteq \mathbb{R}$, where χ_S is the characteristic function of S. Thus,

$$(E^Q(S)\psi)(q) = \begin{cases} \psi(q), & \text{if } q \in S, \\ 0, & \text{if } q \notin S. \end{cases}$$

Consequently, if ψ is a state, then Postulate QM3 gives

$$\begin{aligned} \mu_{\psi,Q}(S) &= \langle \psi, E^Q(S)\psi \rangle = \|E^Q(S)\psi\|^2 \\ &= \int_{\mathbb{R}} \overline{(\chi(S) \cdot \psi)(q)}(\chi(S) \cdot \psi)(q) \, dq \\ &= \int_S \overline{\psi(q)}\psi(q) \, dq, \end{aligned}$$

so

$$\mu_{\psi,Q}(S) = \int_S |\psi(q)|^2 \, dq.$$

With this we arrive at the oft-repeated interpretation of the wave function ψ of a single particle moving in one dimension as the *probability amplitude for position*, meaning that the probability of finding the particle in some Borel subset S of $\mathbb{R}$ is $\int_S |\psi(q)|^2 \, dq$. Stated differently, $|\psi(q)|^2$ is the *probability density for position* measurements. In particular, the probability of finding the particle *somewhere* in $\mathbb{R}$ is $\|\psi\|_{L^2} = 1$.

In a sense we have reversed the historical logic here. The wave function ψ of a particle is often thought to be *defined* as the probability amplitude for the particle's position. Thought of in these terms, our example demonstrates that $(Q\psi)(q) = q\psi(q)$ is the "right" choice for an operator to represent position in quantum mechanics.

Example 6.2.3. For any real constant α we define a unit vector in $L^2(\mathbb{R})$ representing a state by

$$\phi(q) = \frac{1}{\pi^{1/4}} e^{-q^2/2} e^{i\alpha q}.$$

Note that ϕ is in the domain of Q since $\int_{\mathbb{R}} q^2 |\phi(q)|^2 dq = \frac{1}{\pi^{1/2}} \int_{-\infty}^{\infty} q^2 e^{-q^2} dq = \frac{1}{2} < \infty$ (see Example 5.2.3 and Exercise A.0.3.2). Consequently,

$$\langle Q \rangle_\phi = \langle \phi, Q\phi \rangle = \int_{\mathbb{R}} \overline{\phi(q)}(q\phi(q))\, dq = \frac{1}{\pi^{1/2}} \int_{-\infty}^{0} q e^{-q^2} dq + \frac{1}{\pi^{1/2}} \int_{0}^{\infty} q e^{-q^2} dq = 0.$$

In the state ϕ, the expected value of a position measurement is 0. Nevertheless, by Example 6.2.2, the probability that a measurement of position will yield a value in some Borel set $S \subseteq \mathbb{R}$ is

$$\mu_{\phi,Q}(S) = \int_S |\phi(q)|^2 dq = \frac{1}{\pi^{1/2}} \int_S e^{-q^2} dq,$$

which, if S does not have measure zero, is nonzero, although small if S is far from 0. The dispersion is

$$\sigma_\phi^2(Q) = \int_{\mathbb{R}} (\lambda - \langle Q \rangle_\phi)^2 d\langle \phi, E_\lambda^Q \phi \rangle = \int_{\mathbb{R}} \lambda^2 d\langle \phi, E_\lambda^Q \phi \rangle = \|Q\phi\|^2 \quad (\text{see } (6.4))$$

$$= \int_{\mathbb{R}} \overline{q\phi(q)}\, q\phi(q)\, dq = \frac{1}{\pi^{1/2}} \int_{-\infty}^{\infty} q^2 e^{-q^2} dq = \frac{1}{2} \quad (\text{see Exercise A.0.3.2}).$$

Note, by the way, that, since Q has no eigenvalues, the dispersion cannot be zero in *any* state; we will see this again in our discussion of the uncertainty relation in Section 6.3.

Example 6.2.4. Note that the state ϕ in the previous example is, except for a few constants, the ground state ψ_0 of the Hamiltonian H_B for the bosonic harmonic oscillator (Example 5.3.1). We would now like to compute the expected value and dispersion of Q in all of the eigenstates ψ_n, $n = 0, 1, 2, \ldots$, of H_B. Our approach this time will be quite different, however. In particular, we will avoid doing any integrals by exploiting properties of the raising and lowering operators introduced in Example 5.3.1. Recall that these are defined by

$$b = \frac{1}{\sqrt{2m\omega\hbar}} (m\omega Q + iP)$$

and

$$b^\dagger = \frac{1}{\sqrt{2m\omega\hbar}} (m\omega Q - iP).$$

Solving for Q and P gives

$$Q = \sqrt{\frac{\hbar}{2m\omega}}\,(b + b^\dagger)$$

and

$$P = -i\sqrt{\frac{m\omega\hbar}{2}}\,(b - b^\dagger).$$

Thus, for any $n = 0, 1, \ldots,$

$$\langle Q \rangle_{\psi_n} = \langle \psi_n, Q\psi_n \rangle = \sqrt{\frac{\hbar}{2m\omega}}\,\langle \psi_n, (b + b^\dagger)\psi_n \rangle = \sqrt{\frac{\hbar}{2m\omega}}\,[\langle \psi_n, b\psi_n \rangle + \langle \psi_n, b^\dagger \psi_n \rangle].$$

For $n \geq 1$ this becomes

$$\langle Q \rangle_{\psi_n} = \sqrt{\frac{\hbar}{2m\omega}}\,[\langle \psi_n, \sqrt{n}\,\psi_{n-1} \rangle + \langle \psi_n, \sqrt{n+1}\,\psi_{n+1} \rangle],$$

which is zero because $\{\psi_n\}_{n=0}^\infty$ is an orthonormal basis for $L^2(\mathbb{R})$, whereas, for $n = 0$, we obtain

$$\langle Q \rangle_{\psi_0} = \sqrt{\frac{\hbar}{2m\omega}}\,[\langle \psi_0, 0 \rangle + \langle \psi_0, \psi_1 \rangle],$$

which is again zero. Consequently, the expected value of the position operator is zero in every eigenstate:

$$\langle Q \rangle_{\psi_n} = 0, \quad n = 0, 1, \ldots.$$

For the dispersion we have

$$\sigma_{\psi_n}^2(Q) = \|Q\psi_n\|^2 - \langle Q \rangle_{\psi_n} = \|Q\psi_n\|^2 = \langle Q\psi_n, Q\psi_n \rangle = \langle \psi_n, Q^2\psi_n \rangle$$
$$= \frac{\hbar}{2m\omega}\,\langle \psi_n, (b + b^\dagger)^2 \psi_n \rangle$$
$$= \frac{\hbar}{2m\omega}\,[\langle \psi_n, b^2\psi_n \rangle + \langle \psi_n, (b^\dagger)^2\psi_n \rangle + \langle \psi_n, bb^\dagger\psi_n \rangle + \langle \psi_n, b^\dagger b\psi_n \rangle].$$

The first and second terms are zero for exactly the same reason as in the argument for $\langle Q \rangle_{\psi_n}$ above. For the third and fourth terms we note that, on $S(\mathbb{R})$,

$$b^\dagger b + bb^\dagger = 2b^\dagger b + (bb^\dagger - b^\dagger b) = 2b^\dagger b + [b, b^\dagger]_- = 2b^\dagger b + 1 = 2\left(N_B + \frac{1}{2}\right) = \frac{2}{\hbar\omega}H_B.$$

Consequently,

$$\sigma_{\psi_n}^2(Q) = \frac{\hbar}{2m\omega}\,\frac{2}{\hbar\omega}\,[\langle \psi_n, H_B\psi_n \rangle] = \frac{\hbar}{m\omega}\,\frac{\mathcal{E}_n}{\hbar\omega}\,\langle \psi_n, \psi_n \rangle,$$

so

$$\sigma^2_{\psi_n}(Q) = \frac{\hbar}{m\omega}\left(n + \frac{1}{2}\right),\tag{6.5}$$

which becomes arbitrarily large as n increases, that is, as the energy increases.

Example 6.2.5. Note that everything is much easier for the energy H_B itself:

$$\langle H_B \rangle_{\psi_n} = \langle \psi_n, H_B \psi_n \rangle = \mathcal{E}_n \langle \psi_n, \psi_n \rangle = \mathcal{E}_n = \left(n + \frac{1}{2}\right)\hbar\omega,$$

$$\sigma^2_{\psi_n}(H_B) = \|H_B \psi_n\|^2 - \langle H_B \rangle^2_{\psi_n} = \mathcal{E}_n^2 - \mathcal{E}_n^2 = 0.$$

Example 6.2.6. Having interpreted $|\psi(q)|^2$ as the probability density for position measurements in $\mathbb{R}$ we would like to do something similar for momentum. For this we recall that we have already shown that Q and P are unitarily equivalent via the Fourier transform (Example 5.2.7). The idea is to use the Fourier transform $\mathcal{F}$ to move our entire picture from one copy of $L^2(\mathbb{R})$, which physicists call the *position representation*, to another, called the *momentum representation*. The process will perhaps be a bit clearer if we distinguish these two copies of $L^2(\mathbb{R})$ by writing $L^2(\mathbb{R}, dq)$ for the copy in which ψ lives (position representation) and $L^2(\mathbb{R}, dp)$ for the copy in which $\mathcal{F}\psi = \hat{\psi}$ lives (momentum representation). Note that the word *representation* is actually used here in a technical, mathematical sense. When we turn to quantization in Chapter 7 we will see that each of these pictures arises from an "irreducible, unitary, integrable representation of the Heisenberg algebra."

The Fourier transform is therefore regarded as an isometric isomorphism

$$\mathcal{F} : L^2(\mathbb{R}, dq) \to L^2(\mathbb{R}, dp)$$

of L^2 of *position space* onto L^2 of *momentum space*. Any state $\psi \in L^2(\mathbb{R}, dq)$ can then be identified with its image $\mathcal{F}\psi = \hat{\psi} \in L^2(\mathbb{R}, dp)$ and any self-adjoint operator $A :$ $D(A) \to L^2(\mathbb{R}, dq)$ on $L^2(\mathbb{R}, dq)$ can be identified with a self-adjoint operator $\hat{A} = \mathcal{F}A\mathcal{F}^{-1} : \mathcal{F}(D(A)) \to L^2(\mathbb{R}, dp)$ on $L^2(\mathbb{R}, dp)$. In particular, the momentum operator P on $L^2(\mathbb{R}, dq)$ is identified with the operator $\hat{P}$ on $L^2(\mathbb{R}, dp)$ defined by

$$(\hat{P}\hat{\psi})(p) = [\,(\mathcal{F}P\mathcal{F}^{-1})\hat{\psi}\,](p) = \mathcal{F}\left(-i\hbar\frac{d\psi}{dq}\right) = \hbar p \hat{\psi}(p),$$

which is just $\hbar$ times multiplication by the coordinate in $\mathbb{R}$ for $L^2(\mathbb{R}, dp)$. From Example 5.5.2 we know the spectral measure of this operator and we conclude, just as we did there, that

$$\mu_{\hat{\psi},\hat{P}}(S) = \|E^{\hat{P}}(S)\hat{\psi}\|^2 = \int_S |\hbar \hat{\psi}(p)|^2 dp$$

for any Borel set $S \subseteq \mathbb{R}$. But note also that

$$\hat{P} = \mathcal{F}P\mathcal{F}^{-1} \quad \Rightarrow \quad E^{\hat{P}}(S) = \mathcal{F}E^P(S)\mathcal{F}^{-1}$$

$$\Rightarrow \quad \|E^{\hat{P}}(S)\hat{\psi}\|^2 = \|\mathcal{F}E^P(S)\psi\|^2 = \|E^P(S)\psi\|^2 = \mu_{\psi,P}(S),$$

so

$$\mu_{\psi,P}(S) = \int_S |\hbar \hat{\psi}(p)|^2 dp.$$

Thus, $|\hbar\hat{\psi}(p)|^2$ is the *probability density for momentum* measurements in $\mathbb{R}$ and $\hbar\hat{\psi}$ is the *probability amplitude for momentum*.

Example 6.2.7. The arguments from Example 6.2.4 can be repeated essentially *verbatim* to compute the expected value and dispersion of the momentum operator P in all of the eigenstates ψ_n, $n = 0, 1, \ldots$, of H_B, so we will simply record the results. For each $n = 0, 1, \ldots$,

$$\langle P \rangle_{\psi_n} = 0$$

and

$$\sigma^2_{\psi_n}(P) = m\omega\hbar\left(n + \frac{1}{2}\right).$$

For future reference (in Section 6.3) we record a few formulas for the dispersions of the position and momentum observables on $\mathbb{R}$. For Q it follows directly from the second equality in (6.4) that

$$\sigma^2_\psi(Q) = \|(Q - \langle Q\rangle_\psi)\psi\|^2 = \int_\mathbb{R} |(Q\psi)(q) - \langle Q\rangle_\psi \psi(q)|^2 dq = \int_\mathbb{R} (q - \langle Q\rangle_\psi)^2 |\psi(q)|^2 dq. \quad (6.6)$$

For P we claim that

$$\sigma^2_\psi(P) = \sigma^2_{\hat{\psi}}(\hat{P}) = \hbar^2 \int_\mathbb{R} \left(p - \frac{\langle \hat{P}\rangle_{\hat{\psi}}}{\hbar}\right)^2 |\hat{\psi}(p)|^2 dp. \quad (6.7)$$

The first equality in (6.7) follows from

$$\sigma^2_{\hat{\psi}}(\hat{P}) = \|\hat{P}\hat{\psi}\|^2 - \langle \hat{P}\rangle^2_{\hat{\psi}} = \|\mathcal{F}P\mathcal{F}^{-1}\hat{\psi}\|^2 - \langle\hat{\psi}, \hat{P}\hat{\psi}\rangle^2$$

$$= \|\mathcal{F}(P\psi)\|^2 - \langle\mathcal{F}\psi, \mathcal{F}(P\psi)\rangle^2 = \|P\psi\|^2 - \langle\psi, P\psi\rangle^2$$

$$= \sigma^2_\psi(P).$$

The second equality in (6.7) then follows from

$$\sigma_\psi^2(\hat{P}) = \int_R \left|(\hat{P}\psi)(p) - \langle\hat{P}\rangle_{\hat\psi}\hat\psi(p)\right|^2 dp$$

$$= \int_R \left|\hbar p\hat\psi(p) - \langle\hat{P}\rangle_{\hat\psi}\hat\psi(p)\right|^2 dp$$

$$= \hbar^2 \int_R \left(p - \frac{\langle\hat{P}\rangle_{\hat\psi}}{\hbar}\right)^2 \left|\hat\psi(p)\right|^2 dp.$$

We will conclude our initial remarks on Postulate QM3 with a particularly significant special case. Again, we will let ψ denote a unit vector in $\mathcal{H}$ representing some
state. Now, let ϕ be another unit vector in $\mathcal{H}$ representing another state. We have already observed that this second state can be identified with the projection operator P_ϕ
that carries $\mathcal{H}$ onto the one-dimensional subspace of $\mathcal{H}$ spanned by ϕ. But P_ϕ is, in
particular, a self-adjoint operator and we will identify it with an observable. Since it is
not just a self-adjoint operator, but a projection, it corresponds to an observable with
exactly two possible (eigen)values (1 if a measurement determines the system to be in
state ϕ and 0 otherwise). Since P_ϕ is its own spectral decomposition, the probability
that a measurement to determine if the state is ϕ when it is known that the state is ψ
before the measurement is the same as the expected value of the observable and this
is

$$\langle\psi, P_\phi\psi\rangle = \langle\psi, \langle\phi,\psi\rangle\,\phi\rangle = \langle\phi,\psi\rangle\,\langle\psi,\phi\rangle = \overline{\langle\psi,\phi\rangle}\,\langle\psi,\phi\rangle = \left|\langle\psi,\phi\rangle\right|^2.$$

Thus, $|\langle\psi,\phi\rangle|^2$ is the probability of finding the system in state ϕ if it is known to be
in state ψ before the measurement; it is called the *transition probability from state ψ
to state ϕ*. The complex number $\langle\psi,\phi\rangle$ is called the *transition amplitude from ψ to ϕ*.
A case of particular importance to us arises in the following way. Suppose A is an ob-
servable with the property that $\mathcal{H}$ has an orthonormal basis $\{\psi_0, \psi_1, \ldots\}$ of eigenvectors
for A ($A\psi_n = \lambda_n\psi_n$, $n = 0, 1, \ldots$); for example, this is true for the operator H_B of Example 5.3.1. Let ψ be some state and write it $\psi = \sum_{n=0}^\infty \langle\psi_n, \psi\rangle\psi_n$ in terms of the eigenbasis
for A. Then the transition probability from ψ to one of the eigenstates ψ_n of A is just the
squared modulus $|\langle\psi_n, \psi\rangle|^2$ of the ψ_n-component of ψ. Soon we will phrase this in the
following way: $|\langle\psi_n, \psi\rangle|^2$ is the probability that the state ψ will "collapse" to the eigen-
state ψ_n of A when a measurement of A is made. Note that $\sum_{n=0}^\infty |\langle\psi_n, \psi\rangle|^2 = \|\psi\|^2 = 1$.

Our next two postulates deal with the dynamics of a quantum system, that is,
the time evolution of the state. There are two of them (actually, three since we give
two versions of Postulate QM5) because there appear to be two quite different ways
in which the state of the system can change. If you leave the system alone, in particular do not make measurements on it, the state will evolve smoothly in a manner
entirely analogous to the state of a classical system. The first of our two postulates

(Postulate QM4) deals with this scenario. It is probably just what you expect and is, if anything in physics is, universally accepted. The second postulate (Postulate QM5) is quite a different matter. It purports to describe the response of the state to a measurement, that is, to an encounter with an external system. It has been colorfully designated the *collapse of the wave function* and has been a source of controversy since the very early days of quantum mechanics. For this reason we will preface the statement of Postulate QM5 with somewhat more extensive remarks on what motivated it, what it is supposed to mean, some of the attitudes that have been taken toward it and how we will interpret it here.

For the statement of the first of these two postulates we will introduce, somewhat reluctantly, a bit of terminology that is not entirely standard (the reason for our reluctance will be explained shortly). A classical physical system is generally said to be *isolated* if it is not influenced by anything outside itself in the sense that neither matter nor energy can either enter or leave the system. These exist only as idealizations, of course, but the ideal can be achieved to a very high degree of approximation by real physical systems such as, for example, the solar system. It is implicit in this classical notion that the effect of measurements made on the system can be assumed negligible (bouncing a laser beam off of the moon to ascertain its position has negligible effect on the state of the solar system). This is, of course, precisely what one cannot assume of a quantum system. We will use the term *isolated quantum system* for a quantum system that is isolated in the classical sense *and on which no measurements are made*. Naturally, this raises the question of just what counts as a measurement and, as we have pointed out already, this is not a question that, even to this day, has been resolved to everyone's satisfaction. Rather than attempting to describe what a measurement *is* we will adopt our Postulate QM4 below as a definition of what it means to say that measurements are *not* being performed.

We should make just one technical comment before recording Postulate QM4. The state ψ of the system will evolve with time t and we will write the evolving states as $\psi(t)$ or sometimes ψ_t. When the Hilbert space is $L^2(\mathbb{R})$, the wave function ψ is a function of the spatial coordinate q and one writes it as $\psi(q)$. It is common then to write the evolving states as $\psi(q,t)$. This done, it seems natural to write t-derivatives as partial derivatives, but we will see that this requires some care. One must also take care not to get the wrong impression; q and t are not at all on an equal footing here. The time coordinate t enters only as a parameter labeling the states, *not* as an observable like q. In particular, *there is no operator representing t in the formalism of quantum mechanics*. This is not consistent with the spirit of special relativity, which requires that space and time coordinates be treated on an essentially equal footing. This creates some issues for quantum field theory, which is an attempt to reconcile quantum mechanics and special relativity, but this is not our concern at the moment.

Postulate QM4. *Let $\mathcal{H}$ be the Hilbert space of an isolated quantum system. Then there exists a strongly continuous one-parameter group $\{U_t\}_{t\in\mathbb{R}}$ of unitary operators on $\mathcal{H}$,* *called* evolution operators, *with the property that if the state of the system at time t = 0 is ψ(0), then the state at time t is given by*

$$\psi(t) = U_t(\psi(0))$$

for every t ∈ ℝ. By Stone's theorem, Theorem 5.6.2, there is a unique self-adjoint operator H : D(H) → H, called the Hamiltonian of the system, such that $U_t = e^{-itH/\hbar}$ *and therefore*

$$\psi(t) = e^{-itH/\hbar}(\psi(0)).$$

Physically, the Hamiltonian H is identified with the operator representing the total energy of the system.

Note that $\hbar$ is introduced to keep the units consistent with the physical interpretation. Wave functions carry no units (because they are not measurable), so the same must be true of the evolution operators. If H is to represent the energy of the system, then, in our SI units, it would be measured in Joules (J = $m^2(kg)s^{-2}$). The Planck constant has the units of $m^2(kg)s^{-1}$, so $H/\hbar$ has units s^{-1} and, since t is measured in seconds (s), all is well. The minus sign is conventional.

The one-parameter group $\{U_t\}_{t\in\mathbb{R}}$ of unitary operators is the analogue for an isolated quantum system of the classical one-parameter group $\{\phi_t\}_{t\in\mathbb{R}}$ of diffeomorphisms describing the flow of the Hamiltonian vector field in classical mechanics (see (2.49)). That the appropriate analogue of the diffeomorphism ϕ_t is a *unitary* operator U_t deserves some comment. On the surface, the motivation seems clear. A state of the system is represented by a $\psi \in \mathcal{H}$ with $\|\psi\|^2 = 1$, so the same must be true of the evolved states $\psi(t)$, that is, we must have $\langle\psi(t),\psi(t)\rangle = 1$ for all $t \in \mathbb{R}$. Certainly, this will be the case if $\psi(t)$ is obtained from $\psi(0)$ by applying a unitary operator U since $\langle U(\psi(0)), U(\psi(0))\rangle = \langle\psi(0),\psi(0)\rangle = 1$. Note, however, that this is also true if U is anti-unitary since then $\langle U(\psi(0)), U(\psi(0))\rangle = \overline{\langle\psi(0),\psi(0)\rangle} = 1$. Physically, one would probably also wish to assume that the time evolution preserves all transition probabilities $|\langle\psi,\phi\rangle|^2$, but this is also the case for both unitary and anti-unitary operators. Since unitary and anti-unitary operators differ only by composition with the bijection $K : \mathcal{H} \to \mathcal{H}$ that sends ψ to $K\psi = \bar\psi$, one might be tempted to conclude that one choice is as good as the other. Physically, however, matters are not quite so simple (see [Wig3]). Furthermore, it is not so clear that there might not be other possibilities as well. That, in fact, there are no other possibilities is a consequence of a nontrivial result of Wigner [Wig2] that we will describe briefly now. For this we consider the state space $\mathbb{CP}(\mathcal{H})$ of unit rays in $\mathcal{H}$. Note that if $\Psi, \Phi \in \mathbb{CP}(\mathcal{H})$, then one can define $|\langle\Psi,\Phi\rangle|^2 = |\langle\psi,\phi\rangle|^2$ for any $\psi \in \Psi$ and any $\phi \in \Phi$. Then $|\langle\Psi,\Phi\rangle|^2$ is the transition probability from state Ψ to state Φ. Wigner defined a *symmetry* of the quantum system whose Hilbert space is $\mathcal{H}$ to be a bijection $T : \mathbb{CP}(\mathcal{H}) \to \mathbb{CP}(\mathcal{H})$ that preserves transition probabilities in the sense that $|\langle T(\Psi), T(\Phi)\rangle|^2 = |\langle\Psi,\Phi\rangle|^2$

for all $\Psi, \Phi \in \mathbb{CP}(\mathcal{H})$; note that no linearity or continuity assumptions are made. Any unitary or anti-unitary operator U on $\mathcal{H}$ induces a symmetry T_U that carries any representative ψ of Ψ to the representative $U\psi$ of $T_U(\Psi)$. What Wigner proved was that *every* symmetry is induced in this way by a unitary or anti-unitary operator and that anti-unitary operators correspond to discrete symmetries such as time reversal (they are the analogue of reflections in Euclidean geometry). Now we can argue as follows. Suppose that the time evolution is described by an assignment to each $t \in \mathbb{R}$ of a symmetry $\alpha_t : \mathbb{CP}(\mathcal{H}) \to \mathbb{CP}(\mathcal{H})$ and suppose that $t \to \alpha_t$ satisfies $\alpha_{t+s} = \alpha_t \circ \alpha_s$ for all $t, s \in \mathbb{R}$. Then, for any $t \in \mathbb{R}$, $\alpha_t = \alpha_{t/2}^2$. By Wigner's theorem, $\alpha_{t/2}$ is represented by an operator $U_{t/2}$ that is either unitary or anti-unitary. Since the square of an operator that is either unitary or anti-unitary is necessarily unitary, every U_t must be unitary.

In the physics literature Wigner's definition of a symmetry is generally supplemented with the requirement that the operator must commute with the Hamiltonian of the system so that, in particular, it preserves energy levels. There is quite a thorough discussion, from the physicist's point of view, of the applications of such symmetries to concrete quantum mechanical problems in Chapter 4 of [SN].

Note that there is nothing special about $t = 0$ in Postulate QM4. If t_0 is any real number, then $\psi(t_0) = U_{t_0}(\psi(0))$ so $\psi(t + t_0) = U_{t+t_0}(\psi(0)) = U_t(U_{t_0}(\psi(0))) = U_t(\psi(t_0))$ and therefore

$$\psi(t) = \psi((t - t_0) + t_0) = U_{t-t_0}(\psi(t_0)) = e^{-i(t-t_0)H/\hbar}(\psi(t_0)).$$

Thus,

$$U_{t-t_0} = e^{-i(t-t_0)H/\hbar}$$

propagates the state at time t_0 to the state at time t for any $t_0, t \in \mathbb{R}$.

Next let us apply Lemma 5.6.1 to $\psi(t_0)$, which we assume is in the domain of the Hamiltonian H. Then, for any $t \in \mathbb{R}$, $\psi(t) = U_{t-t_0}(\psi(t_0))$ is also in the domain of H by Theorem 5.6.2. Consequently,

$$-(i/\hbar)H(\psi(t_0)) = \lim_{t \to 0} \frac{U_t(\psi(t_0)) - \psi(t_0)}{t}$$
$$= \lim_{t \to t_0} \frac{U_{t-t_0}(\psi(t_0)) - \psi(t_0)}{t - t_0}$$
$$= \lim_{t \to t_0} \frac{\psi(t) - \psi(t_0)}{t - t_0},$$

where the limit is in $\mathcal{H}$. We will write this as $\frac{d\psi(t)}{dt}|_{t=t_0} = -(i/\hbar)H(\psi(t_0))$ or, since t_0 is arbitrary, simply as

$$i\hbar \frac{d\psi(t)}{dt} = H(\psi(t)). \tag{6.8}$$

Equation (6.8) is called the *abstract Schrödinger equation*. We will find that the Hamiltonian operator of a system is generally a differential operator (such as H_B for the harmonic oscillator, or H_0 for the free particle) so that (6.8) is a differential equation that describes the time evolution $\psi(t)$ of the state of an isolated quantum system with Hamiltonian H.

Just to have a few examples to look at, let us once again anticipate the physical interpretations to come in Chapter 7 and have a look at the Schrödinger equation for H_B and H_0. This will also give us the opportunity to introduce the all-important notion of a *propagator*, about which we will have a great deal more to say in Chapter 8. Since both examples are of the same type and this type recurs repeatedly in quantum mechanics we will begin by setting a more general stage.

Our Hilbert space will be $\mathcal{H} = L^2(\mathbb{R})$, so $\psi(q, t)$ is a map from $\mathbb{R}^2$ to $\mathbb{C}$. We will be given a Hamiltonian H that is self-adjoint on some dense linear subspace $\mathcal{D}(H)$ of $L^2(\mathbb{R})$ and we will assume that $\mathcal{D}(H)$ contains $\mathcal{S}(\mathbb{R})$ and that, on $\mathcal{S}(\mathbb{R})$, H is of the form

$$H = H_0 + V(q) = -\frac{\hbar^2}{2m}\Delta + V(q) = -\frac{\hbar^2}{2m}\frac{\partial^2}{\partial q^2} + V(q), \tag{6.9}$$

where $V(q)$ is a real-valued function on $\mathbb{R}$ which acts on $L^2(\mathbb{R})$ as a multiplication operator (thus, for H_0, $V(q) = 0$, and for H_B, $V(q) = (m\omega^2/2)q^2$). Note that we have chosen to write Δ as $\partial^2/\partial q^2$ rather than d^2/dq^2 because we are now viewing ψ as a function of (q, t) rather than as a function of q alone.

Remark 6.2.4. The sum of two unbounded, self-adjoint operators need not be self-adjoint and it is quite a difficult matter to decide what sort of potentials $V(q)$ will give rise to an $H = H_0 + V(q)$ that is self-adjoint. We will have a bit more to say about this in Section 9.2, but for the time being we will simply assume that $V(q)$ is one of them.

Denoting the initial value $\psi(q, 0)$ of the wave function simply $\psi(q)$, the Cauchy problem for the Schrödinger equation (6.9) as it generally appears in treatments of quantum mechanics is

$$i\hbar\frac{\partial\psi(q, t)}{\partial t} = -\frac{\hbar^2}{2m}\frac{\partial^2\psi(q, t)}{\partial q^2} + V(q)\psi(q, t),$$
$$\psi(q, 0) = \psi(q). \tag{6.10}$$

But something just went by a bit too quickly. The abstract Schrödinger equation (6.9) contains in it the t-derivative $d\psi(t)/dt$, which is defined as the limit *in* $\mathcal{H} = L^2(\mathbb{R})$ of the familiar difference quotient. The partial derivative $\partial\psi(q, t)/\partial t$ that appears in the traditional physicist's form (6.10) of the Schrödinger equation is, on the other hand, defined as a limit *in* $\mathbb{C}$ of an equally familiar difference quotient. It is not so clear that these should be the same. Indeed, as a classical partial differential equation (6.10) generally does not make sense in $L^2(\mathbb{R})$ since the elements of $L^2(\mathbb{R})$ need not be differentiable. The best we can hope for is that, in the context of classical solutions to (6.10), the two derivatives coincide so that, in this restricted context, the two versions of the Schrödinger equation describe the same evolution. The following is Exercise 3.1, Chapter IV, of [Prug].

Exercise 6.2.1. Assume that $\psi(q, t)$ is continuously differentiable with respect to t and square integrable with respect to q for every t and that the $L^2(\mathbb{R})$-limit

$$\lim_{\Delta t \to 0} \frac{\psi(q, t + \Delta t) - \psi(q, t)}{\Delta t}$$

exists. Show that the partial derivative $\partial \psi(q, t)/\partial t$ is in $L^2(\mathbb{R})$ for every t and

$$\lim_{\Delta t \to 0} \frac{\psi(q, t + \Delta t) - \psi(q, t)}{\Delta t} = \frac{\partial \psi(q, t)}{\partial t}.$$

Hints: Temporarily denote by $\varphi(q, t)$ the L^2-limit on the left-hand side. Choose a sequence $\Delta t_1, \Delta t_2, \dots$ converging to zero. Then, as $n \to \infty$, the sequence

$$\frac{\psi(q, t + \Delta t_n) - \psi(q, t)}{\Delta t_n}$$

converges in $L^2(\mathbb{R})$ to $\varphi(q, t)$. Consequently, some subsequence converges *pointwise almost everywhere* to $\varphi(q, t)$. Conclude from the assumed t-continuity of $\partial \psi(q, t)/\partial t$ that, as elements of $L^2(\mathbb{R})$, $\varphi(q, t) = \partial \psi(q, t)/\partial t$.

At this point we have assembled enough information to offer something in the way of motivation for calling $P = -i\hbar \frac{d}{dq}$ the "momentum operator" on $\mathbb{R}$. Classically, the momentum of a particle moving in one dimension is the product of its mass m and its velocity. Since a quantum particle has no trajectory in the classical sense one cannot attach any meaning to its classical position or classical velocity at each instant. What one can attach a meaning to are the expected values of position and velocity. We have already argued that $(Q\psi)(q) = q\psi(q)$ is the "right" choice for the position operator. Its expected value in any state ψ is given by

$$\langle Q \rangle_\psi = \langle \psi, Q\psi \rangle = \int_{\mathbb{R}} \overline{\psi}(q) q \psi(q) \, dq.$$

Now we allow the state $\psi = \psi_0$ to evolve in time according to the Schrödinger equation. Writing the state at time t as ψ_t so that $\psi_t(q) = \psi(q, t)$ we inquire as to the time rate of change of the expected value

$$\langle Q \rangle_{\psi_t} = \langle \psi_t, Q\psi_t \rangle = \int_{\mathbb{R}} \overline{\psi}(q, t) q \psi(q, t) dq$$

of position at $t = 0$. Then $\frac{d}{dt} \langle Q \rangle_{\psi_t} |_{t=0}$ is taken to be the expected value of the particle's velocity in state ψ and

$$m \frac{d}{dt} \langle Q \rangle_{\psi_t} \bigg|_{t=0} = m \frac{d}{dt} \int_{\mathbb{R}} \overline{\psi}(q, t) q \psi(q, t) \, dq \bigg|_{t=0}$$

is taken to be the expected value of the particle's momentum in state ψ. The idea is to show that this is precisely $\langle \psi, P\psi \rangle$ and thereby conclude that $P = -i\hbar\frac{d}{dq}$ is the "right" choice for the momentum operator in quantum mechanics. Since our purpose here is motivation we will sweep some technical issues under the rug and assume that $\psi(q,t)$ is continuously differentiable with respect to t and a Schwartz function of q for each t and that we can replace $d\psi_t/dt$ with $\partial\psi(q,t)/\partial t$ in the Schrödinger equation (see Exercise 6.2.1). Now we just compute

$$
\begin{aligned}
\frac{d}{dt}\langle Q \rangle_{\psi_t} &= \frac{d}{dt}\int_R \overline{\psi}(q,t)q\psi(q,t)\,dq \\
&= \int_R q\frac{\partial}{\partial t}[\overline{\psi}(q,t)\psi(q,t)]\,dq \\
&= \int_R q\left[\frac{\partial\overline{\psi}(q,t)}{\partial t}\psi(q,t) + \overline{\psi}(q,t)\frac{\partial\psi(q,t)}{\partial t}\right]dq \\
&= \frac{i\hbar}{2m}\int_R q\left[-\frac{\partial^2\overline{\psi}(q,t)}{\partial q^2}\psi(q,t) + \overline{\psi}(q,t)\frac{\partial^2\psi(q,t)}{\partial q^2}\right]dq,
\end{aligned}
$$

where the last equality follows from the Schrödinger equation and its conjugate; note that the potential terms cancel.

Exercise 6.2.2. Integrate by parts to show that this can be written

$$
\frac{d}{dt}\langle Q \rangle_{\psi_t} = \frac{-i\hbar}{2m}\int_R\left[-\frac{\partial\overline{\psi}(q,t)}{\partial q}\psi(q,t) + \overline{\psi}(q,t)\frac{\partial\psi(q,t)}{\partial q}\right]dq.
$$

Another integration by parts on the first term gives

$$
\begin{aligned}
\frac{d}{dt}\langle Q \rangle_{\psi_t} &= \frac{-i\hbar}{2m}\int_R 2\overline{\psi}(q,t)\frac{\partial\psi(q,t)}{\partial q}\,dq \\
&= \frac{1}{m}\int_R \overline{\psi}(q,t)\left[-i\hbar\frac{\partial}{\partial q}\psi(q,t)\right]dq.
\end{aligned}
$$

At $t = 0$ this gives

$$
m\frac{d}{dt}\langle Q \rangle_{\psi_t}\bigg|_{t=0} = \langle \psi, P\psi \rangle = \langle P \rangle_\psi
$$

and this is what we wanted to show. The general problem of deciding which operators are to represent which observables is a nontrivial one and we will return to it in Chapter 7.

Example 6.2.8. We consider $H_B : \mathcal{D}(H_B) \to L^2(\mathbb{R})$ as in Example 5.4.5 and the Cauchy problem (6.10) for H_B

$$i\hbar\frac{\partial\psi(q,t)}{\partial t} = -\frac{\hbar^2}{2m}\frac{\partial^2\psi(q,t)}{\partial q^2} + (m\omega^2/2)q^2\psi(q,t),$$

$$\psi(q,0) = \psi(q),$$

(6.11)

where $\psi(q)$ is assumed to be in $\mathcal{D}(H_B)$. We know that the solution can be described in terms of the evolution operators by

$$\psi(q,t) = U_t(\psi(q)) = e^{-itH_B/\hbar}(\psi(q)).$$

But $L^2(\mathbb{R})$ has an orthonormal basis $\{\psi_n(q)\}_{n=0}^\infty$ consisting of eigenfunctions of H_B with simple eigenvalues $\mathcal{E}_n$ ($H_B\psi_n = \mathcal{E}_n\psi_n$), so we conclude from (5.39) that this can be written

$$\psi(q,t) = \sum_{n=0}^\infty \langle\psi_n,\psi\rangle e^{-it\mathcal{E}_n/\hbar}\psi_n(q) = \sum_{n=0}^\infty \left(\int_{\mathbb{R}} \overline{\psi_n(x)}\psi(x)dx\right) e^{-it\mathcal{E}_n/\hbar}\psi_n(q),$$

(6.12)

where the convergence is in $L^2(\mathbb{R})$ for each t.

Note that we could omit the conjugation symbol here since, for H_B, the eigenfunctions $\psi_n(x)$ are real-valued. We leave it there because the only property of H_B that played a role in any of this was the fact that $L^2(\mathbb{R})$ has an orthonormal basis consisting of eigenfunctions with simple eigenvalues. This is not as uncommon as one might suppose, but generally the eigenfunctions will be complex-valued. There are various theorems that assert the existence of such an eigenbasis for Hamiltonians of the form $H = H_0 + V(q)$ under various conditions on $V(q)$. This is the case, for example, if $V \in L^\infty_{loc}(\mathbb{R})$ and $\lim_{q\to\infty} V(q) = \infty$; this can be found, for example, in Theorems 7.3 and 7.6 of [Pankov].

The evolution of the wave function for H_B with initial data $\psi(q)$ is completely and explicitly described by (6.12), but even so we would like to play with this just a bit more to motivate an important idea. First let us make the initial time more visible by writing $\psi(x)$ explicitly as $\psi(x,0)$, so that

$$\psi(q,t) = \sum_{n=0}^\infty \left(\int_{\mathbb{R}} \overline{\psi_n(x)}\psi(x,0)dx\right) e^{-it\mathcal{E}_n/\hbar}\psi_n(q).$$

Because of the group property of U_t we can replace $t = 0$ by any $t = t_0$ to obtain

$$\psi(q,t) = \sum_{n=0}^\infty \left(\int_{\mathbb{R}} \overline{\psi_n(x)}\psi(x,t_0)dx\right) e^{-i(t-t_0)\mathcal{E}_n/\hbar}\psi_n(q).$$

For motivational purposes, let us be a bit sloppy here and just assume that we can interchange the summation and integration so that this can be written

$$\psi(q, t) = \int_{\mathbb{R}} K(q, t; x, t_0)\psi(x, t_0)dx, \tag{6.13}$$

where

$$K(q, t; x, t_0) = \sum_{n=0}^{\infty} e^{-i\mathcal{E}_n(t-t_0)/\hbar}\psi_n(q)\overline{\psi_n(x)} = \sum_{n=0}^{\infty} e^{-i(n+\frac{1}{2})\omega(t-t_0)}\psi_n(q)\overline{\psi_n(x)}, \tag{6.14}$$

which is called the *propagator for* H_B, or the *integral kernel for* H_B, or simply the *Schrödinger kernel for* H_B; we will have a great deal more to say about it in Section 7.4 and Chapter 8; in particular, we will exhibit a closed form expression for $K(q, t; x, t_0)$. Moreover, in Chapter 8 we will introduce a precise definition of the propagator for any Hamiltonian and describe Feynman's remarkable representation of it as a *path integral*. For the moment we will content ourselves with a few, admittedly rather vague and unproved hints as to how one should think about it. Let us regard $|\psi(q, t)|^2$ as the probability of locating some particle at position q at time t; in reality, $|\psi(q, t)|^2$ is just a probability density, of course (Example 6.2.2); $\psi(q, t)$ itself will be thought of as the probability amplitude for detecting the particle at location q at time t; similarly for $\psi(x, t_0)$ for any x in $\mathbb{R}$. For q, t and t_0 held fixed, the integral $\int_{\mathbb{R}} K(q, t; x, t_0)\psi(x, t_0)dx$ expresses the amplitude $\psi(q, t)$ as a "sum" over all $x \in \mathbb{R}$ of the products $K(q, t; x, t_0)\psi(x, t_0)$. One thinks of $|K(q, t; x, t_0)|^2$ as the conditional probability that the particle will be detected at q at time t given that it is detected at x at time t_0, so $K(q, t; x, t_0)$ itself is the amplitude for getting from (x, t_0) to (q, t).

We would like to do the same sort of thing for the free particle Hamiltonian H_0. In this case we do not have access to an orthonormal eigenbasis for $L^2(\mathbb{R})$ (there are no eigenfunctions of H_0 in $L^2(\mathbb{R})$) so the procedure is not as straightforward as in Example 6.2.8. Indeed, it seems best to save this for Section 7.3, where we focus entirely on the free particle (see Example 7.3.2).

Moving on to our next "dynamical" postulate, we need to consider the effect of an encounter with some external system. A *measurement* performed on a classical mechanical system is a relatively unambiguous notion. It is an interaction between the system on which the measurement is to be made and some external system ("measuring device") that yields, in one way or another, a value of some classical observable, *but has negligible effect on the state of the measured system.* The distinguishing feature of quantum mechanics is the impossibility of satisfying this last condition. Indeed, it has been argued, particularly in recent years with the increasing prominence of such notions as *entanglement* and *decoherence*, that the very notion of an isolated quantum system upon which measurements are to be made is not meaningful; every physical system is inextricably entangled with and cannot be disentangled from its physical environment (hence our reluctance to introduce the notion of an *isolated quantum system* in the first place). Taking this notion seriously has led to a new perspective on the foundations of quantum mechanics and could potentially resolve many of the conceptual issues associated with the measurement problem that we will soon discuss. For those interested in pursuing these matters further we will only offer the following suggestions. The seminal papers on entanglement are by Einstein, Podolsky and Rosen [EPR], Bell [Bell] and Aspect, Dalibard and Roger [ADR]. For decoherence one might consult the survey article [Schl] and then proceed to the web site http://www.decoherence.de/.

Much of what we need to discuss now is rather problematic, but one thing at least is clear. Whatever reasonable notion of measurement one adopts, there are some things that surely should count as measurements and these have rather startling effects on the state of the system being measured. In our discussion of the two-slit experiments in Section 4.4 we found that any attempt to detect an electron coming through one of the slits obliterates the interference pattern altogether, transforming a state characterized by interference into one in which there is no interference at all. This change in the state is abrupt and discontinuous, occurring at the instant the measurement is made, and cannot be accounted for by the smooth evolution of states described in Postulate QM4. Something new is required and von Neumann formulated what is often called the *projection postulate* or *reduction postulate* or *collapse postulate* to account for this sort of collapse of the state vector. In order to emphasize some of the issues involved we will actually formulate two versions of this collapse hypothesis. The first (Postulate QM5) is the version proposed by von Neumann in [vonNeu]. The second (Postulate QM5′) is an extension of this due to Lüder that includes the first as a special case, but does not follow from the arguments of von Neumann that we will now describe.

Consider first an observable A whose spectrum consists of a discrete set $\sigma(A) = \{\lambda_0, \lambda_1, \ldots\}$ of simple eigenvalues (recall that *discrete* means in the topological sense that each point of the spectrum is isolated in the subspace $\sigma(A)$ of $\mathbb{R}$ and *simple* means that the dimension of each eigenspace is one). A good example to keep in mind is the operator H_B of Example 5.5.4. Assume that $\mathcal{H}$ has an orthonormal basis $\{\psi_0, \psi_1, \ldots\}$ consisting of eigenvectors of A with, say, $A\psi_n = \lambda_n \psi_n$, $n = 0, 1, \ldots$ Let us suppose that the system is in some state ψ and write ψ in terms of the eigenbasis for A as $\psi = \sum_{n=0}^{\infty} \langle \psi_n, \psi \rangle \psi_n$. Now suppose that a measurement of A is made on the system. According to Postulate QM2, the only possible outcome of the measurement is one of the eigenvalues λ_k and this will occur with transition probability $|\langle \psi_k, \psi \rangle|^2$. At this point, von Neumann (page 215 of [vonNeu]) makes a physical assumption that is the source of much of the controversy surrounding our upcoming Postulate QM5. The assumption is that *if this measurement of A on the system yields the value λ_k and is followed "immediately" by another measurement of A on the same system, then the result of the second measurement of A must give the same result λ_k "with certainty."* Since the eigenvalues are simple, the only state for which the outcome of a measurement of A

is λ_k with probability one is ψ_k. Consequently, the *effect of the measurement must be to change the state of the system from* $\sum_{n=0}^{\infty} \langle \psi_n, \psi \rangle \psi_n$ *to* ψ_k. Prior to the measurement the state of the system is some superposition of eigenstates of A which then *collapses* to a single eigenstate when the measurement is performed. It is not altogether clear whether this collapse of the state vector, should you choose to accept it, is to be re-garded as a statement about physics or epistemology. Does the measurement result in a physical change in the quantum system, and therefore in its state, or simply in our knowledge of the system? Each position has its advocates and we are certainly not so presumptuous as to take a stand, but offer the question only as food for thought.

Before opting to accept some variant of this collapse scenario as our Postulate QM5 we need to discuss some of the issues it raises. The most obvious, of course, is that making a measurement on a quantum system is very likely to destroy the system, in which case there is no sense trying to discuss its subsequent state. This is what hap-pens, for example, when we measure the position of an electron in the two-slit ex-periment by letting it collide with a screen (see Remark 6.2.5 for a bit more on this). Henceforth we will bar such destructive measurements from consideration and try to unearth some of the more subtle issues posed by von Neumann's argument.

It is, for example, not entirely clear what is meant by the word "immediately" for the second measurement. Assuming that the measurement is made at time t_0 and the state really does collapse to ψ_k at that instant, then, until another mea-surement is made, Postulate QM4 requires that the new state evolve according to $\psi(t) = e^{-i(t-t_0)H/\hbar}(\psi_k)$. In particular, there is generally no time interval following t_0 during which the state remains ψ_k, so it would seem that a second measurement need not result in λ_k "with certainty." A physicist would respond that what is intended here is a second measurement made after an "infinitesimal" time interval. We will try to make sense of this as a limit statement. We will assume that by "with certainty" von Neumann meant "with probability one" and, as a result, his assumption is that *repeated* second measurements of A when the system is initially in state ψ and the first measurement gives λ_k will yield λ_k with a relative frequency that approaches 1 as the number of repetitions approaches infinity. Now, as the collapsed state evolves smoothly away from ψ_k for $t > t_0$, the probability that the second measurement of A will yield λ_k is, by Postulate QM3, $\mu_{\psi(t),A}(\{\lambda_k\}) = \| E^A(\{\lambda_k\}) \psi(t) \|^2$, which varies con-tinuously with t and therefore approaches 1 as $t \to t_0$. Consequently, the probability that the second measurement of A results in the value λ_k can be made arbitrarily close to 1 by making the measurement sufficiently soon after the collapse. Taking the matter one step further, *repeating the second measurement of A sufficiently often and suffi-ciently soon after the collapse, the relative frequency of the outcome λ_k can be made as close to 1 as desired*. If we are willing to concede, as we must, that nothing in physics is ever measured exactly, then this would seem to be a reasonable interpretation of von Neumann's assumption.

Note that there is one case in which this difficulty associated with the evolution of the collapsed state does not arise. Suppose the observable A we have been considering happens to be the Hamiltonian H itself, that is, the total energy. Then $H\psi_k = \lambda_k\psi_k$, so, by Theorem 5.5.7.3,

$$e^{-i(t-t_0)H/\hbar}(\psi_k) = e^{-i(t-t_0)\lambda_k/\hbar}\psi_k,$$

which differs from ψ_k only in the phase factor $e^{-i(t-t_0)\lambda_k/\hbar}$ and so represents the same state. Normalized eigenvectors for the Hamiltonian are called *stationary states* because they represent states of the system that do not change with time. For these, a second measurement of H on the system should, indeed, yield the same energy as the first with probability one. Of course, for a general Hamiltonian such stationary states (eigenstates) need not exist because, unlike H_B, there need not be any eigenvalues at all.

For observables other than the Hamiltonian, however, there is at least the hint of some tension between the underlying rationale for the collapse hypothesis and the smooth evolution of the state governed by the Schrödinger equation. As evidence that the debate over the collapse of the wave function is still alive and well we offer [CL], the paper ['t Ho2] by a Nobel laureate and the following quote.

> "The dynamics and the postulate of collapse are flatly in contradiction with one another ... the postulate of collapse seems to be right about what happens when we make measurements, and the dynamics seems to be bizarrely wrong about what happens when we make measurements, and yet the dynamics seems to be right about what happens whenever we aren't making measurements."
>
> D. Albert [Albert]

Having sounded the appropriate cautionary notes we now throw caution to the winds and record our first version of the collapse postulate.

Postulate QM5. *Let $\mathcal{H}$ be the Hilbert space of a quantum system with Hamiltonian H and A an observable with a discrete spectrum $\sigma(A) = \{\lambda_0, \lambda_1, \ldots\}$ consisting entirely of simple eigenvalues. Let $\psi_0, \psi_1, \ldots$ be an orthonormal basis for $\mathcal{H}$ with $A\psi_n = \lambda_n\psi_n$, $n = 0, 1, \ldots$. Suppose the system is isolated for $0 \leq t < t_0$ so that its state evolves from some initial state $\psi(0)$ according to $\psi(t) = e^{-itH/\hbar}(\psi(0))$. At $t = t_0$ a measurement of A on the system is made. If the result of the measurement is the eigenvalue λ_k, then the state of the system at time t_0 is ψ_k. For $t > t_0$ the state of the now isolated system evolves according to $\psi(t) = e^{-i(t-t_0)H/\hbar}(\psi_k)$.*

Had the measurement not been made, the state of the system at time t_0 would have been represented by $e^{-it_0H/\hbar}(\psi(0))$. The act of measuring results in a discontinuous jump in the state and the result of the measurement determines what the state jumps to.

Remark 6.2.5. One might also wonder about the physical, that is, experimental evidence that supports the collapse hypothesis. Needless to say, making measurements on a quantum system is a delicate business and making measurements that do not destroy the system in the process is considerably more delicate. Von Neumann (pages 212–214 of [vonNeu]) deduces his hypothesis from a discussion of experiments performed by Compton and Simons on the scattering of photons by electrons. It was not until the 1980s, however, that the technology began to evolve for making successive measurements on a single quantum system. A thorough discussion of the experimental situation through 1992 is available in [BK]. Indicative of the level to which these experimental techniques have evolved is the fact that the 2012 Nobel Prize in Physics was awarded "for ground-breaking experimental methods that enable measuring and manipulation of individual quantum systems" (see http://www.nobelprize.org). At present one can say only that the experimental evidence appears to weigh in on the side of the collapse hypothesis, at least for observables of the type we have been considering. In the end, however, one cannot say that the issue of the collapse of the wave function has been resolved to everyone's satisfaction.

Naturally, not all observables have a discrete spectrum consisting of simple eigenvalues and for these the situation is more tenuous. Begin by considering an observable A that has a discrete spectrum $\sigma(A) = \{\lambda_0, \lambda_1, \ldots\}$ consisting entirely of eigenvalues, but for which the eigenvalues need not be simple. Thus, each λ_n has an eigenspace M_{λ_n} of dimension greater than or equal to one (perhaps countably infinite). Assume that one can still find an orthonormal basis $\{\psi_0, \psi_1, \ldots\}$ for $\mathcal{H}$ consisting of eigenvectors of A. Thus, any state ψ can be written as $\psi = \sum_{n=0}^{\infty} \langle \psi_n, \psi \rangle \psi_n$. It is still true (by Postulate QM2) that a measurement of A in state ψ can only result in one of the eigenvalues of A. Suppose that a measurement is made and the result is λ_k. Invoking von Neumann's argument that a second measurement of A performed "immediately" after the first must also result in λ_k "with certainty," we conclude again that the measurement must collapse the state ψ to a unit eigenvector of A with eigenvalue λ_k. But, if $\dim M_{\lambda_k} > 1$, this does not uniquely determine the collapsed state and von Neumann leaves the matter at this.

"... if the eigenvalue ... is multiple, then the state ... after the measurement is not uniquely determined by the knowledge of the result of the measurement."

von Neumann ([vonNeu], page 218)

Von Neumann did, in fact, have more to say about the case of degenerate eigenvalues, but the end result was not a uniquely determined post-measurement collapsed state.

One can, of course, make a choice of some element in the eigenspace M_{λ_k} to serve as the collapsed state, but no such choice would follow from von Neumann's measurement repeatability assumption alone. Since there appears to be a generally accepted choice in the literature (at least among those who accept the collapse hypothesis at all) we will record a version of this as our Postulate QM5′; it is generally called *Lüder's postulate*.

Postulate QM5′. *Let $\mathcal{H}$ be the Hilbert space of a quantum system with Hamiltonian H and A an observable with a discrete spectrum $\sigma(A) = \{\lambda_0, \lambda_1, \ldots\}$ consisting entirely of (not necessarily simple) eigenvalues. Let $\psi_0, \psi_1, \ldots$ be an orthonormal basis for $\mathcal{H}$ with $A\psi_n = \lambda_n\psi_n$, $n = 0, 1, \ldots$. Suppose the system is isolated for $0 \leq t < t_0$ so that its state evolves from some initial state $\psi(0)$ according to $\psi(t) = e^{-itH/\hbar}(\psi(0))$. Let $\phi = e^{-it_0H/\hbar}(\psi(0))$. At $t = t_0$ a measurement of A on the system is made. If the result of the measurement is the eigenvalue λ_k, then the projection $P_{\lambda_k}\phi = E^A(\{\lambda_k\})\phi$ of ϕ into the eigenspace M_{λ_k} of λ_k is nonzero and the state of the system at time t_0 is represented by the normalized projection of ϕ into M_{λ_k}, that is, by*

$$\psi(t_0) = \frac{P_{\lambda_k}\phi}{\|P_{\lambda_k}\phi\|}.$$

For $t > t_0$ the state of the now isolated system evolves according to

$$\psi(t) = e^{-i(t-t_0)H/\hbar}(\psi(t_0)).$$

A translation of Lüder's paper is available at http://arxiv.org/pdf/quant-ph/0403007v2.pdf, where the rationale behind this choice is described in detail. We will content ourselves with two simple remarks. If $P_{\lambda_k}\phi$ were zero, then ϕ would be orthogonal to the eigenspace M_{λ_k} and so the transition probability from ϕ to any state in M_{λ_k} is zero and collapse to an eigenstate in M_{λ_k} would have probability zero. On the other hand, if $P_{\lambda_k}\phi \neq 0$, collapse onto the normalized projection of ϕ into M_{λ_k} guarantees that Lüder's Postulate QM5′, agrees with von Neumann's Postulate QM5 when the eigenvalues are simple.

For observables with continuous spectrum the situation is even less clear since there need not be any eigenvalues at all and therefore no eigenstates onto which to collapse. Various proposals have been put forth for reasonable versions of the collapse postulate in the presence of a continuous spectrum, but none appears to have been awarded a consensus and so we will content ourselves with a reference to [Srin] for those interested in pursuing the matter.

To introduce our next postulate, let us consider two observables and their corresponding self-adjoint operators A_1 and A_2. Suppose also that A_1 and A_2 happen to commute in the sense that their corresponding spectral measures commute. According to von Neumann's Theorem 5.6.4, both A_1 and A_2 are functions of a single self-adjoint operator, that is, there exists a self-adjoint operator B and two real-valued Borel functions f_1 and f_2 on $\mathbb{R}$ such that $A_i = f_i(B)$, $i = 1, 2$. Assume also that B corresponds to some observable (recall that Postulate QM2 does not ensure this in general, but, barring superselection rules, one usually assumes that every self-adjoint operator corresponds to an observable). Then a measurement of B in any state is, by definition, also a measurement of both $f_1(B)$ and $f_2(B)$ and consequently a measurement of both A_1 and A_2. In this sense one can say that A_1 and A_2 are *simultaneously measurable in any state*. Moreover, since Theorem 5.6.4 applies to arbitrary families of commuting self-adjoint operators we conclude that *any family of observables whose corresponding self-adjoint operators commute can be simultaneously measured in any state* (assuming again that the operator B guaranteed by Theorem 5.6.4 corresponds to some observable).

Now, what about observables that do not commute? Can they also be simultaneously measured in any state? One should take some care with the terminology here. For commuting observables we need only make one measurement of one observable (B) and then we get the measured values of the commuting observables *simultaneously and for free* simply by virtue of the way observables of the form $f(B)$ are defined; there is no question of physically coordinating two different measurements. On the other hand, if A_1 and A_2 do not commute, then there is no *a priori* reason to suppose that we can measure both by simply measuring some third observable and doing a computation. Taken literally, the "simultaneous measurement" of A_1 and A_2 in this case would mean that we must perform two different measurements on the system *at the same time* and this, we claim, raises some issues. How does one ensure that apparatus $\mathcal{A}_1$ for measuring A_1 and apparatus $\mathcal{A}_2$ for measuring A_2 do their measuring at precisely the same instant? We maintain that, since nothing in the laboratory is ever determined exactly, one generally cannot ensure this and for this reason a strict adherence to the literal definition of "simultaneous" is ill-advised.

Remark 6.2.6. We are ignoring relativistic effects here, but one cannot ignore relativity forever and eventually will need to contend also with the *relativity of simultaneity* (see page 23 of [Nab5]).

Once again a physicist will argue that this is no problem provided the two measurements are separated by a time interval that is "infinitesimal" and once again we will interpret this as a statement about sufficiently small time intervals. Unlike our discussion of von Neumann's repeated measurements, however, even if the time interval separating them is very small, there is now the additional issue of which measurement is performed first. Certainly, any interpretation of the "simultaneous measurability" of A_1 and A_2 along these lines must at least require that if the time lapse between them is sufficiently small, then the order in which the measurements actually take place is immaterial. However, since quantum mechanics has nothing to say about the outcome of any individual measurement, a precise formulation of this must be a statement about the relative frequencies of such outcomes, that is, about probabilities. The following arguments are intended only to *motivate* what this precise formulation should look like and how this is related to the commutativity of the corresponding operators. In the end we will state precisely what we are assuming in the form of Postulate QM6.

We consider two observables A_1 and A_2 and two Borel sets S_1 and S_2 in $\mathbb{R}$. Denote the spectral measures of A_1 and A_2 by E^{A_1} and E^{A_2}, respectively, and let ψ be some unit vector in $\mathcal{H}$ representing a state. We intend to compute the joint probability that a measurement of A_1 is in S_1 and a measurement of A_2 is in S_2, first assuming that the A_1 measurement is performed first and then assuming that the A_2 measurement is performed first. Insisting that these two be the same for all states ψ and for all Borel sets $S_1, S_2 \subseteq \mathbb{R}$ will be our way of saying that the order in which the measurements take place is immaterial. We will then show that this alone implies that A_1 and A_2 must commute.

Note that each $E^{A_i}(S_i)$, $i = 1, 2$, being a projection, is an observable with a discrete spectrum of eigenvalues $\{0, 1\}$. Next, observe that

$$\mu_{\psi, A_1}(S_1) = \mu_{\psi, E^{A_1}(S_1)}(\{1\}) = \langle \psi, E^{A_1}(S_1)\psi \rangle = \|E^{A_1}(S_1)\psi\|^2. \tag{6.15}$$

Exercise 6.2.3. Prove (6.15).

This is the probability that a measurement of A_1 in state ψ will yield a value in S_1, that is, the probability that a measurement of $E^{A_1}(S_1)$ in state ψ will have outcome 1. Assuming $E^{A_1}(S_1)\psi \neq 0$, Lüder's postulate applied to $E^{A_1}(S_1)$ implies that when the measurement is made, the state ψ collapses to

$$\varphi = \frac{E^{A_1}(S_1)\psi}{\|E^{A_1}(S_1)\psi\|}.$$

Now, the probability that a measurement of A_2 *in state* φ will yield a value in S_2 is

$$\mu_{\varphi, A_2}(S_2) = \mu_{\varphi, E^{A_2}(S_2)}(\{1\}) = \langle \varphi, E^{A_2}(S_2)\varphi \rangle = \|E^{A_2}(S_2)\varphi\|^2 = \frac{\|E^{A_2}(S_2)E^{A_1}(S_1)\psi\|^2}{\|E^{A_1}(S_1)\psi\|^2}.$$

Assuming that the measurements of $E^{A_1}(S_1)$ and $E^{A_2}(S_2)$ represent independent events, the probability that a measurement of A_1 in state ψ followed "immediately" by a measurement of A_2 in state φ will yield values in S_1 and S_2, respectively, is the product

$$\mu_{\varphi, A_2}(S_2)\, \mu_{\psi, A_1}(S_1) = \|E^{A_2}(S_2)E^{A_1}(S_1)\psi\|^2.$$

The same argument shows that if the measurements are carried out in the reverse order, the result is

$$\|E^{A_1}(S_1)E^{A_2}(S_2)\psi\|^2.$$

In general, there is no reason to suppose that these two joint probabilities are the same. If A_1 and A_2 have the property that they *are* the same for all states ψ and all Borel sets S_1 and S_2 in $\mathbb{R}$, then we will say that the observables A_1 and A_2 are *compatible*.

We show now that, assuming there are no superselection rules, the operators corresponding to compatible observables must commute. The definition of compatibility for A_1 and A_2 implies that for any state ψ and any two Borel sets S_1 and S_2,

$$\langle E^{A_1}(S_1)E^{A_2}(S_2)\psi,\ E^{A_1}(S_1)E^{A_2}(S_2)\psi \rangle = \langle E^{A_2}(S_2)E^{A_1}(S_1)\psi,\ E^{A_2}(S_2)E^{A_1}(S_1)\psi \rangle.$$

Rearranging this using the fact that each $E^{A_i}(S_i)$ is a projection (self-adjoint and idempotent) gives

$$\langle \psi,\ [E^{A_1}(S_1)E^{A_2}(S_2)E^{A_1}(S_1) - E^{A_2}(S_2)E^{A_1}(S_1)E^{A_2}(S_2)]\,\psi \rangle = 0.$$

Exercise 6.2.4. Prove this.

If we now assume that the unit vectors ψ in $\mathcal{H}$ that correspond to states of our quantum system exhaust all of the unit vectors in $\mathcal{H}$ (that is, that there are no superselection rules), then it follows from this that

$$E^{A_1}(S_1)E^{A_2}(S_2)E^{A_1}(S_1) - E^{A_2}(S_2)E^{A_1}(S_1)E^{A_2}(S_2) = 0.$$

From this we obtain

$$
\begin{aligned}
&\left[E^{A_1}(S_1)E^{A_2}(S_2) - E^{A_2}(S_2)E^{A_1}(S_1)\right]^* \left[E^{A_1}(S_1)E^{A_2}(S_2) - E^{A_2}(S_2)E^{A_1}(S_1)\right] \\
&= \left[E^{A_2}(S_2)E^{A_1}(S_1) - E^{A_1}(S_1)E^{A_2}(S_2)\right]\left[E^{A_1}(S_1)E^{A_2}(S_2) - E^{A_2}(S_2)E^{A_1}(S_1)\right] \\
&= E^{A_2}(S_2)E^{A_1}(S_1)E^{A_2}(S_2) - E^{A_1}(S_1)E^{A_2}(S_2)E^{A_1}(S_1)E^{A_2}(S_2) \\
&\quad - E^{A_2}(S_2)E^{A_1}(S_1)E^{A_2}(S_2)E^{A_1}(S_1) + E^{A_1}(S_1)E^{A_2}(S_2)E^{A_1}(S_1) = 0
\end{aligned}
$$

because

$$E^{A_1}(S_1)E^{A_2}(S_2)E^{A_1}(S_1)E^{A_2}(S_2) = E^{A_1}(S_1)E^{A_2}(S_2)E^{A_1}(S_1)$$

and

$$E^{A_2}(S_2)E^{A_1}(S_1)E^{A_2}(S_2)E^{A_1}(S_1) = E^{A_2}(S_2)E^{A_1}(S_1)E^{A_2}(S_2).$$

Exercise 6.2.5. Show that if T is a *bounded* operator on a Hilbert space and $T^*T = 0$, then $T = 0$.

Thus,

$$E^{A_1}(S_1)E^{A_2}(S_2) = E^{A_2}(S_2)E^{A_1}(S_1)$$

for all Borel sets $S_1, S_2 \subseteq \mathbb{R}$ and this is precisely the definition of what it means for the self-adjoint operators A_1 and A_2 to commute. Consequently, A_1 and A_2 can both be written as functions of some self-adjoint operator B, and if B corresponds to an observable (again, no superselection rules), then A_1 and A_2 are simultaneously measurable by measuring B.

A finite family of observables $A_1, \ldots, A_n$ is said to be *compatible* if any two observables in the family are compatible in the sense we have just defined. Postulate QM6 asserts that for finite families of observables, compatibility is to be fully identified with commutativity of the corresponding operators and therefore with simultaneous measurability. Motivated by the joint probability formulas derived above, it also extends the Born–von Neumann formula to include such simultaneously measured values.

Postulate QM6. *Let $\mathcal{H}$ be the Hilbert space of some quantum system and $A_1, \ldots, A_n$ self-adjoint operators on $\mathcal{H}$ corresponding to observables. Then $A_1, \ldots, A_n$ are simultaneously measurable if and only if A_i and A_j commute for all $i, j = 1, \ldots, n$. Moreover, in* *this case, if ψ is any state and $S_1, \ldots, S_n$ are Borel sets in $\mathbb{R}$, then the probability that a simultaneous measurement of the observables $A_1, \ldots, A_n$ will yield values in $S_1, \ldots, S_n$, respectively, is*

$$\| E^{A_1}(S_1) \cdots E^{A_n}(S_n)\psi \|^2 = \langle\, \psi,\, E^{A_1}(S_1) \cdots E^{A_n}(S_n)\psi \,\rangle, \tag{6.16}$$

where $E^{A_1}, \ldots, E^{A_n}$ are the spectral measures of $A_1, \ldots, A_n$, respectively.

Note that since $E^{A_i}(S_i)$ commute, $E^{A_1}(S_1) \cdots E^{A_n}(S_n)$ is a projection and this gives the equality in (6.16).

Because of von Neumann's Theorem 5.6.4, no one argues about the principle that commuting observables should be simultaneously measurable (in virtually any sense in which you might define simultaneously measurable). As one might expect, however, the considerably less obvious converse, that noncommuting observables cannot be simultaneously measured, has not gone unquestioned. Von Neumann presents an argument in favor of this converse quite different from the one we have used here (see pages 223–229 of [vonNeu]), but for the view from the loyal opposition one should consult [PM]. The issue, of course, is that adopting Postulate QM6 essentially *defines* "simultaneous measurability" as commutativity of the corresponding operators and one can argue about how accurately this definition reflects the physical notion of "measuring simultaneously." Mathematical definitions are clear and unambiguous, but physics is rarely like that.

Postulates QM1–QM6 provide a framework on which to build many of the formal aspects of quantum mechanics, but any suggestion that quantum mechanics as a whole is somehow contained in these postulates is wildly false. We will spend some time investigating this formal structure, but when we turn our attention to more concrete issues Postulates QM1–QM6 will need to be supplemented not only with specific information about particular quantum systems, but also with general procedures for implementing the postulates (how does one choose the Hilbert space of the system, or associate operators to its observables, etc.). Furthermore, when we finally come to the quantum mechanical notion of *spin* (in Chapter 10) we will find it necessary to append an additional Postulate QM7 to our list (see Section 10.2).

6.3 Uncertainty relations

We would like to begin by taking a closer look at pairs of observables to which Postulate QM6 does *not* apply. Thus, we consider two self-adjoint operators $A : \mathcal{D}(A) \to \mathcal{H}$ and $B : \mathcal{D}(B) \to \mathcal{H}$ representing observables that are not compatible, that is, do not commute. We know that A and B are not simultaneously measurable in every state.

Remark 6.3.1. The logical negation of "A and B are simultaneously measurable in every state" is "there exist states in which A and B are not simultaneously measurable."

However, we have not defined what it means for observables to be simultaneously measurable in some states, but not in others. There is certainly no obvious reason to exclude such a possibility, but making sense of it would require a refinement of our definitions based on a much more careful look at the measurement process than we have attempted here (see pages 230–231 of [vonNeu]).

Each of the observables A and B has an expected value and a dispersion in every state and we would like to investigate the relationship between the two dispersions. This is based on the following simple lemma.

Lemma 6.3.1. *Let $\mathcal{H}$ be a separable, complex Hilbert space, $A : \mathcal{D}(A) \to \mathcal{H}$ and $B : \mathcal{D}(B) \to \mathcal{H}$ self-adjoint operators on $\mathcal{H}$ and α and β real numbers. Then, for every $\psi \in \mathcal{D}([A, B]_-) = \mathcal{D}(AB) \cap \mathcal{D}(BA)$,*

$$\| (A - \alpha)\psi \|^2 \| (B - \beta)\psi \|^2 \geq \frac{1}{4} | \langle \psi, [A, B]_- \psi \rangle |^2.$$

Proof. Note first that $\mathcal{D}([A - \alpha, B - \beta]_-) = \mathcal{D}([A, B]_-)$ and $[A - \alpha, B - \beta]_- \psi = [A, B]_- \psi$ for every ψ in this domain. Hence, it will suffice to prove the result when $\alpha = \beta = 0$. In this case we just compute

$$| \langle \psi, [A, B]_- \psi \rangle |^2 = | \langle \psi, AB\psi - BA\psi \rangle |^2 = | \langle \psi, AB\psi \rangle - \langle \psi, BA\psi \rangle |^2$$
$$= | \langle A\psi, B\psi \rangle - \langle B\psi, A\psi \rangle |^2 = 4 | \operatorname{Im} \langle A\psi, B\psi \rangle |^2$$
$$\leq 4 \| A\psi \|^2 \| B\psi \|^2. \qquad \square$$

Now, if A and B represent observables, $\psi \in \mathcal{D}([A, B]_-)$ is a unit vector representing a state and we take α and β to be the corresponding expected values of A and B in this state, then, by (6.4), we obtain

$$\sigma_\psi^2(A) \, \sigma_\psi^2(B) \geq \frac{1}{4} | \langle \psi, [A, B]_- \psi \rangle |^2. \tag{6.17}$$

If A and B commute we know that $[A, B]_- \psi = 0$ for every $\psi \in \mathcal{D}([A, B]_-)$ so this last inequality contains no information. Even if A and B do not commute, it is still possible that $[A, B]_- \psi = 0$ for some or all $\psi \in \mathcal{D}([A, B]_-)$, but should it happen that some $\langle \psi, [A, B]_- \psi \rangle \neq 0$, then we have a lower bound on the product of the two dispersions.

Example 6.3.1. We consider the position operator $Q : \mathcal{D}(Q) \to L^2(\mathbb{R})$ and momentum operator $P : \mathcal{D}(P) \to L^2(\mathbb{R})$ on $\mathbb{R}$. We saw in Example 5.6.6 that $\mathcal{S}(\mathbb{R}) \subseteq \mathcal{D}([Q, P]_-)$ and, for every $\psi \in \mathcal{S}(\mathbb{R})$,

$$[Q, P]_- \psi = i\hbar\psi.$$

Consequently, if ψ is a unit vector, $\frac{1}{4} | \langle \psi, [Q, P]_- \psi \rangle |^2 = \frac{1}{4}\hbar^2$, so

$$\sigma_\psi^2(Q) \, \sigma_\psi^2(P) \geq \frac{\hbar^2}{4}. \tag{6.18}$$

Note that, since $\mathcal{S}(\mathbb{R})$ is dense in the domain of $[Q, P]_-$ (in fact, in all of $L^2(\mathbb{R})$) this inequality is satisfied for every state in $\mathcal{D}([Q, P]_-)$. One often sees this written in terms of the *standard deviation* (the nonnegative square root of the dispersion) as

$$\sigma_\psi(Q)\,\sigma_\psi(P) \geq \frac{\hbar}{2}. \tag{6.19}$$

The inequality (6.19) (or (6.18)) is called the *uncertainty relation* for position and momentum.

Example 6.3.2. As a specific example, let ψ_n be one of the eigenstates of H_B. From Examples 6.2.4 and 6.2.7 we have

$$\sigma_{\psi_n}^2(Q)\,\sigma_{\psi_n}^2(P) = \left[\frac{\hbar}{m\omega}\left(n + \frac{1}{2}\right)\right]\left[m\omega\hbar\left(n + \frac{1}{2}\right)\right] = \hbar^2\left(n + \frac{1}{2}\right)^2 \geq \frac{\hbar^2}{4},$$

where equality holds only in the ground state ψ_0.

We have made a point of *not* calling (6.19) the "Heisenberg uncertainty relation," although it is not uncommon to find this name attached to it. We will explain our reluctance to use this terminology shortly, but, for those who would like to see a name attached to such a famous inequality, we might suggest the *Kennard uncertainty relation* in honor of the gentleman who first proved it (see [Kenn]); the more general result (6.17) is due to Robertson (see [Rob]). We should point out also that in 1930 Schrödinger obtained a version of the uncertainty relation that is stronger than (6.17); an annotated translation of Schrödinger's paper is available in [AngBat].

We intend to spend some time sorting out the physical interpretation of the uncertainty relation (6.19), but first we would like to offer another derivation based on the following standard result in Fourier analysis (except for the normalization of the Fourier transform, this is Theorem 1.1 of [FS], which contains a simple proof and a survey of many related results; still more is available in [Fol1]).

Theorem 6.3.2. *Let ψ be any element of $L^2(\mathbb{R}, dq)$ and α and β any two real numbers. Then*

$$\int_{\mathbb{R}} (q - \alpha)^2 |\psi(q)|^2\,dq \int_{\mathbb{R}} (p - \beta)^2 |\hat{\psi}(p)|^2\,dp \geq \frac{\|\psi\|^4}{4}. \tag{6.20}$$

If $\|\psi\| = 1$, then $|\psi(q)|^2 dq$ and $|\hat{\psi}(p)|^2 dp$ are both probability measures on $\mathbb{R}$, the integrals in (6.20) are their variances and the product of these variances is bounded below by $\frac{1}{4}$:

$$\int_{\mathbb{R}} (q - \alpha)^2 |\psi(q)|^2\,dq \int_{\mathbb{R}} (p - \beta)^2 |\hat{\psi}(p)|^2\,dp \geq \frac{1}{4}. \tag{6.21}$$

Intuitively, (6.21) asserts that these probability measures cannot *both* be sharply local-ized at any points α and β in $\mathbb{R}$. In this case we can apply the integral formulas in (6.6) and (6.7) directly to obtain (6.18). We have

$$\sigma_\psi^2(Q)\sigma_\psi^2(P) = \int_{\mathbb{R}} (q - \langle Q \rangle_\psi)^2 |\psi(q)|^2 dq \int_{\mathbb{R}} \left(p - \frac{\langle \hat{P} \rangle_\psi}{\hbar} \right)^2 |\hat{\psi}(p)|^2 dp \geq \frac{\hbar^2}{4}.$$

The first thing we would like to note about these two derivations of $\sigma_\psi(Q)\sigma_\psi(P) \geq \frac{\hbar}{2}$ is that, except for the identification of Q and P with position and momentum opera-tors, there is nothing even remotely resembling physics in either of them. The first is just a special case of a general result on self-adjoint operators, while the second is a special case of an equally general result about functions and their Fourier transforms. Next, it is important to recognize that $\sigma_\psi(Q)\sigma_\psi(P) \geq \frac{\hbar}{2}$ is a *statistical* statement about the standard deviations of a large number of measurements of position and momen-tum on identically prepared systems and, moreover, that each of the standard devi-ations $\sigma_\psi(Q)$ and $\sigma_\psi(P)$ is obtained by making repeated measurement of *just one* of the observables (*not both simultaneously*). Alternatively, one can identify the position and momentum observables in state ψ with their corresponding probability measures $|\psi(q)|^2 dq$ and $|\hat{\psi}(p)|^2 dp$ on $\mathbb{R}$ and regard $\sigma_\psi(Q)\sigma_\psi(P) \geq \frac{\hbar}{2}$ as a limitation on the extent to which *both* measures can be concentrated about their expected values. In any case, the essential observation is that one can infer *nothing* from it regarding the outcome of any single measurement of position and momentum on a system in state ψ. This places the inequality (6.19) in sharp contrast with what is usually referred to as the *Heisenberg uncertainty principle*, described in the following way by Heisenberg him-self.

> "Actually, at every moment the electron has only an inaccurate position and an inaccurate veloc-ity, and between these two inaccuracies there is this uncertainty relation."

Werner Heisenberg

The uncertainty relation to which Heisenberg refers, written in terms of the position and momentum (rather than velocity) of the electron, is

$$\Delta q \, \Delta p \geq \frac{\hbar}{2}, \tag{6.22}$$

where Δq and Δp are called the "uncertainty," or "inaccuracy," in the position and momentum measurements of the electron at some instant. These terms are generally not defined precisely, but only illustrated in various *Gedanken* experiments for the measurement of position and momentum (and, as we will see, they are most certainly *not* the same as the statistical quantities $\sigma_\psi(Q)$ and $\sigma_\psi(P)$). The most famous of these *Gedanken* experiments we will call the *Heisenberg–Bohr microscope*. Here one imag-ines that the position of an electron is determined by "seeing" it in a microscope, that is, by hitting the electron with a photon which then scatters off of the electron and through a lens in which it is visible to an observer.

The quantitative details of the argument involve a bit of optics that we have no need to discuss in detail. However, it is useful to have at least a qualitative appreciation for what is behind the argument, so we will describe a simplified version in which the underlying idea, we hope, is clear, but various numerical factors are swept under the rug by some rough order of magnitude approximations.

Remark 6.3.2. The exposition of an idea should not be construed as an endorsement of that idea. The Heisenberg–Born microscope has probably outlived its usefulness and there is much in the following argument that could be and, indeed, should be and has been criticized. The thrust of the argument is that quantum uncertainty is the result of what has been called the *observer effect*, that is, that any measurement of position necessarily involves a discontinuous and unpredictable effect on momentum; the more delicate the measurement of q, the more pronounced the disturbance of p, thereby ensuring that the product $\Delta q\,\Delta p$ is inevitably bounded below by some universal constant. Quantum uncertainty in the form of (6.19) is rigorously proved and experimentally verified beyond any reasonable doubt, but the same cannot be said of the arguments we embark upon now, nor even of the conclusion to which these arguments lead. For a more sustained critique of Heisenberg's version of the uncertainty principle we refer to [Ozawa] and, especially, [Roz].

The resolving power of the microscope, that is, the minimum distance between two points that the microscope can see as two distinct points, determines the accuracy with which the electron (or any other object) can be located. This resolving power is determined in part by the construction of the microscope (specifically, the angle θ in Figure 6.1), but also by the wavelength λ_{sc} of the scattered photon. Specifically, let us suppose that a photon is scattered off of the electron through an angle ϕ that sends the scattered photon into the lens of the microscope, that is, into the cone in Figure 6.1. Then the resolving power, that is, the "uncertainty" in the position of the electron along the q-axis (parallel to the motion of the incoming photon), is determined by the optics of the microscope to be

$$\Delta q \approx \frac{\lambda_{sc}}{2\sin\theta}.$$

Note that the wavelength λ_{sc} of the scattered photon is not the same as the wavelength λ of the photon impinging upon the electron in Figure 6.1, but the relationship between them is well understood. Briefly, the story is as follows. In 1923, Arthur Compton showed that if one assumes that the scattering of the photon off of the electron is elastic in the sense that (relativistic) energy and momentum are conserved and that the energy of a photon is given by the Einstein relation $E = h\nu = \frac{hc}{\lambda}$ (Section 4.3), then

$$\lambda_{sc} - \lambda = \frac{h}{m_e c}(1 - \cos\phi), \tag{6.23}$$

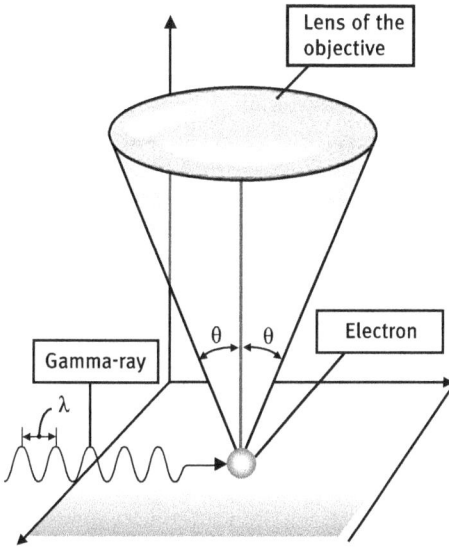

Figure 6.1: Heisenberg–Bohr microscope.

where h is $2\pi\hbar$, m_e is the mass of the electron, c is the speed of light *in vacuo* and ϕ is the angle through which the photon is deflected (one can find this argument in essentially any book on special relativity and, in particular, on pages 88–90 of [Nab5]). The universal constant $\frac{h}{m_e c}$ is called the *Compton wavelength* of the electron and, in SI units (Remark 4.2.1), is

$$\frac{h}{m_e c} \approx 2.4263102175 \times 10^{-12} \, \text{m}. \tag{6.24}$$

Note that the change in wavelength depends only on the scattering angle and not on the incident wavelength. The Compton formula (6.23) has been directly verified experimentally, beginning with experiments conducted by Compton himself, as have the assumptions upon which it is based (for example, that the recoil electron acquires exactly the energy and momentum lost by the incoming photon). A more detailed analysis of this *Compton effect* is available in Section 2.9 of [Bohm].

From all of this we conclude that for a fixed θ, one can decrease the uncertainty Δq arbitrarily by illuminating the electron with photons of small wavelength λ. The electromagnetic spectrum in Figure 4.1 suggests that one would want to choose gamma rays (with, say, $\lambda \approx 10^{-12}$ m) to accomplish this. But by (6.24), if $\lambda \approx 10^{-12}$ m, then λ, λ_{sc} and $\lambda - \lambda_{sc}$ are all of the same order of magnitude and our approximations will use λ for all of them. In particular,

$$\Delta q \approx \frac{\lambda}{2 \sin \theta}.$$

Now, because of their short wavelength gamma rays are quite energetic. According to the Einstein relations (Section 4.3), the energy of a gamma ray of frequency $v = \frac{c}{\lambda}$ is

$E = h\nu = \frac{hc}{\lambda}$ and the magnitude of its momentum is $p = \frac{h\nu}{c} = \frac{h}{\lambda}$. Now we consider the component of the scattered photon's momentum in the q-direction (parallel to the direction of the incoming gamma ray). All we know about the scattered photon is that it enters the field of view somewhere in the angular range between $-\theta$ and θ and so this component is somewhere between $-p \sin \theta$ and $p \sin \theta$. Consequently, the uncertainty in the scattered photon's momentum is $p \sin \theta - (-p \sin \theta) = 2p \sin \theta \approx \frac{2h}{\lambda} \sin \theta$ (the approximation comes from $\lambda_{sc} \approx \lambda$). The Compton effect ensures that the momentum lost by the gamma ray when it scatters is precisely the momentum gained by the electron so the uncertainty in the q-component of the electron's momentum after its collision with the photon is also

$$\Delta p \approx \frac{2h}{\lambda} \sin \theta.$$

Consequently,

$$\Delta q \Delta p \approx \left[\frac{\lambda}{2 \sin \theta} \right] \left[\frac{2h}{\lambda} \sin \theta \right] = h.$$

Since $h > \frac{\hbar}{2}$ this is not as good as $\Delta q \Delta p \geq \frac{\hbar}{2}$, but it is sufficient to indicate what is being claimed here, which is something quite different from our uncertainty relation (6.19). The Heisenberg uncertainty principle (6.22) is a statement about *one* measurement of *two* observables on a *single* quantum system, whereas (6.19) is a statistical statement about *many independent* measurements of these two observables on identical quantum systems.

It is important to clearly make this distinction for several reasons. In the first place, (6.19) is a rigorous mathematical consequence of the most basic assumptions of quantum mechanics, and should it be found (in the laboratory) to be false, one would need to completely rethink the foundations of the subject. On the other hand, (6.22) is not provable (in the mathematical sense) from these basic assumptions alone and our house of cards could conceivably remain standing if it should prove not to be universally valid which, incidentally, it appears not to be (see [Ozawa] and [Roz]).

"The Heisenberg uncertainty principle states that the product of the noise in a position measurement and the momentum disturbance caused by that measurement should be no less than the limit set by Planck's constant, $\hbar/2$, as demonstrated by Heisenberg's thought experiment using a gamma ray microscope. Here I show that this common assumption is false: a universally valid trade-off relation between the noise and the disturbance has an additional correlation term."

– Masanao Ozawa [Ozawa]

"While there is a rigorously proved relationship about uncertainties intrinsic to any quantum system, often referred to as 'Heisenberg's uncertainty principle,' Heisenberg originally formulated his ideas in terms of a relationship between the precision of a measurement and the disturbance it must create. Although this latter relationship is not rigorously proved, it is commonly believed (and taught) as an aspect of the broader uncertainty principle. Here, we experimentally observe a violation of Heisenberg's 'measurement-disturbance relationship,' using weak measurements to characterize a quantum system before and after it interacts with a measurement apparatus."

Rozema *et al.* [Roz]

The Heisenberg uncertainty principle (6.22) has various extensions in physics to pairs of observables other than position and momentum. Essentially, one expects such an uncertainty principle whenever the operators do not commute; we have not introduced these yet, but this occurs, for example, for any two distinct components of angular momentum or spin. However, one also encounters what is called the *time-energy uncertainty principle*, usually written as

$$\Delta t \, \Delta E \geq \frac{\hbar}{2}.$$

This, however, is of a rather different character since, as we have already mentioned, time t is not an observable in quantum mechanics, that is, there is no operator corresponding to t. We would like to say a bit more about this version of the uncertainty principle, but this is most conveniently done by taking a slightly different point of view regarding the formulation of quantum mechanical laws called the "Heisenberg picture," so we will postpone what we have to say until Section 6.4.

6.4 Schrödinger and Heisenberg pictures

Let us begin with a brief synopsis of the picture we have painted thus far of quantum mechanics. A quantum system has associated with it a complex, separable Hilbert space $\mathcal{H}$ and a distinguished self-adjoint operator H, called the Hamiltonian of the system. The states of the system are represented by unit vectors ψ in $\mathcal{H}$ and these evolve in time from an initial state $\psi(0)$ according to $\psi(t) = U_t(\psi(0)) = e^{-itH/\hbar}(\psi(0))$. As a result, the evolving states satisfy the abstract Schrödinger equation $i\hbar \frac{d\psi(t)}{dt} = H(\psi(t))$. Each observable is identified with a self-adjoint operator A that does not change with time. Neither the state vectors ψ nor the observables A are accessible to direct experimental measurement. Rather, the link between the formalism and the physics is contained in the expectation values $\langle A \rangle_\psi = \langle \psi, A\psi \rangle$ and the probability measures $\mu_{\psi,A}(S) = \langle \psi, E^A(S)\psi \rangle$. These contain all of the information that quantum mechanics permits us to know about the system.

We would now like to look at this from a slightly different point of view. As the state evolves, so do the expectation values of any observable. Specifically,

$$\langle A \rangle_{\psi(t)} = \langle \psi(t), A\psi(t) \rangle = \langle U_t(\psi(0)), AU_t(\psi(0)) \rangle = \langle \psi(0), [U_t^{-1}AU_t]\,\psi(0) \rangle,$$

because each U_t is unitary. Now, define a (necessarily self-adjoint) operator

$$A(t) = U_t^{-1}AU_t$$

for each $t \in \mathbb{R}$. Thus,

$$\langle A \rangle_{\psi(t)} = \langle A(t) \rangle_{\psi(0)}$$

for each $t \in \mathbb{R}$. The expectation value of A in the evolved state $\psi(t)$ is the same as the expectation value of the observable $A(t)$ in the initial state $\psi(0)$. Since all of the physics is contained in the expectation values, this presents us with the option of regarding the states as fixed and the observables as evolving in time. From this point of view our quantum system has a *fixed* state ψ and the observables evolve in time from some initial self-adjoint operator $A = A(0)$ according to

$$A(t) = U_t^{-1} A U_t = e^{itH/\hbar} A e^{-itH/\hbar}. \tag{6.25}$$

This is called the *Heisenberg picture* of quantum mechanics to distinguish it from the view we have taken up to this point, which is called the *Schrödinger picture*. Although these two points of view appear to differ from each other rather trivially, the Heisenberg picture occasionally presents some significant advantages and we will now spend a moment seeing what things look like in this picture.

Heisenberg's original formulation of quantum mechanics is generally called *matrix mechanics*, and here the dynamics of a quantum system is defined by time-dependent observables given by infinite matrices which evolve according to the Heisenberg equation (see (6.26) below). We will have a more careful look at the ideas that led Heisenberg to this in Section 7.1. Although Heisenberg, of course, did not view the matter in this way, one can arrive at these matrices by choosing an orthonormal basis for $\mathcal{H}$ contained in $\mathcal{D}(A(t))$ for every t and using it to move from $\mathcal{H}$ to $\ell^2(\mathbb{N})$. The Heisenberg picture is generally regarded as the one most appropriate to quantum field theory.

We should first note that, when A is the Hamiltonian H, Stone's Theorem 5.6.2 implies that each U_t leaves $\mathcal{D}(H)$ invariant and commutes with H, so

$$H(t) = U_t^{-1} H U_t = e^{itH/\hbar} H e^{-itH/\hbar} = H \quad \forall t \in \mathbb{R}.$$

The Hamiltonian is constant in time in the Heisenberg picture. For other observables this is generally not the case, of course, and one would like to have a differential equation describing their time evolution in the same way that the Schrödinger equation describes the time evolution of the states. We will derive such an equation in the case of observables represented by *bounded* self-adjoint operators in the Schrödinger picture. This is a very special case, of course, so we should explain the restriction. In the unbounded case, a rigorous derivation of the equation is substantially complicated by the fact that in the Heisenberg picture, the operators (and therefore their domains) vary with t, so that the usual domain issues also vary with t. Physicists have the good sense to ignore all of these issues and just formally differentiate (6.25), thereby arriving at the very same equation we will derive below. Furthermore, it is not hard to show that if A is unbounded and $A(t) = U_t^{-1} A U_t$, then, for any Borel function f, $f(A(t)) = f(A)(t) = U_t^{-1} f(A) U_t$, so that one can generally study the time evolution of A in terms of the time evolution of the bounded functions of A and these are bounded operators. We will have a bit more to say about the unbounded case after we have proved our theorem. Finally, we recall that, from the point of view of physics, all of the relevant information is contained in the probability measures $\langle \psi, E^A(S)\psi \rangle$ so that, in principle, one requires only the time evolution of the (bounded) projections $E^A(S)$.

The result we need is a special case of Theorem 3.2, Chapter IV, of [Prug], but because of its importance to us we will give the proof here as well.

Theorem 6.4.1. *Let $\mathcal{H}$ be a complex, separable Hilbert space, $H : \mathcal{D}(H) \to \mathcal{H}$ a self-adjoint operator on $\mathcal{H}$ and $U_t = e^{-itH/\hbar}$, $t \in \mathbb{R}$, the one-parameter group of unitary operators determined by H. Let $A : \mathcal{H} \to \mathcal{H}$ be a bounded, self-adjoint operator on $\mathcal{H}$ and define, for each $t \in \mathbb{R}$, $A(t) = U_t^{-1} A U_t$. If ψ and $A(t)\psi$ are in $\mathcal{D}(H)$ for every $t \in \mathbb{R}$, then $A(t)\psi$ satisfies the* Heisenberg equation

$$\frac{dA(t)}{dt}\psi = -\frac{i}{\hbar}[A(t), H]_-\psi, \tag{6.26}$$

where the derivative is the $\mathcal{H}$-limit

$$\frac{dA(t)}{dt}\psi = \lim_{\Delta t \to 0}\left[\frac{A(t + \Delta t) - A(t)}{\Delta t}\psi\right].$$

Remark 6.4.1. Let us simplify the notation a bit and write (6.26) as

$$\frac{dA}{dt} = -\frac{i}{\hbar}[A, H]_-. \tag{6.27}$$

Now compare this with equation (2.65)

$$\frac{df}{dt} = \{f, H\}$$

describing the time evolution of a classical observable in the Hamiltonian picture of mechanics. The analogy is striking and suggests a possible avenue from classical to quantum mechanics, that is, a possible approach to the quantization of classical mechanical systems. The idea is that classical observables should be replaced by self-adjoint operators and Poisson brackets $\{,\}$ by the *quantum bracket*

$$\{,\}_\hbar = -\frac{i}{\hbar}[,]_-.$$

This idea, first proposed by Paul Dirac in his doctoral thesis, is the basis for what is called *canonical quantization* and we will return to it in Chapter 7.

Proof. Adding and subtracting $\frac{U_{t+\Delta t}^{-1} A U_t}{\Delta t}$ to $\frac{A(t+\Delta t)-A(t)}{\Delta t}$ gives

$$\frac{A(t+\Delta t)-A(t)}{\Delta t} = U_{t+\Delta t}^{-1} A \left[\frac{U_{t+\Delta t} - U_t}{\Delta t}\right] + \left[\frac{U_{t+\Delta t}^{-1} - U_t^{-1}}{\Delta t}\right]AU_t.$$

Now, note that

$$\left[\frac{U_{t+\Delta t} - U_t}{\Delta t}\right]\psi = \frac{U_{t+\Delta t}\psi - U_t\psi}{\Delta t} = \frac{U_{\Delta t}(U_t\psi) - (U_t\psi)}{\Delta t} \longrightarrow -\frac{i}{\hbar}HU_t\psi$$

as $\Delta t \to 0$ by Lemma 5.6.1. Since A is bounded and therefore continuous and $U_{t+\Delta t}^{-1}$ is strongly continuous in Δt,

$$U_{t+\Delta t}^{-1} A \left[\frac{U_{t+\Delta t} - U_t}{\Delta t}\right]\psi \longrightarrow -\frac{i}{\hbar}U_t^{-1}AHU_t\psi = -\frac{i}{\hbar}(U_t^{-1}AU_t)H\psi = -\frac{i}{\hbar}A(t)H\psi$$

as $\Delta t \to 0$ (recall that U_t commutes with H by Stone's Theorem 5.6.2).

Exercise 6.4.1. Show similarly that

$$\left[\frac{U_{t+\Delta t}^{-1} - U_t^{-1}}{\Delta t}\right]AU_t\psi \longrightarrow \frac{i}{\hbar}HA(t)\psi$$

as $\Delta t \to 0$.

Combining these two gives

$$\frac{dA(t)}{dt}\psi = -\frac{i}{\hbar}(A(t)H\psi - HA(t)\psi) = -\frac{i}{\hbar}[A(t), H]_-\psi,$$

as required. □

In the physics literature one finds the Heisenberg equation stated quite generally for operators that are perhaps unbounded and with little attention paid to domain issues. Although one cannot rigorously justify this in full generality, it is very often the case that a justification is possible in the cases of physical interest. For example, if A is unbounded, then the domain of $-\frac{i}{\hbar}[A, H]_-$ *may* be quite small, but very often it is not (if A is either the position operator Q or the momentum operator P on $L^2(\mathbb{R})$ and $H = H_B$, then this domain includes all of $S(\mathbb{R})$). If it should happen that $-\frac{i}{\hbar}[A, H]_-$ is essentially self-adjoint on its domain, then, denoting its unique self-adjoint extension also by $-\frac{i}{\hbar}[A, H]_-$, one can *define* the time derivative of A to be this unique self-adjoint extension so that the Heisenberg equation is satisfied by definition. Very often one simply has to supply whatever hypotheses are required to justify a calculation and then check that the hypotheses are satisfied in any case to which the result of the calculation is applied. While such formal calculations are anathema to mathematicians, physics could not get along without them simply because they get to the heart of the matter without distractions. Here is an example (remember that the state ψ does not depend on t in the Heisenberg picture):

$$\frac{d}{dt}\langle A(t)\rangle_\psi = \frac{d}{dt}\langle\psi, A(t)\psi\rangle = \left\langle\psi, \frac{dA(t)}{dt}\psi\right\rangle$$

$$= \left\langle\psi, -\frac{i}{\hbar}[A(t), H]\psi\right\rangle = \left\langle -\frac{i}{\hbar}[A(t), H]\right\rangle_\psi. \tag{6.28}$$

The rate of change of the expectation value of $A(t)$ in state ψ is the expectation value of $-\frac{i}{\hbar}[A(t), H]$ in state ψ; we will return to this in Section 7.4.

We will conclude by returning to an issue we left open in the previous section. There we pointed out that, in addition to the Heisenberg uncertainty principle (6.22) for position and momentum, physicists commonly employ what is called the *time-energy uncertainty principle*, written as

$$\Delta t\, \Delta E \geq \frac{\hbar}{2}. \tag{6.29}$$

We also pointed out that, in addition to all of the subtleties buried in the usual derivation of (6.22) via the Heisenberg microscope, one must now contend with the fact that t is not an observable in quantum mechanics, so it is not represented by an operator. There is more, however. Leaving aside the relativistic prohibition of any universal notion of time, it is not even altogether clear what is meant by t in (6.29) and several different interpretations are possible (this issue is discussed in great detail in [Busch]). We will not attempt to sort out all of these subtleties here, but will offer just one possible interpretation of (6.29).

Remark 6.4.2. There is a reason that physicists believe so strongly that there should be some sort of uncertainty relation involving time and energy. Special relativity requires that various well-known classical physical quantities be merged into a single object in order to ensure relativistic invariance. The most obvious example is the spacetime position 4-vector (x, y, z, t) which combines the classical spatial 3-vector (x, y, z) and the scalar time t-coordinate. Another example has the energy as the time coordinate of a 4-vector whose spatial part is the classical momentum. A physicist will then say, "Well, there you have it; time is to energy as position is to momentum so, quantum mechanically, time and energy should satisfy an uncertainty relation analogous to that for position and momentum." Whether you find this argument persuasive or not is really not the issue since it is intended only to motivate the search for such a relation. As we have already suggested, the difficulty in implementing this search is due in large measure to the fact that time is not an observable in quantum mechanics. A reasonable response to this might be, "Well then, just introduce an operator that represents t." Regrettably, we will find in Chapter 7 that something called the *Stone–von Neumann theorem* prohibits doing this in any physically reasonable way.

Although t is not an observable in quantum mechanics, in the Heisenberg picture observables are functions of t and the idea is to select (appropriately) some observable $A(t)$ evolving from $A(0) = A$ according to the Heisenberg equation (6.26) and let it measure t for us in some way. Fix some state ψ of the system (which, in the Heisenberg picture, does not change with t) and consider the expected values $\langle A(t) \rangle_\psi$ and standard deviations $\sigma_\psi(A(t))$ of the observables $A(t)$ in this state (we must assume that ψ is in the domain of every $A(t)$). Then, by (6.17) and (6.28),

$$\sigma_\psi(A(t))\,\sigma_\psi(H) \geq \frac{1}{2}|\langle \psi, [A(t), H]_- \psi \rangle| = \frac{\hbar}{2}\left|\left\langle \psi, -\frac{i}{\hbar}[A(t), H]_- \psi \right\rangle\right|$$

$$= \frac{\hbar}{2}\left|\left\langle -\frac{i}{\hbar}[A(t), H]_- \right\rangle_\psi\right|,$$

and so

$$\sigma_\psi(A(t))\,\sigma_\psi(H) \geq \frac{\hbar}{2}\left|\frac{d}{dt}\langle A(t) \rangle_\psi\right|.$$

Evaluating at $t = 0$ and assuming that $\frac{d}{dt}\langle A(t) \rangle_\psi|_{t=0} \neq 0$ we can define

$$\Delta t_{\psi,A} = \left|\frac{\sigma_\psi(A)}{\frac{d}{dt}\langle A(t) \rangle_\psi|_{t=0}}\right|,$$

which can be interpreted as (approximately) the time required for the expected value of A to change by an amount equal to one standard deviation, that is, for the statistics of $A(t)$ to change appreciably from that of A. Thus, the "uncertainty" in t is expressed by the average time taken for the expectation of A in state ψ to change by one standard deviation and therefore describes the shortest time scale on which we will be able to notice changes by using the observable A in state ψ. The Hamiltonian represents the total energy, so we let $\Delta E = \sigma_\psi(H)$ and obtain

$$\Delta_{\psi,A}t\,\Delta E \geq \frac{\hbar}{2}.$$

This, of course, depends on the choice of A and ψ, but it is satisfied for any choice of A and ψ satisfying the conditions we just described, so we can let Δt denote the infimum of the $\Delta_{\psi,A}t$ over all such choices and we still have

$$\Delta t\,\Delta E \geq \frac{\hbar}{2}.$$

This then is one possible interpretation of the time-energy uncertainty principle. Whether or not it is the correct interpretation (that is, is actually satisfied by quantum systems) remains to be seen; there is a substantial literature on this and those interested in pursuing the matter might begin with [Busch].

7 Canonical quantization

7.1 $PQ - QP = \frac{h}{2\pi i}$

We have seen in Section 4.3 that Max Planck introduced what we would today call the *quantum hypothesis* in 1900 to explain the observed spectrum of blackbody radiation. Planck did not regard his hypothesis as a new, fundamental principle of physics, but rather as a desperate ploy by which he could arrive at a formula that agreed with experiment. We saw also that, in 1905, Einstein took a different view, arguing that electromagnetic radiation must be regarded as having a dual character. To understand its interference and diffraction one must think of it as a wave phenomenon, but its interaction with matter in, say, the photoelectric effect, can only be accounted for if it is regarded as a stream of particles (light quanta, or photons), each carrying an amount of energy proportional to its frequency. Although Einstein's proposal was initially received with considerable skepticism (not to say, derision), the predictions to which it led were all borne out experimentally, primarily through the work of Robert Millikan. A corresponding proposal that material particles such as electrons might, under certain circumstances, exhibit wave-like behavior did not come until 1924, from Louis de Broglie. In the meantime, however, physicists devoted much effort to investigating the implications of superimposing various "quantization conditions" on classical mechanics for understanding such things as the line spectrum of hydrogen (see Figure 4.3). For example, an electron in an atom was assumed to move along a classical trajectory according to the laws of classical mechanics, but only those orbits satisfying the *Born–Sommerfeld quantization condition*

$$\oint p\,dq = nh, \quad n \in \mathbb{Z}^+, \tag{7.1}$$

were permitted (the integral is over one period of the orbit in phase space). This cobbling together of classical mechanics and *ad hoc* quantization conditions, known as the *old quantum theory*, had its successes, but no underlying logical structure. The goal was simply to take a classical picture (such as the two-body problem in Example 2.2.12) and "quantize" it in some way or another to describe a system that was viewed as analogous to the classical system, but for which classical mechanics failed (such as the hydrogen atom). Clearly, some more systematic procedure would be desirable, but this would have to wait for a more precise understanding of the logical foundations of this new quantum mechanics. This understanding eventually emerged from the work of Heisenberg, Born, Jordan, Dirac and Schrödinger.

In 1925, Heisenberg [Heis1] published a paper the stated intention of which was "to establish a basis for theoretical quantum mechanics founded exclusively upon relationships between quantities which in principle are observable." The paper is notoriously difficult to follow; Nobel laureate Steven Weinberg has referred to it as "pure magic" (see pages 53–54 of [Weinb]). Max Born and Pascual Jordan [BJ], however, saw

https://doi.org/10.1515/9783110751949-007

in the bold vision expressed in Heisenberg's paper a schematic for the logical foundations of the new mechanics. Perhaps the central element in this scheme, which did not appear directly in Heisenberg's paper and which was the key to the canonical quantization procedure later suggested by Dirac, is the identity we have chosen as the title of this section, which, incidentally, is also the epitaph inscribed on the headstone at Max Born's burial site in Göttingen.

As we will see in the next few sections, there is a real sense in which this identity captures the essence of quantum mechanics (in mathematical terms, $[P, Q]_- = \frac{h}{2\pi i}$ is the sole nontrivial relation defining the "Heisenberg Lie algebra"). One should have at least some notion of where it came from and this is what we will try to provide in this introductory section. We will offer only a very crude sketch, but one should keep in mind that even the crudest sketch must rely on Heisenberg's inspiration. Some things are not amenable to mathematical proof and one needs to approach them with an open mind. Some of what follows in this section may look vaguely like mathematics, but it almost certainly is not. In particular, we will pretend that infinite matrices behave exactly like finite matrices and that any infinite series we write down converges as nicely as you might want it to converge. The purpose here is *motivation, not derivation*. For those who would like to follow more closely the ideas in [Heis1] and [BJ] we recommend the two expository papers [AMS] and [FP2].

We will consider a single electron in a periodic orbit about a proton in a hydrogen atom and will focus our attention on the electron. As we mentioned above, the old quantum theory viewed the electron as having a classical trajectory in phase space subject to the laws of classical mechanics, but with the additional constraint that the Bohr–Sommerfeld quantization condition (7.1) must be satisfied. This constraint has the effect of forcing the orbit to lie in one of a discrete set of energy levels (*shells*) around the nucleus which can be labeled by an integer $n = 1, 2, 3, \ldots$ called the *principal quantum number* in such a way that the corresponding energies satisfy $0 < E(1) < E(2) < E(3) < \cdots$. For each fixed quantum number n the periodic classical orbit can be expanded in a Fourier series

$$q(n, t) = \sum_{\alpha=-\infty}^{\infty} q_\alpha(n) e^{i\alpha\omega(n)t}, \tag{7.2}$$

where $\omega(n)$ is the fundamental frequency of the orbit and, because $q(n, t)$ is real, $q_{-\alpha}(n) = \overline{q_\alpha(n)}$. Heisenberg's position, however, was that this classical orbit is unobservable and therefore has no business being built into the foundations of quantum theory. What he needed then was some meaningful quantum analogue of the electron's classical position.

Heisenberg's motivation is often obscure, but he does make quite explicit use of some of those aspects of the old quantum theory that seemed clearly to be pointing in the right direction. Most prominent among these is the *correspondence principle* of Niels Bohr. Roughly, one thinks of this principle as asserting that, in some limiting sense (say, as $h \to 0$), "quantum mechanics reduces to classical mechanics," whatever that is taken to mean. Since this is likely to be interpreted differently in different contexts one should not be surprised to see a number of rather disparate statements all claiming to be the correspondence principle. There are, in fact, at least three commonly accepted interpretations of the correspondence principle (one can find these described succinctly at http://plato.stanford.edu/entries/bohr-correspondence/). The one that is relevant to our sketch of Heisenberg's argument is referred to as the *selection rule interpretation*. One can find a very detailed discussion of both the meaning and the consequences of this version of the correspondence principle in the expository paper [FP1]. For the purposes of our sketch of Heisenberg's argument a few brief comments will suffice.

The measurable quantities associated with an atom are those contained in its line spectrum (Figure 4.3), that is, the frequency and intensity of the spectral lines. These spectral lines arise from photons emitted when an electron "jumps" from one energy level to a lower energy level; the difference in the energies determines the frequency of the emitted photon. The intensity of the spectral line is determined by the probability per unit time for that transition to occur (the more jumps that occur per unit time, the brighter the line). The selection rule interpretation of the correspondence principle asserts that *each allowed transition corresponds to one harmonic component $q_\alpha(n)e^{i\alpha w(n)t}$ of the classical motion, with the transition energy and transition probability corresponding to the harmonic frequency $\alpha w(n)$ and harmonic amplitude $q_\alpha(n)$, respectively*. Now we will try to understand how this led Heisenberg to a quantum analogue of each harmonic component.

We begin by defining a skew-symmetric function $w : \mathbb{Z} \times \mathbb{Z} \to \mathbb{R}$. For any $(n, m) \in \mathbb{Z} \times \mathbb{Z}$, the value $w(n, m)$ will be an integral multiple of $\frac{1}{h}$ that is related to the frequency of the photon associated with the transition $n \to m$ between the energy levels $E(n)$ and $E(m)$. If either n or m is less than or equal to zero, $w(n, m)$ is taken to be zero because there are no energy levels $E(k)$ with $k \leq 0$. The reason for the skew-symmetry can be explained as follows. If $E(n)$ and $E(m)$ are both permissible energy levels and $n > m$, then the transition $n \to m$ is accompanied by the *emission* of a photon with frequency

$$w(n, m) = \frac{1}{h}[E(n) - E(m)].$$

On the other hand, the transition $m \to n$ must be accompanied by the *absorption* of a photon of the same frequency. Thinking of the emission of a photon by the atom as adding energy to the universe outside the atom and the absorption of a photon as subtracting it we are led to define

$$w(m, n) = \frac{1}{h}[E(m) - E(n)] = -w(n, m).$$

Now fix a quantum number $n = 1, 2, 3, \ldots$ and consider one of the harmonic components

$$q_\alpha(n)e^{i\alpha w(n)t} \tag{7.3}$$

of the classical path. Assume for a moment that $0 < \alpha < n$. Then, by the correspondence principle, (7.3) corresponds to the transition

$$n \to n - \alpha$$

from the nth to the $(n - \alpha)$th energy level. Accordingly, we have

$$w(n, n - \alpha) = \frac{1}{\hbar}\left[E(n) - E(n - \alpha)\right] \tag{7.4}$$

for the frequency of the photon emitted during the transition. Heisenberg took $w(n, n - \alpha)$ to be the quantum analogue of the classical frequency $\alpha w(n)$.

Now, if $\alpha = 0$, our skew-symmetry assumption forces $w(n, n - \alpha) = w(n, n) = 0$. Furthermore, if $\alpha \geq n$, then $n - \alpha \leq 0$ and, because there is no energy level $E(n - \alpha)$, we will take $w(n, n - \alpha) = 0 = w(n - \alpha, n)$. Finally, suppose $\alpha < 0$. Then $n - \alpha = n + |\alpha| > n$, so $w(n, n - \alpha) = -w(n - \alpha, n) = \frac{1}{\hbar}[E(n) - E(n - \alpha)]$. Thus, for each fixed quantum number $n = 1, 2, 3, \dots$ we have defined $w(n, m)$ and $w(m, n)$ for all $m \in \mathbb{Z}$. Since $w(n, m) = 0$ whenever $n \leq 0$, this completes the definition of $w : \mathbb{Z} \times \mathbb{Z} \to \mathbb{R}$.

Exercise 7.1.1. Show that for any $n, m, \mu \in \mathbb{Z}$,

$$w(n, \mu) + w(\mu, m) = w(n, m).$$

In particular,

$$w(n, n - \alpha) + w(n - \alpha, n - \beta) = w(n, n - \beta)$$

for all $\alpha, \beta \in \mathbb{Z}$. We will encounter this identity again quite soon.

To obtain Heisenberg's quantum version of the classical harmonic component $q_\alpha(n)e^{i\alpha w(n)t}$ one begins with the replacement

$$e^{i\alpha w(n)t} \to e^{iw(n,n-\alpha)t}.$$

Next Heisenberg introduced quantum analogues $Q(n, n-\alpha)$ of the classical amplitudes $q_\alpha(n)$. These are called complex *transition amplitudes*, or *probability amplitudes* and are assumed to have properties that we now describe. For any $n, m \in \mathbb{Z}$,

$$Q(n, m)$$

is to have the property that $|Q(n, m)|^2$ is a measure of the probability of the transition $n \to m$, whenever such a transition is permissible. If either n or m is less than or equal to zero, $Q(n, m)$ is taken to be zero because at least one of the energy levels $E(n)$ or $E(m)$ does not exist. Heisenberg also introduced a reality condition analogous to $q_{-\alpha}(n) = \overline{q_\alpha(n)}$ by assuming

$$Q(m, n) = \overline{Q(n, m)}.$$

Precisely what these transition amplitudes are in a given context is to be determined by certain differential equations proposed by Heisenberg that are direct analogues of the classical Hamilton equations.

With this Heisenberg's *quantization* of the classical harmonic component $q_\alpha(n)e^{i\alpha\omega(n)t}$ amounts simply to the replacement

$$q_\alpha(n)e^{i\alpha\omega(n)t} \rightarrow Q(n, n - \alpha)e^{i\omega(n,n-\alpha)t}.$$

In particular, the quantization of the harmonic component $q_0(n)$ is just $Q(n, n)$. Evidently, $Q(n, n)$ cannot really be regarded as a transition amplitude (since there is no transition) and so its physical interpretation is not apparent *a priori* and must emerge from the rest of the formalism. Note, however, that each $Q(n, n)$ is real and time-independent (suggesting, perhaps, something conserved). One can learn more about this in [BJ], [BHJ], [AMS] and [FP2].

The harmonic components $q_\alpha(n)e^{i\alpha\omega(n)t}$ are all simply terms in the infinite Fourier series expansion of the electron's classical orbit. Heisenberg wrote that a "similar combination [*that is, sum*] of the corresponding quantum-theoretical quantities seems to be impossible in a unique manner and therefore not meaningful." As an alternative to summing them he simply collected them all together, that is, he suggested that "one may readily regard the ensemble of quantities"

$$\mathbf{Q}(t) = \left\{ Q(n, n - \alpha)e^{i\omega(n,n-\alpha)t} \right\}_{n\in\mathbb{N},\alpha\in\mathbb{Z}}$$

as a representation of the quantum analogue of the classical position $q(t)$ (we no longer write $q(n, t)$ because the "ensemble" $\mathbf{Q}(t)$ contains contributions corresponding to *every* quantum number n).

One should take a moment to appreciate the audacity of this idea. The familiar notion of the classical position of a particle is replaced by an infinite array of complex numbers – *by a physicist, in 1925*. One more comment is worth making at this point. Today we are all trained to view any rectangular array of numbers as a matrix and once we do so the machinery of matrix algebra and matrix analysis is laid before us, free of charge. In 1925, however, Heisenberg did not know what a matrix was and certainly did not have this machinery available to him. We could, of course, ignore this historical anomaly and switch immediately into matrix mode. We feel, however, that this would not only obscure another of Heisenberg's remarkable insights, but would also leap over one of the most interesting parts of this story. For a few moments anyway we will pretend that, with Heisenberg, we have never heard of matrices.

Having decided that the classical notion of the instantaneous position $q(t)$ of an electron in an atom should be replaced in quantum mechanics by the infinite array $\mathbf{Q}(t)$ of complex numbers, it seemed only natural to Heisenberg that the other classical observables associated with the system (momentum, the Hamiltonian, etc.)

should have similar representations. For example, the classical momentum $p(t)$ would be replaced by something of the form

$$\mathbf{P}(t) = \{P(n, n-\alpha)e^{i\omega(n,n-\alpha)t}\}_{n\in\mathbb{N}, \alpha\in\mathbb{Z}}.$$

One might even hazard the guess that if m_e is the mass of the electron, then

$$\mathbf{P}(t) = m_e\dot{\mathbf{Q}}(t) = \{m_e i\omega(n, n-\alpha)Q(n, n-\alpha)e^{i\omega(n,n-\alpha)t}\}_{n\in\mathbb{N}, \alpha\in\mathbb{Z}}.$$

To build a quantum analogue of a Hamiltonian, however, one needs to think a bit more. Classical observables such as the Hamiltonian are *functions* of q and p. Even the classical harmonic oscillator Hamiltonian contains q^2 and p^2, so one must know how to "square" $\mathbf{Q}(t)$, that is, find a quantum analogue $\mathbf{Q}^2(t)$ of $q^2(t)$.

Remark 7.1.1. Do not jump the gun here. Remember that for the moment, we know nothing about matrices and, even if we did, it would not be clear at this stage that matrix multiplication has any physical significance at all in this context. One needs some physical principle that suggests what it "should" mean to square a set of transition amplitudes.

To follow Heisenberg's argument one begins with the corresponding problem as it would be viewed in the old quantum theory. If $q(n, t) = \sum_{\alpha=-\infty}^{\infty} q_\alpha(n)e^{i\alpha\omega(n)t}$, then squaring the Fourier series gives

$$q^2(n, t) = \left(\sum_{\alpha=-\infty}^{\infty} q_\alpha(n)e^{i\alpha\omega(n)t}\right)\left(\sum_{\alpha'=-\infty}^{\infty} q_{\alpha'}(n)e^{i\alpha'\omega(n)t}\right)$$

$$= \sum_{\alpha=-\infty}^{\infty}\sum_{\alpha'=-\infty}^{\infty} q_\alpha(n)q_{\alpha'}(n)e^{i(\alpha\omega(n)+\alpha'\omega(n))t}.$$

Set $\beta = \alpha + \alpha'$ and write this as

$$q^2(n, t) = \sum_{\beta=-\infty}^{\infty} a_\beta(n)e^{i\beta\omega(n)t},$$

where

$$a_\beta(n) = \sum_{\alpha=-\infty}^{\infty} q_\alpha(n)q_{\beta-\alpha}(n).$$

Now rewrite $q^2(n, t)$ once more as

$$q^2(n, t) = \sum_{\beta=-\infty}^{\infty}\sum_{\alpha=-\infty}^{\infty} q_\alpha(n)q_{\beta-\alpha}(n)e^{i[\,\alpha\omega(n)+(\beta-\alpha)\omega(n)\,]t}$$

and read off the harmonic components

$$\sum_{\alpha=-\infty}^{\infty} q_\alpha(n) q_{\beta-\alpha}(n) e^{i[\,\alpha\omega(n)+(\beta-\alpha)\omega(n)\,]t}.$$

The problem then is to define an appropriate quantum analogue $Q^2(n, n-\beta) e^{i\omega(n,n-\beta)t}$ of this.

Note that the Fourier frequencies of the factors in this classical product combine in the simplest possible way in the product (just add them to get $\alpha\omega(n) + (\beta - \alpha)\omega(n)$). Heisenberg then observes that spectral line frequencies do not behave so simply. He is referring to what is called the *Rydberg–Ritz combination rule*. This is an empirical relationship between the frequencies that occur in the line spectrum of any atom, first noted by Ritz in 1908 for hydrogen. It states that the frequency of any line in the spectrum can be expressed as the sum or the difference of the frequencies of two other lines in the spectrum. Mathematically, this takes the form

$$\omega(n, n-\alpha) + \omega(n-\alpha, n-\beta) = \omega(n, n-\beta) \tag{7.5}$$

(compare Exercise 7.1.1). From this Heisenberg concludes that it is "an almost necessary consequence" that the quantum analogue of the harmonic component $\sum_{\alpha=-\infty}^{\infty} q_\alpha(n) q_{\beta-\alpha}(n) e^{i[\,\alpha\omega(n)+(\beta-\alpha)\omega(n)\,]t}$ of $q^2(n, t)$ be given by

$$\sum_{\alpha=-\infty}^{\infty} Q(n, n-\alpha) Q(n-\alpha, n-\beta) e^{i[\,\omega(n,n-\alpha)+\omega(n-\alpha,n-\beta)\,]t}. \tag{7.6}$$

In particular,

$$Q^2(n, n-\beta) = \sum_{\alpha=-\infty}^{\infty} Q(n, n-\alpha) Q(n-\alpha, n-\beta), \tag{7.7}$$

and this is Heisenberg's rule for the multiplication of transition amplitudes.

Confronted with (7.7) it becomes increasingly difficult to go on pretending that we do not know anything about matrices, so it is time to pause and relate the oft-told story of how Heisenberg's new quantum mechanics became *matrix mechanics*. Heisenberg received his PhD in 1923 under the direction of Arnold Sommerfeld in Munich (his topic was *On the Stability and Turbulence of Fluid Flow*). He then moved to Göttingen, became an assistant to Max Born, and completed his Habilitation in 1924. Born and Heisenberg worked on calculating the spectral lines of hydrogen. These calculations, together with what appeared to be fundamental limitations on the applicability of the old quantum theory (to large atoms, for example), led Heisenberg to believe that a thoroughgoing re-evaluation of the logical and mathematical foundations of quantum theory was required. In particular, he felt that quantum theory should be formulated exclusively in terms of quantities that were directly observable (frequencies and intensities of spectral lines rather than classical positions and momenta of electrons).

In June of 1925 Heisenberg suffered a severe allergy attack and left Göttingen for Helgoland (a small island off the coast of Germany in the North Sea that is essentially free of pollen). While there he spent his time climbing, memorizing passages from Goethe and thinking about spectral lines. This last activity culminated in the sort of epiphany that is not granted to mere mortals such as ourselves. Heisenberg quickly wrote down his new vision (I think that is the correct term) of quantum mechanics and the result was the paper [Heis1] that we have been discussing. The strangeness of the ideas in this paper were apparent to Heisenberg, who was reluctant to submit them for publication without first showing them to Born and to his friend Wolfgang Pauli. Born and Pauli, however, were quick to recognize the significance of the paper. Born submitted it to the *Zeitschrift für Physik* and Pauli used the ideas it contained to completely solve the problem of calculating the spectrum of hydrogen.

Born was particularly intrigued with Heisenberg's rule (7.7) for the multiplication of transition amplitudes (for his own recollections, see pages 217–218 of [Born2]). Recalling lectures from his student days by the mathematician Jakob Rosanes, he soon recognized it as nothing other than the (to us) familiar rule for matrix multiplication; as matrices, $(Q^2(n, m)) = (Q(n, m))^2$. Note that, by (7.6), the quantum analogue of $q^2(t)$, as a matrix,

$$\mathbf{Q}^2(t) = \left(\sum_{\alpha=-\infty}^{\infty} Q(n, n - \alpha)Q(n - \alpha, m)e^{i[\, \omega(n,n-\alpha)+\omega(n-\alpha,m)\,]t} \right)$$

is also the matrix square of $\mathbf{Q}(t)$. But note also that, by (7.5), every entry in $\mathbf{Q}^2(t)$ has precisely the same frequency as the corresponding entry of $\mathbf{Q}(t)$. Consequently, these entries also have the same exponential time factors. As a result one generally does not bother to keep track of these time factors, but deals instead only with the transition amplitudes.

Thus, Born has supplied Heisenberg with a ready-made mathematical structure in which to place his view of the foundations of quantum mechanics. You cannot please everyone, however.

> "Yes, I know you are fond of tedious and complicated formalism. You are only going to spoil Heisenberg's physical ideas by your futile mathematics."

> Wolfgang Pauli to Max Born, July 19, 1925

Quantum analogues of the higher powers of $q(t)$ and $p(t)$ can then be defined simply in terms of the corresponding matrix powers. This, in turn, suggests how one might define the quantum version of any classical observable that is a polynomial (or perhaps even a power series) in the classical variables q and p, although Heisenberg himself noted that the noncommutativity of his multiplication rule introduced a "significant difficulty" (we will get back to this soon). From these observations Born and Pascual Jordan (another assistant) reshaped Heisenberg's rather obscure paper into the first cogent and systematic foundation for what would henceforth be known as *matrix mechanics*. This appeared in their paper [BJ], which we will need to follow just a bit further since it contains the first appearance of the identity

$$PQ - QP = \frac{h}{2\pi i},$$

which is the focus of our attention here. Note that, with Q and P interpreted as matrices, the right-hand side of this identity must be thought of as a scalar matrix, that is, a multiple of the identity matrix I. For the sake of clarity we will, for the time being at least, include this explicitly and write

$$PQ - QP = \frac{h}{2\pi i} I. \tag{7.8}$$

The "significant difficulty" to which Heisenberg alluded is clear enough. Classical observables are real-valued functions on phase space and these *commute* under multiplication, whereas matrices do not. Suppose one is trying to write down the quantum analogue of a classical observable which, as a function of q and p, contains the term $q^2 p = qqp = qpq = pq^2$. Replacing q and p with their quantum analogues one has three options for the order in which to write the matrices *and these generally do not give the same result*. The naive quantization procedure we have been hinting at is ambiguous. For some classical observables such as the harmonic oscillator Hamiltonian (which contains only q^2 and p^2), this is not an issue, but in general the difficulty really is significant. Physicists have devised a great many schemes for removing this ambiguity each of which gives rise to what we would call a *quantization procedure*, but different schemes generally give rise to different physical predictions and one can only decide which (if any) gives the correct predictions by consulting the experimentalists. For classical observables that are *quadratic* functions of q and p there is only one apparent ambiguity, that is, QP *versus* PQ. One would therefore like to know something about how they differ, that is, about the commutator $PQ - QP$.

Heisenberg's paper [Heis1] contains a great deal more than we have described so far, but most of this is not our immediate concern ([AMS] contains a detailed analysis of the entire paper). There is one item we cannot ignore, however. Heisenberg postulated differential equations entirely analogous to Hamilton's equations of classical mechanics that should be satisfied by his quantum analogues of position and momentum, but he needed also some reinterpretation of the Bohr–Sommerfeld quantization condition (7.1) to ensure discrete energy levels. The condition he proposed (Equation (16) of [Heis1]) is, in our notation,

$$4\pi m_e \sum_{\alpha=-\infty}^{\infty} Q(n, \alpha) Q(\alpha, n) \omega(\alpha, n) = h, \tag{7.9}$$

where m_e is the mass of the electron and h is Planck's constant.

The path which led Heisenberg to (7.9) is not so easy to follow. Writing out the Born–Sommerfeld condition (7.1) with $p = m_e \dot{q}$ and expanding in Fourier series, he performs a rather odd differentiation with respect to the (discrete) variable n to obtain

$$h = 2\pi m_e \sum_{\alpha=-\infty}^{\infty} \alpha \frac{d}{dn} (\alpha \omega(n) q_\alpha(n) q_{-\alpha}(n)).$$

At this point Heisenberg simply asserts that this equation "has a simple quantum-theoretical reformulation which is related to *Kramers' dispersion theory*" and records his Equation (16) without further ado.

Remark 7.1.2. Today this would generally be known as *Kramers–Heisenberg dispersion theory*. It concerns itself with the scattering of photons by electrons that are bound in an atom and was a topic of great interest in 1925 because the details of atomic structure were by then understood to be intimately related to the emission and absorption of photons. Lacking both the time and the competence to do so properly we will not attempt to describe precisely how Heisenberg was led from dispersion theory to (7.9), but will simply refer those interested in hearing this story to Section III.5 of [MR].

There is, however, one observation we would like to make about (7.9). With

$$\mathbf{Q}(t) = \left(Q(n,m) e^{i\omega(n,m)t} \right)$$

and

$$\mathbf{P}(t) = m_e \dot{\mathbf{Q}}(t) = \left(m_e i\omega(n,m) Q(n,m) e^{i\omega(n,m)t} \right) = \left(P(n,m) e^{i\omega(n,m)t} \right)$$

we compute the products PQ and QP as

$$(PQ)(n,m) = \sum_{\alpha=-\infty}^{\infty} P(n,\alpha) Q(\alpha,m) = m_e i \sum_{\alpha=-\infty}^{\infty} \omega(n,\alpha) Q(n,\alpha) Q(\alpha,m)$$

and, similarly,

$$(QP)(n,m) = m_e i \sum_{\alpha=-\infty}^{\infty} Q(n,\alpha) \omega(\alpha,m) Q(\alpha,m).$$

Now, recalling that $\omega(n,\alpha) = -\omega(\alpha,n)$ we obtain for the diagonal entries $(m = n)$ of the commutator $PQ - QP$

$$(PQ - QP)(n,n) = -2m_e i \sum_{\alpha=-\infty}^{\infty} Q(n,\alpha) Q(\alpha,n) \omega(\alpha,n).$$

Finally, note that Heisenberg's quantization condition (7.9) is precisely the statement that

$$(PQ - QP)(n,n) = \frac{h}{2\pi i}, \quad n = 1, 2, 3, \ldots,$$

that is, that the diagonal entries of the commutator $PQ - QP$ are all equal to $\frac{h}{2\pi i}$. Since the diagonal entries of **PQ** − **QP** are time-independent, these are also all equal to $\frac{h}{2\pi i}$.

If Heisenberg noticed that his quantization condition gave rise to this identity, he did not say so in [Heis1]. Born did notice, however, and conjectured on physical grounds that the off-diagonal entries of $PQ - QP$ must all be zero. Jordan established Born's conjecture by computing the t-derivative of **PQ** − **QP**, using Heisenberg's quantum version of Hamilton's equations to show that the derivative is zero, and then invoking the additional assumption that $w(n, m) \neq 0$ whenever $n \neq m$ to show that the off-diagonal elements are zero (the argument is given in some detail in Section IV of [FP2]).

Whether or not one is willing to take these physical and mathematical arguments of Heisenberg, Born and Jordan seriously is not really the issue here since the resulting identity (7.8) is best regarded as one of the *postulates* of matrix mechanics and its viability should be judged on the basis of the predictions to which matrix mechanics leads. As we emphasized earlier, the goal here was motivation, not derivation. What we hope to have motivated is the underlying algebraic structure of quantum mechanics. In an attempt to unearth a precise definition of this algebraic structure we will rewrite (7.8) just slightly using [,]_ for the matrix commutator, writing $\hbar$ for $h/2\pi$ and appending to it two trivial commutation relations. This results in what are called the *Born–Heisenberg canonical commutation relations*, or *quantum canonical commutation relations*, or simply the *CCR*:

$$[Q, Q]_- = [P, P]_- = 0, \quad [P, Q]_- = -i\hbar I. \tag{7.10}$$

One often sees these relations expressed in the following way. Define the *quantum bracket* $\{ , \}_\hbar$ by

$$\{ , \}_\hbar = -\frac{i}{\hbar}[,]_-.$$

Then (7.10) can be written

$$\{Q, Q\}_\hbar = \{P, P\}_\hbar = 0, \quad \{Q, P\}_\hbar = I. \tag{7.11}$$

In these terms one cannot help but notice the analogy with the $n = 1$ case of the canonical commutation relations (2.66) for classical mechanics, that is,

$$\{q, q\} = \{p, p\} = 0, \quad \{q, p\} = 1. \tag{7.12}$$

The analogy is strengthened by comparing equation (2.65) describing the time evolution of an observable in classical mechanics

$$\frac{df}{dt} = \{f, H\}$$

with the Heisenberg equation (6.27)

$$\frac{dA}{dt} = \{A, H\}_\hbar.$$

Paul Dirac [Dirac1] was the first to suggest that this analogy might provide a method of quantizing classical mechanical systems. His suggestion was simply to find a "reasonable" mapping R from the classical observables $f, g, \ldots$ to the quantum observables $R(f), R(g), \ldots$ (matrices, or operators on a Hilbert space) that sends Poisson brackets to quantum brackets, that is, satisfies

$$\{f, g\} \rightarrow \{R(f), R(g)\}_\hbar$$

and carries (7.12) to (7.11). As it turns out, this is not only more easily said than done, it generally cannot be done at all. Nevertheless, this is what physicists usually mean by *canonical quantization*. We will have a much more careful look at this in the next section.

Finally, note that if we let $C = -i\hbar I$, then, from (7.10), we have

$$[P, P]_- = [Q, Q]_- = [C, C]_- = [P, C]_- = [Q, C]_- = 0, \quad [P, Q]_- = C.$$

If one regards the matrices $\{P, Q, C\}$ as basis elements for the three-dimensional real vector space they span, then these commutation relations show that the commutator provides this vector space with the structure of a (nearly commutative) Lie algebra with C in the center. Of course, in this introductory section we have been rather cavalier about these infinite matrices that are supposed to represent quantum observables, so one cannot claim that this makes any rigorous sense at the moment. In the next section we will attempt to rectify this situation.

7.2 Heisenberg algebras and Heisenberg groups

The physical reasoning and formal manipulations of the preceding section were all intended to simply motivate the rigorous definitions and results that we will describe now. The objective is to formulate a precise notion of the "canonical quantization" of a classical mechanical system and discuss the extent to which it can be realized. In the next two sections we will apply what we learn to the free particle and the harmonic oscillator. We will need to rely rather heavily on basic information about (matrix) Lie groups and Lie algebras and, as we have done previously, will take as our principal references [CM], the lectures by Robert Bryant in [FU], [Hall1], [Knapp], [Nab3], [Sp2] and [Warner].

We begin with the abstract definition of the algebraic structure to which we were led by Heisenberg and Born in the previous section. We will describe a few concrete models of this structure and some of its basic properties and then generalize to accommodate physical systems more complicated than those discussed in Section 7.1.

The *three-dimensional Heisenberg algebra* $\mathfrak{h}_3$ is a three-dimensional, real Lie algebra with a basis $\{X, Y, Z\}$ relative to which the Lie bracket $[\,,\,]$ is determined by

$$[X, Z] = [Y, Z] = 0, \quad [X, Y] = Z. \tag{7.13}$$

In terms of components we have

$$[xX + yY + zZ, x'X + y'Y + z'Z] = (xy' - yx')Z.$$

Exercise 7.2.1. Show that $[A, [B, C]] = 0 \, \forall A, B, C \in \mathfrak{h}_3$ and conclude that the Jacobi identity

$$[A, [B, C]] + [C, [A, B]] + [B, [C, A]] = 0 \quad \forall A, B, C \in \mathfrak{h}_3.$$

is satisfied in $\mathfrak{h}_3$.

Example 7.2.1. It will be useful to have a couple of concrete realizations of $\mathfrak{h}_3$ so we will begin with these.

1. For a classical mechanical system with configuration space $\mathbb{R}$ the phase space is $T^*\mathbb{R} = \mathbb{R}^2$ and the algebra of classical observables is $C^\infty(T^*\mathbb{R})$. This is an infinite-dimensional, real Lie algebra with the Poisson bracket $\{\,,\,\}$ as the Lie bracket. Three particular observables are the coordinate functions q and p and the constant function 1. Consider the linear subspace of $C^\infty(T^*\mathbb{R})$ spanned by $\{q, p, 1\}$. Since

$$\{q, 1\} = \{p, 1\} = 0, \quad \{q, p\} = 1, \tag{7.14}$$

   this is a Lie subalgebra of $C^\infty(T^*\mathbb{R})$ isomorphic to $\mathfrak{h}_3$.

2. The set $\mathfrak{gl}(3; \mathbb{R})$ of all 3×3 real matrices is a Lie algebra under matrix commutator. We consider the subset consisting of those matrices of the form

$$\begin{pmatrix} 0 & x & z \\ 0 & 0 & y \\ 0 & 0 & 0 \end{pmatrix}.$$

Exercise 7.2.2. Show that these form a Lie subalgebra of $\mathfrak{gl}(3; \mathbb{R})$ that is isomorphic to $\mathfrak{h}_3$.

There are a few things worth noting about $\mathfrak{h}_3$. Recall that the center of a Lie algebra $\mathfrak{g}$ consists of all those $B \in \mathfrak{g}$ such that $[A, B] = 0$ for all $A \in \mathfrak{g}$. Certainly, Z is in the center of $\mathfrak{h}_3$, but more is true.

Exercise 7.2.3. Show that the center of $\mathfrak{h}_3$ is the one-dimensional linear subspace spanned by Z.

You have shown that (7.13) implies

$$[[A, B], C] = 0 \qquad (7.15)$$

for all $A, B, C \in \mathfrak{h}_3$. In particular, $\mathfrak{h}_3$ is a nilpotent Lie algebra, but we have in mind a different use for (7.15). Every finite-dimensional Lie algebra $\mathfrak{g}$ is the Lie algebra of some connected Lie group (this is generally known as *Lie's third theorem*). Moreover, there is a *unique* simply connected Lie group whose Lie algebra is $\mathfrak{g}$. This is true, in particular, for $\mathfrak{h}_3$ and we would like to describe this Lie group. For this we will identify $\mathfrak{h}_3$ with the matrix Lie algebra described in Example 7.2.1.2. In general, one gets from a Lie algebra to its Lie group via the exponential map, so we will need to exponentiate matrices of the form

$$xX + yY + zZ = x \begin{pmatrix} 0 & 1 & 0 \\ 0 & 0 & 0 \\ 0 & 0 & 0 \end{pmatrix} + y \begin{pmatrix} 0 & 0 & 0 \\ 0 & 0 & 1 \\ 0 & 0 & 0 \end{pmatrix} + z \begin{pmatrix} 0 & 0 & 1 \\ 0 & 0 & 0 \\ 0 & 0 & 0 \end{pmatrix}.$$

Now, recall that if matrices A and B commute, then $e^{A+B} = e^A e^B$. Since zZ commutes with everything in $\mathfrak{h}_3$, we have

$$e^{xX+yY+zZ} = e^{xX+yY} e^{zZ}.$$

Since $(zZ)^2 = 0$,

$$e^{zZ} = I + zZ = \begin{pmatrix} 1 & 0 & z \\ 0 & 1 & 0 \\ 0 & 0 & 1 \end{pmatrix}.$$

Similarly,

$$e^{xX} = I + xX = \begin{pmatrix} 1 & x & 0 \\ 0 & 1 & 0 \\ 0 & 0 & 1 \end{pmatrix}$$

and

$$e^{yY} = I + yY = \begin{pmatrix} 1 & 0 & 0 \\ 0 & 1 & y \\ 0 & 0 & 1 \end{pmatrix}.$$

Since xX and yY do not commute, e^{xX+yY} takes a bit more work. One can appeal to the *Baker–Campbell–Hausdorff formula*

$$e^A e^B = e^{A+B+\frac{1}{2}[A,B]_- + \frac{1}{12}[A,[A,B]_-]_- - \frac{1}{12}[B,[A,B]_-]_- + \cdots},$$

which, by virtue of (7.15), simplifies in the case of $\mathfrak{h}_3$ to

$$e^A e^B = e^{A+B+\frac{1}{2}[A,B]_-}.$$

There is a very detailed discussion of the Baker–Campbell–Hausdorff formula in Chapter 4 of [Hall1]. Moreover, Theorem 4.1 of [Hall1] contains an independent proof of the special case we need for $\mathfrak{h}_3$. Alternatively, given the simplicity of the matrices, one can verify the following by direct computation:

$$e^{xX} e^{yY} = e^{xX+yY+\frac{1}{2}[xX,yY]_-} = e^{xX+yY+\frac{1}{2}xyZ} = e^{xX+yY} e^{\frac{1}{2}xyZ}.$$

From this we conclude that

$$e^{xX+yY} = e^{xX} e^{yY} e^{-\frac{1}{2}xyZ}.$$

Consequently,

$$e^{yY+xX} = e^{yY} e^{xX} e^{\frac{1}{2}xyZ}.$$

Since $e^{yY+xX} = e^{xX+yY}$, we conclude that

$$e^{xX} e^{yY} = e^{xyZ} e^{yY} e^{xX}. \tag{7.16}$$

Exercise 7.2.4. Put all of this together to show that

$$e^{xX+yY+zZ} = \begin{pmatrix} 1 & x & z+\frac{1}{2}xy \\ 0 & 1 & y \\ 0 & 0 & 1 \end{pmatrix}.$$

Conclude that the exponential map on $\mathfrak{h}_3$ is a bijection onto the set H_3 of all 3×3 matrices of the form

$$\begin{pmatrix} 1 & a & c \\ 0 & 1 & b \\ 0 & 0 & 1 \end{pmatrix}.$$

Exercise 7.2.5. Show that

$$\begin{pmatrix} 1 & a & c \\ 0 & 1 & b \\ 0 & 0 & 1 \end{pmatrix} \begin{pmatrix} 1 & a' & c' \\ 0 & 1 & b' \\ 0 & 0 & 1 \end{pmatrix} = \begin{pmatrix} 1 & a+a' & c+c'+ab' \\ 0 & 1 & b+b' \\ 0 & 0 & 1 \end{pmatrix}$$

and

$$\begin{pmatrix} 1 & a & c \\ 0 & 1 & b \\ 0 & 0 & 1 \end{pmatrix}^{-1} = \begin{pmatrix} 1 & -a & -c+ab \\ 0 & 1 & -b \\ 0 & 0 & 1 \end{pmatrix}.$$

Conclude that, under matrix multiplication, H_3 is a non-Abelian group whose center consists precisely of those elements of the form

$$e^{cZ} = \begin{pmatrix} 1 & 0 & c \\ 0 & 1 & 0 \\ 0 & 0 & 1 \end{pmatrix}.$$

To see that H_3 is a simply connected Lie group, proceed as follows.

Exercise 7.2.6. Define a map from H_3 to $\mathbb{R}^3$ by

$$\begin{pmatrix} 1 & a & c \\ 0 & 1 & b \\ 0 & 0 & 1 \end{pmatrix} \xrightarrow{\phi} (a, \ b, \ c).$$

Regarding H_3 as a topological subspace of the 3×3 real matrices (that is, $\mathbb{R}^9$), show that ϕ is a homeomorphism. In particular, $\phi : H_3 \to \mathbb{R}^3$ is a global chart on H_3 and so H_3 is a differentiable manifold diffeomorphic to $\mathbb{R}^3$.

Exercise 7.2.7. Define a multiplicative structure on $\mathbb{R}^3$ by

$$(x, y, z)(x', y', z') = (x + x', \ y + y', \ z + z' + xy').$$

Show that this defines a group structure on $\mathbb{R}^3$ and that if $\mathbb{R}^3$ is given its usual topology and differentiable structure, it defines a Lie group structure on $\mathbb{R}^3$.

Exercise 7.2.8. Show that when $\mathbb{R}^3$ is provided with the Lie group structure in the previous exercise, $\phi : H_3 \to \mathbb{R}^3$ is a Lie group isomorphism. Conclude that H_3 is the unique simply connected Lie group whose Lie algebra is $\mathfrak{h}_3$.

Exercise 7.2.9. There is another way of describing a Lie group structure on $\mathbb{R}^3$ that is isomorphic to H_3. For this we will denote the elements of $\mathbb{R}^3$ by (x, y, u). Show first that the multiplication defined by

$$(x, y, u)(x', y', u') = \left(x + x', y + y', u + u' + \frac{1}{2}(xy' - x'y) \right)$$

provides $\mathbb{R}^3$ with the structure of a Lie group. Next show that the map $(x, y, z) \to (x, y, u) = (x, y, z - \frac{1}{2}xy)$ satisfies

$$(x + x', y + y', z + z' + xy') \to \left(x + x', y + y', u + u' + \frac{1}{2}(xy' - x'y) \right)$$

and therefore is an isomorphism with the $\mathbb{R}^3$-model of H_3 described in the previous exercises.

Here H_3, thought of either as a group of matrices or as $\mathbb{R}^3$ with either of the group structures just introduced, is called the *three-dimensional Heisenberg group*. There is an interesting way of rephrasing our last view of H_3 that not only makes the definition of the group structure appear a bit less odd, but also suggests a rather elegant generalization. Think of $\mathbb{R}^3$ as $\mathbb{R}^2 \times \mathbb{R} = T^*\mathbb{R} \times \mathbb{R}$ and define a nondegenerate, skew-symmetric, bilinear form ω on $\mathbb{R}^2$ by

$$\omega((x,y),(x',y')) = xy' - x'y.$$

This is just the canonical symplectic form on $T^*\mathbb{R}$ once it has been identified with a bilinear form on $\mathbb{R}^2$ by using the fact that every tangent space to the vector space $\mathbb{R}^2$ is canonically identified with $\mathbb{R}^2$. Then the peculiar group structure we have introduced on $\mathbb{R}^3$ can be thought of as a group structure on $T^*\mathbb{R} \times \mathbb{R}$ defined by

$$(\mathbf{v}, t)(\mathbf{v}', t') = \left(\mathbf{v} + \mathbf{v}', \ t + t' + \frac{1}{2}\omega(\mathbf{v}, \mathbf{v}') \right)$$

for all $\mathbf{v}, \mathbf{v}' \in T^*\mathbb{R}$ and all $t, t' \in \mathbb{R}$. As it happens, one can mimic this definition to associate a "Heisenberg group" and corresponding "Heisenberg algebra" with $V \times \mathbb{R}$ for any finite-dimensional symplectic vector space V (see pages 116–118 of [Berndt]). We will not pursue this in such generality here, but will use the idea to deal with $\mathbb{R}^{2n+1} = \mathbb{R}^{2n} \times \mathbb{R}$. Since there are really no new ideas involved we will simply record the facts.

Let $n \geq 1$ be an integer. The $(2n + 1)$-*dimensional Heisenberg algebra* $\mathfrak{h}_{2n+1}$ is a $(2n + 1)$-dimensional, real Lie algebra with a basis $\{X_1, \ldots, X_n, Y_1, \ldots, Y_n, Z\}$ relative to which the Lie bracket $[\,,\,]$ is determined by

$$[X_i, X_j] = [Y_i, Y_j] = [X_i, Z] = [Y_i, Z] = 0, \quad [X_i, Y_j] = \delta_{ij}Z, \quad i, j = 1, \ldots, n.$$

Example 7.2.2. The corresponding concrete realizations of $\mathfrak{h}_{2n+1}$ are as follows.

1. Let $C^\infty(T^*\mathbb{R}^n) = C^\infty(\mathbb{R}^{2n})$ be the Lie algebra, relative to the Poisson bracket, of classical observables for a mechanical system with configuration space $\mathbb{R}^n$. Then $\mathfrak{h}_{2n+1}$ is isomorphic to the Lie subalgebra of $C^\infty(T^*\mathbb{R}^n)$ generated by $\{q^1, \ldots, q^n, p_1, \ldots, p_n, 1\}$.

2. Let $\mathfrak{gl}(n + 2; \mathbb{R})$ be the Lie algebra of all $(n + 2) \times (n + 2)$ real matrices under the commutator Lie bracket. Then $\mathfrak{h}_{2n+1}$ is isomorphic to the Lie subalgebra of $\mathfrak{gl}(n + 2; \mathbb{R})$ consisting of those matrices of the form

$$\begin{pmatrix} 0 & x^1 & x^2 & \cdots & x^n & z \\ 0 & 0 & 0 & \cdots & 0 & y^1 \\ 0 & 0 & 0 & \cdots & 0 & y^2 \\ \vdots & \vdots & \vdots & \cdots & \vdots & \vdots \\ 0 & 0 & 0 & \cdots & 0 & y^n \\ 0 & 0 & 0 & \cdots & 0 & 0 \end{pmatrix} = \begin{pmatrix} 0 & \mathbf{x} & z \\ \mathbf{0} & 0_n & \mathbf{y} \\ 0 & \mathbf{0} & 0 \end{pmatrix},$$

where 0_n is the $n \times n$ zero matrix, $\mathbf{0}$ is the zero vector in $\mathbb{R}^n$ and $\mathbf{x}$ and $\mathbf{y}$ are arbitrary vectors in $\mathbb{R}^n$ (unless it causes some confusion, we will allow the context to indicate whether the elements of $\mathbb{R}^n$ are to be regarded as row or column vectors).

The simply connected Lie group whose Lie algebra is $\mathfrak{h}_{2n+1}$ is called the $(2n+1)$-*dimensional Heisenberg group* and denoted H_{2n+1}. This can be described in a number of ways. As a matrix group, H_{2n+1} consists precisely of those $(n+2)\times(n+2)$ real matrices of the form

$$
\begin{pmatrix}
1 & a^1 & a^2 & \cdots & a^n & c \\
0 & 1 & 0 & \cdots & 0 & b^1 \\
0 & 0 & 1 & \cdots & 0 & b^2 \\
\vdots & \vdots & \vdots & \cdots & \vdots & \vdots \\
0 & 0 & 0 & \cdots & 1 & b^n \\
0 & 0 & 0 & \cdots & 0 & 1
\end{pmatrix}
=
\begin{pmatrix}
1 & \mathbf{a} & c \\
\mathbf{0} & I_n & \mathbf{b} \\
\mathbf{0} & \mathbf{0} & 1
\end{pmatrix},
$$

where I_n is the $n\times n$ identity matrix. The matrix exponential map is a bijection of $\mathfrak{h}_{2n+1}$ onto H_{2n+1} and is given by

$$
\begin{pmatrix}
0 & \mathbf{x} & z \\
\mathbf{0} & 0_n & \mathbf{y} \\
0 & \mathbf{0} & 0
\end{pmatrix}
\longrightarrow
\begin{pmatrix}
1 & \mathbf{x} & z + \frac{1}{2}\langle \mathbf{x},\mathbf{y}\rangle \\
\mathbf{0} & I_n & \mathbf{y} \\
0 & \mathbf{0} & 1
\end{pmatrix},
$$

where $\langle \mathbf{x},\mathbf{y}\rangle$ is the usual $\mathbb{R}^n$-inner product.

Alternatively, we can identify $\mathbb{R}^{2n+1}$ with $\mathbb{R}^{2n} \times \mathbb{R}$ and define a nondegenerate, skew-symmetric, bilinear form ω on $\mathbb{R}^{2n} = \mathbb{R}^n \times \mathbb{R}^n$ by

$$
\omega(v,v') = \omega((\mathbf{x},\mathbf{y}),(\mathbf{x}',\mathbf{y}')) = \langle \mathbf{x},\mathbf{y}'\rangle - \langle \mathbf{x}',\mathbf{y}\rangle.
$$

Then H_{2n+1} is isomorphic to $\mathbb{R}^{2n} \times \mathbb{R}$ with the group structure defined by

$$
(v,z)(v',z') = \left(v+v', z+z' + \frac{1}{2}\omega(v,v') \right),
$$

that is,

$$
(\mathbf{x},\mathbf{y},z)(\mathbf{x}',\mathbf{y}',z') = \left(\mathbf{x}+\mathbf{x}', \mathbf{y}+\mathbf{y}', z+z' + \frac{1}{2}(\langle \mathbf{x},\mathbf{y}'\rangle - \langle \mathbf{x}',\mathbf{y}\rangle) \right). \tag{7.17}
$$

Before returning to quantum mechanics it is only fair to point out that Heisenberg algebras and Heisenberg groups play decisive roles also in many areas outside of mathematical physics (see, for example, [Howe] and [Fol1] for applications to harmonic analysis).

Now let us see what all of this has to do with Dirac's proposal for quantizing a classical mechanical system. Recall that the idea was to find an "appropriate" mapping from classical observables to quantum observables that sends Poisson brackets to quantum brackets and carries the classical canonical commutation relations (7.12) to the quantum canonical commutation relations (7.11). We will try to write this out a bit more carefully and see what happens. For simplicity we will once again focus on classical systems with one degree of freedom (configuration space $\mathbb{R}$) and then simply record the more or less obvious generalization to higher dimensions.

Remark 7.2.1. We will also temporarily put aside Heisenberg's infinite matrices and return to our previous view of quantum observables as self-adjoint operators on a separable, complex Hilbert space $\mathcal{H}$. As we mentioned previously, one can get back to the matrices simply by choosing an orthonormal basis for $\mathcal{H}$. This essentially establishes the mathematical equivalence of Schrödinger's wave mechanics and Heisenberg's matrix mechanics (see Chapter I, Sections 3 and 4, of [vonNeu] for more on this, or [Casado] for a brief historical survey).

What Dirac is asking us for then is a map R from $C^\infty(T^*\mathbb{R})$ to the self-adjoint operators on some separable, complex Hilbert space $\mathcal{H}$ that satisfies

$$R(\{f,g\}) = -\frac{i}{\hbar}[R(f),R(g)]_-$$

and carries the classical commutation relations (7.12) for position and momentum to the quantum commutation relations (7.11) for the corresponding operators.

Exercise 7.2.10. Define $\pi' = -\frac{i}{\hbar}R$ and show that $R(\{f,g\}) = -\frac{i}{\hbar}[R(f),R(g)]_-$ is equivalent to

$$\pi'(\{f,g\}) = [\pi'(f),\pi'(g)]_-. \tag{7.18}$$

Now, we have already seen that there are all sorts of problems associated with defining the commutator of two self-adjoint operators. If the operators are unbounded the commutator may be defined only at zero. Even if they are bounded the commutator is not self-adjoint, but rather skew-adjoint and one must multiply it by $\pm i$ to get something self-adjoint. Nevertheless, (7.18) at least resembles something familiar. Recall that if $\mathfrak{g}_1$ and $\mathfrak{g}_2$ are two real Lie algebras with brackets $[\,,\,]_1$ and $[\,,\,]_2$, respectively, then a *Lie algebra homomorphism* from $\mathfrak{g}_1$ to $\mathfrak{g}_2$ is a linear map $h : \mathfrak{g}_1 \to \mathfrak{g}_2$ satisfying $h([f,g]_1) = [h(f),h(g)]_2$ for all $f,g \in \mathfrak{g}_1$. If $\mathfrak{g}_2$ is a Lie algebra of operators on some vector space V under commutator, then h is called a *Lie algebra representation* of $\mathfrak{g}_1$ on V. Consequently, *if* the self-adjoint operators on $\mathcal{H}$ formed a Lie algebra (*which they do not*), then π' would be a representation of the classical observables by quantum observables on $\mathcal{H}$.

To make something like this work will require a bit more finesse. We will begin by being somewhat less ambitious. The Heisenberg algebra $\mathfrak{h}_3$ can be identified with the Lie subalgebra of $C^\infty(T^*\mathbb{R})$ spanned by $\{q,p,1\}$ (Example 7.2.1.1) and we will start by looking for an appropriate notion of "representation" only for this Lie algebra; we

will worry later about whether or not such "representations" extend to larger subalgebras of $C^\infty(T^*\mathbb{R})$. We put "representation" in quotes since, for the reasons we have
been discussing, it will be necessary to adapt the definition given above to the infinite-
dimensional context. Eventually, we will opt for the word "realization" instead.

First we should understand what *cannot* be true. The operators in the "representation" certainly cannot act on a finite-dimensional Hilbert space since the images Q, P
and I of q, p and 1 are required to satisfy $QP - PQ = i\hbar I$. In finite dimensions we can take
the trace on both sides, getting zero on the left, but not on the right. In fact, at least
one of Q or P must be unbounded. To see this suppose, to the contrary, that they are
both bounded and satisfy $[Q, P]_- = i\hbar I$ (which now makes sense on all of $\mathcal{H}$). Induction
then gives $[Q, P^n]_- = ni\hbar P^{n-1}$ for any $n \geq 1$. Thus, $n\hbar \|P^{n-1}\| = \|QP^n - P^nQ\| \leq 2\|Q\| \|P^n\|$.
Since P is self-adjoint, it is a normal operator and so $\|P^n\| = \|P\|^n$ for any n (see, for
example, Section 58 of [Simm1]). Consequently, $n\hbar \leq 2\|Q\| \|P\|$ for every n and this is
clearly impossible since the right-hand side is a constant.

The conclusion we draw from all of this is that one simply has to deal with unbounded operators and all of the difficulties presented by their rather problematic
commutators. There are various ways to do this and we will describe one. The idea,
due to Hermann Weyl and based on Theorems 5.6.3 and 5.6.5, is to replace the relations (7.11) with another set of relations involving only bounded (in fact, unitary) operators. We emphasize at the outset, however, that these two sets of relations are *not*
equivalent and we will have to say a bit more about this as we proceed. Before the abstract definitions, however, we will try to get our bearings by looking at an example.

Example 7.2.3. Recall that the position operator $Q : \mathcal{D}(Q) \to L^2(\mathbb{R})$ (Example 5.2.3)
and momentum operator $P : \mathcal{D}(P) \to L^2(\mathbb{R})$ (Example 5.2.4) are defined on the
Schwartz space $\mathcal{S}(\mathbb{R})$ by

$$(Q\psi)(q) = q\psi(q)$$

and

$$(P\psi)(q) = -i\hbar\frac{d}{dq}\psi(q).$$

Note that $\mathcal{S}(\mathbb{R})$ is invariant under both, that is,

$$Q : \mathcal{S}(\mathbb{R}) \to \mathcal{S}(\mathbb{R})$$

and

$$P : \mathcal{S}(\mathbb{R}) \to \mathcal{S}(\mathbb{R}).$$

Furthermore, both Q and P are essentially self-adjoint on $\mathcal{S}(\mathbb{R})$ (Exercise G.2.1). The
commutator $[P, Q]_-$ is well-defined on $\mathcal{S}(\mathbb{R})$ and we have seen that

$$[P, Q]_-\psi(q) = (PQ - QP)\psi(q) = -i\hbar\psi(q) \quad \forall \psi \in \mathcal{S}(\mathbb{R}).$$

Identifying Q with X, P with Y and $i\hbar 1$ with Z we find that the commutation relations for the Heisenberg algebra $\mathfrak{h}_3$ are satisfied on $\mathcal{S}(\mathbb{R})$.

The situation described in this example is essentially the closest one can come to the notion of a "representation" of the Heisenberg algebra $\mathfrak{h}_3$ by unbounded self-adjoint operators, so we are led to formulate the following definition. A *realization* of $\mathfrak{h}_3$ on the separable, complex Hilbert space $\mathcal{H}$ consists of a dense linear subspace $\mathcal{D}$ of $\mathcal{H}$ and two operators Q and P on $\mathcal{H}$ with $\mathcal{D} \subseteq \mathcal{D}(Q)$ and $\mathcal{D} \subseteq \mathcal{D}(P)$ that satisfy:

1. $Q : \mathcal{D} \to \mathcal{D}$ and $P : \mathcal{D} \to \mathcal{D}$,
2. $[P, Q]_{-}\psi = (PQ - QP)\psi = -i\hbar\psi \quad \forall \psi \in \mathcal{D}$ and
3. Q and P are essentially self-adjoint on $\mathcal{D}$.

In this case we say that the unique self-adjoint extensions of Q and P satisfy the *canonical commutation relations*.

The realization of $\mathfrak{h}_3$ described in Example 7.2.3 is called the *Schrödinger realization* of $\mathfrak{h}_3$. We will need to know a bit more about this example.

Example 7.2.4. We consider again the Schrödinger realization of $\mathfrak{h}_3$ described in Example 7.2.3. As is our custom we will denote the unique self-adjoint extensions of $Q : \mathcal{S}(\mathbb{R}) \to L^2(\mathbb{R})$ and $P : \mathcal{S}(\mathbb{R}) \to L^2(\mathbb{R})$ by the same symbols $Q : \mathcal{D}(Q) \to L^2(\mathbb{R})$ and $P : \mathcal{D}(P) \to L^2(\mathbb{R})$. Being self-adjoint, each of these determines a unique strongly continuous one-parameter group of unitary operators on $L^2(\mathbb{R})$, which we will denote by

$$\{U_t\}_{t\in\mathbb{R}} = \{e^{itP}\}_{t\in\mathbb{R}}$$

and

$$\{V_s\}_{s\in\mathbb{R}} = \{e^{isQ}\}_{s\in\mathbb{R}},$$

respectively. We have already seen (Example 5.6.1) that the operator e^{itP} is just translation to the left by $\hbar t$,

$$(e^{itP}\psi)(q) = \psi(q + \hbar t)$$

and (Remark 5.6.2) e^{isQ} is multiplication by e^{isq}:

$$(e^{isQ}\psi)(q) = e^{isq}\psi(q).$$

Now note that for any $\psi \in L^2(\mathbb{R})$,

$$
\begin{aligned}
U_t V_s\psi(q) &= e^{itP}e^{isQ}\psi(q) = e^{itP}(e^{isq}\psi(q)) \\
&= e^{is(q+\hbar t)}\psi(q + \hbar t) = e^{i\hbar ts}e^{isq}\psi(q + \hbar t) \\
&= e^{i\hbar ts}e^{isQ}e^{itP}\psi(q)
\end{aligned}
$$

$$= e^{ihts} V_s U_t \psi(q),$$

so

$$U_t V_s = e^{ihts} V_s U_t$$

on $L^2(\mathbb{R})$.

The thing to note now is that if we had known *only* that these last relations were satisfied, then the commutation relation $[P, Q]_- \psi = -i\hbar\psi$ would have followed from Theorem 5.6.5 for all $\psi \in S(\mathbb{R})$ (indeed, for all $\psi \in \mathcal{D}([P, Q]_-)$), without recourse to Example 7.2.3. Furthermore, Theorem 5.6.3 gives the remaining commutation relations of the Heisenberg algebra, so one can manufacture the Schrödinger realization of $\mathfrak{h}_3$ from relations involving only unitary operators. This is the scenario we will now generalize.

We will be interested in realizations of $\mathfrak{h}_3$ that arise in the manner described in Example 7.2.4. More precisely, suppose $\mathcal{H}$ is a separable, complex Hilbert space and $\{U_t\}_{t\in\mathbb{R}}$ and $\{V_s\}_{s\in\mathbb{R}}$ are two strongly continuous one-parameter groups of unitary operators on $\mathcal{H}$. We say that $\{U_t\}_{t\in\mathbb{R}}$ and $\{V_s\}_{s\in\mathbb{R}}$ satisfy the *Weyl relations* if

$$U_t V_s = e^{ihts} V_s U_t \quad \forall t, s \in \mathbb{R}. \tag{7.19}$$

Remark 7.2.2. Although we have reserved the symbol $\hbar$ for the normalized Planck constant, it is useful now to think of it as representing some (small) positive parameter. It is, after all, an experimentally determined number the value of which not only depends on the choice of units, but is also uncertain to the extent that the result of any measurement is uncertain. More significantly, viewing $\hbar$ as a parameter opens the possibility of taking various limits as $\hbar \to 0^+$ since these should, in some appropriate sense, reproduce the results of classical mechanics (this is the gist of the Bohr correspondence principle).

The following is the corollary to Theorem VIII.14, page 275, of [RS1]; we will sketch the ideas behind the proof shortly.

Theorem 7.2.1. *Let $\{U_t\}_{t\in\mathbb{R}}$ and $\{V_s\}_{s\in\mathbb{R}}$ be two strongly continuous one-parameter groups of unitary operators on the separable, complex Hilbert space $\mathcal{H}$ that satisfy the Weyl relations (7.19). Let $A : \mathcal{D}(A) \to \mathcal{H}$ and $B : \mathcal{D}(B) \to \mathcal{H}$ be the unique self-adjoint operators on $\mathcal{H}$ for which $U_t = e^{itA}$ $\forall t \in \mathbb{R}$ and $V_s = e^{isB}$ $\forall s \in \mathbb{R}$ (Stone's Theorem 5.6.2). Then there exists a dense linear subspace $\mathcal{D} \subseteq \mathcal{H}$ with $\mathcal{D} \subseteq \mathcal{D}(A)$ and $\mathcal{D} \subseteq \mathcal{D}(B)$ and such that:*
1. *$A : \mathcal{D} \to \mathcal{D}$ and $B : \mathcal{D} \to \mathcal{D}$,*
2. *$[A, B]_- \psi = (AB - BA)\psi = -i\hbar\psi$ $\forall \psi \in \mathcal{D}$ and*
3. *A and B are essentially self-adjoint on $\mathcal{D}$.*

The upshot of this is that a pair of strongly continuous one-parameter groups of unitary operators satisfying the Weyl relations will give rise to a realization of $\mathfrak{h}_3$, that is, to a solution to the canonical commutation relations. The question then is, how can one produce such pairs of strongly continuous one-parameter groups of unitary operators? A hint is provided by the identity (7.16) satisfied by the images under the exponential map of the generators $\{X, Y, Z\}$ of the Heisenberg algebra. Changing the notation just a bit we write this as

$$e^{tX} e^{sY} = e^{tsZ} e^{sY} e^{tX}, \quad t, s \in \mathbb{R}.$$

Note that $\{e^{tX}\}_{t\in\mathbb{R}}$ and $\{e^{sY}\}_{s\in\mathbb{R}}$ are both one-parameter subgroups of the Heisenberg group H_3. Now let us suppose we have a group homomorphism $\pi : H_3 \to \mathcal{U}(\mathcal{H})$ from H_3 to the group of unitary operators on some separable, complex Hilbert space $\mathcal{H}$.

Remark 7.2.3. Such a group homomorphism $\pi : H_3 \to \mathcal{U}(\mathcal{H})$ is a "unitary representation of the Heisenberg group." A review of what we need to know about these is contained in Appendix I.

Then

$$\pi(e^{tX}) \pi(e^{sY}) = \pi(e^{tsZ}) \pi(e^{sY}) \pi(e^{tX}), \quad t, s \in \mathbb{R}.$$

Both $\{U_t\}_{t\in\mathbb{R}} = \{\pi(e^{tX})\}_{t\in\mathbb{R}}$ and $\{V_s\}_{s\in\mathbb{R}} = \{\pi(e^{sY})\}_{s\in\mathbb{R}}$ are clearly one-parameter groups of unitary operators on $\mathcal{H}$; whether they are strongly continuous or not will depend on π. Since the center of $\mathfrak{h}_3$ is the one-dimensional subspace spanned by Z and since the exponential map carries $\mathfrak{h}_3$ onto H_3, each $\pi(e^{tsZ})$ commutes with $\pi(g)$ for every $g \in H_3$. For homomorphisms π that are irreducible in a sense defined in Appendix I, we will show that this implies that $\pi(e^{tsZ})$ must be a multiple of the identity (Schur's Lemma I.0.1) and, since it is unitary, it must be a multiple by some complex number of modulus one and we begin to get something that looks like (7.19).

Next we would like to provide a sketch of the ideas that go into the proof of Theorem 7.2.1. As motivation, we first recall that any representation of a matrix Lie group G on a *finite-dimensional* Hilbert space gives rise to a representation of the corresponding Lie algebra $\mathfrak{g}$ simply by differentiation at the identity. More precisely, and more generally, one has the following well-known result (if it is not so well known to you, see Theorem 3.18 of [Hall1]).

Theorem 7.2.2. *Let G and H be matrix Lie groups with Lie algebras $\mathfrak{g}$ and $\mathfrak{h}$, respectively, and suppose $\phi : G \to H$ is a Lie group homomorphism. Then there exists a unique real linear map $\tilde{\phi} : \mathfrak{g} \to \mathfrak{h}$ such that $\phi(e^X) = e^{\tilde{\phi}(X)}$ for every $X \in \mathfrak{g}$. Moreover, $\tilde{\phi}$ satisfies:*
1. $\tilde{\phi}([X, Y]_{\mathfrak{g}}) = [\tilde{\phi}(X), \tilde{\phi}(Y)]_{\mathfrak{h}}$ *for all $X, Y \in \mathfrak{g}$ and*
2. $\tilde{\phi}(X) = \frac{d}{dt}\phi(e^{tX})|_{t=0} = \lim_{t\to 0} \frac{\phi(e^{tX}) - 1_H}{t}$ *for all $X \in \mathfrak{g}$.*

Part 1. of the theorem asserts that $\tilde{\phi}$ is a Lie algebra homomorphism and Part 2. tells us how to compute it from ϕ. If G is connected and simply connected one can show that, conversely, every Lie algebra homomorphism $\mathfrak{g} \to \mathfrak{h}$ is $\tilde{\phi}$ for some Lie group homomor-phism $\phi : G \to H$; this is Theorem 5.33 of [Hall1]. All of this is true, in particular, for *finite-dimensional representations* of G. For representations on infinite-dimensional Hilbert spaces the situation is less simple and we will now sketch the issues involved.

Let G be a matrix Lie group and $\pi : G \to \mathcal{U}(\mathcal{H})$ a strongly continuous, unitary rep-resentation of G on the separable, complex Hilbert space $\mathcal{H}$ (see Appendix I). A vec-tor $v \in \mathcal{H}$ is called a *smooth vector* or C^∞ *vector for π* if

$$g \to \pi(g)v : G \to \mathcal{H}$$

is a C^∞ map from G to $\mathcal{H}$.

Remark 7.2.4. The group G is a matrix Lie group and therefore a finite-dimensional differentiable manifold, but $\mathcal{H}$ is (generally) an infinite-dimensional Hilbert space, so we should say something about what is meant by "C^∞" for a map from G to $\mathcal{H}$. The idea is to regard $\mathcal{H}$ as the simplest example of an infinite-dimensional differentiable manifold, specifically, a Banach manifold modeled on $\mathcal{H}$ with a single, global chart (the identity map on $\mathcal{H}$). Choosing charts on G one can then identify the map $G \to \mathcal{H}$ with a family of maps ("coordinate expressions") from a Euclidean space $\mathbb{R}^n$ ($n = \dim G$) into $\mathcal{H}$ exactly as in the finite-dimensional case. The problem then reduces to defining smoothness for maps between open sets in Banach spaces. There are some minor technical issues due to the infinite dimensionality of the Banach spaces, but the general scheme is exactly as in the finite-dimensional case. For our present purposes we will leave it at this (if you would like to see more details we recommend Chapter 2 of [AMR]).

We let $\mathcal{H}^\infty(\pi)$ denote the set of all smooth vectors for π in $\mathcal{H}$. This is clearly a linear subspace of $\mathcal{H}$ and we claim that it is invariant under $\pi : G \to \mathcal{U}(\mathcal{H})$. To see this, let $v \in \mathcal{H}^\infty(\pi)$ and $g_0 \in G$. We show that $\pi(g_0)v$ is in $\mathcal{H}^\infty(\pi)$, that is, that $g \to \pi(g)(\pi(g_0)v)$ is C^∞. But this is clear since this map is the composition of two maps

$$g \to gg_0 \to \pi(gg_0)v = \pi(g)(\pi(g_0)v),$$

the first of which is smooth because G is a Lie group and the second because $v \in \mathcal{H}^\infty(\pi)$. What is not so obvious, however, is that $\mathcal{H}^\infty(\pi)$ is *dense* in $\mathcal{H}$. This was proved by Gårding in [Går].

Theorem 7.2.3. *Let $\pi : G \to \mathcal{U}(\mathcal{H})$ be a strongly continuous, unitary representation of the Lie group G on the separable, complex Hilbert space $\mathcal{H}$. Then $\mathcal{H}^\infty(\pi)$ is a dense, invariant, linear subspace of $\mathcal{H}$.*

For each X in $\mathfrak{g}$ we define a linear map $d\pi(X) : \mathcal{H}^\infty(\pi) \to \mathcal{H}$ by

$$d\pi(X)v = \lim_{t\to 0} \frac{\pi(e^{tX}) - I}{t}(v) = \lim_{t\to 0} \frac{\pi(e^{tX})v - v}{t} = \frac{d}{dt}\pi(e^{tX})v\Big|_{t=0}, \qquad (7.20)$$

where the limits are in $\mathcal{H}$.

Exercise 7.2.11. Show that the limit in (7.20) exists for every $v \in \mathcal{H}^\infty(\pi)$ and that $d\pi(X)$ depends linearly on $X \in \mathfrak{g}$.

Next we observe that for each $X \in \mathfrak{g}$, $d\pi(X)$ leaves $\mathcal{H}^\infty(\pi)$ invariant, that is, $v \in \mathcal{H}^\infty(\pi) \Rightarrow d\pi(X)v \in \mathcal{H}^\infty(\pi)$. To see this we must show that $g \to \pi(g)d\pi(X)v$ is C^∞. But π is strongly continuous, so

$$\pi(g)d\pi(X)v = \pi(g)\left(\lim_{t\to 0} \frac{\pi(e^{tX})v - v}{t} \right) = \lim_{t\to 0} \frac{\pi(ge^{tX})v - \pi(g)v}{t} = \frac{d}{dt}\pi(ge^{tX})v \Big|_{t=0}.$$

Since $g \to \pi(g)v$ is a C^∞ map, it follows that $g \to \pi(g)d\pi(X)v$ is C^∞ and therefore $d\pi(X)v \in \mathcal{H}^\infty(\pi)$.

Due to the infinite-dimensionality of $\mathcal{H}$ it takes a bit of work, but one can also show that for any $X, Y \in \mathfrak{g}$,

$$d\pi([X, Y]_\mathfrak{g})v = [d\pi(X), d\pi(Y)]_- v = (d\pi(X)d\pi(Y) - d\pi(Y)d\pi(X))v \quad \forall v \in \mathcal{H}^\infty(\pi),$$

or, briefly,

$$d\pi([X, Y]_\mathfrak{g}) = [d\pi(X), d\pi(Y)]_-$$

on $\mathcal{H}^\infty(\pi)$.

Next we show that for any $X \in \mathfrak{g}$ and any positive constant a, the operator

$$ia\, d\pi(X) : \mathcal{H}^\infty(\pi) \to \mathcal{H}$$

is symmetric. To see this note that it clearly suffices to prove the result for $a = 1$. Also note that, because π is a unitary representation,

$$(\pi(e^{tX}))^* = (\pi(e^{tX}))^{-1} = \pi(e^{-tX}).$$

Now we just compute, for any $v, w \in \mathcal{H}^\infty(\pi)$,

$$\left\langle i\, d\pi(X)v, w \right\rangle = \left\langle i \lim_{t\to 0} \frac{\pi(e^{tX}) - I}{t} v, w \right\rangle$$

$$= \lim_{t\to 0} \left\langle i \frac{\pi(e^{tX}) - I}{t} v, w \right\rangle$$

$$= \lim_{t\to 0} \left\langle v, -i \frac{(\pi(e^{tX}))^* - I}{t} w \right\rangle$$

$$= \lim_{t\to 0} \left\langle v, -i \frac{\pi(e^{-tX}) - I}{t} w \right\rangle$$

$$= \lim_{s\to 0} \left\langle v, i \frac{\pi(e^{sX}) - I}{s} w \right\rangle$$

$$= \left\langle v, i\, d\pi(X)w \right\rangle,$$

as required. Note that without the i, $d\pi(X)$ is skew-symmetric.

In the best of all possible worlds we would be able to assert next that the operators $i\,d\pi(X)$ are not merely symmetric, but, in fact, essentially self-adjoint on $\mathcal{H}^{\infty}(\pi)$. Regrettably, Dr. Pangloss was mistaken and things are not so simple. However, Edward Nelson [Nel1] has refined the ideas we have been discussing to prove the following. Given a strongly continuous, unitary representation $\pi : G \to \mathcal{U}(\mathcal{H})$ of a Lie group G on a separable, complex Hilbert space $\mathcal{H}$, there is a dense, linear subspace $\mathcal{D}$ of $\mathcal{H}$ with the following properties:

1. $\mathcal{D} \subseteq \mathcal{H}^{\infty}(\pi)$,
2. $\mathcal{D}$ is invariant under every $i\,d\pi(X)$, that is, $i\,d\pi(X)(\mathcal{D}) \subseteq \mathcal{D}$ for every $X \in \mathfrak{g}$,
3. $i\,d\pi(X)$ is essentially self-adjoint on $\mathcal{D}$ for every $X \in \mathfrak{g}$.

Nelson's procedure was to consider, instead of the smooth vectors $\mathcal{H}^{\infty}(\pi)$ in $\mathcal{H}$ associated with π, what are called *analytic vectors*. By definition, a vector $v \in \mathcal{H}$ is an analytic vector for $\pi : G \to \mathcal{U}(\mathcal{H})$ if the map $g \to \pi(g)v : G \to \mathcal{H}$ is real analytic. Any real Lie group admits a unique real analytic manifold structure for which the group operations are real analytic. In fact, a very famous (and very difficult) theorem of Gleason and Montgomery–Zippin states that a topological group G admits a real-analytic Lie group structure if and only if G is a topological manifold (this is generally regarded as a solution to Hilbert's fifth problem). Even so, there are a number of plausible alternative definitions of an analytic map from G into the Hilbert space $\mathcal{H}$. We will simply record the definition adopted by Nelson (page 579 of [Nel1]). Since G is an analytic manifold it will suffice to define analyticity for a smooth map $u : U \to \mathcal{H}$, where U is an open set in $\mathbb{R}^{n}$ ($n = \dim G$) containing the origin. For any compact set $K \subseteq U$ we let

$$\|u\|_K = \sup_{x \in K} \|u(x)\|_{\mathcal{H}}.$$

We will say that u is *analytic on U* if it is smooth and, for every $x \in U$, there exists an $\epsilon > 0$ such that if K is the closed ball of radius ϵ about x, then

$$\sum_{k=0}^{\infty} \frac{1}{k!} \sum_{1 \le i_1,\dots,i_k \le k} \left\| \frac{\partial}{\partial x^{i_1}} \cdots \frac{\partial}{\partial x^{i_k}} u \right\|_K s_{i_1} \cdots s_{i_k}$$

is absolutely convergent for sufficiently small $s_{i_1}, \dots, s_{i_k}$; for more on analytic vectors, see [Good].

One can also define the notions of smooth vector and analytic vector for a single operator $A : \mathcal{D}(A) \to \mathcal{H}$ on $\mathcal{H}$. Although these are closely related to the corresponding notions for a representation we will save the definitions for Section 9.2, where we will discuss in more detail their relevance to self-adjointness.

Nelson proves that the set $\mathcal{H}^{\omega}(\pi)$ of analytic vectors for π is dense in $\mathcal{H}$ (Theorem 4, Section 8, of [Nel1]). Next he shows how to produce a dense linear subspace $\mathcal{D}$ of $\mathcal{H}$ that contains a dense, invariant set of analytic vectors for every $i\,d\pi(X)$, $X \in \mathfrak{g}$. From this Nelson obtains the essential self-adjointness of each $i\,d\pi(X)$ on $\mathcal{D}$ (this is also Corollary 2, Section X.6, of [RS2]). As usual, we will use the same symbol $i\,d\pi(X)$ to denote the unique self-adjoint extension and will also write $d\pi(X)$ for the unique skew-adjoint extension of $d\pi(X)$. All of the details are available in [Nel1] and we will discuss them no further, but will instead apply what we have learned to the examples of most interest to us, that is, the Heisenberg group H_3 and Heisenberg algebra $\mathfrak{h}_3$.

Let us suppose that $\pi : H_3 \rightarrow \mathcal{U}(\mathcal{H})$ is some strongly continuous, unitary representation of the Heisenberg group on a separable, complex Hilbert space $\mathcal{H}$. The Heisenberg algebra $\mathfrak{h}_3$ is generated by three elements that we have called X, Y and Z subject to the commutation relations $[X,Z] = 0$, $[Y,Z] = 0$ and $[X,Y] = Z$. We consider the corresponding operators $d\pi(X)$, $d\pi(Y)$ and $d\pi(Z)$. These are all defined and essentially skew-adjoint on some common dense, invariant subspace $\mathcal{D}$, where

$$d\pi(X)v = \frac{d}{dx}\pi(e^{xX})v\Big|_{x=0},$$

$$d\pi(Y)v = \frac{d}{dy}\pi(e^{yY})v\Big|_{y=0}$$

and

$$d\pi(Z)v = \frac{d}{dz}\pi(e^{zZ})v\Big|_{z=0}.$$

Now consider the essentially self-adjoint operators on $\mathcal{D}$ defined by

$$i\hbar\,d\pi(X), \quad i\hbar\,d\pi(Y) \quad \text{and} \quad i\hbar\,d\pi(Z)$$

and use the same symbols to denote their unique self-adjoint extensions. On $\mathcal{D}$ we have

$$[\,i\hbar\,d\pi(Y), i\hbar\,d\pi(X)\,]_- = -\hbar^2[\,d\pi(Y), d\pi(X)\,]_- = \hbar^2 d\pi(\,[X,Y]\,) = \hbar^2 d\pi(Z).$$

This much is true for any strongly continuous, unitary representation π of H_3. Now, let us think about $\hbar^2 d\pi(Z)$ for a moment. Suppose we could find a representation π on some Hilbert space that satisfies $d\pi(Z) = -\frac{i}{\hbar}I$. Then

$$[\,i\hbar\,d\pi(Y), i\hbar\,d\pi(X)\,]_- = -i\hbar I,$$

which would amount to a rigorous version of Heisenberg–Born–Jordan–Dirac quantization of a single classical particle moving in one dimension. The task then is to construct such a representation. It should come as no surprise that we will look for a representation that gives rise to the Schrödinger realization of $\mathfrak{h}_3$, that is, we would like to find $\pi : H_3 \rightarrow \mathcal{U}(L^2(\mathbb{R}))$ with $i\hbar\,d\pi(Y) = P$, $i\hbar\,d\pi(X) = Q$ and $i\hbar\,d\pi(Z) = I$. Note that this will be the case if we choose $\pi(e^{xX})$, $\pi(e^{yY})$ and $\pi(e^{zZ})$ as follows:

$$\pi(e^{xX}) = e^{xQ/i\hbar}, \quad \pi(e^{yY}) = e^{yP/i\hbar} \quad \text{and} \quad \pi(e^{zZ}) = e^{zI/i\hbar}.$$

More explicitly, we want

$$[\pi(e^{xX})\psi](q) = e^{xq/i\hbar}\psi(q), \quad [\pi(e^{yY})\psi](q) = \psi(q - y) \quad \text{and}$$
$$[\pi(e^{zZ})\psi](q) = e^{z/i\hbar}\psi(q). \tag{7.21}$$

Exercise 7.2.12. Prove the following group theoretic lemma.

Lemma 7.2.4. *Let G be a group and let $\alpha, \beta, \gamma : \mathbb{R} \to G$ be homomorphisms of the additive group $\mathbb{R}$ into G that satisfy*

$$\alpha(x)\gamma(z) = \gamma(z)\alpha(x), \quad \beta(y)\gamma(z) = \gamma(z)\beta(y) \quad and \quad \alpha(x)\beta(y) = \gamma(xy)\beta(y)\alpha(x)$$

for all $x, y, z \in \mathbb{R}$. Then the map $\pi : H_3 \to G$ defined by

$$\pi(M(x, y, z)) = \gamma(z)\beta(y)\alpha(x),$$

where

$$M(x, y, z) = \begin{pmatrix} 1 & x & z \\ 0 & 1 & y \\ 0 & 0 & 1 \end{pmatrix},$$

is a group homomorphism.

Exercise 7.2.13. Apply Lemma 7.2.4 with $G = \mathcal{U}(L^2(\mathbb{R}))$, $\alpha(x) = e^{xQ/i\hbar}$, $\beta(y) = e^{yP/i\hbar}$ and $\gamma(z) = e^{zI/i\hbar}$ to obtain a group homomorphism $\pi : H_3 \to \mathcal{U}(L^2(\mathbb{R}))$. *Hint:* To verify the required conditions on α, β and γ you will need the Baker–Campbell–Hausdorff formula.

Exercise 7.2.14. Show that the homomorphism $\pi : H_3 \to \mathcal{U}(L^2(\mathbb{R}))$ in Exercise 7.2.13 is strongly continuous. *Hint:* Recall that α, β and γ are strongly continuous one-parameter groups of unitary operators on $L^2(\mathbb{R})$. Show that

$$\|\gamma(z)\beta(y)\alpha(x)\psi - \psi\| \leq \|\alpha(x)\psi - \psi\| + \|\beta(y)\psi - \psi\| + \|\gamma(z)\psi - \psi\|.$$

This depends on the fact that the operators are unitary. Begin by noting that if S and T are any two unitary operators on a Hilbert space $\mathcal{H}$, then one can write $STv - v = STv - Sv + Sv - v$.

Thus, the homomorphism $\pi : H_3 \to \mathcal{U}(L^2(\mathbb{R}))$ in Exercise 7.2.13 is a strongly continuous, unitary representation of H_3 satisfying (7.21). We have therefore succeeded in producing a representation π of the Heisenberg group H_3 whose "infinitesimal version" $d\pi$ reproduces the Schrödinger realization of the Heisenberg algebra $\mathfrak{h}_3$; π is called the *Schrödinger representation of H_3*.

Exercise 7.2.15. Show that the Schrödinger representation π can be written explicitly as

$$(\pi(e^{xX+yY+zZ})\psi)(q) = e^{(xq+z-\frac{1}{2}xy)/i\hbar}\psi(q-y). \tag{7.22}$$

One can show that the smooth vectors for this representation are precisely the Schwartz functions $\mathcal{S}(\mathbb{R}) \subseteq L^2(\mathbb{R})$.

Note that it is *not* the case that every realization of $\mathfrak{h}_3$ arises in this way as $d\pi$ for some representation π of H_3. Those realizations of $\mathfrak{h}_3$ which do arise in this way are said to be *integrable*. To learn more about nonintegrable realizations of $\mathfrak{h}_3$ one can consult [Schm1] and [Schm2].

Next we will use (7.22) to show that the Schrödinger representation of H_3 is irreducible (see Appendix I). For this we will suppose that $\mathcal{H}_0$ is a nonzero, closed, invariant subspace of $L^2(\mathbb{R})$ and show that its orthogonal complement $\mathcal{H}_0^\perp$ is trivial so that $\mathcal{H}_0$ must be all of $L^2(\mathbb{R})$. Select some nonzero $\psi \in \mathcal{H}_0$. Then, for any $\phi \in \mathcal{H}_0^\perp$,

$$\phi \perp \pi(e^{xX+yY})\psi \quad \forall x, y \in \mathbb{R},$$

that is,

$$\int_{\mathbb{R}} e^{(-i/\hbar)xq} e^{(-i/\hbar)x(\frac{y}{2})} \psi(q-y) \overline{\phi(q)} \, dq = 0.$$

Exercise 7.2.16. Use the shift property (Exercise G.2.3) of the Fourier transform $\mathcal{F}_\hbar$ (Remark G.2.1) to show that this can be written

$$\int_{\mathbb{R}} e^{(-i/\hbar)xq} \psi\left(q - \frac{y}{2}\right) \overline{\phi\left(q + \frac{y}{2}\right)} \, dq = 0.$$

We conclude that the Fourier transform of $\psi(q-\frac{y}{2})\overline{\phi(q+\frac{y}{2})}$ is zero for every $y \in \mathbb{R}$. Consequently, $\psi(q-\frac{y}{2})\overline{\phi(q+\frac{y}{2})} = 0$ for almost every $q \in \mathbb{R}$ and for every $y \in \mathbb{R}$. Since the linear transformation $(q, y) \to (X, Y) = (q - \frac{y}{2}, q + \frac{y}{2})$ is invertible, $\psi(X)\overline{\phi(Y)} = 0$ for almost all X and Y. Thus, either ψ or ϕ is the zero element of $L^2(\mathbb{R})$. But ψ is nonzero by assumption, so $\phi = 0 \in L^2(\mathbb{R})$, as required.

Having discovered one strongly continuous, irreducible, unitary representation π of H_3 whose infinitesimal version $d\pi$ gives a realization of $\mathfrak{h}_3$, that is, a solution to the canonical commutation relations, one might wonder if there are others floating around somewhere. Of course, one can always produce such representations that *appear* different on the surface by choosing some unitary operator U of $L^2(\mathbb{R})$ onto itself and replacing each $\pi(g)$ by $U\pi(g)U^{-1}$. This is cheating, however, since, both mathematically and physically, π and $U\pi U^{-1}$ are entirely equivalent (that is, unitarily equivalent). To investigate this question a bit more closely, recall that we pointed out earlier that it is often useful to regard $\hbar$ not as some fixed, universal constant, but rather as a positive parameter that is part of the "input" in the construction of a quantum theory (see Remark 7.2.2). From this point of view it might be best to write the Schrödinger representation as $\pi_\hbar$ rather than simply π. Now note that two different choices of this parameter give rise to representations $\pi_{\hbar_1}$ and $\pi_{\hbar_2}$ that are *really* different, that is, not unitarily equivalent.

Exercise 7.2.17. Show that if $\hbar_1$ and $\hbar_2$ are distinct positive real numbers, then $\pi_{\hbar_1}(e^Z)$ and $\pi_{\hbar_2}(e^Z)$ are not unitarily equivalent operators on $L^2(\mathbb{R})$. *Hint*: Unitarily equivalent operators have the same eigenvalues.

According to Schur's lemma (Theorem I.0.1), any strongly continuous, irreducible, unitary representation of H_3 must send every element e^{zZ} of the center of H_3 to some multiple of the identity by a unit complex number. The various inequivalent Schrödinger representations of H_3 send e^{zZ} to $e^{(-i/\hbar)z}I$ in the center of $\mathcal{U}(L^2(\mathbb{R}))$ and they are distinguished, one from another, simply by the value of $\hbar$. It is a remarkable theorem of Stone and von Neumann that the same statement is true for an irreducible representation of H_3 on any $\mathcal{H}$ and that every such representation is unitarily equivalent to the Schrödinger representation with the same $\hbar$. Indeed, even more is true.

Theorem 7.2.5 (Stone–von Neumann theorem ($n = 1$)). *Let* $\rho \; : \; H_3 \; \rightarrow \; \mathcal{U}(\mathcal{H})$ *be a strongly continuous, unitary representation of the Heisenberg group H_3 on a separable, complex Hilbert space $\mathcal{H}$ with $\rho(e^{zZ}) = e^{(-i/\hbar)z}\mathrm{id}_{\mathcal{H}}$. Then $\mathcal{H}$ is the (finite or countably infinite) direct sum of mutually orthogonal closed subspaces $\mathcal{H}_\alpha$ each of which is invariant under ρ and such that the induced representation $\rho_\alpha \; : \; H_3 \; \rightarrow \; \mathcal{H}_\alpha$ of H_3 on $\mathcal{H}_\alpha$ $(\rho_\alpha(g) = \rho(g)|_{\mathcal{H}_\alpha} \; \forall g \; \in \; H_3)$ is unitarily equivalent to the Schrödinger representation $\pi_\hbar : H_3 \rightarrow \mathcal{U}(L^2(\mathbb{R}))$. In particular, if ρ is irreducible, then ρ is unitarily equivalent to $\pi_\hbar$.*

One often finds the Stone–von Neumann theorem stated in terms of pairs of strongly continuous one-parameter groups of unitary operators that satisfy the Weyl relations. This is entirely equivalent to Theorem 7.2.5, so for the proof we will simply refer to Chapter 14 of [Hall2], Theorem VIII.14, of [RS1] or Theorem 6.4, Chapter IV, of [Prug]. The expository paper [RosenJ] contains a nice synopsis of more recent work related to the Stone–von Neumann theorem (for example, extending it to the fermionic and supersymmetric systems that we will discuss in Chapter 10).

One can interpret the Stone–von Neumann theorem as asserting that for a fixed $\hbar$, there is "really only one" integrable, irreducible solution to the canonical commutation relations and this offers some justification for the physicist's habit of working almost exclusively with the abstract commutation relations themselves and not worrying so much about how they are realized. Of course, it is also true that there is "really only one" separable, infinite-dimensional, complex Hilbert space, but it would be naive to think that it should not matter which model of it one chooses to deal with in a particular context. The same is true of the CCR and we will see advantages to choosing different realizations somewhat later.

Example 7.2.5. Extending everything we have done to n degrees of freedom involves no fundamentally new ideas, so we will simply state the facts. We consider a classical mechanical system with configuration space $\mathbb{R}^n$. Phase space is $T^*(\mathbb{R}^n) = \mathbb{R}^{2n}$ with canonical coordinates $q^1, \ldots, q^n, p_1, \ldots, p_n$ relative to which the classical commutation relations are

$$\{q^j, q^k\} = \{p_j, p_k\} = 0 \quad \text{and} \quad \{q^j, p_k\} = \delta^j_k, \quad \forall j, k = 1, \ldots, n,$$

where $\{ , \}$ is the Poisson bracket. Then $\{q^1, \ldots, q^n, p_1, \ldots, p_n, 1\}$ generate a Lie sub-algebra of the classical observables $C^\infty(T^*(\mathbb{R}^n))$ that is isomorphic to the Heisenberg algebra $\mathfrak{h}_{2n+1}$. The Heisenberg group H_{2n+1} is the unique simply connected Lie group whose Lie algebra is $\mathfrak{h}_{2n+1}$. A *realization* of $\mathfrak{h}_{2n+1}$ on a separable, complex Hilbert space $\mathcal{H}$ consists of a dense linear subspace $\mathcal{D}$ of $\mathcal{H}$ and operators $Q^1, \ldots, Q^n, P_1, \ldots, P_n$ on $\mathcal{H}$ with $\mathcal{D} \subseteq \mathcal{D}(Q^j)$ and $\mathcal{D} \subseteq \mathcal{D}(P_k)$, $j, k = 1, \ldots, n$, that satisfy:

1. $Q^j : \mathcal{D} \to \mathcal{D}$ and $P_k : \mathcal{D} \to \mathcal{D}$ for all $j, k = 1, \ldots, n$,
2. $[Q^j, Q^k]_- \psi = [P_j, P_k]_- \psi = 0$ and $[P_k, Q^j]_- \psi = -i\hbar \delta^j_k \psi$ for all $j, k = 1, \ldots, n$ and for all $\psi \in \mathcal{D}$,
3. $Q^1, \ldots, Q^n, P_1, \ldots, P_n$ are all essentially self-adjoint on $\mathcal{D}$.

In this case we say that the unique self-adjoint extensions of $Q^1, \ldots, Q^n, P_1, \ldots, P_n$, denoted by the same symbols, *satisfy the canonical commutation relations*. If we identify H_{2n+1} with $\mathbb{R}^n \times \mathbb{R}^n \times \mathbb{R}$, where the multiplication is defined by (7.17), then the *Schrödinger representation of H_{2n+1}* on $L^2(\mathbb{R}^n)$ is defined by

$$(\pi_\hbar(\mathbf{x}, \mathbf{y}, z)\psi)(\mathbf{q}) = e^{(\langle \mathbf{x}, \mathbf{q}\rangle + z - \frac{1}{2}\langle \mathbf{x}, \mathbf{y}\rangle)/i\hbar}\, \psi(\mathbf{q} - \mathbf{y}).$$

The infinitesimal version $d\pi_\hbar$ of $\pi_\hbar$ gives the *Schrödinger realization of $\mathfrak{h}_{2n+1}$*:

$$i\hbar \, d\pi_\hbar(X_j)\psi(\mathbf{q}) = Q^j\psi(\mathbf{q}) = q_j\psi(\mathbf{q}),$$

$$i\hbar \, d\pi_\hbar(Y_k)\psi(\mathbf{q}) = P_k\psi(\mathbf{q}) = -i\hbar\frac{\partial\psi}{\partial q^k},$$

$$i\hbar \, d\pi_\hbar(Z)\psi(\mathbf{q}) = \psi(\mathbf{q}),$$

for all $j, k = 1, \ldots, n$ and for all ψ in the Schwartz space. Finally we record the appropriate version of the Stone–von Neumann theorem.

Theorem 7.2.6 (Stone–von Neumann theorem). *Let $\rho : H_{2n+1} \to \mathcal{U}(\mathcal{H})$ be a strongly continuous, unitary representation of the Heisenberg group H_{2n+1} on a separable, complex Hilbert space $\mathcal{H}$ with $\rho(\mathbf{0}, \mathbf{0}, z) = e^{(-i/\hbar)z}\mathrm{id}_\mathcal{H}$. Then $\mathcal{H}$ is the (finite or countably infinite) direct sum of mutually orthogonal closed subspaces $\mathcal{H}_\alpha$ each of which is invariant under ρ and such that the induced representation $\rho_\alpha : H_{2n+1} \to \mathcal{H}_\alpha$ of H_{2n++1} on $\mathcal{H}_\alpha$ ($\rho_\alpha(g) = \rho(g)|_{\mathcal{H}_\alpha} \forall g \in H_{2n+1}$) is unitarily equivalent to the Schrödinger representation $\pi_\hbar : H_{2n+1} \to \mathcal{U}(L^2(\mathbb{R}^n))$. In particular, if ρ is irreducible, then ρ is unitarily equivalent to $\pi_\hbar$.*

At this point we have rather precise information about realizing the classical canonical commutation relations as self-adjoint operators on a Hilbert space and we should pause to ask ourselves how close this has gotten us to Dirac's program for quantizing classical mechanical systems. Reluctantly, we must admit that the answer is, "not very close." Roughly speaking, Dirac asked for a Lie algebra homomorphism from the classical observables to the quantum observables and, at this point, we have managed to do this only for the classical observables that live in the Heisenberg algebra $\mathfrak{h}_3$ and these are all of the form $a + bq + cp$ for $a, b, c \in \mathbb{R}$ (to ease the exposition we will again return to systems with one degree of freedom). Most interesting classical observables (such as the Hamiltonian) are nonlinear functions of q and p and therefore do not live in $\mathfrak{h}_3$. What we need to do then is try to extend our realizations of $\mathfrak{h}_3$ to larger Lie subalgebras of $C^\infty(T^*\mathbb{R})$ that contain the observables we are interested in quantizing. For the classical free particle and the classical harmonic oscillator the Hamiltonians are quadratic functions of q and p, so we will begin by trying to extend just to these. This may seem rather unambitious, but we will see soon that any more ambitious program is doomed to failure.

Begin by considering the linear subspace $\mathcal{P}_2(q, p)$ of $C^\infty(\mathbb{R}^2)$ spanned by $\{1, q, p, q^2, p^2, qp\}$. These are precisely the quadratic classical observables. Computing Poisson brackets gives, in addition to the commutation relations for $\mathfrak{h}_3$,

$$\{qp, p\} = p, \quad \{qp, q\} = -q, \quad \{p^2, q\} = -2p, \quad \{q^2, p\} = 2q \qquad (7.23)$$

and

$$\left\{\frac{q^2}{2}, \frac{p^2}{2}\right\} = qp, \quad \{qp, p^2\} = 2p^2, \quad \{qp, q^2\} = -2q^2. \qquad (7.24)$$

Exercise 7.2.18. Verify all of these.

In particular, $\mathcal{P}_2(q, p)$ is closed under Poisson brackets and is therefore a Lie subalgebra of $C^\infty(\mathbb{R}^2)$. But, according to (7.24), the same is true of the subspace $\mathcal{P}_2^H(q, p)$ spanned by $\{q^2, p^2, qp\}$ consisting of homogeneous quadratic polynomials in q and p. In fact, $\mathcal{P}_2^H(q, p)$ is isomorphic to a very familiar Lie algebra. Recall that $\mathfrak{sl}(2, \mathbb{R})$ denotes the Lie algebra (under matrix commutator) of all 2×2 real matrices with trace zero. It is spanned by

$$e = \begin{pmatrix} 0 & 1 \\ 0 & 0 \end{pmatrix}, \quad f = \begin{pmatrix} 0 & 0 \\ 1 & 0 \end{pmatrix} \quad \text{and} \quad h = \begin{pmatrix} 1 & 0 \\ 0 & -1 \end{pmatrix},$$

which satisfy the commutation relations

$$[e, f]_- = h, \quad [h, e]_- = 2e, \quad [h, f]_- = -2f.$$

Comparing this with (7.24) we find that

$$\frac{p^2}{2} \leftrightarrow e, \quad -\frac{q^2}{2} \leftrightarrow f, \quad qp \leftrightarrow h$$

defines a Lie algebra isomorphism of $\mathcal{P}_2^H(q,p)$ and $\mathfrak{sl}(2,\mathbb{R})$ and we will generally just identify them in this way; $\mathfrak{sl}(2,\mathbb{R})$ is the Lie algebra of the special linear group $SL(2,\mathbb{R})$ consisting of all 2×2 real matrices with determinant one. Despite its seeming simplicity, $SL(2,\mathbb{R})$ and its representations cut a very wide swath in modern mathematics (see, for example, [Lang2]). It is a three-dimensional, noncompact Lie group and all of its nontrivial, irreducible, unitary representations are infinite-dimensional. It is *not* simply connected. Indeed, its fundamental group is $\mathbb{Z}$ and its universal cover is one of that rare breed of finite-dimensional Lie groups that are not matrix groups. It has a double cover $Mp(2,\mathbb{R})$ called the *metaplectic group*. Note that $\mathfrak{h}_3$ is also isomorphic to a Lie subalgebra of $\mathcal{P}_2(q,p)$ and we will now show how $\mathcal{P}_2(q,p)$ can be reconstructed from $\mathfrak{h}_3$ and $\mathfrak{sl}(2,\mathbb{R})$. The result will identify $\mathcal{P}_2(q,p)$ with another well-known Lie algebra.

Remark 7.2.5. We will briefly recall the general notion of the *semi-direct product* of Lie groups; details are available in Section I.15 of [Knapp], although we have adopted a somewhat different notation. Let us begin with two Lie groups H and N and suppose that we are given a smooth left action of H on N by automorphisms, that is, a smooth map $\tau : H \times N \to N$ such that $h \to \tau(h, \cdot)$ is a group homomorphism from H into the group of automorphisms of N. Writing $\tau(h,n) = h \cdot n$ one then has $h_1 \cdot (h_2 \cdot n) = (h_1 h_2) \cdot n$ and $h \cdot (n_1 n_2) = (h \cdot n_1)(h \cdot n_2)$. Then the *semi-direct product*

$$G = H \times_\tau N$$

of H and N determined by τ is the Lie group whose underlying manifold is the product manifold $H \times N$ and whose group operations are defined by

$$(h,n)(h',n') = (hh', n(h \cdot n')),$$
$$1_G = (1_H, 1_N),$$
$$(h,n)^{-1} = (h^{-1}, h^{-1} \cdot n^{-1}).$$

Exercise 7.2.19. Verify the group axioms and show that $G = H \times_\tau N$ is a Lie group.

Example 7.2.6. We take $H = SL(2,\mathbb{R})$ and $N = H_3$. It will be convenient to write the elements of H_3 as $(\binom{x}{y}, z)$ rather than (x,y,z). Now define $\tau : SL(2,\mathbb{R}) \times H_3 \to H_3$ by

$$M \cdot \left(\binom{x}{y}, z \right) = \left(M \binom{x}{y}, z \right)$$

for every $M \in SL(2,\mathbb{R})$. Write the product in H_3 as

$$\left(\binom{x}{y}, z \right)\left(\binom{x'}{y'}, z' \right) = \left(\binom{x+x'}{y+y'}, z + z' + \frac{1}{2}\omega\left(\binom{x}{y}, \binom{x'}{y'} \right) \right),$$

where

$$\omega\left(\binom{x}{y}, \binom{x'}{y'} \right) = xy' - x'y$$

is the canonical symplectic form on $\mathbb{R}^2$.

Exercise 7.2.20. Show that $\omega\left(M\left(\begin{smallmatrix} x \\ y \end{smallmatrix}\right), M\left(\begin{smallmatrix} x' \\ y' \end{smallmatrix}\right)\right) = \omega\left(\left(\begin{smallmatrix} x \\ y \end{smallmatrix}\right), \left(\begin{smallmatrix} x' \\ y' \end{smallmatrix}\right)\right).$

Now we compute

$$
M \cdot \left[\left(\left(\begin{smallmatrix} x \\ y \end{smallmatrix}\right), z\right)\left(\left(\begin{smallmatrix} x' \\ y' \end{smallmatrix}\right), z'\right)\right] = M \cdot \left(\left(\begin{smallmatrix} x+x' \\ y+y' \end{smallmatrix}\right), z+z' + \frac{1}{2}\omega\left(\left(\begin{smallmatrix} x \\ y \end{smallmatrix}\right), \left(\begin{smallmatrix} x' \\ y' \end{smallmatrix}\right)\right)\right)
$$

$$
= \left(M\left(\begin{smallmatrix} x+x' \\ y+y' \end{smallmatrix}\right), z+z' + \frac{1}{2}\omega\left(\left(\begin{smallmatrix} x \\ y \end{smallmatrix}\right), \left(\begin{smallmatrix} x' \\ y' \end{smallmatrix}\right)\right)\right)
$$

$$
= \left(M\left(\begin{smallmatrix} x \\ y \end{smallmatrix}\right) + M\left(\begin{smallmatrix} x' \\ y' \end{smallmatrix}\right), z+z' + \frac{1}{2}\omega\left(M\left(\begin{smallmatrix} x \\ y \end{smallmatrix}\right), M\left(\begin{smallmatrix} x' \\ y' \end{smallmatrix}\right)\right)\right)
$$

$$
= \left(M\left(\begin{smallmatrix} x \\ y \end{smallmatrix}\right), z\right)\left(M\left(\begin{smallmatrix} x' \\ y' \end{smallmatrix}\right), z'\right).
$$

$$
= \left[M \cdot \left(\left(\begin{smallmatrix} x \\ y \end{smallmatrix}\right), z\right)\right]\left[M \cdot \left(\left(\begin{smallmatrix} x' \\ y' \end{smallmatrix}\right), z'\right)\right].
$$

Consequently, each $\tau(M, \cdot)$ is a homomorphism of H_3 and is clearly invertible with inverse $\tau(M^{-1}, \cdot)$. Thus, each $\tau(M, \cdot)$ is a group automorphism of H_3.

Exercise 7.2.21. Show that $M \to \tau(M, \cdot)$ is a group homomorphism of $SL(2, \mathbb{R})$ into the automorphism group $\mathrm{Aut}(H_3)$ of H_3 and that $\tau{:}SL(2, \mathbb{R}) \times H_3 \to H_3$ is smooth.

The corresponding semi-direct product

$$
G^J = SL(2, \mathbb{R}) \times_\tau H_3
$$

is called the *Jacobi group*.

To describe the Lie algebra of a semi-direct product of Lie groups we will need to introduce an analogous "semi-direct product" of Lie algebras (details for this are available in Section I.4 of [Knapp]). Note that, since the underlying manifold of $G = H \times_\tau N$ is the product $H \times N$, the tangent space at the identity in G is just the vector space direct sum $\mathfrak{h} \oplus \mathfrak{n}$ of the Lie algebras of H and N, so the objective is to define an appropriate bracket structure on $\mathfrak{h} \oplus \mathfrak{n}$.

Let $\mathfrak{b}$ be any real Lie algebra. Denote by $\mathrm{End}(\mathfrak{b})$ the group of all Lie algebra homomorphisms of $\mathfrak{b}$. Recall that any $D \in \mathrm{End}(\mathfrak{b})$ satisfying $D([B_1, B_2]_\mathfrak{b}) = [B_1, D(B_2)]_\mathfrak{b} + [D(B_1), B_2]_\mathfrak{b}$ for all $B_1, B_2 \in \mathfrak{b}$ is called a *derivation* of $\mathfrak{b}$. The subset $\mathrm{Der}(\mathfrak{b})$ of $\mathrm{End}(\mathfrak{b})$ consisting of all derivations is itself a Lie algebra under the bracket defined by the commutator $[D, E] = D \circ E - E \circ D$ for all $D, E \in \mathrm{Der}(\mathfrak{b})$. The following is Proposition 1.22 of [Knapp]. For the statement we will identify $\mathfrak{a}$ and $\mathfrak{b}$ with the subspaces of $\mathfrak{a} \oplus \mathfrak{b}$ in which the second, respectively, first coordinate is zero.

Proposition 7.2.7. *Let $\mathfrak{a}$ and $\mathfrak{b}$ be two real Lie algebras and suppose $\pi : \mathfrak{a} \to \mathrm{Der}(\mathfrak{b})$ is a Lie algebra homomorphism. Then there is a unique Lie algebra structure on the vector space direct sum $\mathfrak{g} = \mathfrak{a} \oplus \mathfrak{b}$ satisfying $[A_1, A_2]_\mathfrak{g} = [A_1, A_2]_\mathfrak{a}$ for all $A_1, A_2 \in \mathfrak{a}$, $[B_1, B_2]_\mathfrak{g} =$*

$[B_1, B_2]_{\mathfrak{b}}$ for all $B_1, B_2 \in \mathfrak{b}$ and $[A, B]_{\mathfrak{g}} = \pi(A)(B)$ for all $A \in \mathfrak{a}$ and all $B \in \mathfrak{b}$. Moreover, with this Lie algebra structure, $\mathfrak{a}$ is a Lie subalgebra of $\mathfrak{g}$ and $\mathfrak{b}$ is an ideal in $\mathfrak{g}$.

The Lie algebra $\mathfrak{g}$ described in the proposition is called the *semi-direct product* of the Lie algebras $\mathfrak{a}$ and $\mathfrak{b}$ determined by π and written $\mathfrak{a} \times_\pi \mathfrak{b}$. The idea now is to show that if $G = H \times_\tau N$, then an appropriate choice of $\pi : \mathfrak{h} \to \mathrm{Der}(\mathfrak{n})$ gives $\mathfrak{g} = \mathfrak{h} \times_\pi \mathfrak{n}$. We will simply describe how one must choose π, refer to Proposition 1.124 of [Knapp] for the proof that it works and then write out the example of interest to us.

We consider the Lie group semi-direct product $G = H \times_\tau N$, where $\tau : H \times N \to N$. Fix an $h \in H$. Then $\tau(h, \cdot) : N \to N$ is a Lie group automorphism. Its derivative at the identity 1_N

$$\overline{\tau}(h) = D(\tau(h, \cdot))(1_N) : \mathfrak{n} \to \mathfrak{n}$$

is therefore a Lie algebra isomorphism. This gives a map $\overline{\tau} : H \to \mathrm{GL}(\mathfrak{n})$ from H to the group of invertible linear transformations on the vector space $\mathfrak{n}$. One shows that $\overline{\tau}$ is a smooth group homomorphism

$$\overline{\tau} : H \to \mathrm{Aut}(\mathfrak{n})$$

from H to the automorphism group of $\mathfrak{n}$. The derivative of $\overline{\tau}$ at the identity 1_H, which we denote

$$\pi = D\overline{\tau}\,(1_H),$$

is therefore a linear map from $\mathfrak{h}$ to the Lie algebra of $\mathrm{Aut}(\mathfrak{n})$. Now we note that every element of the Lie algebra of $\mathrm{Aut}(\mathfrak{n})$ is, in fact, a derivation of $\mathfrak{n}$.

Remark 7.2.6. Note that N is a matrix Lie group, and therefore $\mathfrak{n}$ is a Lie algebra of matrices, so $\mathrm{Aut}(\mathfrak{n})$ is also a matrix Lie group. Its Lie algebra is therefore contained in $\mathrm{End}(\mathfrak{n})$. Furthermore, everything in the Lie algebra of $\mathrm{Aut}(\mathfrak{n})$ is $c'(0)$, where $c(t)$ is a curve in $\mathrm{Aut}(\mathfrak{n})$ with $c(0) = \mathrm{id}_\mathfrak{n}$. But $c(t)([X, Y]) = [c(t)X, c(t)Y]$ for each t implies that $c'(0)([X, Y]) = [c'(0)X, Y] + [X, c'(0)Y]$, so $c'(0)$ is a derivation. One can show that, in fact, the Lie algebra of $\mathrm{Aut}(\mathfrak{n})$ is precisely $\mathrm{Der}(\mathfrak{n})$ (this is Proposition 1.120 of [Knapp]).

Consequently,

$$\pi = D\overline{\tau}\,(1_H) : \mathfrak{h} \to \mathrm{Der}(\mathfrak{n}),$$

so we can form the Lie algebra semi-direct product $\mathfrak{h} \times_\pi \mathfrak{n}$, and this is precisely the Lie algebra of $H \times_\tau N$ (this is Proposition 1.124 of [Knapp]). Now we will work out the example of interest to us.

Example 7.2.7. We will return now to the Jacobi group G^J constructed in Example 7.2.6. Recall that $G^J = \mathrm{SL}(2, \mathbb{R}) \times_\tau H_3$, where $\tau : \mathrm{SL}(2, \mathbb{R}) \times H_3 \to H_3$ is given by

$$\tau\left(M, \left(\begin{pmatrix} x \\ y \end{pmatrix}, z \right) \right) = \left(M \begin{pmatrix} x \\ y \end{pmatrix}, z \right).$$

Now fix an $M \in SL(2, \mathbb{R})$ and consider the map $\tau(M, \cdot) : H_3 \to H_3$ given by

$$\left(\begin{pmatrix} x \\ y \end{pmatrix}, z \right) \to \left(M \begin{pmatrix} x \\ y \end{pmatrix}, z \right). \tag{7.25}$$

We need the derivative of this map at $1_{H_3} = \left(\begin{pmatrix} 0 \\ 0 \end{pmatrix}, 0 \right)$. But, as a manifold, H_3 is just $\mathbb{R}^3$ and so $\mathfrak{h}_3$ is also canonically identified with $\mathbb{R}^3$. Furthermore, $\tau(M, \cdot)$ is linear as a map from $\mathbb{R}^3$ to itself, so its derivative, at any point, is the same linear map and we have

$$\overline{\tau}(M) = D(\tau(M, \cdot))(1_{H_3}) = \tau(M, \cdot),$$

that is,

$$\overline{\tau}(M) \left(\begin{pmatrix} x \\ y \end{pmatrix}, z \right) = \left(M \begin{pmatrix} x \\ y \end{pmatrix}, z \right).$$

Consequently, $\overline{\tau}$ is the map on $SL(2, \mathbb{R})$ that carries $M \in SL(2, \mathbb{R})$ onto the map from $\mathfrak{h}_3$ to $\mathfrak{h}_3$ given by (7.25). If we let

$$M = \begin{pmatrix} a & b \\ c & d \end{pmatrix},$$

then the matrix of this map relative to the standard basis

$$e_1 = \left(\begin{pmatrix} 1 \\ 0 \end{pmatrix}, 0 \right), \quad e_2 = \left(\begin{pmatrix} 0 \\ 1 \end{pmatrix}, 0 \right), \quad e_3 = \left(\begin{pmatrix} 0 \\ 0 \end{pmatrix}, 1 \right)$$

is

$$\begin{pmatrix} a & b & 0 \\ c & d & 0 \\ 0 & 0 & 1 \end{pmatrix}.$$

Consequently, $\overline{\tau}$ is the map that sends

$$\begin{pmatrix} a & b \\ c & d \end{pmatrix}$$

in $SL(2, \mathbb{R})$ to the automorphism

$$\begin{pmatrix} a & b & 0 \\ c & d & 0 \\ 0 & 0 & 1 \end{pmatrix}$$

of $\mathfrak{h}_3$.

Exercise 7.2.22. Compute $\pi = D\bar{\tau}(1_{\mathrm{SL}(2,\mathbb{R})})$ and show that it sends $\left(\begin{smallmatrix} \alpha & \beta \\ \gamma & \delta \end{smallmatrix}\right)$ in $\mathfrak{sl}(2,\mathbb{R})$ to the derivation

$$\begin{pmatrix} \alpha & \beta & 0 \\ \gamma & \delta & 0 \\ 0 & 0 & 0 \end{pmatrix}$$

of $\mathfrak{h}_3$.

We can now describe the Lie algebra

$$\mathfrak{g}^J = \mathfrak{sl}(2,\mathbb{R}) \times_\pi \mathfrak{h}_3$$

of the Jacobi group (called, oddly enough, the *Jacobi algebra*); $\mathfrak{g}^J$ is generated by $\{X, Y, Z, e, f, h\}$, where X, Y and Z satisfy the commutation relations of $\mathfrak{h}_3$

$$[X, Z] = [Y, Z] = 0, \quad [X, Y] = Z$$

and e, f and h satisfy the commutation relations of $\mathfrak{sl}(2,\mathbb{R})$

$$[e, f] = h, \quad [h, e] = 2e, \quad [h, f] = -2f.$$

By Proposition 7.2.7, if $A \in \{e, f, h\}$ and $B \in \{X, Y, Z\}$, then

$$[A, B] = \pi(A)(B).$$

For example,

$$[h, Y] = \pi(h)(Y) = \begin{pmatrix} 1 & 0 & 0 \\ 0 & -1 & 0 \\ 0 & 0 & 0 \end{pmatrix} \begin{pmatrix} 0 \\ 1 \\ 0 \end{pmatrix} = \begin{pmatrix} 0 \\ -1 \\ 0 \end{pmatrix} = -Y.$$

Exercise 7.2.23. Show that the only nonzero commutation relations for $\mathfrak{g}^J$ are

$$[e, Y] = X, \quad [f, X] = Y, \quad [h, X] = X, \quad [h, Y] = -Y.$$

Exercise 7.2.24. The Lie algebra $\mathfrak{g}^J$ is often described in other, but equivalent terms. The following exercises are taken from [Berndt].
1. Consider the Lie algebra with basis $\{H, F, G, P, Q, R\}$ and subject to the commutation relations

$$[F, G] = H, \quad [H, F] = 2F, \quad [H, G] = -2G,$$
$$[P, Q] = 2R, \quad [H, Q] = Q, \quad [H, P] = -P,$$
$$[F, P] = -Q, \quad [G, Q] = -P.$$

Show that this Lie algebra is isomorphic to $\mathfrak{g}^J$.

2. Let $\mathcal{P}_2(q,p)$ be the Lie subalgebra of quadratic observables in $C^\infty(T^*\mathbb{R})$ with basis $\{1,q,p,qp,q^2,p^2\}$. Show that the map $\sigma : \mathcal{P}_2(q,p) \to \mathfrak{g}^J$ defined by

$$\sigma(1) = 2R, \quad \sigma(q) = P, \quad \sigma(p) = Q,$$
$$\sigma(qp) = H, \quad \sigma(q^2) = -2G, \quad \sigma(p^2) = 2F$$

is a Lie algebra isomorphism.

We can now view the problem of quantizing the quadratic classical observables in $C^\infty(T^*\mathbb{R})$ as that of extending the Schrödinger realization of the $\mathfrak{h}_3$ Lie subalgebra of $C^\infty(T^*\mathbb{R})$ to the $\mathfrak{g}^J = \mathfrak{sl}(2,\mathbb{R}) \times_\pi \mathfrak{h}_3$ subalgebra.

Remark 7.2.7. According to the Stone–von Neumann theorem, an appropriate realization of $\mathfrak{g}^J$ must restrict to the Schrödinger realization on $\mathfrak{h}_3$ if it is to act irreducibly on the Heisenberg algebra and we take this to be a basic assumption of our quantization procedure. It *is* an assumption, however, and one could certainly conceive of doing without it.

Let us spell this out in more detail. The Schrödinger realization sends 1 to the identity operator on $L^2(\mathbb{R})$ and, on $\mathcal{S}(\mathbb{R}) \subseteq L^2(\mathbb{R})$, is given by

$$q \to Q : (Q\psi)(q) = q\psi(q),$$
$$p \to P : (P\psi)(q) = -i\hbar\frac{d}{dq}\psi(q)$$

and satisfies

$$\{p,q\} \to -\frac{i}{\hbar}[P,Q]_-.$$

What we must do is define appropriate images for q^2, p^2 and qp in such a way that $\{,\} \to -\frac{i}{\hbar}[,]_-$. There is certainly an obvious way to start the process:

$$q^2 \to Q^2 : (Q^2\psi)(q) = q^2\psi(q),$$
$$p^2 \to P^2 : (P^2\psi)(q) = -i\hbar\frac{d}{dq}[(P\psi)(q)] = -\hbar^2\frac{d^2}{dq^2}\psi(q).$$

The element qp presents a problem, however. One might simply try $qp \to QP$. On the other hand, in $C^\infty(T^*\mathbb{R})$, $qp = pq$, so one might just as well try $qp = pq \to PQ$, and these are not the same. This is the infamous *operator ordering problem* of quantization. For quadratic observables the issue is not so serious since we can think of qp as

$$qp = \frac{1}{2}(qp + pq)$$

and take

$$qp \rightarrow \frac{1}{2}(QP + PQ),$$

which is symmetric in Q and P.

We will soon see that this apparent guess is actually forced upon us by the previous assignments, but one should not come to expect this sort of thing since we will see also that for polynomial observables of higher degree, the mathematics does not dictate a "correct" quantization, but only a number of alternatives from which to choose. You will establish the essential self-adjointness of Q^2, P^2 and $\frac{1}{2}(QP + PQ)$ shortly and we will then, as usual, use the same symbols for their unique self-adjoint extensions.

We must, of course, check that these choices preserve the appropriate bracket relations. This is just a little calculus, but worth going through. First note that for any $\psi \in S(\mathbb{R})$,

$$\left[\frac{1}{2}(QP + PQ)\psi \right](q) = \frac{1}{2} \left[-i\hbar q \frac{d}{dq}\psi(q) - i\hbar \left(q \frac{d}{dq}\psi(q) + \psi(q) \right) \right]$$

$$= -i\hbar \left[q \frac{d}{dq}\psi(q) + \frac{1}{2}\psi(q) \right],$$

so we have

$$\frac{1}{2}(QP + PQ) = -i\hbar \left(q \frac{d}{dq} + \frac{1}{2} \right).$$

Since $\{q^2, p^2\} = 4qp$, $\{q^2, qp\} = 2q^2$ and $\{p^2, qp\} = -2p^2$, we must show that, on $S(\mathbb{R})$,

$$-\frac{i}{\hbar}[Q^2, P^2]_- = 4 \left(\frac{1}{2}(QP + PQ) \right),$$

$$-\frac{i}{\hbar} \left[Q^2, \frac{1}{2}(QP + PQ) \right]_- = 2Q^2$$

and

$$-\frac{i}{\hbar} \left[P^2, \frac{1}{2}(QP + PQ) \right]_- = -2P^2.$$

Exercise 7.2.25. Check at least one of these.

Exercise 7.2.26. Show that Q^2, P^2 and $\frac{1}{2}(QP + PQ)$ are all essentially self-adjoint on $S(\mathbb{R})$.

At this point we should be fully prepared to quantize classical polynomial observables up to degree two and we will do some of this in the next two sections. It is not the case, of course, that every classical observable that one would like to quantize is a polynomial of degree two. For example, the Hamiltonian for the Higgs boson contains a quartic term. We should therefore say something about extending the Schrödinger realization of $\mathfrak{h}_3$ beyond the Jacobi algebra $\mathfrak{g}^J$. Needless to say, the operator ordering problems become increasingly severe as the degree increases, but it is not altogether clear that they cannot be resolved. Nevertheless, they cannot, as we will now see.

We would like to briefly describe an example of what the physicists would call a *no-go theorem*. This is basically a statement (sometimes a rigorous theorem) to the effect that something cannot be done. In the case at hand, the (rigorous) theorem goes back to Groenewold [Groe] and Van Hove [VH]. We will discuss only the simplest version of the result and will only sketch the idea of the proof (for more details see [Gotay], [GGT] or Section 5.4 of [Berndt]).

Dirac's proposed quantization scheme asks for a linear map R from $C^\infty(T^*\mathbb{R})$ to the self-adjoint operators on a separable, complex Hilbert space $\mathcal{H}$ that satisfies $R(1) = \mathrm{id}_\mathcal{H}$ and $R(\{f,g\}) = -\frac{i}{\hbar}[R(f), R(g)]_-$. Thus far we have managed to define R only on the $\mathfrak{g}^J$ Lie subalgebra of $C^\infty(T^*\mathbb{R})$ generated by $\{1, q, p, q^2, p^2, qp\}$ and we would now like to know if this map R extends to a larger Lie subalgebra of $C^\infty(T^*\mathbb{R})$. Alas, one has the following negative answer.

Theorem 7.2.8 (Groenewold–Van Hove theorem). *Let $\mathcal{O}$ be a Lie subalgebra of $C^\infty(T^*\mathbb{R})$ that properly contains the Lie subalgebra $P_2(q,p)$ generated by $\{1, q, p, q^2, p^2, qp\}$. Then there does not exist a linear map R from $\mathcal{O}$ to the self-adjoint operators on $L^2(\mathbb{R})$ preserving some fixed dense linear subspace $\mathcal{D} \supseteq S(\mathbb{R})$ and satisfying all of the following:*

$$R(1) = \mathrm{id}_{L^2(\mathbb{R})},$$

$$R(\{f,g\}) = -\frac{i}{\hbar}[R(f), R(g)]_- \quad \forall f,g \in \mathcal{O},$$

$$R(q) = Q \quad [\,(Q\psi)(q) = q\psi(q) \,\forall \psi \in S(\mathbb{R})\,],$$

$$R(p) = P \quad \left[(P\psi)(q) = -i\hbar\frac{d}{dq}\psi(q) \,\forall \psi \in S(\mathbb{R})\right],$$

$$R(q^2) = Q^2 \quad [\,(Q^2\psi)(q) = q^2\psi(q) \,\forall \psi \in S(\mathbb{R})\,],$$

$$R(p^2) = P^2 \quad \left[(P^2\psi)(q) = -\hbar^2\frac{d^2}{dq^2}\psi(q) \,\forall \psi \in S(\mathbb{R})\right].$$

We will describe a few of the ideas behind the relatively simple proof, but will refer to Section 5.4 of [Berndt] for most of the computational details. Let us denote by $P(q,p)$ the Lie subalgebra of $C^\infty(T^*\mathbb{R})$ consisting of all polynomials in q and p (this is actually a Poisson subalgebra of $C^\infty(T^*\mathbb{R})$). We will see that no such mapping R exists even on a Lie subalgebra of $P(q,p)$ larger than $P_2(q,p)$. The first step is to note that $P_2(q,p)$ is actually a *maximal* Lie subalgebra of $P(q,p)$ (Theorem 5.9 of [Berndt]) so that defining R on some subalgebra of $P(q,p)$ that properly contains $P_2(q,p)$ necessarily defines R on all of $P(q,p)$. Now, we will assume that such an R exists and derive a contradiction.

Note that the assumptions we have made about R do not include our earlier "guess" for $R(qp)$. The reason is that, as we will now show, this follows from the rest. Indeed, on $\mathcal{D}$ we have

$$R(\{q,p\}) = R(1) = \mathrm{id}_{L^2(\mathbb{R})} = -\frac{i}{\hbar}[R(q),R(p)]_- = -\frac{i}{\hbar}[Q,P]_-,$$

so

$$QP - PQ = i\hbar\,\mathrm{id}_{L^2(\mathbb{R})}.$$

Next, from $4qp = \{q^2, p^2\}$ we obtain

$$4R(qp) = -\frac{i}{\hbar}[R(q^2),R(p^2)]_- = -\frac{i}{\hbar}[Q^2,P^2]_- = -\frac{i}{\hbar}(Q^2P^2 - P^2Q^2)$$

$$= -\frac{i}{\hbar}(Q(QP)P - P(PQ)Q)$$

$$= -\frac{i}{\hbar}(Q(PQ + i\hbar\,\mathrm{id}_{L^2(\mathbb{R})})P - P(QP - i\hbar\,\mathrm{id}_{L^2(\mathbb{R})})Q)$$

$$= -\frac{i}{\hbar}(QPQP + i\hbar QP - PQPQ + i\hbar PQ)$$

$$= -\frac{i}{\hbar}((PQ + i\hbar\,\mathrm{id}_{L^2(\mathbb{R})})(PQ + i\hbar\,\mathrm{id}_{L^2(\mathbb{R})}) + i\hbar QP - PQPQ + i\hbar PQ)$$

$$= -\frac{i}{\hbar}(2i\hbar PQ - \hbar^2\mathrm{id}_{L^2(\mathbb{R})} + i\hbar QP + i\hbar PQ)$$

$$= -\frac{i}{\hbar}(2i\hbar PQ + i\hbar(QP - PQ) + i\hbar QP + i\hbar PQ) = 2(QP + PQ)$$

and therefore

$$R(qp) = \frac{1}{2}(QP + PQ),$$

as we claimed.

Similar, albeit somewhat more intricate computations show that $R(q^3) = Q^3$. This, together with $q^2p = \frac{1}{6}\{q^3, p^2\}$, gives

$$R(q^2p) = \frac{1}{2}(Q^2P + PQ^2).$$

In the same way, $R(p^3) = P^3$ together with $qp^2 = \frac{1}{6}\{q^2, p^3\}$ gives

$$R(qp^2) = \frac{1}{2}(QP^2 + P^2Q).$$

Now here is the point. Computing two simple Poisson brackets shows that

$$q^2p^2 = \frac{1}{9}\{q^3, p^3\} = \frac{1}{3}\{q^2p, p^2q\},$$

but, with the identities noted above, one obtains

$$R\left(\frac{1}{9}\{q^3, p^3\} \right) = -\frac{i}{9\hbar}[Q^3, P^3]_-$$

$$= -\frac{2}{3}\hbar^2 \mathrm{id}_{L^2(\mathbb{R})} - 2i\hbar QP + Q^2 P^2,$$

whereas

$$R\left(\frac{1}{3}\{q^2 p, p^2 q\} \right) = -\frac{i}{12\hbar}[(Q^2 P + PQ^2), (P^2 Q + QP^2)]$$

$$= -\frac{1}{3}\hbar^2 \mathrm{id}_{L^2(\mathbb{R})} - 2i\hbar QP + Q^2 P^2.$$

Since $R(\frac{1}{3}\{q^2 p, p^2 q\}) \neq R(\frac{1}{9}\{q^3, p^3\})$, we find that our assumptions imply that R must assign two different values to the classical observable $q^2 p^2$ and this is a contradiction. An R such as the one described in Theorem 7.2.8 cannot exist.

This argument does not, of course, imply that it is impossible to quantize quartic polynomials such as $q^2 p^2$ in a manner consistent with the Schrödinger quantization of $P_2(q, p)$. It says only that the assumptions we have made do not uniquely determine the quantization and it is up to us to use whatever additional information is available to make a choice or to adapt our requirements. Needless to say, this is a huge subject and one generally best left to the physicists (a relatively concise synopsis written with both physicists and mathematicians in mind is available in [TAE]). For more on rigorous no-go theorems in quantization one can consult [GGT].

In the next two sections we will apply the quantization map R from the Jacobi algebra $\mathfrak{g}^J \subseteq C^\infty(T^*\mathbb{R})$ to the self-adjoint operators on $L^2(\mathbb{R})$ to the two simplest examples of classical mechanical systems with quadratic Hamiltonians, that is, the free particle and the harmonic oscillator. Needless to say, these are only baby steps toward an understanding of canonical quantization, even in the case of quadratic Hamiltonians. For those who rightly insist on something with more physical substance we recommend [Jaffe] as a first step.

7.3 The free quantum particle

A classical free particle of mass m moving in one dimension has configuration space $\mathbb{R}$ and phase space $T^*(\mathbb{R}) = \mathbb{R}^2$ with coordinates q and (q, p), respectively. The classical Hamiltonian is $\frac{1}{2m} p^2$, which lives in the quadratic Lie subalgebra $\mathfrak{g}^J$ of $C^\infty(T^*\mathbb{R})$ generated by 1, q, p, q^2, p^2 and qp. The quantum phase space is taken to be $L^2(\mathbb{R})$ and the map R from $\mathfrak{g}^J$ to the self-adjoint operators on $L^2(\mathbb{R})$ constructed in Section 7.2 assigns to 1, q, p and $\frac{1}{2m} p^2$ the operators $I = \mathrm{id}_{L^2(\mathbb{R})}, Q, P$ and $H_0 = \frac{1}{2m} P^2$. On the Schwartz space $S(\mathbb{R})$ these are given by $(Q\psi)(q) = q\psi(q)$, $(P\psi)(q) = -i\hbar \frac{d}{dq}\psi(q)$ and $(H_0\psi)(q) = -\frac{\hbar^2}{2m}\frac{d^2}{dq^2}\psi(q)$, respectively, and they are all essentially self-adjoint on $S(\mathbb{R})$.

We recall also that the domain of H_0 is the set of all $\psi \in L^2(\mathbb{R})$ for which $\Delta\psi$ is in $L^2(\mathbb{R})$, where $\Delta\psi$ is the second derivative of ψ thought of as a tempered distribution (Example 5.2.8), and that the spectrum of H_0 is $\sigma(H_0) = [0, \infty)$ (Example 5.4.4). From the latter it follows that, just as in the classical case, the energy of a free quantum particle can assume any nonnegative real value, that is, the energy is *not* "quantized." According to Postulate QM4 of Chapter 6, an initial state $\psi(q, 0)$ of the free particle will evolve in time according to

$$\psi(q, t) = e^{-itH_0/\hbar}\psi(q, 0).$$

The evolution operator $e^{-itH_0/\hbar}$ is given by

$$\mathcal{F}^{-1} Q_{g(p)} \mathcal{F},$$

where $\mathcal{F}$ is the Fourier transform and $Q_{g(p)}$ is multiplication by

$$g(p) = e^{-i\hbar^2 tp^2/2m}$$

(see (5.40)). Just to see how all of this works we will compute a couple of examples. For these we will return to the initial states described in Example 6.2.3 and given by

$$\psi(q, 0) = \frac{1}{\pi^{1/4}} e^{-q^2/2} e^{i\alpha q},$$

where α is any real constant. We showed in Example 6.2.3 that for all of these Q has expected value $\langle Q \rangle_{\psi(q,0)} = 0$ and dispersion $\sigma_{\psi(q,0)}(Q) = \frac{1}{2}$. We will begin with $\alpha = 0$.

Example 7.3.1. We suppose that the initial state of our free particle is $\psi(q, 0) = \frac{1}{\pi^{1/4}} e^{-q^2/2}$ and compute $\psi(q, t) = (\mathcal{F}^{-1} Q_{g(p)} \mathcal{F})\psi(q, 0)$. The Fourier transform of $\psi(q, 0)$ was computed in Example G.2.1:

$$\mathcal{F}(\psi(q, 0)) = \frac{1}{\pi^{1/4}} e^{-p^2/2}.$$

Thus,

$$(Q_{g(p)} \mathcal{F})\psi(q, 0) = \frac{1}{\pi^{1/4}} e^{-\frac{1+i(\hbar^2/m)t}{2}p^2}.$$

For the inverse Fourier transform of this we recall (Example G.2.2) that if $a \in \mathbb{C}$ and $\operatorname{Re}(a) > 0$, then

$$\mathcal{F}(e^{-aq^2/2}) = \frac{1}{\sqrt{a}} e^{-p^2/2a},$$

where the square root has a branch cut along the negative real axis. Consequently,

$$\mathcal{F}^{-1}(e^{-p^2/2a}) = \sqrt{a}\, e^{-aq^2/2}.$$

In the case at hand,

$$a = \frac{1}{1 + i(\hbar^2/m)t},$$

so

$$\psi(q,t) = (\mathcal{F}^{-1} Q_{g(p)} \mathcal{F})\psi(q,0) = \frac{1}{\pi^{1/4}} \frac{1}{\sqrt{1 + i(\hbar^2/m)t}} e^{-\frac{1}{2}(\frac{1}{1+i(\hbar^2/m)t})q^2}.$$

Now we will rewrite this a bit as follows: $\frac{1}{1+i(\hbar^2/m)t} = A + Bi$, where $A = \frac{1}{1+(\hbar^4/m^2)t^2}$ and $B = -\frac{(\hbar^2/m)t}{1+(\hbar^4/m^2)t^2}$. Thus,

$$e^{-\frac{1}{2}(\frac{1}{1+i(\hbar^2/m)t})q^2} = e^{-\frac{1}{2}Bq^2i} e^{-\frac{1}{2}\frac{q^2}{1+(\hbar^4/m^2)t^2}}$$

and

$$\psi(q,t) = \frac{1}{\pi^{1/4}} \frac{1}{\sqrt{1 + i(\hbar^2/m)t}} e^{-\frac{1}{2}Bq^2i} e^{-\frac{1}{2}\frac{q^2}{1+(\hbar^4/m^2)t^2}}.$$

Now, $\sqrt{1 + i(\hbar^2/m)t}$ has modulus $\sqrt[4]{1 + (\hbar^4/m^2)t^2}$. Combine its phase factor with $e^{-\frac{1}{2}Bq^2i}$ and write the result as $e^{i\phi(q,t)}$, where $\phi(q,t)$ is a real-valued function. Then

$$\psi(q,t) = \frac{1}{\sqrt[4]{\pi}} e^{i\phi(q,t)} \frac{1}{\sqrt[4]{1 + (\hbar^4/m^2)t^2}} e^{-\frac{1}{2}\frac{q^2}{1+(\hbar^4/m^2)t^2}}.$$

Consequently,

$$|\psi(q,t)| = \frac{1}{\sqrt[4]{\pi}} \frac{1}{\sqrt[4]{1 + (\hbar^4/m^2)t^2}} e^{-\frac{1}{2}\frac{q^2}{1+(\hbar^4/m^2)t^2}}.$$

We see then that the evolved wave function still peaks at $q = 0$ for every t, but becomes wider and flatter as $t \to \infty$, so that the probability of detecting the particle away from $q = 0$ increases.

Things are a bit different if $\psi(q,0) = \frac{1}{\pi^{1/4}} e^{-q^2/2} e^{i\alpha q}$ with $\alpha \neq 0$.

Exercise 7.3.1. Show that if $\psi(q,0) = \frac{1}{\pi^{1/4}} e^{-q^2/2} e^{i\alpha q}$ with $\alpha > 0$, then

$$|\psi(q,t)| = \frac{1}{\sqrt[4]{\pi}} \frac{1}{\sqrt[4]{1 + (\hbar^4/m^2)t^2}} e^{-\frac{1}{2}\frac{(q-(\alpha\hbar^2/m)t)^2}{1+(\hbar^4/m^2)t^2}}.$$

In this case the initial wave function also peaks at $q = 0$, but the evolving wave functions not only widen and flatten as $t \to \infty$, they also peak at a point that moves to the right with speed $\alpha\hbar^2/m$. The point at which it is most likely to detect the particle is moving along the q-axis.

These last examples are rather atypical, of course, since the initial wave function was chosen in such a way that we could perform all of the required computations explicitly. Next we will look at much more general initial states and try to represent the time evolution in terms of an integral kernel as we did for H_B in (6.13). The results we derive will be critical for understanding the Feynman path integral approach to quantization in Chapter 8.

Example 7.3.2. We consider again the free particle Hamiltonian $H_0 : \mathcal{D}(H_0) \to L^2(\mathbb{R})$ as in Example 5.4.4. It will be useful to start from scratch and look at the Cauchy problem for the corresponding Schrödinger equation which, on $\mathbb{R} \times (0, \infty)$, takes the form

$$i\frac{\partial \psi(q,t)}{\partial t} = -\frac{\hbar}{2m}\frac{\partial^2 \psi(q,t)}{\partial q^2}, \quad (q,t) \in \mathbb{R} \times (0, \infty),$$

$$\lim_{t \to 0^+} \psi(q,t) = \psi_0(q), \quad q \in \mathbb{R}.$$

(7.26)

We begin with a few general observations. First, one cannot help but notice the similarity between the Schrödinger equation in (7.26) and the heat equation we discussed in Example G.3.1. Indeed, if one takes $\alpha = \hbar/2m$ and *formally* makes the change of variable $t \to -it$, then the Schrödinger equation becomes the heat equation (physicists would refer to this formal change of variable as *analytic continuation from physical time to imaginary time*). In Example G.3.1 we were able to express the solution to the heat equation in terms of an integral kernel and this gives us reason to hope. Indeed, throwing caution to the winds one might even conjecture that the Schrödinger kernel for H_0 should be what you get from the heat kernel with $\alpha = \hbar/2m$ by replacing t by it, that is,

$$\sqrt{\frac{m}{2\pi\hbar t i}}\, e^{\, mi(q-x)^2/2\hbar t}.$$

Remarkably enough, this is precisely the kernel we will eventually arrive at (with $\sqrt{i} = e^{\pi i/4} = \frac{1}{\sqrt{2}}(1+i)$), but replacing the formal arguments with rigorous ones will require a little work.

Before getting started on this, however, it will be instructive to digress one more time and consider some simple, but "unphysical" solutions of the Schrödinger equation for H_0. These are analogues of the plane electromagnetic waves we encountered in our discussion of Maxwell's equations in Section 4.2. Specifically, we will consider functions of the form

$$\psi(q,t) = e^{\frac{i}{\hbar}(pq-\omega t)},$$

where ω is a positive constant and p should be regarded as a parameter, different choices giving different functions. Computing a few derivatives gives

$$\frac{\partial \psi}{\partial t} = -\frac{i}{\hbar}\omega\, e^{\frac{i}{\hbar}(pq-\omega t)} = -\frac{i}{\hbar}\omega\, \psi(q,t),$$

$$\frac{\partial \psi}{\partial q} = \frac{i}{\hbar} p \, e^{\frac{i}{\hbar}(pq-\omega t)} = \frac{i}{\hbar} p \psi(q,t) \tag{7.27}$$

and

$$\frac{\partial^2 \psi}{\partial q^2} = -\frac{p^2}{\hbar^2} e^{\frac{i}{\hbar}(pq-\omega t)} = -\frac{p^2}{\hbar^2} \psi(q,t).$$

Substituting into the Schrödinger equation in (7.26) gives

$$\frac{1}{\hbar} \omega \psi(q,t) = \frac{1}{2m\hbar} p^2 \psi(q,t),$$

so $\psi(q,t)$ is a (nontrivial) solution if and only if

$$\omega = \frac{1}{2m} p^2. \tag{7.28}$$

The condition (7.28) is called a *dispersion relation* and if we use it to substitute for ω in $\psi(q,t)$ we obtain the solutions

$$\psi(q,t) = e^{\frac{i}{\hbar}(pq-\frac{t}{2m}p^2)}. \tag{7.29}$$

Clearly, for each fixed t, these functions fail miserably to be in $L^2(\mathbb{R})$ since

$$\int_{\mathbb{R}} |\psi(q,t)|^2 \, dq = \int_{\mathbb{R}} 1 \, dq = \infty.$$

Nevertheless, we will find them to be useful and informative. For instance, it often occurs that an honest state (unit vector in $L^2(\mathbb{R})$) is well approximated over a restricted region of space and time by such a plane wave. Moreover, we will see quite soon that any solution of the Schrödinger equation in (7.26) can be regarded as a (continuous) superposition of such plane waves.

Exercise 7.3.2. Show that for $t > 0$,

$$\int_{\mathbb{R}} e^{\frac{i}{\hbar}(pq-\frac{t}{2m}p^2)} \, dp = (2\pi\hbar) \sqrt{\frac{m}{2\pi t \hbar i}} \, e^{miq^2/2\hbar t}, \tag{7.30}$$

where $\sqrt{i} = e^{\pi i/4} = \frac{1}{\sqrt{2}}(1+i)$. *Hint*: Complete the square in the exponent and use the Gaussian integral

$$\int_{\mathbb{R}} e^{iax^2} \, dx = e^{\text{sgn}(a)\pi i/4} \sqrt{\frac{\pi}{|a|}}, \tag{7.31}$$

where a is a nonzero real number and $\text{sgn}(a)$ is its sign (see (A.4) in Appendix A).

We begin with a few general remarks. It is not uncommon in quantum mechanics to encounter functions which one would like to regard as states, eigenfunctions, etc., but cannot because they do not live in the appropriate Hilbert space $\mathcal{H}$. The plane waves $\psi(q,t)$ defined by (7.29) are such functions. Being solutions to the Schrödinger equation, they look like they should be states, but they are not in $L^2(\mathbb{R})$, so they are not. Physicists would refer to $\psi(q,t)$ as a *nonnormalizable state*, even though it is not really a state at all. Similarly, (7.27) shows that if only they were elements of $L^2(\mathbb{R})$, these plane waves would be eigenfunctions of the momentum operator $P = -i\hbar\frac{\partial}{\partial q}$ with eigenvalues p. These are often referred to as *generalized eigenfunctions* of the momentum operator on $\mathbb{R}$. There are various ways of stepping outside of the Hilbert space and incorporating such functions into a rigorous formalism, one of which we have already gotten a hint of in Appendix G. Recall that if $\mathcal{S}(\mathbb{R})$ is the Schwartz space and $\mathcal{S}'(\mathbb{R})$ is its dual space of tempered distributions, then

$$\mathcal{S}(\mathbb{R}) \subseteq L^2(\mathbb{R}) \subseteq \mathcal{S}'(\mathbb{R}).$$

Note that $\mathcal{S}(\mathbb{R})$ is not a Hilbert space and its Fréchet space topology is strictly finer than the topology it would inherit as a subspace of $L^2(\mathbb{R})$. It is, however, a dense subset of $L^2(\mathbb{R})$ in the norm topology of $L^2(\mathbb{R})$. In turn, $L^2(\mathbb{R})$ is dense in $\mathcal{S}'(\mathbb{R})$. These circumstances qualify $\mathcal{S}(\mathbb{R}) \subseteq L^2(\mathbb{R}) \subseteq \mathcal{S}'(\mathbb{R})$ as an example of what is called a *rigged Hilbert space* or *Gelfand triple*. Note that although the plane waves $\psi(q,t)$ are not in $L^2(\mathbb{R})$, they are certainly in $L^1_{loc}(\mathbb{R})$ and can therefore be regarded as tempered distributions, that is, as elements of $\mathcal{S}'(\mathbb{R})$. Nonnormalizable states and generalized eigenfunctions can be thought of as distributions, living not in $L^2(\mathbb{R})$, but in the larger space $\mathcal{S}'(\mathbb{R})$.

Exercise 7.3.3. The position operator $Q : \mathcal{D}(Q) \rightarrow L^2(\mathbb{R})$ has no eigenfunctions in $L^2(\mathbb{R})$ ($Q\psi = \lambda\psi \Rightarrow (q-\lambda)\psi(q) = 0$ almost everywhere $\Rightarrow \psi = 0 \in L^2(\mathbb{R})$). Explain the sense in which the Dirac delta $\delta_a \in \mathcal{S}'(\mathbb{R})$ is a generalized eigenfunction for Q. *Hint*: Use the Fourier transform.

Now, let us get back to the business of solving (7.26). For the time being we will assume that ψ_0 is smooth with compact support and will look for smooth solutions to the Schrödinger equation on $\mathbb{R} \times (0,\infty)$. Our procedure will be to apply the (spatial) Fourier transform $\mathcal{F}$, solve for $\hat{\psi}(p,t)$ and then apply $\mathcal{F}^{-1}$ to get $\psi(q,t)$. In this way one can only find solutions that actually have Fourier transforms, of course, and we cannot know in advance that there are any. Furthermore, we will assume that there are solutions sufficiently regular that we can differentiate with respect to t under the integral sign to show that

$$\mathcal{F}\left(\frac{\partial\psi}{\partial t}\right) = \frac{\partial\hat{\psi}}{\partial t}.$$

Whether or not these assumptions are justified will be determined by whether or not we find solutions that satisfy them. Applying $\mathcal{F}$ to (7.26) gives

$$\frac{\partial \hat{\psi}}{\partial t} + \frac{i\hbar}{2m} p^2 \hat{\psi} = 0$$

and

$$\lim_{t \to 0^+} \hat{\psi}(p, t) = \hat{\psi}_0(p).$$

The solution to this simple first order, linear initial value problem is

$$\hat{\psi}(p, t) = \hat{\psi}_0(p) e^{-i(\hbar/2m)tp^2}.$$

Now we apply $\mathcal{F}^{-1}$ to obtain

$$\psi(q, t) = \mathcal{F}^{-1}(\hat{\psi}_0(p) e^{-i(\hbar/2m)tp^2}) = \frac{1}{\sqrt{2\pi}} \int_{\mathbb{R}} e^{i(pq - (\hbar/2m)tp^2)} \hat{\psi}_0(p) \, dp. \qquad (7.32)$$

The inverse Fourier transform of $\hat{\psi}_0(p) e^{-i(\hbar/2m)tp^2}$ is given by the integral in (7.32) because ψ_0 is assumed to be smooth with compact support (in particular, Schwartz) and so $\hat{\psi}_0$ is also in $\mathcal{S}(\mathbb{R})$. We should also point out that while the function $\psi(q, t)$ defined by (7.32) clearly satisfies the Schrödinger equation in (7.26), the boundary condition in (7.26) is not at all clear, despite the fact that the corresponding limit for the Fourier transforms is clearly satisfied. We will have more to say about this shortly.

In (7.32) we are asked to compute the inverse Fourier transform of a product of two functions of p and this would lead us to expect $\psi(q, t)$ to be expressed as a convolution (see (G.11)). Naturally, the inverse transform of $\hat{\psi}_0(p)$ is just $\psi_0(q)$, so we need only worry about the inverse transform of $e^{-i(\hbar/2m)tp^2}$. This is something of a problem, however, since $e^{-i(\hbar/2m)tp^2}$ is only in $L^1_{\text{loc}}(\mathbb{R})$, so its inverse Fourier transform exists only as a distribution. Example G.2.2, where we showed that $\mathcal{F}^{-1}(e^{-p^2/2\alpha}) = \sqrt{\alpha}\, e^{-\alpha p^2/2}$ when $\text{Re}(\alpha) > 0$, does not apply directly since, in our present case, $\alpha = -mi/\hbar t$ has real part zero. For this reason, the analysis will be a bit more delicate. For the record, what we intend to prove is that, with $\sqrt{i} = e^{\pi i/4}$,

$$\psi(q, t) = \sqrt{\frac{m}{2\pi\hbar t i}} \int_{\mathbb{R}} e^{mi(q-x)^2/2\hbar t} \, \psi_0(x) \, dx, \qquad (7.33)$$

which we can write as

$$\psi(q, t) = \int_{\mathbb{R}} K(q, t; x, 0) \psi(x, 0) \, dx, \qquad (7.34)$$

where

$$K(q, t; x, 0) = \sqrt{\frac{m}{2\pi\hbar ti}}\, e^{mi(q-x)^2/2\hbar t}. \tag{7.35}$$

Taking the initial condition at $t = t_0$ rather than $t = 0$ one would obtain instead

$$\psi(q, t) = \int_{\mathbb{R}} K(q, t; x, t_0)\psi(x, t_0)\, dx, \tag{7.36}$$

where

$$K(q, t; x, t_0) = \sqrt{\frac{m}{2\pi\hbar(t - t_0)i}}\, e^{mi(q-x)^2/2\hbar(t-t_0)}, \tag{7.37}$$

which is the *propagator*, or *integral kernel for the free particle Hamiltonian H_0*, or simply the *Schrödinger kernel for H_0*. Physicists interpret $|K(q, t; x, t_0)|^2$ as the conditional probability of finding the particle at $q \in \mathbb{R}$ at time t provided it was detected at the point $x \in \mathbb{R}$ at time t_0; $K(q, t; x, t_0)$ itself is interpreted as the probability amplitude for getting from $x \in \mathbb{R}$ at time t_0 to q at time t.

Now we proceed with the proof of (7.33). This will rely rather heavily on material from Appendices A, C and G. As we pointed out above, the result of Example G.2.2 does not apply to $e^{-i(\hbar/2m)tp^2}$. However, see the following exercise.

Exercise 7.3.4. Show that for any $\delta > 0$ and any $t > 0$,

$$\mathcal{F}^{-1}\left(e^{-(\delta+i)(\hbar/2m)tp^2}\right) = \sqrt{\frac{m}{\hbar t(\delta + i)}}\, e^{-mq^2/2\hbar t(\delta+i)},$$

where $\sqrt{}$ refers to the branch of the square root with branch cut along the negative real axis. Conclude that

$$\mathcal{F}^{-1}\left(\hat{\psi}_0(p)e^{-(\delta+i)(\hbar/2m)tp^2}\right) = \sqrt{\frac{m}{2\pi\hbar t(\delta + i)}} \int_{\mathbb{R}} e^{-m(q-x)^2/2\hbar t(\delta+i)}\psi_0(x)\, dx.$$

Exercise 7.3.5. Combine (7.32) and the previous exercise to show that (7.33) will follow if we can prove that

$$\int_{\mathbb{R}} e^{i(pq-(\hbar/2m)tp^2)}\hat{\psi}_0(p)\, dp = \lim_{\delta\to 0^+} \int_{\mathbb{R}} e^{ipq}e^{-(\delta+i)(\hbar/2m)tp^2}\hat{\psi}_0(p)\, dp. \tag{7.38}$$

For the proof of (7.38) we proceed as follows. Fix $t > 0$. We must show that

$$I = \left|\int_{\mathbb{R}} e^{ipq}e^{-(\hbar/2m)tp^2 i}\left(e^{-(\hbar/2m)tp^2\delta} - 1\right)\hat{\psi}_0(p)\, dp\right| \to 0$$

as $\delta \to 0^+$. Let $\epsilon > 0$ be given. Note that $\psi_0 \in S(\mathbb{R}) \Rightarrow \hat{\psi}_0 \in S(\mathbb{R})$ so, in particular, $\| \hat{\psi}_0 \|_{L^1} = \int_{\mathbb{R}} |\hat{\psi}_0(p)| \, dp < \infty$. Thus, for some sufficiently large $R > 0$,

$$
\begin{aligned}
I \le \int_{\mathbb{R}} \left| e^{-(\hbar/2m)tp^2\delta} - 1 \right| |\hat{\psi}_0(p)| \, dp = &\int_{[-R,R]} \left| e^{-(\hbar/2m)tp^2\delta} - 1 \right| |\hat{\psi}_0(p)| \, dp \\
&+ \int_{|p|\ge R} \left| e^{-(\hbar/2m)tp^2\delta} - 1 \right| |\hat{\psi}_0(p)| \, dp \\
\le &\max_{p\in[-R,R]} \left| e^{-(\hbar/2m)tp^2\delta} - 1 \right| \int_{[-R,R]} |\hat{\psi}_0(p)| \, dp \\
&+ \int_{|p|\ge R} |\hat{\psi}_0(p)| \, dp \\
< &\max_{p\in[-R,R]} \left| e^{-(\hbar/2m)tp^2\delta} - 1 \right| \|\hat{\psi}_0\|_{L^1} + \frac{\epsilon}{2}.
\end{aligned}
$$

We fix such an R. Note that all of this is independent of the choice of δ. Now we conclude the proof by showing that

$$
\max_{p\in[-R,R]} \left| e^{-(\hbar/2m)tp^2\delta} - 1 \right|
$$

can be made arbitrarily small by making $\delta > 0$ sufficiently small. But, for any $p \in [-R, R]$ and any $t > 0$,

$$
\begin{aligned}
0 \le tp^2\delta \le tR^2\delta \quad &\Rightarrow \quad -tR^2\delta \le -tp^2\delta \le 0 \\
&\Rightarrow \quad -(\hbar/2m)tR^2\delta \le -(\hbar/2m)tp^2\delta \le 0 \\
&\Rightarrow \quad e^{-(\hbar/2m)tR^2\delta} \le e^{-(\hbar/2m)tp^2\delta} \le 1 \\
&\Rightarrow \quad e^{-(\hbar/2m)tR^2\delta} - 1 \le e^{-(\hbar/2m)tp^2\delta} - 1 \le 0 \\
&\Rightarrow \quad \left| e^{-(\hbar/2m)tp^2\delta} - 1 \right| = 1 - e^{-(\hbar/2m)tp^2\delta} \le 1 - e^{-(\hbar/2m)tR^2\delta} \to 0
\end{aligned}
$$

as $\delta \to 0^+$. This completes the proof of (7.38) and therefore also of (7.33).

At this point we have shown that

$$
\psi(q,t) = \sqrt{\frac{m}{2\pi\hbar t i}} \int_{\mathbb{R}} e^{mi(q-x)^2/2\hbar t} \, \psi_0(x) \, dx
$$

is a solution to the Schrödinger equation

$$
i \frac{\partial \psi(q,t)}{\partial t} = -\frac{\hbar}{2m} \frac{\partial^2 \psi(q,t)}{\partial q^2}
$$

in (7.26), but there remains the issue of the initial condition $\lim_{t\to 0^+} \psi(q,t) = \psi_0(q)$, that is,

$$
\lim_{t\to 0^+} \sqrt{\frac{m}{2\pi\hbar t i}} \int_{\mathbb{R}} e^{mi(q-x)^2/2\hbar t} \, \psi_0(x) \, dx = \psi_0(q). \tag{7.39}
$$

The corresponding result for the heat equation (Exercise G.3.4) was relatively straight-forward, but the i in the exponent here introduces rapid oscillations in the integrand (away from $x = q$) and substantial complications in the analysis of the integral. We will prove (7.39) by applying an important technique for the study of such oscillatory integrals called *stationary phase approximation*. We have included a proof of this in Appendix C, but here we will simply state what we need at the moment. This special case applies to integrals of the form

$$\int_{\mathbb{R}} e^{iTf(x)} g(x)\, dx,$$

where f is a smooth, real-valued function on $\mathbb{R}$ with exactly one nondegenerate critical point $x_0 \in \mathbb{R}$ ($f'(x_0) = 0$ and $f''(x_0) \neq 0$), T is a positive real number and g is smooth with compact support. Then the stationary phase approximation of $\int_{\mathbb{R}} e^{iTf(x)} g(x)dx$ is given by

$$\int_{\mathbb{R}} e^{iTf(x)} g(x)\, dx = \left(\frac{2\pi}{T}\right)^{1/2} e^{\,\mathrm{sgn}(f''(x_0))\,\pi i/4}\, \frac{e^{iTf(x_0)}}{\sqrt{|f''(x_0)|}}\, g(x_0) + O\left(\frac{1}{T^{3/2}}\right) \tag{7.40}$$

as $T \to \infty$. Recall that this means that there exists a constant $M > 0$ and a $T_0 > 0$ such that for all $T \geq T_0$,

$$\left| \int_{\mathbb{R}} e^{iTf(x)} g(x)\, dx - \left(\frac{2\pi}{T}\right)^{1/2} e^{\,\mathrm{sgn}(f''(x_0))\,\pi i/4}\, \frac{e^{iTf(x_0)}}{\sqrt{|f''(x_0)|}}\, g(x_0) \right| \leq M\left(\frac{1}{T^{3/2}}\right).$$

To apply this to the integral $\int_{\mathbb{R}} e^{mi(q-x)^2/2\hbar t} \psi_0(x)\, dx$ we take $T = \frac{1}{t}$, $g(x) = \psi_0(x)$ and $f(x) = m(q-x)^2/2\hbar$. Then $t \to 0^+ \Rightarrow T \to \infty$ and f has exactly one critical point at q, which is nondegenerate because $f''(q) = m/\hbar$. Substituting all of this into (7.40) gives

$$\int_{\mathbb{R}} e^{mi(q-x)^2/2\hbar t} \psi_0(x)\, dx = \sqrt{\frac{2\pi\hbar ti}{m}}\, \psi_0(q) + O(t^{3/2})$$

as $t \to 0^+$, where $\sqrt{i} = e^{\pi i/4}$. Consequently,

$$\sqrt{\frac{m}{2\pi\hbar ti}} \int_{\mathbb{R}} e^{mi(q-x)^2/2\hbar t} \psi_0(x)\, dx = \psi_0(q) + O(t)$$

and this clearly approaches $\psi_0(q)$ as $t \to 0^+$. This completes the proof of (7.39).

Let us summarize what we have done, incorporating the result of Exercise 7.3.2 as we go. If the initial value of the wave function, which we will now denote $\psi(q,0)$, is smooth with compact support, then the time evolution is described by

$$\psi(q,t) = e^{-itH_0/\hbar}(\psi(q,0)) = \int_{\mathbb{R}} K(q,t;x,0)\psi(x,0)\, dx, \tag{7.41}$$

where the Schrödinger kernel is given by

$$K(q, t; x, 0) = \sqrt{\frac{m}{2\pi\hbar t i}} \, e^{mi(q-x)^2/2\hbar t} = \frac{1}{2\pi\hbar} \int_{\mathbb{R}} e^{\frac{i}{\hbar}(p(q-x)-\frac{t}{2m}p^2)} \, dp. \tag{7.42}$$

If the initial condition is given at $t = t_0$ rather than $t = 0$, one has instead

$$\psi(q, t) = e^{-i(t-t_0)H_0/\hbar}(\psi(q, t_0)) = \int_{\mathbb{R}} K(q, t; x, t_0)\psi(x, t_0) \, dx, \tag{7.43}$$

where the Schrödinger kernel is

$$K(q, t; x, t_0) = \sqrt{\frac{m}{2\pi\hbar(t-t_0)i}} \, e^{mi(q-x)^2/2\hbar(t-t_0)} = \frac{1}{2\pi\hbar} \int_{\mathbb{R}} e^{\frac{i}{\hbar}(p(q-x)-\frac{t-t_0}{2m}p^2)} \, dp. \tag{7.44}$$

Thus far we have assumed that the initial data $\psi_0(q) = \psi(q, 0)$ is smooth with compact support, but one can show that equalities (7.41) and (7.43) remain valid as long as ψ_0 is in $L^1(\mathbb{R}) \cap L^2(\mathbb{R})$. For an arbitrary element ψ_0 in $L^2(\mathbb{R})$ these equalities also remain valid provided the integral is interpreted as an integral in the mean, that is, as the following $L^2(\mathbb{R})$-limit:

$$\lim_{M \to \infty} \int_{[-M,M]} K(q, t; x, t_0)\psi(x, t_0) \, dx.$$

Since we really want to "flow" arbitrary L^2 initial states and since this issue will arise again at several points in the sequel, we will provide the proof (for simplicity we will take $t_0 = 0$). We begin by expanding $(q - x)^2 = q^2 + x^2 - 2qx$ to write

$$\sqrt{\frac{m}{2\pi\hbar t i}} \, e^{mi(q-x)^2/2\hbar t}\psi_0(x) = \sqrt{\frac{m}{\hbar t i}} \, e^{miq^2/2\hbar t} \frac{1}{\sqrt{2\pi}} \left[e^{mix^2/2\hbar t}\psi_0(x) \right] e^{-i(mq/\hbar t)x}. \tag{7.45}$$

The function $e^{mix^2/2\hbar t}\psi_0(x)$ is in $L^2(\mathbb{R})$ and so it has a Fourier transform

$$\mathcal{F}(e^{mix^2/2\hbar t}\psi_0(x))$$

that is also in $L^2(\mathbb{R})$. If χ_M denotes the characteristic function of the interval $[-M, M]$, then $e^{mix^2/2\hbar t}\psi_0(x)\chi_M(x)$ is in $L^1(\mathbb{R}) \cap L^2(\mathbb{R})$, so its Fourier transform is

$$\frac{1}{\sqrt{2\pi}} \int_{[-M,M]} \left[e^{mix^2/2\hbar t}\psi_0(x) \right] e^{-iqx} \, dx.$$

Because $e^{mix^2/2\hbar t}\psi_0(x)\chi_M(x)$ converge in $L^2(\mathbb{R})$ to $e^{mix^2/2\hbar t}\psi_0(x)$ as $M \to \infty$, their Fourier transforms converge in $L^2(\mathbb{R})$ to $\mathcal{F}(e^{mix^2/2\hbar t}\psi_0(x))$ as $M \to \infty$, that is,

$$\lim_{M \to \infty} \frac{1}{\sqrt{2\pi}} \int_{[-M,M]} \left[e^{mix^2/2\hbar t}\psi_0(x) \right] e^{-iqx} \, dx = \mathcal{F}(e^{mix^2/2\hbar t}\psi_0(x))(q)$$

in $L^2(\mathbb{R})$. Integrating (7.45) over $[-M, M]$ and taking the L^2-limit as $M \to \infty$ we therefore obtain

$$\lim_{M \to \infty} \int_{[-M,M]} \sqrt{\frac{m}{2\pi\hbar ti}}\, e^{mi(q-x)^2/2\hbar t}\psi_0(x)\, dx = \sqrt{\frac{m}{\hbar ti}}\, e^{miq^2/2\hbar t}\, \mathcal{F}(e^{mix^2/2\hbar t}\psi_0(x))(mq/\hbar t).$$

In particular, this limit (in other words, this integral in the mean) exists. Now, for each $M > 0$ we let $\psi_M = \psi_0\chi_M$. Then, as $M \to \infty$, $\psi_M \to \psi_0$ in $L^2(\mathbb{R})$ and therefore $e^{-itH_0/\hbar}\psi_M \to e^{-itH_0/\hbar}\psi_0$ in $L^2(\mathbb{R})$. Thus,

$$(e^{-itH_0/\hbar}\psi_0)(q) = \lim_{M \to \infty} (e^{-itH_0/\hbar}\psi_M)(q)$$

$$= \lim_{M \to \infty} \int_{[-M,M]} \sqrt{\frac{m}{2\pi\hbar ti}}\, e^{mi(q-x)^2/2\hbar t}\psi_M(x)\, dx$$

$$= \lim_{M \to \infty} \int_{[-M,M]} \sqrt{\frac{m}{2\pi\hbar ti}}\, e^{mi(q-x)^2/2\hbar t}\psi_0(x)\, dx$$

$$= \int_{\mathbb{R}} \sqrt{\frac{m}{2\pi\hbar ti}}\, e^{mi(q-x)^2/2\hbar t}\psi_0(x)\, dx,$$

where the limits are all in $L^2(\mathbb{R})$ and the last integral is to be interpreted as an integral in the mean. Let us summarize all of this in the form of a theorem.

Theorem 7.3.1. *Let $H_0 = -\frac{\hbar^2}{2m}\Delta$ be the free particle Hamiltonian on $L^2(\mathbb{R})$. Then H_0 is self-adjoint on $\mathcal{D}(H_0) = \{\psi \in L^2(\mathbb{R}) : \Delta\psi \in L^2(\mathbb{R})\}$, where Δ is the distributional Laplacian. For any $\psi_0 \in L^2(\mathbb{R})$*

$$(e^{-itH_0/\hbar}\psi_0)(q) = \int_{\mathbb{R}} \sqrt{\frac{m}{2\pi\hbar ti}}\, e^{mi(q-x)^2/2\hbar t}\psi_0(x)\, dx,$$

where, if $\psi_0 \in L^2(\mathbb{R}) - L^1(\mathbb{R})$, the integral must be regarded as an integral in the mean, that is,

$$\int_{\mathbb{R}} \sqrt{\frac{m}{2\pi\hbar ti}}\, e^{mi(q-x)^2/2\hbar t}\psi_0(x)\, dx = \lim_{M \to \infty} \int_{[-M,M]} \sqrt{\frac{m}{2\pi\hbar ti}}\, e^{mi(q-x)^2/2\hbar t}\psi_0(x)\, dx,$$

where the limit is in $L^2(\mathbb{R})$. If $\psi_0(q) = \psi(q, t_0)$ is the state of the free particle at $t = t_0$, then its state at time t is

$$\psi(q, t) = e^{-i(t-t_0)H_0/\hbar}(\psi(q, t_0)) = \int_{\mathbb{R}} K(q, t; x, t_0)\psi(x, t_0)\, dx,$$

where

$$K(q, t; x, t_0) = \sqrt{\frac{m}{2\pi\hbar(t-t_0)i}}\, e^{mi(q-x)^2/2\hbar(t-t_0)} = \frac{1}{2\pi\hbar} \int_{\mathbb{R}} e^{\frac{i}{\hbar}(p(q-x)-\frac{p^2}{2m}(t-t_0))}\, dp.$$

On occasion it will be convenient to alter the notation we have used thus far by replacing (x, t_0) and (q, t) by (q_a, t_a) and (q_b, t_b), respectively, so that

$$K(q_b, t_b; q_a, t_a) = \sqrt{\frac{m}{2\pi\hbar(t_b - t_a)i}}\, e^{mi(q_b - q_a)^2/2\hbar(t_b - t_a)}$$

$$= \frac{1}{2\pi\hbar} \int_{\mathbb{R}} e^{\frac{i}{\hbar}(p(q_b - q_a) - \frac{p^2}{2m}(t_b - t_a))}\, dp. \tag{7.46}$$

We call your attention to the term

$$p(q_b - q_a) - \frac{p^2}{2m}(t_b - t_a),$$

in the exponent. We have seen this and variants of it before and we will see them again so we should take a moment to pay a bit more attention. Our first encounter with such a thing in this section was the expression (7.29) for the (nonnormalizable) plane wave solutions to the Schrödinger equation. These are parametrized by the real number p, so their integral over $-\infty < p < \infty$ can be thought of as a continuous superposition of plane waves of varying frequency $\omega = \frac{1}{2m}p^2$. Next we recall that the classical Lagrangian for a free particle of mass m moving in one dimension is $L(q, \dot{q}) = \frac{1}{2}m\dot{q}^2$. The corresponding canonical momentum is $p = \partial L/\partial \dot{q} = m\dot{q}$ and the Hamiltonian is $H_0(q, p) = \frac{1}{2m}p^2$. Consequently,

$$p\dot{q} - \frac{p^2}{2m} = (m\dot{q})\dot{q} - \frac{1}{2m}(m\dot{q})^2 = \frac{1}{2}m\dot{q}^2,$$

so we can think of $p\dot{q} - \frac{p^2}{2m}$ as simply another way of writing the classical free particle Lagrangian. As a result, for any path α joining $\alpha(t_a) = q_a$ and $\alpha(t_b) = q_b$ in $\mathbb{R}$, the classical action is given by

$$S(\alpha) = \int_{t_a}^{t_b} \left(p\dot{q} - \frac{p^2}{2m} \right) dt.$$

Note now that if p is constant on α (as it is on the classical trajectory in Exercise 2.2.3), we can perform the integrations and obtain

$$S(\alpha) = p(q_b - q_a) - \frac{p^2}{2m}(t_b - t_a),$$

as we did in Exercise 2.2.3. We will see in Chapter 8 that the appearance of the classical action in the propagator is a key ingredient in Feynman's path integral approach to quantization.

It is traditional, and will be convenient, to rephrase some of this by defining $K_0 : \mathbb{R} \times \mathbb{R} \times (0, \infty) \to \mathbb{C}$ by

$$K_0(q, x, t) = K(q, t; x, 0) = \sqrt{\frac{m}{2\pi\hbar ti}}\, e^{mi(q - x)^2/2\hbar t}.$$

Then we can write

$$\psi(q,t) = \int_{\mathbb{R}} K_0(q,x,t)\psi(x,0)\,dx. \tag{7.47}$$

Exercise 7.3.6. Write the one-dimensional heat kernel (G.16) as

$$H^\alpha(q,x,t) = \frac{1}{\sqrt{4\pi\alpha t}}\, e^{-(q-x)^2/4\alpha t}.$$

Here α is required to be a positive real number. Even so, show that by *formally* taking α to be the pure imaginary number $\frac{\hbar}{2m}i$, one turns the heat kernel into the free Schrödinger kernel, that is,

$$H^{\hbar i/2m}(q,x,t) = \sqrt{\frac{m}{2\pi\hbar t i}}\, e^{mi(q-x)^2/2\hbar t} = K_0(q,x,t).$$

Exercise 7.3.7. Show that for each fixed x, $K_0(q,x,t)$ satisfies the free Schrödinger equation

$$i\frac{\partial K_0(q,x,t)}{\partial t} = -\frac{\hbar}{2m}\frac{\partial^2 K_0(q,x,t)}{\partial q^2}$$

on $\mathbb{R} \times (0,\infty)$. Similarly, for each fixed $q \in \mathbb{R}$,

$$i\frac{\partial K_0(q,x,t)}{\partial t} = -\frac{\hbar}{2m}\frac{\partial^2 K_0(q,x,t)}{\partial x^2}$$

on $\mathbb{R} \times (0,\infty)$.

Note that $K_0(q,x,t)$ provides a particularly important solution to the Schrödinger equation in that any other solution can be obtained from it and the initial ($t=0$) data via (7.47). Note also that (7.39) can now be written

$$\lim_{t\to 0^+} \int_{\mathbb{R}} K_0(q,x,t)\psi_0(x)\,dx = \psi_0(q) \tag{7.48}$$

and that this has the following interpretation. If q and t are held fixed, $K_0(q,x,t)$ is certainly an element of $L^1_{\mathrm{loc}}(\mathbb{R})$ and hence can be regarded as a tempered distribution. Since $\psi_0(x)$ can be any element of $S(\mathbb{R})$, (7.48) simply says that *as distributions*

$$\lim_{t\to 0^+} K_0(q,x,t) = \delta(x - q).$$

One often sees this abbreviated in the literature as simply $K_0(q,x,0) = \delta(x-q)$. If we allow ourselves this one small indiscretion we can summarize our discussion by saying that *the kernel $K_0(q,x,t)$ is a (distributional) solution to the free Schrödinger equation* *that is initially the Dirac delta at q.* In the language of partial differential equations one would say that $K(q,x,t)$ is the *fundamental solution* to the free Schrödinger equation.

We will conclude this section by asking you to prove for $K_0(q,x,t)$ a few things you have already proved for the heat kernel (Exercises G.3.5 and G.3.7).

Exercise 7.3.8. Let $q,x,z \in \mathbb{R}$, $s > 0$ and $t > 0$. Show that

$$K_0(q,x,s+t) = \int_{\mathbb{R}} K_0(q,z,t)K_0(z,x,s)\,dz.$$

Hint: See the hint for Exercise G.3.5.

Exercise 7.3.9. Let $k \geq 2$ be an integer, $q,x,z_1,\ldots,z_{k-1} \in \mathbb{R}$ and $t_1,\ldots,t_k > 0$. Show that

$$K_0(q,x,t_1+\cdots+t_k) = \int_{\mathbb{R}^{k-1}} K_0(q,z_1,t_1)K_0(z_1,z_2,t_2)\cdots K_0(z_{k-1},x,t_k)\,dz_1\cdots dz_{k-1}.$$

7.4 The quantum harmonic oscillator

7.4.1 Introduction

Having warmed up on the quantization of the classical free particle in dimension one, we will now turn to a somewhat more challenging example. Recall that the classical harmonic oscillator has configuration space $\mathbb{R}$ and phase space $T^*\mathbb{R} = \mathbb{R}^2$ with coordinates q and (q,p), respectively. The classical Hamiltonian is $\frac{1}{2m}p^2 + \frac{m\omega^2}{2}q^2$, where m and ω are positive constants. The quantum phase space is taken to be $L^2(\mathbb{R})$. Since the Hamiltonian lives in the Jacobi subalgebra $\mathfrak{g}^J$ of the Lie algebra $C^\infty(T^*\mathbb{R})$ of classical observables, we can apply the quantization map R described in Section 7.2 to obtain its quantum analogue

$$H_B = R\left(\frac{1}{2m}p^2 + \frac{m\omega^2}{2}q^2\right) = \frac{1}{2m}P^2 + \frac{m\omega^2}{2}Q^2,$$

which, on $\mathcal{S}(\mathbb{R})$, is given by

$$H_B|_{\mathcal{S}(\mathbb{R})} = -\frac{\hbar^2}{2m}\frac{d^2}{dq^2} + \frac{m\omega^2}{2}q^2.$$

We have already spent a fair amount of time with this operator on $L^2(\mathbb{R})$, but it is worth the effort to review what we know. We have shown (Example 5.3.1) that H_B is essentially self-adjoint on $\mathcal{S}(\mathbb{R})$. This followed from the fact that, on $\mathcal{S}(\mathbb{R})$, it is symmetric and has a discrete set of eigenvalues

$$\mathcal{E}_n = \left(n + \frac{1}{2}\right)\hbar\omega, \quad n = 0,1,2,\ldots,$$

with eigenfunctions $\psi_n(q)$, $n = 0, 1, 2, \ldots$, that live in $S(\mathbb{R})$ and form an orthonormal basis for $L^2(\mathbb{R})$. Specifically,

$$\psi_n(q) = \frac{1}{\sqrt{2^n n!}} \left(\frac{m\omega}{\hbar\pi} \right)^{1/4} e^{-m\omega q^2 / 2\hbar} H_n \left(\sqrt{\frac{m\omega}{\hbar}} \, q \right),$$

where

$$H_n(x) = (-1)^n e^{x^2} \frac{d^n}{dx^n} \left(e^{-x^2} \right)$$

is the nth Hermite polynomial. The eigenvalues $\mathcal{E}_n$ comprise the entire spectrum

$$\sigma(H_B) = \{\mathcal{E}_n\}_{n=0}^{\infty} = \left\{ \left(n + \frac{1}{2} \right) \hbar\omega \right\}_{n=0}^{\infty}$$

of H_B (Example 5.4.5) and all of the eigenspaces are one-dimensional. These eigenvalues are therefore all of the allowed energy levels of the quantum oscillator, so, unlike the free particle, the energy spectrum of the harmonic oscillator is discrete (quantized). The smallest of these eigenvalues is $\mathcal{E}_0 = \frac{1}{2}\hbar\omega$ and the corresponding eigenstate ψ_0 is called the ground state of the oscillator (we emphasize once again that the lowest allowed energy level is *not* zero). The remaining ψ_n, $n = 1, 2, \ldots$, are called excited states. Writing $\psi \in L^2(\mathbb{R})$ as $\psi = \sum_{n=0}^{\infty} \langle \psi_n, \psi \rangle \psi_n$, the domain $\mathcal{D}(H_B)$ of H_B is just the set of ψ for which $\sum_{n=0}^{\infty} \mathcal{E}_n \langle \psi_n, \psi \rangle \psi_n$ converges in $L^2(\mathbb{R})$, that is, for which

$$\sum_{n=0}^{\infty} \mathcal{E}_n^2 \, | \langle \psi_n, \psi \rangle |^2 < \infty.$$

Since 0 is not an eigenvalue, H_B is invertible. Indeed, its inverse is a bounded operator on all of $L^2(\mathbb{R})$ given by

$$H_B^{-1} \phi = \sum_{n=0}^{\infty} \frac{1}{\mathcal{E}_n} \langle \psi_n, \phi \rangle \psi_n$$

(see (5.26)). In Example 5.5.4 it is shown that H_B^{-1} is a compact operator. The evolution operator $e^{-itH_B/\hbar}$ is given by

$$e^{-itH_B/\hbar} \psi = \sum_{n=0}^{\infty} \langle \psi_n, \psi \rangle e^{-(i/\hbar)\mathcal{E}_n t} \psi$$

for any $\psi \in L^2(\mathbb{R})$. The time evolution of an initial state $\psi(q, t_0)$ can be written as

$$\psi(q, t) = \sum_{n=0}^{\infty} \left(\int_{\mathbb{R}} \psi_n(x) \, \psi(x, t_0) dx \right) e^{-(i/\hbar)\mathcal{E}_n(t-t_0)} \psi_n(q),$$

which, at least for sufficiently nice initial data, can be written in terms of an integral kernel as

$$\psi(q, t) = \int_{\mathbb{R}} K(q, t; x, t_0)\psi(x, t_0)dx,$$

where

$$K(q, t; x, t_0) = \sum_{n=0}^{\infty} e^{-(i/\hbar)\mathcal{E}_n(t-t_0)}\psi_n(q)\,\psi_n(x) = \sum_{n=0}^{\infty} e^{-i(n+\frac{1}{2})\omega(t-t_0)}\psi_n(q)\psi_n(x) \qquad (7.49)$$

(see (6.13)). We will expend some effort at the end of this section to obtain a closed form expression for this kernel analogous to the one we found for the free particle in the previous section and will see that here too the classical action will put in an appearance.

The analysis of the quantum harmonic oscillator is greatly illuminated by the introduction of the so-called raising and lowering operators b and $b^\dagger$ defined by

$$b = \frac{1}{\sqrt{2m\omega\hbar}}(m\omega Q + iP)$$

and

$$b^\dagger = \frac{1}{\sqrt{2m\omega\hbar}}(m\omega Q - iP),$$

respectively (our discussion of these began in Section 5.14). These are formal adjoints of each other,

$$\langle b\phi, \psi \rangle = \langle \phi, b^\dagger\psi \rangle \quad \text{and} \quad \langle b^\dagger\psi, \phi \rangle = \langle \psi, b\phi \rangle,$$

and satisfy various algebraic identities, of which we will recall a few. Designating them as raising and lowering operators is motivated by

$$b^\dagger\psi_n = \sqrt{n+1}\,\psi_{n+1}, \quad n = 0, 1, 2, \ldots, \quad \text{and} \quad b\psi_n = \sqrt{n}\,\psi_{n-1}, \quad n = 1, 2, \ldots.$$

On $\mathcal{S}(\mathbb{R})$ we have $[P, Q]_- = -i\hbar$ and it follows from this that

$$[b, b^\dagger]_- = bb^\dagger - b^\dagger b = 1$$

(see (5.16)). Defining the number operator N_B by $N_B = b^\dagger b$ one obtains

$$N_B\psi_n = n\psi_n,$$

$$H_B = \hbar\omega\left(N_B + \frac{1}{2}\right)$$

and various commutation relations such as

$$[N_B, b^\dagger]_- = b^\dagger \quad \text{and} \quad [N_B, b]_- = -b$$

and

$$[H_D, b^\dagger]_- = \hbar\omega b^\dagger \quad \text{and} \quad [H_B, b]_- = -\hbar\omega b.$$

Note that, in particular, $\{N_B, b, b^\dagger, 1\}$ generates a four-dimensional real vector space of operators (*not* self-adjoint operators) on $L^2(\mathbb{R})$ that is closed under $[\,,\,]_-$ on $S(\mathbb{R})$. This motivates the following algebraic definition.

Exercise 7.4.1. The *oscillator algebra* o_4 is a four-dimensional real Lie algebra with a basis $\{N, B_+, B_-, M\}$ subject to the commutation relations

$$[N, B_+] = B_+, \quad [N, B_-] = -B_-, \quad [B_-, B_+] = M, \quad [M, N] = [M, B_+] = [M, B_-] = 0.$$

In particular, M is in the center of o_4.
1. Verify the Jacobi identity for these commutation relations.
2. Define 3×3 matrices

$$D(N) = \begin{pmatrix} 0 & 0 & 0 \\ 0 & 1 & 0 \\ 0 & 0 & 0 \end{pmatrix}, \quad D(B_+) = \begin{pmatrix} 0 & 0 & 0 \\ 0 & 0 & 1 \\ 0 & 0 & 0 \end{pmatrix},$$

$$D(B_-) = \begin{pmatrix} 0 & 1 & 0 \\ 0 & 0 & 0 \\ 0 & 0 & 0 \end{pmatrix}, \quad D(M) = \begin{pmatrix} 0 & 0 & 1 \\ 0 & 0 & 0 \\ 0 & 0 & 0 \end{pmatrix}.$$

   Show that $\{D(N), D(B_+), D(B_-), D(M)\}$ generates a matrix Lie algebra isomorphic to o_4. Whenever convenient we will simply identify o_4 with this matrix Lie algebra.
3. Show that $\{B_+, B_-, M\}$ generates a Lie subalgebra of o_4 isomorphic to the Heisenberg algebra $\mathfrak{h}_3$.
4. Exponentiate a general element $nD(N) + b_+ D(B_+) + b_- D(B_-) + mD(M)$ of this Lie algebra, show that the result is

$$\begin{pmatrix} 1 & b_- e^n & m + b_- b_+ \\ 0 & e^n & b_+ \\ 0 & 0 & 1 \end{pmatrix}$$

   and conclude that the exponential map is a bijection from o_4 onto the set of 3×3 real matrices of the form

$$\begin{pmatrix} 1 & a_{12} & a_{13} \\ 0 & a_{22} & a_{23} \\ 0 & 0 & 1 \end{pmatrix}$$

   with $a_{22} > 0$. Show that these form a group under matrix multiplication.

5. Show that the map that sends

$$\begin{pmatrix} 1 & b_- e^n & m + b_- b_+ \\ 0 & e^n & b_+ \\ 0 & 0 & 1 \end{pmatrix}$$

to (n, b_+, b_-, m) is a bijection onto $\mathbb{R}^4$.

6. The *oscillator group* is the unique simply connected Lie group O_4 with Lie algebra o_4. Identify O_4 with $\mathbb{R}^4$ on which the following multiplicative structure is defined:

$$(n', b'_+, b'_-, m')(n, b_+, b_-, m) = (n' + n,\ b'_+ + b_+ e^{n'},\ b'_- + b_- e^{-n'},$$

$$m' + m - b_- b'_+ e^{-n'}).$$

Finally, we should record, just for reference, how to retrieve the position operator Q and momentum operator P from the raising operator b and the lowering operator $b^\dagger$:

$$Q = \sqrt{\frac{\hbar}{2m\omega}}\, (b^\dagger + b),$$

$$P = i\sqrt{\frac{m\omega\hbar}{2}}\, (b^\dagger - b).$$

There is a very great deal to be said about the quantum harmonic oscillator and, perforce, we cannot say it all here so we should briefly describe what we do intend to say. In the remainder of this section we would like to focus on two issues. The first is a bit of folklore according to which "quantum mechanics reduces to classical mechanics in the limit as $\hbar \to 0$." This is rather vague, of course, but it certainly "should" be true, at least in some moral sense. However, it may not be entirely clear how one would even formulate a precise statement in the hope of being able to prove it. Are certain quantum mechanical "things" supposed to approach various classical "things" as $\hbar \to 0$? If so, what things? Or perhaps the entire classical path of the particle is somehow singled out as $\hbar \to 0$? There are many ways to approach this *classical limit problem* and we will have a look at just a few, including the famous theorem of Ehrenfest, who takes a rather different approach that does not involve letting $\hbar \to 0$. A very thorough and mathematically rigorous discussion of this problem is available in [Lands]. This done we will turn our attention to the derivation of a closed form expression for the Schrödinger kernel of the harmonic oscillator analogous to (7.46) for the free particle. This will again contain the action of the classical trajectory and will reappear in Chapter 8 when we turn to the Feynman path integral for the harmonic oscillator.

7.4.2 The classical limit problem

What kinds of classical and quantum "things" could one reasonably expect to be able to compare in the limit as $\hbar \to 0$? The observables themselves would not seem to be an obvious choice. Classical observables are real-valued functions on phase space so they take a real value at any state (q, p). Quantum observables are operators on a Hilbert space and these certainly do not take real values at a state ψ. Note, however, that in any state ψ any observable A has an expected value $\langle A \rangle_\psi = \langle \psi, A\psi \rangle$ representing, roughly, the expected average of a large number of independent measurements of A in the state ψ and this *is* a real number associated with the quantum state. Moreover, even classically one can measure precise values of observables only "in principle" so that it would seem more physically realistic to deal with probabilities and expected values. We will now carry out one such comparison for the harmonic oscillator.

We consider first the classical oscillator with Hamiltonian $H_{CL} = \frac{1}{2m}p^2 + \frac{m\omega^2}{2}q^2$, where m and ω are two positive constants, the first being the mass and the second the natural frequency of the oscillator. The motion of the oscillator is determined by Hamilton's equations

$$\dot{q} = \frac{\partial H_{CL}}{\partial p} = \frac{1}{m}p,$$

$$\dot{p} = -\frac{\partial H_{CL}}{\partial q} = -m\omega^2 q.$$

These combine to give $\ddot{q}(t) + \omega^2 q(t) = 0$, the general solution to which can be written in the form $q(t) = A \cos(\omega t + \phi)$, where $A \geq 0$ and ϕ are constants determined by

$$A^2 = q(0)^2 + \frac{\dot{q}(0)^2}{\omega^2}$$

and

$$\tan \phi = \frac{q(0)\omega}{\dot{q}(0)}.$$

The motion is periodic, of course, with period $\tau = \frac{2\pi}{\omega}$. The Hamiltonian represents the total energy E of the system and this is conserved in time, so

$$E = H_{CL}(0) = \frac{1}{2m}p(0)^2 + \frac{m\omega^2}{2}q(0)^2 = \frac{m}{2}\dot{q}(0)^2 + \frac{m\omega^2}{2}q(0)^2.$$

From this it follows that

$$E = \frac{m\omega^2}{2}A^2,$$

so the classical motion of the oscillator is constrained to the interval

$$-A = -\sqrt{\frac{2E}{m\omega^2}} \leq q \leq \sqrt{\frac{2E}{m\omega^2}} = A.$$

There is nothing new in any of this, of course, but now we would like to write down the classical probability density function $P_{CL}(q)$ for the position of the oscillator mass (which we will now take to be a point-mass). This is defined by the property that for any closed interval $J \subseteq [-A, A]$,

$$\int_J P_{CL}(q) \, dq$$

is the probability that the mass will be found in the interval J at some randomly chosen instant t. For this we will approximate the probability by sums that can be interpreted as Riemann sums and in such a way that the approximations become better as the partitions become finer. Begin by choosing a partition $[q_0, q_1], [q_1, q_2], \ldots, [q_{n-1}, q_n]$ of J and noting that the probability of finding the mass in J is the sum of the probabilities of finding it in $[q_{i-1}, q_i]$ for $i = 1, 2, \ldots, n$. The probability of finding the mass in $[q_{i-1}, q_i]$ is just the ratio of the time it spends in $[q_{i-1}, q_i]$ during one cycle to the total period of the oscillation. If q_i^* is any point in $[q_{i-1}, q_i]$ and if we denote by $|\dot{q}(q_i^*)|$ the speed of the mass at q_i^*, then the time the mass spends in $[q_{i-1}, q_i]$ during one cycle is approximately

$$\frac{2\Delta q_i}{|\dot{q}(q_i^*)|},$$

where $\Delta q_i = q_i - q_{i-1}$ (once in each direction; hence, the factor of 2). Consequently, the probability of finding the mass in $[q_{i-1}, q_i]$ is approximately

$$\frac{2\frac{\Delta q_i}{|\dot{q}(q_i^*)|}}{\tau} = \frac{\omega}{\pi} \frac{1}{|\dot{q}(q_i^*)|} \Delta q_i.$$

The probability of finding the mass in J is therefore approximately

$$\sum_{i=1}^{n} \frac{\omega}{\pi} \frac{1}{|\dot{q}(q_i^*)|} \Delta q_i,$$

so the probability is given precisely by

$$\int_J \frac{\omega}{\pi} \frac{1}{|\dot{q}(q)|} \, dq.$$

We conclude that

$$P_{CL}(q) = \frac{\omega}{\pi} \frac{1}{|\dot{q}(q)|}.$$

We will put this into a more convenient form by noting that the kinetic energy of the mass is given by $K(q) = \frac{1}{2}m(\dot{q}(q))^2$, but also by $K(q) = E - \frac{m\omega^2}{2}q^2$, so

$$(\dot{q}(q))^2 = \frac{2}{m}\left(E - \frac{m\omega^2}{2}q^2\right) = \frac{2}{m}\left(\frac{m\omega^2}{2}A^2 - \frac{m\omega^2}{2}q^2\right) = \omega^2(A^2 - q^2).$$

The bottom line is

$$P_{CL}(q) = \frac{1}{\pi} \frac{1}{\sqrt{A^2 - q^2}}.$$

Note that

$$\int_{-A}^{A} P_{CL}(q)\, dq = 2 \lim_{T \to A^-} \int_{0}^{T} \frac{1}{\pi} \frac{1}{\sqrt{A^2 - q^2}}\, dq = \frac{2}{\pi} \frac{\pi}{2} = 1,$$

as it should be. Therefore, $P_{CL}(q)$ defines a Borel probability measure on $[-A, A]$ and, if $f(q)$ is any measurable function of the position q, its expected value is

$$\langle f(q) \rangle_E = \int_{-A}^{A} \frac{1}{\pi} \frac{f(q)}{\sqrt{A^2 - q^2}}\, dq \tag{7.50}$$

(the subscript is used to emphasize that the oscillator energy is fixed at E). In particular, the probability of finding the mass in J is the expected value of its characteristic function χ_J with respect to this probability measure.

Now we look at the corresponding quantum system. Here "corresponding" means with the same energy. Classically the energy E of the oscillator can assume any non-negative value, but the energy of the quantum system must be one of the eigenvalues $\mathcal{E}_n = (n + \frac{1}{2})\hbar\omega$, $n = 0, 1, 2, \ldots$, of the Hamiltonian H_B. In order to have a corresponding quantum system at all, we must take E to be one of these. Note, however, that these eigenvalues depend on $\hbar$ and our goal is to take a limit as $\hbar \to 0^+$. In order to take such a limit and at the same time keep the energy fixed we select a sequence $\hbar_0, \hbar_1, \hbar_2, \ldots$ of positive real numbers such that

$$\left(n + \frac{1}{2}\right)\hbar_n \omega = E.$$

Specifically,

$$\hbar_n = \frac{2E}{(2n+1)\omega},$$

so we can accomplish our purpose by taking the limit as $n \to \infty$. The limit of what, you may ask. For each eigenvalue $(n + \frac{1}{2})\hbar_n \omega$ there is precisely one normalized eigenstate

$$\psi_n(q) = \frac{1}{\sqrt{2^n n!}} \left(\frac{m\omega}{\hbar_n \pi}\right)^{1/4} e^{-m\omega q^2 / 2\hbar_n} H_n\left(\sqrt{\frac{m\omega}{\hbar_n}}\, q\right)$$

in which this is the measured energy with probability one. If $f(q)$ is a classical observable we have found its classical expectation value $\langle f(q) \rangle_E$ in (7.50) and we would like to compare this with the limit as $n \to \infty$ of the quantum expectation value of $f(Q)$ in state $\psi_n(q)$ which we will write as

$$\langle f(Q) \rangle_{\psi_n, h_n} = \langle \psi_n, f(Q) \psi_n \rangle.$$

We will write this out explicitly and simplify a bit:

$$\langle f(Q) \rangle_{\psi_n, h_n} = \langle \psi_n, f(Q) \psi_n \rangle = \int_{\mathbb{R}} \overline{\psi_n(q)} f(q) \, \psi_n(q) \, dq = \int_{\mathbb{R}} f(q) \, |\psi_n(q)|^2 \, dq$$

$$= \frac{1}{2^n n!} \sqrt{\frac{m\omega}{\hbar_n \pi}} \int_{\mathbb{R}} f(q) \, e^{-m\omega q^2 / \hbar_n} \, H_n \left(\sqrt{\frac{m\omega}{\hbar_n}} \, q \right)^2 dq$$

$$= \frac{1}{2^n n!} \sqrt{\frac{m\omega}{\hbar_n \pi}} \sqrt{\frac{\hbar_n}{m\omega}} \int_{\mathbb{R}} f \left(\sqrt{\frac{\hbar_n}{m\omega}} \, u \right) e^{-u^2} H_n(u)^2 \, du.$$

Since

$$\sqrt{\frac{\hbar_n}{m\omega}} = \sqrt{\frac{1}{m\omega} \frac{2E}{(2n+1)\omega}} = \sqrt{\frac{2E}{(2n+1)\omega^2}} = \frac{A}{\sqrt{2n+1}},$$

we have

$$\langle f(Q) \rangle_{\psi_n, h_n} = \frac{1}{2^n n! \sqrt{\pi}} \int_{\mathbb{R}} f \left(\frac{Au}{\sqrt{2n+1}} \right) e^{-u^2} H_n(u)^2 \, du. \tag{7.51}$$

What we would like to do is show that, at least for sufficiently nice functions $f(q)$, the quantum expectation value $\langle f(Q) \rangle_{\psi_n, h_n}$ given by (7.51) approaches the classical expectation value $\langle f(q) \rangle_E$ given by (7.50) as $n \to \infty$. We will begin by evaluating the integral

$$\int_{-A}^{A} \frac{1}{\pi} \frac{f(q)}{\sqrt{A^2 - q^2}} \, dq$$

for some simple choices of $f(q)$. Suppose, for example, that $f(q) = \cos(pq)$, where p is a real parameter. Then

$$\int_{-A}^{A} \frac{1}{\pi} \frac{\cos(pq)}{\sqrt{A^2 - q^2}} \, dq = \frac{2}{\pi} \int_{0}^{A} \frac{\cos(pq)}{\sqrt{A^2 - q^2}} \, dq$$

and the substitution $q = A \sin \theta$ turns this into

$$\frac{2}{\pi} \int_{0}^{\pi/2} \cos(pA \sin \theta) \, d\theta.$$

Now we recall that the Bessel function of order zero $J_0(x)$ has the integral representation

$$J_0(x) = \frac{2}{\pi} \int_0^{\pi/2} \cos(x \sin \theta) \, d\theta$$

for all $x \in \mathbb{R}$ (there are proofs of this in Sections 4.7 and 4.9 of [AAR]). We conclude that

$$\langle \cos(pq) \rangle_E = \frac{2}{\pi} \int_0^{\pi/2} \cos(pA \sin \theta) \, d\theta = J_0(pA).$$

Note that for $f(q) = \sin(pq)$, the integral is zero since the integrand is odd. Consequently,

$$\frac{1}{\pi} \int_{-A}^{A} \frac{e^{ipq}}{\sqrt{A^2 - q^2}} \, dq = J_0(pA).$$

For complex-valued functions such as e^{ipq} it will be convenient to write $\langle e^{ipq} \rangle_E$ for the complex number that is the expected value of the real part plus i times the expected value of the imaginary part, so

$$\langle e^{ipq} \rangle_E = J_0(pA).$$

Now suppose $f(q)$ is a Schwartz function. Then its Fourier transform $\hat{f}(p)$ is also a Schwartz function. Write $f(q)$ as

$$f(q) = \mathcal{F}^{-1}(\hat{f}(p))(q) = \frac{1}{\sqrt{2\pi}} \int_{\mathbb{R}} e^{ipq} \hat{f}(p) \, dp$$

and then

$$\langle f(q) \rangle_E = \frac{1}{\pi} \int_{-A}^{A} \frac{f(q)}{\sqrt{A^2 - q^2}} \, dq$$

$$= \frac{1}{\pi} \frac{1}{\sqrt{2\pi}} \int_{-A}^{A} \int_{\mathbb{R}} \frac{e^{ipq}}{\sqrt{A^2 - q^2}} \hat{f}(p) \, dp \, dq.$$

We apply Fubini's theorem to obtain

$$\langle f(q) \rangle_E = \frac{1}{\sqrt{2\pi}} \int_{\mathbb{R}} \left(\frac{1}{\pi} \int_{-A}^{A} \frac{e^{ipq}}{\sqrt{A^2 - q^2}} \, dq \right) \hat{f}(p) \, dp,$$

and hence

$$\langle f(q) \rangle_E = \frac{1}{\sqrt{2\pi}} \int_{\mathbb{R}} J_0(pA)\hat{f}(p)\, dp. \tag{7.52}$$

Now we will start the whole thing over again by looking at $\langle f(Q) \rangle_{\psi_n, h_n}$ when $f(q) = \cos(pq)$. According to (7.51),

$$\langle \cos(pQ) \rangle_{\psi_n, h_n} = \frac{1}{2^n n! \sqrt{\pi}} \int_{\mathbb{R}} \cos\left(\frac{pAu}{\sqrt{2n+1}}\right) e^{-u^2} H_n(u)^2\, du. \tag{7.53}$$

Evaluating integrals of this sort is no mean feat and I will not lie to you; I looked it up. Item number 7.388 (5) on page 806 of [GR] is

$$\int_{\mathbb{R}} \cos(\sqrt{2}\beta u)\, e^{-u^2} H_n(u)^2\, du = 2^n n!\, \sqrt{\pi}\, e^{-\beta^2/2} L_n(\beta^2), \tag{7.54}$$

where $\beta \in \mathbb{R}$ and L_n is the nth Laguerre polynomial

$$L_n(x) = \sum_{k=0}^{n} \binom{n}{k} \frac{(-1)^k}{k!} x^k.$$

With $\beta = \frac{pA}{\sqrt{4n+2}}$ this gives

$$\langle \cos(pQ) \rangle_{\psi_n, h_n} = \exp\left(-\frac{p^2 A^2}{8n+4}\right) L_n\left(\frac{p^2 A^2}{4n+2}\right).$$

Now we need only take the limit of this as $n \to \infty$. The exponential factor clearly approaches one, but for the second factor we must appeal to an old result on the asymptotics of Laguerre polynomials. A special case of Theorem 8.1.3 of [Szegö] states that

$$\lim_{N \to \infty} L_N\left(\frac{x^2}{4N}\right) = J_0(x)$$

uniformly on compact sets. Consequently,

$$\lim_{n \to \infty} L_n\left(\frac{p^2 A^2}{4n+2}\right) = J_0(pA)$$

and therefore

$$\lim_{n \to \infty} \langle \cos(pQ) \rangle_{\psi_n, h_n} = J_0(pA) = \langle \cos(pq) \rangle_E,$$

as we hoped. Since the integrand is odd, $\langle \sin(pQ) \rangle_{\psi_n, h_n} = 0$ for every n, so

$$\lim_{n \to \infty} \langle \sin(pQ) \rangle_{\psi_n, h_n} = 0 = \langle \sin(pq) \rangle_E.$$

Combining these two gives

$$\lim_{n\to\infty} \langle e^{ipQ} \rangle_{\psi_n, \hbar_n} = J_0(pA) = \langle e^{ipq} \rangle_E.$$

Finally, we suppose that $f(q)$ is a Schwartz function. We know that $\langle f(q) \rangle_E$ is given by (7.52) and must show that $\lim_{n\to\infty} \langle f(Q) \rangle_{\psi_n, \hbar_n}$ gives the same result.

Exercise 7.4.2. Show that $\langle f(Q) \rangle_{\psi_n, \hbar_n}$ can be written in the form

$$\langle f(Q) \rangle_{\psi_n, \hbar_n} = \frac{1}{\sqrt{2\pi}} \int_{\mathbb{R}} \hat{f}(p) \left(\frac{1}{2^n n! \sqrt{\pi}} \int_{\mathbb{R}} e^{ipAu/\sqrt{2n+1}} e^{-u^2} H_n(u)^2 \, du \right) dp$$

$$= \frac{1}{\sqrt{2\pi}} \int_{\mathbb{R}} \hat{f}(p) \langle e^{ipQ} \rangle_{\psi_n, \hbar_n} \, dp.$$

Hint: See the argument leading to (7.52).

Exercise 7.4.3. Show that $|\hat{f}(p) \langle e^{ipQ} \rangle_{\psi_n, \hbar_n}| \leq |\hat{f}(p)|$ for every n and every p. *Hint*: Apply the $\beta = 0$ case of (7.54).

Note that, since $\hat{f}(p)$ is a Schwartz function, $|\hat{f}(p)|$ is integrable. Moreover, the sequence $\hat{f}(p) \langle e^{ipQ} \rangle_{\psi_n, \hbar_n}$ converges pointwise to $\hat{f}(p) J_0(pA)$ as $n \to \infty$.

Exercise 7.4.4. Use Lebesgue's dominated convergence theorem to show that

$$\lim_{n\to\infty} \langle f(Q) \rangle_{\psi_n, \hbar_n} = \frac{1}{\sqrt{2\pi}} \int_{\mathbb{R}} J_0(pA) \hat{f}(p) \, dp = \langle f(q) \rangle_E.$$

With this we conclude our admittedly rather modest illustration of what might be meant by (or at least implied by) the assertion that quantum mechanics reduces to classical mechanics as $\hbar \to 0$. We have shown that for the harmonic oscillator, a particularly simple set of quantum expectation values approach the corresponding classical expectation values as $\hbar \to 0$. There are generalizations of this result to more general systems and observables, one of which centers on the notion of a *(canonical) coherent state* (see Section 5.1 of [Lands] for a brief discussion and numerous references).

Examining the behavior of quantum expectation values as $\hbar \to 0$ is not the only possible approach one might take to somehow "retrieving" classical mechanics from quantum mechanics. In 1927, Paul Ehrenfest [Ehren] studied the time evolution of the expectation values of position and momentum for Hamiltonians of the form $-\frac{\hbar^2}{2m}\Delta + V$ and found that they satisfied the classical equations of motion for the position and momentum variables, that is, Hamilton's equations (or, if you prefer, Newton's second law). This is known in physics as *Ehrenfest's theorem* although the argument given by Ehrenfest was not a rigorous proof of a theorem in the mathematical sense. Nevertheless, Ehrenfest's argument is so simple and suggestive that it is well worth describing without worrying too much about technical hypotheses; this done we will briefly consider a rigorous version of the theorem. Whether or not the result can legitimately be regarded as a transition from quantum to classical mechanics is quite another matter, however, and we will have a few words to say about that at the end.

For simplicity we will restrict our discussion to systems with one degree of freedom. For the record we will also recall, from Section 2.3, the classical picture of such systems that we are searching for within the formalism of quantum mechanics. The classical configuration space is $M = \mathbb{R}$ and the phase space is $T^*M = \mathbb{R}^2$ with canonical coordinates (q, p). We are given some classical Hamiltonian $H_{\mathrm{CL}} \in C^\infty(T^*M)$ and the system evolves along the integral curves of the Hamiltonian vector field $X_{H_{\mathrm{CL}}}$. If $f \in C^\infty(T^*M)$ is any classical observable, then its variation along an integral curve of $X_{H_{\mathrm{CL}}}$ is determined by

$$\frac{df}{dt} = \{f, H_{\mathrm{CL}}\},$$

where $\{\,,\,\}$ is the Poisson bracket. In particular, when this is applied to the position observable q and the momentum observable p, one obtains Hamilton's equations

$$\dot{q} = \{q, H_{\mathrm{CL}}\} = \frac{\partial H_{\mathrm{CL}}}{\partial p}$$

and

$$\dot{p} = \{p, H_{\mathrm{CL}}\} = -\frac{\partial H_{\mathrm{CL}}}{\partial q}.$$

Now consider a quantum system with Hilbert space $\mathcal{H} = L^2(\mathbb{R})$. The Hamiltonian H is a self-adjoint operator on $\mathcal{H}$ and we denote its dense domain $\mathcal{D}(H)$. The time evolution of the system is determined by the abstract Schrödinger equation

$$\frac{d\psi(t)}{dt} = -\frac{i}{\hbar} H\psi(t),$$

so that

$$\psi(t) = e^{-itH/\hbar}\psi_0,$$

where $\psi_0 = \psi(0)$ is the initial state and $e^{-itH/\hbar}$ is the one-parameter group of unitary operators determined by Stone's theorem and H. Now let A be some observable, that is, a self-adjoint operator on $\mathcal{H}$ with dense domain $\mathcal{D}(A)$. Under the assumption that $\psi_0 \in \mathcal{D}(A)$ and $\mathcal{D}(A)$ is invariant under $e^{-itH/\hbar}$ for all t we can define the expectation value of A in state $\psi(t)$ for any t by

$$\langle A \rangle_{\psi(t)} = \langle \psi(t), A\psi(t) \rangle = \langle e^{-itH/\hbar}\psi_0, A e^{-itH/\hbar}\psi_0 \rangle.$$

Replacing the Poisson bracket with the quantum bracket, what we would like to assert as the statement of Ehrenfest's theorem is

$$\frac{d}{dt}\langle A \rangle_{\psi(t)} = \left\langle -\frac{i}{\hbar}[A, H]_- \right\rangle_{\psi(t)}. \tag{7.55}$$

The issues here are abundantly clear. Once again we are faced with all of the usual difficulties associated with the commutator of unbounded operators, but now we must even make sense of its expected value in each state $\psi(t)$. Furthermore, there does not appear to be any reason to believe that $\langle A \rangle_{\psi(t)}$ is a differentiable (or, for that matter, even continuous) function of t.

We will address these issues a bit more carefully soon, but let us pretend for a moment that they do not exist and just compute, as Ehrenfest did. We obtain

$$
\begin{aligned}
\frac{d}{dt}\langle A \rangle_{\psi(t)} &= \frac{d}{dt}\langle\, \psi(t), A\psi(t) \,\rangle = \left\langle\, \psi(t), A\frac{d\psi(t)}{dt} \,\right\rangle + \left\langle\, \frac{d\psi(t)}{dt}, A\psi(t) \,\right\rangle \quad \text{(product rule)} \\[2mm]
&= \left\langle\, \psi(t), A\left(-\frac{i}{\hbar}H\psi(t)\right) \,\right\rangle + \left\langle\, -\frac{i}{\hbar}H\psi(t), A\psi(t) \,\right\rangle \quad \text{(Schrödinger equation)} \\[2mm]
&= -\frac{i}{\hbar}\langle\, \psi(t), AH\psi(t) \,\rangle + \frac{i}{\hbar}\langle\, H\psi(t), A\psi(t) \,\rangle \\[2mm]
&= -\frac{i}{\hbar}\langle\, \psi(t), AH\psi(t) \,\rangle + \frac{i}{\hbar}\langle\, \psi(t), HA\psi(t) \,\rangle \quad \text{(self-adjointness of } H\text{)} \\[2mm]
&= -\frac{i}{\hbar}\langle\, \psi(t), (AH - HA)\psi(t) \,\rangle \\[2mm]
&= \left\langle\, \psi(t), -\frac{i}{\hbar}[A, H]_-\psi(t) \,\right\rangle \\[2mm]
&= \left\langle\, -\frac{i}{\hbar}[A, H]_- \,\right\rangle_{\psi(t)}.
\end{aligned}
$$

This little calculation has all sorts of problems and is certainly not a proof, but should at least indicate where Ehrenfest's result might have come from. We would now like to briefly discuss what can be done to turn it into a rigorous theorem (we will be sketching some of the ideas in [FrKo] and [FrSc]).

To make rigorous sense of (7.55) one must first see to it that everything in it is well-defined. Because H and A are two self-adjoint operators on the Hilbert space $\mathcal{H}$, in particular, Stone's theorem guarantees that H generates a unique one-parameter group of unitary operators $e^{-itH/\hbar}$, $t \in \mathbb{R}$; $\psi(t) = e^{-itH/\hbar}\psi_0$ is the time evolution of some initial state ψ_0 which we must assume is in $\mathcal{D}(A)$ in order for $\langle A \rangle_{\psi_0} = \langle \psi_0, A\psi_0 \rangle$ to be defined. Furthermore, since $\langle A \rangle_{\psi(t)} = \langle \psi(t), A\psi(t) \rangle$, each $\psi(t)$ must also be in $\mathcal{D}(A)$, that is, $\mathcal{D}(A)$ must be invariant under $e^{-itH/\hbar}$ for every $t \in \mathbb{R}$. Note that Stone's theorem guarantees that $\mathcal{D}(H)$ is necessarily invariant under every $e^{-itH/\hbar}$.

Dealing with $\langle -\frac{i}{\hbar}[A, H]_- \rangle_{\psi(t)}$ requires a bit more thought due to the problematic nature of $[A, H]_-$ for unbounded operators. Suppose first that we have some ψ that is actually in the domain of $[A, H]_-$. Then

$$\langle \psi, (AH - HA)\psi \rangle = \langle \psi, AH\psi \rangle - \langle \psi, HA\psi \rangle = \langle A\psi, H\psi \rangle - \langle H\psi, A\psi \rangle.$$

Note that this last expression makes sense for any $\psi \in \mathcal{D}(A) \cap \mathcal{D}(H)$, so we can evade the annoying issues associated with the commutator $[A, H]_-$ if we *define*

$$\left\langle -\frac{i}{\hbar}[A, H]_- \right\rangle_\psi = -\frac{i}{\hbar}[\langle A\psi, H\psi \rangle - \langle H\psi, A\psi \rangle]$$

for all $\psi \in \mathcal{D}(A) \cap \mathcal{D}(H)$. Defined in this way, $\langle -\frac{i}{\hbar}[A, H]_- \rangle_{\psi(t)}$ will make sense if $\psi(t) \in \mathcal{D}(A) \cap \mathcal{D}(H)$ for all t, that is, if $\mathcal{D}(A) \cap \mathcal{D}(H)$ is invariant under $e^{-itH/\hbar}$ for every $t \in \mathbb{R}$.

At this stage we know that, provided $\mathcal{D}(A) \cap \mathcal{D}(H)$ is invariant under $e^{-itH/\hbar}$ for every $t \in \mathbb{R}$, both $\langle A \rangle_{\psi(t)}$ and $\langle -\frac{i}{\hbar}[A, H]_- \rangle_{\psi(t)}$ will be well-defined for all t in $\mathbb{R}$. Of course, it is not enough for $\langle A \rangle_{\psi(t)}$ to be well-defined; it must, by (7.55), be (at least) differentiable as a function of t. Perhaps the most interesting part of this story is that our assumption that $\mathcal{D}(A) \cap \mathcal{D}(H)$ is invariant under $e^{-itH/\hbar}$ for every $t \in \mathbb{R}$ is already sufficient to guarantee not only the continuous differentiability of $\langle A \rangle_{\psi(t)}$, but, indeed, also the validity of (7.55). All of this depends on the following rather nontrivial lemma.

Lemma 7.4.1. *Let $\mathcal{H}$ be a separable, complex Hilbert space, $H : \mathcal{D}(H) \to \mathcal{H}$ a self-adjoint operator on $\mathcal{H}$ and $A : \mathcal{D}(A) \to \mathcal{H}$ a symmetric, closed operator on $\mathcal{H}$. Assume that $\mathcal{D}(A) \cap \mathcal{D}(H)$ is invariant under $e^{-itH/\hbar}$ for all $t \in \mathbb{R}$. Then, for any $\psi_0 \in \mathcal{D}(A) \cap \mathcal{D}(H)$,*

$$\sup_{t \in I} \| A e^{-itH/\hbar}\psi_0 \| < \infty$$

for any bounded interval $I \subset \mathbb{R}$.

This is Proposition 2 of [FrSc], where one can find a detailed proof. We will simply show how it is used to prove the following rigorous version of Ehrenfest's theorem.

Theorem 7.4.2 (Ehrenfest's theorem). *Let $\mathcal{H}$ be a separable, complex Hilbert space and let $H : \mathcal{D}(H) \to \mathcal{H}$ and $A : \mathcal{D}(A) \to \mathcal{H}$ be self-adjoint operators on $\mathcal{H}$. Let $e^{-itH/\hbar}$, $t \in \mathbb{R}$, be the one-parameter group of unitary operators generated by H and assume that $\mathcal{D}(A) \cap \mathcal{D}(H)$ is invariant under $e^{-itH/\hbar}$ for all $t \in \mathbb{R}$. For any $\psi_0 \in \mathcal{D}(A) \cap \mathcal{D}(H)$ let $\psi(t) = e^{-itH/\hbar}\psi_0$ for each $t \in \mathbb{R}$ and define*

$$\langle A \rangle_{\psi(t)} = \langle \psi(t), A\psi(t) \rangle$$

for each $t \in \mathbb{R}$. Then $\langle A \rangle_{\psi(t)}$ is a continuously differentiable real-valued function of the real variable t and satisfies

$$\frac{d}{dt}\langle A \rangle_{\psi(t)} = \left\langle -\frac{i}{\hbar}[A, H]_- \right\rangle_{\psi(t)},$$

where

$$\left\langle -\frac{i}{\hbar}[A,H]_- \right\rangle_{\psi(t)} = -\frac{i}{\hbar}[\langle A\psi(t), H\psi(t) \rangle - \langle H\psi(t), A\psi(t) \rangle]$$

for all $t \in \mathbb{R}$.

We will show how this is proved from Lemma 7.4.1 by considering

$$\frac{d}{dt}\langle A\rangle_{\psi(t)} = \frac{d}{dt}\langle \psi(t), A\psi(t) \rangle = \lim_{\Delta t \to 0} \frac{\langle \psi(t+\Delta t), A\psi(t+\Delta t) \rangle - \langle \psi(t), A\psi(t) \rangle}{\Delta t}.$$

Exercise 7.4.5. Show that

$$\frac{\langle \psi(t+\Delta t), A\psi(t+\Delta t) \rangle - \langle \psi(t), A\psi(t) \rangle}{\Delta t} = \left\langle A\psi(t+\Delta t), \frac{\psi(t+\Delta t) - \psi(t)}{\Delta t} \right\rangle$$

$$+ \left\langle \frac{\psi(t+\Delta t) - \psi(t)}{\Delta t}, A\psi(t) \right\rangle. \qquad (7.56)$$

Exercise 7.4.6. Use Lemma 5.6.1 to show that

$$\lim_{\Delta t \to 0} \frac{\psi(t+\Delta t) - \psi(t)}{\Delta t} = -\frac{i}{\hbar} H\psi(t),$$

where the limit is in $\mathcal{H}$.

Consequently, for the limit of the second term in (7.56) we obtain

$$\lim_{\Delta t \to 0} \left\langle \frac{\psi(t+\Delta t) - \psi(t)}{\Delta t}, A\psi(t) \right\rangle = \frac{i}{\hbar}\langle H\psi(t), A\psi(t) \rangle.$$

The first term in (7.56) uses Lemma 7.4.1 and takes just a bit more work. Note first that it will suffice to show that

$$A\psi(t+\Delta t) \to A\psi(t) \quad \text{weakly in } \mathcal{H} \qquad (7.57)$$

as $\Delta t \to 0$ since then the first term in (7.56) approaches

$$-\frac{i}{\hbar}\langle A\psi(t), H\psi(t) \rangle$$

and therefore

$$\lim_{\Delta t \to 0} \frac{\langle \psi(t+\Delta t), A\psi(t+\Delta t) \rangle - \langle \psi(t), A\psi(t) \rangle}{\Delta t}$$

$$= -\frac{i}{\hbar}[\langle A\psi(t), H\psi(t) \rangle - \langle H\psi(t), A\psi(t) \rangle],$$

as required. Our task then is to prove (7.57). We will need to borrow another result from functional analysis. The following is Theorem 1, Chapter V, Section 2, page 126, of [Yosida].

Theorem 7.4.3. *Let $\mathcal{B}$ be a separable, reflexive Banach space (in particular, a separable Hilbert space). Let $\{x_n\}_{n=1}^{\infty}$ be any sequence in $\mathcal{B}$ that is norm bounded. Then there is a subsequence $\{x_{n_k}\}_{k=1}^{\infty}$ of $\{x_n\}_{n=1}^{\infty}$ that converges weakly to some element of $\mathcal{B}$.*

To prove (7.57) we fix a $t \in \mathbb{R}$ and choose an arbitrary sequence $\{\Delta t_n\}_{n=1}^{\infty}$ of real numbers converging to zero. Since $A\psi(t + \Delta t_n) = Ae^{-i(t+\Delta t_n)H/\hbar}\psi_0$ and since we can choose a bounded interval I containing all of the $t + \Delta t_n$, Lemma 7.4.1 implies that $\{A\psi(t + \Delta t_n)\}_{n=1}^{\infty}$ is norm bounded. Theorem 7.4.3 then implies that there is a subsequence $\{A\psi(t + \Delta t_{n_k})\}_{k=1}^{\infty}$ of $\{A\psi(t + \Delta t_n)\}_{n=1}^{\infty}$ that converges weakly to some f in $\mathcal{H}$. We claim that f must be $A\psi(t)$. Since $\mathcal{D}(A)$ is dense in $\mathcal{H}$, it will suffice to show that $\langle f, \phi \rangle = \langle A\psi(t), \phi \rangle$ for every $\phi \in \mathcal{D}(A)$. This is proved by the following calculation, which uses (in order) the weak convergence of $\{A\psi(t + \Delta t_{n_k})\}_{k=1}^{\infty}$, the self-adjointness of A, the continuity of $\psi(t)$ in t, the fact that $\psi(t) \in \mathcal{D}(A)$ for all $t \in \mathbb{R}$ and then self-adjointness again:

$$\langle f, \phi \rangle = \lim_{k \to \infty} \langle A\psi(t + \Delta t_{n_k}), \phi \rangle = \lim_{k \to \infty} \langle \psi(t + \Delta t_{n_k}), A\phi \rangle = \langle \psi(t), A\phi \rangle = \langle A\psi(t), \phi \rangle.$$

Thus, for any sequence $\Delta t_n \to 0$, $\{A\psi(t+\Delta t_n)\}_{n=1}^{\infty}$ contains subsequences that converge weakly and all of these must converge to $A\psi(t)$.

Exercise 7.4.7. Show from this that $A\psi(t + \Delta t)$ converges weakly to $A\psi(t)$ as $\Delta t \to 0$.

This completes the proof of (7.57), so we have shown, modulo Lemma 7.4.1, that $\langle A \rangle_{\psi(t)}$ is a differentiable function of t and satisfies

$$\frac{d}{dt} \langle A \rangle_{\psi(t)} = -\frac{i}{\hbar}[\langle A\psi(t), H\psi(t) \rangle - \langle H\psi(t), A\psi(t) \rangle].$$

Exercise 7.4.8. Show that

$$\frac{d}{dt} \langle A \rangle_{\psi(t)} = \frac{2}{\hbar}\, \mathrm{Im} \langle A\psi(t), H\psi(t) \rangle$$

and use this and what we have proved above to show that $\langle A \rangle_{\psi(t)}$ is *continuously* differentiable as a function of t, thereby completing the proof of Ehrenfest's theorem.

Ehrenfest's theorem is intuitively very appealing and, at first glance, seems to provide a rather direct link between quantum and classical mechanics. One can argue, however, that the link is somewhat illusory. To say that the expected value of the position observable satisfies the classical equation for position does not in any way imply that there is some sort of "particle" traversing a classical path. Even so, in some circumstances it is possible to obtain from Ehrenfest's theorem more convincing quantum-classical associations (see, for example, [SDBS]). Perhaps a more physically persuasive statement might be something like the following. Suppose that a wave function ψ initially (at $t = 0$, say) has an expected value of position that is "close" in some sense to the classical position at $t = 0$, that is, $|\psi(q, 0)|^2$ peaks sharply at this point. Let the system evolve under some Hamiltonian H from $\psi(q,0)$ at $t = 0$ to $\psi(q,T) = e^{-iTH/\hbar}\psi(q,0)$ at $t = T$. Then, as $\hbar \to 0$, $|\psi(q,T)|^2$ will peak sharply at the position of the classical particle at time T. We should emphasize that this is not a rigorous theorem, but one can find a heuristic discussion in terms of path integrals (which we will discuss in Chapter 8) on page 19 of [Schul]. The bottom line here is that the precise relationship between quantum and classical mechanics has not been settled to everyone's (anyone's?) satisfaction and remains a topic of much discussion.

7.4.3 The propagator

We will conclude this section by deriving an explicit, closed form expression for the Schrödinger kernel (propagator) for the harmonic oscillator. There are a number of ways to go about this (see [BB-FF]). Here we will give a direct argument based on the representation we already have available (see (7.49)). The most common procedure in the physics literature is to evaluate Feynman's path integral representation for the propagator and we will do this in Section 8.3.

To ease the notation a bit we will take $t_0 = 0$ and write (7.49) as

$$K(q,t;x,0) = \sum_{n=0}^{\infty} e^{-i(n+\frac{1}{2})\omega t}\psi_n(q)\psi_n(x).$$

Recall from the discussion following (6.14) that $K(q,t;x,0)$ is interpreted as the conditional probability amplitude of finding the particle at q at time t if it was detected at x at time 0. Our computations will show that for fixed x and q, this amplitude acquires a discontinuous phase change as t passes through a value for which ωt is an integer multiple of π. These phase shifts, called *Maslov corrections*, have observable physical effects (see [Horv]). To exhibit this behavior most clearly we will begin by assuming that ωt is *not* an integer multiple of π and let

$$\kappa = \left\lfloor \frac{\omega t}{\pi} \right\rfloor$$

denote the greatest integer less than $\frac{\omega t}{\pi}$. We can then find a unique τ satisfying $0 < \omega \tau < \pi$ with

$$\omega t = \kappa \pi + \omega \tau.$$

With this the propagator becomes

$$K(q,t;x,0) = e^{-i(\frac{\pi}{2})\kappa} \sum_{n=0}^{\infty} e^{-in\kappa\pi} e^{-i(n+\frac{1}{2})\omega\tau}\psi_n(q)\psi_n(x).$$

Now note that the oscillator eigenfunctions $\psi_n(x)$ satisfy $\psi_n(-x) = (-1)^n \psi_n(x)$, so

$$e^{-in\kappa\pi}\psi_n(x) = (-1)^{n\kappa}\psi_n(x) = \left((-1)^k\right)^n\psi_n(x) = \psi_n((-1)^k x),$$

and therefore

$$K(q, t; x, 0) = e^{-i(\frac{\pi}{2})\kappa} \sum_{n=0}^{\infty} e^{-i(n+\frac{1}{2})\omega\tau} \psi_n(q)\psi_n((-1)^k x).$$

Now we rewrite the eigenfunctions as follows. With $u = \sqrt{\frac{m\omega}{\hbar}}\,q$,

$$\psi_n(q) = \frac{1}{\sqrt{2^n n!}}\left(\frac{m\omega}{\hbar\pi}\right)^{1/4} e^{-u^2/2} H_n(u),$$

and with $v = (-1)^k\sqrt{\frac{m\omega}{\hbar}}\,x$,

$$\psi_n((-1)^k x) = \frac{1}{\sqrt{2^n n!}}\left(\frac{m\omega}{\hbar\pi}\right)^{1/4} e^{-v^2/2} H_n(v).$$

Consequently,

$$\psi_n(q)\psi_n((-1)^k x) = \sqrt{\frac{m\omega}{\hbar\pi}}\, e^{-\frac{1}{2}(u^2+v^2)}\frac{H_n(u)H_n(v)}{2^n n!},$$

and therefore

$$K(q, t; x, 0) = \sqrt{\frac{m\omega}{\hbar\pi}}\, e^{-\frac{1}{2}(i\omega\tau+u^2+v^2)}\, e^{-i(\frac{\pi}{2})\kappa} \sum_{n=0}^{\infty} \frac{H_n(u)H_n(v)}{2^n n!}\left(e^{-i\omega\tau}\right)^n. \qquad (7.58)$$

At this point we need to appeal to an old result of Mehler which essentially provides a generating function for $\frac{H_n(u)H_n(v)}{2^n n!}$.

Theorem 7.4.4 (Mehler's formula). *Let H_n be the nth Hermite polynomial and suppose u and v are fixed real numbers. Then for $z \in \mathbb{C}$ with $|z| < 1$,*

$$\sum_{n=0}^{\infty} \frac{H_n(u)H_n(v)}{2^n n!} z^n = \frac{1}{\sqrt{1-z^2}} \exp\left(\frac{2uvz - (u^2 + v^2)z^2}{1-z^2}\right).$$

The usual reference is [Watson], which contains three proofs; another is available in [Iyen]. This not quite good enough for our purposes, however. The radius of convergence of the series expansion in Mehler's formula is 1 and the expansion is obviously not valid at $z = \pm 1$ on the real line. As it happens, it *is* valid everywhere else on the unit circle and this is what we need. This follows at once from another old result in analysis called *Tauber's theorem* (one can find a proof of this in Hardy's classic monograph [Hardy]).

Theorem 7.4.5 (Tauber's theorem). *Suppose* $f(z) = \sum_{n=0}^{\infty} a_n z^n$ *on the open unit disc* $|z| < 1$. *Assume that:*

1. $n a_n \to 0$ *as* $n \to \infty$ *and*
2. *for some fixed* θ, $f(re^{i\theta})$ *approaches a finite limit* L *as* $r \to 1^-$.

Then $\sum_{n=0}^{\infty} a_n (e^{i\theta})^n = L$.

Exercise 7.4.9. Show that Tauber's theorem implies that Mehler's formula is valid for all $z \in \mathbb{C}$ with $|z| \leq 1$ except $z = \pm 1$.

Now recall that $\omega\tau$ satisfies $0 < \omega\tau < \pi$, so $e^{-i\omega\tau} \neq \pm 1$. Thus, we can substitute $e^{-i\omega\tau}$ into Mehler's formula:

$$
\begin{aligned}
\sum_{n=0}^{\infty} \frac{H_n(u)H_n(v)}{2^n n!} (e^{-i\omega\tau})^n \\
= \frac{1}{\sqrt{1 - e^{-2i\omega\tau}}} \exp\left(\frac{2uve^{-i\omega\tau} - (u^2 + v^2)e^{-2i\omega\tau}}{1 - e^{-2i\omega\tau}} \right) \\
= \frac{1}{\sqrt{1 - e^{-2i\omega\tau}}} \exp\left(2uv\frac{e^{-i\omega\tau}}{1 - e^{-2i\omega\tau}} - (u^2 + v^2)\frac{e^{-2i\omega\tau}}{1 - e^{-2i\omega\tau}} \right).
\end{aligned} \tag{7.59}
$$

Exercise 7.4.10. Show that

$$
\frac{1}{\sqrt{1 - e^{-2i\omega\tau}}} = \frac{e^{\frac{1}{2}\omega\tau i}e^{-\frac{\pi}{4}i}}{\sqrt{2 \sin \omega\tau}},
$$

$$
\frac{e^{-i\omega\tau}}{1 - e^{-2i\omega\tau}} = -\frac{i}{2 \sin \omega\tau}
$$

and

$$
\frac{e^{-2i\omega\tau}}{1 - e^{-2i\omega\tau}} = -\frac{i \cos \omega\tau}{2 \sin \omega\tau} - \frac{1}{2}.
$$

Substituting these into (7.59) and simplifying a bit gives

$$
\begin{aligned}
\sum_{n=0}^{\infty} \frac{H_n(u)H_n(v)}{2^n n!} (e^{-i\omega\tau})^n \\
= e^{\frac{1}{2}(\omega\tau i + u^2 + v^2)} \frac{e^{-\frac{\pi}{4}i}}{\sqrt{2 \sin \omega\tau}} \exp\left(\frac{i}{2 \sin \omega\tau}((u^2 + v^2)\cos \omega\tau - 2uv) \right).
\end{aligned} \tag{7.60}
$$

Now note that $\sin \omega\tau = |\sin \omega t| = (-1)^\kappa \sin \omega t$ and $\cos \omega\tau = (-1)^\kappa \cos \omega t$ and recall that $u = \sqrt{\frac{m\omega}{\hbar}}\, q$, $v = (-1)^\kappa \sqrt{\frac{m\omega}{\hbar}}\, x$ and $\kappa = \lfloor \frac{\omega t}{\pi} \rfloor$.

Exercise 7.4.11. Combine (7.60) and (7.59) to obtain the *Feynman–Souriau formula* for the propagator

$$K(q, t; x, 0)$$
$$= \sqrt{\frac{m\omega}{2\pi\hbar |\sin \omega t|}} \, e^{-i(\frac{\pi}{2})(\frac{1}{2} + \lfloor \frac{\omega t}{\pi} \rfloor)} \exp\left(\frac{i}{\hbar} \frac{m\omega}{2 \sin \omega t} [(q^2 + x^2) \cos \omega t - 2qx] \right), \quad (7.61)$$

which is valid whenever ωt is not an integer multiple of π.

As usual, if the initial state is specified at $t = t_0$ rather than $t = 0$, the propagator $K(q, t; x, t_0)$ is given by the Feynman–Souriau formula with t replaced by $T = t - t_0$. As we did for the free particle, we would like to record this for future reference with (q, t) and (x, t_0) replaced by (q_b, t_b) and (q_a, t_a), respectively, Letting $T = t_b - t_a$ we have

$$K(q_b, t_b; q_a, t_a)$$
$$= \sqrt{\frac{m\omega}{2\pi\hbar |\sin \omega T|}} \, e^{-i(\frac{\pi}{2})(\frac{1}{2} + \lfloor \frac{\omega T}{\pi} \rfloor)}$$
$$\times \exp\left(\frac{i}{\hbar} \frac{m\omega}{2 \sin \omega T} [(q_a^2 + q_b^2) \cos \omega T - 2q_a q_b] \right), \quad (7.62)$$

which is valid whenever ωT is not an integer multiple of π.

Exercise 7.4.12. Show that as $\omega \to 0$, the harmonic oscillator propagator (7.62) approaches the free particle propagator (7.46).

A few comments are in order. Feynman first derived a formula for the propagator by evaluating a path integral (see Section 8.3), but his result was less general than (7.61) in that it did not contain the absolute value symbol around $\sin \omega t$ nor the $\lfloor \frac{\omega t}{\pi} \rfloor$ in the exponential factor, that is, it was valid only when $0 < \omega t < \pi$. Souriau was the first to extend Feynman's result to obtain (7.61), but the proof we have given is modeled on [LGM].

The significance of the term $\lfloor \frac{\omega t}{\pi} \rfloor$ in (7.61) is quite clear. For a fixed q and x, when t increases through a value for which ωt is a multiple of π, $\lfloor \frac{\omega t}{\pi} \rfloor$ increases by 1 and $K(q, t; x, 0)$ acquires an additional phase factor of $e^{-i(\frac{\pi}{2})}$. This is the *Maslov correction* referred to earlier. It occurs abruptly as t passes through such a value and, as we mentioned earlier, it has physical effects which have been observed, for example, in optics. Points at which such a Maslov correction occurs are called *caustics* in physics.

Finally, we would like to draw your attention to the term

$$S(q, t; x, 0) = \frac{m\omega}{2 \sin \omega t} [(q^2 + x^2) \cos \omega t - 2qx]$$

in (7.61). You will want to compare this with the result of Exercise 2.2.4 for the action of the classical harmonic oscillator along a solution curve. We will see in Chapter 8 that the appearance of the classical action in the quantum propagator is a key insight into the Feynman path integral approach to quantum mechanics. Just for the record, we will rewrite (7.61) as

$$K(q,t;x,0) = \sqrt{\frac{m\omega}{2\pi\hbar|\sin\omega t|}} \, e^{-i(\frac{\pi}{2})(\frac{1}{2}+\lfloor\frac{\omega t}{\pi}\rfloor)} \, e^{\frac{i}{\hbar}S(q,t;x,0)}.$$

Example 7.4.1. Suppose that at $t = 0$, the initial wave function is the harmonic oscillator ground state

$$\psi_0(q) = \sqrt[4]{\frac{m\omega}{\hbar\pi}} \, e^{-m\omega q^2/2\hbar}. \tag{7.63}$$

Note that

$$|\psi_0(q)|^2 = \sqrt{\frac{m\omega}{\hbar\pi}} \, e^{-m\omega q^2/\hbar}, \tag{7.64}$$

which is just a Gaussian probability distribution centered at $q = 0$. We will compute the time evolution of this state, assuming for simplicity that $0 < \omega t < \pi$.

Exercise 7.4.13. What should the result of this calculation look like?

Now,

$$\psi(q,t) = \int_{\mathbb{R}} K(q,t;x,0)\psi_0(x)\,dx$$

$$= \sqrt[4]{\frac{m\omega}{\hbar\pi}}\sqrt{\frac{m\omega}{2\pi\hbar\sin\omega t}} \, e^{-\pi i/4}$$

$$\times \int_{\mathbb{R}} \exp\left(\frac{im\omega}{2\hbar\sin\omega t}[(q^2+x^2)\cos\omega t - 2qx] - \frac{m\omega}{2\hbar}x^2\right)dx.$$

One can actually evaluate this Gaussian integral by noting that

$$\frac{im\omega}{2\hbar\sin\omega t}[(q^2+x^2)\cos\omega t - 2qx] - \frac{m\omega}{2\hbar}x^2$$

$$= \frac{im\omega e^{i\omega t}}{2\hbar\sin\omega t}[x - e^{-i\omega t}q]^2 - \frac{m\omega}{2\hbar}q^2. \tag{7.65}$$

Indeed,

$$\frac{im\omega e^{i\omega t}}{2\hbar\sin\omega t}[x - e^{-i\omega t}q]^2 - \frac{m\omega}{2\hbar}q^2$$

$$= \frac{im\omega}{2\hbar\sin\omega t}[x^2 e^{i\omega t} - 2qx + e^{-i\omega t}q^2] - \frac{m\omega}{2\hbar}q^2$$

$$= \frac{im\omega}{2\hbar\sin\omega t}[(q^2+x^2)\cos\omega t - 2qx + i(x^2-q^2)\sin\omega t] - \frac{m\omega}{2\hbar}q^2$$

$$= \frac{im\omega}{2\hbar\sin\omega t}[(q^2+x^2)\cos\omega t - 2qx] - \frac{m\omega}{2\hbar}(x^2-q^2) - \frac{m\omega}{2\hbar}q^2$$

$$= \frac{im\omega}{2\hbar \sin \omega t} [(q^2 + x^2) \cos \omega t - 2qx] - \frac{m\omega}{2\hbar} x^2.$$

To compute the integral we write it as follows:

$$\int_{\mathbb{R}} \exp\left(\frac{im\omega}{2\hbar \sin \omega t} [(q^2 + x^2) \cos \omega t - 2qx] - \frac{m\omega}{2\hbar} x^2 \right) dx$$

$$= \int_{\mathbb{R}} \exp\left(\frac{im\omega e^{i\omega t}}{2\hbar \sin \omega t} [x - e^{-i\omega t} q]^2 - \frac{m\omega}{2\hbar} q^2 \right) dx$$

$$= e^{-m\omega q^2/2\hbar} \int_{\mathbb{R}} \exp\left(\frac{im\omega e^{i\omega t}}{2\hbar \sin \omega t} [x - e^{-i\omega t} q]^2 \right) dx$$

$$= e^{-m\omega q^2/2\hbar} \int_{\mathbb{R}} e^{-a(x-b)^2/2} dx$$

$$= \sqrt{\frac{2\pi}{a}} \quad \text{(see (A.6) in Appendix A),}$$

where $\sqrt{}$ means the principal branch of the square root function,

$$a = \frac{m\omega}{\hbar} (1 - i \cot \omega t)$$

and

$$b = e^{-i\omega t} q.$$

Exercise 7.4.14. Show that

$$\sqrt{\frac{2\pi}{a}} = \sqrt{\frac{2\pi\hbar \sin \omega t}{m\omega}} e^{\pi i/4} e^{-i\omega t/2}.$$

Thus,

$$\int_{\mathbb{R}} \exp\left(\frac{im\omega}{2\hbar \sin \omega t} [(q^2 + x^2) \cos \omega t - 2qx] - \frac{m\omega}{2\hbar} x^2 \right) dx$$

$$= e^{-m\omega q^2/2\hbar} \sqrt{\frac{2\pi\hbar \sin \omega t}{m\omega}} e^{\pi i/4} e^{-i\omega t/2}.$$

Putting all of this together we obtain

$$\psi(q,t) = \sqrt[4]{\frac{m\omega}{\hbar\pi}} \sqrt{\frac{m\omega}{2\pi\hbar \sin \omega t}} e^{-\pi i/4} \left(e^{-m\omega q^2/2\hbar} \sqrt{\frac{2\pi\hbar \sin \omega t}{m\omega}} e^{\pi i/4} e^{-i\omega t/2} \right)$$

$$= e^{-i\omega t/2} \psi_0(q).$$

The result of this calculation is no surprise, of course, since the ground state $\psi_0(q)$ is a stationary state of the harmonic oscillator and hence can change only in phase. Nevertheless, the calculation was a nice warm-up for something that *is* a bit surprising and quite interesting.

Exercise 7.4.15. Suppose that at $t = 0$, the initial wave function is the ground state translated to the right by some $a > 0$, that is,

$$\psi_0(q - a) = \sqrt[4]{\frac{m\omega}{\hbar\pi}}\, e^{-m\omega(q-a)^2/2\hbar}.$$

Then

$$|\psi_0(q - a)|^2 = \sqrt{\frac{m\omega}{\hbar\pi}}\, e^{-m\omega(q-a)^2/\hbar}$$

is a Gaussian probability distribution centered at $q = a$. Show that the time evolution is given by

$$\psi(q, t) = \sqrt[4]{\frac{m\omega}{\hbar\pi}}\, e^{-i\omega t/2} \exp\left(-\frac{im\omega}{2\hbar} \left(2aq \sin \omega t - \frac{a^2}{2} \sin 2\omega t \right) \right)$$
$$\cdot \exp\left(-\frac{m\omega}{2\hbar} (q - a \cos \omega t)^2 \right).$$

The phase admittedly evolves in a rather complicated fashion, but note that

$$|\psi(q, t)|^2 = \sqrt{\frac{m\omega}{\hbar\pi}} \exp\left(-\frac{m\omega}{\hbar} (q - a \cos \omega t)^2 \right),$$

and this is simply another Gaussian probability distribution, but centered at

$$q = a \cos \omega t,$$

which oscillates back and forth in precisely the same way as the classical harmonic oscillator. How does one interpret this result physically? *Hint*: The identity you will need in place of (7.65) is

$$\frac{im\omega}{2\hbar \sin \omega t} [\, (q^2 + x^2) \cos \omega t - 2qx\,] - \frac{m\omega}{2\hbar} (x - a)^2$$
$$= \frac{im\omega e^{i\omega t}}{2\hbar \sin \omega t} [\, x - ie^{-i\omega t} (a \sin \omega t - iq) \,]^2 - \frac{m\omega}{2\hbar} (q^2 - 2aqe^{-i\omega t} + a^2 e^{-i\omega t} \cos \omega t),$$

so verify this as well.

As we did for the free particle in Section 7.3, one generally takes the initial state to be specified at $t = 0$ (rather than some $t_0 > 0$) and then suppresses the "0" from the notation. Here we will define

$$K_B(q, x, t) = K(q, t; x, 0).$$

One can then prove analogues of Exercise 7.3.7 and (7.48) to the effect that $K_B(q, x, t)$ is a fundamental solution to the harmonic oscillator equation and also satisfies the group property with respect to t expressed in Exercises 7.3.8 and 7.3.9.

8 Path integral quantization

8.1 Motivation

Quantum mechanics, as we have viewed it so far, is a theory of self-adjoint operators on a Hilbert space and is modeled on the Hamiltonian picture of classical mechanics. Paul Dirac, who viewed Lagrangian mechanics as more fundamental than Hamiltonian, took the first steps toward a Lagrangian formulation of quantum theory in [Dirac3]. Dirac's suggestions were taken up by Richard Feynman in his Princeton PhD thesis (see [Brown]), the result being what is known today as the *Feynman path integral formulation of quantum theory*. Initially, Feynman's approach to quantum mechanics was largely ignored, but it was not long before physicists came around to Feynman's point of view and today path integrals are standard operating procedure in quantum mechanics, and even more so in quantum field theory. Not surprisingly, the literature on the subject is vast. The applications in physics are ubiquitous and the mathematical problem of making some rigorous sense of Feynman's "integrals" (which, as we shall see, are not integrals at all in the usually accepted mathematical sense) has received an enormous amount of attention. As always, our objective here is exceedingly modest. We will try to give some sense of what is behind Feynman's idea and why the idea is so difficult to turn into precise mathematics. We will provide detailed computations of the two simplest cases (the path integral representations of the propagators for the free quantum particle and the harmonic oscillator). This done we will briefly discuss how one might go about dealing with path integrals rigorously. By way of compensation for the modesty of our goals we will try to provide ample references to more serious discussions. Here are a few general sources. One can find Feynman's thesis and a discussion of it in [Brown]; his first published paper on the subject is [Feyn], which is very readable and highly recommended. For a physics-oriented discussion of the path integral and many applications a standard reference is [Schul], but [Fels] might be more congenial for mathematicians. There is a nice, brief survey of various approaches to a rigorous definition of the path integral in [Mazz1] and a great many more details in [AHM], [JL] and [Simon2]; also highly recommended is the paper [Nel3] of Edward Nelson.

We have seen several ways of thinking about the propagator of a quantum mechanical system. It is the integral kernel of the evolution operator. It is also the fundamental solution to the Schrödinger equation. Physically, $K(q_b, t_b; q_a, t_a)$ is the probability amplitude for detecting a particle at q_b at time t_b given that it is known to have been detected at q_a at time t_a. We have already computed two examples explicitly and in the next two sections we will compute them again, but the procedure will be very different indeed.

To understand the rationale behind Feynman's new perspective on quantum mechanics we will begin with a brief recap of our discussion of the two-slit experiment

https://doi.org/10.1515/9783110751949-008

in Section 4.4. The lessons we learned from this experiment are essentially those enumerated by Feynman in Section 1-7, Volume III, of [FLS].

1. Because of the wave-like attributes of particles in quantum mechanics (de Broglie waves) and the resultant interference effects, the probability P that a particular event (such as the arrival of an electron at some location on a screen) will occur is represented as the squared modulus $P = |\psi|^2$ of a complex number ψ called the probability amplitude of the event.

2. When there are several classical alternatives for the way in which the event can occur and no measurements are made on the system, the probability amplitude of the event is the sum of the probability amplitudes for each alternative considered separately. In particular, if there are two possibilities with amplitudes ψ_1 and ψ_2 and probabilities $P_1 = |\psi_1|^2$ and $P_2 = |\psi_2|^2$, then the probability amplitude of the event is $\psi_1 + \psi_2$ and the probability of the event is

$$|\psi_1 + \psi_2|^2 = |\psi_1|^2 + |\psi_2|^2 + 2\,\mathrm{Re}(\psi_1\overline{\psi}_2).$$

The last term represents the effect of interference.

3. When a measurement is made to determine whether one or another of the classical alternatives in 2. is actually taken, the interference is lost and the probability of the event is the sum of the probabilities for each alternative taken separately.

In Section 4.4 the appearance of an interference pattern in the two-slit experiment was explained (qualitatively, at least) by considering the two classical paths from the electron source S, through one of the slits and on to a point X on the screen, viewing each as having a certain probability amplitude and appealing to 2. Feynman's rather remarkable idea in [Brown] and [Feyn] was that not only these two paths, but *every* continuous path from the source to one of the slits and then on to the point on the screen must contribute to the probability amplitude; not just the "reasonable" ones either, but *all* of them, even the crazy ones.

This may sound outlandish at first, but really it is not at all. Indeed, Feynman was led by this seemingly simple experiment to a vastly more general conclusion. His reasoning goes something like this. Consider an electron that is emitted (or just detected) at some point S. The electron wants to get to another point X. What is the probability amplitude that it will be able to do this? If there were a wall W with two slits between S and X, then there would be contributions to the amplitude from (at least) the two piecewise linear paths from S, through one of the slits, and on to X. If the wall had three slits instead of two, we would have to sum three amplitudes instead of two, one for each of the three piecewise linear paths from S to one of the slits and then on to the point X in question. If the wall had n_1 slits we would sum amplitudes over n_1 piecewise linear paths, each consisting of two linear segments. But now suppose that there are two walls W_1 and W_2 instead of just one, the first with n_1 slits and the second with n_2 slits. These determine $n_1 n_2$ piecewise linear paths from S to our point X, each consisting of three linear segments and each with an amplitude that contributes to the total amplitude for getting from S to X. If there are m walls $W_1, W_2, \ldots, W_m$ with $n_1, n_2, \ldots, n_m$ slits, respectively, then we would sum amplitudes over $n_1 n_2 \cdots n_m$ piecewise linear paths from S to X, each with $m + 1$ linear segments. Now for the good part. As $m, n_1, n_2, \ldots, n_m$ become very large, the walls fill up the space between S and X, the slits drill virtually everything out of the walls and the piecewise linear paths acquire more and more, but smaller and smaller linear segments. Since any continuous path from S to X is a uniform limit of such piecewise linear paths, Feynman concludes that, in the limit, there are no walls and no slits, so the electron is moving through empty space and *every continuous path contributes to the amplitude.* Very pretty! Returning to the two-slit experiment itself, everything is exactly the same except there is one fixed, immutable wall W with two slits between S and X and we simply carry out Feynman's argument between S and W and between W and X, that is, we consider only continuous paths that go through one of the two slits in W (the "classical alternatives" in number 2 above).

The question left unresolved by all of this, of course, is *precisely how does each such path "contribute" to the probability amplitude?* The task then would be to assign a probability amplitude to every such continuous path and somehow "add" them. How these amplitudes should be assigned is not so clear, and it is even less clear how one is to "add" so many of them. We will try to supply some general motivation in the remainder of this section and then move on to the explicit examples in Sections 8.2 and 8.3. Not all of us are motivated by the same sort of reasoning, of course. Feynman himself arrived at his path integral by way of physical arguments and an inspired guess due to Dirac. In hindsight, there are other approaches more likely to appeal to mathematicians. It is our feeling that one should see both of these, so we will sketch both. In deference to our intended audience, however, we will begin with the one that is likely to be most congenial to mathematicians (the idea is due to Nelson [Nel3]).

We should be clear on the problem we want to address. We will generally consider only quantum systems with one degree of freedom. The Hilbert space is $\mathcal{H} = L^2(\mathbb{R})$ and we will assume the Hamiltonian is of the form

$$H = H_0 + V = -\frac{\hbar^2}{2m}\Delta + V = -\frac{\hbar^2}{2m}\frac{d^2}{dq^2} + V,$$

where V is some real-valued function on $\mathbb{R}$ that acts on $L^2(\mathbb{R})$ as a multiplication operator. We know that $H_0 = -\frac{\hbar^2}{2m}\Delta$ is defined and self-adjoint on $\mathcal{D}(H_0) = \{\psi \in L^2(\mathbb{R}) : \Delta\psi \in L^2(\mathbb{R})\}$, where Δ is understood in the distributional sense (see Appendix G.2), and that V is defined and self-adjoint on $\mathcal{D}(V) = \{\psi \in L^2(\mathbb{R}) : V\psi \in L^2(\mathbb{R})\}$, where $(V\psi)(q) = V(q)\psi(q)$ for every $q \in \mathbb{R}$. Now, the sum of two *bounded* self-adjoint operators is self-adjoint $((A + B)^* = A^* + B^* = A + B)$, but this is generally not true for unbounded operators where even the "sum" itself could be problematic. On the other hand, we really need $H = H_0 + V$ to be self-adjoint since otherwise it would not give rise to the evolution operator $e^{-itH/\hbar}$ that describes the dynamics (this is Stone's Theorem 5.6.2). An enormous amount of work in quantum mechanics has been and still is devoted to the problem of determining conditions on the potential V sufficient to ensure the self-adjointness of $H_0 + V$. We will have a very brief look at just a few such results in Section 9.2. We point out, however, that the case of real interest to us is the harmonic oscillator potential $V = \frac{m\omega^2}{2}q^2$ and for this we have already proved the self-adjointness of $H_B = H_0 + V$ (see Example 5.3.1). For the remainder of this introduction we will simply *assume that we are dealing with a potential V for which $H_0 + V$ is essentially self-adjoint on $\mathcal{D}(H_0) \cap \mathcal{D}(V)$.* What we *cannot* assume, however, is that H_0 and V commute and this is the first obstacle in our path. Life would be very simple if they did commute for then the evolution operator $e^{-itH/\hbar} = e^{-it(H_0+V)/\hbar}$ would be just $e^{-itH_0/\hbar} e^{-itV/\hbar}$ and this we know all about. What will get us over this obstacle is the famous *Lie–Trotter–Kato product formula*, sometimes called just the *Trotter product formula*.

Theorem 8.1.1 (Lie–Trotter–Kato product formula). *Let A and B be self-adjoint operators on the complex, separable Hilbert space $\mathcal{H}$. If $A + B$ is essentially self-adjoint on $\mathcal{D}(A + B) = \mathcal{D}(A) \cap \mathcal{D}(B)$, then, for every $\psi \in \mathcal{H}$ and each $t \in \mathbb{R}$,*

$$e^{-it\overline{(A+B)}}\psi = \lim_{n\to\infty} \left(\left(e^{-i(\frac{t}{n})A} e^{-i(\frac{t}{n})B} \right)^n \psi \right),$$

where $\overline{(A + B)}$ is the (self-adjoint) closure of $A + B$ and the limit is in $\mathcal{H}$. Stated otherwise, $e^{-it\overline{(A+B)}}$ is the strong limit of $\left(e^{-i(\frac{t}{n})A} e^{-i(\frac{t}{n})B} \right)^n$ as $n \to \infty$. Furthermore, the limit is uniform in t on all compact subsets of $\mathbb{R}$.

The proof of this when either $\mathcal{H}$ is finite-dimensional, or it is infinite-dimensional and A and B are bounded is straightforward (see Theorem VIII.29 of [RS1]). When the operators are unbounded and $A + B$ is actually self-adjoint on $\mathcal{D}(A) \cap \mathcal{D}(B)$ a proof takes a bit more work, but a concise one can be found in Theorem VIII.30 of [RS1] or Appendix B of [Nel3]. Assuming only that $A + B$ is essentially self-adjoint on $\mathcal{D}(A) \cap \mathcal{D}(B)$ necessitates a rather different sort of argument and for this one might want to consult [Ch] or Corollary 11.1.6 of [JL]. There is a generalization of Theorem 8.1.1 in the context of what are called "contractive semigroups of operators" on Banach spaces and their "infinitesimal generators." We will need to discuss this when we consider the Feynman–Kac formula in Section 9.3. What we need to know about semigroups of operators is described in Appendix J.

Our objective is to motivate Feynman's path integral representation for the propagator of the time evolution associated with $H = H_0 + V$. This propagator is the integral kernel of the time evolution operator. More precisely, it is a function $K(q, t; x, t_0)$ defined by the condition that for any fixed t_0, if ψ is the solution to the Schrödinger equation for $H_0 + V$ with initial state $\psi(\cdot, t_0)$, then

$$\psi(q, t) = \left(e^{-(i/\hbar)(t-t_0)(H_0+V)} \psi(\cdot, t_0) \right)(q) = \int_{\mathbb{R}} K(q, t; x, t_0)\psi(x, t_0)\, dx.$$

For the purposes of motivation we will assume that $\psi(x, t_0)$ is very nice, say, a Schwartz function (see Appendix G.2). Intuitively, the amplitude $\psi(q, t)$ for detecting the particle at location q at time t is a (continuous) "sum" of weighted amplitudes $\psi(x, t_0)$ as x varies over all of the possible locations of the particle at time t_0, the weight being the propagator $K(q, t; x, t_0)$. The procedure will be to write out the limit in

$$\psi(q,t) = \lim_{n \to \infty} \left((e^{-(i/\hbar)(\frac{t-t_0}{n})H_0} e^{-(i/\hbar)(\frac{t-t_0}{n})V})^n \ \psi(\cdot, t_0) \right)(q) = \int_{\mathbb{R}} K(q, t; x, t_0)\psi(x, t_0)\, dx$$

explicitly and try to read off $K(q, t; x, t_0)$.

It will streamline the development a bit if we adopt a few notational changes. First, let us fix two real numbers $t_a < t_b$. The time evolution of the system defines a curve in $\mathcal{H} = L^2(\mathbb{R})$, which we will write as $\{\psi_t\}_{t \in \mathbb{R}}$. We will specify the initial state $\psi_{t_a} = \psi(\cdot, t_a)$ at time t_a. For our application of the Lie–Trotter–Kato product formula we will begin by computing, from $(e^{-(i/\hbar)(t_b - t_a)V})\psi(\cdot, t_a) = e^{-(i/\hbar)(t_b - t_a)V(\cdot)}\psi(\cdot, t_a)$ and from (7.46), that

$$\left(e^{-(i/\hbar)(t_b - t_a)H_0} e^{-(i/\hbar)(t_b - t_a)V} \psi(\cdot, t_a) \right)(q_b)$$

$$= \int_{\mathbb{R}} K(q_b, t_b; x, t_a) \left[e^{-(i/\hbar)(t_b - t_a)V(x)}\psi(x, t_a) \right] dx$$

$$= \left(\frac{m}{2\pi i \hbar(t_b - t_a)} \right)^{\frac{1}{2}} \int_{\mathbb{R}} e^{\frac{i}{\hbar}\frac{m}{2}\frac{(q_b - x)^2}{t_b - t_a}} e^{-\frac{i}{\hbar}(t_b - t_a)V(x)}\psi(x, t_a)\, dx$$

$$= \left(\frac{m}{2\pi i \hbar(t_b - t_a)} \right)^{\frac{1}{2}} \int_{\mathbb{R}} e^{\frac{i}{\hbar}[\frac{m}{2}(\frac{q_b - x}{t_b - t_a})^2 - V(x)](t_b - t_a)}\psi(x, t_a)\, dx.$$

Once we get beyond the motivational stage we will want to keep in mind that when ψ_{t_a} is a general element of $L^2(\mathbb{R})$ and not necessarily a Schwartz function, this integral must be thought of as an integral in the mean (see Theorem 7.3.1).

Now we return to the interval $[t_0, t]$. Fix a positive integer n and subdivide the interval into n equal subintervals with endpoints $t_0 < t_1 = t_0 + \Delta t < t_2 = t_1 + \Delta t < \cdots < t_{n-1} < t_n = t_{n-1} + \Delta t = t$, where $\Delta t = \frac{t - t_0}{n}$. Apply the formula we just derived to the interval $[t_a, t_b] = [t_0, t_1]$ and write, instead of x and q_b, q_0 and q_1, respectively. The result is

$$\left(e^{-(i/\hbar)(t_1 - t_0)H_0} e^{-(i/\hbar)(t_1 - t_0)V} \psi(\cdot, t_0) \right)(q_1)$$

$$= \left(\frac{m}{2\pi i \hbar(t_1 - t_0)} \right)^{\frac{1}{2}} \int_{\mathbb{R}} e^{\frac{i}{\hbar}[\frac{m}{2}(\frac{q_1 - q_0}{t_1 - t_0})^2 - V(q_0)](t_1 - t_0)}\psi(q_0, t_0)\, dq_0,$$

or

$$\left(e^{-(i/\hbar)\Delta t H_0} e^{-(i/\hbar)\Delta t V} \psi(\cdot, t_0) \right)(q_1) = \left(\frac{m}{2\pi i \hbar \Delta t} \right)^{\frac{1}{2}} \int_{\mathbb{R}} e^{\frac{i}{\hbar}[\frac{m}{2}(\frac{q_1 - q_0}{\Delta t})^2 - V(q_0)]\Delta t}\psi(q_0, t_0)\, dq_0.$$

Next we compute

$$((e^{-(i/\hbar)\Delta t H_0} e^{-(i/\hbar)\Delta t V})^2 \psi(t_0, \cdot))(q_2)$$

$$= (e^{-(i/\hbar)(t_2-t_1)H_0} e^{-(i/\hbar)(t_2-t_1)V})(e^{-(i/\hbar)(t_1-t_0)H_0} e^{-(i/\hbar)(t_1-t_0)V} \psi(t_0, \cdot))(q_2)$$

$$= \left(\frac{m}{2\pi i \hbar \Delta t}\right)^{\frac{2}{2}} \int_R e^{\frac{i}{\hbar}[\frac{m}{2}(\frac{q_2-q_1}{\Delta t})^2 - V(q_1)]\Delta t} \int_R e^{\frac{i}{\hbar}[\frac{m}{2}(\frac{q_1-q_0}{\Delta t})^2 - V(q_0)]\Delta t} \psi(t_0, q_0) \, dq_0 \, dq_1$$

$$= \left(\frac{m}{2\pi i \hbar \Delta t}\right)^{\frac{2}{2}} \int_R \int_R e^{\frac{i}{\hbar}[[\frac{m}{2}(\frac{q_2-q_1}{\Delta t})^2 - V(q_1)]\Delta t + [\frac{m}{2}(\frac{q_1-q_0}{\Delta t})^2 - V(q_0)]\Delta t]} \psi(t_0, q_0) \, dq_0 \, dq_1$$

$$= \left(\frac{m}{2\pi i \hbar \Delta t}\right)^{\frac{2}{2}} \int_R \int_R e^{\frac{i}{\hbar}\sum_{k=1}^{2}[\frac{m}{2}(\frac{q_k-q_{k-1}}{\Delta t})^2 - V(q_{k-1})]\Delta t} \psi(t_0, q_0) \, dq_0 \, dq_1.$$

Exercise 8.1.1. Continue inductively in this way to arrive at

$$((e^{-(i/\hbar)\Delta t H_0} e^{-(i/\hbar)\Delta t V})^n \psi(t_0, \cdot))(q_n)$$

$$= \left(\frac{m}{2\pi i \hbar \Delta t}\right)^{\frac{n}{2}} \int_R \int_R \cdots \int_R e^{\frac{i}{\hbar}\sum_{k=1}^{n}[\frac{m}{2}(\frac{q_k-q_{k-1}}{\Delta t})^2 - V(q_{k-1})]\Delta t} \psi(t_0, q_0) \, dq_0 \, dq_1 \cdots dq_{n-1}.$$

Now let us define

$$S_n(q_0, q_1, \ldots, q_n; t) = \sum_{k=1}^{n}\left[\frac{m}{2}\left(\frac{q_k - q_{k-1}}{\Delta t}\right)^2 - V(q_{k-1})\right]\Delta t \tag{8.1}$$

and write

$$((e^{-(i/\hbar)\Delta t H_0} e^{-(i/\hbar)\Delta t V})^n \psi(t_0, \cdot))(q_n)$$

$$= \left(\frac{m}{2\pi i \hbar \Delta t}\right)^{\frac{n}{2}} \int_R \cdots \int_R \int_R e^{\frac{i}{\hbar}S_n(q_0, q_1, \ldots, q_n; t)} \psi(t_0, q_0) \, dq_0 \, dq_1 \cdots dq_{n-1}$$

$$= \int_R \left\{\left(\frac{m}{2\pi i \hbar \Delta t}\right)^{\frac{n}{2}} \int_R \cdots \int_R e^{\frac{i}{\hbar}S_n(q_0, q_1, \ldots, q_n; t)} \, dq_1 \cdots dq_{n-1}\right\} \psi(t_0, q_0) \, dq_0.$$

To make contact with our previous notation we set $q_0 = x$ and $q_n = q$ and write this just once more as

$$((e^{-(i/\hbar)\Delta t H_0} e^{-(i/\hbar)\Delta t V})^n \psi_{t_0})(q)$$

$$= \int_R \left\{\left(\frac{m}{2\pi i \hbar \Delta t}\right)^{\frac{n}{2}} \int_R \cdots \int_R \int_R e^{\frac{i}{\hbar}S_n(q, q_1, \ldots, q_{n-1}, x; t)} \, dq_1 \, dq_2 \cdots dq_{n-1}\right\} \psi_{t_0}(x) \, dx.$$

According to the Lie–Trotter–Kato product formula, the sequence of functions of q on the right-hand side converges in L^2 as $n \to \infty$ to the evolved state ψ_t, that is,

$$\psi(q,t) = \lim_{n\to\infty} \int_{\mathbb{R}} \left\{ \left(\frac{m}{2\pi i\hbar\Delta t} \right)^{\frac{n}{2}} \int_{\mathbb{R}} \cdots \int_{\mathbb{R}} e^{\frac{i}{\hbar}S_n(q,q_1,\dots,q_{n-1},x;t)} dq_1 \cdots dq_{n-1} \right\} \psi(x,t_0)\, dx. \quad (8.2)$$

This *suggests* that the expression in the braces should converge to the propagator. The limit

$$\lim_{n\to\infty} \left(\frac{m}{2\pi i\hbar\Delta t} \right)^{\frac{n}{2}} \int_{\mathbb{R}} \cdots \int_{\mathbb{R}} e^{\frac{i}{\hbar}S_n(q,q_1,\dots,q_{n-1},x;t)} dq_1 \cdots dq_{n-1} \quad (8.3)$$

is, by definition, the *Feynman path integral representation for the propagator* $K(q,t;$ $x,t_0)$. The limit in (8.2) is the definition of the *Feynman path integral representation for the wave function* $\psi(q,t)$. We will see in the next two sections that for the free particle and the harmonic oscillator, the limit (8.3) can be computed explicitly and the results agree with the propagators we found in Sections 7.3 and 7.4.

Note, however, that one *cannot* deduce the convergence of the path integral (8.3) to the propagator from the Lie–Trotter–Kato product formula. The reason is that the propagator is the distributional solution to the Schrödinger equation with initial data given by a Dirac delta and the Dirac delta is nowhere to be found in $L^2(\mathbb{R})$. For more on the convergence of the path integral to the propagator one can consult [Fuj1]. Convergence in $L^2(\mathbb{R})$ implies the pointwise almost everywhere convergence of some subsequence, but one might ask whether or not the limit in (8.2) exists in some stronger sense. One can, indeed, prove convergence in various stronger senses, but only for certain restricted classes of potentials V; for more on this, see [Fuj2] and [Fuj3].

The limit in (8.3) does not seem to involve any "paths" and certainly does not look like an "integral" in the usual measure theoretic sense (indeed, we will see in Section 9.3 that it is not). It may seem strange then to refer to it as a "path integral" and to denote it, as physicists generally do, by some variant of a symbol such as

$$\int_{\mathcal{P}_{\mathbb{R}}(q,t;x,t_0)} e^{\frac{i}{\hbar}S(\alpha)} \mathcal{D}q, \quad (8.4)$$

where $\mathcal{P}_{\mathbb{R}}(q,t;x,t_0)$ is the space of continuous paths α in $\mathbb{R}$ joining x at time t_0 with q at time t and $\mathcal{D}q$ is intended to represent the "measure"

$$\mathcal{D}q = \lim_{n\to\infty} \left(\frac{m}{2\pi i\hbar\Delta t} \right)^{\frac{n}{2}} dq_1 \cdots dq_{n-1}. \quad (8.5)$$

There is really nothing sensible to say about this "measure" $\mathcal{D}q$ except that it is purely formal and quite meaningless as it stands. On the other hand, $e^{\frac{i}{\hbar}S(\alpha)}$ is not so obscure. No doubt the expression in (8.1) for $S_n(q_0,\dots,q_n;t)$ looks familiar. If we had a *smooth* path $\alpha : [t_0, t] \to \mathbb{R}$ in $\mathbb{R}$ describing a classical path in a system with Lagrangian $\frac{m}{2}\dot{q}^2 - V(q)$ and if V were *continuous*, then its classical action would be

$$S(\alpha; t) = \int_{t_0}^{t} \left[\frac{m}{2}\dot{\alpha}(s)^2 - V(\alpha(s)) \right] ds.$$

But $t_0 < t_1 = t_0 + \Delta t < t_2 = t_0 + 2\Delta t < \cdots < t_n = t_0 + n\Delta t = t$ is a subdivision of $[t_0, t]$ and if we take $q_k = \alpha(t_k)$, $k = 0, 1, \ldots, n$, then $S_n(q_0, \ldots, q_n; t)$ is just a Riemann sum approximation to $S(\alpha; t)$ and approaches $S(\alpha; t)$ as $n \to \infty$. While the classical action is not defined for a path that is continuous, but not differentiable, the sum in (8.1) is perfectly well-defined for such paths. Intuitively, at least, we can turn this whole business around in the following way. Suppose that $\alpha : [t_0, t] \to \mathbb{R}$ is an arbitrary *continuous* path. For any n we have a partition $t_0 < t_1 = t_0 + \Delta t < t_2 = t_0 + 2\Delta t < \cdots < t_n = t_0 + n\Delta t = t$ of $[t_0, t]$ and we can approximate α by the piecewise linear path with segments joining $(t_{k-1}, q_{k-1}) = (t_{k-1}, \alpha(t_{k-1}))$ and $(t_k, q_k) = (t_k, \alpha(t_k))$ for $k = 1, \ldots, n$. Then $S_n(q_0, \ldots, q_n; t)$ is approximately the action of this polygonal path in a system with Lagrangian $\frac{m}{2}\dot{q}^2 - V(q)$. Since any continuous path can be uniformly approximated arbitrarily well by such piecewise linear paths one should be able to define an "action" functional $S(\alpha)$ on $\mathcal{P}_{\mathbb{R}}(q, t; x, t_0)$.

Since $\mathcal{D}q$, as it was introduced in (8.5), is not a measure, or anything else for that matter, (8.4) certainly cannot be regarded as a Lebesgue integral with respect to $\mathcal{D}q$. This does not preclude the possibility that there exists a measure on $\mathcal{P}_{\mathbb{R}}(q, t; x, t_0)$ relative to which the integral of $e^{\frac{i}{\hbar}S(\alpha)}$ is just the limit in (8.3), and this would certainly be a desirable situation since it would make available all of the rather substantial machinery of Lebesgue theory. Alas, there is no such measure (this is a theorem of Cameron [Cam] that we will discuss in Section 9.3). Nevertheless, the Feynman "integral" is closely related to an honest Lebesgue integral on the space of continuous paths constructed by Norbert Wiener to model the phenomenon of Brownian motion and we will have a bit more to say about this in Section 9.3. For the time being it will be best to think of (8.4) as simply a shorthand notation for the limit (8.3).

Two comments are in order here. First, we note that the original motivation for thinking of the limit (8.3) as some sort of "integral" will become a bit clearer shortly when we sketch Feynman's more physically motivated *sum over histories* approach. Second, one can, in fact, rigorously define measures on infinite-dimensional path spaces such as $\mathcal{P}_{\mathbb{R}}(q, t; x, t_0)$ and we will briefly look at two means of doing this in Section 9.3.

The procedure we have just described for arriving at the Feynman path integral representation of the propagator is not at all the way Feynman thought of it (the Lie–Trotter–Kato product formula did not exist at the time). We will conclude this introduction by briefly sketching the physical ideas that led Feynman to his path integral (although we strongly recommend that the reader go directly to [Feyn] instead). The underlying philosophy has already been described at the beginning of this section. The probabilistic interpretation of the propagator and the fact that quantum mechanics will not permit us to view a particle as following any particular path led Feynman to view $K(q_b, t_b; q_a, t_a)$ as a sum of contributions, one from each continuous path α joining the points in question, that is, one from each *(classically) possible history* of the particle. One might write this symbolically as

$$\sum_{\mathcal{P}_\mathbb{R}(q_b, t_b; q_a, t_a)} K(\alpha),$$

where $K(\alpha)$ is the amplitude assigned to the path $\alpha \in \mathcal{P}_\mathbb{R}(q_b, t_b; q_a, t_a)$. On the other hand, $\mathcal{P}_\mathbb{R}(q_b, t_b; q_a, t_a)$ is rather large, so this might more properly be thought of as a "continuous sum" or "integral"

$$\int_{\mathcal{P}_\mathbb{R}(q_b, t_b; q_a, t_a)} K(\alpha).$$

This is all very well, but how does one assign an amplitude $K(\alpha)$ to every $\alpha \in \mathcal{P}_\mathbb{R}(q_b, t_b; q_a, t_a)$ and, once this is done, how does one actually compute this sum/integral? A proposed answer to the first question comes not from Feynman, but from Dirac [Dirac3], who suggested that the appropriate weight, or amplitude to be assigned to any path α should be determined by its classical action $S(\alpha)$ according to

$$K(\alpha) = e^{\frac{i}{\hbar} S(\alpha)}. \tag{8.6}$$

Dirac's reasoning, described in the first three sections of [Dirac3], is not based on direct physical arguments, but rather on formal analogies between classical Lagrangian mechanics and quantum mechanics that suggested, to Dirac, that the transition amplitude between the states ψ_{t_a} and ψ_{t_b} in quantum mechanics "corresponds to"

$$e^{\frac{i}{\hbar} \int_{t_a}^{t_b} L \, dt}$$

in classical mechanics. We will not pretend to share whatever intuitions led Dirac to this "correspondence," but will simply ask him to speak for himself by referring those who are interested to [Dirac3]. In the end it is probably best to regard (8.6) as the sort of inspired guess that one would expect from Paul Dirac. In hindsight, the recurrence of $e^{\frac{i}{\hbar} S}$ in our calculations of the propagators for the free particle (Section 7.3) and harmonic oscillator (Section 7.4) as well as its emergence from the Lie–Trotter–Kato product formula lend support to the idea, but fundamentally this is just a brilliant insight from a great physicist.

Feynman adopted Dirac's proposal and so his task was to provide an operational definition of

$$\int_{\mathcal{P}_\mathbb{R}(q_b, t_b; q_a, t_a)} e^{\frac{i}{\hbar} S(\alpha)}.$$

Not surprisingly, Feynman's procedure was to approximate and take a limit; it is called *time slicing* in physics. One chooses a partition $t_a = t_0 < t_1 < \cdots < t_{n-1} < t_n = t_b$ of $[t_a, t_b]$ into n subintervals of length $\Delta t = \frac{t_b - t_a}{n}$. Let $q_0 = q_a$ and $q_n = q_b$. Any choice of $q_1, \ldots, q_{n-1}$ in $\mathbb{R}$ gives rise to a piecewise linear path joining (t_{k-1}, q_{k-1}) to (t_k, q_k) for $k = 1, \ldots, n$. Then

$$S_n(q_0, q_1, \ldots, q_{n-1}, q_n) = \sum_{k=1}^{n} \left[\frac{m}{2} \left(\frac{q_k - q_{k-1}}{\Delta t} \right)^2 - V(q_k) \right] \Delta t$$

is approximately the classical action of the piecewise linear path if the Lagrangian is $\frac{m}{2} \dot{q}^2 - V(q)$ and

$$e^{\frac{i}{\hbar} S_n(q_0, q_1, \ldots, q_{n-1}, q_n)}$$

is the contribution this path makes to the propagator. Another choice of $q_1, \ldots, q_{n-1}$ gives another piecewise linear path with another contribution and one obtains the total contribution of all such piecewise linear paths by adding all of these up, that is, by integrating over $-\infty < q_k < \infty$ for each $k = 1, \ldots, n$:

$$\int_{\mathbb{R}} \cdots \int_{\mathbb{R}} e^{\frac{i}{\hbar} S_n(q_a, q_1, \ldots, q_{n-1}, q_b)} \, dq_1 \, \cdots \, dq_{n-1}.$$

Repeating this procedure over and over with larger and larger n and operating under the assumption that any continuous path can be arbitrarily well approximated by such piecewise linear paths with sufficiently many segments, Feynman would like to take the limit of these $(n - 1)$-fold multiple integrals as $n \to \infty$. Realizing full well that the limit will generally not exist, Feynman chooses a normalizing factor of

$$\left(\frac{m}{2\pi i \hbar \Delta t} \right)^{\frac{n}{2}}$$

to multiply the integrals by in order to ensure convergence. Although Feynman offers no hint as to how he arrived at an appropriate factor, one might guess that he chose it in order to guarantee that, in the case of the free particle, his limit not only existed, but gave the right answer for the propagator (we will see this explicitly when we evaluate the free particle path integral in the next section).

Thus, Feynman's *definition* of his path integral is the same as ours; reinserting the mythical $\mathcal{D}q$ it is

$$\int_{\mathcal{P}_{\mathbb{R}}(q_b, t_b; q_a, t_a)} e^{\frac{i}{\hbar} S(\alpha)} \mathcal{D}q = \lim_{n \to \infty} \left(\frac{m}{2\pi i \hbar \Delta t} \right)^{\frac{n}{2}} \int_{\mathbb{R}} \cdots \int_{\mathbb{R}} e^{\frac{i}{\hbar} S_n(q_a, q_1, \ldots, q_{n-1}, q_b)} \, dq_1 \, \cdots \, dq_{n-1},$$

where $\Delta t = \frac{t_b - t_a}{n}$.

One might reasonably ask why anyone would want to write the Feynman integral as an integral if it is not really an integral at all. The answer is, in a sense, psychological. Written as an integral one is inclined to think of it as an integral and to do things with it and to it that one is accustomed to doing with and to integrals (change of variables, integration by parts, stationary phase approximation, and so on). Although these things are in no way justified mathematically by the definition they, quite remarkably, often lead to conclusions that can be verified by other means. This is particularly true in quantum field theory where formal manipulations with path integrals have led to quite extraordinary insights into numerous branches of mathematics seemingly unrelated to quantum field theory. These are, regrettably, beyond our grasp here (for an introduction to the connections between quantum field theory and topology one might begin with [Nash]). We will, however, mention just one rather more mundane instance of this phenomenon because it bears on our discussion of the classical limit problem. If one is willing to take $\int_{\mathcal{P}_{\mathbb{R}}(q_b, t_b; q_a, t_a)} e^{\frac{i}{\hbar} S(\alpha)} \mathcal{D}q$ seriously as an integral, then one cannot help but notice its formal similarity to the oscillatory integrals $\int_{\mathbb{R}} e^{i \, Tf(x)} g(x) \, dx$ considered in Appendix B. For certain of these finite-dimensional oscillatory integrals we found a stationary phase approximation describing their asymptotic behavior as $T \to \infty$. For the path integral this would correspond to the classical limit $\hbar \to 0^+$. Permitting oneself the latitude of believing that the path integral is really an integral and that there is an infinite-dimensional analogue of the stationary phase approximation, a formal application of the approximation to the integral leads, just as in the finite-dimensional case, to the conclusion that the dominant contributions to the path integral come from the stationary points of the action functional and these are just the classical trajectories (see Section 5 of [KM]). In this sense, the $\hbar \to 0^+$ limit of quantum mechanics picks out from among all of the possible paths a particle might follow precisely the one that classical mechanics says it should follow. Physicists have actually taken this a great deal further. There are circumstances in which one can prove (in the finite-dimensional context) that the stationary phase approximation is actually exact. These circumstances are best described in terms of what is called *equivariant cohomology* and hence the results are called *equivariant localization theorems* (see, for example, [BGV]). Formally appropriating these results in the infinite-dimensional context physicists obtain relatively simple closed form expressions for otherwise intractable path integrals. For more on this we will simply refer those interested to [Szabo].

8.2 Path integral for the free quantum particle

This section (and the next) should be regarded as something of a reality check. We have already calculated an explicit formula for the propagator $K(q_b, t_b; q_a, t_a)$ representing the probability amplitude that a free particle will be detected at q_b at time t_b

given that it was detected at q_a at time t_a (Section 7.3). Specifically,

$$K(q_b, t_b; q_a, t_a) = \sqrt{\frac{m}{2\pi i\hbar(t_b - t_a)}}\, e^{\frac{i}{\hbar}\left(\frac{m}{2(t_b - t_a)}(q_b - q_a)^2\right)}.$$

Feynman has assured us that the same propagator can be obtained by "integrating" $e^{\frac{i}{\hbar}S(\alpha)}$ over all continuous paths α in $\mathbb{R}$ from $\alpha(t_a) = q_a$ to $\alpha(t_b) = q_b$ and he has told us precisely what he means by "integrating." We will now compute Feynman's path integral and see if it comes out right. In the next section we will do the same thing for the harmonic oscillator.

To get a feel for how the calculations are done and to re-enforce some of the physical ideas that are behind them we will begin by looking at the approximation to Feynman's integral corresponding to a subdivision of the time interval $[t_a, t_b]$ into just two subintervals:

$$t_0 = t_a < t_1 = \frac{t_a + t_b}{2} < t_b = t_2.$$

Feynman's idea is that every continuous path $\alpha : [t_a, t_b] \to \mathbb{R}$ from $\alpha(t_a) = q_a$ to $\alpha(t_b) = q_b$ contributes to the propagator $K(q_b, t_b; q_a, t_a)$, so this should be true, in particular, for any polygonal path starting at $q_0 = q_a$ at time t_0, going through (t_1, q_1) for some $q_1 \in \mathbb{R}$ and ending at $q_2 = q_b$ at time t_2. Fix such a path and let $\Delta t = \frac{t_b - t_a}{2}$ be the length of each subinterval. Each of the straight line segments is regarded as the classical path of a free particle joining its endpoints and, thought of in this way, each has a classical action given, according to Exercise 2.2.3, by

$$\frac{m}{2\Delta t}(q_k - q_{k-1})^2, \quad k = 1, 2.$$

Thus, the total action associated to the polygonal path is

$$\frac{m}{2\Delta t}\left[(q_1 - q_0)^2 + (q_2 - q_1)^2\right]$$

and the contribution this path makes to the propagator is

$$e^{\frac{i}{\hbar}\frac{m}{2\Delta t}\left[(q_1 - q_0)^2 + (q_2 - q_1)^2\right]}.$$

The collection of all paths of the type we are discussing is obtained by allowing q_1 to vary over $-\infty < q_1 < \infty$ and we must "sum" these up, that is, compute

$$\int_{\mathbb{R}} e^{\frac{i}{\hbar}\frac{m}{2\Delta t}\left[(q_1 - q_0)^2 + (q_2 - q_1)^2\right]}\, dq_1.$$

To turn this and the remaining integrals that we need to do into Gaussian integrals that are evaluated in Appendix A we will need a few algebraic identities. We will prove the first and then leave the rest for you to do in the same way. We will show that for any real numbers x, y and z,

$$(x - y)^2 + (z - x)^2 = 2\left(x - \frac{y + z}{2}\right)^2 + \frac{(y - z)^2}{2} \tag{8.7}$$

(the reason we like this is that x appears only in the first square). To prove this we just complete the square:

$$(x - y)^2 + (z - x)^2 = x^2 - 2xy + y^2 + z^2 - 2xz + x^2$$

$$= 2x^2 - 2x(y + z) + y^2 + z^2$$

$$= 2\left[x^2 - x(y + z) + \frac{y^2}{2} + \frac{z^2}{2}\right]$$

$$= 2\left[x^2 - x(y + z) + \frac{(y + z)^2}{4} - \frac{(y + z)^2}{4} + \frac{y^2}{2} + \frac{z^2}{2}\right]$$

$$= 2\left[\left(x - \frac{y + z}{2}\right)^2 + \frac{y^2}{2} - \frac{y^2 + 2yz + z^2}{4} + \frac{z^2}{2}\right]$$

$$= 2\left[\left(x - \frac{y + z}{2}\right)^2 + \frac{y^2}{4} - \frac{2yz}{4} + \frac{z^2}{4}\right]$$

$$= 2\left(x - \frac{y + z}{2}\right)^2 + \frac{(y - z)^2}{2}.$$

Exercise 8.2.1. Complete the square to show that for any $x, y, z \in \mathbb{R}$,

$$\frac{(x - y)^2}{2} + (z - x)^2 = \frac{3}{2}\left(x - \frac{y + 2z}{3}\right)^2 + \frac{1}{3}(y - z)^2 \tag{8.8}$$

and, in general, for any $n \geq 2$,

$$\frac{(x - y)^2}{n - 1} + (z - x)^2 = \frac{n}{n - 1}\left(x - \frac{y + (n - 1)z}{n}\right)^2 + \frac{1}{n}(y - z)^2. \tag{8.9}$$

Now we use (8.7) and the Gaussian integral

$$\int_{\mathbb{R}} e^{\frac{1}{2}iau^2}\, du = \sqrt{\frac{2\pi i}{a}}, \quad a > 0 \tag{8.10}$$

(see Example A.0.1 in Appendix A), to compute

$$\int_{\mathbb{R}} e^{\frac{i}{\hbar}\frac{m}{2\Delta t}[(q_1 - q_0)^2 + (q_2 - q_1)^2]}\, dq_1 = e^{\frac{i}{\hbar}\frac{m}{2\Delta t}\frac{(q_2 - q_0)^2}{2}}\int_{\mathbb{R}} e^{\frac{i}{\hbar}\frac{m}{2\Delta t}2(q_1 - \frac{q_0 + q_2}{2})^2}\, dq_1$$

$$= e^{\frac{i}{\hbar}\frac{m}{2\Delta t}\frac{(q_2 - q_0)^2}{2}}\frac{1}{\sqrt{2}}\int_{\mathbb{R}} e^{\frac{1}{2}i(\frac{m}{\hbar\Delta t})u^2}\, du$$

$$= e^{\frac{i}{\hbar} \frac{m}{2\Delta t} \frac{(q_2-q_0)^2}{2}} \frac{1}{\sqrt{2}} \sqrt{\frac{2\pi i \hbar \Delta t}{m}}$$

$$= \frac{1}{\sqrt{2}} \left(\frac{2\pi i \hbar \Delta t}{m} \right)^{1/2} e^{\frac{i}{\hbar} \frac{m}{2\Delta t} \frac{(q_b-q_a)^2}{2}}.$$

Let us do this once more. Subdivide $[t_a, t_b]$ into three equal subintervals with end-points

$$t_0 = t_a < t_1 = t_0 + \Delta t < t_2 = t_0 + 2\Delta t < t_b = t_3,$$

where $\Delta t = \frac{t_b - t_a}{3}$. Let $q_0 = q_a$, $q_3 = q_b$, and let $q_1, q_2 \in \mathbb{R}$ be arbitrary. Now consider the polygonal path from (t_0, q_0) to (t_3, q_3) with segments joining (t_0, q_0) and (t_1, q_1), (t_1, q_1) and (t_2, q_2) and (t_2, q_2) and (t_3, q_3). Each of the straight line segments is regarded as the classical path of a free particle joining its endpoints and, thought of in this way, each has a classical action given, according to Exercise 2.2.3, by

$$\frac{m}{2\Delta t} (q_k - q_{k-1})^2, \quad k = 1, 2, 3.$$

Thus, the total action associated to the polygonal path is

$$\frac{m}{2\Delta t} [(q_1 - q_0)^2 + (q_2 - q_1)^2 + (q_3 - q_2)^2]$$

and the contribution this path makes to the propagator is

$$e^{\frac{i}{\hbar} \frac{m}{2\Delta t} [(q_1-q_0)^2+(q_2-q_1)^2+(q_3-q_2)^2]}.$$

Now, the collection of all paths of the type we are discussing is obtained by allowing q_1 and q_2 to vary over $(-\infty, \infty)$ and we must "sum" these up, that is, compute

$$\int_{\mathbb{R}} \int_{\mathbb{R}} e^{\frac{i}{\hbar} \frac{m}{2\Delta t} [(q_1-q_0)^2+(q_2-q_1)^2+(q_3-q_2)^2]} \, dq_1 \, dq_2.$$

For this we use the integral we just evaluated, the algebraic identity (8.8) and the Gaussian (8.10):

$$\int_{\mathbb{R}} \int_{\mathbb{R}} e^{\frac{i}{\hbar} \frac{m}{2\Delta t} [(q_1-q_0)^2+(q_2-q_1)^2+(q_3-q_2)^2]} \, dq_1 \, dq_2$$

$$= \int_{\mathbb{R}} e^{\frac{i}{\hbar} \frac{m}{2\Delta t} (q_3-q_2)^2} \int_{\mathbb{R}} e^{\frac{i}{\hbar} \frac{m}{2\Delta t} [(q_1-q_0)^2+(q_2-q_1)^2]} \, dq_1 \, dq_2$$

$$= \sqrt{\frac{1}{2}} \left(\frac{2\pi i \hbar \Delta t}{m} \right)^{1/2} \int_{\mathbb{R}} e^{\frac{i}{\hbar} \frac{m}{2\Delta t} [\frac{(q_2-q_0)^2}{2} +(q_3-q_2)^2]} \, dq_2$$

$$= \sqrt{\frac{1}{2}} \left(\frac{2\pi i \hbar \Delta t}{m} \right)^{1/2} e^{\frac{i}{\hbar} \frac{m}{2\Delta t} \frac{(q_3-q_0)^2}{3}} \int_{\mathbb{R}} e^{\frac{i}{\hbar} \frac{m}{2\Delta t} \frac{3}{2} (q_2- \frac{q_0+2q_3}{3})^2} \, dq_2$$

$$= \sqrt{\frac{1}{2}} \sqrt{\frac{2}{3}} \left(\frac{2\pi i\hbar\Delta t}{m} \right)^{2/2} e^{\frac{i}{\hbar} \frac{m}{2\Delta t} \frac{(q_3-q_0)^2}{3}}$$

$$= \sqrt{\frac{1}{3}} \left(\frac{2\pi i\hbar(t_b - t_a)}{3m} \right)^{2/2} e^{\frac{i}{\hbar} \left(\frac{m}{2(t_b-t_a)} (q_b-q_a)^2 \right)}.$$

Exercise 8.2.2. Show by induction that for any $n \geq 2$,

$$\int_{\mathbb{R}} \int_{\mathbb{R}} \cdots \int_{\mathbb{R}} e^{\frac{i}{\hbar} \frac{m}{2\Delta t} [(q_1-q_0)^2 + (q_2-q_1)^2 + (q_3-q_2)^2 + \cdots + (q_n-q_{n-1})^2]} \, dq_1 \, dq_2 \, \cdots \, dq_{n-1}$$

$$= \sqrt{\frac{1}{2}} \sqrt{\frac{2}{3}} \cdots \sqrt{\frac{n-1}{n}} \left(\frac{2\pi i\hbar\Delta t}{m} \right)^{\frac{n-1}{2}} e^{\frac{i}{\hbar} \frac{m}{2\Delta t} \frac{(q_n-q_0)^2}{n}}$$

$$= \sqrt{\frac{1}{n}} \left(\frac{2\pi i\hbar(t_b - t_a)}{nm} \right)^{\frac{n-1}{2}} e^{\frac{i}{\hbar} \left(\frac{m}{2(t_b-t_a)} (q_b-q_a)^2 \right)}.$$

Exercise 8.2.3. Show that for any $n \geq 2$,

$$\left(\frac{2\pi i\hbar\Delta t}{m} \right)^{-\frac{n}{2}} \int_{\mathbb{R}} \int_{\mathbb{R}} \cdots \int_{\mathbb{R}} e^{\frac{i}{\hbar} \frac{m}{2\Delta t} [(q_1-q_0)^2 + (q_2-q_1)^2 + (q_3-q_2)^2 + \cdots + (q_n-q_{n-1})^2]} \, dq_1 \, dq_2 \, \cdots \, dq_{n-1}$$

$$= \sqrt{\frac{m}{2\pi i\hbar(t_b - t_a)}} e^{\frac{i}{\hbar} \left(\frac{m}{2(t_b-t_a)} (q_b-q_a)^2 \right)}.$$

According to Feynman's instructions we are to take the limit as $n \to \infty$ of this last expression to obtain his path integral. Since n has disappeared, this is not so hard to do and we have arrived at

$$\int_{\mathcal{P}_{\mathbb{R}}(q_b,t_b;q_a,t_a)} e^{\frac{i}{\hbar} S} \mathcal{D}q = \sqrt{\frac{m}{2\pi i\hbar(t_b - t_a)}} e^{\frac{i}{\hbar} \left(\frac{m}{2(t_b-t_a)} (q_b-q_a)^2 \right)} = K(q_b, t_b; q_a, t_a).$$

In the case of the free particle, at least, the Feynman path integral does, indeed, converge (trivially) to the propagator. We turn next to a somewhat more interesting test case where the limit is not trivial.

8.3 Path integral for the harmonic oscillator

The harmonic oscillator potential is $V(q) = \frac{m\omega^2}{2} q^2$, so the path integral we need to evaluate can be written

$$\int_{\mathcal{P}_{\mathbb{R}}(q_b,t_b;q_a,t_a)} e^{\frac{im}{2\hbar} \int_{t_a}^{t_b} (\dot{q}^2 - \omega^2 q^2) \, dt} \mathcal{D}q$$

$$= \lim_{n \to \infty} \left(\frac{m}{2\pi i\hbar\Delta t} \right)^{\frac{n}{2}} \int_{\mathbb{R}} \cdots \int_{\mathbb{R}} e^{\frac{im}{2\hbar\Delta t} \sum_{k=1}^{n} ((q_k-q_{k-1})^2 - \omega^2 \Delta t^2 q_k^2)} \, dq_1 \, \cdots \, dq_{n-1},$$

where $\Delta t = \frac{t_b - t_a}{n}$, $q_0 = q(t_a) = q_a$ and $q_n = q(t_b) = q_b$. Evaluating this limit is algebraically quite a bit more involved than it was for the free particle in Section 8.2 and the result we are looking for is, as we saw in Section 7.4, also rather more involved. In the hope of minimizing these issues as much as possible we will consider only the case in which the elapsed time interval $T = t_b - t_a$ is small enough to ensure that $0 < \omega T < \pi$. The adjustments required when $\nu\pi < \omega T < (\nu + 1)\pi$, $\nu = 1, 2, \ldots$, are modest and you may want to keep track of them for yourself as we go along. This will leave only the behavior when ωT is an integer multiple of π which, as we found in Section 7.4, involves Maslov corrections and we will briefly discuss this at the end. With these assumptions we can record the result we would like to obtain from the evaluation of the path integral (taken from Section 7.4):

$$\sqrt{\frac{m}{2\pi i \hbar \sin \omega T}} \exp\left(\frac{i}{\hbar} \frac{m\omega}{2 \sin \omega T} [\, (q_a^2 + q_b^2) \cos \omega T - 2q_a q_b \,] \right). \tag{8.11}$$

Just to get in the proper frame of mind we will write out the $n = 3$ term in the sequence of multiple integrals. As motivation for some of the manipulations we mention that the objective is to rewrite the quadratic form $\sum_{k=1}^{n}((q_k - q_{k-1})^2 - \omega^2 \Delta t^2 q_k^2)$ in such a way that:

1. the terms involving only q_0 and q_n are exposed and can be pulled out of the integral and
2. the remaining quadratic form is one to which we can apply a standard Gaussian integration formula from Appendix A.

For the record, the Gaussian integral we propose to apply is the following. Let A be a real, symmetric, nondegenerate $N \times N$ matrix. Then

$$\int_{\mathbb{R}^N} e^{\frac{i}{2}\langle A\mathbf{q},\mathbf{q}\rangle + i\langle \mathbf{p},\mathbf{q}\rangle} \, d^N \mathbf{q} = e^{\frac{N\pi i}{4} - \frac{\nu\pi i}{2}} \frac{\sqrt{(2\pi)^N}}{\sqrt{|\det A|}} e^{-\frac{i}{2}\langle A^{-1}\mathbf{p},\mathbf{p}\rangle}, \tag{8.12}$$

where $\mathbf{q}, \mathbf{p} \in \mathbb{R}^N$, $\langle \, , \, \rangle$ denotes the standard inner product on $\mathbb{R}^N$ and ν is the number of negative eigenvalues of A (see (A.7) of Appendix A).

To rewrite

$$\sum_{k=1}^{3}((q_k - q_{k-1})^2 - \omega^2 \Delta t^2 q_k^2)$$

$$= (q_1 - q_0)^2 + (q_2 - q_1)^2 + (q_3 - q_2)^2 - \omega^2 \Delta t^2 q_1^2 - \omega^2 \Delta t^2 q_2^2 - \omega^2 \Delta t^2 q_3^2$$

appropriately, let $\mathbf{q} = \left(\begin{smallmatrix} q_1 \\ q_2 \end{smallmatrix}\right)$, $\mathbf{p} = \left(\begin{smallmatrix} q_0 \\ q_3 \end{smallmatrix}\right) = \left(\begin{smallmatrix} q_a \\ q_b \end{smallmatrix}\right)$ and

$$A_2 = \begin{pmatrix} 2 - \omega^2 \Delta t^2 & -1 \\ -1 & 2 - \omega^2 \Delta t^2 \end{pmatrix}.$$

A few quick calculations show that

$$\langle A_2\mathbf{q}, \mathbf{q}\rangle = 2q_1^2 - \omega^2\Delta t^2 q_1^2 - 2q_1q_2 + 2q_2^2 - \omega^2\Delta t^2 q_2^2,$$
$$-2\langle \mathbf{p}, \mathbf{q}\rangle = -2q_0q_1 - 2q_2q_3$$

and

$$\langle \mathbf{p}, \mathbf{p}\rangle = q_0^2 + q_3^2,$$

so

$$\langle A_2\mathbf{q}, \mathbf{q}\rangle - 2\langle \mathbf{p}, \mathbf{q}\rangle + \langle \mathbf{p}, \mathbf{p}\rangle$$
$$= (q_1^2 - 2q_0q_1 + q_0^2) + (q_1^2 - 2q_1q_2 + q_2^2) + (q_1^2 - 2q_2q_3 + q_3^2) - \omega^2\Delta t^2 q_1^2 - \omega^2\Delta t^2 q_2^2$$
$$= \sum_{k=1}^{3}\left((q_k - q_{k-1})^2 - \omega^2\Delta t^2 q_k^2\right) + \omega^2\Delta t^2 q_3^2.$$

Consequently,

$$e^{\frac{im}{2\hbar\Delta t}\sum_{k=1}^{3}((q_k-q_{k-1})^2-\omega^2\Delta t^2 q_k^2)} = e^{\frac{im}{2\hbar\Delta t}(q_a^2+q_b^2-\omega^2\Delta t^2 q_b^2)}e^{\frac{im}{2\hbar\Delta t}(\langle A_2\mathbf{q},\mathbf{q}\rangle-2\langle\mathbf{p},\mathbf{q}\rangle)},$$

so

$$\iint_{\mathbb{R}\,\mathbb{R}} e^{\frac{im}{2\hbar\Delta t}\sum_{k=1}^{3}((q_k-q_{k-1})^2-\omega^2\Delta t^2 q_k^2)}\,dq_1\,dq_2$$
$$= e^{\frac{im}{2\hbar\Delta t}(q_a^2+q_b^2-\omega^2\Delta t^2 q_b^2)}\int_{\mathbb{R}^2} e^{\frac{im}{2\hbar\Delta t}(\langle A_2\mathbf{q},\mathbf{q}\rangle-2\langle\mathbf{p},\mathbf{q}\rangle)}d^2\mathbf{q}$$
$$= e^{\frac{im}{2\hbar\Delta t}(q_a^2+q_b^2-\omega^2\Delta t^2 q_b^2)}\int_{\mathbb{R}^2} e^{\frac{i}{2}\langle \frac{m}{\hbar\Delta t}A_2\mathbf{q},\mathbf{q}\rangle+i\langle -\frac{m}{\hbar\Delta t}\mathbf{p},\mathbf{q}\rangle}d^2\mathbf{q}. \qquad (8.13)$$

This integral is just the Gaussian in (8.12) with $A = \frac{m}{\hbar\Delta t}A_2$ and $\mathbf{p}$ replaced by $-\frac{m}{\hbar\Delta t}\mathbf{p}$.

Exercise 8.3.1. Show that

$$\det A = \left(\frac{m}{\hbar\Delta t}\right)^2 \det A_2 = \left(\frac{m}{\hbar\Delta t}\right)^2 [(2-\omega^2\Delta t^2)^2 - 1],$$

$$\left\langle A^{-1}\left(-\frac{m}{\hbar\Delta t}\mathbf{p}\right), -\frac{m}{\hbar\Delta t}\mathbf{p}\right\rangle = \frac{m}{\hbar\Delta t}\langle A_2^{-1}\mathbf{p}, \mathbf{p}\rangle$$

and that the eigenvalues of A are given by

$$\frac{m}{\hbar\Delta t}(2-\omega^2\Delta t^2) \pm \frac{m}{\hbar\Delta t} = \frac{m}{\hbar\Delta t}[(2\pm 1)-\omega^2\Delta t^2]$$
$$= \frac{m}{\hbar\Delta t}(2-\omega^2\Delta t^2) + 2\sqrt{\left(-\frac{m}{\hbar\Delta t}\right)\left(-\frac{m}{\hbar\Delta t}\right)}\cos\left(\frac{k\pi}{2+1}\right),$$

for $k = 1, 2$. The reason for the peculiar expression with the cosine will become clear shortly.

This exercise provides us with *almost* all of the information we would need to evaluate the Gaussian integral in (8.13) using (8.12); almost, but not quite. The problem is the eigenvalues. Their signs depend on the size of $\omega\Delta t$, so the number v_2 of negative eigenvalues is not determined and we need this in (8.12). We will now proceed with the general construction of the nth term in the sequence of multiple integrals defining the path integral. We will find that, *for sufficiently large n*, the number of negative eigenvalues is zero (because we have assumed that $0 < \omega T < \pi$) and this will permit us to do the Gaussian integral and evaluate the limit as $n \to \infty$.

Remark 8.3.1. Before getting started we will need to borrow a result from linear algebra. We consider an $N \times N$ *symmetric, tridiagonal Toeplitz matrix*, that is, one of the form

$$\begin{pmatrix} a & b & 0 & 0 & 0 & \cdots & 0 & 0 \\ b & a & b & 0 & 0 & \cdots & 0 & 0 \\ 0 & b & a & b & 0 & \cdots & 0 & 0 \\ 0 & 0 & b & a & b & \cdots & 0 & 0 \\ \vdots & \vdots & \vdots & \vdots & \vdots & & \vdots & \vdots \\ 0 & 0 & 0 & 0 & 0 & \cdots & a & b \\ 0 & 0 & 0 & 0 & 0 & \cdots & b & a \end{pmatrix},$$

where we assume that a and b are nonzero real numbers. Then there are N distinct real eigenvalues given by

$$\lambda_k = a + 2b \cos\left(\frac{k\pi}{N+1}\right), \quad k = 1, \ldots, N,$$

and a corresponding set of eigenvectors given by

$$V_k = \left(\sin\left(\frac{1 \cdot k\pi}{N+1}\right), \sin\left(\frac{2k\pi}{N+1}\right), \ldots, \sin\left(\frac{Nk\pi}{N+1}\right)\right), \quad k = 1, \ldots, N.$$

Oddly enough, all such matrices have the same eigenvectors. One can find a proof of this and some more general results in [LR].

Now we fix an integer $n \geq 2$ and consider the integral

$$\int_{\mathbb{R}} \cdots \int_{\mathbb{R}} e^{\frac{im}{2\hbar\Delta t} \sum_{k=1}^{n}((q_k - q_{k-1})^2 - \omega^2 \Delta t^2 q_k^2)} \, dq_1 \cdots dq_{n-1},$$

where $\Delta t = \frac{T}{n} = \frac{t_b - t_a}{n}$, $q_0 = q(t_a) = q_a$ and $q_n = q(t_b) = q_b$. The first step is entirely analogous to the $n = 3$ case treated above. We introduce an $(n-1) \times (n-1)$

matrix

$$
A_{n-1} = \begin{pmatrix}
2-\omega^2\Delta t^2 & -1 & 0 & 0 & 0 & \cdots & 0 & 0 \\
-1 & 2-\omega^2\Delta t^2 & -1 & 0 & 0 & \cdots & 0 & 0 \\
0 & -1 & 2-\omega^2\Delta t^2 & -1 & 0 & \cdots & 0 & 0 \\
0 & 0 & -1 & 2-\omega^2\Delta t^2 & -1 & \cdots & 0 & 0 \\
\vdots & \vdots & \vdots & \vdots & \vdots & & \vdots & \vdots \\
0 & 0 & 0 & 0 & 0 & \cdots & 2-\omega^2\Delta t^2 & -1 \\
0 & 0 & 0 & 0 & 0 & \cdots & -1 & 2-\omega^2\Delta t^2
\end{pmatrix}
$$

and define

$$
\mathbf{q} = \begin{pmatrix} q_1 \\ q_2 \\ \vdots \\ q_{n-1} \end{pmatrix} \in \mathbb{R}^{n-1}
$$

and

$$
\mathbf{p} = \begin{pmatrix} q_0 \\ 0 \\ \vdots \\ 0 \\ q_n \end{pmatrix} = \begin{pmatrix} q_a \\ 0 \\ \vdots \\ 0 \\ q_b \end{pmatrix} \in \mathbb{R}^{n-1}.
$$

Exercise 8.3.2. Show that

$$
\langle A_{n-1}\mathbf{q}, \mathbf{q}\rangle - \omega^2\Delta t^2 q_b^2 - 2\langle \mathbf{p}, \mathbf{q}\rangle + q_a^2 + q_b^2 = \sum_{k=1}^{n}\left((q_k - q_{k-1})^2 - \omega^2\Delta t^2 q_k^2\right).
$$

Consequently,

$$
e^{\frac{im}{2\hbar\Delta t}\sum_{k=1}^{n}((q_k-q_{k-1})^2-\omega^2\Delta t^2 q_k^2)} = e^{\frac{im}{2\hbar\Delta t}(q_a^2+q_b^2-\omega^2\Delta t^2 q_b^2)} e^{\frac{im}{2\hbar\Delta t}(\langle A_{n-1}\mathbf{q},\mathbf{q}\rangle - 2\langle \mathbf{p},\mathbf{q}\rangle)},
$$

so

$$
\int_{\mathbb{R}}\cdots\int_{\mathbb{R}} e^{\frac{im}{2\hbar\Delta t}\sum_{k=1}^{n}((q_k-q_{k-1})^2-\omega^2\Delta t^2 q_k^2)}\,dq_1 \cdots dq_{n-1}
$$

$$
= e^{\frac{im}{2\hbar\Delta t}(q_a^2+q_b^2-\omega^2\Delta t^2 q_b^2)}\int_{\mathbb{R}^{n-1}} e^{\frac{im}{2\hbar\Delta t}(\langle A_{n-1}\mathbf{q},\mathbf{q}\rangle - 2\langle \mathbf{p},\mathbf{q}\rangle)}\,d^{n-1}\mathbf{q}
$$

$$
= e^{\frac{im}{2\hbar\Delta t}(q_a^2+q_b^2-\omega^2\Delta t^2 q_b^2)}\int_{\mathbb{R}^{n-1}} e^{\frac{i}{2}\langle \frac{m}{\hbar\Delta t}A_{n-1}\mathbf{q},\mathbf{q}\rangle + i\langle -\frac{m}{\hbar\Delta t}\mathbf{p},\mathbf{q}\rangle}\,d^{n-1}\mathbf{q}.
$$

This integral is just the Gaussian in (8.12) with $A = \frac{m}{\hbar \Delta t} A_{n-1}$ and $\mathbf{p}$ replaced with $-\frac{m}{\hbar \Delta t}\mathbf{p}$. We begin to write out this Gaussian in terms of the number ν_{n-1} of negative eigenvalues of A_{n-1} by noting first that

$$A^{-1}\left(-\frac{m}{\hbar \Delta t}\mathbf{p}\right) = -A_{n-1}^{-1}\mathbf{p},$$

so

$$\left\langle A^{-1}\left(-\frac{m}{\hbar \Delta t}\mathbf{p}\right), -\frac{m}{\hbar \Delta t}\mathbf{p}\right\rangle = \frac{m}{\hbar \Delta t}\langle A_{n-1}^{-1}\mathbf{p}, \mathbf{p}\rangle$$

and

$$\sqrt{|\det A|} = \left(\frac{m}{\hbar \Delta t}\right)^{\frac{n-1}{2}}\sqrt{|\det A_{n-1}|}.$$

Exercise 8.3.3. Use these and (8.12) to show that

$$\int_{\mathbb{R}^{n-1}} e^{\frac{i}{2}\langle \frac{m}{\hbar \Delta t}A_{n-1}\mathbf{q},\mathbf{q}\rangle + i\langle -\frac{m}{\hbar \Delta t}\mathbf{p},\mathbf{q}\rangle} d^{n-1}\mathbf{q}$$

$$= e^{\frac{(n-1)\pi i}{4}} e^{-\frac{\nu_{n-1}\pi i}{2}}\left(\frac{m}{2\pi \hbar \Delta t}\right)^{-\frac{n-1}{2}} |\det A_{n-1}|^{-\frac{1}{2}} e^{-\frac{mi}{2\hbar \Delta t}\langle A_{n-1}^{-1}\mathbf{p},\mathbf{p}\rangle},$$

where ν_{n-1} is the number of negative eigenvalues of A_{n-1}.

Exercise 8.3.4. The definition of the path integral requires that, before taking the limit, we multiply the integral by the normalizing factor $(\frac{m}{2\pi i\hbar \Delta t})^{n/2}$. In preparation for this show that

$$e^{\frac{(n-1)\pi i}{4}}\left(\frac{m}{2\pi i\hbar \Delta t}\right)^{n/2}\left(\frac{m}{2\pi \hbar \Delta t}\right)^{-\frac{n-1}{2}} = \sqrt{\frac{m}{2\pi i\hbar \Delta t}}.$$

Exercise 8.3.5. Put all of this together to obtain

$$\left(\frac{m}{2\pi i\hbar \Delta t}\right)^{\frac{n}{2}}\int_{\mathbb{R}}\cdots\int_{\mathbb{R}} e^{\frac{im}{2\hbar \Delta t}\sum_{k=1}^{n}((q_k-q_{k-1})^2-\omega^2\Delta t^2 q_k^2)} dq_1 \cdots dq_{n-1}$$

$$= e^{-\frac{im}{2\hbar}\omega^2 q_b^2 \Delta t}\left(e^{-\frac{\nu_{n-1}\pi i}{2}}\sqrt{\frac{m}{2\pi i\hbar \Delta t |\det A_{n-1}|}} e^{\frac{im}{2\hbar \Delta t}(q_a^2+q_b^2-\langle A_{n-1}^{-1}\mathbf{p},\mathbf{p}\rangle)}\right). \quad (8.14)$$

Since $\Delta t = \frac{T}{n}$, the first factor clearly approaches 1 as $n \to \infty$, so we need only worry about the second factor. For this we will need some fairly detailed information about:
1. $\det A_{n-1}$,
2. $\langle A_{n-1}\mathbf{p}, \mathbf{p}\rangle$ and
3. ν_{n-1}.

First we will look at the determinant

$$a_{n-1} = \det A_{n-1}.$$

Exercise 8.3.6. Compute the first few of these to show that

$$a_1 = 2 - w^2 \Delta t^2,$$
$$a_2 = (2 - w^2 \Delta t^2)a_1 - 1,$$
$$a_3 = (2 - w^2 \Delta t^2)a_2 - a_1,$$
$$a_4 = (2 - w^2 \Delta t^2)a_3 - a_2$$

and then prove by induction that

$$a_{k-1} = (2 - w^2 \Delta t^2)a_{k-2} - a_{k-3}, \quad k \geq 4.$$

Hint: Expand the determinants by the cofactors of the last row.

Taking $a_0 = 1$ and $a_{-1} = 0$ we have a recurrence relation

$$a_{k-1} = (2 - w^2 \Delta t^2)a_{k-2} - a_{k-3}, \quad k \geq 2,$$

and this is what we must solve. We could now apply general results from the theory of linear, homogeneous, recurrence relations, but let us proceed a bit more directly. Suppose we have found a solution z to the equation

$$z + z^{-1} = 2 - w^2 \Delta t^2$$

(we will actually find some momentarily). Then our recurrence relation can be written as

$$a_{k-1} = (z + z^{-1})a_{k-2} - a_{k-3}, \quad k \geq 2.$$

Now note that $a_l = z^{l+1}$, $l \geq 1$, is a solution to the recurrence relation since

$$(z + z^{-1})a_{k-2} - a_{k-3} = (z + z^{-1})z^{k-1} - z^{k-2} = z^k = a_{k-1}$$

and, similarly, if we let $a_l = (z^{-1})^{l+1} = z^{-l-1}$, then

$$(z + z^{-1})a_{k-2} - a_{k-3} = (z + z^{-1})z^{-k+1} - z^{-k+2} = z^{-k} = a_{k-1}.$$

By the linearity of the recurrence, $a_l = m_1 z^{l+1} + m_2 z^{-l-1}$, $l \geq 1$, is also a solution for any m_1 and m_2. In order to satisfy the initial conditions we set $a_{-1} = 0$, which gives $m_1 + m_2 = 0$ and therefore $a_l = m_1(z^{l+1} - z^{-l-1})$, and then set $a_0 = 1$, giving $m_1(z - z^{-1}) = 1$. In particular, $z \neq \pm 1$, so

$$m_1 = \frac{1}{z - z^{-1}} = -m_2.$$

These initial conditions uniquely determine the solution to our recurrence relation to be

$$a_l = \frac{1}{z - z^{-1}} (z^{l+1} - z^{-l-1}) = \frac{z^{l+1} - z^{-l-1}}{z - z^{-1}}.$$

In particular,

$$\det A_{n-1} = a_{n-1} = \frac{z^n - z^{-n}}{z - z^{-1}}.$$

Now we set about finding explicit solutions to the equation $z + z^{-1} = 2 - w^2 \Delta t^2$. Since $z = 0$ is certainly not a solution, this can be written

$$z^2 - (2 - w^2 \Delta t^2)z + 1 = 0.$$

Remark 8.3.2. For the record we mention that

$$p(z) = z^2 - (2 - w^2 \Delta t^2)z + 1$$

is called the *characteristic polynomial* of our recurrence relation.

An application of the quadratic formula gives

$$z = 1 - \frac{1}{2} w^2 \Delta t^2 \pm \frac{1}{2} w \Delta t \sqrt{w^2 \Delta t^2 - 4}.$$

Now, recall that at the very beginning of all of this we fixed an integer $n \geq 2$ (the number of subintervals into which we partitioned $[t_a, t_b]$). Ultimately, we are interested only in the limit as $n \to \infty$ and so only in large n. Note that, since $\Delta t = \frac{T}{n}$, large n means small $w^2 \Delta t^2$, so eventually the roots we just found will be complex with positive real part. We want to ignore the real roots corresponding to small n, so we will select some positive integer N_0 such that these roots are complex if $n > N_0$. Since n is beginning to play a role now it seems proper to make its appearance explicit by writing $z(n)$ rather than simply z. Thus, for $n > N_0$, our two roots are given by

$$z(n) = 1 - \frac{1}{2} w^2 \Delta t^2 + \frac{1}{2} w \Delta t \sqrt{4 - w^2 \Delta t^2} \, i$$

and its conjugate.

Exercise 8.3.7. Show that $|z(n)|^2 = 1$.

Thus, our two roots lie on the unit circle and are therefore $z(n)$ and $z(n)^{-1}$. We will write these in polar form as

$$z(n) = e^{i\theta(n)} \quad \text{and} \quad z(n)^{-1} = e^{-i\theta(n)}.$$

Note also that, since 1 and −1 are certainly not roots of the characteristic polynomial and $\text{Im}(z(n)) > 0$, we can assume that

$$0 < \theta(n) < \pi \quad \forall n > N_0.$$

Now we can write $\det A_{n-1}$ in the form we were after, that is,

$$\det A_{n-1} = \frac{z(n)^n - z(n)^{-n}}{z(n) - z(n)^{-1}} = \frac{e^{in\theta(n)} - e^{-in\theta(n)}}{e^{i\theta(n)} - e^{-i\theta(n)}} = \frac{\sin n\theta(n)}{\sin \theta(n)}.$$

We will need to know something about the behavior of $\theta(n)$ for large n. For this we first recall that $z(n) + z(n)^{-1}$ is equal to both $e^{i\theta(n)} + e^{-i\theta(n)} = 2 \cos \theta(n)$ and $2 - w^2 \Delta t^2$, so $\cos \theta(n) = 1 - \frac{1}{2} w^2 \Delta t^2$. Moreover, since $0 < \theta(n) < \pi$,

$$\theta(n) = \arccos\left(1 - \frac{1}{2} w^2 \Delta t^2\right) = \arccos\left(1 - \left(\frac{w^2 T^2}{2}\right) n^{-2}\right). \tag{8.15}$$

Our first objective is to show that, with $\theta(n)$ as in (8.15),

$$\theta(n) = w\Delta t + O(n^{-2}) \quad \text{as } n \to \infty,$$

that is, that there exist an integer $N_1 \geq N_0$ and a positive constant M such that

$$|\theta(n) - w\Delta t| \leq \frac{M}{n^2} \quad \forall n \geq N_1.$$

For this it will suffice to show that the real-valued function $\arccos\left(1 - \frac{1}{2} x^2\right)$ of the real variable x is $x + O(x^2)$ as $x \to 0^+$. To do this we would like to apply Taylor's formula with a quadratic remainder at $x = 0$. Unfortunately, this is not differentiable at $x = 0$. Indeed, for $x \neq 0$,

$$\frac{d}{dx} \arccos\left(1 - \frac{1}{2} x^2\right) = \frac{x}{|x|}\left(1 - \frac{x^2}{4}\right)^{-1/2},$$

so at $x = 0$ the right-hand derivative is 1 and the left-hand derivative is −1. To remedy the situation we consider instead

$$f(x) = \begin{cases} \arccos\left(1 - \frac{1}{2} x^2\right), & \text{if } x \geq 0, \\ -\arccos\left(1 - \frac{1}{2} x^2\right), & \text{if } x \leq 0. \end{cases}$$

Then $f(x)$ is differentiable and

$$f'(x) = \left(1 - \frac{x^2}{4}\right)^{-1/2}.$$

Consequently,

$$f''(x) = \frac{x}{4}\left(1 - \frac{x^2}{4}\right)^{-3/2}.$$

Now, by Taylor's formula,

$$f(x) = f(0) + f'(0)x + \frac{\theta''(c)}{2!}x^2 = x + \frac{c}{8}\left(1 - \frac{c^2}{4}\right)^{-3/2}x^2,$$

where c is between 0 and x. But, on $-1 \le c \le 1$, the function $\frac{c}{8}(1 - \frac{c^2}{4})^{-3/2}$ is continuous and therefore bounded by some $K > 0$, so

$$|f(x) - x| \le Kx^2.$$

In particular, if we choose $N_1 \ge N_0$ sufficiently large that $\omega\Delta t = \frac{\omega T}{n} < 1$ whenever $n \ge N_1$, then

$$|\theta(n) - \omega\Delta t| \le \frac{K\omega^2 T^2}{n^2} \quad \forall n \ge N_1,$$

as required.

Next note that $\sin(\omega T + x) = \sin\omega T + O(x)$ as $x \to 0$ implies that $\sin(\omega T + O(n^{-1})) = \sin\omega T + O(O(n^{-1})) = \sin\omega T + O(n^{-1})$ as $n \to \infty$. Thus,

$$\theta(n) = \frac{\omega T}{n} + O(n^{-2}) \quad \Rightarrow \quad n\theta(n) = \omega T + O(n^{-1}) \quad \Rightarrow \quad \sin n\theta(n) = \sin\omega T + O(n^{-1})$$

as $n \to \infty$.

Exercise 8.3.8. Show that

$$\sin\theta(n) = \frac{\omega T}{n} + O(n^{-2})$$

as $n \to \infty$.

But $\sin x = x + O(x^2)$ as $x \to 0$, so $\sin\frac{\omega T}{n} = \frac{\omega T}{n} + O(n^{-2})$ as $n \to \infty$, and therefore

$$\sin\theta(n) = \sin\left(\frac{\omega T}{n}\right) + O(n^{-2})$$

as $n \to \infty$. We will put all of this to use in computing the limits we need for (8.14).

Before moving on to more estimates that we need to compute the limit we will pause for a moment to draw an important conclusion from what we have so far. We will show that for sufficiently large n, A_{n-1} has no negative eigenvalues, so $v_{n-1} = 0$ (this is a consequence of our assumption that $0 < \omega T < \pi$). Begin by fixing some $n \ge N_1$. Since A_{n-1} is an $(n-1) \times (n-1)$ symmetric, tridiagonal Toeplitz matrix, it has $n-1$ distinct real eigenvalues given by

$$\lambda_k = (2 - \omega^2\Delta t^2) - 2\cos\left(\frac{k\pi}{n}\right), \quad k = 1, \ldots, n-1,$$

$$= z(n) + z(n)^{-1} - 2\cos\left(\frac{k\pi}{n}\right), \quad k = 1, \ldots, n-1,$$

$$= 2 \cos \theta(n) - 2 \cos\left(\frac{k\pi}{n}\right), \quad k = 1, \ldots, n-1$$

(see Remark 8.3.1). Now note that $\lambda_k < 0$ if and only if $\cos \theta(n) < \cos\left(\frac{k\pi}{n}\right)$ and, since all of the angles are in $(0, \pi)$, this is the case if and only if

$$\frac{k\pi}{n} < \theta(n).$$

But

$$\frac{k\pi}{n} < \theta(n) \quad \Rightarrow \quad \frac{k\pi}{n} < \omega\Delta t + \frac{M}{n^2} \quad \Rightarrow \quad \frac{k\pi}{n} < \frac{\omega T}{n} + \frac{M}{n^2}$$

$$\Rightarrow \quad k\pi < \omega T + \frac{M}{n} \quad \Rightarrow \quad k < \frac{\omega T}{\pi} + \frac{M}{\pi} \frac{1}{n}$$

$$\Rightarrow \quad k < 1 + \frac{M}{\pi} \frac{1}{n}$$

and this is clearly not possible for arbitrarily large n unless $k = 1$.

Exercise 8.3.9. Show directly that $0 < \omega T < \pi$ implies that for sufficiently large n, $\lambda_1 = (2 - \frac{\omega^2 T^2}{n^2}) - 2 \cos\left(\frac{\pi}{n}\right)$ is positive. *Hint:* Look at a Taylor polynomial for $\cos\left(\frac{\pi}{n}\right)$.

Thus, we can choose an integer $N_2 \geq N_1 \geq N_0$ such that

$$v_{n-1} = 0 \quad \forall n \geq N_2.$$

Exercise 8.3.10. Show that if $v > 0$ is an integer and $v\pi < \omega T < (v+1)\pi$, then, for sufficiently large n, A_{n-1} has precisely v negative eigenvalues.

Now we will get back to some estimates required to compute the limit as $n \to \infty$ of (8.14). We have already shown that

$$\theta(n) = \frac{\omega T}{n} + O(n^{-2}) \quad \text{as } n \to \infty.$$

Exercise 8.3.11. Show that

$$\lim_{n \to \infty} n \sin \frac{\omega T}{n} = \omega T.$$

Now write

$$\frac{n \sin \theta(n)}{T \sin n\theta(n)} = \frac{n \sin \frac{\omega T}{n} + O(n^{-1})}{T \sin \omega T + O(n^{-1})} = \frac{n \sin \frac{\omega T}{n}}{T \sin \omega T + O(n^{-1})} + \frac{O(n^{-1})}{T \sin \omega T + O(n^{-1})}.$$

The second term clearly approaches zero as $n \to \infty$ and, by the previous exercise, the first approaches $\frac{\omega}{\sin \omega T}$, so

$$\lim_{n \to \infty} \frac{n \sin \theta(n)}{T \sin n\theta(n)} = \frac{\omega}{\sin \omega T}.$$

Furthermore, since $\theta(n) = \frac{\omega T}{n} + O(n^{-2})$ and $0 < \omega T < \pi$,

$$\lim_{n \to \infty} \left| \frac{n \sin \theta(n)}{T \sin n\theta(n)} \right| = \frac{\omega}{\sin \omega T}.$$

Recalling that $v_{n-1} = 0$ for $n \geq N_2$ we find that

$$\lim_{n \to \infty} e^{\frac{v_{n-1}\pi i}{2}} \sqrt{\frac{m}{2\pi i \hbar \Delta t \, |\det A_{n-1}|}} = \lim_{n \to \infty} \sqrt{\frac{m}{2\pi i \hbar}} \left| \frac{n \sin \theta(n)}{T \sin n\theta(n)} \right| = \sqrt{\frac{m\omega}{2\pi i \hbar \sin \omega T}}.$$

All that remains for our computation of the limit in (8.14) is to investigate the behavior of

$$e^{\frac{im}{2\hbar \Delta t}(q_a^2 + q_b^2 - \langle A_{n-1}^{-1} \mathbf{p}, \mathbf{p} \rangle)}$$

as $n \to \infty$. We begin with $\langle A_{n-1}^{-1} \mathbf{p}, \mathbf{p} \rangle$. Denoting the entries in the matrix A_{n-1}^{-1} by $B_{i,j}$ we obtain

$$A_{n-1}^{-1} \mathbf{p} = \begin{pmatrix} B_{1,1} & \cdots & B_{1,n-1} \\ B_{2,1} & \cdots & B_{2,n-1} \\ \vdots & & \vdots \\ B_{n-2,1} & \cdots & B_{n-2,n-1} \\ B_{n-1,1} & \cdots & B_{n-1,n-1} \end{pmatrix} \begin{pmatrix} q_a \\ 0 \\ \vdots \\ 0 \\ q_b \end{pmatrix} = \begin{pmatrix} B_{1,1}q_a + B_{1,n-1}q_b \\ B_{2,1}q_a + B_{2,n-1}q_b \\ \vdots \\ B_{n-2,1}q_a + B_{n-2,n-1}q_b \\ B_{n-1,1}q_a + B_{n-1,n-1}q_b \end{pmatrix},$$

and therefore

$$\langle A_{n-1}^{-1} \mathbf{p}, \mathbf{p} \rangle = (B_{1,1}q_a + B_{1,n-1}q_b)q_a + (B_{n-1,1}q_a + B_{n-1,n-1}q_b)q_b.$$

Exercise 8.3.12. Using the fact that A_{n-1}^{-1} is $\frac{1}{\det A_{n-1}}$ times the adjoint of A_{n-1}, show that

$$B_{1,1} = B_{n-1,n-1} = \frac{\sin (n-1)\theta(n)}{\sin n\theta(n)}$$

and

$$B_{1,n-1} = B_{n-1,1} = \frac{\sin \theta(n)}{\sin n\theta(n)}.$$

From this we find that

$$\langle A_{n-1}^{-1} \mathbf{p}, \mathbf{p} \rangle = \frac{\sin (n-1)\theta(n)}{\sin n\theta(n)}(q_a^2 + q_b^2) + \frac{\sin \theta(n)}{\sin n\theta(n)}(2q_a q_b),$$

and therefore

$$q_a^2 + q_b^2 - \langle A_{n-1}^{-1} \mathbf{p}, \mathbf{p} \rangle = \frac{1}{\sin n\theta(n)}((\sin n\theta(n) - \sin (n-1)\theta(n))(q_a^2 + q_b^2)$$
$$- \sin \theta(n)(2q_a q_b)).$$

Exercise 8.3.13. Show that

$$\sin n\theta(n) - \sin (n-1)\theta(n) = \cos n\theta(n) \sin \theta(n) + \sin n\theta(n)(1 - \cos \theta(n)).$$

From this we obtain

$$\frac{im}{2\hbar\Delta t}(q_a^2 + q_b^2 - \langle A_{n-1}^{-1}\mathbf{p}, \mathbf{p}\rangle) = \frac{imn}{2\hbar T \sin n\theta(n)}[[\cos n\theta(n) \sin \theta(n)$$

$$+ \sin n\theta(n)(1 - \cos \theta(n))](q_a^2 + q_b^2) - \sin \theta(n)(2q_a q_b)]$$

$$= \frac{im\omega}{2\hbar}\left[\left(\frac{\cos n\theta(n)}{\sin n\theta(n)}\frac{\sin \theta(n)}{\omega T/n} + \frac{1 - \cos \theta(n)}{\omega T/n}\right)(q_a^2 + q_b^2)\right.$$

$$\left. - \frac{1}{\sin n\theta(n)}\left(\frac{\sin \theta(n)}{\omega T/n}\right)(2q_a q_b)\right].$$

Exercise 8.3.14. Use the various estimates we obtained earlier to show that

$$\lim_{n\to\infty}\frac{im}{2\hbar\Delta t}(q_a^2 + q_b^2 - \langle A_{n-1}^{-1}\mathbf{p}, \mathbf{p}\rangle) = \frac{im\omega}{2\hbar \sin \omega T}((q_a^2 + q_b^2)\cos \omega T - 2q_a q_b),$$

and therefore

$$\lim_{n\to\infty} e^{\frac{im}{2\hbar\Delta t}(q_a^2+q_b^2-\langle A_{n-1}^{-1}\mathbf{p},\mathbf{p}\rangle)} = e^{\frac{i}{\hbar}\frac{m\omega}{\sin \omega T}((q_a^2+q_b^2)\cos \omega T - 2q_a q_b)}.$$

With this we have all of the ingredients required to complete the evaluation of the harmonic oscillator path integral when $0 < \omega T < \pi$.

Exercise 8.3.15. Trace back through all of the results we have obtained in this section to show that, when $0 < \omega T < \pi$,

$$\int_{P_\mathbb{R}(q_b,t_b;q_a,t_a)} e^{\frac{im}{2\hbar}\int_{t_a}^{t_b}(\dot{q}^2-\omega^2 q^2)\,dt}\mathcal{D}q$$

$$= \lim_{n\to\infty}\left(\frac{m}{2\pi i\hbar\Delta t}\right)^{\frac{n}{2}}\int_{\mathbb{R}}\cdots\int_{\mathbb{R}} e^{\frac{im}{2\hbar\Delta t}\sum_{k=1}^{n}((q_k-q_{k-1})^2-\omega^2\Delta t^2 q_k^2)}\,dq_1 \cdots dq_{n-1}$$

$$= \lim_{n\to\infty} e^{-\frac{im}{2\hbar}\omega^2 q_b^2\Delta t}\left(e^{-\frac{v_{n-1}\pi i}{2}}\sqrt{\frac{m}{2\pi i\hbar\Delta t \mid \det A_{n-1}\mid}} e^{\frac{im}{2\hbar\Delta t}(q_a^2+q_b^2-\langle A_{n-1}^{-1}\mathbf{p},\mathbf{p}\rangle)}\right)$$

$$= \sqrt{\frac{m\omega}{2\pi i\hbar \sin \omega T}} e^{\frac{i}{\hbar}\frac{m\omega}{\sin \omega T}((q_a^2+q_b^2)\cos \omega T - 2q_a q_b)}.$$

Mercifully, this agrees with the result (8.11) we obtained from Mehler's formula for the propagator of the quantum harmonic oscillator in Section 7.4.

Exercise 8.3.16. Re-examine the arguments we have just given and make whatever adjustments are required to handle the case in which $\nu\pi < \omega T < (\nu + 1)\pi$ with $\nu > 0$. *Hint*: Use Exercise 8.3.10.

The only issue we have not addressed is the behavior of the path integral as ωT approaches an integral multiple of π. In Section 7.4 we found that the propagator experiences Maslov corrections at these integral multiples of π and that these amount to discontinuous jumps in the phase. To see this behavior in the path integral one can compute the limit, in the distributional sense, of

$$\sqrt{\frac{m\omega}{2\pi i\hbar \, \sin \omega T}} \, e^{\frac{i}{\hbar} \frac{m\omega}{\sin \omega T}((q_a^2+q_b^2)\cos \omega T-2q_aq_b)}$$

as T approaches, for example, 0. The result is simply the Dirac delta $\delta(q_a - q_b)$. We will not carry out this calculation, but if you would like to do so yourself we might suggest (G.12).

9 Sketches of some rigorous results

9.1 Introduction

Let us begin by summarizing Feynman's prescription for evaluating his path integral representation of the propagator $K(q_b, t_b; q_a, t_a)$ for a particle moving along the q-axis from q_a at time t_a to q_b at time t_b under the influence of a Hamiltonian of the form $H = -\frac{\hbar^2}{2m}\Delta + V$, assumed to be self-adjoint on some domain $\mathcal{D}(H)$ in $L^2(\mathbb{R})$. We are told to slice the t-interval $[t_a, t_b]$ into n equal subintervals $[t_{k-1}, t_k]$, $k = 1, \ldots, n$, of length $\Delta t = \frac{t_b - t_a}{n}$, with $t_0 = t_a$ and $t_n = t_b$. Now we let $q_0 = q_a$ and $q_n = q_b$ and take $q_1, \ldots, q_{n-1}$ to be arbitrary real numbers and compute

$$S_n(q_a, q_1, \ldots, q_{n-1}, q_b) = \sum_{k=1}^{n} \left[\frac{m}{2}\left(\frac{q_k - q_{k-1}}{\Delta t} \right)^2 - V(q_{k-1}) \right] \Delta t. \tag{9.1}$$

Next we are to perform the integrations

$$\int_{\mathbb{R}} \cdots \int_{\mathbb{R}} e^{\frac{i}{\hbar} S_n(q_a, q_1, \ldots, q_{n-1}, q_b)} \, dq_1 \ldots dq_{n-1}, \tag{9.2}$$

multiply by the normalizing factor

$$\left(\frac{m}{2\pi i \hbar \Delta t} \right)^{n/2} \tag{9.3}$$

(where $i^{1/2} = e^{\pi i/4}$) and take the limit as $n \to \infty$ to get

$$\int_{\mathcal{P}_{\mathbb{R}}(q_b, t_b; q_a, t_a)} e^{\frac{i}{\hbar} S} \mathcal{D}q = \lim_{n \to \infty} \left(\frac{m}{2\pi i \hbar \Delta t} \right)^{n/2} \int_{\mathbb{R}} \cdots \int_{\mathbb{R}} e^{\frac{i}{\hbar} S_n(q_a, q_1, \ldots, q_{n-1}, q_b)} \, dq_1 \ldots dq_{n-1}. \tag{9.4}$$

The right-hand side is then the *definition* of the left-hand side.

We found in the previous two sections that this worked out admirably in the two simplest cases of the free particle ($V = 0$) and the harmonic oscillator ($V(q) = m\omega^2 q^2/2$), giving us precisely the propagators we had computed by other means earlier. It is an unfortunate fact of life, however, that for even slightly more complicated potentials the explicit computation of the path integral is, at least, orders of magnitude more difficult and, at worst, impossible. Physicists have developed many ingenious schemes for evaluating, or at least approximating, such path integrals (see [Smir]), but this is really not our concern here. We would like to approach this from the other end and look for general theorems that address some of the mathematical issues raised by Feynman's definition. Here are the issues we have in mind.

1. The potentials V with which one must deal are dictated by physics and one cannot simply *assume* that $H = -\frac{\hbar^2}{2m}\Delta + V$ is self-adjoint; one must prove it. If H is not self-adjoint on some domain, then it cannot be regarded as an observable and, more

https://doi.org/10.1515/9783110751949-009

to the point here, does not determine a time evolution $e^{-itH/\hbar}$ at all and so there is no propagator. One would like to see general theorems guaranteeing the self-adjointness of H for certain classes of physically meaningful potentials V.

2. Since

$$\left| e^{\frac{i}{\hbar} S_n(q_a, q_1, \dots, q_{n-1}, q_b)} \right| = 1$$

the exponential in the path integral for the propagator is not Lebesgue integrable on $\mathbb{R}^{n-1}$ and so the meaning of

$$\int_{\mathbb{R}} \cdots \int_{\mathbb{R}} e^{\frac{i}{\hbar} S_n(q_a, q_1, \dots, q_{n-1}, q_b)} \, dq_1 \dots dq_{n-1}$$

is not *a priori* clear.

3. Even assuming that the issue in 2. can be resolved, Feynman defines his path integral as a limit (9.4), but does not specify what sort of limit he has in mind and, of course, limits have the unfortunate habit of not existing when you want them to exist. We would like to see some general results asserting the existence of the limit (in some sense) for various classes of physically reasonable potentials.

4. It would be a fine thing if the path "integral" were really an integral in the Lebesgue sense since this would provide us with an arsenal of very powerful analytical weapons with which to study it. We would therefore like to know if there is a measure on the path space $\mathcal{P}_{\mathbb{R}}(q_b, t_b; q_a, t_a)$ with the property that integrating $e^{\frac{i}{\hbar} S}$ with respect to this measure is the same as evaluating Feynman's limit (9.4).

It goes without saying that all of these issues have received a great deal of attention since 1948 and the literature is not only vast, but technically quite imposing. The following sections can be regarded as nothing more than an appetizer, but we will try to provide sufficient references for those who crave the entire meal.

Before getting started we should point out that much of what we will have to say, particularly in Section 9.2, was proved in order to deal with very specific physical situations and that these generally involve more than one degree of freedom. The classical configuration space of a single particle moving in space, for example, is $\mathbb{R}^3$, not $\mathbb{R}$, so quantization gives rise to the Hilbert space $L^2(\mathbb{R}^3)$, not $L^2(\mathbb{R})$. The hydrogen atom consists of two particles moving in space so its classical configuration space is $\mathbb{R}^3 \times \mathbb{R}^3 = \mathbb{R}^6$ and the corresponding Hilbert space is $L^2(\mathbb{R}^6)$. These additional degrees of freedom can substantially increase the technical issues involved in solving concrete problems, but the corresponding Schrödinger equation is the obvious, natural generalization of (6.10) and we would like to record it here.

We consider n particles with masses $m_1, \dots, m_n$ moving in $\mathbb{R}^3$. For each $k = 1, \dots, n$ we denote by $\mathbf{q}_k$ the position vector of the kth particle and we will label the standard coordinates in $\mathbb{R}^{3n}$ in such a way that

$$\mathbf{q}_1 = (q^1, q^2, q^3), \quad \mathbf{q}_2 = (q^4, q^5, q^6), \quad \ldots, \quad \mathbf{q}_n = (q^{3n-2}, q^{3n-1}, q^{3n}).$$

The potential governing the motion of the particles is a real-valued function $V : \mathbb{R}^{3n} \to \mathbb{R}$ on $\mathbb{R}^{3n}$ and we will write its coordinate expression as $V(\mathbf{q}_1, \ldots, \mathbf{q}_n)$. To each $k = 1, \ldots, n$ we associate a Laplacian

$$\Delta_k = \frac{\partial^2}{(\partial q^{3k-2})^2} + \frac{\partial^2}{(\partial q^{3k-1})^2} + \frac{\partial^2}{(\partial q^{3k})^2},$$

so that the Laplacian Δ on $\mathbb{R}^{3n}$ itself is just the sum of these. The Hamiltonian H is defined on smooth functions in $L^2(\mathbb{R}^{3n})$ by

$$H = -\sum_{k=1}^{n} \frac{\hbar^2}{2m_k} \Delta_k + V. \tag{9.5}$$

The wave function of this system of n particles is written

$$\psi(\mathbf{q}_1, \ldots, \mathbf{q}_n, t)$$

and, if smooth, is assumed to satisfy the *Schrödinger equation*

$$i\hbar \frac{\partial \psi(\mathbf{q}_1, \ldots, \mathbf{q}_n, t)}{\partial t} = \left(-\sum_{k=1}^{n} \frac{\hbar^2}{2m_k} \Delta_k + V \right) \psi(\mathbf{q}_1, \ldots, \mathbf{q}_n, t) \tag{9.6}$$

in the classical sense. For a single particle of mass m we will generally write $\mathbf{q} = (q^1, q^2, q^3)$ and

$$i\hbar \frac{\partial \psi(\mathbf{q}, t)}{\partial t} = \left(-\frac{\hbar^2}{2m} \Delta + V \right) \psi(\mathbf{q}, t).$$

Example 9.1.1. An atom of hydrogen consists of a proton of mass $m_1 = m_p \approx 1.672 \times 10^{-27}$ kg and an electron of mass $m_2 = m_e \approx 9.109 \times 10^{-31}$ kg interacting through a potential that is inversely proportional to the distance between them. More precisely, if $\mathbf{q}_1$ and $\mathbf{q}_2$ denote the position vectors of the proton and electron, respectively, then V is given by *Coulomb's law*

$$V(\mathbf{q}_1, \mathbf{q}_2) = -\frac{1}{4\pi\epsilon_0} \frac{e^2}{\|\mathbf{q}_1 - \mathbf{q}_2\|},$$

where $-e \approx -1.602 \times 10^{-19}$ C is the charge of the electron (so e is the charge of the proton) and ϵ_0 is the vacuum permittivity (see Section 4.2). According to (9.5) the Hamiltonian of this system is

$$H = -\frac{\hbar^2}{2m_p} \Delta_1 - \frac{\hbar^2}{2m_e} \Delta_2 - \frac{1}{4\pi\epsilon_0} \frac{e^2}{\|\mathbf{q}_1 - \mathbf{q}_2\|}. \tag{9.7}$$

The form of the Hamiltonian given in (9.7) is not the most convenient for many purposes. To find something better we will introduce the center of mass coordinates that we first saw in our discussion of the classical two-body problem (Example 2.2.12). Define $\mathbf{r}$ and $\mathbf{R}$ by

$$\mathbf{r} = \mathbf{q}_1 - \mathbf{q}_2$$

and

$$\mathbf{R} = \frac{m_1\mathbf{q}_1 + m_2\mathbf{q}_2}{m_1 + m_2}.$$

Then

$$\mathbf{q}_1 = \mathbf{R} + \frac{m_2\mathbf{r}}{m_1 + m_2}$$

and

$$\mathbf{q}_2 = \mathbf{R} + \frac{m_1\mathbf{r}}{m_1 + m_2}.$$

To express the Hamiltonian in terms of the coordinates $\mathbf{r}$ and $\mathbf{R}$ we let $\Psi(\mathbf{q}_1, \mathbf{q}_2)$ be a function of $\mathbf{q}_1$ and $\mathbf{q}_2$ and $\Phi(\mathbf{r}, \mathbf{R})$ the corresponding function of $\mathbf{r}$ and $\mathbf{R}$, that is,

$$\Phi(\mathbf{r}, \mathbf{R}) = \Psi\left(\mathbf{R} + \frac{m_2\mathbf{r}}{m_1 + m_2}, \mathbf{R} + \frac{m_1\mathbf{r}}{m_1 + m_2}\right)$$

and

$$\Psi(\mathbf{q}_1, \mathbf{q}_2) = \Phi\left(\mathbf{q}_1 - \mathbf{q}_2, \frac{m_1\mathbf{q}_1 + m_2\mathbf{q}_2}{m_1 + m_2}\right).$$

Now introduce a little notation. Write $\mathbf{r} = (r^1, r^2, r^3)$, $\mathbf{R} = (R^1, R^2, R^3)$ and let $r = \|\mathbf{r}\|$. Define operators

$$\Delta_\mathbf{r} = \frac{\partial^2}{(\partial r^1)^2} + \frac{\partial^2}{(\partial r^2)^2} + \frac{\partial^2}{(\partial r^3)^2}$$

and

$$\Delta_\mathbf{R} = \frac{\partial^2}{(\partial R^1)^2} + \frac{\partial^2}{(\partial R^2)^2} + \frac{\partial^2}{(\partial R^3)^2}.$$

Finally, let $M = m_1 + m_2$ be the total mass and $\mu = \frac{m_1 m_2}{m_1 + m_2}$ the so-called *reduced mass*.

Exercise 9.1.1. Show that, in terms of the variables $\mathbf{r}$ and $\mathbf{R}$, the Hamiltonian for the hydrogen atom is given by

$$H = -\frac{\hbar^2}{2M}\Delta_\mathbf{R} - \frac{\hbar^2}{2\mu}\Delta_\mathbf{r} - \frac{e^2}{4\pi\epsilon_0}\frac{1}{r}. \tag{9.8}$$

The real advantage to this form of the Hamiltonian becomes apparent when one confronts the problem of solving the corresponding Schrödinger equation. In terms of the variables **R** and **r** one can separate variables to obtain two equations, one representing the translational motion of the center of mass and the other representing the motion of the two particles relative to each other. From the second of these one computes, for example, the possible energy levels (eigenvalues) of the hydrogen atom. Differences between consecutive energy levels can then be compared directly with the experimentally determined frequencies of the emission lines in the hydrogen spectrum (see Section 4.2). Carrying out these calculations is somewhat tedious, but it is something that everyone should go through. We recommend proceeding in three steps. Begin with a relatively painless undergraduate version of the computations (for example, Sections 2–5 of [Eis]). Proceed then to a more sophisticated, but still physics-oriented treatment in one of the standard graduate texts (say, Chapter IX, Sections I and III, of [Mess1]). Finally, for the rigorous version, consult Chapter II, Section 7, of [Prug].

We should point out that the hydrogen atom, in addition to its significance in physics, is not entirely unrelated to our leitmotif here. As it happens, there is a natural generalization of the harmonic oscillator to dimension four and the Schrödinger equation for the bound states of the hydrogen atom reduces, after a simple change of variable, to the Schrödinger equation for this four-dimensional oscillator. From this one can calculate the energy levels of the hydrogen atom by solving the harmonic oscillator (see [Corn]).

9.2 Self-adjointness of $-\frac{\hbar^2}{2m}\Delta + V$

In this section we would like to have a look at just a few of the many rigorous theorems that have been proved to establish the self-adjointness (or essential self-adjointness) of Schrödinger operators $-\frac{\hbar^2}{2m}\Delta + V$ for various classes of potentials V. We will prove a result strong enough to yield the self-adjointness of the Hamiltonian for the hydrogen atom (Example 9.1.1), but then will be content to state some of the important results and provide references to the proofs. The best general reference is [RS2], Sections X.1–X.6, and its *Notes* to Chapter X. We begin with a famous result of Kato and Rellich that guarantees the self-adjointness of a sufficiently "small" (in some appropriate sense) perturbation of a self-adjoint operator.

Theorem 9.2.1 (Kato–Rellich theorem). *Let $\mathcal{H}$ be a complex, separable Hilbert space, $A : \mathcal{D}(A) \to \mathcal{H}$ a self-adjoint operator on $\mathcal{H}$ and $B : \mathcal{D}(B) \to \mathcal{H}$ a symmetric operator on $\mathcal{H}$ with $\mathcal{D}(A) \subseteq \mathcal{D}(B)$. Suppose there exist real numbers a and b with $a < 1$ such that*

$$\| B\psi \| \le a\| A\psi \| + b\| \psi \|$$

for every $\psi \in \mathcal{D}(A)$. *Then*

$$A + B : \mathcal{D}(A) \to \mathcal{H}$$

is self-adjoint.

We begin with a lemma that is just a rather modest extension of Theorem 5.2.3.

Lemma 9.2.2. *Let* $\mathcal{H}$ *be a complex, separable Hilbert space and* $A : \mathcal{D}(A) \to \mathcal{H}$ *a symmetric operator. Then* A *is self-adjoint if and only if there exists a real number* μ *for which* Image $(A \pm \mu i) = \mathcal{H}$.

Proof. Certainly, if A is self-adjoint, then there is such a μ by Theorem 5.2.3 (namely, $\mu = 1$). Conversely, suppose A is symmetric and there exists a real number μ for which Image $(A \pm \mu i) = \mathcal{H}$. We must show that $\mathcal{D}(A^*) = \mathcal{D}(A)$. Since A is symmetric, $\mathcal{D}(A) \subseteq \mathcal{D}(A^*)$ and $A^*|_{\mathcal{D}(A)} = A$. Thus, we let $\phi \in \mathcal{D}(A^*)$ and show that it is in $\mathcal{D}(A)$. Since Image $(A - \mu i) = \mathcal{H}$, we can select an η in $\mathcal{D}(A)$ with $(A - \mu i)\eta = (A^* - \mu i)\phi$. Since $\mathcal{D}(A) \subseteq \mathcal{D}(A^*)$, $\phi - \eta$ is in $\mathcal{D}(A^*)$. Moreover,

$$(A^* - \mu i)(\phi - \eta) = (A^* - \mu i)\phi - (A^* - \mu i)\eta = (A^* - \mu i)\phi - (A - \mu i)\eta = 0.$$

But Image $(A + \mu i) = \mathcal{H}$, so $0 = $ Image $(A + \mu i)^{\perp} = $ Kernel $(A + \mu i)^* = $ Kernel $(A^* - \mu i)$ and therefore $\phi = \eta$ and, in particular, $\phi \in \mathcal{D}(A)$, as required. $\qquad \square$

Exercise 9.2.1. Show that a symmetric operator $A : \mathcal{D}(A) \to \mathcal{H}$ is self-adjoint if and only if Image $(A \pm \mu i) = \mathcal{H}$ for all real $\mu \neq 0$.

Proof of Kato–Rellich theorem. We intend to apply the previous lemma and show that for sufficiently large real μ, Image $(A + B \pm \mu i) = \mathcal{H}$. First note that, since A is self-adjoint, its spectrum is real so, for any nonzero real μ, $(A + \mu i)^{-1}$ exists and is defined on Image $(A + \mu i)$. But, applying the previous exercise to A, Image $(A + \mu i) = \mathcal{H}$, so $(A + \mu i)^{-1}$ is defined on all of $\mathcal{H}$. We claim that, in fact, $(A + \mu i)^{-1}$ is bounded and satisfies

$$\| (A + \mu i)^{-1} \| \leq \frac{1}{|\mu|}.$$

To see this, first note that for any $\psi \in \mathcal{D}(A)$,

$$\| (A + \mu i)\psi \|^2 = \| A\psi \|^2 + \mu^2 \| \psi \|^2.$$

Indeed,

$$
\begin{aligned}
\| (A + \mu i)\psi \|^2 &= \langle A\psi + \mu i\psi, A\psi + \mu i\psi \rangle \\
&= \langle A\psi, A\psi \rangle + \langle A\psi, \mu i\psi \rangle + \langle \mu i\psi, A\psi \rangle + \langle \mu i\psi, \mu i\psi \rangle \\
&= \| A\psi \|^2 + \mu i \langle A\psi, \psi \rangle - \mu i \langle \psi, A\psi \rangle + \mu^2 \| \psi \|^2
\end{aligned}
$$

$$= \| A\psi \|^2 + \mu i \langle \psi, A\psi \rangle - \mu i \langle \psi, A\psi \rangle + \mu^2 \| \psi \|^2$$
$$= \| A\psi \|^2 + \mu^2 \| \psi \|^2.$$

It follows that

$$\| (A + \mu i)\psi \| \geq |\mu| \, \| \psi \|.$$

Now, for any $\phi \in \mathcal{H}$, applying this to $\psi = (A + \mu i)^{-1}\phi$ gives $\| \phi \| \geq |\mu| \, \| (A + \mu i)^{-1}\phi \|$, that is,

$$\| (A + \mu i)^{-1}\phi \| \leq \frac{1}{|\mu|} \| \phi \|,$$

as required.

Exercise 9.2.2. Show that for any $\phi \in \mathcal{H}$,

$$\| A(A + \mu i)^{-1}\phi \| \leq \| \phi \|.$$

Now, for a and b as specified in the Kato–Rellich theorem we have

$$\| B(A + \mu i)^{-1}\phi \| \leq a \| A(A + \mu i)^{-1}\phi \| + b \| (A + \mu i)^{-1}\phi \|$$

$$\leq a \| \phi \| + b \left(\frac{1}{|\mu|} \| \phi \| \right)$$

and therefore

$$\| B(A + \mu i)^{-1}\phi \| \leq \left(a + \frac{b}{|\mu|} \right) \| \phi \|.$$

Since $a < 1$ we can choose $|\mu|$ sufficiently large to ensure that $a + \frac{b}{|\mu|} < 1$ and therefore

$$\| B(A + \mu i)^{-1} \| < 1.$$

We make use of this in the following way. Write

$$A + B + \mu i = (I + B(A + \mu i)^{-1})(A + \mu i),$$

where I is the identity operator on $\mathcal{H}$. Since

$$\| I - (I + B(A + \mu i)^{-1}) \| < 1,$$

the operator $I + B(A + \mu i)^{-1}$ has a bounded inverse given by the Neumann series

$$\sum_{n=0}^{\infty} (I - (I + B(A + \mu i)^{-1}))^n,$$

where the series converges in the operator norm on $\mathcal{B}(\mathcal{H})$ (see Theorem 2, Section 1, Chapter II, of [Yosida]). In particular, the operator $I + B(A + \mu i)^{-1}$ maps onto $\mathcal{H}$. Since A is self-adjoint, the same is true of $A + \mu i$ and we conclude that

$$\text{Image } (A + B + \mu i) = \text{Image } (I + B(A + \mu i)^{-1})(A + \mu i) = \mathcal{H}.$$

Exercise 9.2.3. Check that exactly the same proof shows that

$$\text{Image }(A + B - \mu i) = \mathcal{H}$$

and conclude that $A + B : \mathcal{D}(A) \to \mathcal{H}$ is self-adjoint. ☐

There is also a version of the Kato–Rellich theorem for essential self-adjointness.

Corollary 9.2.3. *Let $\mathcal{H}$ be a complex, separable Hilbert space, $A : \mathcal{D}(A) \to \mathcal{H}$ an essentially self-adjoint operator on $\mathcal{H}$ and $B : \mathcal{D}(B) \to \mathcal{H}$ a symmetric operator on $\mathcal{H}$ with $\mathcal{D}(A) \subseteq \mathcal{D}(B)$. Suppose there exist real numbers a and b with $a < 1$ such that*

$$\| B\psi \| \le a \| A\psi \| + b \| \psi \|$$

for every $\psi \in \mathcal{D}(A)$. Then

$$A + B : \mathcal{D}(A) \to \mathcal{H}$$

is essentially self-adjoint and

$$\overline{A + B} = \overline{A} + \overline{B}.$$

Proof. Since A is essentially self-adjoint, its closure $\overline{A}$ is its unique self-adjoint extension. Since B is symmetric, it is closable and therefore $\overline{B}$ exists and it too is symmetric. We intend to apply the Kato–Rellich theorem to $\overline{A}$ and $\overline{B}$, so we will first show that $\mathcal{D}(\overline{A}) \subseteq \mathcal{D}(\overline{B})$. Let $\psi \in \mathcal{D}(\overline{A})$. Since $\overline{A}$ is characterized by the fact that its graph $\text{Gr}(\overline{A})$ is the closure in $\mathcal{H} \oplus \mathcal{H}$ of the graph of A, there is a sequence of points ψ_n in $\mathcal{D}(A)$ with $\psi_n \to \psi$ and $A\psi_n \to \overline{A}\psi$. Since

$$\| B(\psi_n - \psi_m) \| \le a \| A(\psi_n - \psi_m) \| + b \| (\psi_n - \psi_m) \|,$$

the sequence of $B\psi_n$ is Cauchy in $\mathcal{H}$ and so $B\psi_n \to \phi$ for some $\phi \in \mathcal{H}$. Then, since B is closed, $\psi \in \mathcal{D}(\overline{B})$ and $\phi = \overline{B}\psi$. In particular, $\mathcal{D}(\overline{A}) \subseteq \mathcal{D}(\overline{B})$. Furthermore, for any $\psi \in \mathcal{D}(\overline{A})$,

$$\| \overline{B}\psi \| = \lim_{n\to\infty} \| B\psi_n \| \le \lim_{n\to\infty} (a\| A\psi_n \| + b\| \psi_n \|) = a\| \overline{A}\psi \| + b\| \psi \|.$$

Thus, $\overline{A}$ and $\overline{B}$ satisfy the hypotheses of the Kato–Rellich theorem and we conclude that

$$\overline{A} + \overline{B} : \mathcal{D}(\overline{A}) \to \mathcal{H}$$

is self-adjoint. In particular, $\overline{A} + \overline{B}$ is a closed extension of $A + B$ and therefore also an extension of $\overline{A + B}$. But $(A + B)\psi_n \to \overline{A}\psi + \overline{B}\psi$, so $(\overline{A + B})\psi = \overline{A}\psi + \overline{B}\psi$, which means that $\overline{A + B}$ is also an extension of $\overline{A} + \overline{B}$, so we conclude that

$$\overline{A + B} = \overline{A} + \overline{B}.$$

In particular, $\overline{A + B}$ is self-adjoint, so $A + B$ is essentially self-adjoint. ☐

In order to apply these results to Schrödinger operators $H = -\frac{\hbar^2}{2m}\Delta + V$ we will need to do two things.

1. We need to establish the self-adjointness of the free particle Hamiltonian $H_0 = -\frac{\hbar^2}{2m}\Delta$ on some appropriate domain $\mathcal{D}(H_0) \supseteq \mathcal{S}(\mathbb{R}^3)$ in $L^2(\mathbb{R}^3)$ and its essential self-adjointness on $\mathcal{S}(\mathbb{R}^3)$. We have already done this for the free particle Hamiltonian on $\mathbb{R}$ in Example 5.2.8 and, as we will see, the proof for $\mathbb{R}^3$ is virtually identical.

Remark 9.2.1. To streamline the exposition a bit we will stick with $\mathbb{R}^3$. Everything we will say is equally true for $\mathbb{R}$ and $\mathbb{R}^2$, but for $\mathbb{R}^N$ with $N \geq 4$ some adjustments are occasionally required. We will point some of these out as we proceed, but for the full story see Chapter X of [RS2].

2. We need to isolate some class of potentials V for which $\mathcal{D}(H_0)$ is contained in the domain $\mathcal{D}(V)$ of the multiplication operator V and for which there exist real numbers $a < 1$ and b such that

$$\| V\psi \| \leq a \| H_0\psi \| + b \| \psi \|$$

for every $\psi \in \mathcal{D}(H_0)$. Our stated objective is to find such a class of potentials large enough to include the Coulomb potential for the hydrogen atom. The condition we eventually decide on may look a bit strange at first sight, so we will use the hydrogen atom to motivate it in the following example.

Example 9.2.1. Writing the Hamiltonian for the hydrogen atom in the form (9.8), the potential is given by

$$V(r) = -\frac{e^2}{4\pi\epsilon_0} \frac{1}{r}.$$

Let χ_1 denote the characteristic function of the closed unit ball $r \leq 1$ in $\mathbb{R}^3$. Then $1 - \chi_1$ is the characteristic function of $r > 1$ and we can write

$$V = V_1 + V_2 = \chi_1 V + (1 - \chi_1)V.$$

Exercise 9.2.4. Show that $V_1 = \chi_1 V$ is in $L^2(\mathbb{R}^3)$ and $V_2 = (1 - \chi_1)V$ is in $L^\infty(\mathbb{R}^3)$.

We intend to show that whenever the potential V is real-valued and can be written as the sum of an element V_1 of $L^2(\mathbb{R}^3)$ and an element V_2 of $L^\infty(\mathbb{R}^3)$, then $H_0 + V$ satisfies the hypotheses of the Kato–Rellich theorem and hence is self-adjoint on $\mathcal{D}(H_0)$.

Remark 9.2.2. We should say a few words about why we need the Kato–Rellich theorem for both self-adjoint and essentially self-adjoint operators and how we intend to use them. A Schrödinger operator is initially defined in the classical sense on some space of differentiable functions (such as $\mathcal{S}(\mathbb{R}^3)$ or $C_0^\infty(\mathbb{R}^3)$). Here it is not self-adjoint, so we would like to find a self-adjoint extension. But this alone is not enough. We would like to be sure that there is *only one* self-adjoint extension since, if there is more than one, then there is more than one contender for the Hamiltonian of the quantum system under consideration and, consequently, for the time evolution of that system. One can see explicitly that different self-adjoint extensions correspond to different physics in Examples 1 and 2, Section X.1, of [RS2]. But suppose we show that H is self-adjoint by showing that the potential satisfies the hypotheses of the Kato–Rellich theorem on $\mathcal{D}(H_0) \subseteq L^2(\mathbb{R}^3)$. Then it certainly also satisfies these conditions on $\mathcal{S}(\mathbb{R}^3)$ and, since H_0 is essentially self-adjoint on $\mathcal{S}(\mathbb{R}^3)$, we conclude from the corollary that $H\,|_{\mathcal{S}(\mathbb{R}^3)}$ is also essentially self-adjoint and hence has a *unique* self-adjoint extension, which must be H. The corollary is a roundabout way of establishing uniqueness.

Now we turn to the first of the two tasks we set ourselves above. For this it will clearly suffice to consider only the Laplace operator $-\Delta$ (we will explain the reason for leaving the minus sign shortly). We begin with a few remarks on the restriction of $-\Delta$ to $\mathcal{S}(\mathbb{R}^3)$. Two integrations by parts show that $-\Delta$ is symmetric on $\mathcal{S}(\mathbb{R}^3)$. Indeed, for ψ and ϕ in $\mathcal{S}(\mathbb{R}^3)$,

$$\langle -\Delta\psi, \phi \rangle = \int_{\mathbb{R}^3} -\Delta\psi(\mathbf{q})\overline{\phi(\mathbf{q})}\, d^3\mathbf{q}$$

$$= \sum_{j=1}^{3} \int_{\mathbb{R}^3} -\partial_j^2\psi(\mathbf{q})\overline{\phi(\mathbf{q})}\, d^3\mathbf{q}$$

$$= \sum_{j=1}^{3} \int_{\mathbb{R}^3} \partial_j\phi(\mathbf{q})\partial_j\overline{\phi(\mathbf{q})}\, d^3\mathbf{q}$$

$$= \sum_{j=1}^{3} \int_{\mathbb{R}^3} -\psi(\mathbf{q})\partial_j^2\overline{\phi(\mathbf{q})}\, d^3\mathbf{q}$$

$$= \langle \psi, -\Delta\phi \rangle.$$

Exercise 9.2.5. Show that for $\psi \in \mathcal{S}(\mathbb{R}^3)$,

$$\langle -\Delta\psi, \psi \rangle \geq 0.$$

We conclude that $-\Delta$ is a positive, symmetric operator on $\mathcal{S}(\mathbb{R}^3)$. The reason this is of interest is that the Friedrichs extension theorem guarantees the existence of a positive, self-adjoint extension of $-\Delta$ (the positivity of $-\Delta$ is the reason for retaining the minus sign). In the case of $-\Delta$ on $\mathcal{S}(\mathbb{R}^3)$ there is, in fact, a *unique* self-adjoint extension, but this does not follow from the Friedrichs theorem, so we will need a more explicit construction. This is done just as it was for $\mathbb{R}$ in Example 5.2.8 by applying the Fourier transform (see Section G.4).

For $\phi \in \mathcal{S}(\mathbb{R}^3)$, it follows from $\mathcal{F}(\partial_\alpha \phi)(\mathbf{p}) = (i\mathbf{p})^\alpha(\mathcal{F}\phi)(\mathbf{p})$ with $\alpha = (0, \ldots, 2, \ldots, 0)$ (2 in the jth slot) that $\mathcal{F}(\partial_j^2 \phi)(\mathbf{p}) = -(p^j)^2\hat{\phi}(\mathbf{p})$ for $j = 1, 2, 3$. Thus,

$$\mathcal{F}(-\Delta\phi)(\mathbf{p}) = \|\mathbf{p}\|^2\hat{\phi}(\mathbf{p}).$$

In particular, for every $\phi \in \mathcal{S}(\mathbb{R}^3)$,

$$-\Delta\phi(\mathbf{q}) = (\mathcal{F}^{-1}Q_{\|\mathbf{p}\|^2}\mathcal{F})\phi(\mathbf{q}),$$

where $Q_{\|\mathbf{p}\|^2}$ is the multiplication operator on $L^2(\mathbb{R}^3)$ defined by $(Q_{\|\mathbf{p}\|^2}\hat{\phi})(\mathbf{p}) = \|\mathbf{p}\|^2\hat{\phi}(\mathbf{p})$. We have noted (in Section G.4) that $\mathcal{F}(\partial_a\psi)(\mathbf{p}) = (i\mathbf{p})^\alpha(\mathcal{F}\psi)(\mathbf{p})$ is true even when ψ is a distribution in $L^2(\mathbb{R}^3)$ provided the derivatives, interpreted in the distributional sense, are also in $L^2(\mathbb{R}^3)$. If $\|\mathbf{p}\|^2\hat{\phi}(\mathbf{p})$ is also in $L^2(\mathbb{R}^3)$, then we can define $-\Delta$ as an operator on $L^2(\mathbb{R}^3)$ that extends $-\Delta$ on $\mathcal{S}(\mathbb{R}^3)$ by

$$-\Delta : \mathcal{D}(-\Delta) \to L^2(\mathbb{R}^3),$$
$$-\Delta\psi = (\mathcal{F}^{-1}Q_{\|\mathbf{p}\|^2}\mathcal{F})\psi,$$

where

$$\mathcal{D}(-\Delta) = \{\psi : \psi \in L^2(\mathbb{R}^3) \text{ and } \|\mathbf{p}\|^2\hat{\psi} \in L^2(\mathbb{R}^3)\}.$$

The domain of $-\Delta$ on $L^2(\mathbb{R}^3)$ can, of course, also be written

$$\mathcal{D}(-\Delta) = \{\psi : \psi \in L^2(\mathbb{R}^3) \text{ and } \Delta\psi \in L^2(\mathbb{R}^3)\}$$

and it is worth pointing out that this is the same as the Sobolev space $H^2(\mathbb{R}^3)$ (see Section G.4). Since $-\Delta = \mathcal{F}^{-1}Q_{\|\mathbf{p}\|^2}\mathcal{F}$ and $\mathcal{F} : L^2(\mathbb{R}^3) \to L^2(\mathbb{R}^3)$ is unitary, $-\Delta$ is unitarily equivalent to $Q_{\|\mathbf{p}\|^2}$.

Exercise 9.2.6. Show that $Q_{\|\mathbf{p}\|^2} : \mathcal{D}(-\Delta) \to L^2(\mathbb{R}^3)$ is self-adjoint. *Hint*: See Exercise 5.2.6 and the discussion preceding it.

According to Lemma 5.2.5, $-\Delta : \mathcal{D}(-\Delta) \to L^2(\mathbb{R}^3)$ is therefore also self-adjoint and this is what we wanted to prove.

Well, that is not quite all we wanted to prove. We now have one self-adjoint extension of $-\Delta|_{\mathcal{S}(\mathbb{R}^3)}$, but to remove any ambiguities we need to know that it is the only one. For this we must show that $-\Delta|_{\mathcal{S}(\mathbb{R}^3)}$ is essentially self-adjoint and therefore has a unique self-adjoint extension. Since the argument is exactly the same as the one given for $\mathcal{S}(\mathbb{R})$ in Section 5.2 we will leave it for you (if you would prefer a little variety, there is another argument in Theorem IX.27 (c) of [RS2]).

Exercise 9.2.7. Show that $-\Delta|_{\mathcal{S}(\mathbb{R}^3)}$ is essentially self-adjoint by proving that

$$\text{Image }(-\Delta|_{\mathcal{S}(\mathbb{R}^3)} \pm i)^\perp$$

are both zero.

Since a positive multiple of $-\Delta$ clearly has all of the properties of $-\Delta$ that we have just established, we have completed the first of the tasks we set for ourselves, that is, the free particle Hamiltonian

$$H_0 = -\frac{\hbar^2}{2m}\Delta$$

on $\mathbb{R}^3$ is self-adjoint on

$$\mathcal{D}(H_0) = \{\psi : \psi \in L^2(\mathbb{R}^3) \text{ and } \|\mathbf{p}\|^2\hat{\psi} \in L^2(\mathbb{R}^3)\}$$
$$= \{\psi : \psi \in L^2(\mathbb{R}^3) \text{ and } \Delta\psi \in L^2(\mathbb{R}^3)\}$$
$$= H^2(\mathbb{R}^3)$$

and essentially self-adjoint on $\mathcal{S}(\mathbb{R}^3)$. Our second objective takes a bit more work, but we will begin by stating precisely what it is we would like to prove.

Theorem 9.2.4. *Let V be a real-valued, measurable function on $\mathbb{R}^3$ that can be written as $V = V_1 + V_2$, where $V_1 \in L^2(\mathbb{R}^3)$ and $V_2 \in L^\infty(\mathbb{R}^3)$. Then*

$$H = H_0 + V = -\frac{\hbar^2}{2m}\Delta + V$$

is essentially self-adjoint on $C_0^\infty(\mathbb{R}^3)$ and self-adjoint on $\mathcal{D}(H_0)$.

Exercise 9.2.8. Show that if $H = H_0 + V = -\frac{\hbar^2}{2m}\Delta + V$ is essentially self-adjoint on $C_0^\infty(\mathbb{R}^n)$, then it is also essentially self-adjoint on $\mathcal{S}(\mathbb{R}^n)$. *Hint*: Use Corollary 5.2.4.

Theorem 9.2.4 remains true if $\mathbb{R}^3$ is everywhere replaced by $\mathbb{R}$ or $\mathbb{R}^2$, but for $\mathbb{R}^n$ with $n \geq 4$ it is not sufficient to assume that V_1 is L^2 (see Theorem X.29 of [RS2]).

By virtue of Example 9.2.1 we conclude from Theorem 9.2.4 that the Hamiltonian of the hydrogen atom is self-adjoint on $\mathcal{D}(H_0)$; we will see shortly that there is a generalization of Theorem 9.2.4 that applies to *any* atom or molecule. For the proof of Theorem 9.2.4 we will need the following lemma.

Lemma 9.2.5. *Let $\psi \in L^2(\mathbb{R}^3)$ be in $\mathcal{D}(H_0)$. Then ψ is in $L^\infty(\mathbb{R}^3)$ and, moreover, for any $a > 0$ there exists a real number b, independent of ψ, such that*

$$\|\psi\|_{L^\infty} \leq a\|H_0\psi\|_{L^2} + b\|\psi\|_{L^2}. \tag{9.9}$$

Proof. We would like to show first that $\hat{\psi} \in L^1(\mathbb{R}^3)$. For this we recall that, by the Hölder inequality (see, for example, Theorem 2.3 of [LL]), the product of two L^2 functions is an L^1 function. Since $\psi \in \mathcal{D}(H_0)$, we know that $\psi \in L^2(\mathbb{R}^3)$, so $\hat{\psi} \in L^2(\mathbb{R}^3)$. Moreover, $\Delta\psi \in L^2(\mathbb{R}^3)$ implies that $\|\mathbf{p}\|^2\hat{\psi} \in L^2(\mathbb{R}^3)$. Consequently, $(1 + \|\mathbf{p}\|^2)\hat{\psi}$ is in $L^2(\mathbb{R}^3)$. But an integration in spherical coordinates shows that

$$\left\| (1 + \|\mathbf{p}\|^2)^{-1} \right\|_{L^2}^2 = \int_{\mathbb{R}^3} \frac{1}{(1 + \|\mathbf{p}\|^2)^2} d^3\mathbf{p} = \pi^2,$$

so $(1 + \|\mathbf{p}\|^2)^{-1}$ is also in $L^2(\mathbb{R}^3)$. Consequently, the product

$$\hat{\psi} = (1 + \|\mathbf{p}\|^2)^{-1}(1 + \|\mathbf{p}\|^2)\hat{\psi}$$

is in $L^1(\mathbb{R}^3)$. In particular, $\|\hat{\psi}\|_{L^1}$ is finite and (by the Hölder inequality again)

$$\|\hat{\psi}\|_{L^1} \leq \left\| (1 + \|\mathbf{p}\|^2)^{-1} \right\|_{L^2} \left\| (1 + \|\mathbf{p}\|^2)\hat{\psi} \right\|_{L^2} \leq \pi \left(\|\hat{\psi}\|_{L^2} + \| \|\mathbf{p}\|^2\hat{\psi}\|_{L^2} \right). \tag{9.10}$$

Exercise 9.2.9. Show that

$$\|\psi\|_{L^\infty} \le (2\pi)^{-3/2}\|\hat\psi\|_{L^1}$$

and conclude that $\psi \in L^\infty(\mathbb{R}^3)$.

Since the Fourier transform is an isometry on $L^2(\mathbb{R}^3)$,

$$\|H_0\psi\|_{L^2} = \|\mathcal{F}(H_0\psi)\|_{L^2} = \frac{\hbar^2}{2m}\,\big\|\,\|\mathbf{p}\|^2\hat\psi\,\big\|_{L^2}$$

and

$$\|\psi\|_{L^2} = \|\hat\psi\|_{L^2},$$

so, for any $a, b \in \mathbb{R}$,

$$a\,\|H_0\psi\|_{L^2} + b\,\|\psi\|_{L^2} = a\left(\frac{\hbar^2}{2m}\right)\big\|\,\|\mathbf{p}\|^2\hat\psi\,\big\|_{L^2} + b\,\|\hat\psi\|_{L^2}.$$

From this and Exercise 9.2.9 we conclude that to prove our lemma it will suffice to show that for any $a' > 0$, there exists a $b' \in \mathbb{R}$ such that

$$\|\hat\psi\|_{L^1} \le a'\,\big\|\,\|\mathbf{p}\|^2\hat\psi\,\big\|_{L^2} + b'\,\|\hat\psi\|_{L^2}.$$

To prove this we fix an arbitrary $r > 0$ and define $\hat\psi_r$ by

$$\hat\psi_r(\mathbf{p}) = r^3\hat\psi(r\mathbf{p}).$$

Exercise 9.2.10. Use the scaling property of the Fourier transform ($\mathcal{F}(\psi(a\mathbf{q})) = \frac{1}{|a|}\hat\psi(\frac{1}{a}\mathbf{p})$ for $a \ne 0$ in $\mathbb{R}$) to show that $\hat\psi_r$ is also the Fourier transform of some element of $\mathcal{D}(H_0)$, so that (9.10) is valid for $\hat\psi_r$ as well, that is,

$$\|\hat\psi_r\|_{L^1} \le \pi\,(\,\|\hat\psi_r\|_{L^2} + \big\|\,\|\mathbf{p}\|^2\hat\psi_r\,\big\|_{L^2}\,). \tag{9.11}$$

Now note that

$$\|\hat\psi_r\|_{L^1} = \int_{\mathbb{R}^3}|\hat\psi_r(\mathbf{p})|\,d^3\mathbf{p} = r^3\int_{\mathbb{R}^3}|\hat\psi(r\mathbf{p})|\,d^3\mathbf{p} = r^3\int_{\mathbb{R}^3}|\hat\psi(\mathbf{p}')|\,|r^{-3}\,d^3\mathbf{p}' = \|\hat\psi\|_{L^1}.$$

Exercise 9.2.11. Show similarly that

$$\|\hat\psi_r\|_{L^2} = r^{3/2}\|\hat\psi\|_{L^2}$$

and

$$\big\|\,\|\mathbf{p}\|^2\hat\psi_r\,\big\|_{L^2} = r^{-1/2}\big\|\,\|\mathbf{p}\|^2\hat\psi\,\big\|_{L^2}.$$

Substituting all of this into (9.11) gives

$$\|\hat{\psi}\|_{L^1} \leq \pi r^{-1/2} \|\|\mathbf{p}\|^2 \hat{\psi}\|_{L^2} + \pi r^{3/2} \|\hat{\psi}\|_{L^2}.$$

Thus, given an $a' > 0$ we choose r so that $\pi r^{-1/2} = a'$ and then, with this value of r, $b' = \pi r^{3/2}$ will satisfy (9.9), so the proof of the lemma is complete. □

Proof of Theorem 9.2.4. We assume that V is a real-valued function on $\mathbb{R}^3$ that can be written as $V = V_1 + V_2$, where $V_1 \in L^2(\mathbb{R}^3)$ and $V_2 \in L^\infty(\mathbb{R}^3)$, and we will apply the Kato–Rellich Theorem 9.2.1, to show that $H = H_0 + V$ is self-adjoint on $\mathcal{D}(H_0)$. Since V, being real-valued, is clearly symmetric as a multiplication operator we must prove two things:
1. $\mathcal{D}(H_0) \subseteq \mathcal{D}(V)$,
2. there exist real numbers a and b with $a < 1$ such that

$$\|V\psi\|_{L^2} \leq a\|H_0\psi\|_{L^2} + b\|\psi\|_{L^2}$$

for all $\psi \in \mathcal{D}(H_0)$.

Note that $\mathcal{D}(V)$ consists of all those ψ in $L^2(\mathbb{R}^3)$ for which $V\psi$ is also in $L^2(\mathbb{R}^3)$. Let ψ be in $\mathcal{D}(H_0)$. In particular, $\psi \in L^2(\mathbb{R}^3)$ and, by Lemma 9.2.5, $\psi \in L^\infty(\mathbb{R}^3)$. Consequently, both $\|\psi\|_{L^2}$ and $\|\psi\|_{L^\infty}$ are finite, so

$$\|V\psi\|_{L^2} = \|V_1\psi + V_2\psi\|_{L^2} \leq \|V_1\psi\|_{L^2} + \|V_2\psi\|_{L^2} \leq \|V_1\|_{L^2}\|\psi\|_{L^\infty} + \|V_2\|_{L^\infty}\|\psi\|_{L^2}$$

implies that $V\psi$ is in $L^2(\mathbb{R}^3)$ and therefore $\mathcal{D}(H_0) \subseteq \mathcal{D}(V)$ as required.

According to Lemma 9.2.5, for *any* $a' > 0$ there is a $b' \in \mathbb{R}$ for which $\|\psi\|_{L^\infty} \leq a'\|H_0\psi\|_{L^2} + b'\|\psi\|_{L^2}$. Thus,

$$\|V\psi\|_{L^2} \leq \|V_1\|_{L^2}\|\psi\|_{L^\infty} + \|V_2\|_{L^\infty}\|\psi\|_{L^2}$$
$$\leq a'\|V_1\|_{L^2}\|H_0\psi\|_{L^2} + (b'\|V_1\|_{L^2} + \|V_2\|_{L^\infty})\|\psi\|_{L^2},$$

so if we choose a' sufficiently small that $a = a'\|V_1\|_{L^2} < 1$ and take $b = b'\|V_1\|_{L^2} + \|V_2\|_{L^\infty}$, then $\|V\psi\|_{L^2} \leq a\|H_0\psi\|_{L^2} + b\|\psi\|_{L^2}$, as required. The argument for the essential self-adjointness of H on $\mathcal{S}(\mathbb{R}^3)$ is described in Remark 9.2.2. □

We will conclude this section by surveying just a few of the many rigorous theorems of this same sort that specify a class of potentials V for which $H = H_0 + V$ is self-adjoint, or at least essentially self-adjoint, on some domain. There is a very detailed account of such results in Sections X.1–X.6 of [RS2] and a guide to even more in the *Notes* to Chapter X of this source. The results we describe will generally assert the *essential self-adjointness* of H on the space of *smooth functions with compact support*. Essential self-adjointness is generally all that matters. For example, we need to be sure of the existence of a unique self-adjoint operator that agrees with $-\frac{\hbar^2}{2m}\Delta + V$ on smooth functions, but the precise domain on which this extension is self-adjoint is very often not relevant (and also very often difficult to determine).

Theorem 9.2.4, from which we obtained the self-adjointness of the Hamiltonian for the hydrogen atom, is but a special case of a very famous result of Kato [Kato] which established the self-adjointness of the Hamiltonian for any atom or molecule. We will describe the Hamiltonian for a neutral (uncharged) atom of atomic number Z. The nucleus therefore contains Z protons and perhaps some neutrons. Being neutral, the atom also has Z electrons. We will denote the mass of the nucleus by M and the mass of each electron by m_e. We will denote the position vector of the nucleus by $\mathbf{q}_0 = (q^1, q^2, q^3)$ and the position vectors of the Z electrons by $\mathbf{q}_1 = (q^4, q^5, q^6), \ldots, \mathbf{q}_Z = (q^{3Z+1}, q^{3Z+2}, q^{3Z+3})$. For each $k = 0, 1, \ldots, Z$ we introduce a Laplacian $\Delta_k = \partial^2_{3k+1} + \partial^2_{3k+2} + \partial^2_{3K+3}$. The potential is determined by the Coulomb interactions between each electron and the nucleus and the Coulomb interactions between each pair of distinct electrons. Specifically, the Hamiltonian is defined on $C^\infty_0(\mathbb{R}^{3Z+3})$ by

$$H = -\frac{\hbar^2}{2M}\Delta_0 - \frac{\hbar^2}{2m_e}\sum_{k=1}^{Z}\Delta_k - \frac{1}{4\pi\epsilon_0}\sum_{k=1}^{Z}\frac{Ze^2}{\|\mathbf{q}_k - \mathbf{q}_0\|} + \frac{1}{4\pi\epsilon_0}\sum_{k=1}^{Z}\sum_{k<j}\frac{e^2}{\|\mathbf{q}_k - \mathbf{q}_j\|} \qquad (9.12)$$

and its essential self-adjointness on $C^\infty_0(\mathbb{R}^{3Z+3})$ is a consequence of the following result of Kato (which is Theorem X.16 of [RS2]).

Theorem 9.2.6 (Kato's theorem). *Let $\{V^k\}_{k=1}^{m}$ be a collection of real-valued functions on $\mathbb{R}^3$ each of which can be written as $V^k = V^k_1 + V^k_2$, where $V^k_1 \in L^2(\mathbb{R}^3)$ and $V^k_2 \in L^\infty(\mathbb{R}^3)$. Let $V^k(\mathbf{y}_k)$ be the multiplication operator on $L^2(\mathbb{R}^{3n})$ obtained by choosing $\mathbf{y}_k$ to be three coordinates of $\mathbb{R}^{3n}$. Finally, let Δ denote the Laplacian on $\mathbb{R}^{3n}$. Then $-\frac{\hbar^2}{2m}\Delta + \sum_{k=1}^{m} V^k(\mathbf{y}_k)$ is essentially self-adjoint on $C^\infty_0(\mathbb{R}^{3n})$.*

Since a great deal of quantum mechanics focuses on the behavior of atoms and molecules, Kato's theorem is arguably the most important self-adjointness result we will see. Even so, not every physically interesting quantum system has a Hamiltonian to which one can apply Kato's theorem to establish self-adjointness. We turn next to one particularly important example.

Very early on in our discussion of the classical harmonic oscillator we emphasized that the potential $V(q) = \frac{1}{2}m\omega^2 q^2$ provides an accurate model of only the *small* displacements of a mass on a spring, or a pendulum, or any system with one degree of freedom near the stable equilibrium point at $q = 0$. When the displacements cannot be regarded as small, but are not "too large" (for example, the spring is not stretched out into a straight piece of wire) one corrects the potential by including additional terms in its Taylor series about the equilibrium point. The simplest example of such a perturbed harmonic oscillator potential would seem to be

$$V(q) = \frac{1}{2}m\omega^2 q^2 + \alpha q^4, \qquad (9.13)$$

where α is a small positive constant.

You are wondering why we jumped over the cubic term. Of course, we did not have to. One can include a multiple of q^3 in the potential, but at the cost of rather seriously distorting the physical situation that one probably has in mind for a "perturbed harmonic oscillator." One would think, for example, that the potential for such a system would be symmetric about the equilibrium point, which it is not with a cubic term. More seriously, including a q^3 term gives a potential that is not bounded from below and so, for large enough displacements, leads to "runaway" solutions which is hardly in keeping with any sort of "oscillator."

A system governed by a potential of the form (9.13) will be referred to as an *anharmonic oscillator*. The natural quantization of such a system leads to a Hamiltonian which, on the smooth functions in $L^2(\mathbb{R})$, is given by

$$H = H_B + \alpha q^4 = -\frac{\hbar^2}{2m}\Delta + \frac{1}{2}m\omega^2 q^2 + \alpha q^4 = -\frac{\hbar^2}{2m}\frac{d^2}{dq^2} + \frac{1}{2}m\omega^2 q^2 + \alpha q^4, \tag{9.14}$$

where $V = \frac{1}{2}m\omega^2 q^2 + \alpha q^4$ acts on $L^2(\mathbb{R})$ as a multiplication operator. The essential self-adjointness of H on $C_0^\infty(\mathbb{R})$ does not follow from the results of Kato, but [RS2] offers five distinct proofs (Section X.2, Example 6; Section X.4, Example 2; Section X.6, Example 5; Section X.9, Example 2; and Section X.10, Example). The simplest of these is just an appeal to the following theorem (Theorem X.28 of [RS2]). We will look at another proof in Example 9.2.2. Needless to say, the theorem also applies to the harmonic oscillator potential $\frac{1}{2}m\omega^2 q^2$ and therefore implies the essential self-adjointness of H_B.

Theorem 9.2.7. *Suppose* $V \in L^2(\mathbb{R}^n)_{\mathrm{loc}}$ *is real-valued and* $V \geq 0$ *pointwise almost everywhere and let* Δ *be the Laplacian on* $\mathbb{R}^n$. *Then* $-\frac{\hbar^2}{2m}\Delta + V$ *is essentially self-adjoint on* $C_0^\infty(\mathbb{R}^n)$.

Recall that $L^2(\mathbb{R}^n)_{\mathrm{loc}}$ consists of all the measurable complex-valued functions on $\mathbb{R}^n$ that are square integrable on compact subsets of $\mathbb{R}^n$. This is certainly true of a continuous function such as $V(q) = \frac{1}{2}m\omega^2 q^2 + \alpha q^4$. Moreover, since $V(q) \geq 0$ for every $q \in \mathbb{R}$ we conclude from the theorem that the anharmonic oscillator Hamiltonian is essentially self-adjoint on $C_0^\infty(\mathbb{R})$. We should also point out that the proof of this theorem is based on a famous distributional inequality called *Kato's inequality* (Theorem X.27 of [RS2]). One can also replace $V \geq 0$ with $V \geq -c$ for any constant $c \geq 0$ so the essential point is that the potential is bounded from below.

One can extend this to handle some potentials that are not bounded from below. Suppose $V : \mathbb{R}^n \to \mathbb{R}$ is written as the difference of its positive and negative parts, that is, $V = V^+ - V^-$, where $V^+(\mathbf{q}) = \max(V(\mathbf{q}), 0)$ and $V^-(\mathbf{q}) = \max(-V(\mathbf{q}), 0)$. Then one can prove the following amalgam of Theorems 9.2.4 and 9.2.7 (this is a special case of Theorem X.29 of [RS2] which also describes the modifications required when $n \geq 4$).

Theorem 9.2.8. *Let* V *be a real-valued function on* $\mathbb{R}^n$, $n = 1, 2, 3$, *and write* $V = V^+ - V^- = \max(V, 0) - \max(-V, 0)$. *If* $V^+ \in L^2(\mathbb{R}^n)_{\mathrm{loc}}$ *and* $V^- = V_1^- + V_2^-$, *where* $V_1^- \in L^2(\mathbb{R}^n)$ *and* $V_2^- \in L^\infty(\mathbb{R}^n)$, *then* $-\frac{\hbar^2}{2m}\Delta + V$ *is essentially self-adjoint on* $C_0^\infty(\mathbb{R}^n)$.

We would like to describe one more result of this sort, primarily because it involves an idea that we have encountered before. In Section 7.2 we introduced the notions of smooth and analytic vectors for a unitary representation. We mentioned also that there were corresponding notions for a single operator $A : D(A) \to \mathcal{H}$ and promised to introduce those notions here and describe their relation to self-adjointness. This is what we will do now.

We consider a densely defined operator $A : D(A) \to \mathcal{H}$ on a complex, separable Hilbert space $\mathcal{H}$. An element ψ of $\mathcal{H}$ is called a *smooth* or C^∞ *vector for A* if it is an element of the domain of every power of A, that is, if

$$\psi \in \bigcap_{n=1}^{\infty} D(A^n).$$

The set of all C^∞ vectors for A will be denoted $C^\infty(A)$. Here is some motivation for the terminology. Suppose $\mathcal{H} = L^2(\mathbb{R}^3)$ and A is the Laplacian Δ. We have seen that any $\psi \in D(\Delta)$ is in the Sobolev space $H^2(\mathbb{R}^3)$. Similar arguments show that any $\psi \in D(\Delta^2)$ is in $H^4(\mathbb{R}^3)$ and, in general, any $\psi \in D(\Delta^n)$ is in $H^{2n}(\mathbb{R}^3)$. Consequently, any $\psi \in \bigcap_{n=1}^{\infty} D(\Delta^n)$ is in every $H^k(\mathbb{R}^3)$. Now one can appeal to a Sobolev embedding theorem according to which any function that is in every Sobolev space $H^k(\mathbb{R}^3)$ is necessarily smooth, that is, in $C^\infty(\mathbb{R}^3)$ (this is Corollary 1.4 of [TaylM]). Thus, $C^\infty(\Delta)$ consists entirely of smooth functions on $\mathbb{R}^3$ (not every smooth function on $\mathbb{R}^3$, of course). All of this is actually true of any elliptic operator on any $L^2(\mathbb{R}^n)$.

Due to the usual domain issues it is entirely possible that $C^\infty(A)$ contains only the zero vector in $\mathcal{H}$, even if A is essentially self-adjoint ([RS2] leaves as Exercise 39 of Chapter X the construction of a self-adjoint operator A with a domain of essential self-adjointness that intersects $D(A^2)$ only in the zero vector). If $A : D(A) \to \mathcal{H}$ is self-adjoint, however, then $C^\infty(A)$ is dense in $D(A)$. This follows from the spectral theorem (if E^A is the spectral measure associated with A and ψ is any element of $\mathcal{H}$, then $E^A(-n, n)\psi$, $n = 1, 2, \ldots$, is a dense set of smooth vectors for A).

For any $\psi \in C^\infty(A)$ one can at least write down the power series

$$\sum_{k=0}^{\infty} \frac{\|A^k \psi\|}{k!} z^k, \tag{9.15}$$

although there is no reason to suppose that it converges for any $z \neq 0$. A vector ψ in $\mathcal{H}$ is said to be an *analytic vector for A* if $\psi \in C^\infty(A)$ and the series (9.15) has a nonzero radius of convergence, which we will denote $r_A(\psi)$. The set of all analytic vectors for A will be denoted $C^\omega(A)$; it is a linear subspace of $\mathcal{H}$, but might well consist of the zero vector alone.

Exercise 9.2.12. Show that if $A : D(A) \to \mathcal{H}$ has an eigenvector ψ, then $\psi \in C^\omega(A)$.

The main result on analytic vectors and essential self-adjointness is due to Nelson [Nel1]. To state it we need a definition. A subset S of the Hilbert space $\mathcal{H}$ is said to be *total* in $\mathcal{H}$ if the set of all finite linear combinations of elements of S is dense in $\mathcal{H}$; for example, any basis for $\mathcal{H}$ is total in $\mathcal{H}$.

Theorem 9.2.9 (Nelson's analytic vector theorem). *Let $A : \mathcal{D}(A) \to \mathcal{H}$ be a symmetric operator on a complex, separable Hilbert space $\mathcal{H}$. If $\mathcal{D}(A)$ contains a total set of analytic vectors for A, then A is essentially self-adjoint.*

The result actually stated in [Nel1] (as Lemma 5.1) is a corollary of this; also see Theorem X.39 of [RS2].

Corollary 9.2.10. *Let $A : \mathcal{D}(A) \to \mathcal{H}$ be a closed, symmetric operator on a complex, separable Hilbert space $\mathcal{H}$. Then A is self-adjoint if and only if $\mathcal{D}(A)$ contains a total set of analytic vectors for A.*

   There are a number of ways to prove Nelson's analytic vector theorem, all of which require a fair amount of work. We will simply suggest four different sources where the proofs have rather different flavors and let the interested reader find one that appeals. One can, of course, consult the proof of Lemma 5.1 in Nelson's paper [Nel1]. Nelson's theorem also appears as Theorem X.39 in [RS2], Theorem 5.6.2 of [BEH] and Theorem 2.31 of [Kant].
   For positive operators Nelson's theorem can be strengthened a bit (recall that an operator $A : \mathcal{D}(A) \to \mathcal{H}$ is said to be *positive* if $\langle \psi, A\psi \rangle \geq 0$ for all $\psi \in \mathcal{D}(A)$). We will say that a vector ψ in $\mathcal{H}$ is *semi-analytic* for A if $\psi \in C^{\infty}(A)$ and the series

$$\sum_{k=0}^{\infty} \frac{\|A^k \psi\|}{(2k)!} z^{2k}$$

has a nonzero radius of convergence. The following is Theorem 1 in [Simon1].

Theorem 9.2.11. *Let $A : \mathcal{D}(A) \to \mathcal{H}$ be a positive, symmetric operator on a complex, separable Hilbert space $\mathcal{H}$. If $\mathcal{D}(A)$ contains a total set of semi-analytic vectors for A, then A is essentially self-adjoint.*

   By way of compensation for not including proofs of these results we would like to work through an interesting example that will establish the essential self-adjointness of the anharmonic oscillator Hamiltonian (9.14) on $\mathcal{S}(\mathbb{R})$ and will also give us another opportunity to make use of the raising and lowering operators for the harmonic oscillator (5.14). This is basically Example 5, Section X.6, of [RS2].

Example 9.2.2 (Anharmonic oscillator Hamiltonian is essentially self-adjoint).
The Hamiltonian for the anharmonic oscillator is given by (9.14), but to keep the arithmetic tolerable we will take $m = 1$, $\omega = 1$ and $\alpha = 4$ and will work in units in which $\hbar = 1$. Consequently, our Hamiltonian is given on $\mathcal{S}(\mathbb{R})$ by

$$H = H_B + 4q^4 = \frac{1}{2}\left(-\frac{d^2}{dq^2} + q^2\right) + 4q^4.$$

We would now like to rewrite H in terms of the raising and lowering operators we introduced for the harmonic oscillator in Example 5.3.1. We will briefly review what we need with the choice of constants we have made. On $\mathcal{S}(\mathbb{R})$ we define

$$b = \frac{1}{\sqrt{2}}(Q + iP)$$

and

$$b^\dagger = \frac{1}{\sqrt{2}}(Q - iP),$$

where Q and P are the position and momentum operators on $\mathcal{S}(\mathbb{R})$. Then $Q = \frac{1}{\sqrt{2}}(b+b^\dagger)$. The number operator is then defined by $N_B = b^\dagger b$ and the Hamiltonian H_B can be written

$$H_B = N_B + \frac{1}{2}.$$

Recall that we used these operators to produce an orthonormal basis for $L^2(\mathbb{R})$ consisting of eigenfunctions $\{\psi_n\}_{n=0}^\infty$ of H_B. Specifically, these were given by the Hermite functions

$$\psi_0(q) = \pi^{-1/4}e^{-q^2/2}$$

and

$$\psi_n(q) = (n!)^{-1/2}b^n\psi_0 = \frac{1}{\sqrt{2^n n!}}\pi^{-1/4}e^{-q^2/2}H_n(q), \quad n = 1, 2, 3, \ldots,$$

where H_n is the nth Hermite polynomial. Our objective is to show that these eigenfunctions of H_B are semi-analytic vectors for H and then appeal to Theorem 9.2.11. Since they lie in $\mathcal{S}(\mathbb{R})$ and not $C_0^\infty(\mathbb{R})$, this explains our decision to show that H is essentially self-adjoint on $\mathcal{S}(\mathbb{R})$ rather than on $C_0^\infty(\mathbb{R})$. Here are a few things we proved in Example 5.3.1 that we will need:

$$b^\dagger\psi_n = \sqrt{n+1}\,\psi_{n+1}, \quad n = 0, 1, 2, \ldots,$$
$$b\psi_0 = 0$$

and

$$b\psi_n = \sqrt{n}\,\psi_{n-1}, \quad n = 1, 2, \ldots.$$

Using these and the fact that $\|\psi_n\| = 1$ for every $n = 0, 1, 2, \ldots$ we find that $\|b^\dagger\psi_n\| = \sqrt{n+1}$ for $n = 0, 1, 2, \ldots$, $\|b\psi_0\| = 0$ and $\|b\psi_n\| = \sqrt{n}$ for $n = 1, 2, \ldots$. If we let $b^\sharp$ denote one of b, $b^\dagger$ or the identity I, then we can conclude from this that

$$\|b^\sharp\psi_n\| \le \sqrt{n+1}, \quad n = 0, 1, 2, \ldots.$$

Similarly, $\|b^\dagger b^\dagger \psi_n\| = \sqrt{n+1}\sqrt{n+2}$, $\|b^\dagger b\psi_0\| = 0$, $\|b^\dagger b\psi_n\| = \sqrt{n}\sqrt{n}$ for $n = 1, 2, \dots$, $\|bb^\dagger\psi_n\| = \sqrt{n+1}\sqrt{n+1}$ for $n = 0, 1, 2, \dots$, $\|bb\psi_0\| = 0$, $\|bb\psi_1\| = 0$ and $\|bb\psi_n\| = \sqrt{n}\sqrt{n-1}$ for $n = 2, 3, \dots$. In particular,

$$\|b^\sharp b^\sharp \psi_n\| \le \sqrt{n+1}\sqrt{n+2}, \quad n = 0, 1, 2, \dots.$$

Exercise 9.2.13. Show that for any $k = 1, 2, \dots,$

$$\|b^\sharp \overset{k}{\dots} b^\sharp \psi_n\| \le \sqrt{n+1} \dots \sqrt{n+k} \le \sqrt{(n+k)!}, \quad n = 0, 1, 2, 3, \dots.$$

Next we rewrite the Hamiltonian H in terms of b and $b^\dagger$ as follows:

$$H = H_B + 4Q^4$$

$$= \left(N_B + \frac{1}{2}\right) + 4Q^4$$

$$= b^\dagger b + \frac{1}{2} + 4\left(\frac{1}{\sqrt{2}}(b + b^\dagger)\right)^4$$

$$= b^\dagger b + \frac{1}{2} + (b + b^\dagger)^4.$$

Expanding $(b+b^\dagger)^4$ (and taking care not to use commutativity) one sees that it consists of 16 terms:

$$bbbb + bbbb^\dagger + bbb^\dagger b + \dots + b^\dagger b^\dagger b^\dagger b^\dagger.$$

Consequently, H contains 18 terms and, for any $k \ge 1$, H^k is a sum of 18^k terms. Each of these terms is a product of the form $c^k b^\sharp \overset{4k}{\dots} b^\sharp$, where $b^\sharp$ is one of b, $b^\dagger$ or I and $0 < c \le 1$. Consequently,

$$\|H^k \psi_n\| \le 18^k \sqrt{n+1} \dots \sqrt{n+4k} \le 18^k \sqrt{(n+4k)!}.$$

Exercise 9.2.14. Prove each of the following.
1. The series $\sum_{k=0}^\infty \frac{\|H^k\psi_n\|}{(2k)!} z^{2k}$ converges on $|z| < (72)^{-1/2}$ for any $n = 0, 1, 2, \dots$. Conclude that each ψ_n is a semi-analytic vector for H. *Hint:* Use the ratio test.
2. H is a positive operator on $\mathcal{S}(\mathbb{R})$.
3. $\{\psi_n\}_{n=0}^\infty$ is total in $L^2(\mathbb{R})$.
4. H is essentially self-adjoint on $\mathcal{S}(\mathbb{R})$.
5. The estimates we have obtained *do not* imply that ψ_n is an analytic vector for H.

9.3 Brownian motion and Wiener measure

We should begin with a little historical perspective.

"There's a model, you should realize,
A paradigm of this that is dancing right before your eyes –
For look well when you let the sun peep in a shuttered room
Pouring forth the brilliance of its beams into the gloom,
And you'll see myriads of motes all moving many ways
Throughout the void and intermingling in the golden rays
As if in everlasting struggle, battling in troops,
Ceaselessly separating and regathering in groups.
From this you can imagine all the motions that take place
Among the atoms that are tossed about in empty space.
For to a certain extent, it is possible for us to trace
Greater things from trivial examples, and discern
In them the trail of knowledge. Another reason you should turn
Your attention to the motes that drift and tumble in the light:
Such turmoil means that there are secret motions, out of sight,
That lie concealed in matter. For you'll see the motes careen
Off course, and then bounce back again, by means of blows unseen,
Drifting now in this direction, now that, on every side.
You may be sure this starts with atoms; they are what provide
The base of this unrest. For atoms are moving on their own,
Then small formations of them, nearest them in scale, are thrown
Into agitation by unseen atomic blows,
And these strike slightly larger clusters, and on and on it goes –
A movement that begins on the atomic level, by slight
Degrees ascends until it is perceptible to our sight,
So that we can behold the dust motes dancing in the sun,
Although the blows that move them can't be seen by anyone."

Lucretius, *On the Nature of Things, Book II, lines 89–141* [Lucr]

And yet, we call it *Brownian motion*; the random movement of minute particles in a liquid or gas as a result of continuous bombardment by the atoms or molecules of the surrounding medium (if you have never seen this you may want to visit https://www.youtube.com/watch?v=cDcprgWiQEY). But this is fair, I suppose, since the botanist Robert Brown made the first careful observations of the phenomenon by watching pollen grains through his microscope in the nineteenth century.

Early on a great many potential mechanisms were proposed to explain these random motions of the particles observed by Brown, from "They're alive!" to the view espoused by Lucretius nearly 2000 years before. The issue was resolved once and for all, in favor of Lucretius, when Einstein took up the problem in a series of five papers between 1905 and 1908. We will have nothing more to say about the history of the problem (if you are interested, consult the first four sections of [Nel4]) and very little to say about the details of Einstein's analysis (all five papers have been translated, with notes in [Ein4]). What we *will* have something to say about is the rigorous path integral reformulation of Brownian motion devised by Norbert Wiener [Wiener] and the rather striking similarity its construction bears to that of the Feynman "integral." We will see that it is *not* possible to modify Wiener's procedure to turn the Feynman integral into an actual (Lebesgue) integral and, indeed, that the Feynman integral simply is not a Lebesgue integral. Nevertheless, the analogy is close enough that Edward Nelson [Nel3] was able to establish a link between the two and we will sketch how this is done in the next section. A much more detailed discussion of all that we will have to say (and a great deal more) is available in [JL].

One should understand that in his first (1905) paper Einstein did not set out explicitly to explain the mechanism responsible for the behavior of Brownian particles. Indeed, at this time Einstein seems to have had access to only very limited information about earlier work on Brownian motion. As to the actual purpose of the paper, we will ask the author to speak for himself.

"In this paper it will be shown that, according to the molecular-kinetic theory of heat, bodies of microscopically visible size suspended in a liquid will perform movements of such magnitude that they can be easily observed in a microscope, on account of the molecular motions of heat. It is possible that the movements to be discussed here are identical with the so-called "Brownian molecular motion"; however, the information available to me regarding the latter is so lacking in precision, that I can form no judgment in the matter. If the movement discussed here can actually be observed (together with the laws relating to it that one would expect to find), then classical thermodynamics can no longer be looked upon as applicable with precision to bodies even of dimensions distinguishable in a microscope: an exact determination of actual atomic dimensions is then possible."

Albert Einstein, *On the Movement of Small Particles Suspended in a Stationary Liquid Demanded by the Molecular-Kinetic Theory of Heat* [Ein4]

In 1905 the existence of atoms and molecules was by no means universally accepted among physicists. The molecular-kinetic theory hypothesizes that such things do exist and that macroscopic properties of fluids such as temperature and pressure can be accounted for by their properties, such as their kinetic energy. The very precise information obtained by Einstein about the motion of particles suspended in the fluid resulting from collisions with these atoms and molecules and their subsequent experimental confirmation validated the molecular-kinetic theory and, by implication, the existence of atoms and molecules.

The experimental confirmation of Einstein's conclusions is largely credited to Jean Perrin [Per] in 1909, who made careful observations of the motion of individual Brownian particles. Perrin was even led to suggest that a typical Brownian path should be represented by a continuous, nowhere differentiable curve and to mention in this regard the famous example of Weierstrass. This is an idealization, of course, since, however small, there is a nonzero time lapse between collisions and during such a time interval the particle should not behave so erratically. Nevertheless, we will see these continuous, nowhere differentiable curves emerge again in the rigorous treatment due to Wiener. The molecular-kinetic theory is, by its very nature, statistical due to the huge number of particles involved (see Exercise 2.4.1 and the comments that follow it). Einstein therefore describes the physical situation with a probability density function $\rho(t, q)$ with $q \in \mathbb{R}$.

Remark 9.3.1. Actually, a probability density function $\rho(t, \mathbf{q})$ with $\mathbf{q} \in \mathbb{R}^3$, but we will restrict ourselves to the one-dimensional situation. One can view this either as an actual Brownian motion taking place on a line or, more realistically, as one coordinate of the three-dimensional Brownian motion.

Then $\rho(t, q)$ is interpreted as the probability density that a Brownian particle is at location q at time t; more precisely, for any measurable subset E of $\mathbb{R}$,

$$\int_E \rho(t, q)\, dq$$

is the probability that the particle is somewhere in E at time t. Rather detailed physical arguments led Einstein to the conclusion that if the particles are not subject to any external forces (gravity, for instance), $\rho(t, q)$ must satisfy the diffusion equation

$$\frac{\partial \rho(t, q)}{\partial t} - D \frac{\partial^2 \rho(t, q)}{\partial q^2} = 0.$$

Moreover, assuming that the particles are spheres of (small) radius a, Einstein argued that the diffusion constant D must be given by

$$D = \frac{\kappa_B T}{6\pi\eta a},$$

where η is the coefficient of viscosity of the fluid, T is the temperature of the fluid (in Kelvin) and κ_B is the Boltzmann constant (see Section 4.3). Let us suppose now that the Brownian particles under observation all originated at some point q_a at the instant t_a. Then the problem we need to solve is

$$\frac{\partial \rho(t, q)}{\partial t} - D \frac{\partial^2 \rho(t, q)}{\partial q^2} = 0, \quad (t, q) \in (t_a, \infty) \times \mathbb{R},$$

$$\lim_{t \to t_a^+} \rho(q, t) = \delta_{q_a}, \quad q \in \mathbb{R}$$

(the limit being in the sense of distributions). But we (or rather, you) have already solved this problem (Example G.3.1 and the exercises therein). Letting

$$H_D(t, q, q_a) = \frac{1}{\sqrt{4\pi D t}} e^{-(q-q_a)^2/4Dt}, \quad (t, q) \in (0, \infty) \times \mathbb{R},$$

denote the heat kernel, that is, the Gaussian distribution with mean q_a and standard deviation $\sqrt{2Dt}$, the solution is

$$H_D(t - t_a, q, q_a) = \frac{1}{\sqrt{4\pi D(t - t_a)}} e^{-(q-q_a)^2/4D(t-t_a)}, \quad (t, q) \in (t_a, \infty) \times \mathbb{R}. \tag{9.16}$$

In particular, for any $-\infty \le \alpha \le \beta \le \infty$, the probability that such a Brownian particle, known to be at q_a at time t_a, will be in $(\alpha, \beta]$ at time $t > t_a$ is given by

$$\int_\alpha^\beta H_D(t - t_a, q, q_a) \, dq = \int_\alpha^\beta \frac{1}{\sqrt{4\pi D(t - t_a)}} e^{-(q-q_a)^2/4D(t-t_a)} \, dq.$$

More generally, if $E \subseteq \mathbb{R}$ is any measurable set, then

$$\int_E H_D(t - t_a, q, q_a) \, dq = \int_E \frac{1}{\sqrt{4\pi D(t - t_a)}} e^{-(q-q_a)^2/4D(t-t_a)} \, dq \qquad (9.17)$$

is the probability that the particle will be in the set E at time t.

Now suppose $t_a < t \le t_b$ and let $t_a = t_0 < t_1 < t_2 < \cdots < t_{n-1} < t_n = t \le t_b$ be a partition of the interval $[t_a, t]$ into n equal subintervals of length $\Delta t = \frac{t - t_a}{n}$ and write $t - t_a = (t_n - t_{n-1}) + \cdots + (t_2 - t_1) + (t_1 - t_0)$.

Exercise 9.3.1. Use Exercise G.3.7 to show that

$$H_D(t - t_a, q, q_a)$$
$$= \int_\mathbb{R} \cdots \int_\mathbb{R} H_D(t_1 - t_0, q_1, q_0) \cdots H_D(t_n - t_{n-1}, q_n, q_{n-1}) dq_1 \cdots dq_{n-1}$$
$$= \left(\frac{1}{4\pi D \Delta t} \right)^{\frac{n}{2}} \int_\mathbb{R} \cdots \int_\mathbb{R} e^{-S_n(q_0, q_1, \ldots, q_{n-1}, q_n; t)} \, dq_1 \ldots dq_{n-1}, \qquad (9.18)$$

where $q_0 = q_a$, $q_n = q$ and

$$S_n(q_0, q_1, \ldots, q_{n-1}, q_n; t) = \frac{1}{4D} \sum_{k=1}^n \left(\frac{q_k - q_{k-1}}{\Delta t} \right)^2 \Delta t.$$

One should compare this with the nth approximation to the Feynman integral in (8.3). There is no potential term since we have not considered Brownian particles subjected to external forces. Otherwise, the analogy seems clear enough, but one must take note of where the analogy breaks down. Simply put, there is no "i" in the exponent in (9.18) so the integrals are not oscillatory, but rather decaying exponentially. We will see that all of the mathematical difficulties associated with the Feynman integral arise from the oscillatory nature of the finite-dimensional integrals in its definition.

There is no particularly compelling reason to insist that the partition $t_a = t_0 < t_1 < t_2 < \cdots < t_{n-1} < t_n = t \le t_b$ of $[t_a, t]$ be uniform, that is, into intervals of the same length. For a general partition one simply has instead

$$H_D(t - t_a, q, q_a)$$
$$= \int_\mathbb{R} \cdots \int_\mathbb{R} H_D(t_1 - t_0, q_1, q_0) \cdots H_D(t_n - t_{n-1}, q_n, q_{n-1}) dq_1 \cdots dq_{n-1}$$

$$= \left[(4\pi D)^n (t_1 - t_0) \cdots (t_n - t_{n-1}) \right]^{-1/2} \int_{\mathbb{R}} \cdots \int_{\mathbb{R}} e^{-\frac{1}{4D} \sum_{k=1}^{n} \frac{(q_k - q_{k-1})^2}{t_k - t_{k-1}}} dq_1 \cdots dq_{n-1}. \qquad (9.19)$$

If, for each $j = 1, \ldots, n$, we have extended real numbers $-\infty \le \alpha_j \le \beta_j \le \infty$, then the probability that a Brownian particle, known to be at q_a at time t_a, will be in $(\alpha_1, \beta_1]$ at time t_1, in $(\alpha_2, \beta_2]$ at time t_2, ... and in $(\alpha_n, \beta_n]$ at time t_n is given by

$$\int_{\alpha_n}^{\beta_n} \cdots \int_{\alpha_1}^{\beta_1} H_D(t_1 - t_0, q_1, q_0) \cdots H_D(t_n - t_{n-1}, q_n, q_{n-1}) dq_1 \cdots dq_n$$

$$= \left[(4\pi D)^n (t_1 - t_0) \cdots (t_n - t_{n-1}) \right]^{-1/2} \int_{\alpha_n}^{\beta_n} \cdots \int_{\alpha_1}^{\beta_1} e^{-\frac{1}{4D} \sum_{k=1}^{n} \frac{(q_k - q_{k-1})^2}{t_k - t_{k-1}}} dq_1 \cdots dq_n. \qquad (9.20)$$

Again more generally, if $E_j \subseteq \mathbb{R}$ is a measurable set for each $j = 1, \ldots, n$, then

$$\int_{E_n} \cdots \int_{E_1} H_D(t_1 - t_0, q_1, q_0) \cdots H_D(t_n - t_{n-1}, q_n, q_{n-1}) dq_1 \cdots dq_n$$

$$= \left[(4\pi D)^n (t_1 - t_0) \cdots (t_n - t_{n-1}) \right]^{-1/2} \int_{E_n} \cdots \int_{E_1} e^{-\frac{1}{4D} \sum_{k=1}^{n} \frac{(q_k - q_{k-1})^2}{t_k - t_{k-1}}} dq_1 \cdots dq_n \qquad (9.21)$$

is the probability that the particle will be in E_1 at time t_1, in E_2 at time t_2, ... and in E_n at time t_n.

Exercise 9.3.2. Show that

$$\int_{\mathbb{R}} \cdots \int_{\mathbb{R}} H_D(t_1 - t_0, q_1, q_0) \cdots H_D(t_n - t_{n-1}, q_n, q_{n-1}) \, dq_1 \cdots dq_n = 1,$$

as it should be.

There are physical assumptions about Brownian motion buried in these formulas. These are encapsulated abstractly in the probabilistic notion of a *Wiener process* (also called a *Brownian motion*). One of these assumptions arises in the following way. Since (9.21) is essentially an iteration of (9.17) in which the individual probabilities are multiplied there is an implicit assumption that the events are independent; getting from E_j to E_{j+1} does not depend on how the particle got from E_{j-1} to E_j. The particle has no memory.

At about this point in our discussion of the Feynman integral we were instructed to take the limit as $n \to \infty$. Wiener, however, had a different idea (in 1923). To state this precisely we will need to introduce a few definitions. We will consider two real numbers $t_a < t_b$ and will denote by $C[t_a, t_b]$ the linear space of continuous real-valued functions $x(t)$ on $[t_a, t_b]$. Supplied with the uniform norm

$$\|x\| = \max_{t_a \le t \le t_b} |x(t)|,$$

$C[t_a, t_b]$ is a real, separable Banach space. Paths of Brownian particles will be represented by elements of $C[t_a, t_b]$ that all start at some fixed point at time t_a and, for simplicity, we will take this fixed point (called q_a above) to be $q = 0$. Thus, the space of Brownian paths *is contained in*

$$C_0[t_a, t_b] = \{x \in C[t_a, t_b] : x(t_a) = 0\}.$$

Since $C_0[t_a, t_b]$ is a closed linear subspace of $C[t_a, t_b]$, it is also a real, separable Banach space when supplied with the uniform norm. In particular, $C_0[t_a, t_b]$ is a topological space and so it has a σ-algebra of Borel sets $\mathcal{B} = \mathcal{B}(C_0[t_a, t_b])$ (see Remark 2.4.1). This is, as always, the σ-algebra generated by the open (or closed) subsets of $C_0[t_a, t_b]$, but there is a more useful description. For any $t_1, \ldots, t_n$ satisfying $t_a = t_0 < t_1 < \cdots < t_{n-1} < t_n \leq t_b$ and any extended real numbers $-\infty \leq \alpha_j \leq \beta_j \leq \infty, j = 1, \ldots, n$, we define

$$I = I(t_1, \ldots, t_n; (\alpha_1, \beta_1] \times \cdots \times (\alpha_n, \beta_n]) = \{x \in C_0[t_a, t_b] : x(t_j) \in (\alpha_j, \beta_j]\}.$$

Sets of this form in $C_0[t_a, t_b]$ are called *cylinder sets* and the collection of all such will be denoted $\mathcal{I}$.

Proposition 9.3.1. *The collection $\mathcal{I}$ of cylinder sets in $C_0[t_a, t_b]$ has the following properties:*
1. *$\emptyset$ and $C_0[t_a, t_b]$ are in $\mathcal{I}$,*
2. *if I_1 and I_2 are in $\mathcal{I}$, then $I_1 \cap I_2$ is in $\mathcal{I}$,*
3. *if I is in $\mathcal{I}$, then $C_0[t_a, t_b] - I$ is a finite disjoint union of elements of $\mathcal{I}$.*

Proof. Part 1. is clear since

$$\emptyset = \{x \in C_0[t_a, t_b] : 1 < x(t_b) \leq 1\}$$

and

$$C_0[t_a, t_b] = \{x \in C_0[t_a, t_b] : -\infty < x(t_b) \leq \infty\}.$$

For 2., let $I_1 = I(t_1, \ldots, t_n; (\alpha_1, \beta_1] \times \cdots \times (\alpha_n, \beta_n])$ and $I_2 = I(s_1, \ldots, s_m; (\gamma_1, \delta_1] \times \cdots \times (\gamma_m, \delta_m])$. Let $\{r_1, \ldots, r_l\} = \{t_1, \ldots, t_n\} \cup \{s_1, \ldots, s_m\}$. If $r_i \in \{t_1, \ldots, t_n\} \cap \{s_1, \ldots, s_m\}$, say $r_i = t_j = s_k$, set $(\mu_i, \nu_i] = (\alpha_j, \beta_j] \cap (\gamma_k, \delta_k]$. If $r_i \in \{t_1, \ldots, t_n\} - \{s_1, \ldots, s_m\}$, say $r_i = t_j$, set $(\mu_i, \nu_i] = (\alpha_j, \beta_j]$ and if $r_i \in \{s_1, \ldots, s_m\} - \{t_1, \ldots, t_n\}$, say $r_i = s_k$, set $(\mu_i, \nu_i] = (\gamma_k, \delta_k]$. Then $I_1 \cap I_2 = I(r_1, \ldots, r_l; (\mu_1, \nu_1] \times \cdots \times (\mu_l, \nu_l])$.

Exercise 9.3.3. Show that $C_0[t_a, t_b] - I(t_1, \ldots, t_n; (\alpha_1, \beta_1] \times \cdots \times (\alpha_n, \beta_n])$ can be written as a finite disjoint union of elements of $\mathcal{I}$. □

Not only is each cylinder set a Borel set in $C_0[t_a, t_b]$, but $\mathcal{B}$ is, in fact, the σ-algebra generated by $\mathcal{I}$. The following is Theorem 3.2.11 of [JL].

Theorem 9.3.2. *The σ-algebra $\sigma(\mathcal{I})$ generated by the cylinder sets in $C_0[t_a, t_b]$ is the σ-algebra $\mathcal{B} = \mathcal{B}(C_0[t_a, t_b])$ of Borel sets in $C_0[t_a, t_b]$.*

Now we can describe Wiener's idea in the following way. Fix a cylinder set $I = I(t_1, \ldots, t_n; (\alpha_1, \beta_1] \times \cdots \times (\alpha_n, \beta_n])$ in $C_0[t_a, t_b]$. Then (9.20), with $q_0 = 0$, assigns to I a probability, that is, the probability that a Brownian particle starting at $q_0 = q_a = 0$ at time t_a will pass through each $(\alpha_j, \beta_j]$ at time t_j for $j = 1, \ldots, n$. Let us regard this probability as a "measure" of the size of I in $C_0[t_a, t_b]$ and denote it $\mathfrak{m}(I)$. Thus, $\mathfrak{m}(I)$ measures the likelihood that the path of a Brownian particle will satisfy the conditions that define I. Wiener set himself the task of constructing a (Lebesgue) measure on some σ-algebra of subsets of $C_0[t_a, t_b]$ containing $\mathcal{I}$ that agrees with $\mathfrak{m}(I)$ for $I \in \mathcal{I}$.

Needless to say, Wiener succeeded admirably. In the intervening years many different constructions of this *Wiener measure* have been devised, all of which involve very substantial technical work. We do not intend to go through the details of any of these, but we will sketch two of them. The first is a very elegant functional analytic argument due to Edward Nelson (Appendix A of [Nel3]). This arrives quickly at a measure with the required properties, but perhaps sacrifices some of the intuitive connection with Brownian motion and the analogy with the Feynman integral. Nevertheless, it is a beautiful argument and worth seeing. The second is entirely analogous to the usual construction of the Lebesgue measure on $\mathbb{R}$, although the technical issues are rather more substantial; Chapter 3 of [JL] fills in many of these details, but not all, so we will supply references for the rest.

Remark 9.3.2. The "sketches" we will offer of the Wiener measure are of the minimalist variety. We will endeavor to state the results precisely and prove a few of the most elementary of them, but there are many deep and technically difficult issues for which we will only provide references. We made the decision that a nodding acquaintance with these ideas was essential for an appreciation of the difficulties inherent in Feynman's path integral, but that the details would probably be of interest only to those who incline toward mathematical analysis.

To prepare for Nelson's construction we should briefly review two items, one a definition from measure theory and the other a theorem from functional analysis (both are discussed in Sections 55–56 of [Hal1] or one can consult Proposition 6, Section 2, Chapter 14 and Theorem 8, Section 3, Chapter 14, of [Roy]). We let X denote a locally compact Hausdorff topological space and $\mathcal{A}$ a σ-algebra on X containing the σ-algebra $\mathcal{B}(X)$ of Borel sets in X. A nonnegative measure μ on the measurable space $(X, \mathcal{A})$ is said to be *regular* if:

1. $\mu(K) < \infty$ for every compact set $K \subseteq X$,
2. $A \in \mathcal{A} \Rightarrow \mu(A) = \inf \{\mu(U) : U \text{ is open in } X \text{ and } A \subseteq U\}$,
3. $U \subseteq X$ open $\Rightarrow \mu(U) = \sup \{\mu(K) : K \text{ is compact in } X \text{ and } K \subseteq U\}$.

The following is one version of the *Riesz representation theorem*, which follows from Theorem 6.19 of [Rud2].

Theorem 9.3.3 (Riesz representation theorem). *Let X be a compact Hausdorff topological space and $C(X)$ its Banach space of continuous, real-valued functions on X (with the uniform norm). Let α be a linear functional on $C(X)$ that is nonnegative ($f \in C(X)$ and $f(x) \geq 0 \, \forall x \in X \Rightarrow \alpha(f) \geq 0$). Then there exists a σ-algebra $\mathcal{A}$ on X containing the σ-algebra $\mathcal{B}(X)$ of Borel sets on X and a regular measure μ on $(X, \mathcal{A})$ such that*

$$\alpha(f) = \int_X f(x) \, d\mu(x)$$

for every $f \in C(X)$. Moreover, the restriction of μ to $\mathcal{B}(X)$ is unique.

Nonnegative linear functionals are often called *positive* linear functionals. Whatever you call them, it is not necessary to assume *a priori* that they are bounded (continuous) for this follows automatically. Indeed, suppose α is nonnegative. Then, for any $f \in C(X)$, $-\|f\| \leq f(x) \leq \|f\| \, \forall x \in X$ and this implies that $-\alpha(1)\|f\| \leq \alpha(f) \leq \alpha(1)\|f\|$. Thus, $|\alpha(f)| \leq \alpha(1)\|f\|$, so α is bounded, that is, continuous.

Exercise 9.3.4. Show from this that

$$\|\alpha\|_{C(X)^*} = \alpha(1),$$

where $C(X)^*$ is the dual of the Banach space $C(X)$.

Exercise 9.3.5. Show that all of this is true for any subalgebra A of $C(X)$ containing 1 and provided with the uniform norm. Specifically, if α is a linear functional on A that is nonnegative ($f \in A$ and $f(x) \geq 0 \, \forall x \in X \Rightarrow \alpha(f) \geq 0$), then α is continuous and $\|\alpha\|_{A^*} = \alpha(1)$, where A^* is the dual of A. Note that A need not be complete, but A^* is always a Banach space (Theorem 4.4.4 of [Fried]).

Nelson considers paths starting at some point $q_0 \in \mathbb{R}$ at time $t_0 = 0$, defined on $[0, \infty)$ and taking values in $\mathbb{R}$ (actually, in $\mathbb{R}^n$, but we will stick to the one-dimensional situation). With essentially obvious modifications, the same procedure works equally well for paths defined on a finite interval $[t_a, t_b]$ and starting at $t_0 = t_a$, but to facilitate a transition to Nelson's paper [Nel3] we will do as Nelson did. When we construct the Wiener measure again using the more familiar procedures from Lebesgue theory we will return to paths defined on $[t_a, t_b]$.

Begin by considering the 1-point compactification $\dot{\mathbb{R}} = \mathbb{R} \cup \{\infty\}$ of $\mathbb{R}$ (see, for example, pages 162–163 of [Simm1]). For each $t \in [0, \infty)$, let $\dot{\mathbb{R}}_t$ be a copy of $\dot{\mathbb{R}}$ and consider the product

$$\Omega = \prod_{t \in [0, \infty)} \dot{\mathbb{R}}_t,$$

provided with the Tychonoff product topology. By the Tychonoff theorem (Theorem A, Section 23, of [Simm1]), Ω is a compact Hausdorff topological space. An element $\omega \in \Omega$ is then a completely arbitrary curve $\omega : [0, \infty) \to \dot{\mathbb{R}}$ in $\dot{\mathbb{R}}$, perhaps discontinuous and perhaps passing through the point at infinity. Inside Ω one finds the subset of continuous curves in $\mathbb{R}$ starting at q_0 at time $t_0 = 0$.

We consider the real Banach space $C(\Omega)$ of continuous, real-valued functions on Ω. Keep in mind that $C(\Omega)$ is also an algebra with unit 1 under pointwise multiplication. Nelson produced measures on Ω by defining, for each fixed $q_0 \in \mathbb{R}$, a nonnegative linear functional α_{q_0} on $C(\Omega)$ and appealing to the Riesz representation theorem. We will define this linear functional first on a subset of $C(\Omega)$ consisting of what are called *finite functions* because their value at any $\omega \in \Omega$ depends on only finitely many of the values $\omega(t)$ taken on by ω. Specifically, suppose $0 = t_0 < t_1 < \cdots < t_n < \infty$ and let

$$F : \prod_{j=1}^{n} \dot{\mathbb{R}}_{t_j} \to \mathbb{R}$$

be a function defined on the corresponding finite product of copies of $\dot{\mathbb{R}}$. Define a function $\varphi = \varphi_{F;t_1,\dots,t_n} : \Omega \to \mathbb{R}$ by

$$\varphi(\omega) = \varphi_{F;t_1,\dots,t_n}(\omega) = F(\omega(t_1), \dots, \omega(t_n)). \tag{9.22}$$

The *evaluation map* $\omega \to (\omega(t_1), \dots, \omega(t_n))$ is just the projection of Ω onto $\prod_{j=1}^{n} \dot{\mathbb{R}}_{t_j}$ and hence is continuous. Consequently, if F is continuous, then so is φ.

Now we consider the set $C_{\text{fin}}(\Omega)$ of all functions on Ω that can be written in the form $\varphi_{F;t_1,\dots,t_n}$ for some $0 = t_0 < t_1 < \cdots < t_n < \infty$ and some *continuous F*. We have just noticed that these are all continuous on Ω, so

$$C_{\text{fin}}(\Omega) \subseteq C(\Omega).$$

There are, of course, a great many things in $C(\Omega)$ that are not in $C_{\text{fin}}(\Omega)$, but you will now show that $C_{\text{fin}}(\Omega)$ is uniformly dense in $C(\Omega)$.

Exercise 9.3.6. Prove each of the following and then appeal to the Stone–Weierstrass theorem (Theorem A, Section 36, of [Simm1]) to conclude that $C_{\text{fin}}(\Omega)$ is uniformly dense in $C(\Omega)$:
1. $C_{\text{fin}}(\Omega)$ is a linear subspace of $C(\Omega)$,
2. $C_{\text{fin}}(\Omega)$ is a subalgebra of $C(\Omega)$ containing the unit element 1,
3. $C_{\text{fin}}(\Omega)$ separates points of Ω in the sense that if $\omega_1 \neq \omega_2$, then there exists a $\varphi \in C_{\text{fin}}(\Omega)$ for which $\varphi(\omega_1) \neq \varphi(\omega_2)$.

With $q_0 \in \mathbb{R}$ fixed, but arbitrary, we define, for any finite function $\varphi = \varphi_{F;t_1,\dots,t_n}$,

$$
\begin{aligned}
\alpha_{q_0}(\varphi) &= \alpha_{q_0}(\varphi_{F;t_1,\dots,t_n}) \\
&= \int_{\dot{\mathbb{R}}} \cdots \int_{\dot{\mathbb{R}}} F(q_1,\dots,q_n) H_D(t_1 - t_0, q_1, q_0) \cdots H_D(t_n - t_{n-1}, q_n, q_{n-1}) dq_1 \cdots dq_n,
\end{aligned}
$$

(9.23)

provided the integral exists ($t_0 = 0$ for us right now, but we have included it in the definition for the sake of symmetry). This requires some interpretation; F is defined on $\prod_{j=1}^{n} \dot{\mathbb{R}}_{t_j}$, so $q_1,\dots,q_n$ are in $\dot{\mathbb{R}}$ and the integrals must be over $\dot{\mathbb{R}}$, which one can identify with the unit circle S^1 by stereographic projection. However, we have not yet attached any meaning to $H_D(t, q, q_0)$ when q is the point at infinity in $\dot{\mathbb{R}}$. We intend to define $H_D(t, \infty, q_0) = 0$ for all $t \geq 0$ and all $q_0 \in \mathbb{R}$. One can either regard this as a statement about the limit of

$$
H_D(t, q, q_0) = \frac{1}{\sqrt{4\pi Dt}} e^{-(q-q_0)^2/4Dt}
$$

as $q \to \infty$ in $\dot{\mathbb{R}}$ or as the physical assumption that the probability density for a Brownian particle to get from any q_0 to ∞ in any finite time is zero. Interpreted in this way the integrand in (9.23) is zero at any point $(q_1,\dots,q_n)$ for which some q_j, $j = 1,\dots,n$, is the point at infinity so that $\alpha_{q_0}(\varphi)$ is completely determined by the values of φ on $\mathbb{R} \times \overset{n}{\cdots} \times \mathbb{R}$ and each integral reduces to an integral over $\mathbb{R}$. Note also that if $E_1,\dots,E_n$ are measurable sets in $\mathbb{R}$ and $F(q_1,\dots,q_n)$ is defined to be 1 if $q_j \in E_j$ for each $j = 1,\dots,n$ and 0 otherwise, then (9.23) reduces to the probability (9.21).

One more remark is in order. We would like α_{q_0} to be a well-defined linear functional on $C_{\text{fin}}(\Omega)$, so one must check that if $\varphi(\omega) = F(\omega(t_1),\dots,\omega(t_n))$ and F does not depend on some given q_j, then the same value for $\alpha_{q_0}(\varphi)$ results if we define φ in terms of the corresponding function of $n - 1$ variables. But the special case of (9.18)

$$
\int_{\mathbb{R}} H_D(t_j - t_{j-1}, q_j, q_{j-1}) H_D(t_{j+1} - t_j, q_{j+1}, q_j) dq_j = H_D(t_{j+1} - t_{j-1}, q_{j+1}, q_{j-1})
$$

(9.24)

allows one to integrate out such a variable q_j and the result follows from this.

We now have a well-defined linear functional α_{q_0} on $C_{\text{fin}}(\Omega)$. Furthermore, $\alpha_{q_0}(1) = 1$ by Exercise 9.3.2 and, since H_D is nonnegative, $\alpha_{q_0}(\varphi) \geq 0$ whenever $\varphi(\omega) \geq 0 \ \forall \omega \in \Omega$. By Exercise 9.3.5, $\|\alpha\|_{C_{\text{fin}}(\Omega)^*} = 1$; α_{q_0} is therefore a nonnegative linear functional of norm 1 on $C_{\text{fin}}(\Omega)$. According to the Hahn–Banach theorem (Theorem 4.8.2 of [Fried]), α_{q_0} has an extension to a bounded linear functional on $C(\Omega)$ of norm 1 and, since $C_{\text{fin}}(\Omega)$ is dense in $C(\Omega)$, this extension is unique and nonnegative. We will continue to denote this unique extension by α_{q_0}.

Now we apply the Riesz representation theorem, Theorem 9.3.3, to obtain a regular measure $\mathfrak{m}_{q_0}$ on some σ-algebra $\mathcal{A}_{q_0}(\Omega)$ on Ω containing the Borel σ-algebra $\mathcal{B}(\Omega)$ on

Ω with the property that

$$\alpha_{q_0}(\varphi) = \int_\Omega \varphi(\omega)\, d\mathfrak{m}_{q_0}(\omega) \quad \forall \varphi \in C(\Omega).$$

Since $\alpha_{q_0}(1) = 1$, $\mathfrak{m}_{q_0}$ is, in fact, a probability measure on Ω. For any fixed q_0, $\mathfrak{m}_{q_0}$ is called a *Wiener measure* on Ω.

Exercise 9.3.7. Show that the Wiener measure of a cylinder set in Ω is the physically correct probability; part of the exercise is to decide what "physically correct probability" means.

One should take careful note of the fact that Nelson's application of the Riesz representation theorem depended in an essential way on the positivity of $H_D(t, q, q_0) = \frac{1}{\sqrt{4\pi Dt}} e^{-(q-q_0)^2/4Dt}$. The Schrödinger kernel looks much like this except that D is complex and so the argument breaks down completely. This does not preclude the possibility that some other argument might produce an analogous (complex) measure appropriate to the Feynman integral, but we will see that a theorem of Cameron [Cam] shows that, in fact, no such measure exists.

This is not exactly what we promised, however. We set out looking for a probability measure on the space of *continuous* paths *in* $\mathbb{R}$, but the paths in Ω are completely arbitrary and are even permitted to pass through ∞. What one would like to show is that although Ω is much too large for our purposes, the Wiener measures concentrate on the subset of Ω consisting of continuous paths that do not pass through ∞ in the sense that this subset has full measure (namely, 1). This, in fact, is where the really hard work begins. The following theorem, originally due to Wiener, is proved by Nelson as Theorem 4 of his paper [Nel3].

Theorem 9.3.4. *Fix $q_0 \in \mathbb{R}$ and let*

$$C([0,\infty), \mathbb{R}) = \{\omega \in \Omega : \omega \text{ is continuous and } \omega(t) \in \mathbb{R} \ \forall t \in [0,\infty)\}.$$

Then $C([0,\infty), \mathbb{R})$ is a Borel set in Ω and $\mathfrak{m}_{q_0}(C([0,\infty), \mathbb{R})) = 1$.

Stated otherwise, the set of points in Ω that correspond to paths that are discontinuous or pass through the point at infinity is, measure theoretically, negligible (has Wiener measure zero). In the language of probability, an element of Ω is almost surely real-valued and continuous.

One can obtain more precise information in the following way. Recall that if $0 < \beta \le 1$, a map $\omega : [0,\infty) \to \mathbb{R}$ is said to be *(locally) Hölder continuous of order β* if, for every $0 < m < \infty$, there exists an $M > 0$ such that

$$\left| \omega(s) - \omega(t) \right| \le M\, |s - t|^\beta \quad \forall s, t \in [0, m].$$

We will denote by Ω_β the set of all elements in Ω that are Hölder continuous of order β. The following result combines Corollary 3.4.4 of [JL] and Theorem 5.2 of [Simon2].

Theorem 9.3.5. *Fix $q_0 \in \mathbb{R}$ and $0 < \beta \leq 1$. Then Ω_β is a Borel set in Ω and*

$$
\mathfrak{m}_{q_0}(\Omega_\beta) = \begin{cases} 1 & \text{if } 0 < \beta < \frac{1}{2}, \\ 0 & \text{if } \frac{1}{2} \leq \beta \leq 1. \end{cases}
$$

A path in Ω is almost surely Hölder continuous of order $\beta < \frac{1}{2}$, but almost surely not Hölder continuous of order $\beta \geq \frac{1}{2}$.

Finally, we address the issue of differentiability. We have seen that Perrin, on the basis of his experimental results, suggested that the path of a Brownian particle might well be modeled by a continuous, nowhere differentiable function of the type first described by Weierstrass. Although this seems rather implausible physically it is remarkable that, in 1933, Paley, Wiener and Zygmund proved that a path in Ω is almost surely a function of this type. The following is Theorem 1.30 of [MP]; we should mention that this reference focuses on the probabilistic definition of a Brownian motion, which we have not emphasized here, so one will need to absorb some additional terminology in order to read the proof. A somewhat less ambitious result (a path in Ω is almost surely differentiable at most on a subset of Lebesgue measure zero) has a more accessible proof and is Theorem 3.4.7 of [JL].

Theorem 9.3.6. *Fix $q_0 \in \mathbb{R}$ and let*

$$
C_{\text{ND}}([0,\infty),\mathbb{R}) = \{\, \omega \in C([0,\infty),\mathbb{R}) : \omega(t) \text{ is nowhere differentiable on } [0,\infty) \,\}.
$$

Then $C_{\text{ND}}([0,\infty),\mathbb{R})$ is a Borel set in Ω and $\mathfrak{m}_{q_0}(C_{\text{ND}}([0,\infty),\mathbb{R})) = 1$.

Now we will turn to the second construction of Wiener measure, a bit less elegant perhaps, but also certain to be more familiar to those who recall the usual construction of the Lebesgue measure on $\mathbb{R}^n$. We will begin with a schematic of the procedure; for the sake of clarity we have included in Appendix F a sketch of some of the basic measure theory that goes into the construction (Sections 1–2, Chapter 12, of [Roy] contain everything we will need).

We begin with the underlying set $X = C_0[t_a, t_b]$ and its semi-algebra $\mathcal{I}$ of cylinder sets. In order for the Carathéodory machinery (see Appendix F) to kick in we need only define a pre-measure $\mathfrak{m}$ on $\mathcal{I}$. The idea is to do this in such a way that for any $I \in \mathcal{I}$, $\mathfrak{m}(I)$ has the appropriate physical interpretation in terms of Brownian motion. This gives us no real choice; we must take

$$
\mathfrak{m}(I) = \mathfrak{m}(I(t_1, \ldots, t_n; (\alpha_1, \beta_1] \times \cdots \times (\alpha_n, \beta_n]))
$$

$$
= \int_{\alpha_n}^{\beta_n} \cdots \int_{\alpha_1}^{\beta_1} H_D(t_1 - t_0, q_1, q_0) \cdots H_D(t_n - t_{n-1}, q_n, q_{n-1}) \, dq_1 \cdots dq_n, \tag{9.25}
$$

where $q_0 = 0$ is the fixed "starting point" for the curves. It will be useful later to have this written as a Lebesgue integral over the rectangle $(\alpha_1, \beta_1] \times \cdots \times (\alpha_n, \beta_n]$. We will write

$$m(I(t_1,\ldots,t_n;(\alpha_1,\beta_1] \times \cdots \times (\alpha_n,\beta_n])) = \int_{(\alpha_1,\beta_1]\times\cdots\times(\alpha_n,\beta_n]} W_n(\mathbf{t},\mathbf{q})\,d^n\mathbf{q}, \qquad (9.26)$$

where $\mathbf{t} = (t_1,\ldots,t_n)$ with $t_a = t_0 < t_1 < \cdots < t_{n-1} < t_n \le t_b$, $\mathbf{q} = (q_1,\ldots,q_n)$, $d^n\mathbf{q}$ stands for Lebesgue measure on $\mathbb{R}^n$ and

$$W_n(\mathbf{t},\mathbf{q}) = H_D(t_1 - t_0, q_1, q_0) \cdots H_D(t_n - t_{n-1}, q_n, q_{n-1}) \qquad (9.27)$$

$$= \left[(4\pi D)^n (t_1 - t_0) \cdots (t_n - t_{n-1})\right]^{-1/2} e^{-\frac{1}{4D}\sum_{k=1}^n \frac{(q_k - q_{k-1})^2}{t_k - t_{k-1}}} \qquad (9.28)$$

and again we recall that $q_0 = 0$.

Before confronting the issue of whether or not this is a pre-measure one should notice that it is not obviously well-defined. The reason is that an element I of $\mathcal{I}$ generally has many different representations as $I(t_1,\ldots,t_n;(\alpha_1,\beta_1] \times \cdots \times (\alpha_n,\beta_n])$ and (9.25) appears to depend on the representation. However, the different representations for a given $I \in \mathcal{I}$ cannot be *too* different; they can differ, one from another, only by the insertion or deletion of subdivision points s for which the corresponding half-open interval is $(-\infty,\infty]$ since it is only these that have no effect on the set of functions in I. To show that $m(I)$ is well-defined we can begin by choosing from among all of these representations of I a "minimal" one $I(t_1,\ldots,t_n;(\alpha_1,\beta_1]\times\cdots\times(\alpha_n,\beta_n])$ in which all such subdivision points have been deleted so that for each $(\alpha_j,\beta_j]$, $j = 1,\ldots,n$, at least one of α_j or β_j is finite. Then one need only show that adding to $t_1,\ldots,t_n$ more subdivision points with "restriction intervals" $(-\infty,\infty]$ does not alter the value of $m(I)$. By induction, it clearly suffices to do this for one point, say, s. Suppose first that $t_0 < s < t_1$ so that the representation for I is

$$I(s,t_1,\ldots,t_n;(-\infty,\infty] \times (\alpha_1,\beta_1] \times \cdots \times (\alpha_n,\beta_n]).$$

Now note that if we use this representation in the definition of $m(I)$, we obtain

$$\int_{\alpha_n}^{\beta_n} \cdots \int_{\alpha_1}^{\beta_1} \int_{-\infty}^{\infty} H_D(s - t_0, v, q_0) H_D(t_1 - s, q_1, v)$$

$$\cdots H_D(t_n - t_{n-1}, q_n, q_{n-1})\,dv\,dq_1 \cdots dq_n.$$

According to (9.24) the first integration (with respect to v) gives

$$\int_{-\infty}^{\infty} H_D(s - t_0, v, q_0) H_D(t_1 - s, q_1, v)\,dv = H_D(t_1 - t_0, q_1, q_0),$$

so this is the same as

$$\int_{\alpha_n}^{\beta_n} \cdots \int_{\alpha_1}^{\beta_1} H_D(t_1 - t_0, q_1, q_0) \cdots H_D(t_n - t_{n-1}, q_n, q_{n-1})\,dq_1 \cdots dq_n,$$

which is what we wanted to show.

Exercise 9.3.8. Show that the same argument works if s is in any of the open intervals (t_{k-1}, t_k), $k = 2, \ldots, n$, so m is well-defined on $\mathcal{I}$. *Hint*: Justify interchanging the order of integration.

Exercise 9.3.9. Show that $m(\emptyset) = 0$, $m(I) > 0$ if $I \neq \emptyset$ and $m(C_0[t_a, t_b]) = 1$.

The good news is that there is only one item left to verify in order to show that m defines a pre-measure on $\mathcal{I}$, that is, countable additivity. One must show that if $I_1, I_2, \ldots$ are pairwise disjoint elements of $\mathcal{I}$ and if $I = \bigsqcup_{k=1}^{\infty} I_k$ is also in $\mathcal{I}$, then $m(I) = \sum_{k=1}^{\infty} m(I_k)$. Once this is done the Carathéodory procedure described in Appendix F produces for us a measure on $C_0[t_a, t_b]$ that takes the "right" values on cylinder sets. The bad news is that countable additivity is by far the deepest and most difficult part of the construction, that proofs are not so easy to find in the literature and that those one can find tend to require a substantial background in stochastic analysis. Our basic source for the material on Wiener measure [JL] does not include a proof, but suggests a few references, among them a relatively short proof on pages 13–14 of [Kal]; one can also consult Theorem 3.1.1 and Appendix A.4 of [GJ] for a proof. We will not attempt to sketch the proof here. However, one can get some sense of why the argument for countable additivity must be rather subtle by noting that it must somehow carefully distinguish curves that are "merely" continuous from those that have derivatives somewhere since, in the end, the set of curves that are differentiable somewhere will have Wiener measure zero and so, in a sense, "do not count." In the functional analytic approach of Nelson that we described earlier one seems to get countable additivity for free from the Riesz representation theorem, but only because the deep issues are shifted to showing that the measure on Ω concentrates on the continuous curves in $\mathbb{R}$.

Modulo this one, admittedly shameless omission on our part we have produced a pre-measure m on the pre-algebra $\mathcal{I}$ of cylinder sets in $C_0[t_a, t_b]$. Carathéodory's procedure then provides a complete measure, which we will also denote m, on a σ-algebra $\mathcal{W}$ of subsets of $C_0[t_a, t_b]$ containing $\sigma(\mathcal{I}) = \mathcal{B}(C_0[t_a, t_b])$; $(C_0[t_a, t_b], \mathcal{W}, m)$ is, in fact, the completion of $(C_0[t_a, t_b], \mathcal{B}(C_0[t_a, t_b]), m|_{\mathcal{B}(C_0[t_a,t_b])})$, so $\mathcal{W}$ consists precisely of sets of the form $A \cup B$, where $A \in \mathcal{B}(C_0[t_a, t_b])$ and B is a subset of some $C \in \mathcal{B}(C_0[t_a, t_b])$ with $m(C) = 0$. The elements of $\mathcal{W}$ are called *Wiener measurable sets*; m itself is called the *Wiener measure* on $C_0[t_a, t_b]$. More accurately, one should say that m is the Wiener measure starting at $q_0 = 0$ since this was built into our definition of m on the cylinder sets in (9.25); one might even prefer to denote it m_0 and use m_{q_0} to indicate the analogous measure built with starting point q_0. Until it becomes necessary to be careful about the distinction, however, we will stick with m.

This construction of Wiener measure is so analogous to the usual construction of the Lebesgue measure on $\mathbb{R}^n$ that we would do well to sound a cautionary note. Wiener measure does *not* share all of the nice properties of Lebesgue measure on $\mathbb{R}^n$. For example, it is not translation invariant. Indeed, one can show that there is no nontrivial translation invariant Borel measure on $C_0[t_a, t_b]$ at all; this is Theorem 3.1.3 of [JL].

Here is another example. Lebesgue measure on $\mathbb{R}^n$ has the property that if M is a measurable set in $\mathbb{R}^n$, then, for any $c \in \mathbb{R}$, $cM = \{cx : x \in M\}$ is also measurable. The analogous statement in $C_0[t_a, t_b]$, however, is false. There exist Wiener measurable sets M in $C_0[t_a, t_b]$ for which $2M = \{2x : x \in M\}$ is not Wiener measurable (see [JS] or Section 4.2 of [JL]).

So far we only know how to compute the Wiener measure of cylinder sets, so we will try to enlarge our arsenal just a bit. Fix $(t_1, \dots, t_n)$ with $t_a = t_0 < t_1 < \cdots < t_n \leq t_b$ and denote this n-tuple

$$\mathbf{t} = (t_1, \dots, t_n).$$

Now define the *evaluation map*

$$\mathrm{ev}_{\mathbf{t}} : C_0[t_a, t_b] \to \mathbb{R}^n$$

by

$$\mathrm{ev}_{\mathbf{t}}(x) = \big(x(t_1), \dots, x(t_n)\big)$$

for every $x \in C_0[t_a, t_b]$. Note that $\mathrm{ev}_{\mathbf{t}}$ is clearly linear and it is bounded because

$$\left\| \mathrm{ev}_{\mathbf{t}}(x) \right\|^2 = x(t_1)^2 + \cdots + x(t_n)^2 \leq n \left(\max_{t_a \leq t \leq t_b} |x(t)| \right)^2 = n \|x\|^2.$$

Thus, $\mathrm{ev}_{\mathbf{t}}$ is continuous and therefore Borel measurable, that is, for any Borel set A in $\mathbb{R}^n$,

$$\mathrm{ev}_{\mathbf{t}}^{-1}(A) = \{x \in C_0[t_a, t_b] : (x(t_1), \dots, x(t_n)) \in A\}$$

is a Borel set in $C_0[t_a, t_b]$. The converse is also true, that is, $\mathrm{ev}_{\mathbf{t}}^{-1}(A) \in \mathcal{B}(C_0[t_a, t_b])$ if and only if $A \in \mathcal{B}(\mathbb{R}^n)$; this is Proposition 3.5.1 of [JL].

In particular, $\mathfrak{m}(\mathrm{ev}_{\mathbf{t}}^{-1}(A))$ is defined and we would like to compute it. The result we will prove is entirely analogous to (9.26). Specifically, we will show that

$$\mathfrak{m}\big(\mathrm{ev}_{\mathbf{t}}^{-1}(A)\big) = \int_A W_n(\mathbf{t}, \mathbf{q}) \, d^n\mathbf{q},$$

where $W_n(\mathbf{t}, \mathbf{q})$ is given by (9.27). The function $W_n(\mathbf{t}, \cdot)$ is positive and satisfies

$$\int_{\mathbb{R}^n} W_n(\mathbf{t}, \mathbf{q}) \, d^n\mathbf{q} = 1.$$

Thus, we can use it to define a probability measure ν_n on $\mathbb{R}^n$ by

$$\nu_n(E) = \int_E W_n(\mathbf{t}, \mathbf{q}) \, d^n\mathbf{q}$$

for every Lebesgue measurable set E in $\mathbb{R}^n$. We wish to compare this with the pushforward measure $(\mathrm{ev}_t)_*(m)$ on $\mathbb{R}^n$ defined by

$$((\mathrm{ev}_t)_*(m))(E) = m(\mathrm{ev}_t^{-1}(E))$$

(see Remark 4.3.3). According to (9.26), the measures ν_n and $(\mathrm{ev}_t)_*(m)$ agree when E is a rectangle of the form $(\alpha_1, \beta_1] \times \cdots \times (\alpha_n, \beta_n]$. But these rectangles generate the Borel sets $\mathcal{B}(\mathbb{R}^n)$ in $\mathbb{R}^n$, so

$$\nu_n = (\mathrm{ev}_t)_*(m) \quad \text{on } \mathcal{B}(\mathbb{R}^n).$$

Consequently, for all $A \in \mathcal{B}(\mathbb{R}^n)$,

$$m(\mathrm{ev}_t^{-1}(A)) = ((\mathrm{ev}_t)_*(m))(A) = \nu_n(A) = \int_A W_n(\mathbf{t}, \mathbf{q})\, d^n\mathbf{q},$$

as required. Note, in particular, that if A has Lebesgue measure zero, then $\mathrm{ev}_t^{-1}(A)$ has Wiener measure zero.

We will show now that, in fact, the same is true for any Lebesgue measurable set in $\mathbb{R}^n$. Denote the σ-algebra of Lebesgue measurable sets in $\mathbb{R}^n$ by $\mathcal{A}_{\mathrm{Leb}}(\mathbb{R}^n)$ and write μ_{Leb} for the Lebesgue measure..

Theorem 9.3.7. *Let* $\mathbf{t} = (t_1, \dots, t_n)$ *be fixed, where* $t_a = t_0 < t_1 < \cdots < t_n \leq t_b$. *Then*

$$\mathrm{ev}_t : (C_0[t_a, t_b], \mathcal{W}) \to (\mathbb{R}^n, \mathcal{A}_{\mathrm{Leb}}(\mathbb{R}^n))$$

defined by

$$\mathrm{ev}_t(x) = (x(t_1), \dots, x(t_n))$$

is measurable and, for any $E \in \mathcal{A}_{\mathrm{Leb}}(\mathbb{R}^n)$,

$$m(\mathrm{ev}_t^{-1}(E)) = \int_E W_n(\mathbf{t}, \mathbf{q})\, d^n\mathbf{q}.$$

Proof. First we show that ev_t is measurable. Let $E \in \mathcal{A}_{\mathrm{Leb}}(\mathbb{R}^n)$. Since $(\mathbb{R}^n, \mathcal{A}_{\mathrm{Leb}}(\mathbb{R}^n), \mu_{\mathrm{Leb}})$ is the completion of $(\mathbb{R}^n, \mathcal{B}(\mathbb{R}^n), \mu_{\mathrm{Leb}}|_{\mathcal{B}(\mathbb{R}^n)})$, E can be written as $E = A \cup B$, where $A \in \mathcal{B}(\mathbb{R}^n)$ and $B \subseteq C$, where C is in $\mathcal{B}(\mathbb{R}^n)$ and has $\mu_{\mathrm{Leb}}(C) = 0$. Now,

$$\mathrm{ev}_t^{-1}(E) = \mathrm{ev}_t^{-1}(A) \cup \mathrm{ev}_t^{-1}(B)$$

and $\mathrm{ev}_t^{-1}(B) \subseteq \mathrm{ev}_t^{-1}(C)$. We have just seen that $\mathrm{ev}_t^{-1}(A)$ and $\mathrm{ev}_t^{-1}(C)$ are in $\mathcal{B}(C_0[t_a, t_b])$ and $m(\mathrm{ev}_t^{-1}(C)) = 0$. But $(C_0[t_a, t_b], \mathcal{W}, m)$ is the completion of $(C_0[t_a, t_b], \mathcal{B}(C_0[t_a, t_b]), m|_{\mathcal{B}(C_0[t_a, t_b])})$. It follows that $\mathrm{ev}_t^{-1}(E)$ is in $\mathcal{W}$, so ev_t is measurable.

Remark 9.3.3. This shows that if $E \subseteq \mathbb{R}^n$ is Lebesgue measurable, then $\mathrm{ev}_t^{-1}(E)$ is Wiener measurable. In fact, the converse is also true, that is, $\mathrm{ev}_t^{-1}(E)$ is Wiener measurable if and only if E is Lebesgue measurable; this is Theorem 3.5.2 of [JL].

Completeness also implies that $\mathrm{ev}_t^{-1}(B)$ is in $\mathcal{W}$ and $\mathfrak{m}(\mathrm{ev}_t^{-1}(B)) = 0$. Consequently,

$$\mathfrak{m}\big(\,\mathrm{ev}_t^{-1}(E)\,\big) = \mathfrak{m}\big(\,\mathrm{ev}_t^{-1}(A)\,\big) = \int_A W_n(\mathbf{t}, \mathbf{q})\, d^n\mathbf{q} = \int_E W_n(\mathbf{t}, \mathbf{q})\, d^n\mathbf{q},$$

as required. ☐

Note that this result can be rephrased by saying that the pushforward measure $(\mathrm{ev}_t)_*(\mathfrak{m})$ on $\mathbb{R}^n$ agrees with the measure on $\mathbb{R}^n$ determined by the density function $W_n(\mathbf{t}, \mathbf{q})$ and the Lebesgue measure.

The Wiener measure of a set of the form $\mathrm{ev}_t^{-1}(E)$ for E Lebesgue measurable in $\mathbb{R}^n$ can be computed as a finite-dimensional integral. We will now show that, more generally, the integral with respect to Wiener measure of a function on $C_0[t_a, t_b]$ that depends only on the values each $x \in C_0[t_a, t_b]$ takes at $\mathbf{t} = (t_1, \ldots, t_n)$ is just a finite-dimensional integral over $\mathbb{R}^n$; as we did in Nelson's construction we will call these *finite functions* in $C_0[t_a, t_b]$. More precisely, we will prove the following.

Theorem 9.3.8. *Let* $\mathbf{t} = (t_1, \ldots, t_n)$ *be fixed, where* $t_a = t_0 < t_1 < \cdots < t_n \leq t_b$, *and let* $f : \mathbb{R}^n \to \mathbb{R}$ *be a Lebesgue measurable function. Define* $\varphi : C_0[t_a, t_b] \to \mathbb{R}$ *by* $\varphi = f \circ \mathrm{ev}_t$, *that is,*

$$\varphi(x) = f\big(x(t_1), \ldots, x(t_n)\big)$$

for all $x \in C_0[t_a, t_b]$. *Then*

$$\int_{C_0[t_a,t_b]} \varphi(x)\, d\mathfrak{m}(x) = \int_{\mathbb{R}^n} f(\mathbf{q}) W_n(\mathbf{t}, \mathbf{q})\, d^n\mathbf{q},$$

where $W_n(\mathbf{t}, \mathbf{q})$ *is given by (9.27) and the equality is interpreted in the strong sense that if either side is defined, whether finite or infinite, then so is the other side and they agree.*

Proof. We have shown that $\mathrm{ev}_t : C_0[t_a, t_b] \to \mathbb{R}^n$ is measurable so we can apply the change of variables formula (4.44) to obtain

$$\int_{C_0[t_a,t_b]} \varphi(x)\, d\mathfrak{m}(x) = \int_{C_0[t_a,t_b]} (f \circ \mathrm{ev}_t)(x)\, d\mathfrak{m}(x) = \int_{\mathbb{R}^n} f(\mathbf{q})\, d((\mathrm{ev}_t)_*(\mathfrak{m}))(\mathbf{q}).$$

But we have also shown that $d((\mathrm{ev}_t)_*(\mathfrak{m}))(\mathbf{q}) = W_n(\mathbf{t}, \mathbf{q})\, d^n\mathbf{q}$, so the result follows. ☐

We should probably compute one or two simple integrals (many more examples can be found in Section 3.3 of [JL] and in [GY]). Let us fix $\mathbf{t} = (t_1)$ with $t_a = t_0 < t_1 \leq t_b$

and write ev_{t_1} for ev_t. Then $ev_{t_1}(x) = x(t_1)$ and, for any measurable function $f : \mathbb{R} \to \mathbb{R}$, $(f \circ ev_{t_1})(x) = f(x(t_1))$ and

$$\int_{C_0[t_a,t_b]} f(x(t_1))\, dm(x) = \int_{\mathbb{R}} f(q_1) W_1(t_1, q_1)\, dq_1 = \int_{\mathbb{R}} f(q_1) H_D(t_1 - t_0, q_1, q_0)\, dq_1.$$

Example 9.3.1. Take f to be the identity map on $\mathbb{R}$, that is, $f(q_1) = q_1$. Then $(f \circ ev_{t_1})(x) = ev_{t_1}(x) = x(t_1)$, so

$$\int_{C_0[t_a,t_b]} x(t_1)\, dm(x) = \int_{\mathbb{R}} q_1 W_1(t_1, q_1)\, dq_1$$

$$= [4\pi D(t_1 - t_0)]^{-1/2} \int_{\mathbb{R}} q_1 e^{-\frac{1}{4D}\frac{(q_1-q_0)^2}{t_1-t_0}}\, dq_1$$

$$= [4\pi D(t_1 - t_0)]^{-1/2} \int_{\mathbb{R}} q_1 e^{-\frac{1}{4D}\frac{q_1^2}{t_1-t_0}}\, dq_1$$

$$= 0$$

because the integrand is odd.

Example 9.3.2. Take f to be $f(q_1) = q_1^2$. Then

$$\int_{C_0[t_a,t_b]} x(t_1)^2\, dm(x) = \int_{\mathbb{R}} q_1^2 W_1(t_1, q_1)\, dq_1$$

$$= [4\pi D(t_1 - t_0)]^{-1/2} \int_{\mathbb{R}} q_1^2 e^{-\frac{1}{4D}\frac{q_1^2}{t_1-t_0}}\, dq_1$$

$$= [4\pi D(t_1 - t_0)]^{-1/2} [4D(t_1 - t_0)]^{3/2} \int_{\mathbb{R}} u^2 e^{-u^2}\, du$$

$$= 2D(t_1 - t_0),$$

where we have used the Gaussian integral

$$\int_{\mathbb{R}} u^2 e^{-u^2}\, du = \frac{\sqrt{\pi}}{2}$$

(see Exercise A.0.3.2 of Appendix A).

These examples may seem a bit artificial, so we would like to try to get something more interesting out of Theorem 9.3.8. For this, however, and for other purposes as well we will need the famous *Hille–Yosida theorem* and a generalization of the Lie–Trotter–Kato product formula (Theorem 8.1.1). These are discussed in Appendix J. It is shown in Example J.0.3 that the heat semigroup $\{T_t\}_{t\geq 0}$ (with $D = 1$) coincides with the semigroup $\{e^{t\Delta}\}_{t\geq 0}$ generated by minus the Laplacian. We will now use this and Theorem 9.3.8 to show that *heat flow can be represented as a Wiener (path) integral*. We will work on the interval $[0, t]$ for some fixed, but arbitrary $t > 0$ and will apply Theorem 9.3.8 to the trivial partition $0 = t_0 < t_1 = t$. For reasons that will become clear momentarily we will use y_1 as the integration variable and write, for any $\psi_0 \in L^2(\mathbb{R})$,

$$(e^{t\Delta}\psi_0)(q) = \int_{\mathbb{R}} (4\pi t)^{-1/2} e^{-(y_0 - y_1)^2/4t} \psi_0(y_1)\, dy_1,$$

where $y_0 = q$. Make the change of variable $q_1 = y_1 - q$ and $q_0 = y_0 - q = q - q = 0$ (which we need to apply Theorem 9.3.8). Then

$$(e^{t\Delta}\psi_0)(q) = \int_{\mathbb{R}} (4\pi t)^{-1/2} e^{(q_0 - q_1)^2/4t} \psi_0(q_1 + q)\, dq_1,$$

where $q_0 = 0$. Now note that, with $\mathbf{t} = (t_1) = (t)$ written simply as t and $\mathbf{q} = (q_1)$ written as q_1,

$$W_1(\mathbf{t}, \mathbf{q}) = W_1(t, q_1) = (4\pi t)^{-1/2} e^{-(q_0 - q_1)^2/4t},$$

so

$$(e^{t\Delta}\psi_0)(q) = \int_{\mathbb{R}} \psi_0(q_1 + q) W_1(t, q_1)\, dq_1 = \int_{C_0[0,t]} (f \circ \mathrm{ev}_{t_1})(x)\, d\mathrm{m}_t(x),$$

where $f(q_1) = \psi_0(q_1 + q)$ and we have written m_t to emphasize the dependence of the Wiener measure on t. Since $t_1 = t$ we finally arrive at

$$(e^{t\Delta}\psi_0)(q) = \int_{C_0[0,t]} \psi_0(x(t) + q)\, d\mathrm{m}_t(x) \qquad (9.29)$$

as the *path integral representation for the heat flow*. Shortly we will describe a very substantial extension of this result which combines Theorem 9.3.8 with the Trotter product formula in order to accommodate a nonzero potential term.

The Wiener measure of a Wiener measurable set $W \in \mathcal{W}$ has, in itself, a nice physical interpretation; $\mathrm{m}(W)$ is the probability that a Brownian path starting at $q_0 = 0$ satisfies whatever conditions define W. Even so, this is not where its real significance lies for us. Recall that we got into this business in the first place because we were interested in whether or not the Feynman "integral" was really an integral and that the reason we cared about Feynman integrals was that they describe the time evolution operators

$$e^{-(i/\hbar)(t - t_0)(H_0 + V)} \qquad (9.30)$$

of a quantum system, either directly (8.2) or by way of the propagator (8.3). We will soon have something to say about whether or not the Feynman "integral" is an integral in the same sense that the Wiener integral is, but first we would like to see that the analogous "evolution question" for the heat equation has an entirely satisfactory and rigorous solution in the context of the Wiener integral. The first step in this direction is the path integral representation (9.29) for $e^{t\Delta}$, but now we would like to include a (sufficiently nice) potential V and consider the evolution operator $e^{t(\Delta-V)}$. The result we will describe is a very special case of the so-called *Feynman–Kac formula*. To keep the argument as simple as possible we will make some wildly extravagant assumptions, but we will mention also a much more general result that is proved in Theorem 12.1.1 of [JL]; there is also a proof fashioned on the Nelson approach to the Wiener integral in Theorem X.68 of [RS2].

We will again work on the interval $[0, t]$ for some fixed, but arbitrary $t > 0$ and will consider an operator of the form $-\Delta + V$ on $L^2(\mathbb{R})$, where $V : \mathbb{R} \to \mathbb{R}$ is a real-valued (potential) function concerning which we will need to make some assumptions. The assumptions will have to be sufficient to guarantee that $-(-\Delta + V) = \Delta - V$ generates a semigroup $e^{t(\Delta-V)}$ of operators on $L^2(\mathbb{R})$ to which we can apply the generalized Trotter product formula in Theorem J.0.5. Now, in Section 9.2 we isolated a number of conditions on V that ensure the self-adjointness of $-\Delta + V$. In particular, by Theorem 9.2.4, if V happens to be *continuous and bounded*, then $-\Delta + V$ is self-adjoint on $\mathcal{D}(\Delta)$ and this is what we will assume for our proof. This is our "wildly extravagant" assumption. Everything we will do can be proved under the much weaker assumption that V satisfies the conditions specified in Theorem 9.2.8; proofs in this case are available in Theorem 12.1.1 of [JL] and Theorem X.68 of [RS2].

We will not assume that V is nonnegative since we saw in Example J.0.4 that boundedness is enough to ensure that

$$e^{t(\Delta-V)}\psi_0 = \lim_{n\to\infty}\left(e^{\frac{t}{n}\Delta}e^{-\frac{t}{n}V}\right)^n\psi_0$$

for every $\psi_0 \in L^2(\mathbb{R})$. Under these assumptions we will derive a *path integral representation for heat flow*. Specifically, we will show that for every $\psi_0 \in L^2(\mathbb{R})$, every $t > 0$ and (Lebesgue) almost every $q \in \mathbb{R}$,

$$\left(e^{t(\Delta-V)}\psi_0\right)(q) = \int_{C_0[0,t]} e^{-\int_0^t V(x(s)+q)\,ds}\psi_0(x(t) + q)\,d\mathfrak{m}_t(x), \tag{9.31}$$

where $\mathfrak{m}_t$ is the Wiener measure on $C_0[0, t] = \{x : [0, t] \to \mathbb{R} : x$ is continuous and $x(0) = 0\}$. Note that if V happens to be zero, then this reduces to (9.29). We begin, as we did for the Feynman integral, by writing out the Trotter products. Fix a $t > 0$ and a $\psi_0 \in L^2(\mathbb{R})$. For reasons that will become clear shortly we will use $y, y_1, \ldots, y_n$ as integration variables. From

$$\left(e^{t\Delta}\psi_0\right)(q) = \int_\mathbb{R} (4\pi t)^{-1/2} e^{-(q-y)^2/4t}\psi_0(y)\,dy,$$

we conclude that

$$(e^{t\Delta/2}e^{-tV/2}\psi_0)(q) = (4\pi(t/2))^{-1/2} \int_{\mathbb{R}} e^{-(q-y_2)^2/4(t/2)} e^{-(t/2)V(y_2)}\psi_0(y_2)\,dy_2.$$

Thus,

$$
\begin{aligned}
&[(e^{t\Delta/2}e^{-tV/2})^2 \psi_0](q)\\
&= [(e^{t\Delta/2}e^{-tV/2})(e^{t\Delta/2}e^{-tV/2}\psi_0)](q)\\
&= (4\pi(t/2))^{-2/2} \int_{\mathbb{R}} e^{-(q-y_1)^2/4(t/2)} e^{-(t/2)V(y_1)} \left[\int_{\mathbb{R}} e^{-(y_1-y_2)^2/4(t/2)} e^{-(t/2)V(y_2)}\psi_0(y_2)\,dy_2 \right] dy_1\\
&= (4\pi(t/2))^{-2/2} \int_{\mathbb{R}}\int_{\mathbb{R}} e^{-\frac{\sum_{k=1}^2 (y_{k-1}-y_k)^2}{4(t/2)}} e^{-(t/2)\sum_{k=1}^2 V(y_k)}\psi_0(y_2)\,dy_2\,dy_1,
\end{aligned}
$$

where $y_0 = q$. Continuing inductively gives

$$
\begin{aligned}
&[(e^{t\Delta/n}e^{-tV/n})^n \psi_0](q)\\
&= (4\pi(t/n))^{-n/2} \int_{\mathbb{R}}\cdots\int_{\mathbb{R}} e^{-\frac{\sum_{k=1}^n (y_{k-1}-y_k)^2}{4(t/n)}} e^{-(t/n)\sum_{k=1}^n V(y_k)}\psi_0(y_n)\,dy_n\cdots dy_1, \quad (9.32)
\end{aligned}
$$

where $y_0 = q$. Now introduce new variables $q_k = y_k - q$ for $k = 0,1,\ldots,n$ so that $q_0 = 0$ and

$$
\begin{aligned}
&[(e^{t\Delta/n}e^{-tV/n})^n \psi_0](q)\\
&= (4\pi(t/n))^{-n/2} \int_{\mathbb{R}}\cdots\int_{\mathbb{R}} e^{-\frac{\sum_{k=1}^n (q_{k-1}-q_k)^2}{4(t/n)}} e^{-(t/n)\sum_{k=1}^n V(q_k+q)}\psi_0(q_n+q)\,dq_n\cdots dq_1. \quad (9.33)
\end{aligned}
$$

Now we will use Theorem 9.3.8 to show that the right-hand side of (9.33) can be written as a Wiener integral. We consider the partition $0 = t_0 < t_1 < \cdots < t_n = t$ of $[0,t]$, where $t_k = kt/n$, $k = 0,1,\ldots,n$. Let $\mathbf{t} = (t_1,\ldots,t_n)$ and $\mathbf{q} = (q_1,\ldots,q_n) \in \mathbb{R}^n$. Then

$$W_n(\mathbf{t},\mathbf{q}) = (4\pi(t/n))^{-n/2} e^{-\frac{\sum_{k=1}^n (q_{k-1}-q_k)^2}{4(t/n)}}.$$

Since ψ_0 is defined almost everywhere on $\mathbb{R}$, the same is true of

$$f(\mathbf{q}) = f(q_1,\ldots,q_n) = e^{-(t/n)\sum_{k=1}^n V(q_k+q)}\psi_0(q_n+q).$$

According to Theorem 9.3.8,

$$\int_{\mathbb{R}^n} f(\mathbf{q})W_n(\mathbf{t},\mathbf{q})\,d^n\mathbf{q} = \int_{C_0[0,t]} (f \circ \mathrm{ev_t})(x)\,dm_t(x),$$

so we conclude that

$$[(e^{t\Delta/n}e^{-tV/n})^n \psi_0](q) = \int_{C_0[0,t]} e^{-(t/n)\sum_{k=1}^n V(x(kt/n)+q)}\psi_0(x(t)+q)\, dm_t(x).$$

We have already seen that our hypotheses concerning V imply that, in $L^2(\mathbb{R})$,

$$\lim_{n\to\infty}(e^{t\Delta/n}e^{-tV/n})^n \psi_0 = e^{t(\Delta-V)}\psi_0.$$

Convergence in $L^2(\mathbb{R})$ implies that some subsequence converges pointwise almost everywhere, so there is a subsequence $\{n_j\}_{j=1}^\infty$ of $\{n\}_{n=1}^\infty$ such that for almost every $q \in \mathbb{R}$,

$$\lim_{j\to\infty}((e^{t\Delta/n_j}e^{-tV/n_j})^{n_j}\psi_0)(q) = (e^{t(\Delta-V)}\psi_0)(q).$$

Now, since every $x \in C_0[0,t]$ is continuous and since we have assumed that V is continuous,

$$\lim_{j\to\infty}\sum_{j=1}^{n_j} V(x(jt/n_j)+q)(t/n_j) = \int_0^t V(x(s)+q)\, ds$$

because the left-hand side is a limit of Riemann sums for the right-hand side. Consequently,

$$\lim_{j\to\infty} e^{-\frac{t}{n_j}\sum_{j=1}^{n_j} V(x(jt/n_j)+q)} = e^{-\int_0^t(V(x(s)+q)\, ds}.$$

Next observe that $x(t) + q = ev_t(x) + q$.

Exercise 9.3.10. Use the fact that ψ_0 is defined almost everywhere on $\mathbb{R}$ and Theorem 9.3.7 to show that $\psi_0(x(t) + q)$ is defined for m_t-almost every $x \in C_0[0,t]$.

Thus, for m_t-almost every $x \in C_0[0,t]$ and almost every $q \in \mathbb{R}$,

$$\lim_{j\to\infty} e^{-(t/n_j)\sum_{j=1}^{n_j} V(x(jt/n_j)+q)}\psi_0(x(t) + q) = e^{-\int_0^t(V(x(s)+q)\, ds}\psi_0(x(t) + q).$$

We can therefore complete the proof of (9.31) by showing that we can take the limit

$$\lim_{j\to\infty}\int_{C_0[0,t]} e^{-(t/n_j)\sum_{j=1}^{n_j} V(x(jt/n_j)+q)}\psi_0(x(t) + q)\, dm_t(x)$$

inside the integral. Since the Wiener integral is an honest Lebesgue integral we are free to apply the dominated convergence theorem (see, for example, Theorem 1.8 of [LL]). But since V is assumed bounded we can let M be a positive constant for which $|V(u)| \le M$ for every $u \in \mathbb{R}$ and then

$$\left|e^{-(t/n_j)\sum_{j=1}^{n_j} V(x(jt/n_j)+q)}\psi_0(x(t) + q)\right| \le e^{(t/n_j)\sum_{j=1}^{n_j} M}\left|\psi_0(x(t) + q)\right| = e^{tM}\left|\psi_0(x(t) + q)\right|.$$

Exercise 9.3.11. Show that

$$\int_{C_0[0,t]} e^{tM} \left| \psi_0(x(t) + q) \right| dm_t(x) < \infty$$

and conclude that the dominated convergence theorem gives

$$\lim_{j \to \infty} \int_{C_0[0,t]} e^{-(t/n_j) \sum_{j=1}^{n_j} V(x(jt/n_j)+q)} \psi_0(x(t) + q) \, dm_t(x)$$

$$= \int_{C_0[0,t]} e^{-\int_0^t V(x(s)+q) \, ds} \psi_0(x(t) + q) \, dm_t(x).$$

Putting all of this together gives (9.31) and we have a path integral representation for the evolution operators $e^{t(\Delta-V)}$ for the heat/diffusion equation on $L^2(\mathbb{R})$ when the potential is continuous and bounded; we mention once more that the same result can be proved with much less restrictive conditions on V and refer those interested to Theorem 12.1.1 of [JL] and Theorem X.68 of [RS2].

All of this is very nice and has had an enormous impact on a wide range of mathematical disciplines (see, for example, [KacM]), but what we were really hoping for was an analogous result for the quantum mechanical time evolution operators

$$e^{it(\Delta-V)}$$

(to stress the similarity with the diffusion operators we have adopted units in which $\hbar = 1$ and taken $t_0 = 0$ and $m = \frac{1}{2}$ in (9.30)). The similarity of the evolution operators certainly tempts one to believe that it should be possible to modify the construction of the Wiener measure to accommodate the quantum evolution. Feynman suspected something of the sort.

> "Some sort of complex measure is being associated with the space of functions x(t). Finite results can be obtained under unexpected circumstances because the measure is not positive everywhere, but the contributions from most of the paths largely cancel out. These curious mathematical problems are sidestepped by the subdivision process. However, one feels as Cavalieri must have felt calculating the volume of a pyramid before the invention of calculus."

Feynman [Feyn], page 8.

Even one of the greatest mathematicians of the twentieth century succumbed to the temptation.

> "It is natural that such a complex measure ... will be just as "good" as Wiener measure. ... A strict proof of this fact does not differ from the corresponding proof for the case of Wiener measure."

Gel'fand and Yaglom [GY], page 58.

Cameron [Cam], however, admonishes us that it is sometimes wise to resist temptation. Let us take a moment to decide what one would want from a "Feynman measure"
and then see what Cameron has to say about it. In Section 8.1 we discussed Feynman's
interpretation of the two-slit experiment and how it led to his notion of a path integral.
The idea was to "weight" each classically possible path for the particle with an amplitude and "add" these amplitudes with a normalizing factor over all such paths. As
we did for Brownian motion a bit earlier we can consider the following special case.
Let $0 = t_0 < t_1 < \cdots < t_{n-1} < t_n = t$ be a uniform partition of $[0, t]$ with $\Delta t = t/n$ and
consider the cylinder set

$$I = I(t_1, \ldots, t_n; (\alpha_1, \beta_1] \times \cdots \times (\alpha_n, \beta_n])$$
$$= \{x \in C_0[0, t] : x(t_j) \in (\alpha_j, \beta_j], j = 1, \ldots, n\}.$$

For a Brownian particle starring at $q_0 = 0$ at $t_0 = 0$ the probability that the particle is
in $(\alpha_k, \beta_k]$ at time t_k for each $k = 1, \ldots, n$ is

$$\int_{\alpha_n}^{\beta_n} \cdots \int_{\alpha_1}^{\beta_1} (4\pi D(t/n))^{-n/2} e^{-\frac{1}{D} \sum_{k=1}^{n} \frac{(q_k - q_{k-1})^2}{4(t/n)}} \, dq_1 \cdots dq_n, \tag{9.34}$$

where $q_0 = 0$. For a free ($V = 0$) quantum particle. Feynman's expression for the
amplitude of a particle passing through $(\alpha_k, \beta_k]$ at time t_k for each $k = 1, \ldots, n$ is

$$\int_{\alpha_n}^{\beta_n} \cdots \int_{\alpha_1}^{\beta_1} (4\pi i(t/n))^{-n/2} e^{i \sum_{k=1}^{n} \frac{(q_k - q_{k-1})^2}{4(t/n)}} \, dq_1 \cdots dq_n, \tag{9.35}$$

where we continue to use units in which $\hbar = 1$ and take $t_0 = 0$ and $m = \frac{1}{2}$. Formally,
at least, (9.35) is just (9.34) with complex "diffusion constant" $D = i$. The construction of the Wiener measure begins with (9.34) which is, by definition, the measure
of $I = I(t_1, \ldots, t_n; (\alpha_1, \beta_1] \times \cdots \times (\alpha_n, \beta_n])$. This definition is fundamentally what is behind the ability of the Wiener integral to represent the evolution operators of the dif-
fusion equation. By the same token, any measure on $C_0[0, t]$ that is to provide a path
integral representation for the time evolution of a quantum particle's probability amplitude must begin by assigning the "correct" (according to Feynman) amplitude to
$I(t_1, \ldots, t_n; (\alpha_1, \beta_1] \times \cdots \times (\alpha_n, \beta_n])$, that is, (9.35); note that this is complex. The existence of such a measure is the question addressed by Cameron.

Remark 9.3.4. To state Cameron's result we should briefly review a few items concerning complex measures (details are available in Section 6.1 of [Rud2]). Let $\mathcal{A}$ be
a σ-algebra on the set X. A *complex measure* on the measurable space $(X, \mathcal{A})$ is a
complex-valued map $v : \mathcal{A} \to \mathbb{C}$ map on $\mathcal{A}$ that satisfies:
1. $v(\emptyset) = 0$,

2.   if $A_k \in \mathcal{A}$ for $k = 1, 2, \ldots$ with $A_{k_1} \cap A_{k_2} = \emptyset \ \forall k_1 \neq k_2$, then

$$v\left(\bigsqcup_{k=1}^{\infty} A_k \right) = \sum_{k=1}^{\infty} v(A_k),$$

where the series is required to converge absolutely.

For example, if μ is an ordinary (positive) measure on $(X, \mathcal{A})$ and φ is an element of $L^1(X, \mu)$, then $v(A) = \int_A \varphi \, d\mu$ defines a complex measure on $(X, \mathcal{A})$. For any complex measure v on $(X, \mathcal{A})$ one defines the map $|v|$ on $\mathcal{A}$ by

$$|v|(A) = \sup \sum |v(A_k)|,$$

where the supremum is taken over all sequences $A_1, A_2, \ldots$ of pairwise disjoint elements of $\mathcal{A}$ with $A = \bigsqcup_{k=1}^{\infty} A_k$. One can show (Theorem 6.2 of [Rud2]) that $|v|$ is an ordinary (positive) measure on $(X, \mathcal{A})$; it is called the *total variation measure* of v. Furthermore (Theorem 6.4 of [Rud2]), $|v|$ is a *finite* measure, that is, $|v|(X) < \infty$; $|v|(X)$ is called the *total variation* of v. This is generally expressed by saying that *any complex measure has finite total variation*. For example, if v is given by $v(A) = \int_A \varphi \, d\mu$ for some $\varphi \in L^1(X, \mathcal{A})$, then the total variation of v is just the L^1 norm of φ.

Cameron [Cam] has shown that there is no complex measure on $C_0[0, t]$ with the property that the measure of every cylinder set $I(t_1, \ldots, t_n; (\alpha_1, \beta_1] \times \cdots \times (\alpha_n, \beta_n])$ is given by

$$\int_{\alpha_n}^{\beta_n} \cdots \int_{\alpha_1}^{\beta_1} (4\pi i(t/n))^{-n/2} e^{i \sum_{k=1}^{n} \frac{(q_k - q_{k-1})^2}{4(t/n)}} \, dq_1 \cdots dq_n.$$

The proof amounts to showing that if such a measure existed, it would have to have infinite total variation, which is not possible (see the previous remark). This is not to say that making rigorous sense of the Feynman integral is hopeless, but only that the most obvious attempt to do so cannot succeed. Many other, less obvious attempts have been made over the years and we will have a few words to say about this in the next section.

9.4 Analytic continuation

The previous section ended on what might be considered a discouraging note. We had hoped to mimic Wiener's construction of his (Lebesgue) measure on $C_0[0, t]$ and the path integral representation it gives rise to for the evolution operators of the diffusion equation in the case of the Schrödinger equation and the quantum time evolution, thereby exhibiting the Feynman "integral" as an actual integral. Cameron, however, has disabused us of the notion that the Feynman integral can be regarded as an integral in the Lebesgue sense. Even so, this is not the end of the story. One would still like to fit the Feynman integral into some rigorous mathematical context even if that context cannot be the familiar Lebesgue theory. Now, one can argue that we already have a perfectly rigorous definition of the Feynman integral. It is defined as a limit, which may or may not exist, and one can simply set out to prove various convergence theorems (see, for example, [Fuj1], [Fuj2] and [Fuj3]). The only obvious issue one might take with this is that, in general, knowing simply that a limit exists does not always tell you a great deal about it. For example, $\sum_{n=1}^{\infty} \frac{1}{n^5}$ certainly converges. The sum of the series has a name; it is the value $\zeta(5)$ of the Riemann zeta function at 5. Is this an irrational number? To date, no one knows. We intend to look in another direction.

The search for its appropriate rigorous context has been ongoing essentially since Feynman introduced his integral. Various approaches have been proposed and one can obtain a brief overview of some of the most successful of these in [Klau] and [Mazz1]; much more detailed discussions and comparisons are available in [JL]. We will simply illustrate the sort of thing that can be done by focusing on just one of these. The approach we will describe was historically the first and is probably the most "obvious" thing to try. The motivation could not be simpler. Let us consider the free Schrödinger equation

$$i\,\frac{\partial \psi(q,t)}{\partial t} = -\frac{\hbar}{2m}\frac{\partial^2 \psi(q,t)}{\partial q^2}.$$

Now *formally* introduce a new variable T, picturesquely referred to as *imaginary time* and defined by

$$T = it.$$

One quick application of the chain rule shows that, in terms of the variables (q, T), the Schrödinger equation becomes

$$\frac{\partial \psi(q, T)}{\partial T} - \frac{\hbar}{2m}\frac{\partial^2 \psi(q, T)}{\partial q^2} = 0.$$

Et voilà, we have the heat equation with diffusion constant $D = \frac{\hbar}{2m}$. On the other hand, the substitution $t = -iT$ formally turns the heat equation with $D = \frac{\hbar}{2m}$ into the Schrödinger equation. This is amusing enough, but since $T = it$ is not a legitimate change of the (real) variable t, it really contains little usable information beyond a hint as to how we might want to proceed.

We have learned a fair amount about the heat equation at this point, even path integral representations for its solutions, and we would like to use the hint provided by the formal substitution $t \to -it$ to build a rigorous bridge from what we know to what we would like to know. Here is the plan. We begin by fixing a potential function $V : \mathbb{R} \to \mathbb{R}$. As we did earlier in our proof of the Feynman–Kac formula we

would like to illustrate the idea with minimal technical difficulties by assuming that
V is *continuous and bounded*. This is much more restrictive than necessary and physically uninteresting, but the idea is the same for the proof of the more general result
that assumes only that V satisfies the conditions specified in Theorem 9.2.8 (see Theorem 13.3.1 of [JL]). We will also return to the system of units in which $\hbar = 1$ and
again take $m = \frac{1}{2}$. Given these assumptions we conclude from Theorem 9.2.4 that
$H = H_0 + V = -\Delta + V$ is self-adjoint on $\mathcal{D}(\Delta)$. Moreover, since we have assumed that V
is continuous and since every $x \in C_0[0, t]$ is continuous by definition, the expression

$$e^{-\int_0^t V(x(s)+q)\, ds}$$

if defined for every x in $C_0[0, t]$ and every $q \in \mathbb{R}$. Now, the idea is to consider the right-
hand side of the Feynman–Kac formula (9.31), for which we will introduce the new
symbol

$$(J_V(t)\psi)(q) = \int_{C_0[0,t]} e^{-\int_0^t V(x(s)+q)\, ds}\psi(x(t) + q)\, d\mathfrak{m}_t(x), \qquad (9.36)$$

where ψ is an arbitrary element of $L^2(\mathbb{R})$. For each $t > 0$ and almost every $q \in \mathbb{R}$
we have, by the Feynman–Kac formula, $(J_V(t)\psi)(q) = (e^{t(\Delta-V)}\psi)(q)$, or, more simply,
$J_V(t) = e^{t(\Delta-V)}$. Our "hint" above suggests that we try to extend $J_V(t)$ as an operator-
valued function of t to the case in which t is pure imaginary. Needless to say, just
making the formal substitution suggested above in the Wiener integral is meaningless,
but we will now see how to do this extension rigorously.

Remark 9.4.1. We would like to perform an analytic continuation much in the spirit of
classical complex analysis (Chapter 16 of [Roy]) except that the mapping we will want
to continue analytically takes values in the Banach space $\mathcal{B}(L^2(\mathbb{R}))$ of bounded linear
operators on $L^2(\mathbb{R})$. We must therefore define what we mean by analyticity in this context. There are three natural choices for such a definition depending on the topology
one chooses for $\mathcal{B}(L^2(\mathbb{R}))$, but, as it happens, these all give rise to the same notion of
analyticity (Sections 3.10–3.14 of [HP] contain the generalization of classical complex
analysis to vector- and operator-valued functions). We will formulate the definition in
the following way. Let D be a domain (connected, open set) in the complex plane $\mathbb{C}$
and suppose f is a mapping from D to $\mathcal{B}(L^2(\mathbb{R}))$. Then f is said to be *analytic* on D if
$z \in D \to \langle \psi, f(z)\psi \rangle$ is an ordinary $\mathbb{C}$-valued analytic map on D for every $\psi \in L^2(\mathbb{R})$.

The first step is to show that, because we have assumed that V is bounded, each
exponential $e^{t(\Delta-V)}$, $t > 0$, is a *bounded* operator on $L^2(\mathbb{R})$; once this is done we can
regard $t \to J_V(t) = e^{t(\Delta-V)}$ as a map from $(0, \infty)$ to $\mathcal{B}(L^2(\mathbb{R}))$ and we can try to analytically continue it to a map from $\mathbb{C}_+ = \{z \in \mathbb{C} : \mathrm{Re}(z) > 0\}$ to $\mathcal{B}(L^2(\mathbb{R}))$. For this analytic
continuation we will actually need, and will now prove, a bit more.

Recall that $-\Delta$ is a positive, self-adjoint operator so $\langle \psi, -\Delta\psi \rangle \geq 0$ for all $\psi \in \mathcal{D}(\Delta)$. Furthermore, since V is real-valued and bounded, $\langle \psi, V\psi \rangle \geq m \|\psi\|^2$, where $m = \inf_{q \in \mathbb{R}} V(q)$. Consequently,

$$\langle \psi, (-\Delta + V)\psi \rangle \geq m \|\psi\|^2.$$

A symmetric operator $A : \mathcal{D}(A) \to \mathcal{H}$ on a Hilbert space $\mathcal{H}$ is said to be *semi-bounded*, or *bounded from below*, if, for some $m \in \mathbb{R}$, $\langle \psi, A\psi \rangle \geq m \|\psi\|^2$ for all $\psi \in \mathcal{D}(A)$; equivalently, $\langle \psi, (m - A)\psi \rangle \leq 0 \ \forall \psi \in \mathcal{D}(A)$.

Lemma 9.4.1. *Let $A : \mathcal{D}(A) \to \mathcal{H}$ be a self-adjoint, semi-bounded operator on the Hilbert space $\mathcal{H}$ with $\langle \psi, A\psi \rangle \geq m \|\psi\|^2 \ \forall \psi \in \mathcal{D}(A)$. Then the spectrum $\sigma(A)$ of A is contained in $[m, \infty)$.*

Proof. Since A is self-adjoint, $\sigma(A) \subseteq \mathbb{R}$. We will show that an $x < m$ cannot be in the spectrum of A. Let $\varepsilon = m - x$. Then $\varepsilon > 0$ and, for any $\psi \in \mathcal{D}(A)$,

$$\|(x - A)\psi\|^2 = \langle (m - A)\psi - \varepsilon\psi, (m - A)\psi - \varepsilon\psi \rangle$$
$$= \varepsilon^2 \|\psi\|^2 + \|(m - A)\psi\|^2 - \varepsilon\langle \psi, (m - A)\psi \rangle - \varepsilon\langle (m - A)\psi, \psi \rangle$$
$$\geq \varepsilon^2 \|\psi\|^2$$

since the last two terms are greater than or equal to zero. From this it follows that $x - A$ is injective and, for any φ in the image of $x - A$,

$$\|(x - A)^{-1}\varphi\| \leq \frac{1}{\varepsilon^2} \|\varphi\|,$$

so $(x - A)^{-1}$ is bounded. Thus, x is in neither the point spectrum nor the continuous spectrum of A. Since A is self-adjoint, its residual spectrum is empty and we conclude that $x \notin \sigma(A)$, so $\sigma(A) \subseteq [m, \infty)$. $\qquad\square$

Applying this to $-\Delta + V$ we find that $\sigma(-\Delta + V) \subseteq [m, \infty)$, where $m = \inf_{q \in \mathbb{R}} V(q)$.

Exercise 9.4.1. Show that this implies that for each fixed $z \in \mathbb{C}_+ = \{z \in \mathbb{C} : \text{Re}(z) \geq 0\}$, the function $f_z : \sigma(-\Delta + V) \to \mathbb{C}$ defined by $f_z(u) = e^{-zu}$ is bounded.

It therefore follows from the functional calculus (Theorem 5.5.7, Part 2.) that the operator

$$e^{-z(-\Delta+V)} = e^{z(\Delta-V)}$$

is bounded for every $z \in \mathbb{C}_+$. In particular, $t \to J_V(t) = e^{t(\Delta - V)}$ is a map from $(0, \infty)$ to $\mathcal{B}(L^2(\mathbb{R}))$. Now, if we can analytically extend this map to pure imaginary values of t, then, by the Feynman–Kac formula, we have extended the Wiener integral on the right-hand side of (9.31). If this does what the Feynman integral is supposed to do (and we will see that it does), then we can regard this extension as a rigorous definition of the Feynman integral.

An analytic continuation is always easier to find if you know in your heart what the extension should be in advance and, in this case, we certainly have a reasonable candidate for the extension of $e^{t(\Delta-V)}$ to $\mathbb{C}_+$, namely, $e^{z(\Delta-V)}$ for $z \in \mathbb{C}_+$. We will show that $z \in \mathbb{C}_+ \rightarrow e^{z(\Delta-V)} \in \mathcal{B}(L^2(\mathbb{R}))$ is analytic, but we will also need a continuity condition on the imaginary axis since it is $e^{it(\Delta-V)}$ that we are really after. We will show that the map $z \rightarrow e^{z(\Delta-V)}$ is:

1. strongly continuous on $\overline{\mathbb{C}}_+$ and
2. analytic on $\mathbb{C}_+$.

The strong continuity is a simple consequence of the functional calculus (specifically, Theorem 5.5.7, Part 5.). To see this, let $\{z_n\}_{n=1}^{\infty}$ be a sequence in $\overline{\mathbb{C}}_+$ converging to z in $\overline{\mathbb{C}}_+$. Then $\{f_{z_n}(u)\}_{n=1}^{\infty} = \{e^{-z_n u}\}_{n=1}^{\infty}$ converges to e^{-zu} for each u in $\sigma(\Delta - V)$.

Exercise 9.4.2. Show that the sequence $\{\|e^{-z_n u}\|_{\infty}\}_{n=1}^{\infty}$ is bounded, where $\|e^{-z_n u}\|_{\infty} = \sup\{|e^{-z_n u}| : u \in \sigma(\Delta - V)\}$.

By Theorem 5.5.7.5, $e^{z_n(\Delta-V)}$ converges strongly to $e^{z(\Delta-V)}$, so $z \rightarrow e^{z(\Delta-V)}$ is strongly continuous on $\overline{\mathbb{C}}_+$.

Next we show that $z \rightarrow e^{z(\Delta-V)}$ is analytic on $\mathbb{C}_+$. According to Remark 9.4.1 we must show that for each fixed ψ in $L^2(\mathbb{R})$,

$$z \in \mathbb{C}_+ \rightarrow \langle \psi, e^{z(\Delta-V)}\psi\rangle \in \mathbb{C}$$

is an analytic complex-valued function of a complex variable. It will suffice to prove this when $\|\psi\| = 1$. Now, since $-\Delta + V$ is self-adjoint on $\mathcal{D}(\Delta)$, it has, by the Spectral Theorem 5.5.6, an associated projection-valued measure E on $\mathbb{R}$ and, from this, a corresponding resolution of the identity $\{E_\lambda\}_{\lambda \in \mathbb{R}}$. Together, E and ψ determine a probability measure $\langle \psi, E\psi\rangle$ which, by (5.34), is concentrated on the spectrum $\sigma(-\Delta + V)$. Moreover, by the functional calculus (Theorem 5.5.7),

$$\langle \psi, e^{-z(-\Delta+V)}\psi\rangle = \int_{\mathbb{R}} e^{-z\lambda}d\langle \psi, E_\lambda\psi\rangle = \int_{m}^{\infty} e^{-z\lambda}d\langle \psi, E_\lambda\psi\rangle,$$

since $\sigma(-\Delta + V) \subseteq [m, \infty)$, where $m = \inf_{q\in\mathbb{R}} V(q)$. Since we have already shown that $z \rightarrow e^{z(\Delta-V)}$ is strongly continuous,

$$z \rightarrow \langle \psi, e^{z(\Delta-V)}\psi\rangle \qquad\qquad (9.37)$$

is a continuous complex-valued function of complex variable. Now, to prove analyticity we will apply Morera's theorem, which we now recall in the form stated in Theorem 10.17 of [Rud2].

Theorem 9.4.2 (Morera's theorem). *Let D be a domain in the complex plane $\mathbb{C}$ and $f : D \to \mathbb{C}$ a continuous function. If*

$$\int_{\partial\Gamma} f(z)dz = 0$$

for every closed triangular region Γ in D, then f is analytic on D ($\partial\Gamma$ is the simple closed contour that is the boundary of Γ).

Thus, we need to show that for any closed triangular region Γ in $\mathbb{C}_+$, the integral

$$\int_{\partial\Gamma} \langle\psi, e^{-z(-\Delta+V)}\psi\rangle \, dz = \int_{\partial\Gamma} \left[\int_m^\infty e^{-z\lambda} d\langle\psi, E_\lambda\psi\rangle \right] dz \tag{9.38}$$

is zero. Note that if we could justify interchanging the order of integration, then, since the Cauchy integral theorem gives

$$\int_{\partial\Gamma} e^{-z\lambda} dz = 0$$

for each λ, the result would follow. For this we need to verify the applicability of Fubini's theorem. To be clear we will record the form of Fubini's theorem that we need. The following is Theorem 8.8 (c) of [Rud2].

Theorem 9.4.3 (Fubini's theorem). *Let $(X, \mathcal{A}_X, \mu_X)$ and $(Y, \mathcal{A}_Y, \mu_Y)$ be σ-finite measure spaces and let $f : X \times Y \to \mathbb{C}$ be a complex-valued L^1 function on the product measure space $(X \times Y, \mathcal{A}_{X\times Y}, \mu_X \times \mu_Y)$. Let $f_x : Y \to \mathbb{C}$ and $f^y : X \to \mathbb{C}$ be given by $f_x(y) = f(x,y)$ and $f^y(x) = f(x,y)$. Then $f_x \in L^1(Y, \mu_Y)$ for μ_X-almost all $x \in X$ and $f^y \in L^1(X, \mu_X)$ for μ_Y-almost all $y \in Y$. Define*

$$\varphi(x) = \int_Y f_x(y) \, d\mu_Y(y)$$

for μ_X-almost all $x \in X$ and

$$\psi(y) = \int_X f^y(x) \, d\mu_X(x)$$

for μ_Y-almost all $y \in Y$. Then $\varphi \in L^1(X, \mu_X)$, $\psi \in L^1(Y, \mu_Y)$ and

$$\int_X \varphi(x) \, d\mu_X(x) = \int_{X\times Y} f(x,y) \, d(\mu_X \times \mu_Y)(x,y) = \int_Y \psi(y) \, d\mu_Y(y).$$

In particular,

$$\int_X \left[\int_Y f_x(y) \, d\mu_Y(y) \right] d\mu_X(x) = \int_Y \left[\int_X f^y(x) \, d\mu_X(x) \right] d\mu_Y(y).$$

The most obvious difficulty we have in applying Fubini's theorem to (9.38) is that we do not (yet) have two measure spaces. The inner integral is fine, being an integral over $[m, \infty)$ with respect to a probability measure on $\mathbb{R}$. The outer integral, however, is a contour integral in the complex plane, so we will need to see if this can be regarded as a Lebesgue integral. This is indeed possible, but to do so one requires some properties of the Lebesgue–Stieltjes integral. In Appendix H.3 it is shown that the contour integral over $\partial \Gamma$ can be written as a sum of Lebesgue–Stieltjes integrals, so (9.38) is a sum of iterated integrals, the first integration being with respect to the probability measure $d\langle \psi, E_\lambda \psi \rangle$ and the second being with respect to some Lebesgue–Stieltjes measure da which is a finite measure on each edge of the triangle. It will therefore suffice to justify the application of Fubini's theorem to each of these integrals. For each of these, Fubini's theorem requires that the integrand be an L^1 function with respect to the product measure. Now, we have already seen that for each fixed z in $\overline{\mathbb{C}}_+$, $e^{-z\lambda}$ is bounded on $[m, \infty)$. Moreover, since the triangle Γ is bounded away from $\mathbb{C}_- = \{z \in \mathbb{C} : \mathrm{Re}(z) < 0\}$, $e^{-z\lambda}$ is also bounded on $\partial \Gamma$ for each fixed λ. Since a bounded, continuous function on a finite measure space is integrable, $e^{-z\lambda}$ is integrable on each product measure space and the result follows. This completes the proof that $z \to \langle \psi, e^{z(\Delta - V)} \psi \rangle$ is analytic on $\mathbb{C}_+$.

We now have an extension of the operator-valued function $J_V(t)$, defined on $t > 0$ by (9.36), to $\overline{\mathbb{C}}_+$ that is strongly continuous on $\overline{\mathbb{C}}_+$ and analytic on $\mathbb{C}_+$. Since analytic functions on $\mathbb{C}_+$ are uniquely determined by their values on $(0, \infty)$ and since continuity then uniquely determines the extension to $\overline{\mathbb{C}}_+$, this extension is the only one with these properties. Furthermore, the extension is given explicitly by $J_V(z) = e^{-z(-\Delta + V)}$ for $z \in \overline{\mathbb{C}}_+$. In particular, for $z = it$, $t \geq 0$, we have

$$ J_V(it) = e^{-it(-\Delta + V)} = e^{it(\Delta - V)}, $$

and this, according to our Postulate QM4 (Section 6.2), is the unitary time evolution operator for a quantum system with Hamiltonian $H = -\Delta + V$ (in units for which $\hbar = 1$ and with $m = \frac{1}{2}$), so, for any initial state ψ, $J_V(t)\psi$ is a solution to the initial value problem for the corresponding abstract Schrödinger equation (5.41). This is, of course, the same result we obtained from the Trotter product formula, which gave rise to the Feynman integral in the first place. Since $J_V(it)$ is the analytic continuation of the Wiener integral in (9.36) to the imaginary axis and since it describes the time evolution of a quantum state, we shall refer to it as the *analytic-in-time operator-valued Feynman integral*.

We should mention once again that in this section we have made extremely restrictive assumptions about the potential V in order to lay bare the underlying ideas, but that the result we have arrived at can be proved under much milder hypotheses (see Theorem 13.3.1 of [JL]).

We will conclude this section by noting that analytic continuation *in time* is not the only approach one might have taken here to build a rigorous bridge between the heat equation and the Schrödinger equation. In his rather remarkable paper [Nel3], Edward Nelson analytically continued *in the mass parameter* and was able to obtain results similar to those we have described here, but for a quite different family of potentials that included some that are highly singular and much closer to the needs of the physicists. For this one should consult [Nel3] directly, but it is also discussed in some detail in Section 13.5 of [JL].

10 Fermionic and supersymmetric harmonic oscillators

10.1 The Stern–Gerlach experiment and spin one-half

Long ago we conceded that our initial foray into the quantum mechanics of particles such as the electron was incomplete in that we consciously suppressed a critical aspect of their behavior known as *spin*. The time has come now to do what we can to remedy this. In truth, we cannot do all that we would like to do because spin is a concept that lives most naturally in the context of relativistic quantum mechanics (specifically, the Dirac equation). Nevertheless, we will try to provide some sense of what this phenomenon is and how the physicists have incorporated it into their mathematical model of the quantum world. In this section we will first briefly describe the famous *Stern–Gerlach experiment*, in which this very strange behavior was first observed. The experiment (first performed in 1922) was *not* originally designed to observe spin. Indeed, the notion of spin was not introduced until 1925 (by Uhlenbeck and Goudsmit). While the historical development of these ideas makes for an interesting story, it is a bit convoluted and, we feel, could only distract us from our purpose here, so we will not discuss it (if you are interested in this sort of thing you might consult Section IV.3 of [MR]).

We should be clear at the outset. There is nothing like quantum mechanical spin in classical physics. This behavior is new, bizarre and wholly quantum mechanical. There is, however, a classical *analogy*. The analogy is inadequate and can be misleading if taken too seriously, but it is the best we can do, so we will begin by describing it.

Imagine a spherical mass m of radius a moving through space on a circular orbit of radius $R \gg a$ about some point O and, at the same time, spinning around an axis through one of its diameters (to a reasonable approximation, the earth does all of this). Due to its orbital motion, the mass has an angular momentum $\mathbf{L} = \mathbf{r} \times (m\mathbf{v}) = \mathbf{r} \times \mathbf{p}$ (see (2.20)), which we will now call its *orbital angular momentum*. The spinning of the mass around its axis contributes additional angular momentum that one calculates by subdividing the spherical region occupied by the mass into subregions, regarding each subregion as a mass in a circular orbit about a point on the axis, approximating its angular momentum, adding all of these and taking the limit as the regions shrink to points. The resulting integral gives the angular momentum due to rotation. This is called the *rotational angular momentum*, is denoted $\mathbf{S}$ and is given by

$$\mathbf{S} = I\boldsymbol{\omega},$$

where I is the moment of inertia of the sphere and $\boldsymbol{\omega}$ is the angular velocity ($\boldsymbol{\omega}$ is along the axis of rotation in the direction determined by the right-hand rule from the direction of the rotation). If the mass is assumed to be uniformly distributed throughout

https://doi.org/10.1515/9783110751949-010

the sphere (in other words, if the sphere has constant density), then an exercise in calculus gives

$$\mathbf{S} = \frac{2}{5}ma^2\boldsymbol{\omega}.$$

The *total angular momentum* of the sphere is $\mathbf{L} + \mathbf{S}$.

Now let us suppose, in addition, that the sphere is charged. Due to its orbital motion the charged sphere behaves like a current loop. As we saw in Section 4.2, Maxwell's equations imply that moving charges give rise to magnetic fields. If we assume that our current loop is very small (or, equivalently, that we are viewing it from a distance) the corresponding magnetic field is that of a *magnetic dipole* (see Sections 14-5 and 34-2, Volume II, of [FLS]). All we need to know about this is that this magnetic dipole is described by a vector $\boldsymbol{\mu}_\mathbf{L}$ called its *orbital magnetic moment* that is proportional to the orbital angular momentum. Specifically,

$$\boldsymbol{\mu}_\mathbf{L} = \frac{q}{2m}\mathbf{L},$$

where q is the charge of the sphere (which can be positive or negative). Similarly, the rotational angular momentum of the charge gives rise to a magnetic field that is also that of a magnetic dipole and is described by a *rotational magnetic moment* $\boldsymbol{\mu}_\mathbf{S}$ given by

$$\boldsymbol{\mu}_\mathbf{S} = \frac{q}{2m}\mathbf{S}. \tag{10.1}$$

The *total magnetic moment* $\boldsymbol{\mu}$ is

$$\boldsymbol{\mu} = \boldsymbol{\mu}_\mathbf{L} + \boldsymbol{\mu}_\mathbf{S} = \frac{q}{2m}(\mathbf{L} + \mathbf{S}).$$

The significance of the magnetic moment $\boldsymbol{\mu}$ of the dipole is that it describes the strength and direction of the dipole field and determines the torque

$$\boldsymbol{\tau} = \boldsymbol{\mu} \times \mathbf{B}$$

experienced by the magnetic dipole when placed in an external magnetic field $\mathbf{B}$. If the magnetic field $\mathbf{B}$ is uniform (that is, constant), then its only effect on the dipole is to force $\boldsymbol{\mu}$ to precess around a cone whose axis is along $\mathbf{B}$ in the same way that the axis of a spinning top precesses around the direction of the earth's gravitational field (see Figure 10.1 and Section 2, Chapter 11, of [Eis]). Note that this precession does not change the projection $\boldsymbol{\mu} \cdot \mathbf{B}$ of $\boldsymbol{\mu}$ along $\mathbf{B}$.

If the $\mathbf{B}$-field is not uniform, however, there will be an additional translational force acting on the mass which, if m is moving through the field, will push it off course. Precisely what this deflection will be depends, of course, on the nature of $\mathbf{B}$, and we will say a bit more about this in a moment.

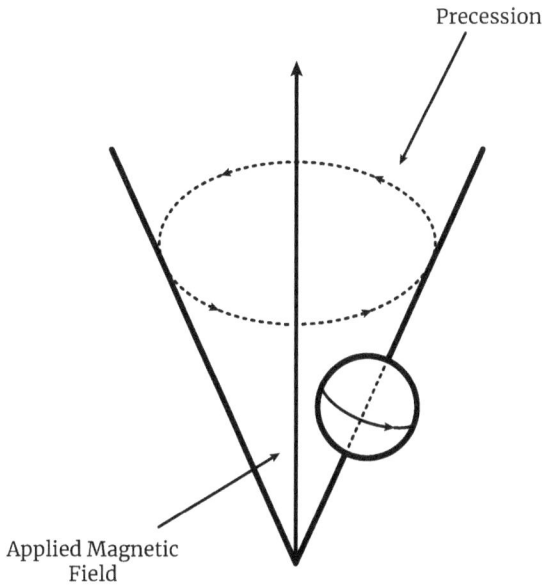

Precession

Applied Magnetic
Field

Figure 10.1: Precession. Reproduced from author's book with the
permission of Springer.

Now we can begin our discussion of the Stern–Gerlach experiment (a schematic is shown in Figure 10.2). We are interested in whether or not the electron has a rotational magnetic moment and, if so, whether or not its behavior is adequately described by classical physics. What we will do is send a certain beam of electrically neutral atoms through a nonuniform magnetic field **B** and then let them hit a photographic plate to record how their paths were deflected by the field. The atoms must be electrically neutral so that the deflections due to the charge (of which we are already aware) do not mask any deflections due to magnetic moments of the atoms. In particular, we cannot do this with free electrons. The atoms must also have the property that any magnetic moment they might have could be due only to a single electron somewhere within it. Stern and Gerlach chose atoms of silver (Ag) which they obtained by evaporating the metal in a furnace and focusing the resulting gas of Ag atoms into a beam aimed at a magnetic field.

Silver is a good choice, but for reasons that are not so apparent. A proper explanation requires some hindsight (not all of the information was available to Stern and Gerlach) as well as some quantum mechanical properties of atoms that we have not discussed here. Nevertheless, it is worth saying at least once since otherwise one is left with all sorts of unanswered questions about the validity of the experiment. So, here it is. The stable isotopes of Ag have 47 electrons, 47 protons and either 60 or 62 neutrons, so, in particular, they are electrically neutral. Since the magnetic moment is inversely proportional to the mass and since the masses of the proton and neutron are each approximately 2000 times the mass of the electron, one can assume that any magnetic moments of the nucleons will have a negligible effect on the magnetic mo-

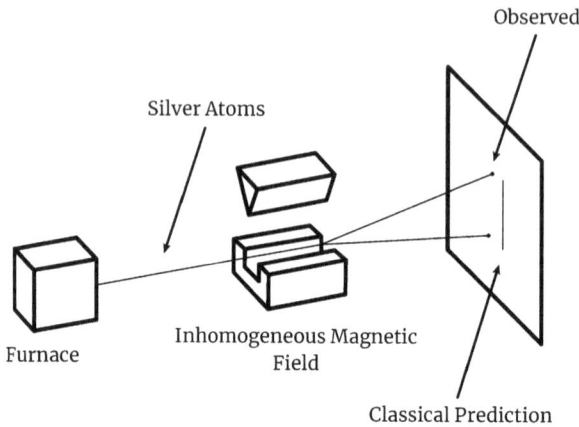

Figure 10.2: Stern–Gerlach experiment. Reproduced from author's book with the permission of Springer.

ment of the atom and can therefore be ignored. Of the 47 electrons, 46 are contained in closed, inner shells (energy levels), and these, it turns out, can be represented as a spherically symmetric cloud with no orbital or rotational angular momentum (this is not at all obvious). The remaining electron is in what is termed the outer 5s-shell, and an electron in an s-state has no orbital angular momentum (again, not obvious). Granting all of this, the only possible source of any magnetic moment for a Ag atom is a rotational angular momentum of its outer 5s-electron. Whatever happens in the experiment is attributable to the electron and the rest of the silver atom is just a package designed to ensure this.

We will first see what the classical picture of an electron with a rotational magnetic moment would lead us to expect in the Stern–Gerlach experiment and will then describe the results that Stern and Gerlach actually obtained (a more thorough, but quite readable account of the physics is available Chapter 11 of [Eis]). For this we will need to be more specific about the magnetic field $\mathbf{B}$ that we intend to send the Ag atoms through. Let us introduce a coordinate system in Figure 10.2 in such a way that the Ag atoms move in the direction of the y-axis and the vertical axis of symmetry of the magnet is along the z-axis, so that the x-axis is perpendicular to the paper. The magnet itself can be designed to produce a field that is nonuniform, but does not vary with y, is predominantly in the z-direction and is symmetric with respect to the yz-plane. The interaction between the neutral Ag atom (with magnetic moment $\boldsymbol{\mu}$) and the nonuniform magnetic field $\mathbf{B}$ provides the atom with a potential energy $-\boldsymbol{\mu} \cdot \mathbf{B}$ so that the atom experiences a force

$$\mathbf{F} = \nabla(\boldsymbol{\mu} \cdot \mathbf{B}) = \nabla(\mu_x B_x + \mu_y B_y + \mu_z B_z). \tag{10.2}$$

For the sort of magnetic field we have just described, $B_y = 0$ and B_z dominates B_x. From this one finds that the translational motion is governed primarily by

$$F_z \approx \mu_z \frac{\partial B_z}{\partial z} \tag{10.3}$$

(see pages 333–334 of [Eis]). The conclusion we draw from this is that the displacements from the intended path of the silver atoms will be in the z-direction (up and down in Figure 10.2) and the forces causing these displacements are proportional to the z-component of the magnetic moment. Of course, different orientations of the magnetic moment $\boldsymbol{\mu}$ among the various Ag atoms will lead to different values of μ_z and therefore to different displacements. Moreover, due to the random thermal effects of the furnace, one would expect that the silver atoms exit with their magnetic moments $\boldsymbol{\mu}$ randomly oriented in space so that their z-components could take on any value in the interval $[-|\boldsymbol{\mu}|, |\boldsymbol{\mu}|]$. As a result, the expectation based on classical physics would be that the deflected Ag atoms will impact the photographic plate at points that cover an entire vertical line segment (see the segment labeled "Classical prediction" in Figure 10.2). Note that writing $q = -e$ for the charge of the electron and $m = m_e$ for its mass, we find

$$F_z \approx \mu_z \frac{\partial B_z}{\partial z} = -\frac{e}{2m_e} S_z \frac{\partial B_z}{\partial z},$$

so that the deflection of an individual Ag atom is a measure of the component S_z of $\mathbf{S}$ in the direction of the magnetic field gradient.

This, however, is not at all what Stern and Gerlach observed. What they found was that the silver atoms arrived at the screen at *only two points*, one above and one the same distance below the y-axis (again, see Figure 10.2). The experiment was repeated with different orientations of the magnet (that is, different choices for the z-axis) and different atoms and *nothing* changed. We seem to be dealing with a very peculiar sort of "vector" $\mathbf{S}$. The classical picture would have us believe that, however it is oriented in space, its projection onto *any* axis is always one of two things. Needless to say, ordinary vectors in $\mathbb{R}^3$ do not behave this way. What we are really being told is that the classical picture is simply *wrong*. The property of electrons that manifests itself in the Stern–Gerlach experiment is in some ways analogous to what one would expect classically of a small charged sphere rotating about some axis, but the analogy can only be taken so far. It is, for example, not possible to make an electron "spin faster (or slower)" to alter the length of its projection onto an axis. This is always the same; it is a characteristic feature of the electron. What we are dealing with is an intrinsic property of the electron that does not depend on its motion (or anything else); for this reason it is often referred to as the *intrinsic angular momentum* of the electron, but, unlike its classical counterpart, it is *quantized*.

Not only the electron, but every particle (elementary particle, atom, molecule, etc.) in quantum mechanics is supplied with some sort of intrinsic angular momentum. Although we will make no serious use of this, we will need some of the terminology, so we will briefly describe the general situation here (for more details see, for example, Chapter 11 of [Eis], or Chapters 14 and 17 of [Bohm]). The basic idea is that these particles exhibit behaviors that mimic what one would expect of angular momentum, but that cannot be accounted for by any motion of the particle. To quantify these behaviors every particle is assigned a *spin quantum number s*. The allowed values of s are

$$0, \frac{1}{2}, 1, \frac{3}{2}, 2, \frac{5}{2}, \ldots, \frac{n-1}{2}, \ldots, \tag{10.4}$$

where $n = 1, 2, 3, 4, 5, \ldots$. Intuitively, one might think of n as the number of dots that appear on the photographic plate if a beam of such particles is sent through a Stern–Gerlach apparatus. According to this scheme an electron has *spin* $\frac{1}{2}$ ($n = 2$). Particles with half-integer spin $\frac{1}{2}, \frac{3}{2}, \frac{5}{2}, \ldots$ are called *fermions*, while those with integer spin $0, 1, 2, \ldots$ are called *bosons*. We will eventually see that fermions and bosons have very different properties and play very different roles in particle physics. Among the fermions, particles of spin $\frac{1}{2}$ are by far the principal players. Indeed, one must look long and hard to find a fermion of higher spin. The best known examples are the so-called Δ *baryons*, which have spin $\frac{3}{2}$, but you dare not blink if you are looking for one of these since their mean lifetime is about 5.63×10^{-24} seconds. Among the bosons, the photon has spin 1 and the very recently observed Higgs boson has spin 0, whereas the conjectured, but not yet observed graviton has spin 2.

We have seen that the classical vector **S** used to describe the rotational angular momentum does not travel well into the quantum domain, where it simply does not behave the way one expects a vector to behave. Nevertheless, it is still convenient to collect together the quantities S_x, S_y and S_z, measured, for example, by a Stern–Gerlach apparatus aligned along the x-, y- and z-axes, and refer to the triple

$$\mathbf{S} = (S_x, S_y, S_z)$$

as the *spin vector*. Quantum theory decrees that for a particle with spin quantum number s, the only allowed values for the "components" S_x, S_y and S_z are

$$-s\hbar, \; -(s-1)\hbar, \; \ldots, (s-1)\hbar, \; s\hbar.$$

In particular, for a spin $\frac{1}{2}$ particle such as the electron there are only two possible values, for example,

$$S_z = \pm\frac{\hbar}{2},$$

and these correspond to the two dots in our Stern–Gerlach experiment. As in the classical case one can associate a *spin magnetic moment* $\boldsymbol{\mu_S}$ to each spin vector **S**, but the classical definition requires an adjustment. For the electron this is given by

$$\boldsymbol{\mu_S} = g_e\left(\frac{-e}{2m_e}\right)\mathbf{S}, \tag{10.5}$$

where $-e$ is the charge of the electron, m_e is the mass of the electron and g_e is a dimensionless constant called the *electron spin g-factor*. As it happens, g_e is the most accurately measured constant in physics, with a value of approximately

$$g_e \approx 2.00231930419922 \pm \left(1.5 \times 10^{-12}\right)$$

(see [OHUG] for more on the measurement of g_e). The Dirac equation predicts a value of $g_e = 2$ and the corrections are accounted for by quantum electrodynamics. This is one of the reasons that physicists have such confidence in quantum electrodynamics, and quantum field theory in general, despite the fact that they do not have the sort of rigorous mathematical foundations that mathematicians would like to see.

With this synopsis of the general situation behind us we will return the particular case of spin $\frac{1}{2}$ and, still more particularly, to the electron. We know that the classical picture of the electron as a tiny spinning ball cannot describe what is actually observed, so we must look for another picture that can do this. We pointed out at the beginning of this section that the correct picture is to be found in relativistic quantum mechanics and the Dirac equation. What we will describe now is the nonrelativistic precursor of Dirac's theory due to Wolfgang Pauli [Pauli1].

Whatever this picture is, it must be a quantum mechanical one, so we are looking for a Hilbert space $\mathcal{H}$ and some self-adjoint operators on it to represent the observables S_x, S_y and S_z. Previously we represented the state of the electron by a wave function $\psi(x, y, z)$ that is in $L^2(\mathbb{R}^3)$, but we now know that the state of a spin $\frac{1}{2}$ particle must depend on more than just x, y, and z since these alone cannot tell us which of the two paths an electron is likely to follow in a Stern–Gerlach apparatus; we say "likely to" because we can no longer hope to know more than probabilities. What we would like to do is isolate some appropriate notion of the "spin state" of the particle that will provide us with the information we need to describe these probabilities. Now, we know that the only possible values of S_z are $\pm\frac{\hbar}{2}$. By analogy with the classical situation one might view this as saying that the spin vector **S** can only be either "up" or "down," but nothing in-between. This suggests that we consider wave functions

$$\psi(x, y, z, \sigma) \tag{10.6}$$

that depend on x, y, z and an additional discrete variable σ that can take only two values, say, $\sigma = 1$ and $\sigma = 2$ (or, if you prefer, $\sigma = $ up and $\sigma = $ down). Then $|\psi(x, y, z, 1)|^2$ would represent the probability density for locating the electron at (x, y, z) with $S_z = \frac{\hbar}{2}$ and similarly $|\psi(x, y, z, 2)|^2$ is the probability density for locating the electron at (x, y, z) with $S_z = -\frac{\hbar}{2}$. Stated this way it sounds a little strange, but note that this is precisely the same as describing the state of the electron with *two functions* $\psi_1(x, y, z) = \psi(x, y, z, 1)$ and $\psi_2(x, y, z) = \psi(x, y, z, 2)$ and this is what we will do. Specifically, we will identify the wave function of a spin $\frac{1}{2}$ particle with a (column) vector

$$\begin{pmatrix} \psi_1(x, y, z) \\ \psi_2(x, y, z) \end{pmatrix},$$

where ψ_1 and ψ_2 are in $L^2(\mathbb{R}^3)$ and

$$\int_{\mathbb{R}^3} \left(|\psi_1(x, y, z)|^2 + |\psi_2(x, y, z)|^2 \right) d\mu = 1$$

because the probability of finding the electron somewhere with either $S_z = \frac{\hbar}{2}$ or $S_z = -\frac{\hbar}{2}$ is 1. The Hilbert space is therefore $\mathcal{H} = L^2(\mathbb{R}^3) \oplus L^2(\mathbb{R}^3)$.

Now we must isolate self-adjoint operators on $\mathcal{H}$ to represent the observables S_x, S_y and S_z. Since these observables represent an intrinsic property of a spin $\frac{1}{2}$ particle, independent of x, y and z, we will want the operators to act only on the spin coordinates 1 and 2 and the action should be constant in (x, y, z). Thus, we are simply looking for 2×2 complex, self-adjoint (that is, Hermitian) matrices. Since the only possible observed values are $\pm\frac{\hbar}{2}$, these must be the eigenvalues of each matrix. There are, of course, many such matrices floating around, and we must choose three of them. The motivation for our choice is based on the following exercise and the fact that S_x, S_y and S_z correspond to measurements made along the directions of an *oriented, orthonormal basis for* $\mathbb{R}^3$.

Exercise 10.1.1. Denote by $\mathcal{R}^3$ the set of all 2×2 complex, Hermitian matrices with trace zero.

1. Show that every $X \in \mathcal{R}^3$ can be uniquely written as

$$X = \begin{pmatrix} x^3 & x^1 - ix^2 \\ x^1 + ix^2 & -x^3 \end{pmatrix} = x^1\sigma_1 + x^2\sigma_2 + x^3\sigma_3,$$

where x^1, x^2 and x^3 are real numbers and

$$\sigma_1 = \begin{pmatrix} 0 & 1 \\ 1 & 0 \end{pmatrix}, \quad \sigma_2 = \begin{pmatrix} 0 & -i \\ i & 0 \end{pmatrix}, \quad \sigma_3 = \begin{pmatrix} 1 & 0 \\ 0 & -1 \end{pmatrix}$$

are the so-called *Pauli spin matrices*.

2. Show that, with the operations of matrix addition and (real) scalar multiplication, $\mathcal{R}^3$ is a three-dimensional, real vector space and $\{\sigma_1, \sigma_2, \sigma_3\}$ is a basis. Consequently, $\mathcal{R}^3$ is linearly isomorphic to $\mathbb{R}^3$. Furthermore, defining an orientation on $\mathcal{R}^3$ by decreeing that $\{\sigma_1, \sigma_2, \sigma_3\}$ is an oriented basis, the map $X \to (x^1, x^2, x^3)$ is an orientation preserving isomorphism when $\mathbb{R}^3$ is given its usual orientation.

3. Show that

$$\sigma_1\sigma_2 = i\sigma_3, \quad \sigma_2\sigma_3 = i\sigma_1, \quad \sigma_3\sigma_1 = i\sigma_2, \quad \sigma_1\sigma_2\sigma_3 = iI,$$

where I is the 2×2 identity matrix.

4. Show that σ_1, σ_2, σ_3 satisfy the following *commutation relations*:

$$[\sigma_1, \sigma_2]_- = 2i\sigma_3, \quad [\sigma_2, \sigma_3]_- = 2i\sigma_1, \quad [\sigma_3, \sigma_1]_- = 2i\sigma_2. \tag{10.7}$$

5. Show that σ_1, σ_2, σ_3 satisfy the following *anticommutation relations*:

$$[\sigma_i, \sigma_j]_+ = 2\delta_{ij}I, \quad i, j = 1, 2, 3, \tag{10.8}$$

where δ_{ij} is the Kronecker delta and $[A, B]_+ = AB + BA$ is the anticommutator.

6. Show that if $X = x^1\sigma_1 + x^2\sigma_2 + x^3\sigma_3$ and $Y = y^1\sigma_1 + y^2\sigma_2 + y^3\sigma_3$, then

$$\frac{1}{2}[X, Y]_+ = (x^1y^1 + x^2y^2 + x^3y^3)I.$$

Conclude that if one defines an inner product $\langle X, Y\rangle_{\mathcal{R}^3}$ on $\mathcal{R}^3$ by

$$\frac{1}{2}[X, Y]_+ = \langle X, Y\rangle_{\mathcal{R}^3} I,$$

then $\{\sigma_1, \sigma_2, \sigma_3\}$ is an *oriented, orthonormal basis for* $\mathcal{R}^3$ and $\mathcal{R}^3$ is isometric to $\mathbb{R}^3$. We will refer to $\mathcal{R}^3$ as the *spin model* of $\mathbb{R}^3$.

7. Regard the matrices σ_1, σ_2, σ_3 as linear operators on $\mathbb{C}^2$ (as a two-dimensional, complex vector space with its standard Hermitian inner product) and show that each of these operators has eigenvalues ± 1 with normalized, orthogonal eigenvectors given as follows:

$$\sigma_1 : \quad \frac{1}{\sqrt{2}}\begin{pmatrix}1\\1\end{pmatrix}, \quad \frac{1}{\sqrt{2}}\begin{pmatrix}1\\-1\end{pmatrix},$$

$$\sigma_2 : \quad \frac{1}{\sqrt{2}}\begin{pmatrix}1\\i\end{pmatrix}, \quad \frac{1}{\sqrt{2}}\begin{pmatrix}1\\-i\end{pmatrix},$$

$$\sigma_3 : \quad \begin{pmatrix}1\\0\end{pmatrix}, \quad \begin{pmatrix}0\\1\end{pmatrix}.$$

This spin model of $\mathbb{R}^3$ contains a great deal of useful information and we will return to it shortly, but first we will use the Pauli spin matrices to define the operators corresponding to the spin components S_x, S_y and S_z. Ordinarily we would use the same symbols to denote the corresponding operators, but it will be much more convenient at this point to denote the operators S_1, S_2 and S_3 and also to opt for coordinates x^1, x^2 and x^3 rather than x, y and z. Specifically, we define

$$S_1 = \frac{\hbar}{2}\sigma_1, \quad S_2 = \frac{\hbar}{2}\sigma_2 \quad \text{and} \quad S_3 = \frac{\hbar}{2}\sigma_3. \tag{10.9}$$

Each of these is clearly Hermitian, and it follows from Exercise 10.1.1.7 that each has eigenvalues $\pm\frac{\hbar}{2}$. Note that, in terms of S_1, S_2 and S_3, the commutation relations (10.7) become

$$[S_1, S_2]_- = i\hbar S_3, \quad [S_2, S_3]_- = i\hbar S_1, \quad [S_3, S_1]_- = i\hbar S_2. \tag{10.10}$$

In particular, these operators do not commute and so, according to our Postulate QM6, no pair of them is simultaneously measurable. This we know imposes uncertainty relations on the measurements of the various spin components. Let us just see what one of these looks like by applying (6.17) to S_3 and S_1 for an electron whose state is described by the two-component wave function

$$\psi = \begin{pmatrix} \psi_1 \\ \psi_2 \end{pmatrix}.$$

We have

$$\sigma_\psi(S_3)\,\sigma_\psi(S_1) \geq \frac{1}{2}\,|\,\langle \psi,\, [S_3, S_1]_-\psi \rangle\,| = \frac{1}{2}\,|\,\langle \psi,\, i\hbar S_2\psi \rangle\,|$$

$$= \frac{1}{2}\,|\,\langle \psi,\, i(\hbar^2/2)\sigma_2\psi \rangle\,|$$

$$= \frac{\hbar^2}{4}\,\left|\,\left\langle \begin{pmatrix} \psi_1 \\ \psi_2 \end{pmatrix},\, \begin{pmatrix} 0 & -i \\ i & 0 \end{pmatrix}\begin{pmatrix} \psi_1 \\ \psi_2 \end{pmatrix} \right\rangle\,\right|$$

$$= \frac{\hbar^2}{4}\,\left|\,\left\langle \begin{pmatrix} \psi_1 \\ \psi_2 \end{pmatrix},\, \begin{pmatrix} -i\psi_2 \\ i\psi_1 \end{pmatrix} \right\rangle\,\right|$$

$$= \frac{\hbar^2}{4}\,|\,\overline{\psi}_1(-i\psi_2) + \overline{\psi}_2(i\psi_1)\,|$$

$$= \frac{\hbar^2}{4}\,|\,2\,\mathrm{Im}\,(\overline{\psi}_1\psi_2)\,|.$$

Thus,

$$\sigma_\psi(S_3)\,\sigma_\psi(S_1) \geq \frac{\hbar^2}{2}\,|\,\mathrm{Im}\,(\overline{\psi}_1\psi_2)\,|.$$

Exercise 10.1.2. Find an expression for the expectation value $\langle S_3 \rangle_\psi$ of S_3 in the state

$$\psi = \begin{pmatrix} \psi_1 \\ \psi_2 \end{pmatrix}.$$

We should point out that Pauli's motivation for the introduction of the specific operators S_1, S_2 and S_3 was not the same as what we have described here. He too was looking for 2×2, complex Hermitian matrices with eigenvalues $\pm\frac{\hbar}{2}$, but instead of looking for a basis for the copy $\mathcal{R}^3$ of $\mathbb{R}^3$ he insisted that the matrices he was after satisfy the commutation relations (10.10). His reason was that these are precisely the same as the commutation relations satisfied by the components of the quantized (orbital) angular momentum (see, for example, Sections 14.2 and 14.3 of [Bohm]) and he sought to keep spin angular momentum and orbital angular momentum on the same formal footing since, classically, they really are the same thing.

Having decided to model the spin state of an electron by a two-component wave function, Pauli [Pauli1] then proposed a Hamiltonian to describe the interaction of this "spinning" electron with an external electromagnetic field and wrote down a corresponding "Schrödinger equation," now generally known as the *Pauli equation*. The solutions to the equation accurately describe the behavior of a nonrelativistic spin $\frac{1}{2}$

charged particle in an electromagnetic field. The equation, however, has two short-comings. The most obvious is that, in general, one cannot always ignore relativistic effects in particle physics. More fundamentally, perhaps, is that in Pauli's treatment of the spin of the electron is put in by hand and one would prefer to see it arise of its own accord from some more fundamental hypothesis (such as relativistic invariance). Both of these issues were beautifully resolved by Dirac [Dirac2] in 1928 (for a brief taste of how this was done we might suggest Section 2.4 of [Nab4]). We will not pursue these matters any further here, but will instead turn to a quite remarkable property of spin $\frac{1}{2}$ particles that will lead us unexpectedly into topology and back to the discussion in Example 2.2.13. One of our objectives is to get some idea of what spinors are and what they have to do with spin.

The result of the Stern–Gerlach experiment is completely independent of the direction in $\mathbb{R}^3$ onto which one chooses to project the spin vector and hence is independent of the choice of oriented, orthonormal basis giving rise to the coordinates x^1, x^2 and x^3. The Pauli Hamiltonian and the corresponding equations of motion therefore need to be invariant under such a change of coordinates, that is, invariant under the action of rotation group SO(3) on the coordinates. This is also true in classical mechanics, of course, and in this context we have actually dealt with rotational invariance when we discussed symmetries of Lagrangians in Section 2.2. Classically the situation is somewhat more straightforward and it is important to understand why, so we will begin with a review of the classical picture.

The mathematical objects used to describe the physical quantities that appear in classical physics (that is, scalars, vectors and tensors) all have perfectly explicit, well-defined transformation laws under a change of coordinates, so one need only substitute these into the basic equations and check that they retain the same form ($\mathbf{F} = m\mathbf{A}$ is a vector equation precisely because both forces and accelerations transform in the same way and masses remain constant). These transformation laws are defined rigorously in a way that we will now describe.

Remark 10.1.1. We will need a few facts about the representations of SO(3). These are described in a bit more detail in Section 2.4 of [Nab4], but for the full story one can consult [Gel].

Recall that a *representation* of the group SO(3) on a finite-dimensional vector space $\mathcal{V}$ is a group homomorphism $T : \mathrm{SO}(3) \to \mathrm{GL}(\mathcal{V})$ of SO(3) into the group of invertible linear operators on $\mathcal{V}$ (which can be identified with a group of invertible matrices once a basis for $\mathcal{V}$ is chosen). In particular,

$$T(R_1 R_2) = T(R_1)T(R_2) \quad \forall R_1, R_2 \in \mathrm{SO}(3),$$

$T(\mathrm{id}_{\mathbb{R}^3}) = \mathrm{id}_{\mathcal{V}}$ and $T(R^{-1}) = T(R)^{-1}$ for every $R \in \mathrm{SO}(3)$. The representation T is said to be *irreducible* if there is no nontrivial linear subspace of $\mathcal{V}$ that is invariant under $T(R)$ for every $R \in \mathrm{SO}(3)$; every representation of SO(3) can be built from irreducible representations. The elements of $\mathcal{V}$ are called *carriers* of the representation and these are used to represent the various physical and geometrical objects of interest. For each $R \in SO(3)$, $T(R) : \mathcal{V} \to \mathcal{V}$ describes how each carrier transforms under the rotation R. With a choice of basis each carrier is represented by a set of numbers, that is, components, and each $T(R)$ is represented by a matrix which describes the transformation law for these components under the rotation R. For a real scalar such as the mass one takes $\mathcal{V} = \mathbb{R}$ and $T(R) = \mathrm{id}_{\mathbb{R}}$ for every $R \in SO(3)$ so that the value is the same in every coordinate system. For a vector V in $\mathbb{R}^3$ such as the momentum, we take $\mathcal{V} = \mathbb{R}^3$ and $T(R) = R$ for every $R \in SO(3)$ because vectors, by definition, transform in the same way as the coordinates. Specifically, suppose $R = (R^i{}_j)_{i,j=1,2,3} \in SO(3)$ and $\hat{x}^i = R^i{}_j x^j$, $i = 1, 2, 3$, are the rotated coordinates (here we sum over $j = 1, 2, 3$). Then the components of V in the two coordinate systems are related by

$$\hat{V}^i = R^i{}_j V^j, \quad i = 1, 2, 3.$$

Vector *fields* transform as vectors at each point, that is, if we write $R^{-1} = (R_i{}^j)_{i,j=1,2,3}$ so that $x^j = R_i{}^j \hat{x}^i$, then

$$\hat{V}^i(\hat{x}^1, \hat{x}^2, \hat{x}^3) = (R^i{}_j V^j)(R_k{}^1 \hat{x}^k, R_k{}^2 \hat{x}^k, R_k{}^3 \hat{x}^k), \quad i = 1, 2, 3.$$

For a second rank tensor field S (such as the stress tensor in continuum mechanics), $\mathcal{V} = \mathbb{R}^9$ and $T(R) = R \otimes R$ is the tensor product of the matrix R with itself, so the nine components are related by

$$\hat{S}^{ij}(\hat{x}^1, \hat{x}^2, \hat{x}^3) = (R^i{}_m R^j{}_n S^{mn})(R_k{}^1 \hat{x}^k, R_k{}^2 \hat{x}^k, R_k{}^3 \hat{x}^k), \quad i, j = 1, 2, 3,$$

where we sum over $k, m, n = 1, 2, 3$.

Two representations $T_1 : SO(3) \to GL(\mathcal{V}_1)$ and $T_2 : SO(3) \to GL(\mathcal{V}_2)$ are *equivalent* if there exists an invertible linear transformation $P : \mathcal{V}_1 \to \mathcal{V}_2$ of $\mathcal{V}_1$ onto $\mathcal{V}_2$ for which

$$T_2 = P \circ T_1 \circ P^{-1}.$$

This is the case if and only if there exist bases for $\mathcal{V}_1$ and $\mathcal{V}_2$ relative to which the matrices of $T_1(R)$ and $T_2(R)$ are the same for every $R \in SO(3)$, that is, if and only if they differ only in a choice of coordinates.

The point to all of this is that the representations of $SO(3)$ determine transformation laws and each of these expresses a certain type of rotational invariance in the sense that if the components of two quantities of the same type are equal in one coordinate system, then the components in any rotated coordinate system are also equal because they transform in the same way.

Let us return now to Pauli's two-component electron. To establish some sort of rotational invariance it would seem that what we need is some representation of $SO(3)$ on the two-dimensional complex vector space $\mathbb{C}^2$ of pairs

$$\psi = \begin{pmatrix} \psi_1 \\ \psi_2 \end{pmatrix}$$

of complex numbers. The good news is that all of the representations of SO(3) are known. The bad news is that (up to equivalence, that is, up to a change of coordinates) there is only one irreducible representation of SO(3) on $\mathbb{C}^2$, namely, the one that sends every $R \in$ SO(3) to the 2×2 identity matrix. This representation leaves every element of $\mathbb{C}^2$ fixed for every rotation R and we claim that this clearly will not do for our purposes. The reason is simple. Since ψ_1 and ψ_2 represent the probability amplitudes for detecting the electron with spin up and spin down, respectively, a rotation that reverses the direction of the z-axis must interchange ψ_1 and ψ_2 (up to phase),

$$\begin{pmatrix} \psi_1 \\ \psi_2 \end{pmatrix} \rightarrow e^{i\phi} \begin{pmatrix} \psi_2 \\ \psi_1 \end{pmatrix},$$

and the one representation we have available does not do this.

All is not lost, however. Indeed, a moment's thought should make it clear that our classical picture of rotational invariance has missed an essential feature of quantum mechanics. Wave functions are determined only up to phase, so, in particular, $\pm T(R)\psi$ describe precisely the same state and so the transformation law/representation is determined only up to sign. Physicists are inclined to say that what we need is not a representation of SO(3), but rather a "two-valued representation" of SO(3)

$$R \rightarrow \pm T(R).$$

This really makes no sense, of course, since functions are never "two-valued." Nevertheless, there is a perfectly rigorous construction that will allow us to make sense of the underlying idea. To describe this we will need to exploit a remarkable relationship between SO(3) and the group SU(2) of 2×2 unitary matrices with determinant 1, specifically, that SU(2) is the universal double covering group of SO(3). Everything we will need is proved in Appendix A of [Nab3], but we will sketch a few of the ideas to provide some intuition. SU(2) and SO(3) are both Lie groups (see Section 5.8 of [Nab3]). The claim is that there exists a smooth, surjective group homomorphism

$$\text{Spin} : \text{SU}(2) \rightarrow \text{SO}(3)$$

with kernel $\pm \begin{pmatrix} 1 & 0 \\ 0 & 1 \end{pmatrix}$ and with the property that each point in SO(3) has an open neighborhood V whose inverse image under Spin is a disjoint union of two open sets in SU(2), each of which is mapped diffeomorphically onto V by Spin (this last property is the meaning of "double cover"). In particular, SO(3) is isomorphic to SU(2) / $\mathbb{Z}_2$ and SU(2) is *locally* diffeomorphic to SO(3). Since SU(2) is homeomorphic to the 3-sphere S^3 (Theorem 1.1.4 of [Nab3]), this implies that SO(3) is homeomorphic to real, projective 3-space $\mathbb{RP}^3$ (Section 1.2 of [Nab3]). With this we can record the fundamental groups of SU(2) and SO(3) (see pages 117–119 and Theorem 2.4.5 of [Nab3]):

$$\pi_1(\text{SU}(2)) \cong 1 \quad (\text{SU}(2) \text{ is simply connected}),$$
$$\pi_1(\text{SO}(3)) \cong \mathbb{Z}_2.$$

Let us take a moment to see where the map Spin comes from. The most efficient way to do this is to identify $\mathbb{R}^3$ with $\mathcal{R}^3$ as in Exercise 10.1.1. Now note that for any $U \in \mathrm{SU}(2)$ and any $X \in \mathcal{R}^3$, $UX\overline{U}^T = UXU^{-1}$ is also in $\mathcal{R}^3$.

Exercise 10.1.3. Prove this.

Consequently, for each $U \in \mathrm{SU}(2)$ we can define a map

$$R_U : \mathcal{R}^3 \to \mathcal{R}^3$$

by

$$R_U(X) = UX\overline{U}^T = UXU^{-1} \quad \forall X \in \mathcal{R}^3.$$

Exercise 10.1.4. Show that R_U is an orthogonal transformation of $\mathcal{R}^3$ for each $U \in \mathrm{SU}(2)$.

In particular, the determinant of each R_U is either 1 or -1.

Exercise 10.1.5. Show that $\mathrm{SU}(2)$ consists precisely of those 2×2 complex matrices of the form

$$\begin{pmatrix} \alpha & \beta \\ -\overline{\beta} & \overline{\alpha} \end{pmatrix},$$

where $|\alpha|^2 + |\beta|^2 = 1$. Also show that if $\alpha = a + bi$ and $\beta = c + di$, then the matrix of R_U relative to the oriented, orthonormal basis $\{\sigma_1, \sigma_2, \sigma_3\}$ for $\mathcal{R}^3$ is

$$\begin{pmatrix} a^2 - b^2 - c^2 + d^2 & 2ab + 2cd & -2ac + 2bd \\ -2ab + 2cd & a^2 - b^2 + c^2 - d^2 & 2ad + 2bc \\ 2ac + 2bd & 2bc - 2ad & a^2 + b^2 - c^2 - d^2 \end{pmatrix}.$$

Note that the determinant of this matrix is a continuous, real-valued function on $\mathrm{SU}(2)$.

Exercise 10.1.6. Show that each R_U has determinant 1. *Hint*: Continuity.

Consequently, we can define Spin : $\mathrm{SU}(2) \to \mathrm{SO}(3)$ by $\mathrm{Spin}(U) = R_U$ for each $U \in \mathrm{SU}(2)$.

Exercise 10.1.7. Prove that Spin is a smooth group homomorphism with kernel

$$\pm \begin{pmatrix} 1 & 0 \\ 0 & 1 \end{pmatrix}.$$

The proof of surjectivity requires a bit more work and an appeal to Theorem 2.2.2. It turns out that for any element R of $\mathrm{SO}(3)$, expressed in the form described in Theorem 2.2.2, one can simply write down an element U of $\mathrm{SU}(2)$ for which $\mathrm{Spin}(\pm U) = R$

(details are available on page 398 of [Nab3]). That SU(2) is a double cover for SO(3) follows from the fact that S^3 is a double cover for $\mathbb{RP}^3$ (pages 64–65 of [Nab3]).

Now suppose $T : \mathrm{SO}(3) \to \mathrm{GL}(\mathcal{V})$ is a representation of SO(3). Composing with Spin gives a group homomorphism

$$\tilde{T} = T \circ \mathrm{Spin} : \mathrm{SU}(2) \to \mathrm{GL}(\mathcal{V}) \tag{10.11}$$

from SU(2) to GL($\mathcal{V}$), that is, a representation of SU(2) on $\mathcal{V}$. Thus, every representation of SO(3) gives rise to a representation of SU(2). The converse is not true, however. Specifically, a given representation $\tilde{T} : \mathrm{SU}(2) \to \mathrm{GL}(\mathcal{V})$ of SU(2) will clearly descend to a representation of SO(3) if and only if $\tilde{T}(-U) = \tilde{T}(U)$ for every $U \in \mathrm{SU}(2)$. The representations of SU(2) that *do not* satisfy this condition, but instead satisfy $\tilde{T}(-U) = -\tilde{T}(U)$ for every $U \in \mathrm{SU}(2)$ are what the physicists mean when they refer to *two-valued representations* of SO(3), although they are not representations of SO(3) at all, of course. There certainly is a representation of SU(2) on $\mathbb{C}^2$ that satisfies this condition, namely, the one that sends every element of SU(2) to itself. Up to equivalence this is, in fact, the only irreducible one and it is traditionally denoted $D^{\frac{1}{2}}$:

$$D^{\frac{1}{2}} : \mathrm{SU}(2) \to \mathrm{GL}(\mathbb{C}^2),$$

$$D^{\frac{1}{2}}(U) = U \quad \forall U \in \mathrm{SU}(2).$$

The carriers $\left(\begin{smallmatrix} \psi_1 \\ \psi_2 \end{smallmatrix}\right) \in \mathbb{C}^2$ of this representation are called *two-component spinors*. There is a great deal of beautiful mathematics hidden in this apparently simple idea of a two-component spinor. For a synopsis of some of its connections with Clifford algebras and Hopf bundles from the point of view of spin $\frac{1}{2}$ physics we recommend [Socol]. For the full story of Clifford algebras and spinors in general see [LM].

Exercise 10.1.8. Show that

$$U = \begin{pmatrix} \cos \frac{\pi}{2} & -i \sin \frac{\pi}{2} \\ -i \sin \frac{\pi}{2} & \cos \frac{\pi}{2} \end{pmatrix}$$

is in SU(2) and Spin($\pm U$) is a rotation about the x-axis *through π (not $\pi/2$)* and therefore reverses the direction of the z-axis. Then note that

$$D^{\frac{1}{2}}(U) \begin{pmatrix} \psi_1 \\ \psi_2 \end{pmatrix} = e^{-\frac{\pi}{2}i} \begin{pmatrix} \psi_2 \\ \psi_1 \end{pmatrix}$$

so that, up to phase, $D^{\frac{1}{2}}(U)$ just reverses the spin components ψ_1 and ψ_2 as Stern–Gerlach insists that it should. *Hint:* For Spin($\pm U$) use Exercise 10.1.4.

Note that there is an interesting doubling of angles under Spin in the previous exercise. This is a characteristic feature of Spin and arises because $\mathrm{Spin}(U)(X) = UX\overline{U}^T$

essentially "squares" U. It is also the feature that will lead us to one of the most remarkable properties of spin $\frac{1}{2}$ particles (and fermions in general). To uncover this property we will make considerable use of the discussion in Example 2.2.13, so we suggest a review of this before you proceed.

As we have learned (Example 2.2.13), the physical process of rotating an object in space is modeled mathematically by a continuous curve $\alpha : [t_0, t_1] \to SO(3)$ in $SO(3)$. For example,

$$\begin{pmatrix} 1 & 0 & 0 \\ 0 & \cos t & -\sin t \\ 0 & \sin t & \cos t \end{pmatrix}$$

is a rotation through t radians about the x-axis, so the curve $\alpha : [0, 2\pi] \to SO(3)$ defined by this matrix for $0 \le t \le 2\pi$ defines a continuous rotation about the x-axis through $360°$. This is a loop at the 3×3 identity matrix $I_{3\times3}$ in $SO(3)$ and hence it determines an element $[\alpha]$ of the fundamental group $\pi_1(SO(3))$. On the other hand, the curve $\alpha^2 : [0, 4\pi] \to SO(3)$ defined by the same matrix for $0 \le t \le 4\pi$ (α traversed twice) defines a rotation about the x-axis through $720°$ and determines the element $[\alpha^2] = [\alpha]^2$ in $\pi_1(SO(3))$. How does the wave function of a spin $\frac{1}{2}$ particle respond to these two rotations? To answer this we recall that, since Spin : $SU(2) \to SO(3)$ is a covering space, curves in the covered space $SO(3)$ lift uniquely to curves in the covering space $SU(2)$ once an initial point is selected (see Corollary 1.5.13 of [Nab3]). Since α begins at $I_{3\times3}$ and $\mathrm{Spin}(\pm I_{2\times2}) = I_{3\times3}$, there is a unique curve $\tilde{\alpha} : [0, 2\pi] \to SU(2)$ with $\mathrm{Spin} \circ \tilde{\alpha} = \alpha$ and $\tilde{\alpha}(0) = I_{2\times2}$.

Exercise 10.1.9. Show that $\tilde{\alpha}(t)$ is given by

$$\begin{pmatrix} \cos \frac{t}{2} & -i \sin \frac{t}{2} \\ -i \sin \frac{t}{2} & \cos \frac{t}{2} \end{pmatrix},$$

where $0 \le t \le 2\pi$. Similarly, the unique lift of α^2 starting at $I_{2\times2}$ is given by this same matrix with $0 \le t \le 4\pi$. Note that the lift of α begins at $I_{2\times2}$ and ends at $-I_{2\times2}$, whereas the lift of α^2 begins and ends at $I_{2\times2}$, passing through $-I_{2\times2}$ along the way.

The response of the two-component spinor wave function $\psi = \begin{pmatrix} \psi_1 \\ \psi_2 \end{pmatrix}$ to these rotations is described by $D^{\frac{1}{2}}$ applied to the points along the lifted curves in $SU(2)$. In particular,

$$D^{\frac{1}{2}}(\tilde{\alpha}(0)) \begin{pmatrix} \psi_1 \\ \psi_2 \end{pmatrix} = D^{\frac{1}{2}}(I_{2\times2}) \begin{pmatrix} \psi_1 \\ \psi_2 \end{pmatrix} = \begin{pmatrix} \psi_1 \\ \psi_2 \end{pmatrix},$$

whereas

$$D^{\frac{1}{2}}(\tilde{\alpha}(2\pi)) \begin{pmatrix} \psi_1 \\ \psi_2 \end{pmatrix} = D^{\frac{1}{2}}(-I_{2\times2}) \begin{pmatrix} \psi_1 \\ \psi_2 \end{pmatrix} = -\begin{pmatrix} \psi_1 \\ \psi_2 \end{pmatrix},$$

so a rotation through $360°$ reverses the sign of the wave function.

Exercise 10.1.10. Show that a rotation through $720°$ returns the wave function to its original value.

We have therefore found a physical system (namely, a spin $\frac{1}{2}$ particle) whose wave function is changed by a rotation through $360°$, but returns to its original value after a rotation through $720°$. It certainly does seem strange to our macroscopically conditioned brains that a $360°$ rotation and a $720°$ rotation could result in "something different," but this is quantum mechanics, after all, where strange is the order of the day. Alleviating this feeling of strangeness to some degree is really the point of Dirac's ingenious scissors experiment, which we described in Example 2.2.13.

Note, however, that mathematically the difference between α and α^2 is not at all mysterious if we keep in mind that $\pi_1(SO(3)) \cong \mathbb{Z}_2$, but SU(2) is simply connected. The loop α^2 in SO(3) lifts to a loop in SU(2) which must be null-homotopic by simple connectivity. Pushing the null-homotopy down to SO(3) by Spin shows that α^2 itself is null-homotopic. However, α cannot be null-homotopic since its lift to SU(2) is a *path* from $I_{2\times 2}$ to $-I_{2\times 2}$ and not a loop at all (see the homotopy lifting theorem, Theorem 2.4.1 of [Nab3]). Representing $\mathbb{Z}_2$ as the multiplicative group $\{[-1], [1]\}$ of integers *mod* 2 we can write all of this symbolically as $[\alpha] = [-1] \in \pi_1(SO(3))$, but $[\alpha^2] = [\alpha]^2 = [-1]^2 = [1] \in \pi_1(SO(3))$.

The homotopy type of a rotation (thought of as a curve in SO(3)) determines the effect of the rotation on the wave function of a spin $\frac{1}{2}$ particle. This is interesting enough, but one might wonder whether or not this change in the wave function resulting from a 2π rotation is actually observable, that is, whether or not it has any physical consequences. After all, $\pm\psi$ represent the same *state* of the particle, so perhaps all of this is just a peculiarity of the mathematical model and not physics at all. For some thoughts on this, see [AS].

10.2 Anticommutation relations and the fermionic harmonic oscillator

The quantum systems that we have examined thus far have all arisen in essentially the same way. One begins with a familiar classical system and chooses some appropriate quantization scheme to build a "corresponding" quantum system. Generally, one then checks that the quantum system approaches the classical system in some limiting sense (say, as $\hbar \to 0$). We have gotten away with this so far only because we have conscientiously ignored the existence of particles with half-integer spin and, in particular, the fact that the electron has spin $\frac{1}{2}$. Having come face-to-face with the Stern–Gerlach experiment and the bizarre reaction of spin $\frac{1}{2}$ particles to rotations, we should probably not continue to do this. There is a problem, however. Since nothing in the classical world behaves like a fermion, we have nothing to quantize! The quantum mechanics of a fermion system must be built from scratch without reference to any "corresponding" classical system. This is a big deal because fermion systems exist in great abundance and are, in a sense, responsible for everything that we see and experience in the world around us. Every atom is built from spin $\frac{1}{2}$ particles (electrons, protons and neutrons and even the quarks from which protons and neutrons are built). The arrangement of the corresponding elements in the periodic table as well as all of their chemical properties are explicable only in terms of the quantum theory of fermions. Even neutron stars and white dwarf stars can exist only because fermions are subject to what is called the Pauli exclusion principle; indeed, it is thanks to this principle that *we* exist to see and experience the world around us (see [Lieb]). In this section we will describe the standard operating procedure in quantum theory for dealing with fermions and illustrate the procedure by building a fermionic analogue of the bosonic harmonic oscillator.

The standard operating procedure to which we referred above is easy to state ("change the commutators $[\,,\,]_-$ to anticommutators $[\,,\,]_+$"), but not so easy to motivate or justify. One can simply say, as one often does in quantum mechanics, that the proof of the pudding is in the eating, so one must simply be patient and see what consequences can be drawn and whether or not they jibe with the experimental facts. This is fine, but rather unsatisfying. After all, someone had to actually think of trying it and presumably had reasons for doing so.

In this particular case, that person was Pascual Jordan. In 1927 Jordan was troubled by what he perceived to be an inconsistency between, on the one hand, the canonical quantization procedure of Dirac (Section 7.2) in which one represents classical (Poisson bracket) commutation relations as operators on a Hilbert space and, on the other hand, the nature of the intrinsic angular momentum (spin) of an electron. Very roughly, the idea goes something like this. Dirac's program would arrive at the quantum description of the electron with spin by quantizing the classical angular momentum of a charged spinning sphere. This it would do by identifying canonically conjugate coordinates for the classical problem and representing them as self-adjoint operators on a Hilbert space. Now, classically these conjugate coordinates consist of a projection of the magnetic moment onto some axis, say, S_z, and a corresponding angular coordinate ϕ_z. What bothered Jordan was that there is no meaningful (that is, measurable) quantum analogue of ϕ_z since, as far as anyone can tell, the electron behaves like a structureless, point particle, so, in particular, it has no marker on it that would make such an angle measurable. But now recall from the previous section that, according to Pauli, the components S_i, $i = 1, 2, 3$, of the spin vector are given by $S_i = \frac{\hbar}{2}\sigma_i$, $i = 1, 2, 3$, and that $\sigma_1, \sigma_2, \sigma_3$ satisfy $[\sigma_i, \sigma_j]_+ = 2\delta_{ij} I_{2\times2}$. It follows that

$$[S_i, S_j]_+ = \frac{\hbar^2}{2} \delta_{ij} I_{2\times2}, \quad i, j = 1, 2, 3.$$

Now we will define two new operators a and $a^\dagger$ that should be compared with (5.14) and (5.15):

$$a = \frac{1}{\hbar}(S_3 + iS_1) \quad \text{and} \quad a^\dagger = \frac{1}{\hbar}(S_3 - iS_1).$$

Exercise 10.2.1. Show that

$$[a, a]_+ = [a^\dagger, a^\dagger]_+ = 0 \quad \text{and} \quad [a, a^\dagger]_+ = I_{2\times2} \tag{10.12}$$

and note that one obtains the same result if S_1 is replaced by S_2 in the definitions of a and $a^\dagger$.

The relations in (10.12) are, of course, strikingly similar to the canonical commutation relations, but involve anticommutators rather than commutators. From this one can at least imagine how anticommutation relations might play a role in the description of fermions and what might have led Jordan to his insight. However, the real story behind this lies in much greater depths, specifically, in the so-called *spin-statistics theorem* of quantum field theory. Even a precise statement of this result would take us very far outside of our comfort zone here, but we would feel remiss if we did not at least try to provide some sense of what it is all about.

For those who are quite properly dissatisfied with the brief sketch that follows we can suggest the following sources for more detailed discussions. The spin-statistics theorem is, in fact, a rigorous mathematical theorem in axiomatic quantum field theory. The standard reference for the rather demanding proof is [SW], which also provides a schematic of the Wightman axioms for quantum field theory, but not a great deal in the way of physical motivation. The original, less rigorous, but more physically based argument is due to Wolfgang Pauli [Pauli2]. A great deal of effort has been expended in the search for a simpler proof of the result, but without much success. Regarding these one can consult [DS2] and [Wight] and the book [DS1], which also contains some historical perspective and excerpts from many of the seminal papers.

We should emphasize that being a rigorous theorem in axiomatic quantum field theory should not lead one to assume that the conclusion drawn from the spin-statistics theorem is an incontestable physical fact. The proof is based on assumptions (the Wightman axioms) about the essential prerequisites of any quantum field theory and these have certainly not gone unquestioned. Furthermore, quantum field theory is a *relativistic* theory and quantum mechanics is not so, at our level (the level of quantum mechanics) the spin-statistics theorem is not a theorem at all and if we want to make use of its conclusion (and we do) this must be introduced as an addition to our list of Postulates QM1–QM6 (Section 6.2). This is what we will do, but we will need to build up to it slowly.

The first order of business is to take note of yet one more peculiarity of quantum mechanics. In both classical and quantum mechanics any two electrons are *identical* in the sense that they have precisely the same characteristic properties (classically, their mass and charge and quantum mechanically their mass, charge and spin).

Classical physics, however, allows one to *distinguish* even identical particles by simply keeping track of their trajectories in space (electron number 1 is here and electron number 2 is there). In quantum mechanics particles do not have trajectories, only
wave functions, so this option is not available and one is led to the conclusion that any
two electrons are *indistinguishable*. Intuitively, this means that the *state* of a system
consisting of more than one electron remains the same if two of the electrons are in-
terchanged (we will make this more precise in a moment). The same is true of protons
and neutrons and, indeed, of any identical quantum particles.

In order to make more precise the notion of indistinguishable particles as well as
to formulate a version of the spin-statistics theorem we will need to say something
about how quantum mechanics associates a Hilbert space to a system consisting of
more than one particle. This has not come up in our discussions yet and the procedure does not follow from our list of Postulates QM1–QM6, so we will need a new one.
We will also need some basic facts about tensor products of Hilbert spaces. We have
outlined this in Appendix K and will use the notation and terminology established
there.

We consider a quantum system S consisting of N particles in space. The Hilbert
space associated to each particle is $\mathcal{H} = L^2(\mathbb{R}^3)$. Now consider the tensor product
$\otimes^N \mathcal{H} = \otimes^N L^2(\mathbb{R}^3)$, one factor for each particle. The particles are said to be *identical*
if the action of the symmetric group S_N on $\otimes^N L^2(\mathbb{R}^3)$ leaves the *state of the system* invariant, that is, if

$$\psi(\mathbf{x}_{\sigma(1)}, \ldots, \mathbf{x}_{\sigma(N)}) = R(\sigma)\psi(\mathbf{x}_1, \ldots, \mathbf{x}_N) \quad \forall \sigma \in S_N,$$

where each $R(\sigma)$ is a complex number of modulus one (phase factor). Note that R is a
homomorphism of S_N to the group of unit complex numbers, of which there are only
two, namely, $R(\sigma) = 1 \ \forall \sigma \in S_N$ and $R(\sigma) = \text{sgn } \sigma \ \forall \sigma \in S_N$. We conclude that quantum
particles fall into two types, namely, those for which a system of N such particles has a
wave function that is symmetric (states are in $\otimes_S^N L^2(\mathbb{R}^3)$) and those for which a system
of N such particles has a wave function that is antisymmetric (states are in $\otimes_A^N L^2(\mathbb{R}^N)$).
The final postulate we add to our list (Postulates QM1–QM6) identifies these two types
and will serve as our version of the spin-statistics theorem.

Postulate QM7. *Let S be a quantum system consisting of N identical particles in space.
The Hilbert space of each individual particle is $L^2(\mathbb{R}^3)$. Then the states of the system S
are in $\otimes^N L^2(\mathbb{R}^3)$ and are either:*
1. *symmetric (that is, in $\otimes_S^N L^2(\mathbb{R}^3)$), in which case the particles are bosons (integer
   spin), or*
2. *antisymmetric (that is, in $\otimes_A^N L^2(\mathbb{R}^3)$), in which case the particles are fermions (half-
   integer spin).*

The most remarkable aspect of this postulate is the relationship it establishes between
the intrinsic angular momentum (spin) of a particle and the behavior of a system of

such particles (statistics). It would be a very good thing to be able to offer some sort of intuitive motivation for this relationship, but, as the saying goes, "I got nothing!" Indeed, it seems that no one has anything to offer in this regard. Here is the oft-quoted comment of Richard Feynman.

> "We apologize for the fact that we cannot give you an elementary explanation. An explanation has been worked out by Pauli from complicated arguments of quantum field theory and relativity. He has shown that the two must necessarily go together, but we have not been able to find a way of reproducing his arguments on an elementary level. It appears to be one of the few places in physics where there is a rule which can be stated very simply, but for which no one has found a simple and easy explanation. The explanation is deep down in relativistic quantum mechanics. This probably means that we do not have a complete understanding of the fundamental principle involved."

> Richard Feynman ([FLS], Volume III, 4-3)

We should point out a particularly important consequence of Postulate QM7. Suppose S consists of N identical fermions (say, electrons). Assume that at some instant, one of the particles is in a state represented by ψ_1, another is in a state represented by $\psi_2, \ldots,$ and so on. Then the state of the system is represented by $\Psi = \psi_1 \otimes \psi_2 \otimes \cdots \otimes \psi_N \in \otimes_A^N L^2(\mathbb{R}^3)$. What happens if two of the fermions are in the same state, that is, $\psi_i = \psi_j$ for some $1 \le i < j \le N$? If σ_{ij} is the permutation that switches i and j and leaves everything else alone (which is odd), then $\sigma_{ij} \cdot \Psi = -\Psi$ by antisymmetry. But $\psi_i = \psi_j$ clearly implies that $\sigma_{ij} \cdot \Psi = \Psi$ and these cannot both be true unless $\Psi = 0 \in \otimes_A^N L^2(\mathbb{R}^3)$, which is not a state at all. From this we obtain the famous *Pauli exclusion principle*.

In a system of identical fermions, no two of the fermions can be in the same state.

As we mentioned earlier, this principle is responsible not only for the existence of the world we see around us, but for our existence as well. Historically, it was not derived from the spin-statistics theorem (which it predated) and, indeed, it was not "derived" from anything. Pauli proposed the exclusion principle (for electrons in an atom) as an *ad hoc* hypothesis which he found could explain a huge number of otherwise mysterious, but incontestable experimental facts. Even after establishing the spin-statistics theorem and being awarded the Nobel Prize, Pauli admitted to an uneasy feeling about the logical status of the exclusion principle.

> "Already in my original paper I stressed the circumstance that I was unable to give a logical reason for the exclusion principle or to deduce it from more general assumptions. I had always the feeling, and I still have it today, that this is a deficiency."

> Wolfgang Pauli ([Pauli3])

Finally, we must try to see what all of this has to do with commutation and anticommutation relations. This will take us somewhat outside of our element since it

Figure 10.3: Pair production.

involves a technique, known as *second quantization*, that is more specifically geared toward quantum field theory. For this reason we will provide only a brief sketch (more details are available in Section 4.5 of [Fol3] and Sections 12.1 and 12.2 of [BEH]).

The first construction we will require is not particularly well motivated by what we have done previously, but is easily explained. In elementary particle physics the interactions that take place generally do not leave the number of particles fixed. For example, in Figure 10.3 a gamma ray (which, being a photon, is not visible) enters from below and, at the vertex joining the two spirals, interacts with a nearby atom, decays into an electron–positron pair (the spirals) and ejects an electron from the atom (the third track leaving the vertex). Shortly thereafter the ejected electron emits a gamma ray (again, not visible) which then also decays into an electron–positron pair with straighter tracks (the < above of the vertex).

Interactions of this sort are described in quantum field theory and this is not within our purview, but we will borrow the mathematical device used there to associate a Hilbert space to a system with a possibly varying number of particles. We will restrict our attention to systems of identical fermions, although the procedure is much more general than this (see Section 4.5 of [Fol3] and Sections 12.1 and 12.2 of [BEH]).

We begin with the Hilbert space $\mathcal{H} = L^2(\mathbb{R}^3)$ of a single fermion; this we will refer to as the *one-particle Hilbert space* and will also write it as $\otimes_A^1 \mathcal{H}$. For any $N \geq 1$, the antisymmetric Nth tensor power $\otimes_A^N \mathcal{H}$ of $\mathcal{H}$ (see Appendix K) will be called the *N-particle Hilbert space*. Also set $\otimes_A^0 \mathcal{H} = \mathbb{C}$ and refer to this as the *zero-particle Hilbert space*. The *algebraic direct sum* of the vector spaces $\otimes_A^N \mathcal{H}$

$$\bigoplus_{\text{alg}} \{\otimes_A^N \mathcal{H} : N = 0, 1, 2, \ldots\} = \mathbb{C} \oplus_{\text{alg}} \mathcal{H} \oplus_{\text{alg}} \otimes_A^2 \mathcal{H} \oplus_{\text{alg}} \cdots$$

is called the *finite particle space*. An element of the finite particle space has only finitely many nonzero coordinates, each one of which is an N-particle state in $\otimes_A^N \mathcal{H}$ for some N. For each $N \geq 0$ we will identify $\otimes_A^N \mathcal{H}$ with the subspace of the direct sum in which all of the coordinates are zero except the Nth. On the other hand, the *Hilbert space direct sum* of $\otimes_A^N \mathcal{H}$ is called the *antisymmetric*, or *fermionic Fock space* of $\mathcal{H}$ and is denoted

$$\mathcal{F}_A(\mathcal{H}) = \bigoplus \{\otimes_A^N \mathcal{H} : N = 0, 1, 2, \ldots\} = \mathbb{C} \oplus \mathcal{H} \oplus \otimes_A^2 \mathcal{H} \oplus \cdots.$$

Recall that if $\mathcal{H}_0, \mathcal{H}_1, \mathcal{H}_2, \ldots$ are Hilbert spaces, then their Hilbert space direct sum $\mathcal{H} = \bigoplus_{n=0}^{\infty} \mathcal{H}_n$ is the linear subspace of the vector space direct product $\prod_{n=0}^{\infty} \mathcal{H}_n$ consisting of all sequences $(x_0, x_1, x_2, \ldots)$ with $x_n \in \mathcal{H}_n$ for each $n = 0, 1, 2, \ldots$ and $\sum_{n=0}^{\infty} \|x_n\|_{\mathcal{H}_n}^2 < \infty$ and with the inner product defined by

$$\langle (x_n)_{n=0}^{\infty}, (y_n)_{n=0}^{\infty} \rangle_{\mathcal{H}} = \sum_{n=0}^{\infty} \langle x_n, y_n \rangle_{\mathcal{H}_n}.$$

The algebraic direct sum of the vector spaces $\mathcal{H}_0, \mathcal{H}_1, \mathcal{H}_2, \ldots$ is naturally identified with a dense linear subspace of $\bigoplus_{n=0}^{\infty} \mathcal{H}_n$. We can therefore identify the elements of $\mathcal{F}_A(L^2(\mathbb{R}^3))$ with sequences

$$(\psi_0, \psi_1(\mathbf{x}_1), \psi_2(\mathbf{x}_1, \mathbf{x}_2), \ldots, \psi_N(\mathbf{x}_1, \mathbf{x}_2, \ldots, \mathbf{x}_N), \ldots),$$

where

$$|\psi_0|^2 + \|\psi_1(\mathbf{x}_1)\|_{L^2(\mathbb{R}^3)}^2 + \cdots + \|\psi_N(\mathbf{x}_1, \mathbf{x}_2, \ldots, \mathbf{x}_N)\|_{L^2(\mathbb{R}^3 \times \overset{N}{\cdots} \times \mathbb{R}^3)}^2 + \cdots < \infty$$

and each of the functions is antisymmetric with respect to permutations of its arguments (up to a set of measure zero).

Now we would like to define a number of operators on the Fock space $\mathcal{F}_A(\mathcal{H})$ analogous to the raising and lowering operators for the bosonic harmonic oscillator. The difference is that, whereas for the oscillator these operators raised and lowered the number of energy quanta, we now think of them as creating and annihilating *particles*, that is, mapping $\otimes_A^N \mathcal{H} \to \otimes_A^{N+1} \mathcal{H}$ and $\otimes_A^N \mathcal{H} \to \otimes_A^{N-1} \mathcal{H}$, respectively. Begin by looking in the algebraic direct sum $\bigoplus_{\text{alg}}(\otimes_A^N \mathcal{H})$ of the particle spaces. For each $\phi \in \mathcal{H}$ we define $a(\phi)^\dagger : \otimes_A^N \mathcal{H} \to \otimes_A^{N+1} \mathcal{H}$ to be the linear extension to $\otimes_A^N \mathcal{H}$ of the map defined as follows. If $N = 0$, then $a(\phi)^\dagger(1) = \phi$, and if $N \geq 1$, then

$$a(\phi)^\dagger(\psi_1 \otimes_A \cdots \otimes_A \psi_N) = \sqrt{N+1} \,(\phi \otimes_A \psi_1 \otimes_A \cdots \otimes_A \psi_N).$$

This defines a linear map on all of $\bigoplus_{\text{alg}}(\otimes_A^N \mathcal{H})$. Next define $a(\phi) : \otimes_A^N \mathcal{H} \to \otimes_A^{N-1} \mathcal{H}$ for $N \geq 1$ to be the linear extension to $\otimes_A^N \mathcal{H}$ of the map defined as follows. If $N = 1$, then $a(\phi)(\psi) = \langle \phi, \psi \rangle_{\mathcal{H}}$, and if $N \geq 2$,

$$a(\phi)(\psi_1 \otimes_A \cdots \otimes_A \psi_N) = \frac{1}{\sqrt{N}} \sum_{j=1}^{N} (-1)^{j-1} \langle \phi, \psi_j \rangle_{\mathcal{H}} (\psi_1 \otimes_A \cdots \hat{\psi}_j \cdots \otimes_A \psi_N),$$

where the hat $\hat{\ }$ indicates that ψ_j is omitted. To cover $N = 0$ we can simply take $\otimes_A^{-1} \mathcal{H}$ to be the vector space containing only 0 and let $a(\phi) : \otimes_A^0 \mathcal{H} \to \otimes_A^{-1} \mathcal{H}$ be the map that sends everything to 0. The linear maps $a(\phi)$ and $a(\phi)^\dagger$ are formal adjoints of each other with respect to the inner product introduced above, that is, on $\bigoplus_{\text{alg}}(\otimes_A^N \mathcal{H})$,

$$\langle a(\phi)\Psi, \Phi \rangle = \langle \Psi, a(\phi)^\dagger \Phi \rangle.$$

Exercise 10.2.2. Verify the following special case:

$$\langle a(\phi)(\psi_1 \otimes_A \psi_2 \otimes_A \psi_3), \varphi_1 \otimes_A \varphi_2 \rangle = \langle \psi_1 \otimes_A \psi_2 \otimes_A \psi_3, a(\phi)^\dagger(\varphi_1 \otimes_A \varphi_2) \rangle.$$

Hint: Show that both sides are equal to

$$\frac{1}{\sqrt{3}} \frac{1}{2!} [\, \langle \psi_1, \phi \rangle \langle \psi_2, \varphi_1 \rangle \langle \psi_3, \varphi_2 \rangle - \langle \psi_1, \phi \rangle \langle \psi_2, \varphi_2 \rangle \langle \psi_3, \varphi_1 \rangle - \langle \psi_2, \phi \rangle \langle \psi_1, \varphi_1 \rangle \langle \psi_3, \varphi_2 \rangle$$
$$+ \langle \psi_2, \phi \rangle \langle \psi_1, \varphi_2 \rangle \langle \psi_3, \varphi_1 \rangle + \langle \psi_3, \phi \rangle \langle \psi_1, \varphi_1 \rangle \langle \psi_2, \varphi_2 \rangle - \langle \psi_3, \phi \rangle \langle \psi_1, \varphi_2 \rangle \langle \psi_2, \varphi_1 \rangle \,].$$

Exercise 10.2.3. Show that on $\bigoplus_{\text{alg}}(\otimes_A^N \mathcal{H})$:
1. $[a(\phi_1)^\dagger, a(\phi_2)^\dagger]_+ = 0$ and
2. $[a(\phi_1), a(\phi_2)]_+ = 0$.

Next we show that on $\bigoplus_{\text{alg}}(\otimes_A^N \mathcal{H})$,

$$[a(\phi_1), a(\phi_2)^\dagger]_+ = \langle \phi_1, \phi_2 \rangle I, \tag{10.13}$$

where I is the identity operator on $\mathcal{H}$. For this we note first that

$$a(\phi_1)a(\phi_2)^\dagger(\psi_1 \otimes_A \cdots \otimes_A \psi_N)$$
$$= \sqrt{N+1}\, a(\phi_1)(\phi_2 \otimes_A \psi_1 \otimes_A \cdots \otimes_A \psi_N)$$
$$= \langle \phi_1, \phi_2 \rangle (\psi_1 \otimes_A \cdots \otimes_A \psi_N) + \sum_{j=1}^{N} (-1)^j \langle \phi_1, \psi_j \rangle \phi_2 \otimes_A \psi_1 \otimes_A \cdots \hat{\psi}_j \cdots \otimes_A \psi_N.$$

But also

$$a(\phi_2)^\dagger a(\phi_1)(\psi_1 \otimes_A \cdots \otimes_A \psi_N)$$
$$= a(\phi_2)^\dagger \left(\frac{1}{\sqrt{N}} \sum_{j=1}^{N} (-1)^{j-1} \langle \phi_1, \psi_j \rangle \psi_1 \otimes_A \cdots \hat{\psi}_j \cdots \otimes_A \psi_N \right)$$

$$= \sum_{j=1}^{N}(-1)^{j-1}\langle\phi_1,\psi_j\rangle\phi_2\otimes_A\psi_1\otimes_A\cdots\hat{\psi}_j\cdots\otimes_A\psi_N.$$

Adding these two gives

$$[a(\phi_1),a(\phi_2)^\dagger]_+(\psi_1\otimes_A\cdots\otimes_A\psi_N) = \langle\phi_1,\phi_2\rangle(\psi_1\otimes_A\cdots\otimes_A\psi_N),$$

as required.

Exercise 10.2.4. Use (10.13) to show that on $\bigoplus_{\text{alg}}(\otimes_A^N\mathcal{H})$,

$$\|a(\phi)\Psi\|^2 + \|a(\phi)^\dagger\Psi\|^2 = \|\phi\|^2\|\Psi\|^2$$

and conclude from this that $a(\phi)$ and $a(\phi)^\dagger$ extend uniquely to *bounded* operators on the Fock space $\mathcal{F}_A(\mathcal{H})$.

We will use the same symbols $a(\phi)$ and $a(\phi)^\dagger$ for these extensions and will refer to them as the *annihilation* and *creation* operators on $\mathcal{F}_A(\mathcal{H})$, respectively. For systems of identical fermions they play roles analogous to the lowering and raising operators we introduced for the bosonic harmonic oscillator.

Finally, suppose $\{e_1,e_2,\ldots\}$ is an orthonormal basis for $\mathcal{H}$ and define operators a_i and $a_i^\dagger$ by

$$a_i = a(e_i) \quad\text{and}\quad a_i^\dagger = a(e_i)^\dagger, \quad i = 1,2,\ldots.$$

Then (10.13) and the identities in Exercise 10.2.3 give

$$[a_i,a_j]_+ = 0, \quad [a_i^\dagger,a_j^\dagger]_+ = 0 \quad\text{and}\quad [a_i,a_j^\dagger]_+ = \delta_{ij}I, \quad i,j = 1,2,\ldots. \tag{10.14}$$

These, or (10.13) and the identities in Exercise 10.2.3 that gave rise to them, are generally referred to as the *canonical anticommutation relations*.

Our objective here has been to suggest that, just as bosonic systems (like the bosonic harmonic oscillator) are characterized by commutation relations, fermionic systems are characterized by anticommutation relations. This is the key to the construction of the so-called "fermionic harmonic oscillator," to which we turn now.

The construction is based on the picture of the bosonic oscillator in terms of its lowering (b) and raising ($b^\dagger$) operators, so a review of this material in Section 5.3 would probably be in order. In particular, one should recall their commutation relations

$$[b,b]_- = [b^\dagger,b^\dagger]_- = 0 \quad\text{and}\quad [b,b^\dagger]_- = 1. \tag{10.15}$$

There is also a corresponding bosonic number operator

$$N_B = b^\dagger b$$

in terms of which the bosonic oscillator Hamiltonian can be written

$$H_B = \frac{1}{2}\hbar\omega[b^\dagger,b]_+ = \hbar\omega\left(N_B + \frac{1}{2}\right).$$

All of the essential physical information about the bosonic oscillator (for example, the spectrum of the Hamiltonian H_B) is contained in any algebra of operators satisfying these relations. Adopting an algebraic point of view, one could *identify* the bosonic harmonic oscillator with this algebra and this choice of Hamiltonian. This is the path we will follow in the fermionic case.

According to the "standard operating procedure" described at the beginning of this section our course would seem clear. To define a fermionic analogue of the bosonic harmonic oscillator we are simply to change the commutators in (10.15) to anticommutators:

$$[f,f]_+ = [f^\dagger,f^\dagger]_+ = 0 \quad \text{and} \quad [f,f^\dagger]_+ = 1. \tag{10.16}$$

This raises some issues, however. In the bosonic case, b and $b^\dagger$ were defined to be operators on $L^2(\mathbb{R})$ obtained from the canonical quantization of the Poisson bracket commutation relations for the classical harmonic oscillator. As a result one can, modulo the usual domain difficulties, make sense of the commutators. As we have gone to some lengths to emphasize, however, there are no classical fermionic systems and therefore nothing to quantize, so it is not at all clear where $f, f^\dagger$, 0 and 1 are supposed to live and what meaning is to be attached to $[\,,\,]_+$ in (10.16). For this reason we will need to begin in a somewhat more abstract algebraic setting.

An *algebra with involution* is a complex algebra $\mathcal{A}$ with multiplicative unit $1_\mathcal{A}$ on which is defined a conjugate-linear map $A \mapsto A^\dagger : \mathcal{A} \to \mathcal{A}$, called an *involution*, that satisfies $(A^\dagger)^\dagger = A$ and $(AB)^\dagger = B^\dagger A^\dagger$ for all $A, B \in \mathcal{A}$. An element A of $\mathcal{A}$ is said to be *self-adjoint* if $A^\dagger = A$. Since $\mathcal{A}$ is an algebra, one can define the commutator and anticommutator on $\mathcal{A} \times \mathcal{A}$ by $[A,B]_- = AB - BA$ and $[A,B]_+ = AB + BA$.

An obvious example of an algebra with involution is the algebra $\mathcal{B}(\mathcal{H})$ of bounded operators on a complex Hilbert space $\mathcal{H}$ with the involution † taken to be the Hilbert space adjoint *. If $\mathcal{A}$ is an arbitrary algebra with involution and $\mathcal{H}$ is a complex Hilbert space, then a linear map $\pi : \mathcal{A} \to \mathcal{B}(\mathcal{H})$ satisfying $\pi(AB) = \pi(A)\pi(B)$, $\pi(1_\mathcal{A}) = \mathrm{id}_\mathcal{H}$ and $\pi(A^\dagger) = \pi(A)^*$ for all $A, B \in \mathcal{A}$ is called a *representation* of $\mathcal{A}$ on $\mathcal{H}$. The representation π is said to be *faithful* if it is injective so that distinct elements of $\mathcal{A}$ are represented by distinct operators. One can also consider representations of $\mathcal{A}$ by unbounded operators on $\mathcal{H}$, but we will have no need to do so here. Indeed, we will actually require only finite-dimensional Hilbert spaces.

It is common to define the algebras with involution of interest in physics in terms of generators and relations with the relations specified as commutation or anticommutation relations. One then establishes the existence of such an algebra by finding a faithful representation of the generators and relations. For example, we would like to identify the *fermionic harmonic oscillator algebra* with an algebra with involution containing elements $\{f, f^\dagger, 0, 1\}$, with 0 being the additive identity and 1 the multiplicative identity and subject to the anticommutation relations

$$[f, f]_+ = [f^\dagger, f^\dagger]_+ = 0 \quad \text{and} \quad [f, f^\dagger]_+ = 1. \tag{10.17}$$

In fact, it is quite easy to construct a concrete representation of this algebra in terms of spin operators. Specifically, we take the Hilbert space $\mathcal{H}$ to be $\mathbb{C}^2$ and define Hermitian operators on $\mathbb{C}^2$ with respect to the standard basis $e_0 = \left(\begin{smallmatrix} 0 \\ 1 \end{smallmatrix}\right)$ and $e_1 = \left(\begin{smallmatrix} 1 \\ 0 \end{smallmatrix}\right)$ for $\mathbb{C}^2$ by

$$S_1 = \frac{\hbar}{2}\sigma_1, \quad S_2 = \frac{\hbar}{2}\sigma_2 \quad \text{and} \quad S_3 = \frac{\hbar}{2}\sigma_3,$$

where σ_1, σ_2 and σ_3 are the Pauli spin matrices (see Exercise 10.1.1). These satisfy

$$[S_i, S_j]_+ = \frac{\hbar^2}{2}\delta_{ij}1, \quad i, j = 1, 2, 3,$$

where we denote by 1 the identity operator on $\mathbb{C}^2$. Now introduce operators

$$f = \frac{1}{\hbar}(S_1 + iS_2)$$

and

$$f^\dagger = \frac{1}{\hbar}(S_1 - iS_2).$$

Exercise 10.2.5. Show that $f^\dagger$ is the adjoint f^* of f on $\mathbb{C}^2$ and then verify

$$[f, f]_+ = [f^\dagger, f^\dagger]_+ = 0 \quad \text{and} \quad [f, f^\dagger]_+ = 1.$$

The algebra of 2×2 complex matrices generated by f, $f^\dagger$ and 1 is a concrete faithful representation of the fermionic harmonic oscillator algebra.

Exercise 10.2.6. Show that, with respect to the standard basis $e_0 = \left(\begin{smallmatrix} 0 \\ 1 \end{smallmatrix}\right)$ and $e_1 = \left(\begin{smallmatrix} 1 \\ 0 \end{smallmatrix}\right)$ for $\mathbb{C}^2$,

$$f = \begin{pmatrix} 0 & 0 \\ 1 & 0 \end{pmatrix} \quad \text{and} \quad f^\dagger = \begin{pmatrix} 0 & 1 \\ 0 & 0 \end{pmatrix},$$

so that $fe_0 = 0$, $fe_1 = e_0$, $f^\dagger e_0 = e_1$ and $f^\dagger e_1 = e_0$.

We will refer to $f^\dagger$ and f as the *fermionic creation* and *annihilation operators*, respectively. Pursuing the analogy with the bosonic oscillator we introduce also the *fermionic number operator* by

$$N_F = f^\dagger f.$$

Note that this is self-adjoint and

$$N_F^2 = (f^\dagger f)^2 = (1 - ff^\dagger)^2 = 1 - ff^\dagger - ff^\dagger + ff^\dagger ff^\dagger$$
$$= 1 - ff^\dagger - ff^\dagger + f(1 - ff^\dagger)f^\dagger = 1 - ff^\dagger - (ff)(f^\dagger f^\dagger) = 1 - ff^\dagger$$
$$= N_F,$$

so it is also idempotent, that is, N_F is a projection. Consequently, N_F is an observable and its spectrum consists of just the two eigenvalues 0 and 1. The only possible observed values of N_F are 0 and 1, which is another reflection of the Pauli exclusion principle (at most one fermion per state).

Exercise 10.2.7. Show that

$$N_F = \begin{pmatrix} 1 & 0 \\ 0 & 0 \end{pmatrix},$$

so that $N_F e_0 = 0 \cdot e_0$ and $N_F e_1 = 1 \cdot e_1$.

In particular, the eigenspaces of N_F corresponding to the eigenvalues 0 and 1 are Span $\{e_0\}$ and Span $\{e_1\}$, respectively. The states in Span $\{e_0\}$ are said to be *unoccupied*, while those in Span $\{e_1\}$ are *occupied*.

Exercise 10.2.8. Define two new operators on $\mathbb{C}^2$ by

$$f_1 = f^\dagger + f \quad \text{and} \quad f_2 = -i(f^\dagger - f).$$

Show that the anticommutation relations (10.17) are equivalent to

$$[f_i, f_j]_+ = 2\delta_{ij}1, \quad i, j = 1, 2. \tag{10.18}$$

These are the defining relations for the *complex Clifford algebra* $\text{Cl}(2, \mathbb{C})$. We will not pursue this any further here, but for those who are interested, the standard introduction to Clifford algebras is Chapter 1 of [LM].

Taking the analogy with the bosonic oscillator one step further we will introduce a Hamiltonian H_F for the fermionic oscillator by

$$H_F = \frac{1}{2}\hbar\omega[f^\dagger, f]_- = \hbar\omega\left(N_F - \frac{1}{2}\right). \tag{10.19}$$

We will call the fermionic harmonic oscillator algebra together with this Hamiltonian the *fermionic harmonic oscillator*. Note that, unlike H_B, which is symmetric under the interchange $b \leftrightarrow b^\dagger$, H_F is antisymmetric under $f \leftrightarrow f^\dagger$.

Exercise 10.2.9. Show H_F is self-adjoint and that the spectrum of H_F is $\{-\frac{1}{2}\hbar\omega, \frac{1}{2}\hbar\omega\}$ so that these are the only two allowed energy levels for the fermionic harmonic oscillator.

At the risk of being tedious we would like to emphasize once again that there is no classical mechanical system that one can quantize to obtain the fermionic harmonic oscillator, or any other fermionic system; spin does not exist in classical physics and therefore neither do fermions. That is not to say, however, that one cannot construct something "like" a classical system and a procedure not unlike canonical quantization that will give rise to such systems. Precisely why one would want to do this might not be so clear. For the moment it will suffice to say that the effort is well rewarded, particularly when one begins the search for a framework in which bosons and fermions can be seen from a more unified point of view. This framework is called *supersymmetry*, and we will have an ever-so-brief encounter with it in the next section.

Carrying out this program, however, will require a rather abrupt paradigm shift. The obstacles (both mathematical and psychological) that will confront us are not unlike those faced by mathematicians wanting to solve polynomial equations before a rigorous construction of complex numbers. The intuitive solution was simple and clear; introduce a new "number" i with the formal property that $i^2 = -1$ and write the solutions to your equation as "numbers" that look like $a + bi$. Eventually this program was carried out with enough rigor to satisfy mathematicians, but this step was not entirely trivial and it did not come first. In a similar vein, if we put aside our mathematical scruples for a moment we might even write something like

$$[\,\theta_1 A_1, \theta_2 A_2\,]_- = \theta_1 \theta_2 (A_1 A_2) - \theta_2 \theta_1 (A_2 A_1)$$

$$= \begin{cases} \theta_1 \theta_2\,[A_1, A_2]_-, & \text{if } \theta_1 \text{ and } \theta_2 \text{ commute,} \\ \theta_1 \theta_2\,[A_1, A_2]_+, & \text{if } \theta_1 \text{ and } \theta_2 \text{ anticommute,} \end{cases}$$

where θ_1 and θ_2 are "numbers," assuming that we have somehow made sense of what it means for two "numbers" to anticommute. Granting for a moment that this is possible, one should be able to treat commutators and anticommutators (that is, bosons and fermions) from a more unified point of view.

The notion that one might use "anticommuting numbers" to build rigorous "pseudo-classical" models of fermions was first proposed in [Martin]. Subsequently, the underlying mathematical structures required to do this were extensively developed (see, for example, [Ber1]). We will have a bit more to say about the issues involved in constructing a unified point of view for bosons and fermions in the next section. Here we will limit ourselves to a few remarks on what a pseudo-classical analogue of a fermion might look like (a more detailed discussion is available in Chapter 7 of [Takh]).

The first order of business, of course, is to decide just what an "anticommuting number" is supposed to be. Fortuitously, objects of just the sort required have been well known in mathematics since the nineteenth century; they are called *generators of a Grassmann algebra*. Grassmann algebras can be viewed in a number of different ways (there is a very detailed treatment in Chapter 1, Part I, of [Ber2]). The most direct is in terms of generators and relations. A *Grassmann algebra* with *n generators* is an associative, unital, complex algebra $\mathrm{Gr}(n)$ for which there are generators $\theta_1, \ldots, \theta_n$ subject to the relations

$$\theta_i \theta_j + \theta_j \theta_i = 0, \quad i, j = 1, \ldots, n. \tag{10.20}$$

In particular, each of the generators is nilpotent in $\mathrm{Gr}(n)$; specifically,

$$\theta_i^2 = 0, \quad i = 1, \ldots, n.$$

Note that $\mathrm{Gr}(n)$ can be realized concretely in several ways. For example, if $A_{\mathbb{C}}(\theta_1, \ldots, \theta_n)$ is the associative, unital, complex algebra freely generated by $\theta_1, \ldots, \theta_n$ and J is the two-sided ideal generated by elements of the form $\theta_i \theta_j + \theta_j \theta_i$, then $\mathrm{Gr}(n)$ is the quotient $A_{\mathbb{C}}(\theta_1, \ldots, \theta_n)/J$. On the other hand, if $V_{\mathbb{C}}(\theta_1, \ldots, \theta_n)$ is the complex vector space freely generated by $\theta_1, \ldots, \theta_n$, then $\mathrm{Gr}(n)$ is isomorphic to the exterior algebra $\bigwedge V_{\mathbb{C}}(\theta_1, \ldots, \theta_n)$ of $V_{\mathbb{C}}(\theta_1, \ldots, \theta_n)$, that is, the antisymmetric part of the tensor algebra. From our present point of view the best way to think of $\mathrm{Gr}(n)$ is as the algebra $\mathbb{C}[\theta_1, \ldots, \theta_n]$ of complex polynomials in the anticommuting variables $\theta_1, \ldots, \theta_n$, any element α of which can be written uniquely in the form

$$\alpha = c_0 + c_1 \theta_1 + \cdots + c_n \theta_n + c_{12} \theta_1 \theta_2 + \cdots$$
$$+ c_{i_1 i_2 \cdots i_k} \theta_{i_1} \theta_{i_2} \cdots \theta_{i_k} + \cdots + c_{12 \cdots n} \theta_1 \theta_2 \cdots \theta_n, \tag{10.21}$$

where $1 \le i_1 < i_2 < \cdots < i_k \le n$ for $k = 1, \ldots, n$ and $c_0, c_1, c_{12}, \ldots, c_{12 \cdots n}$ are all in $\mathbb{C}$. There are no polynomials of higher degree because the generators are nilpotent.

Exercise 10.2.10. Show that the dimension of $\mathrm{Gr}(n)$ as a complex vector space is 2^n.

The decomposition (10.21) provides $\mathrm{Gr}(n)$ with a grading

$$\mathrm{Gr}(n) = \bigoplus_{k=0}^{n} \mathrm{Gr}^k(n), \tag{10.22}$$

where $\mathrm{Gr}^0(n) = \mathbb{C}$ and, for $1 \le k \le n$, $\mathrm{Gr}^k(n)$ consists of those elements that are homogeneous of degree k, that is, linear combinations of terms of the form $\theta_{i_1} \theta_{i_2} \cdots \theta_{i_k}$ with $1 \le i_1 < i_2 < \cdots < i_k \le n$. We will denote the degree of a homogeneous element α by $|\alpha|$. Note that the multiplication in $\mathrm{Gr}(n)$ satisfies

$$\mathrm{Gr}^k(n) \cdot \mathrm{Gr}^l(n) \subseteq \mathrm{Gr}^{k+l}(n),$$

where $\mathrm{Gr}^{k+l}(n) = 0$ if $k + l > n$.

Exercise 10.2.11. Show that if $\alpha = \theta_{i_1} \theta_{i_2} \cdots \theta_{i_k}$ and $\beta = \theta_{j_1} \theta_{j_2} \cdots \theta_{j_l}$, then

$$\alpha\beta = (-1)^{kl} \beta\alpha,$$

so for any homogeneous elements α and β,

$$\alpha\beta = (-1)^{|\alpha| \, |\beta|} \beta\alpha.$$

If we write the additive group of integers modulo 2 as $\mathbb{Z}_2 = \{\mathbf{0}, \mathbf{1}\}$, then (10.22) provides $\mathrm{Gr}(n)$ with a $\mathbb{Z}_2$-grading

$$\mathrm{Gr}(n) = \mathrm{Gr}^{\mathbf{0}}(n) \oplus \mathrm{Gr}^{\mathbf{1}}(n),$$

where $\mathrm{Gr}^{\mathbf{0}}(n)$ is the direct sum of the $\mathrm{Gr}^k(n)$ with k even and $\mathrm{Gr}^{\mathbf{1}}(n)$ is the direct sum of the $\mathrm{Gr}^k(n)$ with k odd. The elements of $\mathrm{Gr}^{\mathbf{0}}(n)$ are said to be *even*, while those in $\mathrm{Gr}^{\mathbf{1}}(n)$ are said to be *odd*. Even elements commute with either even elements or odd elements, but two odd elements anticommute. Of course, an element α of $\mathrm{Gr}(n)$ need not be either even or odd, but it can be written uniquely as the sum $\alpha = \alpha^{\mathbf{0}} + \alpha^{\mathbf{1}}$ of its even and odd parts.

Next we will define an involution on $\mathrm{Gr}(n)$, called *complex conjugation*, as follows. If $\alpha \in \mathrm{Gr}(n)$ is given by (10.21), then

$$\alpha \mapsto \overline{\alpha} = \overline{c}_0 + \overline{c}_1 \theta_1 \cdots + \overline{c}_{i_1 i_2 \cdots i_k} \theta_{i_k} \cdots \theta_{i_2} \theta_{i_1} + \cdots + \overline{c}_{12 \cdots n} \theta_n \cdots \theta_2 \theta_1$$

(note the reversal of the θ-factors).

Exercise 10.2.12. Show that this does, in fact, define an involution on $\mathrm{Gr}(n)$.

We will say that an element α in $\mathrm{Gr}(n)$ is *real* if $\overline{\alpha} = \alpha$ (these are the self-adjoint elements) and *imaginary* if $\overline{\alpha} = -\alpha$. Thus, each generator θ_i is a real element of $\mathrm{Gr}(n)$.

Exercise 10.2.13. Show that $\theta_{i_1} \theta_{i_2} \cdots \theta_{i_k}$ is real if and only if $\frac{k(k-1)}{2}$ is an even integer and it is imaginary otherwise.

Finally we observe that, being a complex vector space of dimension 2^n, $\mathrm{Gr}(n)$ admits a standard Hermitian inner product $\langle \, , \, \rangle_{\mathrm{Gr}(n)}$ that can be defined by simply decreeing that the basis

$$\{1, \theta_1, \ldots, \theta_n, \theta_1 \theta_2, \ldots, \theta_{i_1} \theta_{i_2} \cdots \theta_{i_k}, \ldots, \theta_1 \theta_2 \cdots \theta_n\},$$

with $1 \le i_1 < i_2 < \cdots < i_k \le n$, is an orthonormal basis and extending $\langle \, , \, \rangle_{\mathrm{Gr}(n)}$ so that it is linear in the second slot and conjugate linear in the first slot. In particular,

$$\langle \theta_{i_1} \theta_{i_2} \cdots \theta_{i_k}, \theta_{j_1} \theta_{j_2} \cdots \theta_{j_l} \rangle_{\mathrm{Gr}(n)} = \delta_{kl} \delta_{i_1 j_1} \delta_{i_2 j_2} \cdots \delta_{i_k j_k}.$$

With this, $\mathrm{Gr}(n)$ becomes a finite-dimensional Hilbert space and therefore a Banach space with the corresponding norm $\| \; \|_{\mathrm{Gr}(n)}$. The results of the following two exercises will not be called upon here, but are worth seeing.

Exercise 10.2.14. Show that $\|1\|_{\mathrm{Gr}(n)} = 1$ and $\| \alpha \beta \|_{\mathrm{Gr}(n)} \le \| \alpha \|_{\mathrm{Gr}(n)} \| \beta \|_{\mathrm{Gr}(n)}$, so that $\mathrm{Gr}(n)$ is a Banach algebra. *Hint*: For the definition and a very nice, elementary discussion of Banach algebras see Chapter Twelve of [Simm1].

Exercise 10.2.15. Show that $\| \overline{\alpha} \alpha \|_{\mathrm{Gr}(n)} = \| \alpha \|^2_{\mathrm{Gr}(n)}$, so that $\mathrm{Gr}(n)$ is a B^*-algebra. *Hint*: See Section 72 of [Simm1].

We can now return to fermionic systems. We should be clear on how we are interpreting the Grassmann algebra $Gr(n)$, which we will now identify with the algebra $\mathbb{C}[\theta_1, \ldots, \theta_n]$ of complex polynomials in the anticommuting variables $\theta_1, \ldots, \theta_n$. The generators $\theta_1, \ldots, \theta_n$ are thought of as the fermionic analogues of the ordinary (commuting) variables $q^1, \ldots, q^k, p_1, \ldots, p_k$ in the phase space of a classical mechanical system although we need not assume that n is even. The polynomials in $Gr(n)$ can be thought of as functions of $\theta_1, \ldots, \theta_n$, so the real (that is, self-adjoint) elements of $Gr(n)$ correspond to the fermionic analogue of classical observables.

Note that, unlike $q^1, \ldots, q^k, p_1, \ldots, p_k$, the fermionic variables $\theta_1, \ldots, \theta_n$ are not coordinates "in" anything. There is no fermionic analogue of the classical phase space, only of its algebra of functions. That a mathematical object can be completely characterized by some algebra of functions on it is an idea that goes back to a beautiful result of Banach and Stone which reconstructs any compact Hausdorff topological space X as the maximal ideal space of its algebra $C(X)$ of continuous complex-valued functions (see Section 74 of [Simm1]). This idea has evolved to an extraordinary level of depth, primarily through the work of Alain Connes on *noncommutative geometry* (see [Connes]).

On the other hand, $Gr(n)$ is a finite-dimensional, complex Hilbert space. We will conclude this section by describing a "Schrödinger-like" representation of the canonical anticommutation relations

$$[a_i, a_j]_+ = 0, \quad [a_i^\dagger, a_j^\dagger]_+ = 0 \quad \text{and} \quad [a_i, a_j^\dagger]_+ = \delta_{ij}I, \quad i,j = 1, 2, \ldots, n, \qquad (10.23)$$

by multiplication and differentiation operators on $Gr(n)$. Note that the canonical anticommutation relations in (10.14) have $i, j = 1, 2, \ldots$, but (10.23) has $i, j = 1, 2, \ldots, n$. It is possible to define the Grassmann algebra on an infinite number of anticommuting variables $\theta_1, \theta_2, \ldots$ and extend what we are about to do, but this requires more analytical work and we will simply refer those interested to [Tuyn].

The calculus of anticommuting variables was initiated and developed by Berezin, but we will require only the most elementary parts of this (see [Ber2] for a thorough discussion). For each generator θ_i we define a linear operator

$$\Theta_i : Gr(n) \to Gr(n)$$

on $Gr(n)$ that multiplies on the left by θ_i, that is,

$$\Theta_i \alpha = \theta_i \alpha.$$

For example, $\Theta_2(\theta_1\theta_2\theta_3) = -\Theta_2(\theta_2\theta_1\theta_3) = -\theta_2^2\theta_1\theta_3 = 0$. Next we define the *left partial differentiation operator*

$$\frac{\partial^L}{\partial \theta_i} : Gr(n) \to Gr(n)$$

on $\mathrm{Gr}(n)$. It is enough to define the operator on basis elements and extend linearly. The idea is quite simple. If there is no θ_i present, the result is zero. If there is a θ_i present, move it all the way to the left using anticommutativity and then drop θ_i. For example,

$$\frac{\partial^L}{\partial\theta_2}(\theta_1\theta_2\theta_3) = -\frac{\partial^L}{\partial\theta_2}(\theta_2\theta_1\theta_3) = -\theta_1\theta_3.$$

In general,

$$\frac{\partial^L}{\partial\theta_i}(\theta_{i_1}\cdots\theta_{i_k}) = \sum_{l=1}^{k}(-1)^{l-1}\delta_{ii_l}\theta_{i_1}\cdots\hat{\theta}_{i_l}\cdots\theta_{i_k},$$

where, as usual, the hat $\hat{\ }$ indicates that θ_{i_l} has been omitted. Similarly, we define the *right partial differentiation operator*

$$\frac{\partial^R}{\partial\theta_i} : \mathrm{Gr}(n) \to \mathrm{Gr}(n)$$

by moving θ_i all the way to the right, that is,

$$(\theta_{i_1}\cdots\theta_{i_k})\frac{\partial^R}{\partial\theta_i} = \sum_{l=1}^{k}(-1)^{k-l}\delta_{ii_l}\theta_{i_1}\cdots\hat{\theta}_{i_l}\cdots\theta_{i_k}.$$

Exercise 10.2.16. Prove the following graded versions of the product rule. For homogeneous elements α and β of $\mathrm{Gr}(n)$,

$$\frac{\partial^L}{\partial\theta_i}(\alpha\beta) = \frac{\partial^L\alpha}{\partial\theta_i}\beta + (-1)^{|\alpha|}\alpha\frac{\partial^L\beta}{\partial\theta_i}$$

and

$$(\alpha\beta)\frac{\partial^R}{\partial\theta_i} = \alpha\left(\beta\frac{\partial^R}{\partial\theta_i}\right) + (-1)^{|\beta|}\left(\alpha\frac{\partial^R}{\partial\theta_i}\right)\beta.$$

Hint: It is enough to do this for basis elements.

First we will show that, with respect to the inner product $\langle\,,\,\rangle_{\mathrm{Gr}(n)}$, the operators Θ_i and $\frac{\partial^L}{\partial\theta_i}$ are adjoints of each other. Since $\mathrm{Gr}(n)$ is finite-dimensional, there are no domain issues, so we need only show that $\langle\frac{\partial^L}{\partial\theta_i}\alpha, \beta\rangle_{\mathrm{Gr}(n)} = \langle\alpha, \Theta_i\beta\rangle_{\mathrm{Gr}(n)}$. Moreover, it is enough to prove this for basis elements, so we must show that

$$\left\langle \frac{\partial^L}{\partial\theta_i}(\theta_{i_1}\cdots\theta_{i_k}), \theta_{j_1}\cdots\theta_{j_l} \right\rangle_{\mathrm{Gr}(n)} \tag{10.24}$$

is equal to

$$\langle \theta_{i_1}\cdots\theta_{i_k}, \Theta_i(\theta_{j_1}\cdots\theta_{j_l}) \rangle_{\mathrm{Gr}(n)}. \tag{10.25}$$

Suppose first that (10.24) is nonzero. Then, in particular, $\theta_i = \theta_{i_m}$ for some $m = 1, \ldots, k$. Thus,

$$\frac{\partial^L}{\partial \theta_i}(\theta_{i_1} \cdots \theta_{i_k}) = (-1)^{m-1} \theta_{i_1} \cdots \theta_{i_{m-1}} \theta_{i_{m+1}} \cdots \theta_{i_k}.$$

Since (10.24) is nonzero, $\theta_{i_1} \cdots \theta_{i_{m-1}} \theta_{i_{m+1}} \cdots \theta_{i_k}$ must be equal to $\theta_{j_1} \cdots \theta_{j_l}$, so (10.24) is equal to $(-1)^{m-1}$. Moreover,

$$\Theta_i(\theta_{j_1} \cdots \theta_{j_l}) = \Theta_i(\theta_{i_1} \cdots \theta_{i_{m-1}} \theta_{i_{m+1}} \cdots \theta_{i_k})$$
$$= \theta_{i_m} \theta_{i_1} \cdots \theta_{i_{m-1}} \theta_{i_{m+1}} \cdots \theta_{i_k}$$
$$= (-1)^{m-1} \theta_{i_1} \cdots \theta_{i_{m-1}} \theta_{i_m} \theta_{i_{m+1}} \cdots \theta_{i_k},$$

and therefore (10.25) is also equal to $(-1)^{m-1}$, as required.

Exercise 10.2.17. Show that if (10.24) is equal to zero, then (10.25) is also equal to zero and thereby complete the proof that

$$\Theta_i = \left(\frac{\partial^L}{\partial \theta_i} \right)^*.$$

Finally, we will prove that the operators Θ_i and $\frac{\partial^L}{\partial \theta_i}$, $i = 1, \ldots, n$, provide an irreducible representation of the canonical anticommutation relations (10.23) on $\mathrm{Gr}(n)$. First we show that

$$[\Theta_i, \Theta_j]_+ = 0, \quad \left[\frac{\partial^L}{\partial \theta_i}, \frac{\partial^L}{\partial \theta_j} \right]_+ = 0 \quad \text{and} \quad \left[\Theta_i, \frac{\partial^L}{\partial \theta_j} \right]_+ = \delta_{ij} I, \quad i, j = 1, \ldots, n,$$

where I is the identity operator on $\mathrm{Gr}(n)$.

Exercise 10.2.18. Prove the first two of these.

Now consider

$$\left[\Theta_i, \frac{\partial^L}{\partial \theta_j} \right]_+ (\theta_{i_1} \cdots \theta_{i_k}) = \Theta_i \frac{\partial^L}{\partial \theta_j}(\theta_{i_1} \cdots \theta_{i_k}) + \frac{\partial^L}{\partial \theta_j} \Theta_i(\theta_{i_1} \cdots \theta_{i_k}).$$

Suppose first that $i = j$. If θ_i is not among the factors of $\theta_{i_1} \cdots \theta_{i_k}$, then the first term is zero and the second term is $\theta_{i_1} \cdots \theta_{i_k}$. If θ_i is among the factors of $\theta_{i_1} \cdots \theta_{i_k}$, then the second term is zero and the first term is $\theta_{i_1} \cdots \theta_{i_k}$. In either case, the result is proved. Now suppose $i \neq j$. If θ_j is among the factors of $\theta_{i_1} \cdots \theta_{i_k}$, then the two terms differ by a sign, so the sum is zero. If θ_j is not among the factors of $\theta_{i_1} \cdots \theta_{i_k}$, then both terms are zero and, again, the result is proved.

All that remains is to show that this representation of the canonical anticommutation relations by the operators Θ_i and $\frac{\partial^L}{\partial \theta_i}$, $i = 1, \ldots, n$, is irreducible, that is, if V is a linear subspace of $\mathrm{Gr}(n)$ that is invariant under all of these operators, then V is either the subspace consisting only of 0, or it is all of $\text{Gr}(n)$. First we note that if $\pi : \text{Gr}(n) \to \text{Gr}(n)$ is any linear map that commutes with all of the Θ_i and $\frac{\partial^L}{\partial \theta_i}$, then π must be a complex multiple of the identity operator on $\text{Gr}(n)$. Indeed,

$$\frac{\partial^L}{\partial \theta_i}(\pi(1)) = \pi\left(\frac{\partial^L}{\partial \theta_i}(1)\right) = \pi(0) = 0$$

for all $i = 1, \ldots, n$ implies that $\pi(1) = c \cdot 1$ for some $c \in \mathbb{C}$. But $\pi(\theta_i) = \pi(\Theta_i(1)) = \Theta_i(\pi(1)) = c \Theta_i(1) = c\theta_i$. Now, if $i_1 < i_2$, then

$$\pi(\theta_{i_1} \theta_{i_2}) = \pi(\Theta_{i_1}(\theta_{i_2})) = \Theta_{i_1}(\pi(\theta_{i_2})) = \Theta_{i_1}(c\theta_{i_2}) = c\Theta_{i_1}(\theta_{i_2}) = c(\theta_{i_1} \theta_{i_2}).$$

Continuing inductively gives

$$\pi(\theta_{i_1} \theta_{i_2} \cdots \theta_{i_k}) = c(\theta_{i_1} \theta_{i_2} \cdots \theta_{i_k})$$

and from this it follows that $\pi = cI$.

Now suppose that V is a linear subspace of $\text{Gr}(n)$ that is invariant under all of the operators Θ_i and $\frac{\partial^L}{\partial \theta_i}$, $i = 1, \ldots, n$.

Exercise 10.2.19. Show that the orthogonal complement $V^\perp$ of V with respect to the inner product $\langle \, , \, \rangle_{\text{Gr}(n)}$ is also invariant under all of the operators Θ_i and $\frac{\partial^L}{\partial \theta_i}$, $i = 1, \ldots, n$.

Let $\pi_V : \text{Gr}(n) = V \oplus V^\perp \to V$ be the orthogonal projection onto V. We claim that π_V commutes with all of the operators Θ_i and $\frac{\partial^L}{\partial \theta_i}$, $i = 1, \ldots, n$. To see this let $v + v^\perp \in \text{Gr}(n)$ with $v \in V$ and $v^\perp \in V^\perp$. Then $\Theta_i(v + v^\perp) = \Theta_i(v) + \Theta_i(v^\perp)$ with $\Theta_i(v) \in V$ and $\Theta_i(v^\perp) \in V^\perp$. Thus, $\pi_V(\Theta_i(v + v^\perp)) = \Theta_i(v) = \Theta_i(\pi_V(v + v^\perp))$, so $\pi_V \Theta_i = \Theta_i \pi_V$.

Exercise 10.2.20. Show that $\pi_V \frac{\partial^L}{\partial \theta_i} = \frac{\partial^L}{\partial \theta_i} \pi_V$.

Consequently, $\pi_V = cI$ for some $c \in \mathbb{C}$. Thus, Kernel (π_V) is either $\{0\}$ (if $c \neq 0$) or $\text{Gr}(n)$ (if $c = 0$). But Kernel $(\pi_V) = V^\perp$, so V is either $\text{Gr}(n)$ or $\{0\}$, as required.

There is an important aspect of the calculus of anticommuting variables called *Berezin integration* that we have not discussed here. The basics can be found in Section 2.3, Chapter 7, of [Takh]; for the whole story see [Ber1], [Ber2] and [Tuyn]. Section 4 of [Takh] describes the analogue of the Feynman path integral for anticommuting variables, which we will also not discuss here.

10.3 $N = 2$ supersymmetry and the harmonic oscillator

Supersymmetry (SUSY) is an idea that was born in the 1970s in physics. It soon became the darling of the particle physics community due to the promise it held for resolving many fundamental questions not addressed by the standard model of elementary particles and, at the same time, spawned an entirely new branch of mathematics, also called supersymmetry. We have seen that bosons and fermions behave quite differently and appear to require very different sorts of theoretical models for their description. SUSY postulates the existence of a new type of quantum symmetry operator that interchanges bosons and fermions and thereby permits one to view these two seemingly different sorts of animals from a more unified perspective. The idea is clearly very appealing, but even its most ardent proponents concede that its Achilles heel appears to be that the proposal is based on philosophical and aesthetic convictions rather than experimental facts. Here is how the situation was viewed in 1985 by a physicist whose work did much to advance the cause of supersymmetry.

> "Modern science shares with both the Greek and earlier philosophies the conviction that the observed universe is founded on simple underlying principles which can be understood and elaborated through disciplined intellectual endeavor. By the Middle Ages, this conviction had, in Christian Europe, become stratified into a system of Natural Philosophy that entirely and consciously ignored the realities of the physical world and based all its insights on thought, and Faith, alone. The break with the medieval tradition occurred when the scientific revolution of the 16th and 17th centuries established an undisputed dominance in the exact sciences of fact over idea, of observation over conjecture, and of practicality over aesthetics. Experiment and observation were established as the ultimate judge of theory. Modern particle physics, in seeking a single unified theory of all elementary particles and their fundamental interactions, appears to be reaching the limits of this process and finds itself forced, in part and often very reluctantly, to revert for guidelines to the "medieval" principles of symmetry and beauty."

Martin Sohnius [Sohn]

This may seem a rather peculiar attitude for theoretical physics, but it is not without precedent. Paul Dirac believed very strongly that mathematical beauty was not simply a useful criterion for, but a decisive guide to physical laws.

> "... it is more important to have beauty in one's equations than to have them fit experiment. ... If there is not complete agreement between the results of one's work and experiment, one should not allow oneself to be too discouraged, because the discrepancy may well be due to minor features that are not properly taken into account and that will get cleared up with further developments of the theory."

Paul Dirac [Dirac4]

Dirac's philosophical and aesthetic convictions served him extraordinarily well, the most obvious example being the relativistic equation for spin $\frac{1}{2}$ particles to which they guided him. The equation predicted the existence of an "antiparticle" for the electron which, at the time, was nowhere to be found in nature. It was not long, however, before Carl Anderson found the tracks of these so-called "positrons" in his cloud chamber, thereby vindicating Dirac's equation in a rather spectacular way. Supersymmetry makes an analogous prediction. Every elementary particle should have an associated "superpartner" (the hypothetical superpartner of the electron has been christened the "selectron"). None of these supersymmetric pairs is to be found in the currently known elementary particle zoo, so one must try to produce them as the result of collisions in particle accelerators. However, the masses of the superpartners are predicted to be quite large, so enormous amounts of energy would be required to produce them ($E = mc^2$). Until quite recently such energies were beyond the reach of even the largest particle accelerators. This changed, however, with the construction of the Large Hadron Collider (LHC) at CERN in Switzerland, which is capable of producing the required energies. For adherents to the SUSY philosophy, the news has not been good; none of the presumably accessible superpartners has been detected. This is quite a serious problem for particle physics which, for nearly 45 years, has pinned its hopes on supersymmetry (one can read more about this "crisis" in [LS]). Of course, the new branch of mathematics, also known as supersymmetry, does not really care. The mathematics is either beautiful, or it is not, either useful, or not, and it happens to be both. Whatever the eventual fate of SUSY in physics, the subject is still worth knowing something about.

Before getting started we should be clear on what we can, and cannot, deliver here. Whether one has in mind the physics or the mathematics, supersymmetry is a vast and complex subject and we cannot pretend to offer anything even remotely resembling an introduction to either. Although a bit dated now, [Sohn] is a highly regarded survey of the physics. For the mathematics one might consult [Tuyn], [Ber2] and [Vara] or, for a different approach, [Rogers]. Our very modest goal here is to describe the simplest possible system that exhibits supersymmetry and try to place it within some general context by defining what is called $N = 2$ *supersymmetry*. This done we will briefly describe a familiar mathematical structure that serves as a model of $N = 2$ supersymmetry and as the starting point of Edward Witten's extremely influential paper [Witt2] on Morse theory.

What we would like to do is combine the bosonic and fermionic oscillators into a single quantum system and then describe symmetry operators on this new system that interchange bosonic and fermionic states. In order to have all of the components assembled before us we will begin with a synopsis of what we have already done (see Table 10.1).

The combined system is called the *supersymmetric harmonic oscillator* or *SUSY harmonic oscillator* and its Hilbert space is taken to be

$$\mathcal{H}_S = \mathcal{H}_B \otimes \mathcal{H}_F = L^2(\mathbb{R}) \otimes \mathbb{C}^2.$$

The first order of business is to transfer all of the relevant operators on $\mathcal{H}_B$ and $\mathcal{H}_F$ to the tensor product. To do this we note the following general construction. Suppose $\mathcal{H}_1$ and $\mathcal{H}_2$ are two complex, separable Hilbert spaces and T_1 and T_2 are operators on $\mathcal{H}_1$ and $\mathcal{H}_2$, respectively. If T_1 and T_2 are bounded, then one defines a bounded operator

$$T_1 \otimes T_2 : \mathcal{H}_1 \otimes \mathcal{H}_2 \to \mathcal{H}_1 \otimes \mathcal{H}_2$$

Table 10.1: Bosonic and Fermionic Harmonic Oscillators

Bosonic harmonic oscillator

Hilbert space: $\mathcal{H}_B = L^2(\mathbb{R})$
Lowering operator: $b = \frac{1}{\sqrt{2m\omega\hbar}}(m\omega Q + iP)$
Raising operator: $b^\dagger = \frac{1}{\sqrt{2m\omega\hbar}}(m\omega Q - iP)$
Commutation relations: $[b,b]_- = [b^\dagger,b^\dagger]_- = 0$, $[b,b^\dagger]_- = 1$
Number operator: $N_B = b^\dagger b$
Hamiltonian: $H_B = \frac{1}{2}\hbar\omega[b^\dagger,b]_+ = \hbar\omega(N_B + \frac{1}{2})$
Orthonormal basis: $\psi_0 = (\frac{m\omega}{\hbar\pi})^{1/4} e^{-m\omega q^2/2\hbar}$, $\psi_n = \frac{1}{n!}(b^\dagger)^n\psi_0$, $n = 1,2,\ldots$
$N_B\psi_n = n\psi_n$, $n = 0,1,2,\ldots$
$H_B\psi_n = \mathcal{E}_n\psi_n = (n + \frac{1}{2})\hbar\omega\psi_n$, $n = 0,1,2,\ldots$

Fermionic harmonic oscillator

Hilbert space: $\mathcal{H}_F = \mathbb{C}^2$
Annihilation operator: $f = \frac{1}{\hbar}(S_1 + iS_2)$
Creation operator: $f^\dagger = \frac{1}{\hbar}(S_1 - iS_2)$
Anticommutation relations: $[f,f]_+ = [f^\dagger,f^\dagger]_+ = 0$, $[f,f^\dagger]_+ = 1$
Number operator: $N_F = f^\dagger f$
Hamiltonian: $H_F = \frac{1}{2}\hbar\omega[f^\dagger,f]_- = \hbar\omega(N_F - \frac{1}{2})$
Orthonormal basis: $e_0 = \binom{0}{1}$, $e_1 = \binom{1}{0}$
$N_F e_0 = 0 \cdot e_0$, $N_F e_1 = 1 \cdot e_1$
$H_F e_0 = -\frac{1}{2}\hbar\omega e_0$, $H_F e_1 = \frac{1}{2}\hbar\omega e_1$

as follows. On the algebraic tensor product $\mathcal{H}_1 \otimes_{alg} \mathcal{H}_2$ one defines $(T_1 \otimes T_2)(\phi_1 \otimes \phi_2) = (T_1\phi_1) \otimes (T_2\phi_2)$ and then extends by linearity. Since $\mathcal{H}_1 \otimes_{alg} \mathcal{H}_2$ is dense in $\mathcal{H}_1 \otimes \mathcal{H}_2$, boundedness gives a unique bounded extension to all of $\mathcal{H}_1 \otimes \mathcal{H}_2$. If the operators are unbounded, then one can define $T_1 \otimes T_2$ at least on the algebraic tensor product $\mathcal{D}(T_1) \otimes_{alg} \mathcal{D}(T_2)$, which is again dense in $\mathcal{H}_1 \otimes \mathcal{H}_2$. We will require only the case in which one of the two operators is the identity. Since we will have a number of such identity operators floating around we will abandon the rather nondescript 1 in favor of $\mathrm{id}_{\mathcal{H}_B}$, $\mathrm{id}_{\mathcal{H}_F}$ and $\mathrm{id}_{\mathcal{H}_S}$. We will also eschew the usual custom of using the same symbol for an operator on $\mathcal{H}_1$ or $\mathcal{H}_2$ and the induced operator on the tensor product obtained by tensoring with the identity. Now define the *number operator* N_S on $\mathcal{H}_S$ by

$$N_S = (N_B \otimes \mathrm{id}_{\mathcal{H}_F}) + (\mathrm{id}_{\mathcal{H}_B} \otimes N_F)$$

and the *Hamiltonian* H_S by

$$H_S = (H_B \otimes \mathrm{id}_{\mathcal{H}_F}) + (\mathrm{id}_{\mathcal{H}_B} \otimes H_F).$$

Exercise 10.3.1. Prove each of the following, where the products indicate compositions:

1. $N_B \otimes \mathrm{id}_{\mathcal{H}_F} = (b^\dagger \otimes \mathrm{id}_{\mathcal{H}_F})(b \otimes \mathrm{id}_{\mathcal{H}_F})$,

2. $\mathrm{id}_{\mathcal{H}_B} \otimes N_F = (\mathrm{id}_{\mathcal{H}_B} \otimes f^\dagger)(\mathrm{id}_{\mathcal{H}_B} \otimes f)$,
3. $H_S = \hbar\omega \left[(b^\dagger \otimes \mathrm{id}_{\mathcal{H}_F})(b \otimes \mathrm{id}_{\mathcal{H}_F}) + (\mathrm{id}_{\mathcal{H}_B} \otimes f^\dagger)(\mathrm{id}_{\mathcal{H}_B} \otimes f) \right] = \hbar\omega N_S$.

The next few items on the agenda will be easier to write if we introduce a bit of notation. If T is an operator on $\mathcal{H}$ and T has an eigenvalue λ we will write the corresponding eigenspace $E_\lambda(T)$. Note that the operator $\mathrm{id}_{\mathcal{H}_B} \otimes N_F$ on $\mathcal{H}_B \otimes \mathcal{H}_F$ is bounded and given by $(\mathrm{id}_{\mathcal{H}_B} \otimes N_F)(\psi \otimes (a_0 e_0 + a_1 e_1)) = \psi \otimes (a_1 e_1)$.

Exercise 10.3.2. Show that $\mathrm{id}_{\mathcal{H}_B} \otimes N_F$ is self-adjoint and satisfies $(\mathrm{id}_{\mathcal{H}_B} \otimes N_F)^2 = \mathrm{id}_{\mathcal{H}_B} \otimes N_F$ so that it is an orthogonal projection. Conclude that its spectrum consists of just the two eigenvalues $\lambda = 0, 1$ and show that $E_0(\mathrm{id}_{\mathcal{H}_B} \otimes N_F) = \mathcal{H}_B \otimes E_0(N_F)$ and $E_1(\mathrm{id}_{\mathcal{H}_B} \otimes N_F) = \mathcal{H}_B \otimes E_1(N_F)$.

An essential ingredient in what we would like to do here is a unitary involution τ on $\mathcal{H}_S$ defined by

$$\tau = \mathrm{id}_{\mathcal{H}_S} - 2(\mathrm{id}_{\mathcal{H}_B} \otimes N_F).$$

We show that τ is, indeed, a unitary involution as follows. We let φ be an element of $\mathcal{H}_S$ and compute

$$
\begin{aligned}
\tau^2\varphi &= \tau(\varphi - 2(\mathrm{id}_{\mathcal{H}_B} \otimes N_F)\varphi) \\
&= \varphi - 2(\mathrm{id}_{\mathcal{H}_B} \otimes N_F)\varphi - 2(\mathrm{id}_{\mathcal{H}_B} \otimes N_F)(\varphi - 2(\mathrm{id}_{\mathcal{H}_B} \otimes N_F)\varphi) \\
&= \varphi - 4(\mathrm{id}_{\mathcal{H}_B} \otimes N_F)\varphi + 4(\mathrm{id}_{\mathcal{H}_B} \otimes N_F)^2\varphi \\
&= \varphi,
\end{aligned}
$$

so τ is an involution:

$$\tau^2 = \mathrm{id}_{\mathcal{H}_S}$$

Exercise 10.3.3. Show that $\tau = \mathrm{id}_{\mathcal{H}_S} - 2(\mathrm{id}_{\mathcal{H}_B} \otimes N_F)$ satisfies $\langle \tau\varphi, \tau\varphi \rangle = \langle \varphi, \varphi \rangle$ for every $\varphi \in \mathcal{H}_S$ and conclude that τ is a unitary operator on $\mathcal{H}_S$.

Since τ is an involution, its spectrum consists of just the two eigenvalues $\lambda = 1, -1$. Furthermore, since $\tau\varphi = \varphi \Leftrightarrow \varphi - 2(\mathrm{id}_{\mathcal{H}_B} \otimes N_F)\varphi = \varphi \Leftrightarrow (\mathrm{id}_{\mathcal{H}_B} \otimes N_F)\varphi = 0$,

$$E_1(\tau) = E_0(\mathrm{id}_{\mathcal{H}_B} \otimes N_F) = \mathcal{H}_B \otimes E_0(N_F).$$

Similarly,

$$E_{-1}(\tau) = E_1(\mathrm{id}_{\mathcal{H}_B} \otimes N_F) = \mathcal{H}_B \otimes E_1(N_F).$$

Since τ is unitary, $\mathcal{H}_S$ has an orthogonal decomposition as the direct sum of these eigenspaces. Note that $E_0(N_F)$ and $E_1(N_F)$ are both isomorphic to $\mathbb{C}$, so $E_1(\tau)$ and $E_{-1}(\tau)$ are both copies of $\mathcal{H}_B$. We will denote them

$$\mathcal{H}_B^+ = E_1(\tau) = \{\varphi \in \mathcal{H}_S : \tau\varphi = \varphi\}$$

and

$$\mathcal{H}_B^- = E_{-1}(\tau) = \{\varphi \in \mathcal{H}_S : \tau\varphi = -\varphi\},$$

so that

$$\mathcal{H}_S = \mathcal{H}_B^+ \oplus \mathcal{H}_B^-.$$

The states in $\mathcal{H}_B^+$ will be referred to as *bosonic*, while those in $\mathcal{H}_B^-$ are *fermionic*.

Remark 10.3.1. If we had not already used up the symbols $\mathcal{H}_B$ and $\mathcal{H}_F$ for the Hilbert spaces of the bosonic and fermionic oscillators, we would have called these subspaces $\mathcal{H}_B$ and $\mathcal{H}_F$ instead of $\mathcal{H}_B^+$ and $\mathcal{H}_B^-$, as is more customary. When we generalize in a few moments and do not begin with the harmonic oscillators, we will adopt this more common notation.

Exercise 10.3.4. Show that if φ is in $\mathcal{H}_B^+$, then $H_S\varphi$ is also in $\mathcal{H}_B^+$, and if φ is in $\mathcal{H}_B^-$, then $H_S\varphi$ is also in $\mathcal{H}_B^-$.

Consequently, we can define operators

$$H_S^+ : \mathcal{H}_B^+ \to \mathcal{H}_B^+$$

and

$$H_S^- : \mathcal{H}_B^- \to \mathcal{H}_B^-$$

by restricting H_S. It is often convenient to think of H_S as a diagonal matrix acting on column vectors of states:

$$H_S = \begin{pmatrix} H_S^+ & 0 \\ 0 & H_S^- \end{pmatrix} : \begin{matrix} \mathcal{H}_B^+ \\ \oplus \\ \mathcal{H}_B^- \end{matrix} \to \begin{matrix} \mathcal{H}_B^+ \\ \oplus \\ \mathcal{H}_B^- \end{matrix}.$$

The Hamiltonian H_S, as you have just shown, preserves the bosonic and fermionic subspaces of $\mathcal{H}_S$. Now we will introduce two operators $D^\pm$ that switch them. Specifically, we define

$$D^+\varphi = \sqrt{\hbar\omega}\,(b \otimes \mathrm{id}_{\mathcal{H}_F})(\mathrm{id}_{\mathcal{H}_B} \otimes f^\dagger)$$

and

$$D^-\varphi = \sqrt{\hbar\omega}\,(b^\dagger \otimes \mathrm{id}_{\mathcal{H}_F})(\mathrm{id}_{\mathcal{H}_B} \otimes f).$$

Evaluating each of these at $\varphi = \psi \otimes A \in \mathcal{H}_S$ gives

$$D^+\varphi = D^+(\psi \otimes A) = \sqrt{\hbar\omega}\,(b\psi) \otimes (f^\dagger A)$$

and

$$D^-\varphi = D^-(\psi \otimes A) = \sqrt{\hbar\omega}\,(b^\dagger\psi) \otimes (fA).$$

Exercise 10.3.5. Prove each of the following:
1. $\varphi \in \mathcal{H}_B^+ \Rightarrow D^+\varphi \in \mathcal{H}_B^-$,
2. $\varphi \in \mathcal{H}_B^- \Rightarrow D^-\varphi \in \mathcal{H}_B^+$.

Thus,

$$D^+|_{\mathcal{H}_B^+} : \mathcal{H}_B^+ \to \mathcal{H}_B^-$$

and

$$D^-|_{\mathcal{H}_B^-} : \mathcal{H}_B^- \to \mathcal{H}_B^+.$$

Now note that

$$
\begin{aligned}
D^+D^-\varphi &= D^+(\sqrt{\hbar\omega}\,(b^\dagger\psi) \otimes (fA)) \\
&= \hbar\omega\,((b \otimes \mathrm{id}_{\mathcal{H}_F})(\mathrm{id}_{\mathcal{H}_B} \otimes f^\dagger))((b^\dagger\psi) \otimes (fA)) \\
&= \hbar\omega\,(b \otimes \mathrm{id}_{\mathcal{H}_F})((b^\dagger\psi) \otimes (f^\dagger fA)) \\
&= \hbar\omega\,(bb^\dagger\psi) \otimes (N_F A)
\end{aligned}
$$

and, similarly,

$$D^-D^+\varphi = \hbar\omega\,(N_B\psi) \otimes (ff^\dagger A).$$

Exercise 10.3.6. Show that

$$[D^+, D^-]_+ = D^+D^- + D^-D^+ = \hbar\omega\,(N_B \otimes \mathrm{id}_{\mathcal{H}_F} + \mathrm{id}_{\mathcal{H}_B} \otimes N_F) = H_S.$$

Next we compute

$$D^+D^+\varphi = \sqrt{\hbar\omega}\,D^+((b\psi) \otimes (f^\dagger A)) = \hbar\omega\,(bb\psi) \otimes (f^\dagger f^\dagger A) = \hbar\omega\,(bb\psi) \otimes 0 = 0$$

and, similarly,

$$D^-D^-\varphi = 0.$$

It follows from these that

$$(D^+ + D^-)^2 = D^+D^+ + D^+D^- + D^-D^+ + D^-D^- = 0 + H_S + 0 = H_S.$$

For future reference and in the hope that it might look vaguely familiar we will summarize a few of the identities we have just derived:

$$
\begin{aligned}
D^+D^+ &= 0, \\
D^-D^- &= 0, \qquad\qquad\qquad\qquad (10.26)
\end{aligned}
$$

$$\left(D^+ + D^-\right)^2 = H_S.$$

If these do not look familiar you might want to browse through Chapter 6 of [Warner] or Chapter 2 of [Jost] on Hodge theory. We will return to this shortly.

Soon we will distill from the example we are now investigating the essential ingredients of what is called $N = 2$ *supersymmetry*. There are three of them, the first two of which (a Hilbert space $\mathcal{H}_S$ of states and a unitary involution τ) we have already seen. The third is an operator Q whose square is taken to be the Hamiltonian of the system. We have just seen what this must be in our example. Define

$$Q = D^+ + D^-$$

so that

$$Q^2 = H_S.$$

From this it follows at once that Q commutes with the Hamiltonian so that

$$[Q, H_S]_- = 0.$$

Exercise 10.3.7. Show that
1. $\tau D^+ = -D^+\tau$ and
2. $\tau D^- = -D^-\tau$

and conclude that

$$[Q, \tau]_+ = 0.$$

Exercise 10.3.8. Show that for $\varphi_1, \varphi_2 \in \mathcal{H}_S$,

$$\langle\, \varphi_1, Q\varphi_2 \,\rangle = \langle\, Q\varphi_1, \varphi_2 \,\rangle.$$

At this point we have seen enough of the SUSY oscillator to motivate the general definition of $N = 2$ supersymmetry, but there is one last item we would like to discuss. Note that $\mathcal{H}_B$ has an orthonormal basis $\{\psi_{n_b}\}_{n_b=0}^\infty$ of eigenstates for H_B with energy $\hbar\omega\,(n_b + \tfrac{1}{2})$:

$$H_B\psi_{n_b} = \hbar\omega\left(n_b + \frac{1}{2}\right)\psi_{n_b}, \quad n_b = 0, 1, 2, \ldots.$$

The ground state is ψ_0 and has energy $\tfrac{1}{2}\hbar\omega$. All of the eigenspaces are one-dimensional. Similarly, $\mathcal{H}_F$ has an orthonormal basis $\{e_{n_f}\}_{n_f=0,1}$ of eigenstates for H_F with energy $\hbar\omega\,(n_f - \tfrac{1}{2})$:

$$H_F e_{n_f} = \hbar\omega\left(n_f - \frac{1}{2}\right)e_{n_f}, \quad n_f = 0, 1.$$

The ground state is e_0 and has energy $-\frac{1}{2}\hbar\omega$. All of the eigenspaces are one-dimensional. Therefore, $\mathcal{H}_S$ has an orthonormal basis consisting of all

$$\psi_{n_b} \otimes e_{n_f}, \quad n_b = 0, 1, 2, \ldots, \quad n_f = 0, 1.$$

Each $\psi_{n_b} \otimes e_{n_f}$ is an eigenstate for H_S with energy $\hbar\omega(n_b + n_f)$ because

$$H_S(\psi_{n_b} \otimes e_{n_f}) = (H_B \otimes \mathrm{id}_{\mathcal{H}_F})(\psi_{n_b} \otimes e_{n_f}) + (\mathrm{id}_{\mathcal{H}_B} \otimes H_F)(\psi_{n_b} \otimes e_{n_f})$$

$$= \hbar\omega\left(n_b + \frac{1}{2}\right)\psi_{n_b} \otimes e_{n_f} + \hbar\omega\left(n_f - \frac{1}{2}\right)\psi_{n_b} \otimes e_{n_f}$$

$$= \hbar\omega\,(n_b + n_f)\,\psi_{n_b} \otimes e_{n_f}.$$

The ground state is $\psi_0 \otimes e_0$ and has energy 0. While the ground state is unique, we will show now that all of the remaining eigenvalues are degenerate (the eigenspaces have dimension greater than 1). To see this, fix an $n_b \geq 1$. Now note that

$$\psi_{n_b} \otimes e_0 \quad \text{and} \quad \psi_{n_b-1} \otimes e_1$$

have the same energy $\hbar\omega n_b$ and

$$\psi_{n_b} \otimes e_1 \quad \text{and} \quad \psi_{n_b+1} \otimes e_0$$

both have energy $\hbar\omega(n_b + 1)$. Also,

$$\psi_0 \otimes e_1 \quad \text{and} \quad \psi_1 \otimes e_0$$

both have energy $\hbar\omega$. One can phrase these results rather suggestively as follows. The simultaneous creation of one bosonic quantum of energy and annihilation of one fermionic quantum of energy (or *vice versa*) leaves the total energy unchanged.

There is much more one could say about the SUSY oscillator (see, for example, Sections 2.1–2.5 of [Bagchi]), but we would now like to abstract the essential features of this example in the form of a definition. We will say that a triple

$$(\mathcal{H}_S, \tau, Q)$$

consisting of a complex, separable Hilbert space $\mathcal{H}_S$, a unitary involution τ on $\mathcal{H}_S$ and a self-adjoint operator Q on $\mathcal{H}_S$ that anticommutes with τ is an instance of $N = 2$ *supersymmetry*. The supersymmetric harmonic oscillator is one such and we will see another soon. First we would like to develop a few of the elementary consequences of the definition. We will pursue this only far enough that we can say a few words about what Edward Witten [Witt2] has called "the most important question about a supersymmetric theory."

Since τ is a unitary involution, its spectrum consists only of the two eigenvalues ±1 and $\mathcal{H}_S$ has an orthogonal decomposition into the direct sum of the corresponding eigenspaces $E_{\pm1}(\tau)$. We define the *bosonic* and *fermionic* subspaces of $\mathcal{H}_S$ by

$$\mathcal{H}_B = E_1(\tau)$$

and

$$\mathcal{H}_F = E_{-1}(\tau),$$

respectively, so that

$$\mathcal{H}_S = \mathcal{H}_B \oplus \mathcal{H}_F.$$

The *Hamiltonian* of $(\mathcal{H}_S, \tau, Q)$ is defined by

$$H_S = Q^2.$$

There are domain issues here, of course, but these need to be resolved in each example separately. The same is true of much of what follows. Note that since Q is self-adjoint,

$$\langle H_S\psi_1, \psi_2 \rangle = \langle \psi_1, H_S\psi_2 \rangle,$$

and since Q anticommutes with τ,

$$\tau H_S = \tau Q^2 = (\tau Q)(Q) = (-Q\tau)Q = -Q(\tau Q) = -Q(-Q\tau) = Q^2\tau = H_S\tau,$$

so

$$[H_S, \tau]_- = 0.$$

Now, if $\varphi \in \mathcal{H}_B$, then $\tau\varphi = \varphi$, so

$$H_S\varphi = H_S(\tau\varphi) = \tau H_S\varphi,$$

so

$$\tau H_S\varphi = \tau^2 H_S\varphi = H_S\varphi,$$

and therefore $H_S\varphi$ is also in $\mathcal{H}_B$. Thus,

$$H_S(\mathcal{H}_B) \subseteq \mathcal{H}_B.$$

Exercise 10.3.9. Prove each of the following:
1. $H_S(\mathcal{H}_F) \subseteq \mathcal{H}_F$,
2. $Q(\mathcal{H}_B) \subseteq \mathcal{H}_F$,

3. $Q(\mathcal{H}_F) \subseteq \mathcal{H}_B$.

Thus, H_S preserves the bosonic and fermionic subspaces, but Q reverses them. We can therefore define operators

$$H_B = H_S|_{\mathcal{H}_B} : \mathcal{H}_B \to \mathcal{H}_B,$$
$$H_F = H_S|_{\mathcal{H}_F} : \mathcal{H}_F \to \mathcal{H}_F,$$
$$D^+ = Q|_{\mathcal{H}_B} : \mathcal{H}_B \to \mathcal{H}_F,$$
$$D^- = Q|_{\mathcal{H}_F} : \mathcal{H}_F \to \mathcal{H}_B.$$

Then

$$H_B = D^- D^+$$

and

$$H_F = D^+ D^-.$$

Exercise 10.3.10. Show that for $\psi \in \mathcal{H}_B$ and $\varphi \in \mathcal{H}_F$,

$$\langle \psi, D^- \varphi \rangle = \langle D^+ \psi, \varphi \rangle.$$

Exercise 10.3.11. Prove each of the following:
1. Kernel (H_B) = Kernel (D^+),
2. Kernel (H_F) = Kernel (D^-),
3. Kernel (H_S) = Kernel (Q) = Kernel $(D^+) \oplus$ Kernel (D^-).

It is sometimes convenient to write H_S and Q as 2×2 matrices of operators acting on column vectors of states:

$$H_S = \begin{pmatrix} H_B & 0 \\ 0 & H_F \end{pmatrix} = \begin{pmatrix} D^- D^+ & 0 \\ 0 & D^+ D^- \end{pmatrix} : \begin{matrix} \mathcal{H}_B \\ \oplus \\ \mathcal{H}_F \end{matrix} \to \begin{matrix} \mathcal{H}_B \\ \oplus \\ \mathcal{H}_F \end{matrix},$$

$$Q = \begin{pmatrix} 0 & D^- \\ D^+ & 0 \end{pmatrix} : \begin{matrix} \mathcal{H}_B \\ \oplus \\ \mathcal{H}_F \end{matrix} \to \begin{matrix} \mathcal{H}_B \\ \oplus \\ \mathcal{H}_F \end{matrix}.$$

Finally, we will define operators Q_1 and Q_2, called *supercharges*, or *generators of the supersymmetry*. The fact that there are two of them accounts for the "$N = 2$" in "$N = 2$ supersymmetry." Specifically, we set

$$Q_1 = Q$$

and

$$Q_2 = iQ\tau.$$

Both of these reverse the bosonic and fermionic subspaces of $\mathcal{H}_S$ and satisfy

$$\text{Kernel } Q_1 = \text{Kernel } Q_2 = \text{Kernel } Q = \text{Kernel } H_S,$$
$$\langle Q_1\psi_1, \psi_2 \rangle = \langle Q\psi_1, \psi_2 \rangle = \langle \psi_1, Q\psi_2 \rangle = \langle \psi_1, Q_1\psi_2 \rangle$$

and

$$\langle Q_2\psi_1, \psi_2 \rangle = \langle iQ\tau\psi_1, \psi_2 \rangle = -i\langle \psi_1, \tau Q\psi_2 \rangle = i\langle \psi_1, Q\tau\psi_2 \rangle = \langle \psi_1, Q_2\psi_2 \rangle.$$

Moreover,

$$Q_1^2 = Q^2 = H_S$$

and

$$Q_2^2 = -(Q\tau)(Q\tau) = -Q(\tau Q)\tau = Q(Q\tau)\tau = Q^2\tau^2 = Q^2 = H_S,$$

so

$$Q_1^2 = Q_2^2 = H_S.$$

Exercise 10.3.12. Show that

$$[Q_1, Q_2]_+ = 0.$$

We can write these last few identities as

$$[Q_i, Q_j]_+ = 2\delta_{ij}H_S, \quad i, j = 1, 2.$$

Exercise 10.3.13. Show that Q_1 and Q_2 both commute with the Hamiltonian, that is,

$$[H_S, Q_i]_- = 0, \quad i = 1, 2.$$

The seemingly endless barrage of identities that you are being subjected to actually has a serious purpose. We are in the process of uncovering a very fundamental notion in supersymmetry. To fully expose it, however, will require one more identity and, for this, a few definitions. Operators such as H_S that preserve the bosonic and fermionic subspaces of $\mathcal{H}_S$ are said to be *even* and to have *degree* $0 \in \mathbb{Z}_2$. Operators such as Q_1 and Q_2 that reverse the bosonic and fermionic subspaces of $\mathcal{H}_S$ are said to be *odd* and to have *degree* $1 \in \mathbb{Z}_2$. If an operator A is either even or odd we will say that it is *homogeneous* and we will write its degree as $|A| \in \mathbb{Z}_2$. Since $\mathcal{H}_S = \mathcal{H}_B \oplus \mathcal{H}_F$, any operator can be written as the sum of an even operator and an odd operator.

As we have seen, some operators satisfy commutation relations and some satisfy anticommutation relations. To express all of these various relations in a uniform way we will define the *supercommutator* of two homogeneous operators A and B on $\mathcal{H}_S$ by

$$[A, B]_S = AB - (-1)^{|A||B|} BA$$

and then extend to all operators by decreeing that $[\,,\,]_S$ should be bilinear. Note that products and sums of degrees are computed in $\mathbb{Z}_2 = \{\mathbf{0}, \mathbf{1}\}$ and $(-1)^{\mathbf{0}} = 1$, while $(-1)^{\mathbf{1}} = -1$. If A and B are both odd, then $[A, B]_S = [A, B]_+$, while if either A or B is even, $[A, B]_S = [A, B]_-$. Note that it follows immediately from this that

$$[A, B]_S = -(-1)^{|A||B|}[B, A]_S.$$

The final identity we need is called the *super Jacobi identity* and it states that

$$(-1)^{|A||C|}\,[A, [B, C]_S]_S + (-1)^{|C||B|}\,[C, [A, B]_S]_S + (-1)^{|B||A|}\,[B, [C, A]_S]_S = 0.$$

To prove this we need to write out

$$
\begin{aligned}
[A, [B, C]_S]_S &= [A, BC - (-1)^{|B||C|}CB]_S = [A, BC]_S + (-1)^{|B||C|}[A, CB]_S \\
&= ABC - (-1)^{|A||BC|}BCA - (-1)^{|B||C|}\left(ACB - (-1)^{|A||CB|}CBA\right) \\
&= ABC - (-1)^{|A||B|+|A||C|}BCA - (-1)^{|B||C|}\left(ACB - (-1)^{|A||C|+|A||B|}CBA\right).
\end{aligned}
$$

Consequently,

$$
\begin{aligned}
(-1)^{|A||C|}[A, [B, C]_S]_S = {}&(-1)^{|A||C|}ABC - (-1)^{|A||B|}BCA \\
&- (-1)^{|A||C|+|B||C|}ACB + (-1)^{|A||B|+|B||C|}CBA.
\end{aligned}
$$

Changing the names gives

$$
\begin{aligned}
(-1)^{|C||B|}[C, [A, B]_S]_S = {}&(-1)^{|C||B|}CAB - (-1)^{|C||A|}ABC \\
&- (-1)^{|C||B|+|A||B|}CBA + (-1)^{|C||A|+|A||B|}BAC
\end{aligned}
$$

and

$$
\begin{aligned}
(-1)^{|B||A|}[B, [C, A]_S]_S = {}&(-1)^{|B||A|}BCA - (-1)^{|B||C|}CAB \\
&- (-1)^{|B||A|+|C||A|}BAC + (-1)^{|B||C|+|C||A|}ACB.
\end{aligned}
$$

Adding these last three, one sees that everything cancels on the right-hand side.

Here is what we have shown. Let $\mathfrak{g} = \mathfrak{g}_0 \oplus \mathfrak{g}_1$, where $\mathfrak{g}_0$ is the vector space freely generated by the even operator H_S and $\mathfrak{g}_1$ is the vector space freely generated by the two odd operators Q_1 and Q_2. The elements of $\mathfrak{g_0}$ are even and of degree **0**, while those of $\mathfrak{g_1}$ are odd and of degree **1**. For homogeneous elements A and B in $\mathfrak{g}$, define

$$[A, B]_S = AB - (-1)^{|A|\,|B|} BA$$

and extend by bilinearity. Then $[Q_i, Q_j]_S = 2\delta_{ij} H_S$, $i, j = 1, 2$, and $[H_S, Q_i]_S = 0$, $i = 1, 2$, imply that

$$[\,\mathfrak{g_i},\,\mathfrak{g_j}\,]_S \subseteq \mathfrak{g_{i+j}}, \quad \mathbf{i}, \mathbf{j} \in \mathbb{Z}_2, \tag{10.27}$$

and for homogeneous elements we have

$$[A, B]_S = -(-1)^{|A|\,|B|} [B, A]_S \tag{10.28}$$

and

$$(-1)^{|A|\,|C|} [A, [B, C]_S]_S + (-1)^{|C|\,|B|} [C, [A, B]_S]_S + (-1)^{|B|\,|A|} [B, [C, A]_S]_S = 0. \tag{10.29}$$

Let us define a *Lie superalgebra* to be a $\mathbb{Z}_2$-graded vector space $\mathfrak{g} = \mathfrak{g_0} \oplus \mathfrak{g_1}$ with a bilinear map $[\,,\,]_S : \mathfrak{g} \oplus \mathfrak{g} \to \mathfrak{g}$ satisfying (10.27) and, for homogeneous elements of $\mathfrak{g}$, (10.28) and (10.29). We see then that $N = 2$ supersymmetry gives rise to a natural Lie superalgebra.

Lie superalgebras are sometimes called super Lie algebras, but the terminology can be misleading since a Lie superalgebra is not a Lie algebra at all. Note, however, that the restriction of $[\,,\,]_S$ to $\mathfrak{g_0} \times \mathfrak{g_0}$ is an ordinary Lie bracket and therefore $\mathfrak{g_0}$ is a Lie algebra. For the applications of supersymmetry to particle physics one must proceed in the other direction, that is, one must begin with a given Lie algebra $\mathfrak{g_0}$ (such as the so-called Poincaré algebra) and construct a Lie superalgebra for which $\mathfrak{g_0}$ is the even part. Such Lie superalgebras are regarded as the infinitesimal generators of supersymmetries in the same way that Lie algebras are regarded as infinitesimal generators of symmetries in mechanics. A proper introduction to this requires very sophisticated physical ideas that are beyond our level here (one might begin with the preface and introduction to [Sohn]). The rigorous study of Lie superalgebras was initiated by Kac in [KacV].

Before leaving the general subject of $N = 2$ supersymmetry and proceeding to our final example we would like to say a few words about a fundamental problem in supersymmetry that has had a profound impact on both physics and mathematics. Recall that the Hamiltonian for the supersymmetric harmonic oscillator has a ground state $\psi_0 \otimes e_0$ with energy 0 and that this ground state is unique. Now, in any $N = 2$ supersymmetric theory the Hamiltonian is, by definition, a square ($H_S = Q^2$), so it follows from self-adjointness that its spectrum $\sigma(H_S)$ is contained in $[0, \infty)$. In general, 0 may or may not be in the spectrum and, if it is, it may or may not be an eigenvalue. We will let

$$E_0 = \inf \sigma(H_S) \tag{10.30}$$

and we will consider only the case in which E_0 is actually an eigenvalue of H_S so that there exists a $\psi_0 \in \mathcal{H}_S$ with $\|\psi_0\| = 1$ and

$$H_S\psi_0 = E_0\psi_0. \tag{10.31}$$

Then ψ_0 is called a *ground state* of ($\mathcal{H}_S$, τ, Q). Note that it follows from this that

$$H_S(Q\psi_0) = Q(H_S\psi_0) = Q(E_0\psi_0) = E_0(Q\psi_0). \tag{10.32}$$

Thus, $Q\psi_0$ is also an eigenstate of H_S with energy E_0. Let us think about this physically for a second. We think of ψ_0 as the state of some particle, either boson or fermion, with ground state energy E_0, and $Q\psi_0$ is the state of another particle, either fermion or boson because Q reverses these, with the same ground state energy E_0. The particles corresponding to ψ_0 and $Q\psi_0$ are called *superpartners* and quantum field theory has something to say about these superpartners. Specifically, if the ground state energy $E_0 = 0$, then ψ_0 *and* $Q\psi_0$ *must have the same mass*; this is not at all obvious, of course.

The problem with this is that bosons and fermions with the same mass have simply not been observed, and if they exist, they *should* have been observed, since lots of bosons and fermions have been produced in particle accelerators and the mass/energy is the only impediment to the production of particles in accelerators. The conclusion we draw from this is that a realistic supersymmetric theory cannot have a ground state energy $E_0 = 0$. There is some terminology in physics used to describe what is going on here.

An $N = 2$ supersymmetry ($\mathcal{H}_S$, τ, Q) for which $E_0 = 0$ is said to be *unbroken*; if $E_0 > 0$, then the supersymmetry ($\mathcal{H}_S$, τ, Q) is said to be *spontaneously broken*. We have just seen that a realistic supersymmetric theory must be spontaneously broken and this simply amounts to the requirement that the equation $H_S\psi = 0$ has no non-trivial solutions in $\mathcal{H}_S$ or, equivalently (Exercise 10.3.11), the equation

$$Q\psi = 0$$

has no nontrivial solutions in $\mathcal{H}_S$. Stated otherwise, in a spontaneously broken super-symmetric theory zero is not an eigenvalue of Q.

Remark 10.3.2. Group theory provides a context in which *symmetry breaking* in physics can be defined precisely and studied. We will not pursue this at all and will say only that if supersymmetry (symmetry between bosons and fermions) is a real physical symmetry, then the symmetry must have been *broken* at some point in the past and is no longer visible to us because we do not observe bosons and fermions of the same mass. This is analogous to the discovery made by Pierre Curie that, beyond a certain critical temperature T_c, ferromagnetic materials lose their magnetic properties because the alignment of the magnetic moments of the atoms is destroyed by thermal agitation. For $T > T_c$ the ground state (state of minimal energy) has a rotational symmetry in the sense that there is no preferred direction in space, but for

$T < T_c$ this symmetry is destroyed. Note that the rotational symmetry can be restored by raising the temperature, that is, the ground state energy. It has been suggested that supersymmetry was visible in the very early universe shortly after the Big Bang and could become visible again if we had access to accelerators with sufficiently high energies; so far this does not appear to be working out.

How could one show that $Q\psi = 0$ has no nontrivial solutions in $\mathcal{H}_S$? Generally, this is a difficult problem. Typically, Q is some differential operator and we are asking about its smallest eigenvalue. In some circumstances one can obtain a lower bound on the energy eigenvalues, and if this happens to be positive, then 0 cannot be an eigenvalue and $Q\psi = 0$ can have no nontrivial solutions. As a rule, however, such direct estimates are generally inaccessible. Witten [Witt2] proposed an indirect method that can sometimes provide an answer to our question and we will just briefly describe the idea.

We have Q written as

$$Q = \begin{pmatrix} 0 & D^- \\ D^+ & 0 \end{pmatrix} : \begin{matrix} \mathcal{H}_B \\ \oplus \\ \mathcal{H}_F \end{matrix} \rightarrow \begin{matrix} \mathcal{H}_B \\ \oplus \\ \mathcal{H}_F \end{matrix}$$

and will now focus our attention on $D^+ : \mathcal{H}_B \to \mathcal{H}_F$. Then $D^- : \mathcal{H}_F \to \mathcal{H}_B$ is the adjoint of D^+ (Exercise 10.3.10). We will assume that D^+ is a Fredholm operator. Recall that a densely defined, closed operator $T : \mathcal{H}_1 \to \mathcal{H}_2$ between separable, complex Hilbert spaces is *Fredholm* if it has closed range and both Kernel T and Kernel T^* are finite-dimensional. We can then define the *Fredholm index* of D^+ by

$$\text{ind } D^+ = \dim (\text{Kernel } D^+) - \dim (\text{Kernel } D^-).$$

Now, it follows from Exercise 10.3.11 that

$$\dim (\text{Kernel } Q) = \dim (\text{Kernel } D^+) + \dim (\text{Kernel } D^-),$$

so, in particular, if ind $D^+ \neq 0$, then $\dim (\text{Kernel } Q) \neq 0$ as well, so zero is an eigenvalue of Q and the supersymmetry is unbroken. In the context of supersymmetry, ind D^+ is generally called the *Witten index* and computing it for a given supersymmetric theory is a problem of considerable interest. For those with an interest in pursuing further some specific examples that arise in physics we might suggest [Witt1] or, for a mathematically rigorous treatment, [JLL]. We will say no more about these, but will move on to Section 10.4 and describe one more example of an $N = 2$ supersymmetry ($\mathcal{H}_S, \tau, Q$).

10.4 $N = 2$ supersymmetry and Hodge theory

The example we have in mind in this section arises, not in physics, but in mathematics. The idea behind the example is quite simple, but formulating it all precisely enough to fit our rigorous definition of $N = 2$ supersymmetry draws upon quite a surprising amount of machinery. Since some of this material is likely to exceed the level of preparedness we have heretofore assumed of our generic reader we will spend a bit more time introducing the ideas as we proceed rather than relegating them to various appendices. Good sources for much of what we need are [Warner] and [Jost].

We begin with a compact, connected, oriented, smooth, n-dimensional manifold X with a Riemannian metric $\mathbf{g}$. Recall that $\mathbf{g}$ assigns to each tangent space $T_x(X)$ a positive definite inner product $\mathbf{g}_x = \langle \, , \, \rangle_x$ and that these vary smoothly from point to point in the sense that if $\mathbf{V}$ and $\mathbf{W}$ are smooth vector fields on X, then $x \in X \mapsto \langle \mathbf{V}(x), \mathbf{W}(x) \rangle_x \in \mathbb{R}$ is a smooth real-valued function on X. We will denote by $\Omega^p(X)$, $p = 0, 1, \ldots, n$, the $C^\infty(X; \mathbb{R})$-module of real-valued p-forms on X. Our construction will begin with the de Rham complex

$$0 \rightarrow \Omega^0(X) \overset{d_0}{\rightarrow} \Omega^1(X) \overset{d_1}{\rightarrow} \Omega^2(X) \overset{d_2}{\rightarrow} \cdots$$
$$\overset{d_{p-1}}{\rightarrow} \Omega^p(X) \overset{d_p}{\rightarrow} \Omega^{p+1}(X) \overset{d_{p+1}}{\rightarrow} \cdots \overset{d_{n-1}}{\rightarrow} \Omega^n(X) \rightarrow 0, \tag{10.33}$$

where each d_p is the exterior derivative acting on p-forms so that $d_{p+1}d_p = 0$ for each $p = 0, 1, \ldots, n - 2$. The de Rham cohomology groups with real coefficients are defined for $0 \le p \le n$ by

$$H^p(X; \mathbb{R}) = \text{Kernel} \, (d_p) / \text{Image} \, (d_{p-1}).$$

In particular, for $p = 0$ and $p = n$,

$$H^0(X; \mathbb{R}) = \text{Kernel} \, (d_0), \quad H^n(X; \mathbb{R}) = \Omega^n(X) / \text{Image} \, (d_{n-1}).$$

However, since we are trying to build a *complex* Hilbert space $\mathcal{H}_S$ we will actually be interested in complex-valued differential forms. These are obtained simply by tensoring (over $\mathbb{R}$) each $\Omega^p(X)$ with $\mathbb{C}$ thought of as a two-dimensional real vector space.

Exercise 10.4.1. As a reminder, we should review a few facts about the complexification of a real vector space $\mathcal{V}$. For this we regard $\mathbb{C}$ as a two-dimensional real vector space and define $\mathcal{V}^\mathbb{C} = \mathcal{V} \otimes \mathbb{C}$, where the tensor product is over $\mathbb{R}$. If $\{e_1, \ldots, e_n\}$ is a basis for $\mathcal{V}$ and if we take $\{1, i\}$ as a basis for $\mathbb{C}$, then $\{e_1 \otimes 1, \ldots, e_n \otimes 1, e_1 \otimes i, \ldots, e_n \otimes i\}$ is a basis for $\mathcal{V}^\mathbb{C}$ as a real vector space.
1. Show that $\mathcal{V}^\mathbb{C}$ becomes a complex vector space if one defines complex scalar multiplication by $\alpha(v \otimes \beta) = v \otimes (\alpha\beta)$ for all $v \in \mathcal{V}$ and all $\alpha, \beta \in \mathbb{C}$.
2. Show that any element v of $\mathcal{V}^\mathbb{C}$ can be written as $v = v_1 \otimes 1 + v_2 \otimes i$, where $v_1, v_2 \in \mathcal{V}$, and henceforth adopt the usual notational convention and write this simply as $v_1 + v_2 i$.
3. Show that if $a_1 + a_2 i \in \mathbb{C}$ and $v_1 + v_2 i \in \mathcal{V}^\mathbb{C}$, then

$$(a_1 + a_2 i)(v_1 + v_2 i) = (a_1 v_1 - a_2 v_2) + (a_1 v_2 + a_2 v_1)i.$$

4. Show that the complex dimension of $\mathcal{V}^{\mathbb{C}}$ is equal to the real dimension of $\mathcal{V}$.
5. Let $\langle\,,\,\rangle$ be a positive definite inner product on $\mathcal{V}$ and define $\langle\,,\,\rangle^{\mathbb{C}} : \mathcal{V}^{\mathbb{C}} \times \mathcal{V}^{\mathbb{C}} \to \mathbb{C}$ by

$$\langle v_1 + v_2 i,\, w_1 + w_2 i \rangle^{\mathbb{C}} = \langle v_1, w_1 \rangle + \langle v_2, w_2 \rangle + i\left(\langle v_1, w_2 \rangle - \langle v_2, w_1 \rangle\right).$$

Show that $\langle\,,\,\rangle^{\mathbb{C}}$ is a Hermitian inner product on $\mathcal{V}^{\mathbb{C}}$, complex linear in the *second* slot and conjugate linear in the first.
6. Apply this construction to $\Omega^p(X)$ and show that the elements of $\Omega^p(X)^{\mathbb{C}} = \Omega^p(X) \otimes \mathbb{C}$ can be regarded as differential forms with complex-valued coefficients.

One defines d_p on $\Omega^p(X) \otimes \mathbb{C}$ by computing exterior derivatives of real and imaginary parts. Then the entire apparatus of de Rham theory goes through without change for

$$0 \to \Omega^0(X) \otimes \mathbb{C} \xrightarrow{d_0} \Omega^1(X) \otimes \mathbb{C} \xrightarrow{d_1} \Omega^2(X) \otimes \mathbb{C} \xrightarrow{d_2} \cdots$$

$$\xrightarrow{d_{p-1}} \Omega^p(X) \otimes \mathbb{C} \xrightarrow{d_p} \Omega^{p+1}(X) \otimes \mathbb{C} \xrightarrow{d_{p+1}} \cdots \xrightarrow{d_{n-1}} \Omega^n(X) \otimes \mathbb{C} \to 0. \qquad (10.34)$$

For example, the de Rham cohomology groups with complex coefficients are defined for $0 \le p \le n$ by

$$H^p(X; \mathbb{C}) = \text{Kernel}\,(d_p)/\,\text{Image}\,(d_{p-1}).$$

In particular, for $p = 0$ and $p = n$,

$$H^0(X; \mathbb{C}) = \text{Kernel}\,(d_0), \quad H^n(X; \mathbb{C}) = \Omega^n(X)/\,\text{Image}\,(d_{n-1}).$$

We will summarize just those parts of the apparatus that we will need.

The elements of $\Omega^0(X)$ are simply smooth, real-valued functions on X. Smooth vector fields on X can also be regarded as functions on X, specifically, as sections of the tangent bundle TX and smooth 1-forms on X, that is, the elements of $\Omega^1(X)$, can be identified with sections of the cotangent bundle T^*X. As it happens, the elements of any $\Omega^p(X)$ can be described in a similar way (indeed, this is how p-forms are defined in Section 2.14 of [Warner]). Since we will soon find this point of view particularly fruitful, we will pause for a moment to briefly describe the general context in which it is done (details are available in Section 1.5 of [Jost] and in many other places as well, for example, Volume I of [Sp2]).

Note that TX and T^*X are manifolds built by supplying a differentiable structure to the disjoint union of the vector spaces $T_x(X)$ and $T_x^*(X)$, for $x \in X$, respectively, and for each there is a natural smooth projection π onto X (see Appendix D). The general context we have in mind is contained in the following definitions. Let $\mathbb{F}$ denote either $\mathbb{R}$ or $\mathbb{C}$ and k a positive integer. Then a *k-dimensional smooth $\mathbb{F}$-vector bundle* over the manifold X consists of a smooth manifold E and a smooth map $\pi : E \to X$ of E onto X such that the following conditions are satisfied:

1. Each of the *fibers* $\pi^{-1}(x_0)$ for $x_0 \in X$ has the structure of a k-dimensional vector space over $\mathbb{F}$.
2. (*Local triviality*) For each $x_0 \in X$ there exists an open neighborhood U of x_0 in X and a diffeomorphism $\Phi : \pi^{-1}(U) \to U \times \mathbb{F}^k$ such that each of the following is satisfied:
   a. $\pi^1 \circ \Phi = \pi$, where $\pi^1 : U \times \mathbb{F}^k \to U$ is the projection onto the first factor;
   b. for each $x \in U$ the map $\Phi_x : \pi^{-1}(x) \to \mathbb{F}^k$ defined by $\Phi_x = \pi^2 \circ \Phi|_{\pi^{-1}(x)}$ is an $\mathbb{F}$-vector space isomorphism, where $\pi^2 : U \times \mathbb{F}^k \to \mathbb{F}^k$ is the projection onto the second factor.

Here k is called the *fiber dimension* of the vector bundle and the pair (U, Φ) is called a *local trivialization* of the bundle. By shrinking U if necessary, one can (and we will) always assume that U is a coordinate neighborhood for the manifold X. A *section* of the vector bundle is a smooth map $s : X \to E$ satisfying $\pi \circ s = \mathrm{id}_X$. Thus, a section selects an element in the vector space $\pi^{-1}(x)$ for each $x \in X$ and the selections vary smoothly with x. The *trivial k-dimensional $\mathbb{F}$-vector bundle* over X is simply the product $E = X \times \mathbb{F}^k$ with $\pi : X \times \mathbb{F}^k \to X$ being just the projection onto the first factor. Note that smooth $\mathbb{F}$-valued functions on X can be regarded as sections of the trivial bundle $X \times \mathbb{F}$ by simply identifying the function with its graph.

All of the usual operations by which one produces new vector spaces from given vector spaces (subspaces, duals, direct sums, tensor products, exterior powers, etc.) have direct analogues for vector bundles which just apply the vector space operations to the fibers. In particular, one can construct the pth exterior power of the cotangent bundle T^*X, generally denoted $\wedge^p(T^*X)$. Smooth p-forms on X are then identified with sections of $\wedge^p(T^*X)$. Similarly, one can complexify the fibers of T^*X and form the pth exterior power to obtain a vector bundle denoted $\wedge^p(T^*X \otimes \mathbb{C})$. The complex-valued p-forms on X are just sections of $\wedge^p(T^*X \otimes \mathbb{C})$.

This point of view is fruitful because it allows us to think of the the de Rham complexes (10.33) and (10.34) as sequences of differential operators on sections of vector bundles, and as we will see, this is not only useful for us, but also suggests the possibility of a vast generalization of what we will describe here.

The fibers of an $\mathbb{R}$-vector bundle are isomorphic copies of some real vector space $\mathcal{V}$. If $\mathcal{V}$ has a positive definite inner product $\langle \, , \, \rangle$, then this will induce a positive definite inner product $\langle \, , \, \rangle_x$ on each fiber $\pi^{-1}(x)$. In general, a *fiber metric* on an $\mathbb{R}$-vector bundle $\pi : E \to X$ is an assignment of a positive definite inner product $\langle \, , \, \rangle_x$ to each fiber $\pi^{-1}(x)$ that varies smoothly with x in the sense that $x \mapsto \langle s_1(x), s_2(x) \rangle_x$ is a smooth real-valued function on X for any sections s_1 and s_2. Local triviality and a partition of unity argument (see Remark C.0.1) imply that these always exist. If $E = TX$, this is just a Riemannian metric. A *fiber metric* on a $\mathbb{C}$-vector bundle $\pi : E \to X$ is an assignment of a Hermitian inner product $\langle \, , \, \rangle_x$ to each fiber $\pi^{-1}(x)$ that varies smoothly with x in the sense that $x \mapsto \langle s_1(x), s_2(x) \rangle_x$ is a smooth complex-valued function on X for any sections s_1 and s_2.

Since X is a compact, oriented n-manifold with a Riemannian metric $\mathbf{g}$, there is a standard procedure for supplying each $\Omega^p(X)$ with a positive definite inner product which we will briefly sketch. The metric and the orientation determine a unique volume form $\mathrm{vol}_\mathbf{g} \in \Omega^n(X)$ and any element of $\Omega^n(X)$ can be uniquely written as some smooth, real-valued function on X times $\mathrm{vol}_\mathbf{g}$. Moreover, the metric and orientation also determine, for each $p = 0, 1, \ldots, n$, an isomorphism

$$* : \Omega^p(X) \to \Omega^{n-p}(X),$$

called the Hodge star operator; the image of $\beta \in \Omega^p(X)$ under this isomorphism is denoted $*\beta$. Consequently, for $\alpha, \beta \in \Omega^p(X)$, $\alpha \wedge *\beta$ is in $\Omega^n(X)$, so

$$\alpha \wedge *\beta = \langle \alpha, \beta \rangle \, \mathrm{vol}_\mathbf{g}$$

for some smooth, real-valued function $\langle \alpha, \beta \rangle$ on X called the pointwise inner product of α and β. One then obtains an inner product on $\Omega^p(X)$ by integrating over X:

$$\int_X \alpha \wedge *\beta = \int_X \langle \alpha, \beta \rangle \, \mathrm{vol}_\mathbf{g}, \quad \alpha, \beta \in \Omega^p(X).$$

Applying Exercise 10.4.1.5 to $\langle \, , \, \rangle$ on each fiber, one obtains a Hermitian fiber metric on $\wedge^p(T^*X \otimes \mathbb{C})$ which we will also write simply as $\langle \, , \, \rangle$ rather than $\langle \, , \, \rangle^{\mathbb{C}}$ since we will not have occasion to use the real-valued function again. Integrating then gives an L^2-inner product on $\Omega^p(X) \otimes \mathbb{C}$:

$$\langle \alpha, \beta \rangle_{L^2} = \int_X \langle \alpha, \beta \rangle \, \mathrm{vol}_\mathbf{g}, \quad \alpha, \beta \in \Omega^p(X) \otimes \mathbb{C}.$$

The completion of $\Omega^p(X) \otimes \mathbb{C}$ with respect to this inner product is called the space of L^2-sections of $\wedge^p(T^*X \otimes \mathbb{C})$, or simply the space of L^2-forms of degree p on X, and we will denote it

$$L^2(\Omega^p(X) \otimes \mathbb{C}).$$

The elements of $L^2(\Omega^p(X) \otimes \mathbb{C})$ are, as usual, equivalence classes of sections that differ only on a set of measure zero, the measure being the one determined by the metric volume form $\mathrm{vol}_\mathbf{g}$ on X. This is a Hilbert space, but, to be clear, it is not the Hilbert space of the example we are in the process of constructing.

Because we are interested primarily in the construction of a complex Hilbert space $\mathcal{H}_S$ for our example of an $N = 2$ supersymmetry we will henceforth restrict our attention to the complex case. With respect to the L^2-inner products on the smooth forms, each $d_{p-1} : \Omega^{p-1}(X) \otimes \mathbb{C} \to \Omega^p(X) \otimes \mathbb{C}$ has a formal adjoint $\delta_p : \Omega^p(X) \otimes \mathbb{C} \to \Omega^{p-1}(X) \otimes \mathbb{C}$ defined by the condition that for $\alpha \in \Omega^{p-1}(X) \otimes \mathbb{C}$ and $\beta \in \Omega^p(X) \otimes \mathbb{C}$,

$$\langle d_{p-1}\alpha, \beta \rangle_{L^2} = \langle \alpha, \delta_p \beta \rangle_{L^2}.$$

We have

$$\Omega^{p-1}(X) \otimes \mathbb{C} \xrightarrow{d_{p-1}} \Omega^p(X) \otimes \mathbb{C} \xrightarrow{d_p} \Omega^{p+1}(X) \otimes \mathbb{C},$$

$$\Omega^{p-1}(X) \otimes \mathbb{C} \xleftarrow{\delta_p} \Omega^p(X) \otimes \mathbb{C} \xleftarrow{\delta_{p+1}} \Omega^{p+1}(X) \otimes \mathbb{C}.$$

Exercise 10.4.2. Show that $\delta_{p-1}\delta_p = 0$ for $p = 1, \ldots, n$.

It follows from basic properties of the Hodge star operator and Stokes' theorem that

$$\delta_p = (-1)^{n(p+1)+1} * d_{n-p} *$$

(see Proposition 6.2 of [Warner]). Now define, for each $p = 0, 1, \ldots, n$, the *Hodge Laplacian* (also called the *Laplace–Beltrami operator*)

$$\Delta_p : \Omega^p(X) \otimes \mathbb{C} \rightarrow \Omega^p(X) \otimes \mathbb{C}$$

as follows. For $p = 0$,

$$\Delta_0 = \delta_1 d_0,$$

and for $p = n$,

$$\Delta_n = d_{n-1}\delta_n,$$

while for $1 \le p \le n - 1$,

$$\Delta_p = \delta_{p+1} d_p + d_{p-1} \delta_p.$$

Note that for $\alpha, \beta \in \Omega^p(X) \otimes \mathbb{C}$,

$$\begin{aligned}
\langle \Delta_p \alpha, \beta \rangle_{L^2} &= \langle \delta_{p+1} d_p \alpha + d_{p-1} \delta_p \alpha, \beta \rangle_{L^2} \\
&= \langle \delta_{p+1} d_p \alpha, \beta \rangle_{L^2} + \langle d_{p-1} \delta_p \alpha, \beta \rangle_{L^2} \\
&= \langle d_p \alpha, d_p \beta \rangle_{L^2} + \langle \delta_p \alpha, \delta_p \beta \rangle_{L^2} \\
&= \langle \alpha, \delta_{p+1} d_p \beta \rangle_{L^2} + \langle \alpha, d_{p-1} \delta_p \beta \rangle_{L^2} \\
&= \langle \alpha, \Delta_p \beta \rangle_{L^2},
\end{aligned}$$

so Δ_p is formally self-adjoint with respect to $\langle\,,\,\rangle_{L^2}$ for $1 \le p \le n - 1$; the same is true of Δ_0 and Δ_n. A p-form α in the kernel of Δ_p ($\Delta_p \alpha = 0 \in \Omega^p(X) \otimes \mathbb{C}$) is said to be *Hodge harmonic*.

Exercise 10.4.3. Show that $\alpha \in \Omega^p(X) \otimes \mathbb{C}$ is Hodge harmonic if and only if it is closed ($d_p \alpha = 0$) and coclosed ($\delta_p \alpha = 0$). *Hint*: One direction is obvious; for the other, compute $\langle \Delta_p \alpha, \alpha \rangle_{L^2}$.

Remark 10.4.1. Although we will not make any serious use of them we would feel re-miss if we did not mention two quite deep results that lie at the heart of Hodge theory (these are proved in great detail in Chapter 6 of [Warner]).

Theorem 10.4.1 (Hodge decomposition theorem). *Let X be a compact, connected, ori-ented, smooth, Riemannian n-manifold and $0 \le p \le n$ an integer. Then the space of Hodge harmonic p-forms is finite-dimensional and $\Omega^p(X) \otimes \mathbb{C}$ admits an L^2-orthogonal direct sum decomposition*

$$\Omega^p(X) \otimes \mathbb{C} = \text{Image } \Delta_p \oplus \text{Kernel } \Delta_p.$$

Consequently, the equation $\Delta_p \alpha = \beta$ has a solution in $\Omega^p(X) \otimes \mathbb{C}$ if and only if β is orthog-onal to the space of Hodge harmonic p-forms.

Corollary 10.4.2. *Each de Rham cohomology class on a compact, oriented, Riemannian manifold contains a unique Hodge harmonic representative.*

We should note in passing that, by the corollary, the pth de Rham cohomology group of a compact, oriented, Riemannian n-manifold X is isomorphic to the space of Hodge harmonic p-forms on X for each $p = 0, 1, \ldots, n$. In particular, the Euler char-acteristic $\chi(X)$ of X, which is defined as the alternating sum of the dimensions of the cohomology groups, is also given by

$$\sum_{p=0}^{n} (-1)^p \dim (\text{Kernel } \Delta_p). \tag{10.35}$$

But the Euler characteristic is a *topological* invariant of a compact manifold and this last formula expresses it in terms of *analytic* data (the number of independent solu-tions to the partial differential equations $\Delta_p \alpha = 0$ on $\Omega^p(X) \otimes \mathbb{C}$ for $p = 0, 1, \ldots, n$). The expression in (10.35) is called the *analytic index* of the de Rham complex (10.34) and it just so happens to be a topological invariant of the underlying manifold. There is a vast generalization of this scenario due to Atiyah and Singer. We will have nothing further to say about this other than to suggest [LM] or [Palais1] as sources for those interested in pursuing it.

Next we need to consolidate all of the spaces of forms $\Omega^p(X) \otimes \mathbb{C}$ into a single object. We do this by forming their vector space direct sum:

$$\Omega^*(X) \otimes \mathbb{C} = \bigoplus_{p=0}^{n} \Omega^p(X) \otimes \mathbb{C}.$$

We will identify each $\Omega^p(X) \otimes \mathbb{C}$ with a subspace of $\Omega^*(X) \otimes \mathbb{C}$ and write the elements of $\Omega^*(X) \otimes \mathbb{C}$ as sums of elements of these subspaces. We note that $\Omega^*(X) \otimes \mathbb{C}$ is not only a complex vector space, but also a $C^\infty(X; \mathbb{C})$-module, and we will extend the Hermi-tian inner products on the $\Omega^p(X) \otimes \mathbb{C}$ to $\Omega^*(X) \otimes \mathbb{C}$ by declaring the distinct summands

$\Omega^p(X) \otimes \mathbb{C}$ to be mutually orthogonal. Furthermore, we will define linear transformations

$$d : \Omega^*(X) \otimes \mathbb{C} \to \Omega^*(X) \otimes \mathbb{C},$$
$$\delta : \Omega^*(X) \otimes \mathbb{C} \to \Omega^*(X) \otimes \mathbb{C}$$

and

$$\Delta : \Omega^*(X) \otimes \mathbb{C} \to \Omega^*(X) \otimes \mathbb{C}$$

by

$$d \,|_{\Omega^p(X) \otimes \mathbb{C}} = d_p,$$
$$\delta \,|_{\Omega^p(X) \otimes \mathbb{C}} = \delta_p$$

and

$$\Delta \,|_{\Omega^p(X) \otimes \mathbb{C}} = \Delta_p,$$

respectively. The object of real interest, however, is the linear transformation

$$d + \delta : \Omega^*(X) \otimes \mathbb{C} \to \Omega^*(X) \otimes \mathbb{C}.$$

Note that for $\alpha \in \Omega^p(X) \otimes \mathbb{C}$,

$$(d + \delta)\alpha = d_p \alpha + \delta_p \alpha \in [\Omega^{p+1}(X) \otimes \mathbb{C}] \oplus [\Omega^{p-1}(X) \otimes \mathbb{C}].$$

Exercise 10.4.4. Show that $d + \delta$ is formally self-adjoint and for $\alpha \in \Omega^p(X) \otimes \mathbb{C}$,

$$(d + \delta)^2 \alpha = \Delta_p \alpha.$$

The result of this exercise can be written simply as $(d + \delta)^2 = \Delta$. Let us record this along with two other identities that we have seen:

$$d^2 = 0,$$
$$\delta^2 = 0, \tag{10.36}$$
$$(d + \delta)^2 = \Delta.$$

These should now be compared with (10.26).

Next define a linear map $\mathcal{T} : \Omega^*(X) \otimes \mathbb{C} \to \Omega^*(X) \otimes \mathbb{C}$ by

$$\mathcal{T} \,|_{\Omega^p(X) \otimes \mathbb{C}} = (-1)^p,$$

that is, $\mathcal{T}\alpha = (-1)^p \alpha$ for all $\alpha \in \Omega^p(X) \otimes \mathbb{C}$. Then $\mathcal{T}$ clearly preserves the Hermitian inner product on $\Omega^*(X) \otimes \mathbb{C}$, satisfies $\mathcal{T}^2 = \mathrm{id}_{\Omega^*(X) \otimes \mathbb{C}}$ and anticommutes with

$d + \delta$. Consequently, $\mathcal{T}$ has precisely two eigenvalues, 1 and –1, and $\Omega^*(X) \otimes \mathbb{C}$ has an orthogonal decomposition into the direct sum of the corresponding eigenspaces. The eigenspace $E_1(\mathcal{T})$ consists precisely of the forms with even degree, while $E_{-1}(\mathcal{T})$ consists of those with odd degree, so we obtain a $\mathbb{Z}_2$-grading

$$\Omega^*(X) \otimes \mathbb{C} = [\Omega^0(X) \otimes \mathbb{C}] \oplus [\Omega^1(X) \otimes \mathbb{C}],$$

where

$$\Omega^0(X) \otimes \mathbb{C} = \bigoplus_{p \equiv 0 \bmod 2} \Omega^p(X) \otimes \mathbb{C}$$

and

$$\Omega^1(X) \otimes \mathbb{C} = \bigoplus_{p \equiv 1 \bmod 2} \Omega^p(X) \otimes \mathbb{C}.$$

This should all be sounding very familiar, so we will pause for a moment to compare what we have at this point with what we actually want. Here is what we want:

$$
\begin{aligned}
\text{Hilbert space:} \quad & \mathcal{H}_S \\
\text{Unitary involution:} \quad & \tau \\
\text{Supercharge:} \quad & Q \\
\text{Hamiltonian:} \quad & H_S = Q^2
\end{aligned}
$$

And here is what we have:

$$
\begin{aligned}
& \Omega^*(X) \otimes \mathbb{C} \\
& \mathcal{T} \\
& d + \delta \\
& \Delta = (d + \delta)^2.
\end{aligned}
$$

Now, $\Omega^*(X) \otimes \mathbb{C}$ is a Hermitian inner product space, $\mathcal{T}$ is a unitary involution on this inner product space and $d + \delta$ is formally self-adjoint and anticommutes with $\mathcal{T}$, but $\Omega^*(X) \otimes \mathbb{C}$ is *not* a Hilbert space because it is not complete. To produce an $N = 2$ supersymmetry we must complete $\Omega^*(X) \otimes \mathbb{C}$ to a Hilbert space and extend the operators $\mathcal{T}$ and $d + \delta$ to the completion while preserving the desired properties of these operators. Not surprisingly, this requires some analytic work.

We mentioned earlier that it will be useful to think of the smooth differential forms on X as sections of vector bundles. The reason is that one can then regard any exterior derivative operator as a first order differential operator on these sections; now we will see why this is useful. We begin with a few general definitions. Let $\pi_E : E \to X$ and $\pi_F : F \to X$ be two complex vector bundles over the compact, connected, oriented, Riemannian n-manifold X with fiber dimensions k and l, respectively, and each equipped with a Hermitian fiber metric, denoted $\langle\,,\,\rangle_E$ and $\langle\,,\,\rangle_F$, respectively. We will denote the $C^\infty(X;\mathbb{C})$-modules of smooth sections of the vector bundles by $\Gamma(E)$ and $\Gamma(F)$, respectively; in particular, these are complex vector spaces. A complex-linear map

$$D : \Gamma(E) \rightarrow \Gamma(F)$$

is called a *(linear) differential operator* of *order m* if, roughly speaking, *"it locally looks like an mth order differential operator on $\mathbb{R}^n$."* To make this precise we select an open coordinate neighborhood U on X with coordinates $x^1, \ldots, x^n$ and with the property that both E and F have local trivializations $\Phi_E : \pi_E^{-1}(U) \rightarrow U \times \mathbb{C}^k$ and $\Phi_F : \pi_F^{-1}(U) \rightarrow U \times \mathbb{C}^l$ over U. On U the sections of E and F can be identified with elements of $C^\infty(U;\mathbb{C}^k)$ and $C^\infty(U;\mathbb{C}^l)$, respectively, and what we require is that for every $f \in C^\infty(U;\mathbb{C}^k)$, $Df \in C^\infty(U;\mathbb{C}^l)$ is of the form

$$(Df)(x) = \sum_{|\alpha|\leq m} A^\alpha(x)(\partial_\alpha f)(x), \qquad (10.37)$$

where α is a multi-index (see Appendix G.4), f is a column vector of smooth, complex-valued functions on U, $\partial_\alpha f$ is computed entrywise and $A^\alpha(x)$ is some $k \times l$ matrix of smooth, complex-valued functions on U with $A^\alpha \neq 0$ for some α with $|\alpha| = m$. A change of coordinates shows that this definition is independent of the choice of coordinates, but it is also possible to give an invariant definition of linear differential operators in terms of jet bundles (see Section 3, Chapter IV, of [Palais1]).

Example 10.4.1. For our purposes the most important examples are the exterior differentiation operators

$$d_p : \Omega^p(X) \otimes \mathbb{C} \rightarrow \Omega^{p+1}(X) \otimes \mathbb{C},$$

which are differential operators of order 1. Just to get an idea of how the notation works we will write it out explicitly when $p = 1$ and $n = 3$. Thus, we choose an arbitrary coordinate neighborhood U in X with coordinates x^1, x^2, x^3 and above which the exterior bundles $\wedge^1(T^*X \otimes \mathbb{C})$ and $\wedge^2(T^*X \otimes \mathbb{C})$ are both trivial. On U we identify the sections in $\Gamma(\wedge^1(T^*X \otimes \mathbb{C})) = \Omega^1(X) \otimes \mathbb{C}$ and $\Gamma(\wedge^2(T^*X \otimes \mathbb{C})) = \Omega^2(X) \otimes \mathbb{C}$ with elements of $C^\infty(U;\mathbb{C}^3)$ and $C^\infty(U;\mathbb{C}^3)$, respectively, by identifying forms with their components relative to the standard bases $\{dx^1, dx^2, dx^3\}$ and $\{dx^1 \wedge dx^2, dx^2 \wedge dx^3, dx^1 \wedge dx^3\}$. Since $\alpha = a_1 dx^1 + a_2 dx^2 + a_3 dx^3$ implies $d_1\alpha = (\partial_1 a_2 - \partial_2 a_1)dx^1 \wedge dx^2 + (\partial_2 a_3 - \partial_3 a_2)dx^2 \wedge dx^3 + (\partial_1 a_3 - \partial_3 a_1)dx^1 \wedge dx^3$, we have

$$d_1 \begin{pmatrix} a_1 \\ a_2 \\ a_3 \end{pmatrix} = \begin{pmatrix} \partial_1 a_2 - \partial_2 a_1 \\ \partial_3 a_2 - \partial_2 a_3 \\ \partial_1 a_3 - \partial_3 a_1 \end{pmatrix}$$

$$
= \begin{pmatrix} 0 & 1 & 0 \\ 0 & 0 & 0 \\ 0 & 0 & 1 \end{pmatrix} \begin{pmatrix} \partial_1 a_1 \\ \partial_1 a_2 \\ \partial_1 a_3 \end{pmatrix} + \begin{pmatrix} -1 & 0 & 0 \\ 0 & 0 & 1 \\ 0 & 0 & 0 \end{pmatrix} \begin{pmatrix} \partial_2 a_1 \\ \partial_2 a_2 \\ \partial_2 a_3 \end{pmatrix}
$$

$$
+ \begin{pmatrix} 0 & 0 & 0 \\ 0 & -1 & 0 \\ -1 & 0 & 0 \end{pmatrix} \begin{pmatrix} \partial_3 a_1 \\ \partial_3 a_2 \\ \partial_3 a_3 \end{pmatrix}
$$

$$
= \begin{pmatrix} 0 & 1 & 0 \\ 0 & 0 & 0 \\ 0 & 0 & 1 \end{pmatrix} \partial_1 \begin{pmatrix} a_1 \\ a_2 \\ a_3 \end{pmatrix} + \begin{pmatrix} -1 & 0 & 0 \\ 0 & 0 & 1 \\ 0 & 0 & 0 \end{pmatrix} \partial_2 \begin{pmatrix} a_1 \\ a_2 \\ a_3 \end{pmatrix}
$$

$$
+ \begin{pmatrix} 0 & 0 & 0 \\ 0 & -1 & 0 \\ -1 & 0 & 0 \end{pmatrix} \partial_3 \begin{pmatrix} a_1 \\ a_2 \\ a_3 \end{pmatrix}.
$$

Although a bit messier to write out explicitly it should be clear that the same sort of decomposition occurs for any $d_p : \Omega^p(X) \otimes \mathbb{C} \to \Omega^{p+1}(X) \otimes \mathbb{C}$, so these are, indeed, first order differential operators.

To obtain the Hilbert space $\mathcal{H}_S$ one needs to extend our discussion of the Sobolev spaces in Appendix G.4 from complex-valued functions on $\mathbb{R}^n$ to sections of complex vector bundles. This can be done in a number of ways and we will sketch one of them (a much more detailed discussion from a more general point of view is available in Chapters IX and X of [Palais1]). Intuitively, the idea is simple enough. Locally, a section of $\pi_E : E \to X$ is a smooth map from an open set U in $\mathbb{R}^n$ to $\mathbb{C}^k$ and for these we have defined Sobolev norms, so we would like to piece these together with a partition of unity (see Remark C.0.1) to get a Sobolev norm on the sections of E, and then complete the space of sections with respect to this norm to get a Hilbert space. Here are a few more details.

We will consider an arbitrary complex vector bundle $\pi_E : E \to X$ with Hermitian fiber metric $\langle\,,\,\rangle_E$ over a compact, connected, oriented, Riemannian n-manifold X. We begin by constructing a specific type of finite open cover for X. We will denote by $B^n(r)$ the open ball of radius $r > 0$ about the origin in $\mathbb{R}^n$ and by $\overline{B}^n(r)$ its closure in $\mathbb{R}^n$. At each point x_0 in X we can choose a chart $\varphi_{x_0} : U_{x_0} \to B^n(\frac{3}{2})$ whose image is $B^n(\frac{3}{2})$. Thus, $\varphi_{x_0} : U_{x_0} \to B^n(\frac{3}{2})$ is a diffeomorphism. In particular, U_{x_0} is contractible, so the bundle $\pi_E : E \to X$ is trivial over U_{x_0} and $\pi_E^{-1}(U_{x_0})$ can be identified with $U_{x_0} \times \mathbb{C}^k$.

Remark 10.4.2. Any vector bundle over a contractible space is trivial, but this is certainly not obvious. We will not supply a proof, but will simply send those interested in seeing one to Section 3, Chapter II, of [Osborn].

Denote by $y = (y^1, \ldots, y^n)$ the coordinates supplied to U_{x_0} by φ_{x_0}. These are also coordinates on $V_{x_0} = \varphi_{x_0}^{-1}(B^n(1)) \subseteq U_{x_0}$ and on $W_{x_0} = \varphi_{x_0}^{-1}(B^n(\frac{1}{\sqrt{2}})) \subseteq V_{x_0} \subseteq U_{x_0}$. On V_{x_0}

we make the change of coordinates

$$y = (y^1, \ldots, y^n) \rightarrow x = (x^1, \ldots, x^n) = \frac{1}{\sqrt{1 - \|y\|^2}} (y^1, \ldots, y^n)$$

to get a new chart ϕ_{x_0} on V_{x_0} and W_{x_0}. Note that $\phi_{x_0}(V_{x_0}) = \mathbb{R}^n$ and $\phi_{x_0}(W_{x_0}) = B^n(1)$. All of this can be done for any $x_0 \in X$, so $\{W_{x_0} : x_0 \in X\}$ is an open cover for X. By compactness, we can select a finite subcover $\{W_1, \ldots, W_N\}$ together with open sets $V_1, \ldots, V_N$ and $U_1, \ldots, U_N$ in X satisfying $W_i \subseteq V_i \subseteq \overline{V}_i \subseteq U_i$ for $i = 1, \ldots, N$ and each of the following:

1. $\pi_E : E \rightarrow X$ is trivial over U_i for each $i = 1, \ldots, N$,
2. there are charts $\phi_i : V_i \rightarrow \mathbb{R}^n$ with $\phi_i(V_i) = \mathbb{R}^n$ and $\phi_i(W_i) = B^n(1)$ for $i = 1, \ldots, N$.

Now we select a smooth partition of unity $\{\chi_1, \ldots, \chi_N\}$ on X subordinate to $\{W_1, \ldots, W_N\}$, that is, a family of smooth functions $\chi_i : X \rightarrow [0,1]$ on X with supports satisfying supp $\chi_i \subseteq W_i$ for each $i = 1, \ldots, N$ and, for each $x \in X$, $\sum_{i=1}^N \chi_i(x) = 1$.

Now consider a section $s : X \rightarrow E$ of E. Since $\sum_{i=1}^N \chi_i(x) = 1$ for each $x \in X$, we can write s as $s = \sum_{i=1}^N s_i$, where $s_i = \chi_i s$ for each $i = 1, \ldots, N$. In the coordinate trivializations we have chosen each s_i can be identified with a smooth, $\mathbb{C}^k$-valued function on $\mathbb{R}^n$ with compact support contained in the unit ball $B^n(1)$. For these we have Sobolev K-norms (see Appendix G.4), so we can define the Sobolev K-norm of s by

$$\|s\|_{H^K} = \sum_{i=1}^N \|s_i\|_{H^K}.$$

The completion of $\Gamma(E)$ with respect to this norm is a Hilbert space that we will denote

$$L_K^2(E).$$

In particular, the smooth sections $\Gamma(E)$ are dense in every $L_K^2(E)$. Clearly, we made a great many choices in arriving at this definition (trivializations, coordinates, partitions of unity), but one can show that different choices give rise to equivalent norms and therefore to the same $L_K^2(E)$. This is proved in Section 2, Chapter III, of [LM].

We now have available a plethora of Hilbert spaces of sections of our vector bundle $\pi_E : E \rightarrow X$ which we can arrange in a chain of dense inclusions (compare (G.22))

$$\cdots \subseteq L_K^2(E) \subseteq \cdots \subseteq L_2^2(E) \subseteq L_1^2(E) \subseteq L_0^2(E) = L^2(E).$$

These Sobolev spaces have many properties that make them ideal arenas in which to study partial differential equations. Although we will need relatively few of these it seems worthwhile to enumerate some of the most basic properties just to get a sense of how nice these spaces really are. For the proofs of these and for a great deal more as well we refer you to any one of our principal references, that is, [Warner], [LM]

or [Palais1]. Each of these sources also contains a great deal of important information about *elliptic operators*, which is perhaps the most interesting part of the story,
but which we will not consider here.

We let $\pi_E : E \to X$ and $\pi_F : F \to X$ be two complex vector bundles with fiber
dimensions k and l, respectively, over the compact, connected, oriented, Riemannian
n-manifold X.

1. Every differential operator

$$D : \Gamma(E) \to \Gamma(F)$$

   of order m extends to a *bounded* linear operator

$$D_K : L^2_K(E) \to L^2_{K-m}(F)$$

   for every $K \geq m$.

2. The Hilbert space adjoint of the extension D_K of D is the extension of the formal
   L^2-adjoint D^* of D:

$$(D_K)^* = (D^*)_{K-m}.$$

3. Let $C^0(E)$ denote the continuous sections of E. Then, for $K > \frac{n}{2}$,

$$L^2_K(E) \subseteq C^0(E)$$

   in the sense that every equivalence class in $L^2_K(E)$ has a continuous representative.

4. Let $C^r(E)$ denote the r times continuously differentiable sections of E. Then, for
   $K > \frac{n}{2} + r$,

$$L^2_K(E) \subseteq C^r(E)$$

   in the sense that every equivalence class in $L^2_K(E)$ has an r times continuously
   differentiable representative. Thus, by choosing K sufficiently large we can ensure
   any desired degree of differentiability for the elements of $L^2_K(E)$. In particular, a
   section that is in $L^2_K(E)$ for every $K = 1, 2, \ldots$ is smooth.

5. The inclusion $L^2_{K+1}(E) \subseteq L^2_K(E)$ is compact in the sense that a sequence that is
   bounded in $L^2_{K+1}(E)$ has a subsequence that converges in $L^2_K(E)$.

Now, finally we can specialize all of this machinery to construct our example of an $N =$
2 supersymmetry ($\mathcal{H}_S$, τ, Q). The procedure is virtually identical to that for $\Omega^*(X) \otimes \mathbb{C}$,
but with smooth objects replaced by L^2 objects. Begin by considering the complex
vector bundle $\wedge^p(T^*X \otimes \mathbb{C})$. Then $\Omega^p(X) \otimes \mathbb{C} = \Gamma(\wedge^p(T^*X \otimes \mathbb{C}))$. The exterior derivative
$d_p : \Gamma(\wedge^p(T^*X \otimes \mathbb{C})) \to \Gamma(\wedge^{p+1}(T^*X \otimes \mathbb{C}))$ is a differential operator of order 1, so it
extends to a bounded operator

$$d_p : L^2_1(\wedge^p(T^*X \otimes \mathbb{C})) \to L^2(\wedge^{p+1}(T^*X \otimes \mathbb{C}))$$

that we will also denote d_p. The Hilbert space adjoint of this operator is the extension of the formal L^2 adjoint $\delta_{p+1} : \Gamma(\wedge^{p+1}(T^*X \otimes \mathbb{C})) \to \Gamma(\wedge^p(T^*X \otimes \mathbb{C}))$ of d_p and will also be denoted

$$\delta_{p+1} : L^2\left(\wedge^{p+1}(T^*X \otimes \mathbb{C})\right) \to L_1^2\left(\wedge^p(T^*X \otimes \mathbb{C})\right).$$

Note that, since $L_1^2\left(\wedge^p(T^*X \otimes \mathbb{C})\right)$ is a dense linear subspace of $L^2\left(\wedge^p(T^*X \otimes \mathbb{C})\right)$, we can regard d_p as an unbounded operator on $L^2\left(\wedge^p(T^*X \otimes \mathbb{C})\right)$. Now define $\mathcal{H}_S$ to be the Hilbert space direct sum

$$\mathcal{H}_S = \bigoplus_{p=0}^{n} L^2\left(\wedge^p(T^*X \otimes \mathbb{C})\right)$$

of the L^2 p-forms on X for $p = 0, 1, \ldots, n$. Define operators d and δ on $\mathcal{H}_S$ by $d|_{L^2(\wedge^p(T^*X \otimes \mathbb{C}))} = d_p$ and $\delta|_{L^2(\wedge^p(T^*X \otimes \mathbb{C}))} = \delta_p$. Then let

$$Q = d + \delta : \mathcal{H}_S \to \mathcal{H}_S$$

and define $\tau : \mathcal{H}_S \to \mathcal{H}_S$ by

$$\tau|_{L^2(\wedge^p(T^*X \otimes \mathbb{C}))} = (-1)^p,$$

that is, $\tau\alpha = (-1)^p \alpha$ for all $\alpha \in L^2(\wedge^p(T^*X \otimes \mathbb{C}))$. Then τ is a unitary involution on $\mathcal{H}_S$ that commutes with Q, so we have produced all of the required ingredients for $(\mathcal{H}_S, \tau, Q)$. The corresponding Hamiltonian is $H_S = Q^2 = (d + \delta)^2 = d\delta + \delta d$, which is the extension of the Hodge Laplacian to $\bigoplus_{p=0}^n L_2^2 (\wedge^p(T^*X \otimes \mathbb{C}))$ regarded as an unbounded operator on $\mathcal{H}_S$.

We will conclude with the recommendation that you proceed from here directly to the remarkable paper [Witt2] in which this example of an $N = 2$ supersymmetry drawn from Hodge theory led Edward Witten to a new view of Morse theory and opened the door to the extraordinary impact that physics has had on topology in the past four decades.

A Gaussian integrals

The purpose of this appendix is to evaluate just those Gaussian integrals that we have need of in the body of the text. Most of these are integrals over $\mathbb{R}$ of some real or complex quadratic exponential. Some of these must be regarded as improper Riemann integrals and some can also be regarded as Lebesgue integrals. Although the relationship between these two is no doubt familiar, it seems prudent to begin by establishing some notation to distinguish them and reviewing some of the basic facts. This material can be found in most books treating measure and integration ([Apos], in particular, has everything we will need).

Let $[a, b]$ be a compact interval in $\mathbb{R}$ and f a real- or complex-valued function on $[a, b]$. Then the Riemann integral $\int_a^b f(x)dx$ exists if and only if f is continuous almost everywhere. In this case, the Lebesgue integral $\int_{[a,b]} f d\mu$ also exists and the two are equal (we will use μ for the Lebesgue measure). If f is Riemann integrable on $[a, b]$ for all $b \geq a$ and if the limit $\lim_{b\to\infty} \int_a^b f(x)dx$ exists, then the improper Riemann integral of f over $[a, \infty)$ is defined by

$$\int_a^\infty f(x)dx = \lim_{b\to\infty} \int_a^b f(x)dx.$$

The improper Riemann integral $\int_{-\infty}^a f(x)dx$ is defined analogously. If $\int_{-\infty}^0 f(x)dx$ and $\int_0^\infty f(x)dx$ both exist, then one also defines

$$\int_{\mathbb{R}} f(x)dx = \int_{-\infty}^\infty f(x)dx = \int_{-\infty}^0 f(x)dx + \int_0^\infty f(x)dx.$$

If $\int_{-\infty}^\infty f(x)dx$ exists, then so does the limit $\lim_{b\to\infty} \int_{-b}^b f(x)dx$ and these two are equal. However, the limit $\lim_{b\to\infty} \int_{-b}^b f(x)dx$ can exist even when the improper integral $\int_{-\infty}^\infty f(x)dx$ does not (for example, when $f(x) = x$). When it exists, the limit $\lim_{b\to\infty} \int_{-b}^b f(x)dx$ is called the Cauchy principal value of $\int_{-\infty}^\infty f(x)dx$. The following is Theorem 10.31 of [Apos].

Theorem A.0.1. *Suppose f is defined on $[a, \infty)$ and Lebesgue integrable on $[a, b]$ for every $b \geq a$. Suppose also that there exists a positive constant M with the property that $\int_{[a,b]} |f| \, d\mu \leq M$ for all $b \geq a$. Then f is Lebesgue integrable on $[a, \infty)$, $\lim_{b\to\infty} \int_{[a,b]} f d\mu$ exists and*

$$\int_{[a,\infty)} f d\mu = \lim_{b\to\infty} \int_{[a,b]} f d\mu.$$

There is an analogous result for functions defined on $(-\infty, a]$. Finally, we record the following analogue of Theorem A.0.1 for Riemann integrable functions (which is Theorem 10.33 of [Apos]); there is, of course, a corresponding result for $(-\infty, a]$.

https://doi.org/10.1515/9783110751949-011

Theorem A.0.2. *Suppose f is defined on $[a, \infty)$ and Riemann integrable on $[a, b]$ for every $b \geq a$. Suppose also that there exists a positive constant M with the property that $\int_a^b |f(x)| \, dx \leq M$ for all $b \geq a$. Then the improper Riemann integrals of f and $|f|$ both exist on $[a, \infty)$. Moreover, f is Lebesgue integrable on $[a, \infty)$ and*

$$\int_{[a,\infty)} f d\mu = \int_a^\infty f(x) dx.$$

Now we can begin computing the integrals we need. The first is essentially the example (or exercise) that one finds in every calculus book in the world. We have

$$\int_{-\infty}^\infty e^{-x^2} dx = \sqrt{\pi} = \int_{\mathbb{R}} e^{-x^2} d\mu(x). \tag{A.1}$$

To prove this we will show first that $\int_{-\infty}^\infty e^{-x^2} dx$ exists so that we can compute it as the Cauchy principal value. Note first that for $x > 1$, $0 < e^{-x^2} < xe^{-x^2}$, so if $b > 1$,

$$\int_1^b e^{-x^2} dx < \int_1^b xe^{-x^2} dx = \frac{1}{2}(e^{-1} - e^{-b^2}) < \frac{1}{2e}.$$

Consequently,

$$\int_0^b e^{-x^2} dx < M,$$

where

$$M = \int_0^1 e^{-x^2} dx + \frac{1}{2e}.$$

In particular, $\int_0^\infty e^{-x^2} dx$ exists.

Exercise A.0.1. Show that $\int_{-\infty}^0 e^{-x^2} dx$ exists.

We conclude that $\int_{-\infty}^\infty e^{-x^2} dx$ exists and therefore agrees with its Cauchy principal value, that is,

$$\int_{-\infty}^\infty e^{-x^2} dx = \lim_{b \to \infty} \int_{-b}^b e^{-x^2} dx.$$

To compute this we temporarily let $I(b) = \int_{-b}^{b} e^{-x^2} dx$ and compute

$$I(b)^2 = \left(\int_{-b}^{b} e^{-x^2} dx \right) \left(\int_{-b}^{b} e^{-y^2} dy \right)$$

$$= \int_{-b}^{b} \left(\int_{-b}^{b} e^{-y^2} dy \right) e^{-x^2} dx$$

$$= \int_{-b}^{b} \int_{-b}^{b} e^{-(x^2+y^2)} dy dx$$

$$= \iint_{[-b,b]\times[-b,b]} e^{-(x^2+y^2)} d\mu(x,y),$$

where, in the last equality, we have used Fubini's theorem to turn the iterated integral into a double integral over the square. Now we will bound this double integral above and below as follows. Let R_1 be the disc of radius b about the origin in $\mathbb{R}^2$ and R_2 the disc of radius $\sqrt{2}\,b$ about the origin. Then R_1 is inscribed in the square $[-b,b] \times [-b,b]$ and R_2 is circumscribed about the square. Since $e^{-(x^2+y^2)}$ is positive we have

$$\iint_{R_1} e^{-(x^2+y^2)} d\mu(x,y) \leq I(b)^2 \leq \iint_{R_2} e^{-(x^2+y^2)} d\mu(x,y),$$

which, in polar coordinates, gives

$$\int_0^{2\pi} \int_0^{b} e^{-r^2} r\, dr\, d\theta \leq I(b)^2 \leq \int_0^{2\pi} \int_0^{\sqrt{2}\,b} e^{-r^2} r\, dr\, d\theta$$

and therefore

$$\pi(1 - e^{-b^2}) \leq I(b)^2 \leq \pi(1 - e^{-2b^2}).$$

Thus, $\lim_{b\to\infty} I(b)^2 = \pi = (\lim_{b\to\infty} I(b))^2$, so $\lim_{b\to\infty} I(b) = \sqrt{\pi}$, that is,

$$\int_{-\infty}^{\infty} e^{-x^2} dx = \sqrt{\pi}.$$

Exercise A.0.2. Show that

$$\int_{\mathbb{R}} e^{-x^2} d\mu(x) = \sqrt{\pi}.$$

Exercise A.0.3. Prove each of the following:

1. $\int_{-\infty}^{\infty} xe^{-x^2} dx = \int_{\mathbb{R}} xe^{-x^2} d\mu(x) = 0$,

2. $\int_{-\infty}^{\infty} x^2 e^{-x^2} dx = \int_{\mathbb{R}} x^2 e^{-x^2} d\mu(x) = \frac{\sqrt{\pi}}{2}$, (*Hint*: Integrate by parts.)

3. for any $a > 0$,

$$\int_{-\infty}^{\infty} e^{-ax^2/2} dx = \int_{\mathbb{R}} e^{-ax^2/2} d\mu(x) = \sqrt{\frac{2\pi}{a}},$$

4. for $a > 0, b, c \in \mathbb{R}$,

$$\int_{-\infty}^{\infty} e^{-ax^2/2+bx+c} dx = \int_{\mathbb{R}} e^{-ax^2/2+bx+c} d\mu(x) = \sqrt{\frac{2\pi}{a}} \, e^{\frac{b^2}{2a}+c}.$$

(*Hint*: Complete the square.)

All of the examples we have seen so far are integrals that can be regarded either as improper Riemann integrals or Lebesgue integrals. This is not true of the next example.

Example A.0.1. We will prove that for $a > 0$,

$$\int_{-\infty}^{\infty} e^{iax^2/2} dx = \sqrt{\frac{2\pi}{a}} \, e^{\pi i/4} = \sqrt{\frac{2\pi i}{a}} \quad (a > 0), \tag{A.2}$$

where we have taken the value of $\sqrt{i}$ to be $e^{\pi i/4}$. Note that, since $|e^{iax^2/2}| = 1$ is not Lebesgue integrable over $\mathbb{R}$, neither is $e^{iax^2/2}$. Thus, we will need to examine $\int_{-\infty}^{0} e^{iax^2/2} dx$ and $\int_{0}^{\infty} e^{iax^2/2} dx$ separately. Integrals of this sort are often handled by relating them to contour integrals for functions of a complex variable. Although we intend to take a different approach shortly it is worth the effort to see how this technique works, so this is what we will do here. Note first that

$$\int_{0}^{\infty} e^{iax^2/2} dx = \sqrt{\frac{2}{a}} \int_{0}^{\infty} e^{ix^2} dx,$$

so we need only evaluate $\int_{0}^{\infty} e^{ix^2} dx$. For this we will consider the contour integral

$$\int_{C} e^{iz^2} dz,$$

where C is the closed (slice of pizza) contour consisting of the following three segments (R is an arbitrary positive real number):

$$C_1 : z_1(x) = x, \quad 0 \le x \le R,$$

$$C_R : z_R(\theta) = Re^{i\theta}, \quad 0 \le \theta \le \frac{\pi}{4}.$$

and $-C_2$, where

$$C_2 : z_2(x) = e^{\pi i/4} x, \quad 0 \le x \le R.$$

Since e^{iz^2} is analytic on all of $\mathbb{C}$, the Cauchy integral theorem implies that $\int_C e^{iz^2} dz = 0$, so

$$
\begin{aligned}
0 &= \int_{C_1} e^{iz^2} dz + \int_{C_R} e^{iz^2} dz - \int_{C_2} e^{iz^2} dz \\
&= \int_0^R e^{ix^2} dx + \int_{C_R} e^{iz^2} dz - e^{\pi i/4} \int_0^R e^{-x^2} dx.
\end{aligned}
$$

Exercise A.0.4. Compute the contour integrals over C_1 and C_2 and verify the second equality.

Taking the limit as $R \to \infty$ gives

$$\int_0^\infty e^{ix^2} dx = e^{\pi i/4} \left(\frac{\sqrt{\pi}}{2} \right) + \lim_{R \to \infty} \int_{C_R} e^{iz^2} dz.$$

Next we will show that $\lim_{R \to \infty} \int_{C_R} e^{iz^2} dz = 0$, so that

$$\int_0^\infty e^{ix^2} dx = e^{\pi i/4} \left(\frac{\sqrt{\pi}}{2} \right)$$

and therefore

$$\int_0^\infty e^{iax^2/2} dx = \sqrt{\frac{2}{a}} e^{\pi i/4} \left(\frac{\sqrt{\pi}}{2} \right) = \frac{1}{2} \sqrt{\frac{2\pi}{a}} e^{\pi i/4}.$$

For the limit $\lim_{R \to \infty} \int_{C_R} e^{iz^2} dz$ we will need what is called *Jordan's inequality*.

Exercise A.0.5. Show that if $0 \le \phi \le \pi/2$ and $a > 0$, then

$$\int_0^{\pi/2} e^{-a \sin \phi} d\phi < \frac{\pi}{2a}.$$

Hint: Show that $\sin \phi \ge \frac{2}{\pi} \phi$ for $0 \le \phi \le \pi/2$.

Now we observe that, on C_R,

$$\left| e^{iz^2} \right| = \left| e^{iR^2 e^{2\theta i}} \right| = e^{-R^2 \sin 2\theta}$$

and therefore

$$\left| \int_{C_R} e^{iz^2} dz \right| = \left| \int_0^{\pi/4} e^{iR^2 e^{2\theta i}} (iRe^{i\theta}) d\theta \right| \leq R \int_0^{\pi/4} e^{-R^2 \sin 2\theta} d\theta$$

$$= \frac{R}{2} \int_0^{\pi/2} e^{-R^2 \sin \phi} d\phi < \frac{\pi}{4R}.$$

Consequently, $\lim_{R \to \infty} \int_{C_R} e^{iz^2} dz = 0$ and this completes the proof of

$$\int_0^\infty e^{ix^2} dx = e^{\pi i/4} \left(\frac{\sqrt{\pi}}{2} \right).$$

As we noted earlier this gives us

$$\int_0^\infty e^{iax^2/2} dx = \frac{1}{2} \sqrt{\frac{2\pi}{a}} e^{\pi i/4}.$$

Exercise A.0.6. Show that for $a > 0$,

$$\int_0^\infty \cos\left(\frac{ax^2}{2}\right) dx = \int_0^\infty \sin\left(\frac{ax^2}{2}\right) dx = \frac{1}{2} \sqrt{\frac{\pi}{a}}.$$

These are called *Fresnel integrals*.

Exercise A.0.7. Show that for $a > 0$,

$$\int_{-\infty}^0 e^{iax^2/2} dx = \frac{1}{2} \sqrt{\frac{2\pi}{a}} e^{\pi i/4}$$

and conclude that

$$\int_{-\infty}^\infty e^{iax^2/2} dx = \sqrt{\frac{2\pi}{a}} e^{\pi i/4} = \sqrt{\frac{2\pi i}{a}} \quad (a > 0),$$

as required by (A.2).

Exercise A.0.8. Show that if a is any nonzero real number, then

$$\int_{-\infty}^\infty e^{iax^2/2} dx = \sqrt{\frac{2\pi}{|a|}} e^{\text{sgn}(a)\pi i/4} = \sqrt{\frac{2\pi i}{a}} \quad (a \in \mathbb{R}, a \neq 0), \tag{A.3}$$

where $\text{sgn}(a) = 1$ if $a > 0$, $\text{sgn}(a) = -1$ if $a < 0$ and $\sqrt{i} = e^{\pi i/4}$.

Example A.0.2. Now we would like to describe a different approach that will yield a more general result. We begin by recalling a result from complex analysis. We consider the improper Riemann integral $\int_a^\infty f(x,z)\,dx$, where $a \in \mathbb{R}$ and $f(x,z)$ is a function of the real variable x in $[a,\infty)$ and z is a complex number in some domain D of the complex plane $\mathbb{C}$. We will assume that:

1. the integral converges for each fixed value of $z \in D$,
2. for each fixed $x \in [a,\infty)$, $f(x,z)$ is an analytic function of z on D,
3. $\frac{\partial f(x,z)}{\partial z}$ is a continuous function of $(x,z) \in [a,\infty) \times D$,
4. $\int_a^\infty \frac{\partial f(x,z)}{\partial z}\,dx$ converges uniformly on D. Note that uniform convergence means that there exists a function $M(x)$ such that $|\frac{\partial f(x,z)}{\partial z}| \le M(x)$ for all $z \in D$ and $\int_a^\infty M(x)\,dx$ converges.

It follows from these assumptions that $\int_a^\infty f(x,z)\,dx$ is an analytic function of z on D (this is 5.32 of [WW]). There is an analogous theorem for $\int_{-\infty}^a f(x,z)\,dx$ and therefore also for $\int_{-\infty}^\infty f(x,z)\,dx$.

Now we consider the function $f(x,z) = e^{-zx^2/2}$, where $x \in [0,\infty)$ and z is in the right half $(\mathrm{Re}(z) > 0)$ of the complex plane. Then, since

$$e^{-zx^2/2} = e^{-\mathrm{Re}(z)x^2/2}e^{-i\,\mathrm{Im}(z)x^2/2},$$

we have $|e^{-zx^2/2}| = e^{-\mathrm{Re}(z)x^2/2}$. According to Exercise A.0.3.3, $\int_0^\infty e^{-\mathrm{Re}(z)x^2/2}\,dx$ converges. Consequently, $\int_0^\infty e^{-zx^2/2}\,dx$ converges uniformly on $\mathrm{Re}(z) > 0$.

Exercise A.0.9. Let $f(x,z) = e^{-zx^2/2}$, where $x \in [0,\infty)$ and z is in the right half $(\mathrm{Re}(z) > 0)$ of the complex plane. Show that $\int_0^\infty \frac{\partial f(x,z)}{\partial z}\,dx$ converges uniformly on $\mathrm{Re}(z) > 0$.

From this it follows that $\int_0^\infty e^{-zx^2/2}\,dx$ is an analytic function of z on $\mathrm{Re}(z) > 0$. Similarly, $\int_{-\infty}^0 e^{-zx^2/2}\,dx$ is an analytic function of z on $\mathrm{Re}(z) > 0$ and consequently the same is true of $\int_{-\infty}^\infty e^{-zx^2/2}\,dx$. Changing the z to an a we conclude that $\int_{-\infty}^\infty e^{-ax^2/2}\,dx$ is an analytic function of a on $\mathrm{Re}(a) > 0$. Now note that if a is a positive real number, then Exercise A.0.3.3 gives the value of this integral as $\sqrt{\frac{2\pi}{a}}$. On $\mathrm{Re}(a) > 0$ the principal branch of the square root function $\sqrt{\ }$ (branch cut along the negative real axis) is analytic and gives the positive square root of a positive real number. Consequently, $\sqrt{\frac{2\pi}{a}}$ is an analytic function of a on $\mathrm{Re}(a) > 0$ that agrees with $\int_{-\infty}^\infty e^{-ax^2/2}\,dx$ on the positive real axis. As a result, they must agree everywhere on $\mathrm{Re}(a) > 0$ (Corollary to Theorem 10.18 of [Rud2]), so

$$\int_{-\infty}^\infty e^{-ax^2/2}\,dx = \sqrt{\frac{2\pi}{a}} \quad (a \in \mathbb{C}, \mathrm{Re}(a) > 0), \tag{A.4}$$

where $\sqrt{\ }$ is the principal branch of the square root function.

Exercise A.0.10. Show that (A.4) is also true if a is on the imaginary axis. Specifically, show that

$$\int_{-\infty}^{\infty} e^{-aix^2/2} dx = \sqrt{\frac{2\pi}{\alpha}}\, e^{-\pi i/4} \quad (\alpha > 0)$$

and

$$\int_{-\infty}^{\infty} e^{-aix^2/2} dx = \sqrt{\frac{2\pi}{|\alpha|}}\, e^{\pi i/4} \quad (\alpha < 0).$$

Show also that this can be written as

$$\int_{-\infty}^{\infty} e^{\operatorname{sgn}(\alpha)\,|\alpha| i x^2/2} dx = \sqrt{\frac{2\pi}{|\alpha|}}\, e^{\operatorname{sgn}(\alpha)\pi i/4}. \tag{A.5}$$

Hint: Exercise A.0.6

Next note that if b is a *real* number, then the simple substitution $x \to x + b$ gives

$$\int_{-\infty}^{\infty} e^{-a(x-b)^2/2} dx = \int_{-\infty}^{\infty} e^{-ax^2/2} dx = \sqrt{\frac{2\pi}{a}} \quad (a \in \mathbb{C}, \operatorname{Re}(a) > 0). \tag{A.6}$$

When b is complex this substitution makes no sense since x is real and $x+b$ is complex. Nevertheless, (A.6) *is still true when b is complex.* Although we will forego the details here one can see this by carrying out a contour integral argument analogous to that in Example A.0.1. Specifically, one shows that for any fixed a with $\operatorname{Re}(a) > 0$, the contour integral of $e^{-a(z-b)^2/2}$ over the curve $z(x) = x$, $-\infty < x < \infty$ (which we want), is the same as its contour integral over $z(x) = x + b$, $-\infty < x < \infty$ (which we know is $\sqrt{\frac{2\pi}{a}}$). This can be done by integrating around the parallelogram with vertices $-R$, R, $R + b$ and $-R + b$, applying the Cauchy integral theorem and showing that, as $R \to \infty$, the contributions from the two nonhorizontal sides go to zero.

All of the examples we have seen thus far have been one-dimensional integrals, but in Section 8.3 we also need an n-dimensional Gaussian integral (see (8.12)). The general context for such integrals is as follows. We are given some real, symmetric, nonsingular, $N \times N$ matrix $A = (A_{ij})_{i,j=1,\dots,N}$. We write the usual inner product on $\mathbb{R}^N$ as $\langle \mathbf{x}, \mathbf{J} \rangle = \sum_{i=1}^{N} x^i J^i$. Then A determines a quadratic form on $\mathbb{R}^N$ given by $\langle \mathbf{x}, A\mathbf{x} \rangle = \sum_{i,j=1}^{N} A_{ij} x^i x^j$. Writing $d^N \mathbf{x} = dx^1 \cdots dx^N$, the result we require is

$$\int_{\mathbb{R}^N} e^{i\langle \mathbf{x}, A\mathbf{x}\rangle/2 + i\langle \mathbf{x}, \mathbf{J}\rangle} d^N \mathbf{x} = e^{N\pi i/4 - v\pi i/2} \sqrt{\frac{(2\pi)^N}{|\det A|}}\, e^{-i\langle A^{-1}\mathbf{J}, \mathbf{J}\rangle} \tag{A.7}$$

for all $\mathbf{J} \in \mathbb{R}^N$, where v is the number of negative eigenvalues of A. We will see where this comes from in a moment, but as a warm-up we will first prove something simpler.

Example A.0.3. Suppose in addition that A is positive definite (its N distinct real eigenvalues are positive). We will show that

$$\int_{\mathbb{R}^N} e^{i\langle \mathbf{x}, A\mathbf{x}\rangle/2} d^N\mathbf{x} = e^{N\pi i/4}\sqrt{\frac{(2\pi)^N}{\det A}}. \tag{A.8}$$

Since A is a real symmetric matrix we can find some orthogonal matrix S with determinant 1 for which $SAS^{-1} = D$ is the diagonal matrix whose diagonal entries are the distinct real eigenvalues $a_1, \ldots, a_N$ of A. Since A is positive definite, all of these eigenvalues are positive, so $\det A = a_1 \cdots a_N$ is also positive. Now make the change of variable $\mathbf{y} = S\mathbf{x}$. Then

$$\langle \mathbf{x}, A\mathbf{x}\rangle = \langle S\mathbf{x}, S(A\mathbf{x})\rangle = \langle \mathbf{y}, D(S\mathbf{x})\rangle = \langle \mathbf{y}, D\mathbf{y}\rangle = a_1(y^1)^2 + \cdots + a_N(y^N)^2.$$

Since $d^N\mathbf{x} = |\det S| \, d^N\mathbf{y} = d^N\mathbf{y} = dy^1 \cdots dy^N$ we have

$$\int_{\mathbb{R}^N} e^{i\langle \mathbf{x}, A\mathbf{x}\rangle/2} d^N\mathbf{x} = \int_{\mathbb{R}} \cdots \int_{\mathbb{R}} e^{ia_1(y^1)^2/2 + \cdots + ia_N(y^N)^2/2} dy^1 \cdots dy^N$$

$$= \int_{\mathbb{R}} \cdots \int_{\mathbb{R}} e^{ia_1(y^1)^2/2} \cdots e^{ia_N(y^N)^2/2} dy^1 \cdots dy^N$$

$$= \left(\int_{\mathbb{R}} e^{ia_1(y^1)^2/2} dy^1\right) \cdots \left(\int_{\mathbb{R}} e^{ia_N(y^N)^2/2} dy^N\right)$$

$$= \left(\sqrt{\frac{2\pi}{a_1}} e^{\pi i/4}\right) \cdots \left(\sqrt{\frac{2\pi}{a_N}} e^{\pi i/4}\right) \quad \text{(Exercise A.0.10)}$$

$$= e^{N\pi i/4}\sqrt{\frac{(2\pi)^N}{\det A}},$$

as required.

Exercise A.0.11. Show that if A is not necessarily positive definite (but still real, symmetric and nonsingular), then

$$\int_{\mathbb{R}^N} e^{i\langle \mathbf{x}, A\mathbf{x}\rangle/2} d^N\mathbf{x} = e^{N\pi i/4 - v\pi i/2}\sqrt{\frac{(2\pi)^N}{|\det A|}},$$

where v is the number of negative eigenvalues of A. *Hint:* Exercise A.0.10 and note that $N - 2v$ is the signature $\text{sgn}(A)$ of A, that is, the number of positive eigenvalues minus the number of negative eigenvalues.

Now let us try the same sort of thing for $\int_{\mathbb{R}^N} e^{i\langle \mathbf{x}, A\mathbf{x}\rangle/2 + i\langle \mathbf{x}, \mathbf{J}\rangle} d^N\mathbf{x}$. We assume only that A is a real, symmetric, nonsingular, $N \times N$ matrix and denote by v the number of negative eigenvalues of A and by $\text{sgn}(A) = N - 2v$ the signature of A; $\mathbf{J} \in \mathbb{R}^N$ is arbitrary. We select an orthogonal matrix $S = (S_{ij})_{i,j=1,\ldots,N}$ with determinant 1 such that $SAS^{-1} = D$ is a diagonal matrix with diagonal entries $a_1, \ldots, a_N$ that are the distinct real nonzero eigenvalues of A. Making the change of variable $\mathbf{y} = S\mathbf{x}$, one has, exactly as in the previous example, $\langle \mathbf{x}, A\mathbf{x} \rangle = a_1(y^1)^2 + \cdots + a_N(y^N)^2$. In addition,

$$\langle \mathbf{x}, \mathbf{J} \rangle = \langle S^{-1}\mathbf{y}, \mathbf{J} \rangle = \langle S^T\mathbf{y}, \mathbf{J} \rangle = \langle \mathbf{y}, S\mathbf{J} \rangle = b^1 y^1 + \cdots + b^N y^N,$$

where we have written

$$SJ = \begin{pmatrix} S_{1j}J^j \\ \vdots \\ S_{Nj}J^j \end{pmatrix} = \begin{pmatrix} b^1 \\ \vdots \\ b^N \end{pmatrix}.$$

Then

$$\int_{\mathbb{R}^N} e^{i\langle \mathbf{x}, A\mathbf{x} \rangle/2 + i\langle \mathbf{x}, \mathbf{J} \rangle} d^N\mathbf{x} = \int_{\mathbb{R}^N} e^{i[a_1(y^1)^2/2 + b^1 y^1]} \cdots e^{i[a_N(y^N)^2/2 + b^N y^N]} d^N\mathbf{y}$$

$$= \left(\int_{\mathbb{R}} e^{i[a_1(y^1)^2/2 + b^1 y^1]} dy^1 \right) \cdots \left(\int_{\mathbb{R}} e^{i[a_N(y^N)^2/2 + b^N y^N]} dy^N \right). \qquad (A.9)$$

Now, for any $a \neq 0$ and any $b \in \mathbb{R}$ we complete the square to obtain

$$\int_{\mathbb{R}} e^{i[ay^2/2 + by]} dy = e^{-ib^2/2a} \int_{\mathbb{R}} e^{ia(y + \frac{b}{a})^2/2} dy = e^{-ib^2/2a} \int_{\mathbb{R}} e^{iau^2/2} du$$

$$= e^{-ib^2/2a} \begin{cases} \sqrt{\frac{2\pi}{|a|}} e^{-\pi i/4}, & \text{if } a < 0, \\ \sqrt{\frac{2\pi}{a}} e^{\pi i/4}, & \text{if } a > 0. \end{cases}$$

Applying this to each of the factors in (A.9) and keeping in mind that $a_1, \ldots, a_N$ are the eigenvalues of A we obtain

$$\int_{\mathbb{R}^N} e^{i\langle \mathbf{x}, A\mathbf{x} \rangle/2 + i\langle \mathbf{x}, \mathbf{J} \rangle} d^N\mathbf{x} = \sqrt{\frac{(2\pi)^N}{|\det A|}} e^{\operatorname{sgn}(A)\pi i/4} \left[e^{-i\sum_{j=1}^N \frac{(b^j)^2}{2a_j}} \right].$$

Next we note that

$$\sum_{j=1}^N \frac{(b^j)^2}{2a_j} = \frac{1}{2} \langle SJ, D^{-1}SJ \rangle = \frac{1}{2} \langle J, S^T D^{-1}SJ \rangle = \frac{1}{2} \langle J, (S^{-1}D^{-1}S)J \rangle$$

$$= \frac{1}{2} \langle J, (S^{-1}(SA^{-1}S^{-1})S)J \rangle = \frac{1}{2} \langle J, A^{-1}J \rangle.$$

With this and $\operatorname{sgn}(A) = N - 2\nu$ we therefore obtain

$$\int_{\mathbb{R}^N} e^{i\langle \mathbf{x}, A\mathbf{x} \rangle/2 + i\langle \mathbf{x}, \mathbf{J} \rangle} d^N\mathbf{x} = e^{N\pi i/4 - \nu\pi i/2} \sqrt{\frac{(2\pi)^N}{|\det A|}} e^{-i\langle A^{-1}J, J \rangle},$$

as required.

B Morse lemma

The result we would like to prove in this appendix is not particularly difficult, but it has an enormous number of beautiful ramifications. The subject that evolves from it is called *Morse theory*, but, regrettably, this is not our subject here. Our only excuse for including a proof (and we would have jumped at any excuse) is that the result, called the *Morse lemma*, is a key ingredient in the proof of the stationary phase approximation to which we will turn in Appendix C. For those who would like to see more we suggest the following. One of the simpler applications of the Morse lemma is the proof of Reeb's elegant topological characterization of the n-dimensional sphere S^n and one can find this in Section 5-12 of [Nab2]. Milnor's beautiful book [Milnor] is everyone's recommended source for a deeper introduction to Morse theory. You will certainly also want to begin browsing through [Bott] and, if you incline toward physics and/or supersymmetry, the very influential paper [Witt2] by Witten. But now to the task at hand.

Theorem B.0.1. *Let U be an open ball centered at $(a^1, \ldots, a^n)$ in $\mathbb{R}^n$ and $f : U \to \mathbb{R}$ a smooth, real-valued function on U.*

1. *There exist smooth, real-valued functions $f_1, \ldots, f_n$ on U satisfying*

$$f(x^1, \ldots, x^n) = f(a^1, \ldots, a^n) + x^1 f_1(x^1, \ldots, x^n) + \cdots + x^n f_n(x^1, \ldots, x^n)$$

$$= f(a^1, \ldots, a^n) + \sum_{i=1}^{n} x^i f_i(x^1, \ldots, x^n)$$

for all $(x^1, \ldots, x^n) \in U$ and

$$f_i(a^1, \ldots, a^n) = \frac{\partial f}{\partial x^i}(a^1, \ldots, a^n), \quad i = 1, \ldots, n.$$

2. *If f has a critical point at $(a^1, \ldots, a^n)$, that is, if $\frac{\partial f}{\partial x^i}(a^1, \ldots, a^n) = 0$ for each $i = 1, \ldots, n$, then there exist smooth functions $f_{ij}, i, j = 1, \ldots, n$, on U such that*

$$f(x^1, \ldots, x^n) = f(a^1, \ldots, a^n) + \sum_{i,j=1}^{n} x^i x^j f_{ij}(x^1, \ldots, x^n)$$

and

$$f_{ij}(a^1, \ldots, a^n) = \frac{\partial^2 f}{\partial x^i \partial x^j}(a^1, \ldots, a^n), \quad i, j = 1, \ldots, n.$$

3. *(Morse lemma) If f has a critical point at $(a^1, \ldots, a^n)$ that is nondegenerate, that is, for which the Hessian matrix*

$$\left(\frac{\partial^2 f}{\partial x^i \partial x^j}(a^1, \ldots, a^n) \right)_{i,j=1,\ldots,n}$$

https://doi.org/10.1515/9783110751949-012

is nonsingular, then there is a diffeomorphism $\varphi : U \to \varphi(U) \subseteq \mathbb{R}^n$,

$$\varphi(x^1,\ldots,x^n) = (y^1,\ldots,y^n),$$

of U onto an open neighborhood $\varphi(U)$ of $(0,\ldots,0)$ in $\mathbb{R}^n$ such that $\varphi(a^1,\ldots,a^n) = (0,\ldots,0)$ and

$$(f \circ \varphi^{-1})(y^1,\ldots,y^n) = f(a^1,\ldots,a^n) - (y^1)^2 - \cdots - (y^l)^2 + (y^{l+1})^2 + \cdots + (y^n)^2 \quad \text{(B.1)}$$

for $(y^1,\ldots,y^n) \in \varphi(U)$, where $0 \le l \le n$ is an integer. Moreover, l is the same for any such diffeomorphism.

Note that $f \circ \varphi^{-1}$ is just f expressed in terms of the coordinates $(y^1,\ldots,y^n)$ on U determined by φ and we will generally write this simply as $f(y^1,\ldots,y^n)$. The content of the Morse lemma is that any smooth function is, locally, near a nondegenerate critical point, quadratic in some coordinates.

Proof. We will begin by asking you to simplify the arithmetic just a bit.

Exercise B.0.1. Show that it will suffice to prove the theorem when $(a^1,\ldots,a^n) = (0,\ldots,0)$ and $f(0,\ldots,0) = 0$.

Consequently, we will assume that f is smooth on the open ball U centered at $(0,\ldots,0)$ in $\mathbb{R}^n$ and $f(0,\ldots,0) = 0$. Note that for any $(x^1,\ldots,x^n)$ in U, $t(x^1,\ldots,x^n) = (tx^1,\ldots,tx^n)$ is also in U for every $0 \le t \le 1$, so we can define

$$f_i(x^1,\ldots,x^n) = \int_0^1 \frac{\partial f}{\partial x^i}(tx^1,\ldots,tx^n)\,dt.$$

These are clearly smooth on U and

$$\sum_{i=1}^n x^i f_i(x^1,\ldots,x^n) = \sum_{i=1}^n x^i \int_0^1 \frac{\partial f}{\partial x^i}(tx^1,\ldots,tx^n)\,dt$$

$$= \int_0^1 \sum_{i=1}^n x^i \frac{\partial f}{\partial x^i}(tx^1,\ldots,tx^n)\,dt$$

$$= \int_0^1 \frac{d}{dt} f(tx^1,\ldots,tx^n)\,dt$$

$$= f(tx^1,\ldots,tx^n)|_0^1$$

$$= f(x^1,\ldots,x^n).$$

Moreover,

$$f_i(0,\ldots,0) = \int_0^1 \frac{\partial f}{\partial x^i}(0,\ldots,0)\, dt = \frac{\partial f}{\partial x^i}(0,\ldots,0), \quad i = 1,\ldots,n,$$

and this completes the proof of 1.

To prove 2. we assume, in addition, that $\frac{\partial f}{\partial x^i}(0,\ldots,0) = 0$ for each $i = 1,\ldots,n$ and apply the argument in 1. to $\frac{\partial f}{\partial x^i}$ as follows:

$$
\begin{aligned}
f(x^1,\ldots,x^n) &= \sum_{j=1}^n x^j f_j(x^1,\ldots,x^n)\\[2mm]
&= \sum_{j=1}^n x^j \int_0^1 \frac{\partial f}{\partial x^i}(tx^1,\ldots,tx^n)\, dt\\[2mm]
&= \sum_{j=1}^n x^j \int_0^1 \sum_{i=1}^n x^i \left(\int_0^1 \frac{\partial^2 f}{\partial x^i \partial x^j}(stx^1,\ldots,stx^n)\, ds \right) dt\\[2mm]
&= \sum_{i,j=1}^n x^i x^j \left(\int_0^1 \int_0^1 \frac{\partial^2 f}{\partial x^i \partial x^j}(stx^1,\ldots,stx^n)\, ds\, dt \right).
\end{aligned}
$$

Now we let

$$f_{ij}(x^1,\ldots,x^n) = \int_0^1 \int_0^1 \frac{\partial^2 f}{\partial x^i \partial x^j}(stx^1,\ldots,stx^n)\, ds\, dt.$$

Again, each f_{ij} is clearly smooth on U and

$$f_{ij}(0,\ldots,0) = \frac{\partial^2 f}{\partial x^i \partial x^j}(0,\ldots,0),$$

so the proof of 2. is complete.

For the proof of 3. we increment our assumptions once more and assume that the critical point at $(0,\ldots,0)$ is nondegenerate, that is, that the Hessian matrix

$$\left(\frac{\partial^2 f}{\partial x^i \partial x^j}(0,\ldots,0) \right)_{i,j=1,\ldots,n}$$

is nonsingular. We begin with the result from 2. We have

$$f(x^1,\ldots,x^n) = \sum_{i,j=1}^n x^i x^j f_{ij}(x^1,\ldots,x^n), \tag{B.2}$$

where

$$f_{ij}(0,\ldots,0) = \frac{\partial^2 f}{\partial x^i \partial x^j}(0,\ldots,0). \tag{B.3}$$

Exercise B.0.2. Show that we can assume that the functions f_{ij} satisfy

$$f_{ji} = f_{ij} \tag{B.4}$$

for all $i, j = 1, \ldots, n$. More precisely, show that the functions $f'_{ij} = \frac{1}{2}(f_{ij} + f_{ji})$ have a nonsingular Hessian at $(0, \ldots, 0)$ and satisfy (B.2), (B.3) and (B.4).

Now we consider the quadratic form $\sum_{i,j=1}^{n} a_{ij}x^i x^j$ on $\mathbb{R}^n$, where $a_{ij} = f_{ij}(0, \ldots, 0)$ for all $i, j = 1, \ldots, n$. By assumption, the matrix $(a_{ij})_{i,j=1,\ldots,n}$ is nonsingular. In particular, it is not identically zero, that is, some a_{ij} is nonzero. It follows from this that some nonsingular linear transformation of $\mathbb{R}^n$ will provide new coordinates $\hat{x}^1, \ldots, \hat{x}^n$ in terms of which the quadratic form is given by $\sum_{i,j=1}^{n} \hat{a}_{ij}\hat{x}^i \hat{x}^j$, where $\hat{a}_{11} \neq 0$.

Remark B.0.1. This is a standard result about nonzero, real quadratic forms (see, for example, the Lemma in Section 8, Chapter IX, of [BM]), but we will sketch the proof. Suppose first that some coefficient on the diagonal is nonzero, say, a_{ii}. Consider the nonsingular linear transformation that interchanges x^i and x^1, that is, $\hat{x}^1 = x^i$, $\hat{x}^i = x^1$ and $\hat{x}^j = x^j$ for $j \neq 1, i$. Then $\hat{a}_{11} = a_{ii} \neq 0$. Suppose, on the other hand, that $a_{ii} = 0$ for each $i = 1, \ldots, n$, but $a_{ij} \neq 0$ for some $i \neq j$. The nonsingular linear transformation that interchanges x^i and x^1 as well as x^j and x^2 gives $\hat{a}_{12} = a_{ij} \neq 0$, so we might as well assume at the outset that $a_{12} \neq 0$. Then the terms in the quadratic form involving x^1 and x^2 are

$$a_{11}(x^1)^2 + 2a_{12}x^1 x^2 + a_{22}(x^2)^2 = 2a_{12}x^1 x^2$$

(recall that we are assuming $a_{ii} = 0$ for all $i = 1, \ldots, n$). Now define new coordinates by $x^1 = \hat{x}^1 + \hat{x}^2$, $x^2 = \hat{x}^1 - \hat{x}^2$ and $x^i = \hat{x}^i$ for $i = 3, \ldots, n$. Then the terms in the quadratic form involving $\hat{x}^1$ and $\hat{x}^2$ are

$$2a_{12}x^1 x^2 = 2a_{12}(\hat{x}^1 + \hat{x}^2)(\hat{x}^1 - \hat{x}^2) = 2a_{12}(\hat{x}^1)^2 - 2a_{12}(\hat{x}^2)^2,$$

so $\hat{a}_{11} = 2a_{12} \neq 0$, as required.

Rewriting $f(x^1, \ldots, x^n) = \sum_{i,j=1}^{n} x^i x^j f_{ij}(x^1, \ldots, x^n)$ in terms of the coordinates $\hat{x}^1, \ldots, \hat{x}^n$ determined by this nonsingular linear transformation gives

$$f(\hat{x}^1, \ldots, \hat{x}^n) = \sum_{i,j=1}^{n} \hat{x}^i \hat{x}^j h_{ij}(\hat{x}^1, \ldots, \hat{x}^n),$$

where the functions $h_{ij}(\hat{x}^1, \ldots, \hat{x}^n)$ are smooth and, as above, we can assume that $h_{ji} = h_{ij}$ for all $i, j = 1, \ldots, n$. Moreover, computing second derivatives gives

$$h_{ij}(0, \ldots, 0) = \frac{\partial^2 f}{\partial \hat{x}^i \partial \hat{x}^j}(0, \ldots, 0),$$

and, by construction,

$$h_{11}(0, \ldots, 0) \neq 0.$$

Exercise B.0.3. Show that if A is the nonsingular linear transformation that carries $(x^1, \ldots, x^n)$ to $(\hat{x}^1, \ldots, \hat{x}^n)$, then

$$\left(\frac{\partial^2 f}{\partial \hat{x}^i \partial \hat{x}^j}(0, \ldots, 0) \right)_{i,j=1,\ldots,n} = A^T \left(\frac{\partial^2 f}{\partial x^i \partial x^j}(0, \ldots, 0) \right)_{i,j=1,\ldots,n} A,$$

and conclude that $(h_{ij}(0, \ldots, 0))_{i,j=1,\ldots,n}$ is nonsingular.

Now, since h_{11} is smooth and nonzero at $(0, \ldots, 0)$ it must be nonzero on some neighborhood of $(0, \ldots, 0)$ and on that neighborhood

$$| h_{11} |^{1/2}$$

is smooth and nonzero. We will now make a coordinate transformation on some neighborhood of $(0, \ldots, 0)$ that will alter only $\hat{x}^1$. To see where it comes from we recommend the following exercise.

Exercise B.0.4. Complete the square to show that

$$h_{11}(\hat{x}^1)^2 + 2 \sum_{j=2}^{n} h_{1j} \hat{x}^1 \hat{x}^j = \pm \left(| h_{11} |^{1/2} \left(\hat{x}^1 + \sum_{j=2}^{n} \frac{h_{1j}}{h_{11}} \hat{x}^j \right) \right)^2 - h_{11} \left(\sum_{j=2}^{n} \frac{h_{1j}}{h_{11}} \hat{x}^j \right)^2.$$

Now define $(u^1, \ldots, u^n)$ by

$$\begin{cases} u^1 = | h_{11} |^{1/2} (\hat{x}^1 + \sum_{j=2}^{n} \frac{h_{1j}}{h_{11}} \hat{x}^j), \\ u^2 = \hat{x}^2, \\ \vdots \\ u^n = \hat{x}^n. \end{cases}$$

This is a smooth map from some neighborhood of $(0, \ldots, 0)$ in $\mathbb{R}^n$ to $\mathbb{R}^n$ carrying $(0, \ldots, 0)$ to $(0, \ldots, 0)$ and its Jacobian is

$$\begin{pmatrix} \frac{\partial u^1}{\partial \hat{x}^1} & \frac{\partial u^1}{\partial \hat{x}^2} & \cdots & \frac{\partial u^1}{\partial \hat{x}^n} \\ 0 & 1 & \cdots & 0 \\ \vdots & \vdots & & \vdots \\ 0 & 0 & \cdots & 1 \end{pmatrix}.$$

The determinant of the Jacobian is therefore

$$\frac{\partial u^1}{\partial \hat{x}^1} = | h_{11} |^{1/2} \frac{\partial}{\partial \hat{x}^1} \left(\hat{x}^1 + \sum_{j=2}^{n} \frac{h_{1j}}{h_{11}} \hat{x}^j \right) + \frac{\partial | h_{11} |^{1/2}}{\partial \hat{x}^1} \left(\hat{x}^1 + \sum_{j=2}^{n} \frac{h_{1j}}{h_{11}} \hat{x}^j \right),$$

which, at $(0, \ldots, 0)$, is

$$\frac{\partial u^1}{\partial \hat{x}^1}(0, \ldots, 0) = | h_{11}(0, \ldots, 0) |^{1/2} \neq 0.$$

By the inverse function theorem the map is therefore a diffeomorphism of some neighborhood of $(0, \ldots, 0)$ in $\mathbb{R}^n$ onto another neighborhood of $(0, \ldots, 0)$ in $\mathbb{R}^n$ and therefore $(u^1, \ldots, u^n)$ are coordinates on some neighborhood of $(0, \ldots, 0)$ in $\mathbb{R}^n$.

Exercise B.0.5. Show that f is given in terms of the coordinates $u^1, \ldots, u^n$ by

$$f(u^1, \ldots, u^n) = \pm(u^1)^2 + \sum_{i,j=2}^{n} u^i u^j g_{ij}(u^1, \ldots, u^n)$$

for some smooth functions $g_{ij}(u^1, \ldots, u^n)$, $i, j = 2, \ldots, n$.

Now we would like to continue inductively, applying the same argument to $\sum_{i,j=2}^{n} u^i u^j g_{ij}(u^1, \ldots, u^n)$. To do so we must show that the matrix $(g_{ij}(0, \ldots, 0))_{i,j=2,\ldots,n}$ is nonsingular. But we know that the Hessian of f at $(0, \ldots, 0)$, computed in *any* coordinates, is nonsingular and

$$\left(\frac{\partial^2 f}{\partial u^i \partial u^j}(0, \ldots, 0) \right)_{i,j=1,\ldots,n} = \begin{pmatrix} \pm 2 & 0 & \cdots & & 0 \\ 0 & & & & \\ \vdots & & (g_{ij}(0, \ldots, 0))_{i,j=2,\ldots,n} & \\ 0 & & & & \end{pmatrix},$$

so $(g_{ij}(0, \ldots, 0))_{i,j=2,\ldots,n}$ must be nonsingular as well. As before, we can assume that $g_{ji} = g_{ij}$ for all $i, j = 2, \ldots, n$ and $g_{22}(0, \ldots, 0) \neq 0$. Defining $v^1, \ldots, v^n$ by

$$\begin{cases} v^1 = u^1, \\ v^2 = |g_{22}|^{1/2} \left(u^2 + \sum_{j=3}^{n} \frac{g_{2j}}{g_{22}} u^j \right), \\ v^2 = u^3, \\ \vdots \\ v^n = u^n \end{cases}$$

gives

$$\sum_{i,j=2}^{n} u^i u^j g_{ij}(u^1, \ldots, u^n) = \pm(v^2)^2 + \sum_{i,j=3,\ldots,n} v^i v^j p_{ij}(v^1, \ldots, v^n)$$

and hence

$$f(v^1, \ldots, v^n) = \pm(v^1)^2 \pm (v^2)^2 + \sum_{i,j=3,\ldots,n} v^i v^j p_{ij}(v^1, \ldots, v^n)$$

on some neighborhood of $(0, \ldots, 0)$ in $\mathbb{R}^n$.

Exercise B.0.6. Explicitly carry out the induction argument required to obtain coordinates $y^1, \ldots, y^n$ on some neighborhood of $(0, \ldots, 0)$ in $\mathbb{R}^n$ relative to which $f(y^1, \ldots, y^n) = \pm(y^1)^2 \pm \cdots \pm (y^n)^2$.

Renumbering the coordinates if necessary we can write

$$f(y^1,\ldots,y^n) = -(y^1)^2 - \cdots - (y^l)^2 + (y^{l+1})^2 + \cdots + (y^n)^2$$

for some integer $0 \le l \le n$.

All that remains is to show that if $z^1,\ldots,z^n$ are local coordinates at $(0,\ldots,0)$ in $\mathbb{R}^n$ for which

$$f(z^1,\ldots,z^n) = -(z^1)^2 - \cdots - (z^m)^2 + (z^{m+1})^2 + \cdots + (z^n)^2,$$

then $m = l$.

Exercise B.0.7. Compute the Hessian of f at the critical point $(0,\ldots,0)$ in both coordinate systems and show that these matrices are similar.

According to Sylvester's law of inertia (Theorem 6-z_3 of [Her]) these matrices must have the same signature (number of positive eigenvalues minus the number of negative eigenvalues), so $n - 2l = n - 2m$, and therefore $m = l$. ☐

We will conclude with a definition and a few remarks. If U is an open set in $\mathbb{R}^n$ and $f : U \to \mathbb{R}$ is a smooth, real-valued function, then f is said to be a *Morse function* if all of its critical points are nondegenerate. Although this sounds like a rather restrictive condition, there is a sense in which Morse functions are the common state of affairs. The following is Lemma 5-22 of [Nab2].

Theorem B.0.2. *Let U be an open set in $\mathbb{R}^n$ and $g : U \to \mathbb{R}$ an arbitrary smooth function. For each $\mathbf{b} = (b^1,\ldots,b^n)$ in $\mathbb{R}^n$ define a smooth function $g_{\mathbf{b}} : U \to \mathbb{R}$ by*

$$g_{\mathbf{b}}(x^1,\ldots,x^n) = g(x^1,\ldots,x^n) + b^1 x^1 + \cdots + b^n x^n.$$

Then the set of all $\mathbf{b} \in \mathbb{R}^n$ for which $g_{\mathbf{b}}$ fails to be a Morse function has (Lebesgue) measure zero in $\mathbb{R}^n$.

Intuitively, almost every linear perturbation of any smooth function is Morse. Note also that the critical points of a Morse function f are *isolated* in the sense that each one has an open neighborhood that contains no other critical points (since a sum or difference of squares has only one critical point, any open set on which f has the form $f(a^1,\ldots,a^n) - (y^1)^2 - \cdots - (y^l)^2 + (y^{l+1})^2 + \cdots + (y^n)^2$ contains only the critical point at $(y^1,\ldots,y^n) = (0,\ldots,0)$). The integer l is called the *index* of f at the critical point. If $l = 0$, then f has a relative minimum at $(a^1,\ldots,a^n)$; if $l = n$, then f has a relative maximum at $(a^1,\ldots,a^n)$. In general, l specifies the number of independent directions in which f decreases at $(a^1,\ldots,a^n)$. Finally, although we will have no need of this, we point out that the Morse lemma, being purely local, generalizes at once to smooth functions defined on finite-dimensional manifolds; there is also a version for smooth functions on Banach spaces (see [Palais2]).

C Stationary phase approximation

Recall that in Example 7.3.2 we found an integral representation for the solution to the Cauchy problem

$$i\frac{\partial\psi(q,t)}{\partial t} = -\frac{\hbar}{2m}\frac{\partial^2\psi(q,t)}{\partial q^2}, \quad (q,t) \in \mathbb{R} \times (0,\infty),$$

$$\lim_{t\to 0^+}\psi(q,t) = \psi_0(q), \quad q \in \mathbb{R}$$

for the free Schrödinger equation with initial data $\psi_0(q)$ (assumed to be smooth with compact support). Briefly, the procedure was to take the Fourier transform to obtain the solution

$$\psi(q,t) = \sqrt{\frac{m}{2\pi\hbar ti}} \int_{\mathbb{R}} e^{mi(q-x)^2/2\hbar t}\,\psi_0(x)\,dx$$

to the Schrödinger equation and then deal with the problem of showing that this approached $\psi_0(q)$ as $t \to 0^+$. Because of the oscillatory nature of the integral, this limit was not at all straightforward and we had to appeal to what is called its *stationary phase approximation*. Our objective in this appendix is to provide the proof of this. Since such oscillatory integrals arise with great regularity in classical as well as quantum physics and can only rarely be evaluated explicitly, the result we will prove is a well-worn part of any physicist's toolkit (see, for example, [GS3] for applications to geometric optics).

We will begin, as we did in Example 7.3.2, by noting (i) that the oscillatory integral that was under consideration there is of the general form

$$\int_{\mathbb{R}} e^{iTf(x)}g(x)\,dx, \tag{C.1}$$

where $T = \frac{1}{t}$, $f(x) = m(q-x)^2/2\hbar$ and $g(x) = \psi_0(x)$ and (ii) that we were interested in its asymptotic behavior as $T \to \infty$. We will now assume that both $f(x)$ and $g(x)$ are smooth and that $g(x)$ has compact support. The intuitive rationale behind the approximation we are looking for goes something like this. The exponential factor $e^{iTf(x)}$ is oscillatory. Near a point x_0 where $f'(x_0) \neq 0$ the rate at which it is oscillating (that is, the frequency) changes with x and, as T gets large, these oscillations become more and more rapid near x_0. One might expect that these very rapid oscillations near x_0 with varying frequencies will (approximately) cancel (real and imaginary parts are positive as often as they are negative) so that the dominant contributions to the integral as $T \to \infty$ will come from neighborhoods of points with $f'(x_0) = 0$ where $f(x)$ is nearly constant; we will state this precisely and prove it soon. The problem then is to compute the contributions due to the critical (that is, stationary) points of $f(x)$. If the critical point x_0 is degenerate ($f''(x_0) = 0$), then the behavior of $f(x)$ nearby can be

https://doi.org/10.1515/9783110751949-013

very complicated and it is difficult to say anything in general. On the other hand, if x_0 is a nondegenerate critical point of $f(x)$, then the Morse lemma (Theorem B.0.1.3) implies that, nearby, f is quadratic in some coordinates and one might hope to compute the contribution as a Gaussian integral. For this reason, we will assume throughout that f is a Morse function (Appendix B); note that, by virtue of Theorem B.0.2, this is not as serious a restriction as it might seem. The general result we will prove gives the following stationary phase approximation in the special case of a Morse function $f : \mathbb{R} \to \mathbb{R}$ with exactly one nondegenerate critical point at x_0:

$$\int_{\mathbb{R}} e^{iTf(x)} g(x)\, dx = \left(\frac{2\pi}{T}\right)^{1/2} e^{\operatorname{sgn}(f''(x_0))\,\pi i/4}\, \frac{e^{iTf(x_0)}}{\sqrt{|f''(x_0)|}}\, g(x_0) + O\!\left(\frac{1}{T^{3/2}}\right) \tag{C.2}$$

as $T \to \infty$. Recall that this means that there exist constants $M > 0$ and $T_0 > 0$ such that for all $T \geq T_0$,

$$\left| \int_{\mathbb{R}} e^{iTf(x)} g(x)\, dx - \left(\frac{2\pi}{T}\right)^{1/2} e^{\operatorname{sgn}(f''(x_0))\,\pi i/4}\, \frac{e^{iTf(x_0)}}{\sqrt{|f''(x_0)|}}\, g(x_0) \right| \leq M\!\left(\frac{1}{T^{3/2}}\right).$$

Although we will give the proof shortly it might be instructive to see a quick, informal calculation that gives some sense of where the terms in the approximation come from. For this we will take $g(x) = 1$, but one should keep in mind that this does not have compact support, so the general result we will prove is not applicable; nevertheless, the following rough computation is illuminating. Thus, we will consider the integral

$$\int_{\mathbb{R}} e^{iTf(x)}\, dx,$$

where $f(x)$ has a nondegenerate critical point at x_0. We approximate $f(x)$ by its second Taylor polynomial

$$f(x) \approx f(x_0) + \frac{1}{2} f''(x_0)(x - x_0)^2$$

at x_0 and then

$$\int_{\mathbb{R}} e^{iTf(x)}\, dx \approx \int_{\mathbb{R}} e^{iT[f(x_0) + \frac{1}{2} f''(x_0)(x - x_0)^2]}\, dx.$$

But

$$\int_{\mathbb{R}} e^{iT[f(x_0) + \frac{1}{2} f''(x_0)(x - x_0)^2]}\, dx = e^{iTf(x_0)} \int_{\mathbb{R}} e^{iTf''(x_0)(x - x_0)^2/2}\, dx$$

$$= e^{iTf(x_0)} \int_{\mathbb{R}} e^{iTf''(x_0)x^2/2}\, dx$$

$$= e^{iTf(x_0)} \int_{\mathbb{R}} e^{iT \, \mathrm{sgn}(f''(x_0)) \, |f''(x_0)| \, x^2/2} \, dx$$

$$= \left(\frac{2\pi}{T}\right)^{1/2} e^{\mathrm{sgn}(f''(x_0)) \, \pi i /4} \, \frac{e^{iTf(x_0)}}{\sqrt{|f''(x_0)|}},$$

where the last equality follows from the Gaussian integral (A.5).

For the remainder of this appendix we will consider the following situation. We consider the Morse function $f : \mathbb{R}^n \to \mathbb{R}$ with finitely many critical points $\mathbf{p}_1, \ldots, \mathbf{p}_N$, $g : \mathbb{R}^n \to \mathbb{R}$ is a smooth function with compact support and $T > 0$ is a real number. We will consider the integral

$$\int_{\mathbb{R}^n} e^{iTf(\mathbf{x})} g(\mathbf{x}) \, d^n \mathbf{x}$$

and will investigate its asymptotic behavior as $T \to \infty$. The first order of business is to confirm our earlier intuitive suspicion that the noncritical points of f do not contribute to the limit $\lim_{T \to \infty} \int_{\mathbb{R}^n} e^{iTf(\mathbf{x})} g(\mathbf{x}) \, d^n \mathbf{x}$. Specifically, we will let $U = \mathbb{R}^n - \{\mathbf{p}_1, \ldots, \mathbf{p}_N\}$ and will show that the integral

$$\int_U e^{iTf(\mathbf{x})} g(\mathbf{x}) \, d^n \mathbf{x}$$

approaches zero as $T \to \infty$. In fact, we will show that for any $m = 1, 2, \ldots,$

$$\int_U e^{iTf(\mathbf{x})} g(\mathbf{x}) \, d^n \mathbf{x} = O(T^{-m}) \quad \text{as } T \to \infty.$$

Recall that this means that for each such m, there exist positive constants $M(m)$ and $T_0(m)$ such that for all $T \geq T_0(m)$,

$$\left| \int_U e^{iTf(\mathbf{x})} g(\mathbf{x}) \, d^n \mathbf{x} \right| \leq \frac{M(m)}{T^m}.$$

To prove this we note that the gradient of f defines a nonvanishing, smooth vector field

$$V = \nabla f = \sum_{j=1}^n \frac{\partial f}{\partial x^j} \frac{\partial}{\partial x^j}$$

on U. Moreover,

$$V[e^{iTf(\mathbf{x})}] = \sum_{j=1}^n \frac{\partial f}{\partial x^j} e^{iTf(\mathbf{x})} (iT) \frac{\partial f}{\partial x^j} = iTe^{iTf(\mathbf{x})} \|V\|^2.$$

Defining the smooth vector field W by

$$W = \frac{1}{T\|V\|^2} V,$$

we can write this as

$$e^{iTf(\mathbf{x})} = -iW\left[e^{iTf(\mathbf{x})}\right].$$

Thus,

$$\int_U e^{iTf(\mathbf{x})} g(\mathbf{x})\, d^n\mathbf{x} = -i\int_U W\left[e^{iTf(\mathbf{x})}\right]g(\mathbf{x})\, d^n\mathbf{x} = i\int_U e^{iTf(\mathbf{x})} W[g(\mathbf{x})]\, d^n\mathbf{x},$$

where we have integrated by parts componentwise in W and used the fact that $g(\mathbf{x})$ has compact support. Continuing this computation we obtain

$$\int_U e^{iTf(\mathbf{x})} g(\mathbf{x})\, d^n\mathbf{x} = i\int_U e^{iTf(\mathbf{x})}\left(\frac{1}{T\|V\|^2} V[g(\mathbf{x})]\right) d^n\mathbf{x} = \frac{1}{T}\left(i\int_U e^{iTf(\mathbf{x})} h(\mathbf{x})\, d^n\mathbf{x}\right),$$

where $h(\mathbf{x}) = \frac{1}{\|V\|^2} V[g(\mathbf{x})]$. Note that $h(\mathbf{x})$ also has compact support. We find then that

$$\left|\int_U e^{iTf(\mathbf{x})} g(\mathbf{x})\, d^n\mathbf{x}\right| \le \frac{\int_U |h(\mathbf{x})|\, d^n\mathbf{x}}{T}$$

for all $T > 0$, so

$$\int_U e^{iTf(\mathbf{x})} g(\mathbf{x})\, d^n\mathbf{x} = O(T^{-1}) \quad \text{as } T \to \infty.$$

Exercise C.0.1. Use the fact that $h(\mathbf{x})$ also has compact support to repeat the same argument and show that

$$\int_U e^{iTf(\mathbf{x})} g(\mathbf{x})\, d^n\mathbf{x} = O(T^{-2}) \quad \text{as } T \to \infty$$

and then continue inductively to obtain

$$\int_U e^{iTf(\mathbf{x})} g(\mathbf{x})\, d^n\mathbf{x} = O(T^{-m}) \quad \text{as } T \to \infty$$

for every $m \ge 1$. In particular

$$\lim_{T \to \infty} \int_U e^{iTf(\mathbf{x})} g(\mathbf{x})\, d^n\mathbf{x} = 0. \tag{C.3}$$

Now we must handle the critical points. Choose disjoint open balls $U_1, \ldots, U_N$ centered at $\mathbf{p}_1, \ldots, \mathbf{p}_N$, respectively, and sufficiently small that on each one f takes the form (B.1) specified by the Morse lemma. Also let $U_{N+1} = \mathbb{R}^n - \{\mathbf{p}_1, \ldots, \mathbf{p}_N\}$. Then $\{U_1, \ldots, U_N, U_{N+1}\}$ is an open cover of $\mathbb{R}^n$. Now we choose a partition of unity $\{\varphi_1, \ldots, \varphi_N, \varphi_{N+1}\}$ subordinate to this open cover.

Remark C.0.1. We recall that this means the following. Each $\varphi_j, j = 1, \ldots, N, N + 1$, is a smooth function from $\mathbb{R}^n$ to $[0, 1]$ with the following properties:
1. The support supp (φ_j) of φ_j is contained in U_j for each $j = 1, \ldots, N, N + 1$.
2. For each $\mathbf{x} \in \mathbb{R}^n$, $\sum_{j=1}^{N+1} \varphi_j(\mathbf{x}) = 1$.

The existence of such a partition of unity is not obvious, but is proved, for example, in Theorem 3-11 of [Sp1], Appendix 3, Volume I, of [KN], or Corollary 3.1.5 of [Nab4].

Thus, we can write $g(\mathbf{x}) = \sum_{j=1}^{N+1} g(\mathbf{x})\varphi_j(\mathbf{x})$ and so

$$\int_{\mathbb{R}^n} e^{iTf(\mathbf{x})}g(\mathbf{x}) \, d^n\mathbf{x} = \sum_{j=1}^{N+1} \int_{U_j} e^{iTf(\mathbf{x})}g(\mathbf{x})\varphi_j(\mathbf{x}) \, d^n\mathbf{x}$$

$$= \sum_{j=1}^{N} \int_{U_j} e^{iTf(\mathbf{x})}g(\mathbf{x})\varphi_j(\mathbf{x}) \, d^n\mathbf{x} + \int_{U_{N+1}} e^{iTf(\mathbf{x})}g(\mathbf{x})\varphi_{N+1}(\mathbf{x}) \, d^n\mathbf{x}. \qquad \text{(C.4)}$$

Observe that each $g(\mathbf{x})\varphi_j(\mathbf{x})$ has compact support, so we already know that the last integral above is $O(T^{-m})$ as $T \to \infty$ for every $m = 1, 2, \ldots$. Now, fix some $j = 1, \ldots, N$ and, for convenience, write $U_j = U$, $\varphi_j = \varphi$ and $\mathbf{p}_j = \mathbf{p}$. Also let $g(\mathbf{x})\varphi(\mathbf{x}) = h(\mathbf{x})$. Then $h(\mathbf{x})$ has compact support and we will consider the integral

$$\int_U e^{iTf(\mathbf{x})}h(\mathbf{x}) \, d^n\mathbf{x}.$$

Recall that U is an open ball centered at $\mathbf{p}$ and we have assumed that it was chosen sufficiently small that there are local coordinates $y^1, \ldots, y^n$ on U relative to which

$$f(\mathbf{y}) = f(\mathbf{p}) + [-(y^1)^2 - \cdots - (y^l)^2 + (y^{l+1})^2 + \cdots + (y^n)^2]/2 = f(\mathbf{p}) + Q(\mathbf{y})/2,$$

where $Q(\mathbf{y}) = -(y^1)^2 - \cdots - (y^l)^2 + (y^{l+1})^2 + \cdots + (y^n)^2$. Let $(\frac{\partial \mathbf{x}}{\partial \mathbf{y}})$ denote the Jacobian of the coordinate transformation $\mathbf{x} = \mathbf{x}(\mathbf{y})$. Then

$$\int_U e^{iTf(\mathbf{x})}h(\mathbf{x}) \, d^n\mathbf{x} = \int_U e^{iTf(\mathbf{y})}h(\mathbf{y}) \left|\det\left(\frac{\partial \mathbf{x}}{\partial \mathbf{y}}\right)\right| d^n\mathbf{y}.$$

The function $h(\mathbf{y}) \, | \det(\frac{\partial \mathbf{x}}{\partial \mathbf{y}}) |$ is also smooth with compact support contained in U. We extend it to all of $\mathbb{R}^n$ by taking it to be zero on $\mathbb{R}^n - U$. Call this extension $F(\mathbf{y})$ and write it as

$$F(\mathbf{y}) = F(\mathbf{p}) + y^1 F_1(\mathbf{y}) + \cdots + y^n F_n(\mathbf{y}),$$

where $F_1, \ldots, F_n$ are smooth and have compact support. Then

$$\int_U e^{iTf(\mathbf{x})}h(\mathbf{x}) \, d^n\mathbf{x} = F(\mathbf{p}) \int_{\mathbb{R}^n} e^{iTf(\mathbf{y})} \, d^n\mathbf{y} + \sum_{k=1}^{n} \int_{\mathbb{R}^n} e^{iTf(\mathbf{y})}y^k F_k(\mathbf{y}) \, d^n\mathbf{y}. \qquad \text{(C.5)}$$

Now we will examine each of these integrals separately. Begin by evaluating $F(\mathbf{p})$.

Exercise C.0.2. Show that $\varphi(\mathbf{p}) = 1$.

Thus,

$$F(\mathbf{p}) = h(\mathbf{p}) \left| \det\left(\frac{\partial \mathbf{x}}{\partial \mathbf{y}}\right)(\mathbf{p}) \right| = g(\mathbf{p}) \left| \det\left(\frac{\partial \mathbf{x}}{\partial \mathbf{y}}\right)(\mathbf{p}) \right|.$$

To evaluate $|\det(\frac{\partial \mathbf{x}}{\partial \mathbf{y}})(\mathbf{p})|$ we note that, since $\mathbf{p}$ is a critical point of f, the Hessian of f at $\mathbf{p}$ computed in the $\mathbf{y}$-coordinates is related to the Hessian of f at $\mathbf{p}$ computed in the $\mathbf{x}$-coordinates by

$$\left(\frac{\partial^2 f}{\partial y^i \partial y^j}(\mathbf{p})\right) = \left(\frac{\partial \mathbf{x}}{\partial \mathbf{y}}(\mathbf{p})\right)^T \left(\frac{\partial^2 f}{\partial x^i \partial x^j}(\mathbf{p})\right)\left(\frac{\partial \mathbf{x}}{\partial \mathbf{y}}(\mathbf{p})\right).$$

Taking the determinant on both sides gives

$$1 = \left| \det\left(\frac{\partial \mathbf{x}}{\partial \mathbf{y}}\right)(\mathbf{p}) \right|^2 |\det H(f(\mathbf{x}))(\mathbf{p})|,$$

where we have written $H(f(\mathbf{x}))$ for the Hessian $(\frac{\partial^2 f}{\partial x^i \partial x^j})$ of f in the $\mathbf{x}$-coordinates. Consequently,

$$\left| \det\left(\frac{\partial \mathbf{x}}{\partial \mathbf{y}}\right)(\mathbf{p}) \right| = \frac{1}{\sqrt{|\det H(f(\mathbf{x}))(\mathbf{p})|}},$$

so

$$F(\mathbf{p}) = \frac{g(\mathbf{p})}{\sqrt{|\det H(f(\mathbf{x}))(\mathbf{p})|}}.$$

Exercise C.0.3. Show that

$$\int_{\mathbb{R}^n} e^{iTf(\mathbf{y})} \, d^n \mathbf{y} = \left(\frac{2\pi}{T}\right)^{n/2} e^{\pi i \, \mathrm{sgn}(H(f(\mathbf{x}))(\mathbf{p}))/4} e^{iTf(\mathbf{p})}$$

and therefore

$$F(\mathbf{p}) \int_{\mathbb{R}^n} e^{iTf(\mathbf{y})} \, d^n \mathbf{y} = \left(\frac{2\pi}{T}\right)^{n/2} e^{\pi i \, \mathrm{sgn}(H(f(\mathbf{x}))(\mathbf{p}))/4} \frac{e^{iTf(\mathbf{p})}}{\sqrt{|\det H(f(\mathbf{x}))(\mathbf{p})|}} g(\mathbf{p}).$$

This takes care of the first term in (C.5). Note, in particular, that this term is $O(T^{-n/2})$ as $T \to \infty$. Keep in mind, however, that all of this has been done for $U = U_j$, $\varphi = \varphi_j$, and $\mathbf{p} = \mathbf{p}_j$ for some fixed $j = 1, \ldots, N$, so what we will need in (C.4) is the sum

$$\sum_{j=1}^{N} \int_{U_j} e^{iTf(\mathbf{x})} g(\mathbf{x}) \varphi_j(\mathbf{x}) \, d^n \mathbf{x}$$

$$= \sum_{j=1}^{N} \left(\frac{2\pi}{T}\right)^{n/2} e^{\pi i \, \mathrm{sgn}(H(f(\mathbf{x}))(\mathbf{p}_j))/4} \frac{e^{iTf(\mathbf{p}_j)}}{\sqrt{|\det H(f(\mathbf{x}))(\mathbf{p}_j)|}} g(\mathbf{p}_j). \tag{C.6}$$

These are all $O(t^{-n/2})$.

All that remains is to consider the integrals

$$\sum_{k=1}^{n} \int_{\mathbb{R}^n} e^{iTf(\mathbf{y})} y^k F_k(\mathbf{y}) \, d^n \mathbf{y}$$

in (C.5). Fix some $k = 1, \ldots, n$ and consider the integral

$$\int_{\mathbb{R}^n} e^{iTf(\mathbf{y})} y^k F_k(\mathbf{y}) \, d^n \mathbf{y}.$$

Compute

$$\frac{\partial}{\partial y^k}\left(e^{iTf(\mathbf{y})}\right) = e^{iTf(\mathbf{y})}(iT)\frac{\partial f}{\partial y^k} = e^{iTf(\mathbf{y})}(iT)\frac{\partial}{\partial y^k}\left(f(\mathbf{p}) + Q(\mathbf{y})/2\right) = (\pm i)Te^{iTf(\mathbf{y})}y^k.$$

Thus,

$$e^{iTf(\mathbf{y})}y^k = (\mp i)\left(\frac{1}{T}\right)\frac{\partial}{\partial y^k}\left(e^{iTf(\mathbf{y})}\right),$$

so

$$\int_{\mathbb{R}^n} e^{iTf(\mathbf{y})}y^k F_k(\mathbf{y}) \, d^n \mathbf{y} = \frac{1}{T}(\mp i)\int_{\mathbb{R}^n} \frac{\partial}{\partial y^k}\left(e^{iTf(\mathbf{y})}\right)F_k(\mathbf{y}) \, d^N \mathbf{y}$$

$$= \frac{1}{T}(\pm i)\int_{\mathbb{R}^n} e^{iTf(\mathbf{y})}\frac{\partial F_k}{\partial y^k} \, d^n \mathbf{y},$$

where, for the last equality, we integrated by parts and used the fact that F_k has compact support. Since $\frac{\partial F_k}{\partial y^k}$ has compact support this last integral has exactly the same form as the integral $\int_{\mathbb{R}^n} e^{iTf(\mathbf{x})}g(\mathbf{x}) \, d^n \mathbf{x}$ with which we began, that is, we can write

$$\int_{\mathbb{R}^n} e^{iTf(\mathbf{y})}y^k F_k(\mathbf{y}) \, d^n \mathbf{y} = \frac{1}{T}(\pm i)\int_{\mathbb{R}^n} e^{iTf(\mathbf{y})}\phi(\mathbf{y}) \, d^n \mathbf{y}$$

and repeat the entire argument to obtain

$$\int_{\mathbb{R}^n} e^{iTf(\mathbf{y})}y^k F_k(\mathbf{y}) \, d^n \mathbf{y} = \frac{1}{T}(\pm i)\left[\Phi(\mathbf{p})\int_{\mathbb{R}^n} e^{iTf(\mathbf{y})} \, d^n \mathbf{y} + \sum_{k=1}^{n} e^{iTf(\mathbf{y})}y^k \Phi_k(\mathbf{y}) \, d^n \mathbf{y}\right].$$

We have seen that the first term inside the bracket is $O(T^{-n/2})$, so with the extra factor of $\frac{1}{T}$, the first term on the right-hand side is $O(T^{-n/2-1})$ and for the rest we pick up yet another factor of $\frac{1}{T}$. Continuing we find that the lowest order terms are the $O(T^{-n/2})$ terms in (C.6). Consequently,

$$\int_{\mathbb{R}^n} e^{iTf(\mathbf{x})}g(\mathbf{x}) \, d^n \mathbf{x}$$

$$= \sum_{j=1}^{N}\left(\frac{2\pi}{T}\right)^{n/2} e^{\pi i \, \mathrm{sgn}(H(f(\mathbf{x}))(\mathbf{p}_j))/4} \frac{e^{iTf(\mathbf{p}_j)}}{\sqrt{|\det H(f(\mathbf{x}))(\mathbf{p}_j)|}} g(\mathbf{p}_j) + O\left(\frac{1}{T^{n/2+1}}\right). \tag{C.7}$$

The first term on the right-hand side of (C.7) is the *stationary phase approximation* to the integral on the left-hand side. Note that when $n = 1$ and there is exactly one nondegenerate critical point at x_0, this reduces to (C.2).

This is really all we need here, but we would like to conclude with a few remarks. We have already pointed out the uses made of this approximation in dealing with oscillatory integrals that cannot be evaluated explicitly. What we have not pointed out yet is the remarkable fact that there are circumstances in which the stationary phase approximation is *exact*, that is, the sum on the right-hand side of (C.7) is equal to the integral on the left, so that the error terms vanish. Precisely when this occurs involves some rather deep topological issues and the phenomenon itself is best viewed from the perspective of localization theorems in equivariant cohomology. For those who would like to learn more about this we suggest [Kirwan], [DH], [Atiyah], [AB] and [BGV]. Needless to say, from the point of view of quantum theory it would be very desirable if some sort of stationary phase approximation could be established in the infinite-dimensional context and applied to Feynman path "integrals." Better yet, one would like analogues of those finite-dimensional equivariant localization theorems that guarantee the exactness of the approximation. From the rigorous mathematical point of view this is quite a nontrivial problem (see, however, Section 10.3 of [AHM]). Physicists are undeterred by this, of course, as one can see from, for example, [Szabo].

D Tangent and cotangent bundles

Here we would simply like to establish the notation we will be using for the tangent and cotangent bundles of a smooth manifold. The general definition of a vector bundle, of which these are particular examples, appears in Section 10.4.

We will view TM in the following way. As a set, TM consists of all pairs (p, v_p), where $p \in M$ and v_p is in the tangent space $T_p(M)$ to M at p. There is a natural projection $\pi : TM \to M$ of TM onto M defined by $\pi(p, v_p) = p$. The topology and manifold structure of TM are defined as follows. Let (U, ϕ) be a chart on M, where $U \subseteq M$ is an open set and ϕ is a homeomorphism of U onto the open set $\phi(U)$ in $\mathbb{R}^n$. Denote the coordinate functions of (U, ϕ) by $q^1, \ldots, q^n$ and their coordinate velocity vector fields by $\partial_{q^1}, \ldots, \partial_{q^n}$. Then, for $p \in U$ and $v_p \in T_p(M)$,

$$v_p = v_p(q^1)\partial_{q^1}|_p + \cdots + v_p(q^n)\partial_{q^n}|_p = v_p(q^i)\partial_{q^i}|_p,$$

where the summation convention is used in the last equality. We will often write this without the references to p as

$$v = v(q^1)\partial_{q^1} + \cdots + v(q^n)\partial_{q^n} = v(q^i)\partial_{q^i}.$$

Now let $\tilde{U} = \pi^{-1}(U) \subseteq TM$ and define $\tilde{\phi} : \tilde{U} \to \phi(U) \times \mathbb{R}^n \subseteq \mathbb{R}^n \times \mathbb{R}^n = \mathbb{R}^{2n}$ by

$$\tilde{\phi}(p, v_p) = (q^1(p), \ldots, q^n(p), \dot{q}^1(v_p), \ldots, \dot{q}^n(v_p)),$$

where

$$\dot{q}^i(v_p) = v_p(q^i), \ i = 1, \ldots, n.$$

Now, a subset $\mathcal{U}$ of TM is said to be *open* in TM if, for every chart (U, ϕ) on M, $\tilde{\phi}(\tilde{U} \cap \mathcal{U})$ is open in $\mathbb{R}^{2n}$. This determines a topology on TM and, relative to this topology, $(\tilde{U}, \tilde{\phi})$ are charts with coordinate functions $(q^1, \ldots, q^n, \dot{q}^1, \ldots, \dot{q}^n)$. These overlap smoothly and hence determine a differentiable structure for TM. Coordinates of this type that arise from charts on M are called *natural coordinates* on TM (physicists also call them *generalized coordinates*). The tangent bundle of $M = \mathbb{R}^n$ is, as a topological space and as a differentiable manifold, just the product $M \times \mathbb{R}^n = \mathbb{R}^n \times \mathbb{R}^n$, but generally tangent bundles are not topological products. For example, the tangent bundle TS^2 of the 2-sphere S^2 cannot be the topological product $S^2 \times \mathbb{R}^2$ since, if it were, then S^2 would admit a continuous, nonvanishing vector field and this would violate a classical (and rather deep) theorem in topology (see Theorem 16.5 of [Gre]). Any tangent bundle is, however, *locally* a product since $\tilde{\phi} : \tilde{U} \to \phi(U) \times \mathbb{R}^n \subseteq \mathbb{R}^n \times \mathbb{R}^n$ is a homeomorphism. A *(global) section* of the tangent bundle TM is a smooth map $s : M \to TM$ for which $\pi \circ s = \mathrm{id}_M$. Such a section therefore picks out a tangent vector $s(p) \in T_p(M)$ at each $p \in M$ and these tangent vectors vary smoothly from point to point in M. One can therefore identify a section of TM with a smooth vector field on M.

https://doi.org/10.1515/9783110751949-014

The definition of T^*M is very much like the definition of TM. As a set, T^*M consists of all pairs (p, η_p), where $p \in M$ and η_p is an element of the dual $T_p^*(M)$ of the tangent space $T_p(M)$ at p, that is, η_p is a real-valued, linear function on $T_p(M)$. There is a natural projection $\pi : T^*M \to M$ of T^*M onto M defined by $\pi(p, \eta_p) = p$ (unless it is likely to cause confusion we will use the same symbol for the projection on both TM and T^*M). If (U, ϕ) is a chart on M with coordinate functions $q^1, \ldots, q^n$, then the basis for $T_p^*(M)$ dual to the basis $\partial_{q^1}|_p, \ldots, \partial_{q^n}|_p$ for $T_p(M)$ is given by the coordinate differentials $dq^1|_p, \ldots, dq^n|_p$ at p. Thus, for $p \in M$ and $\eta_p \in T_p^*(M)$,

$$\eta_p = \eta_p(\partial_{q^1}|_p)dq^1|_p + \cdots + \eta_p(\partial_{q^n}|_p)dq^n|_p,$$

which we will be inclined to write simply as

$$\eta = \eta(\partial_{q^1})dq^1 + \cdots + \eta(\partial_{q^n})dq^n = \eta(\partial_{q^i})dq^i,$$

where the summation convention is used in the last equality. Define $(\tilde{U}, \tilde{\phi})$ by $\tilde{U} = \pi^{-1}(U) \subseteq T^*M$ and $\tilde{\phi} : \tilde{U} \to \phi(U) \times \mathbb{R}^n$, where

$$\tilde{\phi}(p, \eta_p) = (q^1(p), \ldots, q^n(p), \eta_p(\partial_{q^1}|_p), \ldots, \eta_p(\partial_{q^n}|_p)).$$

Now a subset $\mathcal{U}$ of T^*M is said to be *open* in T^*M if, for every chart (U, ϕ) on M, $\tilde{\phi}(\tilde{U} \cap \mathcal{U})$ is open in $\phi(U) \times \mathbb{R}^n$. This determines a topology on T^*M and, relative to this topology, $(\tilde{U}, \tilde{\phi})$ are charts on T^*M. These overlap smoothly and hence determine a differentiable structure for T^*M. Coordinates associated with these charts are called *natural coordinates* on T^*M and are denoted

$$(q^1, \ldots, q^n, \xi_1, \ldots, \xi_n). \tag{D.1}$$

Thus, the coordinate function ξ_i on T^*M is given by $\xi_i(p, \eta_p) = \eta_p(\partial_{q^i}|_p)$. The cotangent bundle $T^*\mathbb{R}^n$ of $\mathbb{R}^n$ is just the product manifold $\mathbb{R}^n \times \mathbb{R}^n$, but, as for tangent bundles, T^*M is generally not just a topological product space. A *(global)section* of the cotangent bundle T^*M is a smooth map $s : M \to T^*M$ for which $\pi \circ s = \mathrm{id}_M$. Such a section therefore picks out a covector $s(p) \in T_p^*(M)$ at each $p \in M$, and these covectors vary smoothly from point to point in M. One can therefore identify a section of T^*M with a smooth 1-form on M.

E Poisson and wave equations

In our derivations of the Lorenz and Coulomb gauges (Section 4.2) we required the existence of smooth solutions to the Poisson equation

$$\nabla^2 u = g$$

and the wave equation

$$\nabla^2 u - \frac{1}{c^2}\frac{\partial^2 u}{\partial t^2} = g$$

on $\mathbb{R}^3$ when the right-hand side g is smooth. These existence and regularity results are classical and can be deduced from various general theorems on linear partial differential equations; standard sources for this are [Evans] and [TaylM]. We will briefly describe the results available in two other sources that lead directly to the specific theorems we need and which may be somewhat more accessible.

The paper [Ros] contains an elementary proof of the following existence theorem for (in particular) the Poisson equation (and, as a bonus, the Malgrange–Ehrenpreis theorem on the existence of fundamental solutions for arbitrary constant coefficient linear partial differential equations). We begin by simply stating the result and will then try to clarify with a few remarks. Recall that $L^2_{\mathrm{loc}}(\mathbb{R}^n)$ consists of all Lebesgue measurable functions on $\mathbb{R}^n$ that are square integrable on every compact subset of $\mathbb{R}^n$. A constant coefficient linear differential operator D of order m on $\mathbb{R}^n$ is of the form

$$D = \sum_{|\alpha|\leq m} A^\alpha \partial_\alpha,$$

where α is a multi-index (see Appendix G.4) and each A^α a constant.

Theorem E.0.1. *Let D be a (nonzero) constant coefficient linear differential operator on $\mathbb{R}^n$. Then for every $g \in L^2_{\mathrm{loc}}(\mathbb{R}^n)$ there exists a $u \in L^2_{\mathrm{loc}}(\mathbb{R}^n)$ such that $Du = g$.*

A few remarks are in order. For the Poisson equation, D is just the Laplacian on $\mathbb{R}^n$. Since the elements of $L^2_{\mathrm{loc}}(\mathbb{R}^n)$ need not be differentiable, the sense in which $u \in L^2_{\mathrm{loc}}(\mathbb{R}^n)$ is a solution to $Du = g$ cannot be the usual, classical one ("compute the derivatives, plug into the equation and get an identity"). What is meant here is that u is a *weak solution* to $Du = g$. We will have much more to say about this when the need arises, but just to be clear we will state explicitly what is intended here. As motivation note that if u were a smooth, classical solution to $Du = g$ with g smooth, then it would certainly be the case that $\langle Du, \varphi \rangle = \langle g, \varphi \rangle$ for any smooth function φ with compact support on $\mathbb{R}^n$, where $\langle\,,\,\rangle$ is the $L^2(\mathbb{R}^n)$ inner product. Performing a few integrations by parts using the fact that φ has compact support then gives $\langle u, D^*\varphi \rangle = \langle g, \varphi \rangle$ for any such φ, where D^* is the adjoint of the differential operator D; for the Poisson equation, D^* is also the Laplacian (it is worth doing the integrations by parts in this case

https://doi.org/10.1515/9783110751949-015

just to see this). Having shifted all of the differentiations to φ we can now define a *weak solution* of $Du = g$ to be an element of $L^2_{loc}(\mathbb{R}^n)$ that satisfies $\langle u, D^*\varphi \rangle = \langle g, \varphi \rangle$ for any smooth function φ with compact support on $\mathbb{R}^n$. Certainly, any smooth g is in $L^2_{loc}(\mathbb{R}^n)$, so we are guaranteed the existence of an $L^2_{loc}(\mathbb{R}^n)$ solution u. For a general D one can do no better than this, but for the Poisson equation there are *elliptic regularity* results available which imply that a weak solution u is necessarily smooth (that part of elliptic regularity that is directly relevant to the Poisson equation is covered concisely in Chapter 10 of [LL]). Consequently, we get the result we need for the Poisson equation.

For the wave equation, which is hyperbolic rather than elliptic, we will state an existence and regularity result for a broader class of equations and for which there is a self-contained proof available at [Holz]. The following is Theorem 5.1 from this source. Here $\square$ is used to denote the *d'Alembertian*, or *wave operator*

$$\square = \nabla^2 - \frac{1}{c^2}\frac{\partial^2}{\partial t^2},$$

where c is a positive constant and we write $x = (x^1, \dots, x^n) \in \mathbb{R}^n$ and $(t, x) \in \mathbb{R}^{n+1}$.

Theorem E.0.2. *Let u_0, u_1 be in $C^\infty(\mathbb{R}^n)$ and let $b^1, \dots, b^n, c$ and g all be in $C^\infty(\mathbb{R}^{n+1})$. Then there exists a unique $u \in C^\infty(\mathbb{R}^{n+1})$ satisfying*

$$\square u + \sum_{k=1}^{n} b^k(t,x)\frac{\partial u}{\partial x^k} + c(t,x)u + g(t,x) = 0,$$

$$u(0,x) = u_0(x),$$

$$\frac{\partial u}{\partial t}(0,x) = u_1(x).$$

In particular, the wave equation

$$\nabla^2 u - \frac{1}{c^2}\frac{\partial^2 u}{\partial t^2} = g$$

has lots of smooth solutions if g is smooth.

F Carathéodory procedure

The following review is intended only to establish some notation and terminology and to recall a few basic results that we will need in the construction of the Wiener measure in Section 9.3 (see Sections 1–2, Chapter 12, of [Roy] for details).

Let X denote some nonempty set and $\mathcal{P}(X)$ its power set (the collection of all subsets of X). For any subset $S \subseteq \mathcal{P}(X)$ containing $\emptyset$, a *pre-measure* on S is a nonnegative, extended real-valued function $\mu_S : S \rightarrow [0, \infty]$ on S that satisfies the following two conditions (we use $\bigsqcup$ for disjoint unions):

1. $\mu_S(\emptyset) = 0$,
2. if $S_k \in S$ for $k = 1, 2, \ldots$ with $S_{k_1} \cap S_{k_2} = \emptyset \, \forall \, k_1 \neq k_2$ and if $\bigsqcup_{k=1}^{\infty} S_k$ is in S, then

$$\mu_S\left(\bigsqcup_{k=1}^{\infty} S_k \right) = \sum_{k=1}^{\infty} \mu_S(S_k).$$

A *semi-algebra* on X is a collection $\mathcal{I} \subseteq \mathcal{P}(X)$ of subsets of X that contains $\emptyset$ and X, is closed under finite intersections and has the property that if A is in $\mathcal{I}$, then $X - A$ can be written as a finite disjoint union of elements of $\mathcal{I}$. Note that, by Proposition 9.3.1, the collection $\mathcal{I}$ of cylinder sets in $C_0[t_a, t_b]$ is a semi-algebra. It is true, but not at all obvious that for a fixed q_0, the probabilities assigned to cylinder sets by (9.20) define a pre-measure on $\mathcal{I}$.

An *algebra* $\mathcal{A}$ on X is a collection of subsets of X that contains $\emptyset$ and X and has the property that $A, B \in \mathcal{A} \Rightarrow A \cup B \in \mathcal{A}$ and $A - B \in \mathcal{A}$. Denote by $\mathcal{A}(\mathcal{I}) \subseteq \mathcal{P}(X)$ the algebra generated by the semi-algebra $\mathcal{I}$; $\mathcal{A}(\mathcal{I})$ is just the intersection of all the algebras in $\mathcal{P}(X)$ containing $\mathcal{I}$. Any pre-measure $\mu_{\mathcal{I}}$ on $\mathcal{I}$ extends in a natural way to a pre-measure $\mu_{\mathcal{A}(\mathcal{I})}$ on $\mathcal{A}(\mathcal{I})$; see Proposition 9 and Exercise 5, Section 2, Chapter 12, of [Roy].

An algebra on X that is closed under countable unions is called a σ-*algebra* on X. The σ-algebra generated by $S \subseteq \mathcal{P}(X)$ is the intersection of all the σ-algebras on X containing S and will be denoted $\sigma(S)$. If $\mathcal{I}$ is a semi-algebra and $\mathcal{A}(\mathcal{I})$ is the algebra it generates, then $\sigma(\mathcal{A}(\mathcal{I})) = \sigma(\mathcal{I})$. A pre-measure μ on some σ-algebra $\mathcal{A}$ in X is called a *measure* on the *measurable space* $(X, \mathcal{A})$, or simply a measure on X if the σ-algebra is understood. The elements of $\mathcal{A}$ are called μ-*measurable*, or simply *measurable*, sets. The triple $(X, \mathcal{A}, \mu)$ is called a *measure space*. We should recall also that a measure space $(X, \mathcal{A}, \mu)$ is said to be *complete* if $\mathcal{A}$ contains all subsets of sets of μ-measure zero, that is, if $A \in \mathcal{A}$ with $\mu(A) = 0$ and $B \subseteq A$, then $B \in \mathcal{A}$ (and so $\mu(B) = 0$). For any measure space $(X, \mathcal{A}, \mu)$ one can find a complete measure space $(X, \mathcal{A}_0, \mu_0)$ with $\mathcal{A} \subseteq \mathcal{A}_0$, $\mu_0|_{\mathcal{A}} = \mu$ and $\mathcal{A}_0 = \{A \cup B : A \in \mathcal{A} \text{ and } B \subseteq C \text{ with } C \in \mathcal{A} \text{ and } \mu(C) = 0\}$. Then $(X, \mathcal{A}_0, \mu_0)$ is called the *completion* of $(X, \mathcal{A}, \mu)$; see Proposition 4, Section 1, Chapter 11, of [Roy]. For example, denoting the σ-algebra of Lebesgue measurable sets in $\mathbb{R}^n$ by $\mathcal{A}_{\text{Leb}}(\mathbb{R}^n)$ and writing μ_{Leb} for the Lebesgue measure, $(\mathbb{R}^n, \mathcal{A}_{\text{Leb}}(\mathbb{R}^n), \mu_{\text{Leb}})$ is

https://doi.org/10.1515/9783110751949-016

the completion $(\mathbb{R}^n, \mathcal{B}(\mathbb{R}^n), \mu_{\mathrm{Leb}}|_{\mathcal{B}(\mathbb{R}^n)})$, where $\mathcal{B}(\mathbb{R}^n)$ is the σ-algebra of Borel sets in $\mathbb{R}^n$ (the σ-algebra generated by the open, or closed, sets in $\mathbb{R}^n$).

Our problem is to describe how to get from a pre-measure on a semi-algebra on X to a measure on some σ-algebra on X. The standard procedure for doing this is due to Carathéodory and we will review it briefly. Assume that we are given a pre-measure μ on the semi-algebra $\mathcal{I}$ and that we have extended it to a pre-measure on the algebra $\mathcal{A}(\mathcal{I})$ generated by $\mathcal{I}$. For simplicity we will denote this extended pre-measure by $\mu : \mathcal{A}(\mathcal{I}) \rightarrow [0, \infty]$ also. Now let $T \in \mathcal{P}(X)$ denote an *arbitrary* subset of X. An $\mathcal{A}(\mathcal{I})$-*cover* of T is a sequence $A_1, A_2, \ldots$ of elements of $\mathcal{A}(\mathcal{I})$ with $T \subseteq \bigcup_{k=1}^{\infty} A_k$. For a given T there are generally many of these; we define

$$\mu^*(T) = \inf \left\{ \sum_{k=1}^{\infty} \mu(A_k) \right\},$$

where the infimum is over all $\mathcal{A}(\mathcal{I})$-covers $\{A_k\}_{k=1}^{\infty}$ of T. The function

$$\mu^* : \mathcal{P}(X) \rightarrow [0, \infty]$$

thus defined is called an *outer measure* on X. What this means is that μ^* has the following properties (see Lemma 4, Section 2, Chapter 12, of [Roy]):

1. $\mu^*(\emptyset) = 0$,
2. $A \subseteq B \Rightarrow \mu^*(A) \leq \mu^*(B), \forall A, B \in \mathcal{P}(X)$,
3. $\mu^*(\bigcup_{k=1}^{\infty} A_k) \leq \sum_{k=1}^{\infty} \mu^*(A_k), \forall A_1, A_2, \ldots \in \mathcal{P}(X)$.

To turn the countable subadditivity in 3. into countable additivity Carathéodory restricts attention to those subsets of X that "split every other set additively," as Halmos puts it (page 44 of [Hal1]). More precisely, we will say that a subset M of X is μ^*-*measurable* if, for every $T \in \mathcal{P}(X)$,

$$\mu^*(T) = \mu^*(T \cap M) + \mu^*(T \cap (X - M)).$$

One can then show that the collection $\sigma(\mu^*)$ of all μ^*-measurable sets is a σ-algebra on X and that $\mu^*|_{\sigma(\mu^*)}$ is a complete measure on $(X, \sigma(\mu^*))$; see Theorem 1, Section 1, Chapter 12, of [Roy]. Furthermore, $\sigma(\mathcal{I}) = \sigma(\mathcal{A}(\mathcal{I})) \subseteq \sigma(\mu^*)$ and $\mu^*|_{\sigma(\mathcal{I})} = \mu$; see Lemma 5, Section 2, Chapter 12, of [Roy]. We will call $\mu^*|_{\sigma(\mathcal{I})}$ the *measure induced by the pre-measure* μ and we will use the same symbol μ for both.

G Schwartz space, Fourier transform, distributions and Sobolev spaces

G.1 Introduction

Eventually, we will need to review, and even generalize, the basic properties of the Fourier transform on $\mathbb{R}^n$, but for the moment we will consider only the $n = 1$ case. We will provide a fairly extensive synopsis, but, again, only a synopsis. All of this material is accessible in many sources, for example, Sections IX.1 and IX.2 of [RS2], Sections 4.3.1 and 5.2.1 of [Evans], Sections 2-1 and 2-2 of [SW], Sections 1–3, Chapter VI, of [Yosida] and Chapters 5–7 of [LL].

G.2 Dimension 1

The *Schwartz space* $\mathcal{S}(\mathbb{R})$ of rapidly decreasing smooth, complex-valued functions on $\mathbb{R}$ is defined by

$$\mathcal{S}(\mathbb{R}) = \left\{ f \in C^\infty(\mathbb{R}) : \forall k, n = 0, 1, 2, \ldots, \sup_{q \in \mathbb{R}} \left| q^k f^{(n)}(q) \right| < \infty \right\},$$

where $f^{(n)}(q)$ denotes the nth derivative of $f(q)$ and $f^{(0)}(q) = f(q)$. Thus, the elements of $\mathcal{S}(\mathbb{R})$ are smooth, complex-valued functions on $\mathbb{R}$ which, together with all of their derivatives, decrease in modulus more rapidly than the reciprocal of any polynomial as $q \to \pm\infty$. Examples include such things as $f(q) = q^m e^{-q^2/2}$ for any integer m. Moreover, $\mathcal{S}(\mathbb{R})$ is an infinite-dimensional complex vector space. It also has the topological structure of a *Fréchet space*. What this means is the following (one can find a useful discussion of Fréchet spaces with lots of examples in [Ham]). For each $k, n = 0, 1, 2, \ldots$ we define a semi-norm on $\mathcal{S}(\mathbb{R})$ by

$$\| f \|_{k,n} = \sup_{q \in \mathbb{R}} \left| q^k f^{(n)}(q) \right|.$$

These are semi-norms and not norms because $\| f \|_{k,n} = 0$ does not, in general, imply $f = 0$; for example, if f is a constant function and $n > 0$. These determine a complete metric ρ on $\mathcal{S}(\mathbb{R})$ defined by

$$\rho(f, g) = \sum_{k,n=0}^{\infty} 2^{-k-n} \frac{\| f - g \|_{k,n}}{1 + \| f - g \|_{k,n}}$$

and with the property that for any sequence $\{f_j\}_{j=1}^{\infty}$ in $\mathcal{S}(\mathbb{R})$, $\rho(f, f_j) \to 0$ if and only if $\| f - f_j \|_{k,n} \to 0$ for all k and n. The *Fréchet topology* on $\mathcal{S}(\mathbb{R})$ is the metric topology determined by ρ. Note that convergence in this topology is very restrictive. For example, the sequence $\{\frac{1}{j} e^{-q^2/j^2}\}_{j=1}^{\infty}$ converges uniformly to zero on all of $\mathbb{R}$, but does not

https://doi.org/10.1515/9783110751949-017

converge in $S(\mathbb{R})$ because, for example,

$$\left\|\frac{1}{j}e^{-q^2/j^2}\right\|_{1,0} = \sup_{q\in\mathbb{R}}\left|\frac{q}{j}e^{-q^2/j^2}\right| = \max_{x\in\mathbb{R}}\left|xe^{-x^2}\right| = \frac{1}{\sqrt{2e}}.$$

Note that $C_0^\infty(\mathbb{R})$ is contained in $S(\mathbb{R})$; relative to the Fréchet space topology it is, in fact, dense in $S(\mathbb{R})$.

Exercise G.2.1. Show that the position operator Q (Example 5.2.3) and momentum operator P (Example 5.2.4) are both essentially self-adjoint on the Schwartz space $S(\mathbb{R})$.

On $S(\mathbb{R})$ we define the *Fourier transform*

$$\mathcal{F} : S(\mathbb{R}) \rightarrow S(\mathbb{R})$$

as follows. For $f \in S(\mathbb{R})$, $\mathcal{F}f = \hat{f}$ is the complex-valued function of a real variable p defined by

$$(\mathcal{F}f)(p) = \hat{f}(p) = \frac{1}{\sqrt{2\pi}} \int_{\mathbb{R}} e^{-ipq}f(q)dq. \tag{G.1}$$

Remark G.2.1. We should point out that there are numerous alternative conventions in the literature for the definition of the Fourier transform, most of which differ from each other only in the choice of various constants (we have adopted the convention of [RS2]). In harmonic analysis, for example, it is common to define $\hat{f}(p) = \int_{\mathbb{R}} e^{-2\pi ipq}f(q)dq$, whereas in quantum mechanics one often sees $\hat{f}(p) = \frac{1}{\sqrt{2\pi\hbar}} \int_{\mathbb{R}} e^{-ipq/\hbar}f(q)dq$. We will, in fact, use this last convention when it seems prudent to arrive at the formulas one encounters in the physics literature (in which case we will write $\mathcal{F}_\hbar$ for $\mathcal{F}$). The essential features of the transform are the same for all of these variants and one need only make a change of variable to keep track of where the constants appear or disappear. The crucial property of all of these definitions is that *the Fourier transform of a Schwartz function of q is a Schwartz function of p* (the reason this is crucial will become clear when we extend the Fourier transform, first to $L^2(\mathbb{R})$ and then to tempered distributions).

In fact, $\mathcal{F}$ is a linear bijection of $S(\mathbb{R})$ onto $S(\mathbb{R})$; it is also continuous with respect to the Fréchet topology. We will compute just one simple example that we will need later.

Example G.2.1. For any positive real number α we define $f \in S(\mathbb{R})$ by $f(q) = e^{-\alpha q^2/2}$. Then

$$(\mathcal{F}f)(p) = \hat{f}(p) = \frac{1}{\sqrt{2\pi}} \int_{\mathbb{R}} e^{-ipq}e^{-\alpha q^2/2}dq = \frac{1}{\sqrt{2\pi}} \int_{\mathbb{R}} e^{-\alpha q^2/2-ipq}dq.$$

Making the change of variable $t = \sqrt{\frac{\alpha}{2}}\, q$ and then completing the square in the exponent gives

$$\hat{f}(p) = \frac{1}{\sqrt{2\pi}}\sqrt{\frac{2}{\alpha}} \int_{\mathbb{R}} e^{-(t+ip/\sqrt{2\alpha})^2} e^{-p^2/2\alpha} dt$$

$$= \frac{e^{-p^2/2\alpha}}{\sqrt{\alpha\pi}} \int_{\mathbb{R}} e^{-(t+ip/\sqrt{2\alpha})^2} dt.$$

To evaluate this last integral we will appeal to the Cauchy integral theorem to conclude that for every $R > 0$, $\int_{C_R} e^{-z^2} dz = 0$, where C_R is the counterclockwise boundary of the closed rectangle $[-R, R] \times [0, p/\sqrt{2\alpha}]$. Thus,

$$0 = \int_{-R}^{R} e^{-t^2} dt + i \int_{0}^{p/\sqrt{2\alpha}} e^{-(R+i\tau)^2} d\tau - \int_{-R}^{R} e^{-(t+ip/\sqrt{2\alpha})^2} dt - i \int_{0}^{p/\sqrt{2\alpha}} e^{-(-R+i\tau)^2} d\tau.$$

As $R \to \infty$ the second and fourth integrals clearly go to zero, so this reduces to

$$\int_{\mathbb{R}} e^{-(t+ip/\sqrt{2\alpha})^2} dt = \int_{\mathbb{R}} e^{-t^2} dt = \sqrt{\pi}$$

(for the last equality see (A.1) in Appendix A). Consequently,

$$\hat{f}(p) = \frac{e^{-p^2/2\alpha}}{\sqrt{\alpha}}. \tag{G.2}$$

With a bit more work one can show in a similar way that this formula is also true when α is complex with positive real part, provided $\sqrt{\alpha}$ is taken to be the branch of the square root with branch cut along the negative real axis. There is another argument giving this same result that we will describe in Example G.2.2.

Since $\mathcal{F}: S(\mathbb{R}) \to S(\mathbb{R})$ is a bijection, it has an inverse

$$\mathcal{F}^{-1}: S(\mathbb{R}) \to S(\mathbb{R}),$$

and this turns out to be easy to describe. For $g \in S(\mathbb{R})$, $\mathcal{F}^{-1}g = \check{g}$ is given by

$$(\mathcal{F}^{-1}g)(q) = \check{g}(q) = \frac{1}{\sqrt{2\pi}} \int_{\mathbb{R}} e^{ipq} g(p) dp. \tag{G.3}$$

Note that $\mathcal{F}^{-1}$ is also Fréchet continuous.

It is worth taking a moment to write out the equality $f = \mathcal{F}^{-1}(\mathcal{F}f)$ for $f \in S(\mathbb{R})$ explicitly in terms of (G.1) and (G.3),

$$f(q) = \frac{1}{\sqrt{2\pi}} \int_{\mathbb{R}} e^{ipq} \hat{f}(p) dp = \int_{\mathbb{R}} \left(\frac{1}{2\pi} \int_{\mathbb{R}} e^{-ipq} f(q) dq \right) e^{ipq} dp, \tag{G.4}$$

and note its similarity to the Fourier series expansion

$$f(\theta) = \sum_{n\in\mathbb{Z}} \left(\frac{1}{2\pi} \int_{S^1} e^{-in\phi} f(\phi)d\phi \right) e^{in\theta}$$

of a 2π-periodic function on $\mathbb{R}$ or, equivalently, a function on the circle S^1. Both of these are, in fact, instances of the same phenomenon, one discrete and one continuous. One thinks of the value $\hat{f}(p)$ of the Fourier transform of f at $p \in \mathbb{R}$ as the pth component of $f(q)$ in the integral ("continuous sum") decomposition of f given by (G.4). Motivated by many of the common applications (such as vibrating strings), one often views q intuitively as a coordinate in physical space, while p is a frequency. Then (G.4) is a frequency decomposition of f with $\hat{f}(p)$ quantifying the "amount" of the frequency p contained in $f(q)$.

Every element of $S(\mathbb{R})$ is also an element of $L^2(\mathbb{R})$, so we can identify $S(\mathbb{R})$ with a linear subspace (*not* a topological subspace) of $L^2(\mathbb{R})$; it is, in fact, dense in $L^2(\mathbb{R})$. The *Plancherel theorem* then gives $\langle \hat{f}, \hat{g} \rangle = \langle f, g \rangle\ \forall f, g \in S(\mathbb{R})$, so $\mathcal{F}$ extends to a unique *unitary* map, also denoted $\mathcal{F}$, of $L^2(\mathbb{R})$ onto itself:

$$\mathcal{F} : L^2(\mathbb{R}) \to L^2(\mathbb{R}).$$

This extension is also referred to as the *Fourier transform*, or sometimes the *Fourier–Plancherel transform*, and its inverse

$$\mathcal{F}^{-1} = \mathcal{F}^* : L^2(\mathbb{R}) \to L^2(\mathbb{R})$$

will still be called the *inverse Fourier transform*. For $f \in L^1(\mathbb{R}) \cap L^2(\mathbb{R})$, $\hat{f}$ can be computed from the same formula (G.1) as for $S(\mathbb{R})$, but for a general element of $L^2(\mathbb{R})$ this integral need not exist. One way to describe $\hat{f}$ for an arbitrary element of $L^2(\mathbb{R})$ is as an L^2-limit of Fourier transforms of functions in $L^1(\mathbb{R}) \cap L^2(\mathbb{R})$. Specifically, for every $n \geq 1$ we define $f_n(q) = f(q)\chi_n(q)$, where χ_n is the characteristic function of the interval $[-n, n]$. Then $f_n \in L^1(\mathbb{R}) \cap L^2(\mathbb{R})$ and the monotone convergence theorem implies that, in $L^2(\mathbb{R})$, $f_n \to f$ as $n \to \infty$. In particular, $\{f_n\}_{n=1}^\infty$ is Cauchy in $L^2(\mathbb{R})$, so, by Parseval's theorem, $\{\hat{f}_n\}_{n=1}^\infty$ is also Cauchy in $L^2(\mathbb{R})$ and therefore it converges in $L^2(\mathbb{R})$. Computing $\hat{f}_n$ from (G.1) one has the following L^2-limit for $\mathcal{F}f = \hat{f}$:

$$\mathcal{F}f = \hat{f} = \lim_{n\to\infty} \hat{f}_n.$$

We should point out that, although the (Lebesgue) integral that defines the Fourier transform on $S(\mathbb{R})$ and $L^1(\mathbb{R})$ generally does not exist for $f \in L^2(\mathbb{R})$, it is nevertheless the case that the Fourier transform of such an f can be computed as an *improper* integral. Indeed,

$$(\mathcal{F}f)(p) = \hat{f}(p) = \lim_{M\to\infty} \frac{1}{\sqrt{2\pi}} \int_{[-M,M]} e^{-ipq} f(q)dq, \qquad (G.5)$$

where the limit is in $L^2(\mathbb{R})$. In light of this it is not uncommon to hear it said that the Fourier transform of an $L^2(\mathbb{R})$ function can be written in the form (G.1), provided the integral is interpreted in the sense of (G.5). We will bow to the tradition of using the same (Lebesgue) integral symbol to denote the function defined by this L^2-limit. However, since we will face this same issue at several points in the text we would like to at least have some terminology to distinguish this new notion of "integral." We will define it here just for functions on $\mathbb{R}$ and will generalize when the need arises. Let g be an element of $L^2(\mathbb{R})$ and let $k : \mathbb{R} \times \mathbb{R} \to \mathbb{C}$ be a function with the following properties. For any $M > 0$,

1. $k(\cdot, p) \in L^1([-M, M])$ for almost every $p \in \mathbb{R}$,
2. $\int_{[-M,M]} k(q, p)\, dq$ is in $L^2(\mathbb{R})$ as a function of p.

Then we say that g is the *integral in the mean*, or the *mean-square integral* of k if

$$\lim_{M \to \infty} \int_{[-M,M]} k(q, \cdot)\, dq = g(\cdot),$$

where the limit is in $L^2(\mathbb{R})$, that is, if

$$\lim_{M \to \infty} \left\| g(\cdot) - \int_{[-M,M]} k(q, \cdot)\, dq \right\|_{L^2} = 0.$$

In this case we will abuse notation a bit and still write

$$g(p) = \int_{\mathbb{R}} k(q, p)\, dq.$$

Much of the significance $\mathcal{F}$ and $\mathcal{F}^{-1}$ resides in the fact that, on the differentiable elements of $L^2(\mathbb{R})$, they convert differentiation into multiplication and multiplication into differentiation. Specifically, for $f \in \mathcal{S}(\mathbb{R})$, integration by parts gives all of the following:

$$\left(\frac{df}{dq} \right)^{\wedge}(p) = ip\hat{f}(p), \tag{G.6}$$

$$\left(qf(q) \right)^{\wedge}(p) = i\frac{d\hat{f}}{dp}, \tag{G.7}$$

$$\left(p\hat{f}(p) \right)^{\vee}(q) = -i\frac{df}{dq}, \tag{G.8}$$

$$\left(\frac{d\hat{f}}{dp} \right)^{\vee}(q) = -iqf(q). \tag{G.9}$$

Exercise G.2.2. Prove (G.6), (G.7), (G.8) and (G.9).

We will see many examples of how these are used, but here is a simple one.

Example G.2.2. We will give another derivation of the Fourier transform of $f(q) = e^{-aq^2/2}$ (see Example G.2.1), but this time we will allow α to be complex with positive real part. Compute the derivative of $f(q)$ to obtain $f'(q) + \alpha q f(q) = 0$. Taking the Fourier transform then gives $ip\hat{f}(p) + ia\hat{f}'(p) = 0$. Solving the first order equation for $\hat{f}(p)$ gives $\hat{f}(p) = \hat{f}(0)e^{-p^2/2\alpha}$. Now, by definition, $\hat{f}(0) = \frac{1}{\sqrt{2\pi}}\int_{\mathbb{R}} e^{-aq^2/2} dq$, and if α is real and positive, this is just $\frac{1}{\sqrt{\alpha}}$. But we saw in Example A.0.2 that this formula is still valid if α is complex with positive real part provided $\sqrt{\alpha}$ is the principal value of the square root. Consequently, under these assumptions we obtain, as promised earlier, the same formula (G.2)

$$\hat{f}(p) = \frac{e^{-p^2/2\alpha}}{\sqrt{\alpha}}$$

for the Fourier transform.

We will also need to know what the Fourier transform does to a product of two functions in $S(\mathbb{R})$. What it does is surprisingly simple. If $f, g \in S(\mathbb{R})$, then fg is also in $S(\mathbb{R})$ and the *convolution* of f and g is the function $f * g \in S(\mathbb{R})$ defined by

$$(f * g)(t) = \int_{\mathbb{R}} f(t - \tau)g(\tau) \, d\tau = \int_{\mathbb{R}} f(\tau)g(t - \tau) \, d\tau = (g * f)(t).$$

Then one can show that

$$(fg)^\wedge = \frac{1}{\sqrt{2\pi}} \hat{f} * \hat{g} \tag{G.10}$$

and

$$(f * g)^\wedge = \sqrt{2\pi} \hat{f} \hat{g} \tag{G.11}$$

(see Theorem IX.3 of [RS2]).

Exercise G.2.3. Prove the following *shift properties* of the Fourier transform. For any $f \in S(\mathbb{R})$ and a in $\mathbb{R}$,

$$\mathcal{F}(f(q - a)) = e^{-iap}\hat{f}(p)$$

and

$$\mathcal{F}(e^{iaq}f(q)) = \hat{f}(p - a).$$

Exercise G.2.4. Prove the following *scaling property* of the Fourier transform. For any $f \in S(\mathbb{R})$ and $a \neq 0$ in $\mathbb{R}$,

$$\mathcal{F}(f(aq)) = \frac{1}{|a|} \hat{f}\left(\frac{1}{a}p\right).$$

The Fourier transform also extends to a class of objects much more general than functions in $L^2(\mathbb{R})$ and we will need to deal with this. The topological dual of $S(\mathbb{R})$ is the linear space $S'(\mathbb{R})$ of all complex-valued linear functionals on $S(\mathbb{R})$ that are continuous with respect to the Fréchet topology. An element T of $S'(\mathbb{R})$ is called a *tempered distribution* on $\mathbb{R}$; the Schwartz functions on which they operate to produce complex numbers are then called *test functions*. Every ψ in $L^1_{loc}(\mathbb{R})$ (the complex-valued measurable functions on $\mathbb{R}$ that are integrable on compact subsets of $\mathbb{R}$) gives rise to a tempered distribution T_ψ by defining

$$T_\psi[f] = \int_\mathbb{R} f(q)\psi(q)dq$$

for every $f \in S(\mathbb{R})$. Note that the integral exists due to the fact that ψ is locally integrable and f decays rapidly as $q \to \pm\infty$. Distributions of this type are said to be *regular*, whereas all other tempered distributions are called *singular*. One generally does not bother to distinguish between the function $\psi \in L^1_{loc}(\mathbb{R})$ and the tempered distribution T_ψ. In particular, since $L^2(\mathbb{R}) \subseteq L^1_{loc}(\mathbb{R})$, every element of $L^2(\mathbb{R})$ gives rise to a tempered distribution, so we can identify $L^2(\mathbb{R})$ with a *subset* of $S'(\mathbb{R})$:

$$S(\mathbb{R}) \subseteq L^2(\mathbb{R}) \subseteq S'(\mathbb{R}).$$

With this in mind one often allows oneself such abuses of terminology as the phrase "distributions in $L^2(\mathbb{R})$." We should point out that the dual of a Fréchet space is generally not a Fréchet space, so $S'(\mathbb{R})$ does not come equipped with a ready-made topology. However, if one defines sequential convergence in $S'(\mathbb{R})$ pointwise on $S(\mathbb{R})$, that is, $T_j \to T$ if and only if $T_j[f] \to T[f]$ in $\mathbb{C}$ for every $f \in S(\mathbb{R})$, then every element of $S'(\mathbb{R})$ is the limit of a sequence in $L^2(\mathbb{R})$. Here is an example.

Example G.2.3. Fix $a \in \mathbb{R}$ and define the *Dirac delta* at a, denoted

$$\delta_a : S(\mathbb{R}) \to \mathbb{C},$$

by

$$\delta_a[f] = f(a)$$

for every $f \in S(\mathbb{R})$. Then δ_a is clearly linear on $S(\mathbb{R})$. To show that δ_a is continuous on $S(\mathbb{R})$ we suppose that $f_j \to f$ in $S(\mathbb{R})$, that is, $\|f - f_j\|_{k,n} \to 0$ for all $k, n = 0, 1, 2, \ldots$. We must show that $\delta_a[f_j] \to \delta_a[f]$ in $\mathbb{C}$, that is, $f_j(a) \to f(a)$. But this is precisely the statement that $\|f - f_j\|_{0,0} \to 0$. Consequently, δ_a is a tempered distribution.

Next we show that δ_a is the limit of a sequence of distributions in $L^2(\mathbb{R})$. For this we must find a sequence of elements ψ_j of $L^2(\mathbb{R})$ with $T_{\psi_j}[f] \to \delta_a[f] \ \forall f \in S(\mathbb{R})$. There are many ways to do this, but the simplest choice is to let ψ_j be $j/2$ times the characteristic function of the interval $[a - \frac{1}{j}, a + \frac{1}{j}]$:

$$\psi_j = \frac{j}{2} \chi_{[a-\frac{1}{j}, a+\frac{1}{j}]}.$$

Now let $\epsilon > 0$ be given and choose j so that $|f(q) - f(a)| < \epsilon$ for all $q \in [a - \frac{1}{j}, a + \frac{1}{j}]$. Then

$$\left| \int_{\mathbb{R}} f(q)\psi_j(q)dq - f(a) \right| = \left| \frac{j}{2} \int_{a-\frac{1}{j}}^{a+\frac{1}{j}} f(q)dq - \frac{j}{2} \int_{a-\frac{1}{j}}^{a+\frac{1}{j}} f(a)dq \right|$$

$$= \left| \frac{j}{2} \int_{a-\frac{1}{j}}^{a+\frac{1}{j}} (f(q) - f(a))dq \right|$$

$$\leq \frac{j}{2} \int_{a-\frac{1}{j}}^{a+\frac{1}{j}} |f(q) - f(a)| \, dq$$

$$< \frac{j}{2} \epsilon \left(\left(a + \frac{1}{j} \right) - \left(a - \frac{1}{j} \right) \right) = \epsilon,$$

as required. As we mentioned above, this is by no means the only way to represent the Dirac delta as a limit in $\mathcal{S}'(\mathbb{R})$ of a family of functions (thought of as distributions). For future reference we will record just one more. Specifically, one can show that in $\mathcal{S}'(\mathbb{R})$,

$$\lim_{t \to 0^+} \frac{1}{\sqrt{2\pi i t}} e^{i(q-a)^2/2t} = \delta_a. \tag{G.12}$$

One final remark on the Dirac delta is in order. We note that δ_a is an element of $\mathcal{S}'(\mathbb{R})$, but is not T_ψ for any ψ in $L^1_{\text{loc}}(\mathbb{R})$. Nevertheless, it is common, particularly in the physics literature, to write it as if it were by introducing a fictional object called the Dirac delta "function" $\delta(q - a)$ and writing

$$\int_{\mathbb{R}} f(q)\delta(q - a)dq = f(a)$$

rather than $\delta_a[f] = f(a)$. One should understand, however, that this is just notation and should not to be mistaken for what it looks like, that is, the integral of $f(q)\delta(q-a)$ with respect to Lebesgue measure. Alternatively, one can regard $\delta(q - a)dq$ as a name for the point measure at a on $\mathbb{R}$, in which case $\int_{\mathbb{R}} f(q)\delta(q - a)dq$ is the integral of $f(q)$ with respect to this measure.

Although tempered distributions are, from the point of view of ordinary calculus, rather singular objects, one can extend the notions of derivative and Fourier transform to them by shifting these operations to the test functions. Specifically, we define the *distributional derivative* of $T \in S'(\mathbb{R})$ to be the tempered distribution T' defined by

$$T'[f] = -T[f'].$$

Note that if $T = T_\psi$, where ψ is a Schwartz function, then $\psi' \in L^1_{loc}(\mathbb{R})$, and for any $f \in S(\mathbb{R})$, integration by parts gives

$$T_\psi[f'] = \int_\mathbb{R} f'(q)\psi(q)dq = -\int_\mathbb{R} f(q)\psi'(q)dq = -T_{\psi'}[f],$$

so the minus sign in the definition is to ensure that $T'_\psi[f] = T_{\psi'}[f]$ when ψ is in $S(\mathbb{R})$.

In the case of regular distributions for which the distributional derivative is also regular we would like to rephrase this slightly. Suppose then that $\psi \in L^1_{loc}(\mathbb{R})$ and T_ψ is the corresponding regular distribution. Assume that T'_ψ is also regular, so that $T'_\psi = T_\varphi$ for some $\varphi \in L^1_{loc}(\mathbb{R})$. Then, for every $f \in S(\mathbb{R})$,

$$\int_\mathbb{R} f'(q)\psi(q)\, dq = -\int_\mathbb{R} f(q)\varphi(q)\, dq.$$

Note the resemblance to the integration by parts formula with φ playing the role of the derivative of ψ. It is not hard to see that a $\varphi \in L^1_{loc}(\mathbb{R})$ with this property (if it exists) must be unique and we will call it the *weak derivative* of $\psi \in L^1_{loc}(\mathbb{R})$. This provides a natural generalization of the usual notion of derivative that applies to functions that need not have derivatives in the usual sense. To see how things work out in practice we will compute a simple example.

Example G.2.4. Define $\psi : \mathbb{R} \to \mathbb{R}$ by

$$\psi(x) = \begin{cases} 0 & \text{if } x \le 0, \\ 2x & \text{if } 0 \le x \le \frac{1}{2}, \\ 2 - 2x & \text{if } \frac{1}{2} \le x \le 1, \\ 0 & \text{if } x \ge 1. \end{cases}$$

Note that $\psi(x)$ is certainly in $L^1_{loc}(\mathbb{R})$ and, in the usual sense, it is not differentiable at $x = 0, \frac{1}{2}, 1$. We will show that $\psi(x)$ has a weak derivative $L^1_{loc}(\mathbb{R})$ represented by a function $\varphi(x)$ that is defined arbitrarily at $x = 0, \frac{1}{2}, 1$ and elsewhere is given by

$$\varphi(x) = \begin{cases} 0 & \text{if } x < 0, \\ 2 & \text{if } 0 < x < \frac{1}{2}, \\ -2 & \text{if } \frac{1}{2} < x < 1, \\ 0 & \text{if } x > 1. \end{cases}$$

Now we must show that for every $\phi \in S(\mathbb{R})$, $\int_{\mathbb{R}} \psi(x)\phi'(x)\,dx = -\int_{\mathbb{R}} \varphi(x)\phi(x)\,dx$. For this we just compute

$$
\int_{\mathbb{R}} \varphi(x)\phi(x)\,dx = 2\int_0^{1/2} \phi(x)\,dx - 2\int_{1/2}^1 \phi(x)\,dx
$$

$$
= 2\left\{-\int_0^{1/2} x\phi'(x)\,dx + x\phi(x)\Big|_0^{1/2}\right\}
$$

$$
- 2\left\{-\int_{1/2}^1 (x-1)\phi'(x)\,dx + (x-1)\phi(x)\Big|_{1/2}^1\right\}
$$

$$
= -\int_0^{1/2} 2x\phi'(x)\,dx - \int_{1/2}^1 (2-2x)\phi'(x)\,dx
$$

$$
= -\int_{\mathbb{R}} \psi(x)\phi'(x)\,dx.
$$

Higher order derivatives are defined inductively so that $T^{(n)}[f] = (-1)^n T[f^{(n)}]$. Similarly, we define the *Fourier transform* and *inverse Fourier transform* of T to be the distributions $\mathcal{F}T = \hat{T}$ and $\mathcal{F}^{-1}T = \check{T}$ given by

$$
(\mathcal{F}T)[f] = T[\mathcal{F}f]
$$

and

$$
(\mathcal{F}^{-1}T)[f] = T[\mathcal{F}^{-1}f].
$$

Then $\mathcal{F}$ and $\mathcal{F}^{-1}$ are inverse bijections of $S'(\mathbb{R})$ onto itself.

Note that if $\psi \in S(\mathbb{R}) \subseteq L^2(\mathbb{R})$, then its derivative ψ' can be written as $\psi' = (\mathcal{F}^{-1} \circ Q_{ip} \circ \mathcal{F})\psi$, where Q_{ip} is the operator that multiplies by ip, and this is also in $S(\mathbb{R})$. If ψ is in $L^2(\mathbb{R})$, but not in $S(\mathbb{R})$, then we can regard ψ as a tempered distribution and its distributional derivative is still obtained by applying $\mathcal{F}^{-1} \circ Q_{ip} \circ \mathcal{F}$. This is a distribution, but need not be an L^2-distribution. Indeed, this will be the case if and only if ψ satisfies $ip\hat{\psi}(p) \in L^2(\mathbb{R})$. For such $\psi \in L^2(\mathbb{R})$ the distributional derivative is again an element of $L^2(\mathbb{R})$ and we will call it the L^2-*derivative* of ψ.

Similar remarks apply to higher order derivatives. In particular, as an operator on $L^2(\mathbb{R})$, the second derivative, or *one-dimensional Laplacian*

$$
\Delta = \frac{d^2}{dq^2} = \mathcal{F}^{-1} \circ Q_{-p^2} \circ \mathcal{F}
$$

is defined on

$$
\mathcal{D}(\Delta) = \{\psi \in L^2(\mathbb{R}) : Q_{-p^2}\hat{\psi} \in L^2(\mathbb{R})\}.
$$

It follows from Lemma 5.2.5 that Δ is self-adjoint on this domain.

Example G.2.5. From the definition of the Dirac delta it is clear that

$$\delta_a^{(n)}[f] = (-1)^n f^{(n)}(a)$$

for any $n = 0, 1, 2, \ldots$. Furthermore,

$$(\mathcal{F}\delta_a)[f] = \delta_a[\mathcal{F}f] = \delta_a[\hat{f}] = \hat{f}(a) = \int_{\mathbb{R}} f(q)\frac{e^{-iaq}}{\sqrt{2\pi}}\,dq.$$

As we mentioned earlier, when a distribution such as $\mathcal{F}\delta_a$, takes values that are given by integration next to some $L^1_{loc}(\mathbb{R})$ function it is common to identify the distribution with the function so that one is likely to see the result of this example written

$$\mathcal{F}\delta_a = \frac{e^{-iap}}{\sqrt{2\pi}}.$$

Similarly, regarding e^{iaq} as a distribution,

$$\mathcal{F}(e^{iaq}) = \sqrt{2\pi}\,\delta_a.$$

To prove this one simply computes from (G.4) that

$$\mathcal{F}(e^{iaq})[f] = (e^{iaq})[\mathcal{F}f] = (e^{iaq})[\hat{f}] = \int_{\mathbb{R}} \hat{f}(p)e^{iap}\,dp = \sqrt{2\pi}f(a) = \sqrt{2\pi}\,\delta_a[f].$$

We will extend all of this material to functions on $\mathbb{R}^n$ in Section G.4, but first we will take an important detour to consider the heat kernel on $\mathbb{R}$.

G.3 Heat kernel

We would like to illustrate the role played by Fourier transforms and distributions in partial differential equations. This will be a recurring theme for us in the text, but for the moment we will be content with a very important example that exhibits all of the essential features and to which we will return when we discuss such things as "propagators" and "path integrals." What we intend to do is compute the one-dimensional heat kernel and then, through a sequence of exercises, describe a number of its most important properties. This will serve as a warm-up for the somewhat more involved case of the Schrödinger kernel and will also lay the foundation for our discussion of the path integral formulation of quantum mechanics, which we take up in Chapter 8.

Example G.3.1. We let D denote a positive real number and consider the following initial value problem for the *one-dimensional heat equation* (also called the *one-dimensional diffusion equation*):

$$\frac{\partial\psi(t,x)}{\partial t} - D\frac{\partial^2\psi(t,x)}{\partial x^2} = 0, \quad (t,x) \in (0,\infty) \times \mathbb{R}, \tag{G.13}$$

$$\lim_{t\to 0^+} \psi(t,x) = \psi_0(x), \quad x \in \mathbb{R}. \tag{G.14}$$

We will begin by assuming that the initial data $\psi_0(x)$ is in the Schwartz space $\mathcal{S}(\mathbb{R})$ and will look for a solution that is also in $\mathcal{S}(\mathbb{R})$ for each $t \in (0, \infty)$ and for which $\psi(t, x)$ and $\partial\psi(t, x)/\partial t$ are continuous. The procedure will be to apply the Fourier transform $\mathcal{F}$ (with respect to x) to both sides of the heat equation, solve the resulting equation for $\hat\psi(t, p)$ and then apply the inverse transform $\mathcal{F}^{-1}$ to get $\psi(t, x)$. Our assumptions ensure that one can differentiate under the integral sign to obtain

$$\left(\frac{\partial\psi(t, x)}{\partial t} \right)^{\wedge} = \frac{\partial\hat\psi(t, p)}{\partial t}$$

and hence applying $\mathcal{F}$ to (G.13) gives

$$\frac{\partial\hat\psi(t, p)}{\partial t} + Dp^2\hat\psi(t, p) = 0.$$

The initial condition becomes

$$\lim_{t \to 0^+} \hat\psi(t, p) = \hat\psi_0(p).$$

The solution to this simple first order initial value problem is

$$\hat\psi(t, p) = \hat\psi_0(p)e^{-Dtp^2}.$$

Therefore, $\psi(t, x)$ is the inverse Fourier transform of the product of $\hat\psi_0(p)$ and e^{-Dtp^2} and this, by (G.11), is $\frac{1}{\sqrt{2\pi}}$ times the convolution of $\psi_0(x)$ and the inverse transform of e^{-Dtp^2}. We have already computed the latter, so we obtain the following explicit formula for a solution to our initial value problem for the heat equation:

$$\psi(t, x) = \frac{1}{\sqrt{4\pi Dt}} \int_{\mathbb{R}} e^{-(x-y)^2/4Dt} \psi_0(y)\, dy. \tag{G.15}$$

Note that we have used the indefinite article *a* rather than the definite article *the*. Solutions to the heat equation on $\mathbb{R}$ with given initial data need not be unique. This phenomenon was first investigated by Tychonoff [Tych].

Now define a map $H_D : (0, \infty) \times \mathbb{R} \times \mathbb{R} \to \mathbb{R}$ by

$$H_D(t, x, y) = H_D^t(x, y) = \frac{1}{\sqrt{4\pi Dt}} e^{-(x-y)^2/4Dt}. \tag{G.16}$$

Then

$$\psi(t, x) = \int_{\mathbb{R}} H_D(t, x, y)\psi_0(y)\, dy. \tag{G.17}$$

Note that $\psi(t, x)$ is just the convolution product of ψ_0 and $K_D^t(x) = \frac{1}{\sqrt{4\pi Dt}} e^{-x^2/4Dt}$:

$$\psi(t, x) = (K_D^t * \psi_0)(x).$$

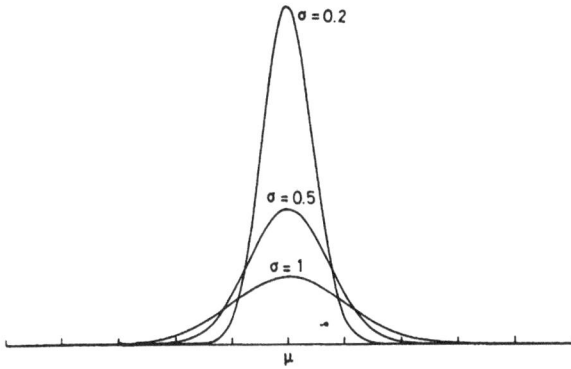

Figure G.1: Gaussian distribution.

Also note that, since the Gaussian (or normal) distribution with mean μ and standard deviation σ is defined by

$$\frac{1}{\sigma\sqrt{2\pi}}e^{-(x-\mu)^2/2\sigma^2},$$

the function $H(t,x,y)$ can be regarded, for each fixed $t > 0$ and $y \in \mathbb{R}$, as the normal distribution with mean y and standard distribution $\sqrt{2Dt}$. As $t \to 0^+$, $\sigma \to 0$ and the distribution peeks more and more sharply at the mean (see Figure G.1). Intuitively, one would say that $H_D(t,x,y)$ approaches the Dirac delta at y as $t \to 0^+$. You will prove this shortly.

In physics the positive constant D is called the *diffusion constant* and knowing how it is determined by the basic parameters of a physical system is an important problem. For most mathematical purposes, however, it does not play a significant role and one can take $D = 1$. For any fixed value of $D > 0$ we will call (G.16) the *(one-dimensional) heat kernel*. The study of the heat kernel and its generalizations to higher-dimensional manifolds and manifolds with boundary plays a significant role in partial differential equations, geometry and mathematical physics (see, for example, [RosenS]). Much of what we will have to say about it and its analogue for the Schrödinger equation rests on the properties we explore now in a sequence of exercises.

Exercise G.3.1. Hold $y \in \mathbb{R}$ fixed. Show that $H_D(t,x,y)$ is a solution to the heat equation

$$\frac{\partial}{\partial t} H_D(t,x,y) - D\frac{\partial^2}{\partial x^2} H_D(t,x,y) = 0.$$

Since $H_D(t,x,y)$ is symmetric in x and y, it is also true that

$$\frac{\partial}{\partial t} H_D(t,x,y) - D\frac{\partial^2}{\partial y^2} H_D(t,x,y) = 0$$

when $x \in \mathbb{R}$ is held fixed.

Exercise G.3.2. Hold $y \in \mathbb{R}$ and $t > 0$ fixed. Show that

$$\int_{\mathbb{R}} H_D(t, x, y) \, dx = 1.$$

Conclude that $H_D(t, x, y)$ determines a family of probability measures $\mu_D^{t,y}$ on $\mathbb{R}$ parametrized by $(t, y) \in (0, \infty) \times \mathbb{R}$ and defined by

$$\mu_D^{t,y}(M) = \int_M H_D(t, x, y) \, dx$$

for any Lebesgue measurable set M in $\mathbb{R}$.

We will have very little to say about heat flow here. However, the heat/diffusion equation arises also in the context of *Brownian motion* and this will play an important role in our discussion of Feynman's path integral in Section 9.3. We will reserve a more detailed discussion for later, but will mention at this point that the probability measures $\mu_D^{t,y}$ that we have just introduced have the following physical interpretation. Suppose that a particle undergoing Brownian motion in $\mathbb{R}$ is to be found at $y \in \mathbb{R}$ when $t = 0$. Then, for any measurable set M in $\mathbb{R}$, $\mu_D^{t,y}(M)$ is the probability that the particle will be found in M at time t. In this case the constant D is determined in a very specific way by the particles and the fluid in which the motion is taking place.

Exercise G.3.3. Fix $y \in \mathbb{R}$ and regard $t > 0$ as a parameter. Then Exercise G.3.2 gives a one-parameter family of probability measures $\{\mu_D^{t,y} : t > 0\}$. Each of these can be regarded as a tempered distribution whose value at any $f \in S(\mathbb{R})$ is

$$\mu_D^{t,y}[f] = \int_{\mathbb{R}} f(x) \, d\mu_D^{t,y}(x) = \int_{\mathbb{R}} f(x) H_D(t, x, y) \, dx.$$

Show that as $t \to 0^+$, these distributions converge in $S'(\mathbb{R})$ to the Dirac delta δ_y at y, that is, for any $f \in S(\mathbb{R})$,

$$\lim_{t \to 0^+} \int_{\mathbb{R}} f(x) H_D(t, x, y) \, dx = \delta_y[f] = f(y).$$

Hint: For any $\epsilon > 0$,

$$\int_{\mathbb{R}} f(x) H_D(t, x, y) \, dx - f(y) = \int_{[y-\epsilon, y+\epsilon]} (f(x) - f(y)) H_D(t, x, y) \, dx$$

$$+ \int_{\mathbb{R} - [y-\epsilon, y+\epsilon]} (f(x) - f(y)) H_D(t, x, y) \, dx.$$

One can express the content of Exercises G.3.1 and G.3.3 by saying that for a fixed y, the heat kernel $H_D(t, x, y)$ is the *fundamental solution* to the heat equation, that is, it is a solution to $\frac{\partial}{\partial t} H_D(t,x,y) - D\frac{\partial^2}{\partial x^2} H_D(t,x,y) = 0$ on $(0,\infty) \times \mathbb{R}$ satisfying the initial condition $H_D(0,x,y) = \delta_y$, where this last statement is understood in the sense that for each fixed y, $\mu_D^{t,y}$ converge as distributions to δ_y as $t \to 0^+$; equivalently, the $L^2(\mathbb{R})$ functions $H_D(t,x,y)$ of x, regarded as distributions, converge to δ_y as $t \to 0^+$. Somewhat more generally, for any $t_0 \in \mathbb{R}$, $H_D(t-t_0,x,y)$ satisfies

$$\frac{\partial H_D(t-t_0,x,y)}{\partial t} - D\frac{\partial^2 H_D(t-t_0,x,y)}{\partial x^2} = 0, \quad (t,x) \in (t_0,\infty) \times \mathbb{R},$$

$$\lim_{t \to t_0^+} H_D(t-t_0,x,y) = \delta_y, \quad x \in \mathbb{R}.$$

Exercise G.3.4. Show that the solution (G.15) to the heat equation (G.13) satisfies the initial condition $\psi(0,x) = \psi_0(x)$ in the sense that $\lim_{t\to 0^+} \psi(t,x) = \psi_0(x)$ for every $x \in \mathbb{R}$.

Exercise G.3.5. Let $x,y,z \in \mathbb{R}$, $s > 0$ and $t > 0$. Show that

$$H_D(t+s,x,y) = \int_{\mathbb{R}} H_D(t,x,z) H_D(s,z,y)\, dz.$$

Hint: Note that

$$\frac{(x-z)^2}{4t} + \frac{(z-y)^2}{4s} = \frac{s+t}{4st}\left(z - \frac{sx+ty}{s+t}\right)^2 + \frac{(x-y)^2}{4(s+t)}.$$

Now use the Gaussian integral

$$\int_{\mathbb{R}} e^{iax^2/2}\, dx = e^{\operatorname{sgn}(a)\pi i/4} \sqrt{\frac{2\pi}{|a|}},$$

where a is a nonzero real number and $\operatorname{sgn}(a)$ is its sign (see (A.3) in Appendix A).

Exercise G.3.6. For each $t > 0$ define a map T_t on $L^2(\mathbb{R})$ by

$$(T_t u)(x) = \int_{\mathbb{R}} H_D(t,x,y)\, u(y)\, dy$$

for every $u \in L^2(\mathbb{R})$. Take T_0 to be the identity map on $L^2(\mathbb{R})$. Show that each T_t is a bounded linear operator on $L^2(\mathbb{R})$ and conclude from Exercise G.3.5 that

$$T_{t+s} = T_t T_s, \quad \forall t, s \geq 0.$$

This qualifies $\{T_t\}_{t\geq 0}$ as a *semigroup of operators* on $L^2(\mathbb{R})$. Show that $\{T_t\}_{t\geq 0}$ is *strongly continuous* in the sense that

$$t \to 0^+ \text{ in } \mathbb{R} \text{ implies } T_t u \to u \text{ in } L^2(\mathbb{R}) \ \forall u \in L^2(\mathbb{R}).$$

Remark G.3.1. We will discuss such strongly continuous semigroups of operators in more detail in Appendix J.

Show also that the semigroup $\{T_t\}_{t\geq 0}$ is *contractive* in the sense that the operator norm $\|T_t\|$ of each T_t satisfies

$$\|T_t\| \leq 1.$$

Hint: For this last part you will need *Young's inequality*, which we will state in the following form. Suppose p, q and r are integers that satisfy $1 \leq p, q, r < \infty$ and

$$\frac{1}{p} + \frac{1}{q} = \frac{1}{r} + 1.$$

If $f \in L^p(\mathbb{R})$ and $g \in L^q(\mathbb{R})$, then

$$\|f * g\|_{L^r} \leq \|f\|_{L^p} \|g\|_{L^q},$$

where $f * g$ is the convolution product of f and g.

We will have much more to say about strongly continuous semigroups of operators in Appendix J. For the moment we would simply like to introduce a bit of traditional notation that will be explained somewhat later. We will call $\{T_t\}_{t\geq 0}$ the *heat semigroup* and write T_t symbolically as

$$T_t = e^{tD\Delta},$$

where Δ is the one-dimensional Laplacian. This exponential notation can *sometimes* be identified with an actual exponential function of an operator, but for the time being it is best to think of it simply as notation. With it we can write (G.17) as

$$\psi(t, x) = e^{tD\Delta}\psi_0(x).$$

Exercise G.3.7. Let $k \geq 2$ be an integer, $x, y, z_1, \ldots, z_{k-1} \in \mathbb{R}$ and $t_1, \ldots, t_k > 0$. Show that

$$H_D(t_1 + \cdots + t_k, x, y) = \int_{\mathbb{R}}\int_{\mathbb{R}} \cdots \int_{\mathbb{R}} H_D(t_1, x, z_1) H_D(t_2, z_1, z_2)$$

$$\cdots H_D(t_k, z_{k-1}, y)\, dz_1\, dz_2 \cdots dz_{k-1}.$$

G.4 Dimension n

Let $\mathbb{N}$ denote the set of nonnegative integers and $\mathbb{N}^N = \mathbb{N}\times \overset{N}{\cdots} \times\mathbb{N}$ the set of N-tuples of nonnegative integers. An element $\alpha = (\alpha_1, \ldots, \alpha_N)$ of $\mathbb{N}^N$ will be called a *multi-index*.

For each such multi-index α we write $|\alpha|$ for the sum $\alpha_1 + \cdots + \alpha_N$. If $\mathbf{q} = (q^1, \ldots, q^N) \in \mathbb{R}^N$ and ϕ is a smooth real- or complex-valued function on $\mathbb{R}^N$, we will denote by $\partial_\alpha \phi$ the partial derivative

$$(\partial_\alpha \phi)(\mathbf{q}) = \left(\frac{\partial}{\partial q^1} \right)^{\alpha_1} \cdots \left(\frac{\partial}{\partial q^N} \right)^{\alpha_N} \phi(\mathbf{q}).$$

If $\alpha = (0, \ldots, 0)$, then $\partial_\alpha \phi = \phi$. If $\alpha = (1, 0, 0, \ldots, 0, 0), (0, 1, 0, \ldots, 0, 0), \ldots, (0, 0, 0, \ldots, 0, 1)$, we will write $\partial_\alpha \phi$ as $\partial_1 \phi, \partial_2 \phi, \ldots, \partial_N \phi$ so that $\partial_k \phi = \partial \phi / \partial q^k$ for $k = 1, 2, \ldots, N$. We will write $\mathbf{q}^\alpha$ for the monomial

$$\mathbf{q}^\alpha = (q^1)^{\alpha_1} \cdots (q^N)^{\alpha_N}.$$

The *Schwartz space* $\mathcal{S}(\mathbb{R}^N)$ consists of all smooth, complex-valued functions ϕ on $\mathbb{R}^N$ for which

$$\sup_{\mathbf{q} \in \mathbb{R}^N} | \mathbf{q}^\alpha (\partial_\beta \phi)(\mathbf{q}) | < \infty$$

for all multi-indices α and β. These are the functions which, together with all of their partial derivatives, decay more rapidly than the reciprocal of any polynomial in $q^1, \ldots, q^N$ as $\|\mathbf{q}\| \to \infty$. Examples include such things as

$$\phi(\mathbf{q}) = Q(\mathbf{q}) e^{-c \| \mathbf{q} - \mathbf{q}_0 \|^2},$$

where $c > 0$, $\mathbf{q}_0 \in \mathbb{R}^N$ and $Q(\mathbf{q}) = \sum_{|\alpha| \le d} a_\alpha \mathbf{q}^\alpha$ is a polynomial on $\mathbb{R}^N$.

On $\mathcal{S}(\mathbb{R}^N)$ we can define a countable family of semi-norms

$$\|\phi\|_{\alpha, \beta} = \sup_{\mathbf{q} \in \mathbb{R}^N} | \mathbf{q}^\alpha (\partial_\beta \phi)(\mathbf{q}) |,$$

parametrized by pairs of multi-indices $\alpha, \beta \in \mathbb{N}^N$. Although each $\|\phi\|_{\alpha, \beta}$ is only a semi-norm, the family of all such has the property that $\|\phi\|_{\alpha, \beta} = 0 \; \forall \alpha, \beta \in \mathbb{N}^N \Rightarrow \phi = 0$, so these combine to give a metric

$$\rho(\phi_1, \phi_2) = \sum_{\alpha, \beta \in \mathbb{N}^N} \frac{1}{2^{|\alpha| + |\beta|}} \frac{\|\phi_1 - \phi_2\|_{\alpha, \beta}}{1 + \|\phi_1 - \phi_2\|_{\alpha, \beta}}$$

that is, moreover, complete (Cauchy sequences converge). We supply $\mathcal{S}(\mathbb{R}^N)$ with the topology determined by this metric and $\mathcal{S}(\mathbb{R}^N)$ thereby becomes a *Fréchet space* (again we refer to [Ham] for a very thorough discussion of Fréchet spaces).

The complex-valued, linear functionals on $\mathcal{S}(\mathbb{R}^N)$ that are continuous with respect to this Fréchet topology are called *tempered distributions* on $\mathbb{R}^N$ and the linear space of all such is denoted $\mathcal{S}'(\mathbb{R}^N)$. The elements of $\mathcal{S}(\mathbb{R}^N)$ are called *test functions*. Every

ψ in $L^1_{loc}(\mathbb{R}^N)$ (complex-valued, measurable functions on $\mathbb{R}^N$ that are integrable on compact subsets of $\mathbb{R}^N$) gives rise to a tempered distribution T_ψ by defining

$$T_\psi[\phi] = \int_{\mathbb{R}^N} \phi(\mathbf{q})\psi(\mathbf{q})\, d^N\mathbf{q}$$

for every $\phi \in S(\mathbb{R}^N)$. As in the $N = 1$ case we will often simply identify T_ψ and ψ. Since $L^2(\mathbb{R}^N) \subseteq L^1_{loc}(\mathbb{R}^N)$, every L^2 function on $\mathbb{R}^N$ gives rise to a tempered distribution in this way. Distributions of the form T_ψ for some $\psi \in L^1_{loc}(\mathbb{R}^N)$ are called *regular distributions*, while all of the others are called *singular distributions*. An example of a singular distribution is the *Dirac delta* at $\mathbf{a} \in \mathbb{R}^N$, denoted $\delta_\mathbf{a}$ and defined by

$$\delta_\mathbf{a}[\phi] = \phi(\mathbf{a}) \quad \forall \phi \in S(\mathbb{R}^N).$$

Sequential convergence in $S'(\mathbb{R}^N)$ is defined pointwise on $S(\mathbb{R}^N)$, that is, a sequence $\{T_n\}$ in $S'(\mathbb{R}^N)$ converges to T in $S'(\mathbb{R}^N)$ if and only if $\{T_n(\phi)\}$ converges in $\mathbb{C}$ to $T(\phi)$ for every $\phi \in S(\mathbb{R}^N)$. For any multi-index α, the *αth distributional derivative* of a distribution T is defined by

$$\partial_\alpha T[\phi] = (-1)^{|\alpha|} T[\partial_\alpha \phi]$$

for every $\phi \in S(\mathbb{R}^N)$. If $T = T_\psi$ for some $\psi \in L^1_{loc}(\mathbb{R}^N)$ and if $\partial_\alpha T_\psi$ is also regular so that there exists a (necessarily unique) element $\partial_\alpha^w \psi$ of $L^1_{loc}(\mathbb{R}^N)$ with $\partial_\alpha T = T_{\partial_\alpha^w \psi}$, then $\partial_\alpha^w \psi$ is called the *αth weak derivative* of ψ. This is characterized by

$$\int_{\mathbb{R}^N} \psi(\mathbf{q})(\partial_\alpha \phi)(\mathbf{q})\, d^N\mathbf{q} = (-1)^{|\alpha|} \int_{\mathbb{R}^N} (\partial_\alpha^w \psi)(\mathbf{q})\phi(\mathbf{q})\, d^N\mathbf{q}$$

for every ϕ in $S(\mathbb{R}^N)$ (see Example G.2.4).

It is not uncommon in the literature to drop the w and use the same symbol for ordinary and weak derivatives. There is no real harm in this since the two coincide when both make sense. Also note that if we adhere to the convention of identifying an $L^1_{loc}(\mathbb{R}^N)$ function with the corresponding regular distribution, then distributional derivatives and weak derivatives are also identified, *provided the weak derivative exists* (the distributional derivative always exists, but it need not be regular).

For $\alpha = (1,0,0,\ldots,0,0),(0,1,0,\ldots,0,0),\ldots,(0,0,0,\ldots,0,1)$ the weak derivatives $\partial_\alpha^w \psi$ are written $\partial_1^w \psi, \partial_2^w \psi, \ldots, \partial_N^w \psi$. Thus, for each $k = 1,\ldots,N$,

$$\int_{\mathbb{R}^N} \psi(\mathbf{q})\,(\partial_k \phi)(\mathbf{q})\, d^N\mathbf{q} = -\int_{\mathbb{R}^N} (\partial_k^w \psi)(\mathbf{q})\,\phi(\mathbf{q})\, d^N\mathbf{q}$$

for every $\phi \in S(\mathbb{R}^N)$. Similarly we will write $\partial_{k_1}^w \partial_{k_2}^w \psi$ for $\partial_\alpha^w \psi$ when α has 1 in the k_1 and k_2 slots and 0 elsewhere; the order is immaterial because of the corresponding result for smooth functions (Theorem 2-5 of [Sp1]). The *weak gradient of ψ* is the N-tuple

$$\nabla^w \psi = (\partial_1^w \psi, \partial_2^w \psi, \ldots, \partial_N^w \psi),$$

provided each $\partial_k^w \psi$ exists.

Exercise G.4.1. Suppose $\psi_1, \psi_2 \in L^1_{loc}(\mathbb{R}^n)$ and the weak derivatives $\partial^w_\alpha \psi_1$ and $\partial^w_\alpha \psi_2$ exist. Let $c_1, c_2 \in \mathbb{C}$. Show that $\partial^w_\alpha (c_1\psi_1 + c_2\psi_2)$ exists and is given by $\partial^w_\alpha (c_1\psi_1 + c_2\psi_2) = c_1 \partial^w_\alpha \psi_1 + c_2 \partial^w_\alpha \psi_2$.

Weak derivatives share many of the other basic properties of ordinary derivatives and we will record those we require as the need arises (also see Theorem 1, Section 5.2.3, of [Evans]). The particular use we would like to make of weak derivatives at the moment is the description of certain Hilbert spaces that will play an essential role when we start looking for domains of differential operators on $L^2(\mathbb{R}^N)$.

We define the *Sobolev space* $H^1(\mathbb{R}^N)$ as follows. As a set, $H^1(\mathbb{R}^N)$ is the subset of $L^1_{loc}(\mathbb{R}^N)$ consisting of those elements that are in $L^2(\mathbb{R}^N)$ and for which the first order weak derivatives exist and are also in $L^2(\mathbb{R}^N)$, that is,

$$H^1(\mathbb{R}^N) = \{\psi \in L^1_{loc}(\mathbb{R}^N) : \psi, \partial^w_k \psi \in L^2(\mathbb{R}^N), \ k = 1, 2, \ldots, N\}.$$

The inner product on $H^1(\mathbb{R}^N)$ is taken to be

$$\langle \psi_1, \psi_2 \rangle_{H^1} = \langle \psi_1, \psi_2 \rangle_{L^2} + \langle \partial^w_1 \psi_1, \partial^w_1 \psi_2 \rangle_{L^2} + \cdots + \langle \partial^w_N \psi_1, \partial^w_N \psi_2 \rangle_{L^2}, \qquad \text{(G.18)}$$

so the corresponding norm is

$$\|\psi\|^2_{H^1} = \|\psi\|^2_{L^2} + \sum_{k=1}^N \|\partial^w_k \psi\|^2_{L^2}. \qquad \text{(G.19)}$$

With this inner product, $H^1(\mathbb{R}^N)$ is a Hilbert space (Theorem 2, Section 5.2, of [Evans]). Relative to the norm topology determined by (G.19), the smooth functions on $\mathbb{R}^N$ are dense. Indeed, one can show that the set $C^\infty_0(\mathbb{R}^N)$ of smooth functions with compact support is dense in $H^1(\mathbb{R}^N)$ relative to the $H^1(\mathbb{R}^N)$-norm (Theorem 7.6 of [LL]). In fact, $H^1(\mathbb{R}^N)$ is often defined to be the Hilbert space completion of $C^\infty_0(\mathbb{R}^N)$ relative to the inner product (G.18); see Section 3.3 of [Fried].

It is worth noting that the integration by parts formula

$$\int_{\mathbb{R}^N} \psi(\mathbf{q}) \, (\partial_k \phi)(\mathbf{q}) \, d^N\mathbf{q} = - \int_{\mathbb{R}^N} (\partial^w_k \psi)(\mathbf{q}) \, \phi(\mathbf{q}) \, d^N\mathbf{q}$$

for every $\phi \in S(\mathbb{R}^N)$, which is essentially the definition of the weak derivatives, generalizes to an integration by parts formula on $H^1(\mathbb{R}^N)$: If ψ and ϕ are *both* in $H^1(\mathbb{R}^N)$, then

$$\int_{\mathbb{R}^N} \psi(\mathbf{q}) \, (\partial^w_k \phi)(\mathbf{q}) \, d^N\mathbf{q} = - \int_{\mathbb{R}^N} (\partial^w_k \psi)(\mathbf{q}) \, \phi(\mathbf{q}) \, d^N\mathbf{q}.$$

This is Theorem 7.7 of [LL].

Next define the *Sobolev space* $H^2(\mathbb{R}^N)$ to be the subset of $L^1_{\text{loc}}(\mathbb{R}^N)$ consisting of those elements that are in $L^2(\mathbb{R}^N)$ and for which the first and second order weak derivatives exist and are in $L^2(\mathbb{R}^N)$, that is,

$$H^2(\mathbb{R}^N) = \{\psi \in L^1_{\text{loc}}(\mathbb{R}^N) : \psi, \partial^w_k \psi, \partial^w_{k_1}\partial^w_{k_2}\psi \in L^2(\mathbb{R}^N), k, k_1, k_2 = 1, 2, \dots, N\}.$$

Note that $H^2(\mathbb{R}^N)$ is also a Hilbert space with inner product

$$\langle \psi_1, \psi_2 \rangle_{H^2} = \langle \psi_1, \psi_2 \rangle_{L^2} + \sum_{k=1}^{N} \langle \partial^w_k \psi_1, \partial^w_k \psi_2 \rangle_{L^2} + \sum_{k_1=1}^{N}\sum_{k_2=1}^{N} \langle \partial^w_{k_1}\partial^w_{k_2}\psi_1, \partial^w_{k_1}\partial^w_{k_2}\psi_2 \rangle_{L^2} \quad \text{(G.20)}$$

and corresponding norm

$$\|\psi\|^2_{H^2} = \|\psi\|^2_{L^2} + \sum_{k=1}^{N} \|\partial^w_k \psi\|^2_{L^2} + \sum_{k_1=1}^{N}\sum_{k_2=1}^{N} \|\partial^w_{k_1}\partial^w_{k_2}\psi\|^2_{L^2} \quad \text{(G.21)}$$

(Theorem 2, Section 5.2, of [Evans]). As for $H^1(\mathbb{R}^N)$, one can also define $H^2(\mathbb{R}^N)$ to be the Hilbert space completion of $C^\infty_0(\mathbb{R}^N)$ with respect to the inner product defined by (G.20).

We will need only $H^1(\mathbb{R}^N)$ and $H^2(\mathbb{R}^N)$, but for integers $K \geq 3$, the Sobolev spaces $H^K(\mathbb{R}^N)$ are defined analogously, so, *as sets*,

$$\cdots \subseteq H^K(\mathbb{R}^N) \subseteq \cdots \subseteq H^2(\mathbb{R}^N) \subseteq H^1(\mathbb{R}^N) \subseteq L^2(\mathbb{R}^N), \quad \text{(G.22)}$$

although each of these has a different inner product. Much more refined information about these inclusions and about the degree of regularity one can expect of the elements of a given Sobolev space can be obtained from the so-called *Sobolev inequalities*, which are discussed in detail in Chapter 5 of [Evans]. We mention also that for $\mathbb{C}^k$-valued functions, the Sobolev norms are defined to be the sum of the Sobolev norms of the coordinate functions and one thereby obtains Sobolev spaces of $\mathbb{C}^k$-valued functions. We will use this to generalize the Sobolev norms to sections of complex vector bundles in Section 10.4.

The *Fourier transform* of $\phi \in S(\mathbb{R}^N)$ is defined by

$$(\mathcal{F}\phi)(\mathbf{p}) = \hat{\phi}(\mathbf{p}) = \frac{1}{(2\pi)^{N/2}} \int_{\mathbb{R}^N} e^{-i\mathbf{q}\cdot\mathbf{p}}\phi(\mathbf{q})\, d^N\mathbf{q}, \quad \text{(G.23)}$$

where $\mathbf{q} \cdot \mathbf{p} = \sum_{i=1}^{N} q^i p^i$ if $\mathbf{q} = (q^1, \dots, q^N) \in \mathbb{R}^N$ and $\mathbf{p} = (p^1, \dots, p^N) \in \mathbb{R}^N$.

One should really think of the $\mathbb{R}^N$ in which $\mathbf{q}$ lives and the $\mathbb{R}^N$ in which $\mathbf{p}$ lives as distinct. One often identifies the $\mathbf{p}$ copy of $\mathbb{R}^N$ with the dual of the $\mathbf{q}$ copy of $\mathbb{R}^N$. In this case it would be more proper to subscript the components of $\mathbf{p}$ as $(p_1, \dots, p_N)$ and think of $\mathbf{q} \cdot \mathbf{p} = \sum_{i=1}^{N} q^i p_i$ as the natural pairing on $\mathbb{R}^N \times (\mathbb{R}^N)^*$ rather than the usual inner product on $\mathbb{R}^N$.

We have already mentioned (Remark G.2.1) that when $N = 1$, there are many variants of this definition, differing one from another by various constants. These same variants are in use when $N > 1$, but here there is even more flexibility. Indeed, one does not alter the essential features of the Fourier transform by taking $\mathbf{q} \cdot \mathbf{p}$ to be not the usual inner product on $\mathbb{R}^n$, but any nondegenerate, symmetric, bilinear form on $\mathbb{R}^N$.

Exercise G.4.2. Let A be an $N \times N$, symmetric, positive definite matrix. Show that the Fourier transform of

$$\phi(\mathbf{q}) = e^{-\frac{1}{2}\mathbf{q} \cdot A\mathbf{q}}$$

is given by

$$\hat{\phi}(\mathbf{p}) = \frac{1}{\sqrt{\det A}} e^{-\frac{1}{2}\mathbf{p} \cdot A^{-1}\mathbf{p}}.$$

Hint: The $N = 1$ case is Example G.2.1. For $N > 1$ choose an orthogonal matrix B such that

$$B^T A B = \Lambda = \operatorname{diag}(\lambda_1, \ldots, \lambda_N).$$

In the integral defining the Fourier transform make the changes of variable $\mathbf{q} = B\tilde{\mathbf{q}}$ and $\mathbf{p} = B\tilde{\mathbf{p}}$ and apply the $N = 1$ case N times.

The Fourier transform of a Schwartz function of $\mathbf{q}$ is a Schwartz function of $\mathbf{p}$. Indeed, the mapping $\mathcal{F} : S(\mathbb{R}^N) \to S(\mathbb{R}^N)$ that sends ϕ to $\mathcal{F}\phi = \hat{\phi}$ is a (Fréchet) continuous, linear, bijection with a continuous inverse $\mathcal{F}^{-1} : S(\mathbb{R}^N) \to S(\mathbb{R}^N)$ given by

$$(\mathcal{F}^{-1}\phi)(\mathbf{q}) = \check{\phi}(\mathbf{q}) = \frac{1}{(2\pi)^{N/2}} \int_{\mathbb{R}^N} e^{i\mathbf{q} \cdot \mathbf{p}} \phi(\mathbf{p}) \, d^N \mathbf{p}. \tag{G.24}$$

This is called the Fourier inversion theorem and is Theorem IX.1 of [RS2].

All of the familiar properties of the Fourier transform on $\mathbb{R}$ have analogues on $\mathbb{R}^N$. Here are a few of the most commonly used. For any $\phi, \phi_1, \phi_2 \in S(\mathbb{R}^N)$, any multi-index α, any $a \neq 0$ in $\mathbb{R}$ and any $\mathbf{a} \in \mathbb{R}^N$,

1. $\mathcal{F}(\partial_\alpha \phi)(\mathbf{p}) = (i\mathbf{p})^\alpha (\mathcal{F}\phi)(\mathbf{p})$,
2. $\mathcal{F}((-i\mathbf{q})^\alpha \phi)(\mathbf{p}) = \partial_\alpha (\mathcal{F}\phi)(\mathbf{p})$,
3. $\mathcal{F}^{-1}(\partial_\alpha \phi)(\mathbf{q}) = (-i\mathbf{q})^\alpha (\mathcal{F}^{-1}\phi)(\mathbf{q})$,
4. $\mathcal{F}^{-1}((i\mathbf{p})^\alpha \phi)(\mathbf{q}) = \partial_\alpha (\mathcal{F}^{-1}\phi)(\mathbf{q})$,
5. $\mathcal{F}(\phi(\mathbf{q} - \mathbf{a})) = e^{-i\mathbf{a} \cdot \mathbf{p}} \hat{\phi}(\mathbf{p})$,
6. $\mathcal{F}(e^{i\mathbf{a} \cdot \mathbf{q}} \phi(\mathbf{q})) = \hat{\phi}(\mathbf{p} - \mathbf{a})$,
7. $\mathcal{F}(\phi(a\mathbf{q})) = \frac{1}{|a|} \hat{\phi}(\frac{1}{a}\mathbf{p})$,

8. $\mathcal{F}(\phi_1\phi_2)(\mathbf{p}) = (2\pi)^{-N/2}(\hat{\phi}_1 * \hat{\phi}_2)(\mathbf{p})$, where the convolution product $\hat{\phi}_1 * \hat{\phi}_2$ is defined by

$$(\hat{\phi}_1 * \hat{\phi}_2)(\mathbf{p}) = \int_{\mathbb{R}^N} \hat{\phi}_1(\mathbf{p} - \mathbf{y})\hat{\phi}_2(\mathbf{y}) \, d^N\mathbf{y}.$$

Furthermore, $\mathcal{F} : S(\mathbb{R}^N) \to S(\mathbb{R}^N)$ preserves the $L^2(\mathbb{R}^N)$ norm, that is,

$$\int_{\mathbb{R}^N} |\phi(\mathbf{q})|^2 d^N\mathbf{q} = \int_{\mathbb{R}^N} |\hat{\phi}(\mathbf{p})|^2 d^N\mathbf{p}$$

for every $\phi \in S(\mathbb{R}^N)$ (Corollary to Theorem IX.1 of [RS2]). Since $S(\mathbb{R}^N)$ is dense in $L^2(\mathbb{R}^N)$ and $\mathcal{F}$ carries $S(\mathbb{R}^N)$ onto $S(\mathbb{R}^N)$, this implies that $\mathcal{F}$ extends by continuity to a unitary operator of $L^2(\mathbb{R}^N)$ onto itself, which we will continue to denote

$$\mathcal{F} : L^2(\mathbb{R}^N) \to L^2(\mathbb{R}^N)$$

(this is called the *Plancherel theorem*). We will continue to refer to $\mathcal{F} : L^2(\mathbb{R}^N) \to L^2(\mathbb{R}^N)$ as the Fourier transform, although it is often called the *Fourier–Plancherel transform*. Note that $\mathcal{F}^{-1} : S(\mathbb{R}^N) \to S(\mathbb{R}^N)$ extends to the $L^2(\mathbb{R}^N)$-adjoint of $\mathcal{F} : L^2(\mathbb{R}^N) \to L^2(\mathbb{R}^N)$. For ϕ in $L^1(\mathbb{R}^N) \cap L^2(\mathbb{R}^N)$, $\mathcal{F}\phi$ is computed from the integral (G.23), but for an element of $L^2(\mathbb{R}^N)$ that is not Lebesgue integrable on $\mathbb{R}^N$ this integral will not converge. As in the one-dimensional case one can compute $\mathcal{F}\phi$ either as a limit in $L^2(\mathbb{R}^N)$ of the Fourier transforms of a sequence of functions in $L^1(\mathbb{R}^N) \cap L^2(\mathbb{R}^N)$ converging to ϕ or as

$$(\mathcal{F}\phi)(\mathbf{p}) = \hat{\phi}(\mathbf{p}) = \lim_{M \to \infty} \frac{1}{(2\pi)^{N/2}} \int_{[-M,M]^N} e^{-i\mathbf{q}\cdot\mathbf{p}} \phi(\mathbf{q}) \, d^N\mathbf{q},$$

where the limit is in $L^2(\mathbb{R}^N)$. Also as in the one-dimensional case, the L^2-limit on the right-hand side is generally written simply $\frac{1}{(2\pi)^{N/2}} \int_{\mathbb{R}^N} e^{-i\mathbf{q}\cdot\mathbf{p}} \phi(\mathbf{q}) \, d^N\mathbf{q}$, but this is a new use of the integral sign and we would like to extend the terminology to make the distinction.

Let g be an element of $L^2(\mathbb{R}^N)$ and let $k : \mathbb{R}^N \times \mathbb{R}^N \to \mathbb{C}$ be a function with the following properties. For any $M > 0$,
1. $k(\cdot, \mathbf{p}) \in L^1([-M, M]^N)$ for almost every $\mathbf{p} \in \mathbb{R}^N$,
2. $\int_{[-M,M]^N} k(\mathbf{q}, \mathbf{p}) \, d^N\mathbf{q}$ is in $L^2(\mathbb{R}^N)$ as a function of $\mathbf{p}$.

Then we say that g is the *integral in the mean*, or the *mean-square integral* of k if

$$\lim_{M \to \infty} \int_{[-M,M]^N} k(\mathbf{q}, \cdot) \, d^N\mathbf{q} = g(\cdot),$$

where the limit is in $L^2(\mathbb{R}^N)$, that is, if

$$\lim_{M\to\infty}\left\| g(\cdot) - \int_{[-M,M]^N} k(\mathbf{q},\cdot)\, d^N\mathbf{q} \right\|_{L^2} = 0.$$

In this case we will abuse notation a bit and still write

$$g(p) = \int_{\mathbb{R}^N} k(\mathbf{q},\mathbf{p})\, d^N\mathbf{q}.$$

The Fourier transform actually extends to all $\phi \in L^1(\mathbb{R}^N)$ by (G.23), but $\hat{\phi}$ will, in general, only be in the space $C^0_\infty(\mathbb{R}^N)$ of continuous functions that vanish at infinity ($|\hat{\phi}(\mathbf{p})| \to 0$ as $\|\mathbf{p}\| \to \infty$). This is the so-called *Riemann–Lebesgue lemma* (see Theorem IX.7 of [RS2]). Moreover, $\mathcal{F}$ maps $L^1(\mathbb{R}^N)$ into, but not onto $C^0_\infty(\mathbb{R}^N)$.

Just as in the one-dimensional case, the Fourier transform and its inverse extend beyond $L^2(\mathbb{R}^N)$ to the tempered distributions $\mathcal{S}'(\mathbb{R}^N) \supseteq L^2(\mathbb{R}^N)$ via the definitions

$$(\mathcal{F}T)[\phi] = \hat{T}[\phi] = T[\mathcal{F}\phi] = T[\hat{\phi}] \quad \forall \phi \in \mathcal{S}(\mathbb{R}^N)$$

and

$$(\mathcal{F}^{-1}T)[\phi] = \check{T}[\phi] = T[\mathcal{F}^{-1}\phi] = T[\check{\phi}] \quad \forall \phi \in \mathcal{S}(\mathbb{R}^N).$$

For example,

$$\mathcal{F}\delta_{\mathbf{a}} = \frac{1}{(2\pi)^{N/2}}\, e^{-i\mathbf{a}\cdot\mathbf{p}},$$

where the function on the right-hand side is identified with the corresponding regular distribution.

Both $\mathcal{F}: \mathcal{S}'(\mathbb{R}^N) \to \mathcal{S}'(\mathbb{R}^N)$ and $\mathcal{F}^{-1}: \mathcal{S}'(\mathbb{R}^N) \to \mathcal{S}'(\mathbb{R}^N)$ are still linear bijections. Moreover, if T is any tempered distribution and f is any function on $\mathbb{R}^N$ that does not grow too rapidly (more precisely, has the property that if ϕ is a Schwartz function, then $f\phi$ is also a Schwartz function), then one can define a distribution fT by

$$(fT)[\phi] = T[f\phi].$$

This is certainly the case if f is a polynomial on $\mathbb{R}^n$. With this definition one can show that properties 1.–4. above are still valid when ϕ is taken to be a distribution in $L^2(\mathbb{R}^N)$, provided the distributional derivatives are also in $L^2(\mathbb{R}^N)$. Thus, even for distributions, the Fourier transform takes derivatives to products, which is essentially its *raison d'être*. For example, if ψ is in $L^2(\mathbb{R}^N)$ and if $\partial_k^w\psi$ exists and is in $L^2(\mathbb{R}^N)$ for each $k = 1,\dots,N$, then

$$\mathcal{F}(\partial_k^w\psi)(\mathbf{p}) = i\,(p^k)\,\hat{\psi}(\mathbf{p}),$$

where $\mathbf{p} = (p^1, \ldots, p^k, \ldots, p^N)$. But $\mathcal{F}$ is an isometry on $L^2(\mathbb{R}^N)$, so

$$\| \partial_k^w \psi \|_{L^2}^2 = \int_{\mathbb{R}^N} (p^k)^2 |\hat{\psi}(\mathbf{p})|^2 \, d^N\mathbf{p}.$$

Consequently,

$$\| \psi \|_{L^2}^2 + \sum_{k=1}^N \| \partial_k^w \psi \|_{L^2}^2 = \int_{\mathbb{R}^N} (1 + \|\mathbf{p}\|^2) |\hat{\psi}(\mathbf{p})|^2 \, d^N\mathbf{p},$$

and we conclude that if ψ is in $H^1(\mathbb{R}^N)$, then $(1 + \|\mathbf{p}\|^2)^{\frac{1}{2}} \hat{\psi}(\mathbf{p})$ is in $L^2(\mathbb{R}^N)$ and

$$\| \psi \|_{H^1} = \| (1 + \|\mathbf{p}\|^2)^{\frac{1}{2}} \hat{\psi}(\mathbf{p}) \|_{L^2}.$$

For functions in $L^2(\mathbb{R}^N)$ it is also true that, conversely, if $(1 + \|\mathbf{p}\|^2)^{\frac{1}{2}} \hat{\psi}(\mathbf{p})$ is in $L^2(\mathbb{R}^N)$, then ψ is in $H^1(\mathbb{R}^N)$ (Theorem 7.9 of [LL]). There are analogous Fourier characterizations of all of the Sobolev spaces (Theorem 7, Section 5.8.4, of [Evans]).

H Stieltjes integrals

H.1 Introduction

In the body of the text we have a number of occasions to draw on properties of the Riemann–Stieltjes and Lebesgue–Stieltjes integrals, and we will provide a brief description of what we need here. One can find thorough discussions in Chapter 7 of [Apos], Chapter III of [RiSz.N], Section 1.5 of [Fol2] and Chapter III of [Saks].

H.2 Riemann–Stieltjes integrals

Begin with a closed, bounded interval $[a, b]$ in $\mathbb{R}$ and a real-valued function α : $[a, b] \to \mathbb{R}$ of bounded variation; recall that α is of bounded variation if and only if it is the difference of two nondecreasing functions (see Section 4 of [RiSz.N] or Section 3.5 of [Fol2]). Then, for any continuous, real-valued function f on $[a, b]$, the *Riemann–Stieltjes integral* of f with respect to α is denoted

$$\int_a^b f(\tau)d\alpha(\tau)$$

and is defined to be the limit of the sums

$$\sum_{k=1}^n f(\tau_k^*)[\alpha(\tau_k) - \alpha(\tau_{k-1})] = \sum_{k=1}^n f(\tau_k^*)\Delta\alpha(\tau_k), \tag{H.1}$$

as $\max(\tau_k - \tau_{k-1}) \to 0$, where

$$a = \tau_0 < \tau_1 < \cdots < \tau_n = b$$

and

$$\tau_{k-1} < \tau_k^* \le \tau_k.$$

That the limit exists and is independent of the choice of the partitions and the points τ_k^* is Theorem 7.27 of [Apos]. If $\alpha(\tau) = \tau$ this is simply the ordinary Riemann integral. Moreover, one can often evaluate Riemann–Stieltjes integrals by converting them to ordinary Riemann integrals. For example, if α is strictly increasing on $[a, b]$ and α' exists and is Riemann integrable on $[a, b]$, then f is integrable with respect to α on $[a, b]$ if and only if $f\alpha'$ is Riemann integrable on $[a, b]$ and, in this case,

$$\int_a^b f(\tau)d\alpha(\tau) = \int_a^b f(\tau)\alpha'(\tau)d\tau. \tag{H.2}$$

https://doi.org/10.1515/9783110751949-018

In many cases of interest, $\int_a^b f(\tau)d\alpha(\tau)$ can also be regarded as the Lebesgue integral of f with respect to a certain measure μ_α on $\mathbb{R}$ determined by α. For example, if α is continuous and increasing, μ_α is determined by the requirement that for any $a \leq t_1 < t_2 \leq b$, the measure of any of the intervals $[t_1, t_2]$, $[t_1, t_2)$, $(t_1, t_2]$ or (t_1, t_2) is $\alpha(t_2) - \alpha(t_1)$. This is quite convenient since it makes such things as the dominated convergence theorem available for use with Stieltjes integrals. If α and f are complex-valued functions on $[a, b]$, the first of bounded variation and the second continuous, then the definition is exactly the same, but the products in (H.1) are interpreted as products of complex numbers. Doing the arithmetic, one finds that if $\alpha = \alpha_1 + i\alpha_2$ and $f = f_1 + if_2$, then

$$\int_a^b f(\tau)d\alpha(\tau) = \left(\int_a^b f_1(\tau)d\alpha_1(\tau) - \int_a^b f_2(\tau)d\alpha_2(\tau) \right)$$
$$+ i \left(\int_a^b f_2(\tau)d\alpha_1(\tau) + \int_a^b f_1(\tau)d\alpha_2(\tau) \right).$$

Various methods of extending these definitions to a wider class of functions f on $[a, b]$, but with α still of bounded variation are discussed in Sections 56 and 57 of [RiSz.N]. Clearly, one can define *improper Riemann–Stieltjes integrals* by the same limits that are used to define improper Riemann integrals but, just as clearly, these limits need not exist.

H.3 Lebesgue–Stieltjes integrals

What we would like to do first is describe *all* of the regular Borel measures on $\mathbb{R}$. We will make use of some of the material described in Appendix F. Note that, with obvious modifications, the procedure we will describe works equally well for any interval in $\mathbb{R}$. We begin with the collection of all subsets of $\mathbb{R}$ of the form $(a, b]$, (a, ∞) or $\emptyset$, where $-\infty \leq a < b < \infty$. The collection of all finite disjoint unions of such sets forms an algebra $\mathcal{A}$ and the σ-algebra generated by $\mathcal{A}$ coincides with the Borel σ-algebra $\mathcal{B}(\mathbb{R})$ of $\mathbb{R}$. Now fix some function $\alpha : \mathbb{R} \to \mathbb{R}$ that is nondecreasing and right-continuous ($\lim_{\tau \to \tau_0^+} \alpha(\tau) = \alpha(\tau_0)$ for each $\tau_0 \in \mathbb{R}$). Note that if we happen to start with a finite Borel measure μ on $\mathbb{R}$, then its distribution function $\alpha(\tau) = \mu(-\infty, \tau]$ is nondecreasing and right-continuous. We are basically going to reverse this process. Now define μ_α by $\mu_\alpha(\emptyset) = 0$, and

$$\mu_\alpha \left(\bigsqcup_{k=1}^n (a_k, b_k] \right) = \sum_{k=1}^n \left(\alpha(b_k) - \alpha(a_k) \right)$$

for any finite family of pairwise disjoint intervals $(a_k, b_k]$, $k = 1, \ldots, n$. Then μ_α is a pre-measure on $\mathcal{A}$ (Proposition 1.15 of [Fol2]). As we saw in Appendix F, μ_α then determines an outer measure μ_α^* on the power set $\mathcal{P}(\mathbb{R})$ of $\mathbb{R}$. The μ_α^*-measurable sets form a σ-algebra $\sigma(\mu_\alpha^*)$ on $\mathbb{R}$ containing the σ-algebra $\mathcal{B}(\mathbb{R})$ generated by $\mathcal{A}$ and the restriction of μ_α^* to $\sigma(\mu_\alpha^*)$ is a complete measure which, as usual, we will also denote simply μ_α. This measure is called the *Lebesgue–Stieltjes measure* on $\mathbb{R}$ determined by α; it is, in fact, a regular measure on $\mathbb{R}$ (Theorem 1.18 of [Fol2]). If $\alpha(\tau) = \tau$, then this is just the ordinary Lebesgue measure on $\mathbb{R}$. We should point out that it is also common to refer to the regular Borel measure $\mu_\alpha|_{\mathcal{B}(\mathbb{R})}$ as the Lebesgue–Stieltjes measure and to denote it μ_α as well. The following combines Theorems 1.16 and 1.18 of [Fol2].

Theorem H.3.1. *Let $\alpha : \mathbb{R} \to \mathbb{R}$ be any nondecreasing, right-continuous function on $\mathbb{R}$. Then there exists a unique regular Borel measure μ_α on $\mathbb{R}$ such that $\mu_\alpha(a, b] = \alpha(b) - \alpha(a)$ for all $a < b$ in $\mathbb{R}$. If β is another nondecreasing, right-continuous function, then $\mu_\alpha = \mu_\beta$ if and only if $\alpha - \beta$ is a constant. Conversely, if μ is a regular Borel measure on $\mathbb{R}$ and if we define*

$$
\alpha(\tau) = \begin{cases} \mu(0, \tau], & \text{if } \tau > 0, \\ 0, & \text{if } \tau = 0, \\ -\mu(-\tau, 0], & \text{if } \tau < 0, \end{cases}
$$

then α is nondecreasing and right-continuous and $\mu = \mu_\alpha$.

If we write $\lim_{\tau \to a^-} \alpha(\tau)$ as $\alpha(a^-)$ and $\lim_{\tau \to b^-} \alpha(\tau)$ as $\alpha(b^-)$, then one can check each of the following (this is Exercise 28 in [Fol2]):
1. $\mu_\alpha(\{a\}) = \alpha(a) - \alpha(a^-)$,
2. $\mu_\alpha[a, b) = \alpha(b^-) - \alpha(a^-)$,
3. $\mu_\alpha[a, b] = \alpha(b) - \alpha(a^-)$,
4. $\mu_\alpha(a, b) = \alpha(b^-) - \alpha(a)$.

The integral associated with the Lebesgue–Stieltjes measure μ_α is, naturally enough, called the *Lebesgue–Stieltjes integral* associated with α. If E is a μ_α-measurable set and f is a μ_α-integrable real-valued function on $\mathbb{R}$, then the integral is generally denoted $\int_E f \, d\mu_\alpha$, $\int_E f(\tau) \, d\mu_\alpha(\tau)$ or, more commonly, $\int_E f \, d\alpha$ or $\int_E f(\tau) \, d\alpha(\tau)$; if E is an interval one generally opts for $\int_a^b$ rather than $\int_E$.

The Lebesgue–Stieltjes integral can be defined for functions α that are of bounded variation and right-continuous by using the fact that any such α can be written as the difference $\alpha = v_1 - v_2$ of two functions that are nondecreasing (in fact, increasing) and right-continuous (see Section 3.5 of [Fol2] or Section 4 of [RiSz.N]) and then defining

$$
\int_E f(\tau) \, d\alpha(\tau) = \int_E f(\tau) \, dv_1(\tau) - \int_E f(\tau) \, dv_2(\tau).
$$

One can choose v_1 and v_2 in the following way. Define $v_1(a) = 0$ and, for $a < \tau \le b$, let $v_1(\tau)$ be the total variation of α on $(a, \tau]$. Then v_1 is increasing, as is $v_2 = \alpha - v_1$. It can then be shown that α is right-continuous at τ in $[a, b]$ if and only if v_1 is right-continuous at τ.

For particularly nice functions there are simple computational formulas for Lebesgue–Stieltjes integrals. For example, if f is a bounded Borel function on $[a, b]$ and α is absolutely continuous (Remark 5.2.1), then

$$\int_a^b f(\tau)\, d\alpha(\tau) = \int_a^b f(\tau)\alpha'(\tau)\, d\tau,$$

where the right-hand side is an ordinary Lebesgue integral. In particular, since a continuously differentiable function on $[a, b]$ is absolutely continuous, this is true when $f \in C^0[a, b]$ and $\alpha \in C^1[a, b]$; in this case the Lebesgue–Stieltjes integral agrees with the Riemann–Stieltjes integral.

As we did for the Riemann–Stieltjes integral, we now define the Lebesgue–Stieltjes integral when α and f are complex-valued. If $\alpha = \alpha_1 + i\alpha_2$ and $f = f_1 + if_2$, where the first is of bounded variation and the second is a bounded Borel function, then we set

$$\int_a^b f(\tau)\, d\alpha(\tau) = \left(\int_a^b f_1(\tau)\, d\alpha_1(\tau) - \int_a^b f_2(\tau)\, d\alpha_2(\tau) \right)$$
$$+ i\left(\int_a^b f_2(\tau)\, d\alpha_1(\tau) + \int_a^b f_1(\tau)\, d\alpha_2(\tau) \right).$$

With this we can describe an example that we will need to make use of in Section 9.4. We consider a *curve* $\gamma : [a, b] \rightarrow \mathbb{C}$ in the complex plane, which we will assume is in $C^1[a, b]$, and a continuous, complex-valued function F defined on some open set $U \subseteq \mathbb{C}$ containing the image $\gamma[a, b]$ of γ. Then $F \circ \gamma : [a, b] \rightarrow \mathbb{C}$ is continuous and the Lebesgue–Stieltjes integral of $F \circ \gamma$ with respect to γ is

$$\int_a^b (F \circ \gamma)\, d\gamma = \int_a^b F(\gamma(\tau))\, \gamma'(\tau)\, d\tau,$$

where the multiplication in the integrand on the right-hand side means complex multiplication. This is, of course, just what is ordinarily called the *contour integral* of F along γ and denoted

$$\int_\gamma F(z)\, dz.$$

One handles piecewise C^1-curves in the usual way by integrating over each piece and adding. That is, if $\gamma_k : [a_k, b_k] \rightarrow \mathbb{C}$ is a C^1-curve for each $k = 1, \ldots, n$ with $a = a_1 < b_1 = a_2 < b_2 = a_3 < \cdots < b_{n-1} = a_n < b_n = b$ and $\gamma_k(b_k) = \gamma_{k+1}(a_{k+1})$ for $k = 1, \ldots, n-1$ and if $\sum_{k=1}^{n} \gamma_k : [a, b] \to \mathbb{C}$ is the curve that agrees with γ_k on $[a_k, b_k]$ for each $k = 1, \ldots, n$, then

$$\int\limits_{\sum_{k=1}^{n} \gamma_k} F(z)\, dz = \sum_{k=1}^{n} \int\limits_{\gamma_k} F(z)\, dz = \sum_{k=1}^{n} \int\limits_{a_k}^{b_k} (F \circ \gamma_k)\, d\gamma_k.$$

The point we want to take out of all of this is the following. *A contour integral in the complex plane over a piecewise C^1-curve can be written as a sum of Lebesgue–Stieltjes integrals.*

I Unitary representations and Schur's lemma

Here we will briefly review the general notion of a strongly continuous, unitary representation of a Lie group and provide a proof of the infinite-dimensional version of Schur's lemma that we require for our discussion of the Heisenberg group H_3 in Section 7.2.

We let G be a matrix Lie group (closed subgroup of some general linear group $GL(n, \mathbb{R})$ or $GL(n, \mathbb{C})$) and denote its identity element 1_G or simply 1 if no confusion will arise. Let $\mathcal{H}$ be a separable, complex Hilbert space (either finite- or infinite-dimensional) and $\mathcal{U}(\mathcal{H})$ the group of unitary operators on $\mathcal{H}$. A *strongly continuous, unitary representation of G on $\mathcal{H}$* is a group homomorphism

$$\pi : G \to \mathcal{U}(\mathcal{H})$$

such that for each fixed $v \in \mathcal{H}$, the map

$$g \to \pi(g)v : G \to \mathcal{H}$$

is continuous in the norm topology of $\mathcal{H}$, that is,

$$g \to g_0 \quad \text{in } G \quad \Rightarrow \quad \|\pi(g)v - \pi(g_0)v\| \to 0 \quad \text{in } \mathbb{R}.$$

The representation is said to be *trivial* if it sends every $g \in G$ to the identity operator $\mathrm{id}_{\mathcal{H}} = I$ on $\mathcal{H}$.

Exercise I.0.1. We saw in Remark 3.3.2 that for one-parameter groups of unitary operators, strong continuity is equivalent to a number of apparently weaker assumptions. Since these are simply unitary representations of the Lie group $\mathbb{R}$ under addition, one might hope that something similar is true for unitary representations in general. This is what you will prove here. Let G be a matrix Lie group, $\mathcal{H}$ a separable, complex Hilbert space and $\pi : G \to \mathcal{U}(\mathcal{H})$ a group homomorphism. Show that the following are equivalent:

1. π is strongly continuous,
2. π is weakly continuous, that is, for all $u, v \in \mathcal{H}$,

$$g \to g_0 \quad \text{in } G \quad \Rightarrow \quad \langle \pi(g)u, v \rangle \to \langle \pi(g_0)u, v \rangle \quad \text{in } \mathbb{C},$$

3. for each $u \in \mathcal{H}$, the map $g \in G \to \langle \pi(g)u, u \rangle \in \mathbb{C}$ is continuous at e.

Hint: For 3. $\Rightarrow$ 1. show that $\| \pi(g)u - \pi(g_0)u \|^2 = 2\|u\|^2 - 2\,\mathrm{Re}\,\langle \pi(g_0^{-1}g)u, u \rangle \le |\,\|u\|^2 - \langle \pi(g_0^{-1}g)u, u \rangle\,|$.

A linear subspace $\mathcal{H}_0$ of $\mathcal{H}$ is said to be *invariant* under $\pi : G \to \mathcal{U}(\mathcal{H})$ if $\pi(g)(\mathcal{H}_0) \subseteq \mathcal{H}_0$ for every $g \in G$. The zero subspace $\{0\}$ and $\mathcal{H}$ itself are always invariant. If $\pi :$

https://doi.org/10.1515/9783110751949-019

$G \to \mathcal{U}(\mathcal{H})$ is nontrivial and if $\{0\}$ and $\mathcal{H}$ are the only *closed* invariant subspaces, then $\pi : G \to \mathcal{U}(\mathcal{H})$ is said to be *irreducible*. If there are closed invariant subspaces other than $\{0\}$ and $\mathcal{H}$, then the representation is said to be *reducible*. In a way we will describe in Section 7.2, any reducible unitary representation of the Heisenberg group H_3 can be built from irreducible unitary representations, so we will concern ourselves primarily with the latter. Also in Section 7.2 we will describe all irreducible, unitary representations of H_3, modulo a certain equivalence relation that we now introduce. Two unitary representations $\pi_1 : G \to \mathcal{U}(\mathcal{H}_1)$ and $\pi_2 : G \to \mathcal{U}(\mathcal{H}_2)$ of G are said to be *unitarily equivalent* if there exists a unitary equivalence $U : \mathcal{H}_1 \to \mathcal{H}_2$ of $\mathcal{H}_1$ onto $\mathcal{H}_2$ such that

$$U\pi_1(g) = \pi_2(g)U \quad \forall g \in G,$$

that is,

$$\pi_2(g) = U\pi_1(g)U^{-1} \quad \forall g \in G.$$

Another item that will play a role in our discussion of the Heisenberg group is the following infinite-dimensional version of *Schur's lemma*. The proof is a nice application of the spectral theorem, so we will provide the details. A more general version of Schur's lemma is proved in Appendix 1 of [Lang3].

Theorem I.0.1 (Schur's lemma). *Let G be a matrix Lie group, $\mathcal{H}$ a separable, complex Hilbert space and $\pi : G \to \mathcal{U}(\mathcal{H})$ a strongly continuous unitary representation of G. Then $\pi : G \to \mathcal{U}(\mathcal{H})$ is irreducible if and only if the only bounded operators $A : \mathcal{H} \to \mathcal{H}$ that commute with every $\pi(g)$*

$$\pi(g)A = A\pi(g) \quad \forall g \in G$$

are those of the form $A = cI$, where c is a complex number and I is the identity operator on $\mathcal{H}$.

Proof. Suppose first that the only bounded operators that commute with every $\pi(g)$ are constant multiples of the identity. We will show that the representation is irreducible. Let $\mathcal{H}_0$ be a closed subspace of $\mathcal{H}$ that is invariant under every $\pi(g)$.

Exercise I.0.2. Show that the orthogonal complement $\mathcal{H}_0^\perp$ of $\mathcal{H}_0$ is also invariant under every $\pi(g)$.

Now, let $P : \mathcal{H} \to \mathcal{H}_0$ be the orthogonal projection onto $\mathcal{H}_0$.

Exercise I.0.3. Show that $\pi(g)P = P\pi(g)$ for every $g \in G$.

According to our assumption, P is a constant multiple of the identity. Being a projection, $P^2 = P$, so the constant is either 0 or 1. Thus, $\mathcal{H}_0 = P(\mathcal{H})$ is either 0 or $\mathcal{H}$, as required.

Now, for the converse we will assume that $\pi : G \to \mathcal{U}(\mathcal{H})$ is irreducible and that $A : \mathcal{H} \to \mathcal{H}$ is a bounded operator that commutes with every $\pi(g)$. Let $A^* : \mathcal{H} \to \mathcal{H}$ denote the adjoint of A (also a bounded operator on $\mathcal{H}$).

Exercise I.0.4. Show that A^* also commutes with every $\pi(g)$.

Note that $\frac{1}{2}(A + A^*)$ and $\frac{i}{2}(A - A^*)$ are both self-adjoint and both commute with every $\pi(g)$. Moreover,

$$A = \frac{1}{2}(A + A^*) + \frac{1}{i}\left[\frac{i}{2}(A - A^*)\right].$$

Consequently, it will be enough to prove that bounded *self-adjoint* operators that commute with every $\pi(g)$ must be constant multiples of the identity. Accordingly, we may assume that A is self-adjoint. Then, by the Spectral Theorem 5.5.6, A has associated with it a unique spectral measure E^A. Moreover, since A commutes with every $\pi(g)$, so does $E^A(S)$ for any Borel set $S \subseteq \mathbb{R}$ (Theorem 5.5.7). From this it follows that each closed linear subspace $E^A(S)(\mathcal{H})$ is invariant under $\pi : G \to \mathcal{U}(\mathcal{H})$. But, by irreducibility, this means that

$$E^A(S)(\mathcal{H}) = 0 \quad \text{or} \quad E^A(S)(\mathcal{H}) = \mathcal{H}$$

for every Borel set S in $\mathbb{R}$.

Since A is bounded there exist $a_1 < b_1$ in $\mathbb{R}$ such that if $S \cap [a_1, b_1] = \emptyset$, then $E^A(S) = 0$. In particular, $E^A([a_1, b_1]) = I$. Write

$$[a_1, b_1] = \left[a_1, \frac{a_1 + b_1}{2}\right] \cup \left[\frac{a_1 + b_1}{2}, b_1\right].$$

Now note that if $E^A(\{\frac{a_1+b_1}{2}\}) = I$, then the spectral theorem gives $A = \frac{a_1+b_1}{2}I$ and we are done. Otherwise, E^A must be I on one of the intervals and 0 on the other. Denote by $[a_2, b_2]$ the interval on which it is I. Applying the same argument to $[a_2, b_2]$ we either prove the result (at the midpoint) or we obtain an interval $[a_3, b_3]$ of half the length of $[a_2, b_2]$ on which E^A is I. Continuing inductively, we either prove the result in a finite number of steps or we obtain a nested sequence $[a_1, b_1] \supseteq [a_2, b_2] \supseteq [a_3, b_3] \supseteq \cdots$ of intervals whose lengths approach zero and for which $E^A([a_i, b_i]) = I$ for every $i = 1, 2, 3, \ldots$. By the Cantor intersection theorem (Theorem C, page 73, of [Simm1]), $\bigcap_{i=1}^{\infty}[a_i, b_i] = \{c\}$ for some $c \in \mathbb{R}$. Since $E^A(\mathbb{R} - [a_i, b_i]) = 0$ for each i, $0 = E^A(\bigcup_{i=1}^{\infty}(\mathbb{R} - [a_i, b_i])) = E^A(\mathbb{R} - \bigcap_{i=1}^{\infty}[a_i, b_i]) = E^A(\mathbb{R} - \{c\})$. Thus, $E^A(\{c\}) = I$, so again we have $A = cI$. $\qquad\square$

J Semigroups of operators

Let $\mathcal{E}$ denote a Banach space and, for each $t \geq 0$, let $T_t : \mathcal{E} \to \mathcal{E}$ be a bounded linear operator on $\mathcal{E}$. If $\{T_t\}_{t \geq 0}$ satisfies

1. $T_0 = \mathrm{id}_{\mathcal{E}}$,
2. $T_{t+s} = T_t T_s$, $\forall t, s \geq 0$, and
3. for each $x \in \mathcal{E}$,

$$t \to T_t x : [0, \infty) \to \mathcal{E}$$

is continuous,

then $\{T_t\}_{t \geq 0}$ is called a *strongly continuous semigroup of operators*, or a C^0-*semigroup of operators* on $\mathcal{E}$. The semigroup of operators associated with the heat flow on $\mathbb{R}$ (Exercise G.3.6) is a nice example to keep in mind. We will now describe what is essentially the simplest example; this uses some basic properties of the exponential of a bounded operator, all of which are proved essentially as they are for the matrix exponential function (Chapter IX, Section 6, of [Yosida] contains all of the details).

Example J.0.1. Let $\mathcal{E}$ be a Banach space and $A : \mathcal{E} \to \mathcal{E}$ a bounded linear operator on $\mathcal{E}$. For each $t \geq 0$ define $T_t : \mathcal{E} \to \mathcal{E}$ by

$$T_t = e^{tA} = \sum_{n=0}^{\infty} \frac{(tA)^n}{n!}.$$

Since $\|A\| < \infty$ the series converges, for each fixed t, in the Banach space $\mathcal{B}(\mathcal{E})$ of bounded operators on $\mathcal{E}$ to a bounded operator. Clearly, $T_0 = \mathrm{id}_{\mathcal{E}}$, and since tA and sA commute, $e^{(t+s)A} = e^{tA}e^{sA}$, so $T_{t+s} = T_t T_s$. Because A is bounded we actually have a much stronger continuity condition than the definition requires. Indeed, since

$$\|T_t - \mathrm{id}_{\mathcal{E}}\| = \left\| \sum_{n=1}^{\infty} \frac{(tA)^n}{n!} \right\| \leq \sum_{n=1}^{\infty} \frac{t^n \|A\|^n}{n!} = e^{t\|A\|} - 1$$

and $e^{t\|A\|} - 1 \to 0$ as $t \to 0^+$, $t \to e^{tA}$ is actually continuous as a map into $\mathcal{B}(\mathcal{E})$.

Exercise J.0.1. Show that for any $x \in \mathcal{E}$,

$$\lim_{t \to 0^+} \frac{e^{tA}(x) - x}{t} = Ax$$

and then that

$$\lim_{h \to 0} \frac{e^{(t+h)A}(x) - e^{tA}(x)}{h} = Ae^{tA}(x).$$

https://doi.org/10.1515/9783110751949-020

Write the second of these as

$$\frac{d}{dt}T_t(x) = AT_t(x)$$

and recall that

$$T_0(x) = x.$$

Since every tangent space to a vector space can be identified with that same vector space, even in the infinite-dimensional case, we can think of the operator A as defining a vector field on $\mathcal{E}$ whose value at $x \in \mathcal{E}$ is $Ax \in T_x(\mathcal{E})$. This suggests regarding $T_t(x)$ as the integral curve of the vector field on $\mathcal{E}$ represented by A that starts at $T_0(x) = x$. This is what we would like to generalize.

One can show, from the uniform boundedness theorem (Theorem 4.5.1 of [Fried], or Theorem III.9 of [RS1]), that for any C^0-semigroup $\{T_t\}_{t\geq0}$ of operators on a Banach space $\mathcal{E}$, there exist an $\omega \in \mathbb{R}$ and an $M \geq 0$ such that for each $t \geq 0$, $\|T_t\| \leq Me^{\omega t}$ (see page 246 of [RS2], page 418 of [Lax] or page 232 of [Yosida]). We will use this result only to motivate the following definitions. If ω is a real number, then a C^0-semigroup $\{T_t\}_{t\geq0}$ is said to be ω-contractive if $\|T_t\| \leq e^{\omega t}$ for each $t \geq 0$; $\{T_t\}_{t\geq0}$ is contractive if it is 0-contractive, that is, if $\|T_t\| \leq 1$ for each $t \geq 0$. As a rule, results proved for contractive semigroups have relatively straightforward extensions to the general case (see Section X.8 of [RS2]).

Let $\{T_t\}_{t\geq0}$ be a C^0-semigroup of operators on a Banach space $\mathcal{E}$. We introduce an operator A, called the *infinitesimal generator* of $\{T_t\}_{t\geq0}$, as follows. The domain of A is

$$\mathcal{D}(A) = \left\{x \in \mathcal{E} : \lim_{t\to0^+} \frac{T_t x - x}{t} \text{ exists in } \mathcal{E}\right\}.$$

Then $A : \mathcal{D}(A) \to \mathcal{E}$ is given, at each $x \in \mathcal{D}(A)$, by

$$Ax = \lim_{t\to0^+} \frac{T_t x - x}{t}.$$

The following is Theorem 2, Section 7.4.1, of [Evans], Theorem 4, Section 34.1, of [Lax] and the Proposition in Section X.8 of [RS2].

Theorem J.0.1. *Let $\{T_t\}_{t\geq0}$ be a C^0-semigroup of operators on a Banach space $\mathcal{E}$ and $A : \mathcal{D}(A) \to \mathcal{E}$ its infinitesimal generator. Then $\mathcal{D}(A)$ is a dense linear subspace of $\mathcal{E}$ and A is a closed linear operator on $\mathcal{D}(A)$.*

Remark J.0.1. Recall that A is *closed* if, whenever $x_n \in \mathcal{D}(A)$ for $n = 1, 2, \ldots, x_n \to x$ in $\mathcal{E}$ and $Ax_n \to y$ in $\mathcal{E}$, then $x \in \mathcal{D}(A)$ and $Ax = y$.

Although the following result is proved in [Evans], [Lax] and [Yosida], it is the key to isolating an appropriate notion of "flow" in the infinite-dimensional context, so we will record the simple proof here as well. We will assume that the semigroup is contractive to simplify the proofs of 3. and 4., but the result is true in general (see Theorem 2, Section IX.3, of [Yosida]). In the statement of the theorem the derivative of $t \to T_t x$, for $x \in \mathcal{E}$, is defined to be the following limit in $\mathcal{E}$, provided the limit exists:

$$\frac{d}{dt} T_t x = \lim_{h \to 0} \frac{T_{t+h} x - T_t x}{h}.$$

Theorem J.0.2. *Let $\{T_t\}_{t \geq 0}$ be a (contractive) C^0-semigroup of operators on a Banach space $\mathcal{E}$ and $A : \mathcal{D}(A) \to \mathcal{E}$ its infinitesimal generator. Let x be in $\mathcal{D}(A)$. Then:*

1. *$T_t x$ is in $\mathcal{D}(A)$ for all $t \geq 0$,*
2. *$AT_t x = T_t Ax$ for all $t \geq 0$,*
3. *the map $t \to T_t x$ is continuously differentiable on $t > 0$ and*
4. *$\frac{d}{dt} T_t x = AT_t x$ for all $t > 0$.*

Proof. Since $x \in \mathcal{D}(A)$, the limit $\lim_{s \to 0^+} \frac{T_s x - x}{s}$ exists in $\mathcal{E}$ and is, by definition, Ax. Then, for any $t \geq 0$,

$$\frac{T_s(T_t x) - T_t x}{s} = \frac{T_{s+t} x - T_t x}{s} = \frac{T_{t+s} x - T_t x}{s} = \frac{T_t(T_s x) - T_t x}{s} = T_t \left(\frac{T_s x - x}{s} \right).$$

Since T_t is bounded (continuous), $\lim_{s \to 0^+} \frac{T_s(T_t x) - T_t x}{s}$ exists in $\mathcal{E}$ and therefore $T_t x \in \mathcal{D}(A)$, so 1. is proved. Moreover, taking the limit as $t \to 0^+$ on both sides gives $AT_t x = T_t Ax$, so 2. is proved as well. We will prove 3. and 4. by showing that for any $t > 0$,

$$\lim_{h \to 0} \frac{T_{t+h} x - T_t x}{h} = AT_t x$$

for every $x \in \mathcal{D}(A)$. By 2. it is enough to show that $\lim_{h \to 0} \frac{T_{t+h} x - T_t x}{h} = T_t Ax$ and for this we will examine the limits as $h \to 0^+$ and $h \to 0^-$ separately. Let $h > 0$ and note that

$$\frac{T_{t+h} x - T_t x}{h} - T_t Ax = T_t \left(\frac{T_h x - x}{h} - Ax \right).$$

From the definition of Ax and the continuity of T_t, this approaches zero as $h \to 0^+$. For the limit as $h \to 0^-$ we again assume h is positive (and sufficiently small) and consider

$$\frac{T_t x - T_{t-h} x}{h} - T_t Ax = T_{t-h} \left(\frac{T_h x - x}{h} - Ax \right) + (T_{t-h} Ax - T_t Ax).$$

Since $\{T_t\}_{t \geq 0}$ is assumed contractive, $\|T_{t-h}\| \leq 1$ and we conclude that

$$\left\| \frac{T_t x - T_{t-h} x}{h} - T_t Ax \right\| \leq \left\| \frac{T_h x - x}{h} - Ax \right\| + \|T_{t-h} Ax - T_t Ax\|.$$

Both of the terms on the right-hand side approach zero as $h \to 0^+$. This shows that $\lim_{h \to 0^-} \frac{T_{t+h} x - T_t x}{h} = AT_t x$ and completes the proof of differentiability and 4. Continuity of the derivative follows from 4. and the strong continuity of $\{T_t\}_{t \geq 0}$. □

If we regard the infinitesimal generator A as a vector field on $\mathcal{D}(A)$ with values in $\mathcal{E}$, then, motivated by 4., we call $\{T_t\}_{t\geq0}$ the *flow* of A. Also motivated by 4. is the traditional notation for the semigroup generated by A, that is,

$$T_t = e^{tA}.$$

This notation is very suggestive and convenient. For example, $T_t T_s = T_{t+s}$ becomes $e^{tA}e^{sA} = e^{(t+s)A}$ and $\frac{d}{dt}T_t x = AT_t x$ becomes $\frac{d}{dt}e^{tA}x = Ae^{tA}x$. However, one should keep in mind that it is only under certain circumstances that e^{tA} is literally the exponential of an operator; this is true, for example, if the infinitesimal generator A happens to be a bounded operator (Example J.0.1) and we will mention one other instance of this in a moment.

Typically, one is not given a flow (semigroup of operators) and asked to find the vector field that gives rise to it (its infinitesimal generator). Rather, one is given a vector field and would like to know that a flow exists. The crucial question then is, given an unbounded operator/vector field A, how can one know that it is the infinitesimal generator for some C^0-semigroup of operators? This is the question addressed by the Hille–Yosida theorem, to which we now turn.

We already know that the infinitesimal generator A of any C^0-semigroup $\{T_t\}_{t\geq0}$ of operators on a Banach space $\mathcal{E}$ is a densely defined, closed operator on $\mathcal{E}$. In fact, any such A has two additional properties and, remarkably enough, these two characterize infinitesimal generators among the densely defined, closed operators. To describe these two properties we recall that $\lambda \in \mathbb{C}$ is in the resolvent set $\rho(A)$ of the closed operator A if and only if $\lambda - A : \mathcal{D}(A) \to \mathcal{E}$ is one-to-one and onto and that it follows from this that the resolvent operator $R_\lambda(A) = (\lambda - A)^{-1} : \mathcal{E} \to \mathcal{D}(A)$ is bounded (Theorem, Section VIII.1, of [Yosida]). One can then show that if A is the infinitesimal generator of an ω-contractive semigroup of operators on a Banach space, then:
1. $(\omega, \infty) \subseteq \rho(A)$ and
2. $\|R_\lambda(A)\| = \|(\lambda - A)^{-1}\| \leq \frac{1}{\lambda-\omega} \ \forall \lambda > \omega.$

The proof of this amounts to writing $R_\lambda(A)$ as a Laplace transform

$$R_\lambda(A)x = \int_0^\infty e^{-\lambda s} T_s x \, ds$$

and estimating the integral using $\|T_t\| \leq e^{\omega t}$ (see Theorem 3(ii), Section 7.4.1, of [Evans], Section X.8 of [RS2], Section 34.1 of [Lax] or Section IX.3 of [Yosida]). Remarkably, these two properties alone characterize the infinitesimal generators of ω-contractive semigroups of operators on a Banach space among the densely defined, closed operators.

Theorem J.0.3 (Hille–Yosida theorem). *Let $\mathcal{E}$ be a Banach space and $A : \mathcal{D}(A) \to \mathcal{E}$ a densely defined, closed operator on $\mathcal{E}$. Then A is the infinitesimal generator of an ω-contractive semigroup of operators on $\mathcal{E}$ for some $\omega \in \mathbb{R}$ if and only if*

1. $(\omega, \infty) \subseteq \rho(A)$ and
2. $\|R_\lambda(A)\| = \|(\lambda - A)^{-1}\| \le \frac{1}{\lambda - \omega}$ $\forall \lambda > \omega$.

The contractive case of the Hille–Yosida theorem is Theorem X.47a of [RS2] and Theorem 7, Section 34.2, of [Lax]. The extension to arbitrary C^0-semigroups is generally called the Hille–Yosida–Phillips theorem and is Theorem X.47b of [RS2]. A more general result for locally convex, sequentially complete topological vector spaces appears as the Theorem in Section IX.7 of [Yosida]. Here is a consequence of Hille–Yosida that we will need.

Theorem J.0.4. *Let $\mathcal{H}$ be a Hilbert space and $T : D(A) \to \mathcal{H}$ an operator on $\mathcal{H}$ that is self-adjoint and positive ($\langle T\psi, \psi \rangle \ge 0$ $\forall \psi \in D(T)$). Then $-T$ generates a contractive C^0-semigroup of operators on $\mathcal{H}$.*

Proof. Since T is self-adjoint, so is $-T$. Since T is positive, its spectrum $\sigma(T)$ is contained in $[0, \infty)$, so $\sigma(-T) \subseteq (-\infty, 0]$. Consequently, $(0, \infty) \subseteq \rho(-T)$ and condition 1. of the Hille–Yosida theorem is satisfied. Now suppose $\lambda > 0$ and consider the (bounded) resolvent operator $(\lambda - (-T))^{-1} = (\lambda + T)^{-1}$. For any $\psi \in D(T) = D(-T)$,

$$\|(\lambda + T)\psi\| \, \|\psi\| \ge |\langle (\lambda + T)\psi, \psi \rangle| = \lambda\|\psi\|^2 + \langle T\psi, \psi \rangle \ge \lambda\|\psi\|^2,$$

so

$$\|(\lambda + T)\psi\| \ge \lambda\|\psi\| \quad \forall \psi \in D(T) = D(\lambda + T).$$

Applying this to $(\lambda + T)^{-1}\phi$ for any $\phi \in \mathcal{H}$ gives

$$\|(\lambda + T)^{-1}\phi\| \le \frac{1}{\lambda}\|\phi\|,$$

so

$$\|(\lambda + T)^{-1}\| \le \frac{1}{\lambda}.$$

This verifies condition 2. of the Hille–Yosida theorem, so we conclude that $-T$ is the infinitesimal generator for a contractive C^0-semigroup of operators on $\mathcal{H}$ which we will denote

$$e^{-tT}.$$
□

We mention that in this case the semigroup operator e^{-tT} really is the exponential function of the operator $-tT$ in the sense of the functional calculus. That is, if

$$T = \int_{[0,\infty)} \lambda \, dE_\lambda$$

is the spectral decomposition of T, then

$$e^{-tT} = \int_{[0,\infty)} e^{-t\lambda} \, dE_\lambda.$$

Example J.0.2. Let $V : \mathbb{R} \to \mathbb{R}$ be a nonnegative, measurable function. Then the corresponding multiplication operator on $L^2(\mathbb{R})$, which we will also denote V, is self-adjoint and positive on $\mathcal{D}(V) = \{\phi \in L^2(\mathbb{R}) : V\phi \in L^2(\mathbb{R})\}$. Consequently, $-V$ generates a contractive C^0-semigroup e^{-tV} on $L^2(\mathbb{R})$.

Example J.0.3. We have shown that the operator $-\Delta : \mathcal{D}(\Delta) \to L^2(\mathbb{R})$ is self-adjoint and positive, so we conclude from Theorem J.0.4 that Δ generates a contractive semigroup

$$e^{t\Delta}$$

of operators on $L^2(\mathbb{R})$. We would like show that this semigroup coincides with the heat semigroup of Exercise G.3.6 (with $D = 1$).

Remark J.0.2. There is another proof in Example 2, Section 6, Chapter IX, of [Yosida].

Let ψ_0 be an arbitrary element of $L^2(\mathbb{R})$ and define

$$\psi(t, x) = (e^{t\Delta}\psi_0)(t, x).$$

Part 4. of Theorem J.0.2 implies that

$$\frac{d}{dt}\psi(t, x) = \Delta\psi(t, x)$$

and, moreover,

$$\lim_{t\to 0^+} \psi(t, x) = \psi_0(x).$$

Furthermore, by 3. of Theorem J.0.2, $\psi(t, x)$ is continuously differentiable with respect to t and, by definition, $\psi(t, x)$ is in $L^2(\mathbb{R})$ as a function of x for each t. Appealing to Exercise 6.2.1 we find that

$$\frac{d}{dt}\psi(t, x) = \frac{\partial\psi(t, x)}{\partial t},$$

so $\psi(t, x)$ is a solution to the heat equation (with $D = 1$) and satisfies the initial condition $\lim_{t\to 0^+} \psi(t, x) = \psi_0(x)$. The heat semigroup also has the property that it carries ψ_0 onto a solution to this initial value problem, but, as we have already mentioned, solutions to an initial value problem for the heat equation need not be unique so we can not yet infer that the two semigroups agree. We will approach this somewhat differently.

Denote the heat semigroup by $\{T_t\}_{t\geq 0}$ and denote its infinitesimal generator by A. We will first show that the operators A and Δ agree on the Schwartz space $\mathcal{S}(\mathbb{R}) \subseteq L^2(\mathbb{R})$. Thus, we let ψ_0 be in $\mathcal{S}(\mathbb{R})$. Then

$$(T_t\psi_0)(x) = \int_\mathbb{R} \frac{1}{\sqrt{4\pi t}}e^{-(x-y)^2/4t}\psi_0(y)\,dy = (K^t * \psi_0)(x),$$

where

$$K^t(x) = \frac{1}{\sqrt{4\pi t}} e^{-x^2/4t}.$$

Exercise J.0.2. Compute Fourier transforms and show that

$$\mathcal{F}[T_t\psi_0](p) = e^{-tp^2} \mathcal{F}[\psi_0](p).$$

From this we obtain

$$\mathcal{F}\left[\frac{T_t\psi_0 - \psi_0}{t}\right](p) = \frac{e^{-tp^2} - 1}{t} \mathcal{F}[\psi_0](p)$$

and therefore

$$\frac{T_t\psi_0 - \psi_0}{t} = \mathcal{F}^{-1}\left[\frac{e^{-tp^2} - 1}{t} \mathcal{F}[\psi_0](p)\right].$$

Taking the limit as $t \to 0^+$ then gives

$$A\psi_0 = \mathcal{F}^{-1}[-p^2\mathcal{F}[\psi_0](p)] = \Delta\psi_0.$$

We conclude then that the infinitesimal generator A of the heat semigroup agrees with the Laplacian Δ on $\mathcal{S}(\mathbb{R})$. It follows that e^{tA} and $e^{t\Delta}$ agree on $\mathcal{S}(\mathbb{R})$. Now let ψ be an arbitrary element of $L^2(\mathbb{R})$. Choose a sequence $\{\psi_n\}_{n=1}^{\infty}$ in $\mathcal{S}(\mathbb{R})$ converging to ψ in $L^2(\mathbb{R})$. Then, for any n,

$$e^{tA}\psi - e^{t\Delta}\psi = (e^{tA} - e^{t\Delta})(\psi - \psi_n).$$

Now, $e^{tA} - e^{t\Delta}$ is a bounded operator on $L^2(\mathbb{R})$ with norm, say, M. Thus, *for any n,*

$$\|e^{tA}\psi - e^{t\Delta}\psi\| \leq M\|\psi - \psi_n\|,$$

so $e^{tA}\psi = e^{t\Delta}\psi$ in $L^2(\mathbb{R})$ and we have shown that the semigroup generated by the Laplacian is the heat semigroup.

We mentioned earlier that the Trotter product formula has a generalization to the context of operator semigroups. It is now time to formulate the result we had in mind. The following is Theorem 9, Appendix B, of [Nel3].

Theorem J.0.5. *Let $A : \mathcal{D}(A) \to \mathcal{E}$ and $B : \mathcal{D}(B) \to \mathcal{E}$ be linear operators on the Banach space $\mathcal{E}$ that are infinitesimal generators of contractive semigroups on $\mathcal{E}$. Suppose that $A + B : \mathcal{D}(A) \cap \mathcal{D}(B) \to \mathcal{E}$ also generates a contractive semigroup on $\mathcal{E}$. Then, for all $\varphi \in \mathcal{E}$,*

$$e^{t(A+B)}\varphi = \lim_{n\to\infty} \left(e^{\frac{t}{n}A} e^{\frac{t}{n}B}\right)^n \varphi.$$

It is actually enough to assume that the closure of $(A + B)|_{\mathcal{D}(A) \cap \mathcal{D}(B)}$ generates a contractive semigroup, but the proof requires more work; see Theorem 11.1.4 of [JL], where this result is obtained from a still more general product formula due to Chernoff.

Example J.0.4. Let $V : \mathbb{R} \to \mathbb{R}$ be a bounded, measurable function on $\mathbb{R}$ and let $M > 0$ be such that $|V(x)| \le M$ $\forall x \in \mathbb{R}$. We do not assume that V is nonnegative. However, $V + M : \mathbb{R} \to \mathbb{R}$ is bounded, measurable and nonnegative. Using the same symbols to denote the multiplication operators on $L^2(\mathbb{R})$ determined by V, M and $V + M$, we find that all of these are bounded and self-adjoint and the operator $V + M$ is also positive on $L^2(\mathbb{R})$. Consequently, $-(V+M)$ generates a contractive semigroup of operators on $L^2(\mathbb{R})$ and each of the semigroup operators is just multiplication by the ordinary exponential $e^{-t(V+M)}$. Moreover, each of the multiplication operators e^{-tV} and e^{-tM} is defined and bounded and $e^{-t(V+M)} = e^{-tV}e^{-tM}$. Now, consider also the positive, self-adjoint operator $-\Delta : \mathcal{D}(\Delta) \to L^2(\mathbb{R})$ on $L^2(\mathbb{R})$. Then Δ generates the heat semigroup $e^{t\Delta}$ on $L^2(\mathbb{R})$. Using the fact that the multiplication operator e^{-tM} commutes with every operator on $L^2(\mathbb{R})$ we compute, for any integer $n \ge 1$ and any $\varphi \in L^2(\mathbb{R})$,

$$e^{-tM}\left(e^{\frac{t}{n}\Delta}e^{-\frac{t}{n}V}\right)^n\varphi = \left(e^{-\frac{t}{n}M}e^{\frac{t}{n}\Delta}e^{-\frac{t}{n}V}\right)^n\varphi = \left(e^{\frac{t}{n}\Delta}e^{-\frac{t}{n}(V+M)}\right)^n\varphi.$$

Since $V + M$ is bounded, Theorem 9.2.4 implies that $-\Delta + (V + M)$ is self-adjoint on $\mathcal{D}(\Delta)$ and, since $V + M$ is nonnegative, $-\Delta + (V + M)$ is a positive operator. Consequently, $\Delta - (V + M)$ also generates a contractive semigroup on $L^2(\mathbb{R})$, so Theorem J.0.5 implies that

$$\lim_{n\to\infty}\left(e^{\frac{t}{n}\Delta}e^{-\frac{t}{n}(V+M)}\right)^n\varphi = e^{t(\Delta-(V+M))}\varphi = e^{-tM}e^{t(\Delta-V)}\varphi.$$

Consequently,

$$e^{-tM}\lim_{n\to\infty}\left(e^{\frac{t}{n}\Delta}e^{-\frac{t}{n}V}\right)^n\varphi = e^{-tM}e^{t(\Delta-V)}\varphi,$$

which gives

$$e^{t(\Delta-V)}\varphi = \lim_{n\to\infty}\left(e^{\frac{t}{n}\Delta}e^{-\frac{t}{n}V}\right)^n\varphi.$$

The thing to note is that, although Theorem J.0.5 does not apply directly to $\Delta - V$, the conclusion we have arrived at is precisely the same.

K Hilbert space tensor products

We will require some of the basic properties of tensor products of Hilbert spaces. Everything we need is treated in detail in Chapter II, Sections 6.3–6.5, and Chapter IV, Section 4.5, of [Prug], but one can also consult Section II.4 of [RS1] for a different, but equivalent approach. We will just summarize the items we require. For this we let $\mathcal{H}_1, \ldots, \mathcal{H}_N$ be complex, separable Hilbert spaces with inner products $\langle\,,\,\rangle_1, \ldots, \langle\,,\,\rangle_N$, respectively. Then $\mathcal{H}_1, \ldots, \mathcal{H}_N$ are, in particular, complex vector spaces, so they have an algebraic tensor product $\mathcal{H}_1 \otimes_{\text{alg}} \cdots \otimes_{\text{alg}} \mathcal{H}_N$ (Chapter II, Section 6.4, of [Prug]) consisting of all finite linear combinations of elements of the form $\psi_1 \otimes \cdots \otimes \psi_N$, where $\psi_i \in \mathcal{H}_i$ for each $i = 1, \ldots, N$. Define a complex-valued bilinear map $\langle\,,\,\rangle$ on $\mathcal{H}_1 \otimes_{\text{alg}} \cdots \otimes_{\text{alg}} \mathcal{H}_N$ by

$$\langle \psi_1 \otimes \cdots \otimes \psi_N, \phi_1 \otimes \cdots \otimes \phi_N \rangle = \langle \psi_1, \phi_1 \rangle_1 \cdots \langle \psi_N, \phi_N \rangle_N.$$

This defines an inner product on $\mathcal{H}_1 \otimes_{\text{alg}} \cdots \otimes_{\text{alg}} \mathcal{H}_N$ (Theorem 6.8, Chapter II, of [Prug]) and the *Hilbert space tensor product* $\mathcal{H}_1 \otimes \cdots \otimes \mathcal{H}_N$ of $\mathcal{H}_1, \ldots, \mathcal{H}_N$ is defined to be the completion of $\mathcal{H}_1 \otimes_{\text{alg}} \cdots \otimes_{\text{alg}} \mathcal{H}_N$ with respect to this inner product. Note that $\mathcal{H}_1 \otimes \cdots \otimes \mathcal{H}_N$ is separable and if $\{e_n^i\}_{n=1}^\infty$ is an orthonormal basis for $\mathcal{H}_i$ for each $i = 1, \ldots, N$, then

$$\left\{ e_{n_1}^1 \otimes \cdots \otimes e_{n_N}^N : 1 \leq n_1, \ldots, n_N < \infty \right\}$$

is an orthonormal basis for $\mathcal{H}_1 \otimes \cdots \otimes \mathcal{H}_N$ (Chapter II, Theorem 6.10, of [Prug]).

If $\mathcal{H}_1, \ldots, \mathcal{H}_N$ are all the same Hilbert space $\mathcal{H}$, then we will write the tensor product as $\otimes^N \mathcal{H}$ and call it the *Nth tensor power of $\mathcal{H}$*. Note that the symmetric group S_N of permutations of $\{1, \ldots, N\}$ acts naturally on $\otimes^N \mathcal{H}$ by defining

$$\sigma \cdot (\psi_1 \otimes \cdots \otimes \psi_N) = \psi_{\sigma(1)} \otimes \cdots \otimes \psi_{\sigma(N)}$$

for every $\sigma \in S_N$ and then extending to all of $\otimes^N \mathcal{H}$ by linearity and the density of the algebraic tensor product in $\otimes^N \mathcal{H}$. We define two closed, linear subspaces of $\otimes^N \mathcal{H}$ as follows:

$$\otimes_S^N \mathcal{H} = \{ \Psi \in \otimes^N \mathcal{H} : \sigma \cdot \Psi = \Psi \ \forall \sigma \in S_N \}$$

is called the *Nth symmetric tensor power of $\mathcal{H}$* and

$$\otimes_A^N \mathcal{H} = \{ \Psi \in \otimes^N \mathcal{H} : \sigma \cdot \Psi = (\text{sgn } \sigma) \Psi \ \forall \sigma \in S_N \},$$

where sgn σ is the sign of the permutation σ, is the *Nth antisymmetric tensor power of $\mathcal{H}$*. These subspaces of $\otimes^N \mathcal{H}$ are orthogonal with respect to the inner product defined above and one defines the projections $\pi_S : \otimes^N \mathcal{H} \to \otimes_S^N \mathcal{H}$ and $\pi_A : \otimes^N \mathcal{H} \to \otimes_A^N \mathcal{H}$ to be the extensions of

$$\pi_S(\psi_1 \otimes \cdots \otimes \psi_N) = \frac{1}{N!} \sum_{\sigma \in S_N} \psi_{\sigma(1)} \otimes \cdots \otimes \psi_{\sigma(N)}$$

https://doi.org/10.1515/9783110751949-021

and

$$\pi_A(\psi_1 \otimes \cdots \otimes \psi_N) = \frac{1}{N!} \sum_{\sigma \in S_N} (\text{sgn } \sigma) \, \psi_{\sigma(1)} \otimes \cdots \otimes \psi_{\sigma(N)}.$$

One generally writes

$$\pi_S(\psi_1 \otimes \cdots \otimes \psi_N) = \psi_1 \otimes_S \cdots \otimes_S \psi_N$$

and

$$\pi_A(\psi_1 \otimes \cdots \otimes \psi_N) = \psi_1 \otimes_A \cdots \otimes_A \psi_N.$$

Then $\otimes_S^N(\mathcal{H})$ and $\otimes_A^N(\mathcal{H})$ are the closed linear spans of $\psi_1 \otimes_S \cdots \otimes_S \psi_N$ and $\psi_1 \otimes_A \cdots \otimes_A \psi_N$, respectively.

The particular examples of most interest to us arise in the following way. Let $(X_1, \mathcal{A}_1, \mu_1)$ and $(X_2, \mathcal{A}_2, \mu_2)$ be two measure spaces and consider the corresponding Hilbert spaces $L^2(X_1, \mathcal{A}_1, \mu_1)$ and $L^2(X_2, \mathcal{A}_2, \mu_2)$. If $f \in L^2(X_1, \mathcal{A}_1, \mu_1)$ and $g \in L^2(X_2, \mathcal{A}_2, \mu_2)$, then $f(x_1)g(x_2)$ represents an element of $L^2(X_1 \times X_2, \mathcal{A}_1 \times \mathcal{A}_2, \mu_1 \times \mu_2)$ that we denote $f \cdot g$, that is,

$$(f \cdot g)(x_1, x_2) = f(x_1)g(x_2) \quad \forall (x_1, x_2) \in X_1 \times X_2.$$

With this we can define a linear map from the algebraic tensor product $L^2(X_1, \mathcal{A}_1, \mu_1) \otimes_{\text{alg}} L^2(X_2, \mathcal{A}_2, \mu_2)$ to $L^2(X_1 \times X_2, \mathcal{A}_1 \times \mathcal{A}_2, \mu_1 \times \mu_2)$ by

$$\sum_{k=1}^{n} a_k f_k \otimes g_k \rightarrow \sum_{k=1}^{n} a_k f_k \cdot g_k.$$

According to Theorem 6.9, Chapter II, of [Prug], this map extends uniquely to a unitary equivalence of the Hilbert space tensor product $L^2(X_1, \mathcal{A}_1, \mu_1) \otimes L^2(X_2, \mathcal{A}_2, \mu_2)$ onto $L^2(X_1 \times X_2, \mathcal{A}_1 \times \mathcal{A}_2, \mu_1 \times \mu_2)$. In particular, with Lebesgue measure understood, we can identify, for any $n = 1, 2, \ldots$,

$$L^2(\mathbb{R}^n) \otimes L^2(\mathbb{R}^n) \cong L^2(\mathbb{R}^n \times \mathbb{R}^n) \cong L^2(\mathbb{R}^{2n})$$

and, by induction, for any $N = 1, 2, \ldots$,

$$\otimes^N L^2(\mathbb{R}^n) \cong L^2(\mathbb{R}^n \times \overset{N}{\cdots} \times \mathbb{R}^n) \cong L^2(\mathbb{R}^{Nn}).$$

Thus, we can identify elements of $\otimes^N L^2(\mathbb{R}^n)$ with (equivalence classes of) square integrable functions $\psi(\mathbf{x}_1, \ldots, \mathbf{x}_N)$, where $\mathbf{x}_1, \ldots, \mathbf{x}_N \in \mathbb{R}^n$. The elements of $\otimes_S^N L^2(\mathbb{R}^n)$ then correspond to functions that are (up to a set of measure zero) symmetric under permutations of their variables,

$$\psi(\mathbf{x}_{\sigma(1)}, \ldots, \mathbf{x}_{\sigma(N)}) = \psi(\mathbf{x}_1, \ldots, \mathbf{x}_N) \quad \forall \sigma \in S_N,$$

while those in $\otimes_A^N L^2(\mathbb{R}^n)$ are antisymmetric,

$$\psi(\mathbf{x}_{\sigma(1)}, \ldots, \mathbf{x}_{\sigma(N)}) = (\text{sgn } \sigma) \, \psi(\mathbf{x}_1, \ldots, \mathbf{x}_N) \quad \forall \sigma \in S_N.$$

Bibliography

[Abra]   Abraham, R., *Foundations of Mechanics*, W. A. Benjamin, New York, NY, 1967.

[AMR]   Abraham, R., J. E. Marsden and T. Ratiu, *Manifolds, Tensor Analysis, and Applications*, Second edition, Springer, New York, NY, 2001.

[AS]   Aharonov, Y. and L. Susskind, Observability of the sign change of spinors under 2π rotations, Phys. Rev., 158, 1967, 1237–1238.

[AMS]   Aitchison, I. J. R., D. A. MacManus and T. M. Snyder, Understanding Heisenberg's "Magical" paper of July 1925: a new look at the computational details, Am. J. Phys., 72, 2004, 1370–1379.

[Albert]   Albert, D., *Quantum Mechanics and Experience*, Harvard University Press, Cambridge, MA, 1992.

[AHM]   Albeverio, S. A., R. J. Hoegh and S. Mazzucchi, *Mathematical Theory of Feynman Path Integrals: An Introduction*, Second edition, Springer, New York, NY, 2008.

[AAR]   Andrews, G. E., R. Askey and R. Roy, *Special Functions*, Cambridge University Press, Cambridge, England, 2000.

[AngBat]   Angelow, A. and M.-C. Batoni, About Heisenberg's uncertainty relation (by E. Schrödinger). Annotated by A. Angelow and M.-C. Batoni. (Accessed July 10, 2021, at http://arxiv.org/abs/quant-ph/9903100.)

[Apos]   Apostol, T. M., *Mathematical Analysis*, Second edition, Addison-Wesley Publishing Co., Reading, MA, 1974.

[Arn1]   Arnol'd, V. I., *Ordinary Differential Equations*, MIT Press, Cambridge, MA, 1973.

[Arn2]   Arnol'd, V. I., *Mathematical Methods of Classical Mechanics*, Second edition, Springer, New York, NY, 1989.

[ADR]   Aspect, A., J. Dalibard and G. Roger, Experimental test of Bell's inequalities using time-varying analyzers, Phys. Rev. Lett., 49, 1982, 1804–1807.

[Atiyah]   Atiyah, M. F., Circular symmetry and stationary phase approximation, Astérisque, 131, 1985, 43–59.

[AB]   Atiyah, M. F. and R. Bott, The moment map and equivariant cohomology, Topology, 23, 1984, 1–28.

[AdeP]   Atkins, P. and J. de Paula, *Physical Chemistry*, Eighth edition, Oxford University Press, Oxford, England, 2006.

[BPLB]   Bach, R., D. Pope, S-H. Liou and H. Batelaan, Controlled double-slit electron diffraction, New J. Phys., 15, 2013. (Accessed July 10, 2021, at http://iopscience.iop.org/1367-2630/15/3/033018/media.)

[Bagchi]   Bagchi, B. K., *Supersymmetry in Quantum and Classical Mechanics*, Chapman and Hall/CRC, Boca Raton, FL, 2001.

[Bal]   Ballentine, L. E., The statistical interpretation of quantum mechanics, Rev. Mod. Phys., 42, 1970, 358–381.

[BB-FF]   Barone, F. A., H.Boschi-Filho and C. Farina, Three methods for calculating the Feynman propagator, Am. J. Phys., 71, 2003, 483–491.

[BarRac]   Barut, A. O. and R. Raczka, *Theory of Group Representations and Applications*, Polish Scientific Publishers, Warszawa, Poland, 1977.

[Bell]   Bell, J. S., On the Einstein, Podolsky, Rosen paradox, Physics 1, 1964, 195–200.

[Ber1]   Berezin, F. A., *The Method of Second Quantization*, Pure and Applied Physics, Academic Press, New York, NY, 1966.

[Ber2]   Berezin, F. A., *Introduction to Superanalysis*, Edited by A. A. Kirillov, D. Reidel Publishing Company, Dordrecht, Holland, 1987.

[BS]   Berezin, F. A. and M. A. Shubin, *The Schrödinger Equation*, Kluwer Academic Publishers, Dordrecht, The Netherlands, 1991.

https://doi.org/10.1515/9783110751949-022

[BGV]   Berline, N., E. Getzler and M. Vergne, *Heat Kernels and Dirac Operators*, Springer, New York, NY, 2004.

[Berndt]   Berndt, R., *An Introduction to Symplectic Geometry*, Amer. Math. Soc., Providence, RI, 2001.

[BP]   Bernstein, H. J. and A. V. Phillips, Fiber bundles and quantum theory, Sci. Am., 245, 1981, 94–109.

[BM]   Birkhoff, G. and S. Mac Lane, *A Survey of Modern Algebra*, Third edition, Macmillan, New York, NY, 1971.

[BG]   Bishop, R. L. and S. Goldberg, *Tensor Analysis on Manifolds*, Dover Publications, Inc., Mineola, NY, 1980.

[BD]   Bjorken, J. D. and S. D. Drell, *Relativistic Quantum Fields*, McGraw-Hill Book Company, New York, NY, 1965.

[BEH]   Blank, J., P. Exner and M. Havlicek, *Hilbert Space Operators in Quantum Physics*, Second edition, Springer, New York, NY, 2010.

[Bleeck]   Bleecker, D., *Gauge Theory and Variational Principles*, Addison-Wesley Publishing Company, Reading, MA, 1981.

[Bohm]   Bohm, D., *Quantum Theory*, Dover Publications, Inc., Mineola, NY, 1989.

[BR]   Bohr, N. and L. Rosenfeld, Zur Frage der Meßbarkeit der Electromagnetischen Feldgrößen, Danske Mat.-Fys. Meddr. XII, 8, 1933, 1–65.

[Bolker]   Bolker, E. D., The spinor spanner, Am. Math. Mon., 80, 1973, 977–984.

[Born1]   Born, M., Nobel lecture: the statistical interpretations of quantum mechanics. (Accessed May 8, 2021, at https://www.nobelprize.org/uploads/2018/06/born-lecture.pdf.).

[Born2]   Born, M., *My Life: Recollections of a Nobel Laureate*, Charles Scribner's Sons, New York, NY, 1975.

[BJ]   Born, M. and P. Jordan, Zur Quantenmechanik, Z. Phys., 34, 1925, 858–888. (English translation available in [VdW].).

[BHJ]   Born, M., W. Heisenberg and P. Jordan, Zur Quantenmechanik II, Z. Phys., 35, 1926, 557–615. (English translation available in [VdW].).

[Bott]   Bott, R., Lectures on Morse theory, old and new, Bull. Am. Math. Soc., 7, 2, 1982, 331–358.

[BK]   Braginsky, V. B. and F. Ya. Khalili, *Quantum Measurement*, Cambridge University Press, Cambridge, England, 1992.

[BC]   Brown, J. W. and R. V. Churchill, *Fourier Series and Boundary Value Problems*, Seventh edition, McGraw-Hill, Boston, MA, 2008.

[Brown]   Brown, L. M. (Editor), *Feynman's Thesis: A New Approach to Quantum Theory*, World Scientific, Singapore, 2005.

[Busch]   Busch, P., The time-energy uncertainty relation. (Accessed May 9, 2021, at http://arxiv.org/pdf/quant-ph/0105049v3.pdf.).

[Cam]   Cameron, R. H., The Ilstow and Feynman integrals, J. Anal. Math., 10, 1962/63, 287–361.

[Car1]   Cartan, H., *Differential Forms*, Houghton-Mifflin, Boston, MA, 1970.

[Car2]   Cartan, H., *Differential Calculus*, Kershaw Publishing, London, England, 1971.

[Casado]   Casado, C. M. M., A brief history of the mathematical equivalence of the two quantum mechanics, Lat. Am. J. Phys. Educ., 2, 2008, 152–155.

[CI]   Chamizo, F. and H. Iwaniec, On the sphere problem, Rev. Mat. Iberoam., 11, 1995, 417–429.

[Ch]   Chernoff, P. R., Note on product formulas for operator semigroups, J. Funct. Anal., 2, 1968, 238–242.

[ChM1]   Chernoff, P. R. and J. E. Marsden, Some basic properties of infinite dimensional Hamiltonian systems, Colloq. Intern., 237, 1976, 313–330. (Accessed May 9, 2021, at http://authors.library.caltech.edu/20406/1/ChMa1976.pdf.).

[ChM2]   Chernoff, P. R. and J. E. Marsden, Some remarks on Hamiltonian systems and quantum mechanics, in *Problems of Probability Theory, Statistical Inference, and Statistical Theories of Science*, 35–43, edited by Harper and Hooker, D. Reidel Publishing Company, Dordrecht, Holland, 1976.

[CL]   Cini, M. and J. M. Lévy-Leblond, *Quantum Theory without Reduction*, Taylor and Francis, Boca Raton, FL, 1990.

[Connes] Connes, A., *Noncommutative Geometry*, Academic Press, San Diego, CA, 1994. (Accessed May 9, 2021, at http://www.alainconnes.org/en/downloads.php.).

[Corn]   Cornish, F. H. J., The hydrogen atom and the four-dimensional harmonic oscillator, J. Phys. A, Math. Gen., 17, 1984, 323–327.

[CM]   Curtis, W. D. and F. R. Miller, *Differential Manifolds and Theoretical Physics*, Academic Press, Inc., Orlando, FL, 1985.

[Del]   Deligne, P., *et al.*, Editors, *Quantum Fields and Strings: A Course for Mathematicians*, Volumes 1–2, Amer. Math. Soc., Providence, RI, 1999.

[deMJS]   de Muynck, W. M., P. A. E. M. Janssen and A. Santman, Simultaneous measurement and joint probability distributions in quantum mechanics, Found. Phys., 9, 1979, 71–122.

[DH]   Duistermaat, J. J. and G. J. Heckman, On the variation in the cohomology of the symplectic form of the reduced phase space, Invent. Math., 69, 259–268.

[Dirac1] Dirac, P. A. M., The fundamental equations of quantum mechanics, Proc. R. Soc. Lond. A, 109, 1925, 642–653.

[Dirac2] Dirac, P. A. M., The quantum theory of the electron, Proc. R. Soc. Lond. A, 117, 1928, 610–624.

[Dirac3] Dirac, P. A. M., The Lagrangian in quantum mechanics, Phys. Z. Sowjetunion, 3, 1, 1933, 64–72.

[Dirac4] Dirac, P. A. M., The evolution of the physicist's picture of nature, Sci. Am., 208, 1963, 45–53.

[DS1]   Duck, I. and E. C. G. Sudarshan, *Pauli and the Spin-Statistics Theorem*, World Scientific, Singapore, 1997.

[DS2]   Duck, I. and E. C. G. Sudarshan, Toward an understanding of the spin-statistics theorem, Am. J. Phys., 64, 1998, 284–303.

[DSII]   Dunford, N. and J. S. Schwartz, *Linear Operators: Part II, Spectral Theory*, John Wiley and Sons, New York, NY, 1963.

[Ehren]   Ehrenfest, P., Bemerkung über die angenäherte Gültigkeit der klassischen Mechanik innerhalb der Quantenmechanik, Z. Phys., 45, 1927, 455–457.

[Ein1]   Einstein, A., Über einen die Erzeugung und Verwandlung des Lichtes betreffenden heuristischen Gesichtspunkt, Ann. Phys., 17, 1905, 132–148. (English translation available in [ter H].).

[Ein2]   Einstein, A., Zur Elektrodynamik bewegter Körper, Ann. Phys., 17, 10, 1905, 891–921. (English translation available in [Ein3].).

[Ein3]   Einstein, A., et al., *The Principle of Relativity*, Dover Publications, Inc., Mineola, NY, 1952.

[Ein4]   Einstein, A., *Investigations on the Theory of the Brownian Movement*, Edited with notes by R. Fürth, Translated by A. D. Cowper, Dover Publications, Inc., Mineola, NY, 1956.

[EPR]   Einstein, A., B. Podolsky and N. Rosen, Can the quantum mechanical description of reality be considered complete?, Phys. Rev., 41, 1935, 777–780.

[Eis]   Eisberg, R. M., *Fundamentals of Modern Physics*, John Wiley and Sons, New York, NY, 1967.

[Emch]   Emch, G. G., *Mathematical and Conceptual Foundations of 20th-Century Physics*, North Holland, Amsterdam, The Netherlands, 1984.

[Evans]   Evans, L. C., *Partial Differential Equations*, Second edition, Amer. Math. Soc., Providence, RI, 2010.

[Fadell]   Fadell, E., Homotopy groups of configuration spaces and the string problem of Dirac, Duke Math. J., 29, 1962, 231–242.

[FP1]   Fedak, W. A. and J. J. Prentis, Quantum jumps and classical harmonics, Am. J. Phys. 70, 2002, 332–344.

[FP2]   Fedak, W. A. and J. J. Prentis, The 1925 Born and Jordan paper "On quantum mechanics", Am. J. Phys. 77, 2009, 128–139.

[Fels]   Felsager, B., *Geometry, Particles, and Fields*, Springer, New York, NY, 1998.

[Fermi]   Fermi, E., *Thermodynamics*, Dover Publications, Mineola, NY, 1956.

[FLS]   Feynman, R. P., R. B. Leighton and M. Sands, *The Feynman Lectures on Physics, Vol. I–III*, Addison-Wesley, Reading, MA, 1964.

[Feyn]   Feynman, R. P., Space-time approach to non-relativistic quantum mechanics, Rev. Mod. Phys., 20, 1948, 367–403.

[Flamm]   Flamm, D., Ludwig Boltzmann—a pioneer of modern physics. (Accessed May 9, 2021, at https://arxiv.org/abs/physics/9710007.).

[Fol1]   Folland, G. B., *Harmonic Analysis in Phase Space*, Princeton University Press, Princeton, NJ, 1989.

[Fol2]   Folland, G. B., *Real Analysis: Modern Techniques and their Applications*, Second edition, John Wiley and Sons, New York, NY, 1999.

[Fol3]   Folland, G. B., *Quantum Field Theory: A Tourist Guide for Mathematicians*, Amer. Math. Soc., Providence, RI, 2008.

[FS]   Folland, G. B. and A. Sitaram, The uncertainty principle: a mathematical survey, J. Fourier Anal. Appl., 3, 1997, 207–238.

[FU]   Freed, D. S. and K. K. Uhlenbeck, Editors, *Geometry and Quantum Field Theory*, Amer. Math. Soc., Providence, RI, 1991.

[Fried]   Friedman, A., *Foundations of Modern Analysis*, Dover Publications, Inc., Mineola, NY, 1982.

[FrKo]   Friesecke, G. and M. Koppen, On the Ehrenfest Theorem of Quantum Mechanics, J. Math. Phys., 50, 2009. (Accessed May 9, 2021, at http://arxiv.org/pdf/0907.1877.pdf.).

[FrSc]   Friesecke, G. and B. Schmidt, A sharp version of Ehrenfest's theorem for general self-adjoint operators, Proc. R. Soc. A, 466, 2010, 2137–2143. (Accessed May 9, 2021, at http://arxiv.org/pdf/1003.3372.pdf.).

[Fuj1]   Fujiwara, D., A construction of the fundamental solution for the Schrödinger equations, Proc. Jpn. Acad., Ser. A, Math. Sci., 55, 1979, 10–14.

[Fuj2]   Fujiwara, D., On the nature of convergence of some Feynman path integrals I, Proc. Jpn. Acad., Ser. A, Math. Sci., 55, 1979, 195–200.

[Fuj3]   Fujiwara, D., On the nature of convergence of some Feynman path integrals II, Proc. Jpn. Acad., Ser. A, Math. Sci., 55, 1979, 273–277.

[Gaal]   Gaal, S. A., *Linear Analysis and Representation Theory*, Springer, New York, NY, 1973.

[Går]   Gårding, L., Note on continuous representations of lie groups, Proc. Natl. Acad. Sci. USA, 33, 1947, 331–332.

[Gel]   Gel'fand, I. M., *Representations of the Rotation and Lorentz Groups and their Applications*, Martino Publishing, Mansfield Centre, CT, 2012.

[GN]   Gel'fand, I. M. and M. A. Naimark, On the embedding of normed rings into the ring of operators in Hilbert space, Mat. Sb., 12, 1943, 197–213.

[GS]   Gel'fand, I. M. and G. E. Shilov, *Generalized Functions, Volume I*, Academic Press, Inc., New York, NY, 1964.

[GY]   Gel'fand, I. M. and A. M. Yaglom, Integration in functional spaces and its applications in quantum physics, J. Math. Phys., 1, 1960, 48–69.

[GHL]   Gallot, S., D. Hulin and J. Lafontaine, *Riemannian Geometry*, Third edition, Springer, New York, NY, 2004.

[GJ]   Glimm, J. and A. Jaffe, *Quantum Physics: A Functional Integral Point of View*, Second edition, Springer, New York, NY, 1987.

[Gold]   Goldstein, H., *Classical Mechanics*, Second edition, Addison-Wesley, Reading, MA, 1980.

[Good]   Goodman, R. W., Analytic and entire vectors for representations of lie groups, Trans. Am. Math. Soc., 143, 1969, 55–76.

[Gotay]   Gotay, M. J., On the Groenewold–Van Hove problem for $\mathbb{R}^{2n}$, J. Math. Phys., 40, 1999, 2107–2116.

[GGT]   Gotay, M. J., H. B. Grundling and G. M. Tuynman, Obstruction results in quantization theory, J. Nonlinear Sci., 6, 1996, 469–498.

[GR]   Gradshteyn, I. S. and I. M. Ryzhik, *Table of Integrals, Series, and Products*, Seventh edition, Academic Press, Burlington, MA, 2007.

[Graf]   Grafakos, L., *Classical Fourier Analysis*, Second edition, Springer, New York, NY, 2008.

[Gre]   Greenberg, M., *Lectures on Algebraic Topology*, W. A. Benjamin, New York, NY, 1967.

[Grein]   Greiner, W., *Relativistic Quantum Mechanics: Wave Equations*, Third edition, Springer-Verlag, Berlin, 2000.

[Groe]   Groenewold, H. J., On the principles of elementary quantum mechanics, Physica (Amsterdam), 12, 1946, 405–460.

[GS1]   Guillemin, V. and S. Sternberg, *Symplectic Techniques in Physics*, Cambridge University Press, Cambridge, England, 1984.

[GS2]   Guillemin, V. and S. Sternberg, *Supersymmetry and Equivariant de Rham Theory*, Springer, New York, NY, 1999.

[GS3]   Guillemin, V. and S. Sternberg, *Geometric Asymptotics*, Revised edition, Amer. Math. Soc., Providence, RI, 1999.

[HK]   Hafele, J. C. and R. E. Keating, Around-the-world atomic clocks: observed relativistic time gains, Science, 177, 168–170.

[Hall1]   Hall, B. C., *Lie Groups, Lie Algebras, and Representations: An Elementary Introduction*, Springer, New York, NY, 2003.

[Hall2]   Hall, B. C., *Quantum Theory for Mathematicians*, Springer, New York, NY, 2013.

[Hal1]   Halmos, P. R., *Measure Theory*, Springer, New York, NY, 1974.

[Hal2]   Halmos, P. R., Foundations of probability theory, Am. Math. Mon., 51, 1954, 493–510.

[Hal3]   Halmos, P. R., What does the spectral theorem say?, Am. Math. Mon., 70, 1963, 241–247.

[Ham]   Hamilton, R. S., The inverse function theorem of Nash and Moser, Bull. Am. Math. Soc., 7, 1982, 65–222.

[Hardy]   Hardy, G. H., *Divergent Series*, Clarendon Press, Oxford, England, 1949.

[Heis1]   Heisenberg, W., Über quantentheoretische Umdeutung kinematischer und mechanischer Beziehungen, Z. Phys., 33, 1925, 879–893. (English translation available in [VdW]).

[Heis2]   Heisenberg, W., *The Physical Principles of the Quantum Theory*, Dover Publications, Inc., Mineola, NY, 1949.

[Her]   Herstein, I. N., *Topics in Algebra*, Second edition, John Wiley and Sons, New York, NY, 1975.

[HP]   Hille, E. and R. S. Phillips, *Functional Analysis and Semi-Groups*, Amer. Math. Soc., Providence, RI, 1957.

[Holz]   Holzegel, G., PDE lecture notes, Week10. (Accessed May 9, 2021, at https://www.imperial.ac.uk/people/g.holzegel.).

[Horv]   Horvathy, P. A., The Maslov correction in the semiclassical Feynman integral. (Accessed May 9, 2021, at http://arxiv.org/abs/quant-ph/0702236.).

[Howe]   Howe, R., On the role of the Heisenberg group in harmonic analysis, Bull. Am. Math. Soc., 3, 1980, 821–843.

[Iyen]   Iyengar, K. S. K., A new proof of Mehler's formula and other theorems on hermitian polynomials, Proc. Indian Acad. Sci., Sect. A, Phys. Sci., 10, 3, 1939, 211–216.

[Jaffe]   Jaffe, A., Supplementary notes on canonical quantization and application to a charged particle in a magnetic field. (Accessed May 9, 2021, at http://stuff.mit.edu/afs/athena/course/8/8.06/spring08/handouts.shtml.).

[JLL]   Jaffe, A., A. Lesniewski and M. Lewenstein, Ground state structure in supersymmetric quantum mechanics, Ann. Phys., 178, 1987, 313–329.

[Jamm]   Jammer, M., *The Conceptual Development of Quantum Mechanics*, Tomash Publishers, Los Angeles, CA, 1989.

[JL]   Johnson, G. W. and M. L. Lapidus, *The Feynman Integral and Feynman's Operational Calculus*, Oxford Science Publications, Clarendon Press, Oxford, England, 2000.

[JS]   Johnson, G. W. and D. L. Skoug, Scale-invariant measurability in Wiener space, Pac. J. Math., 83, 1979, 157–176.

[Jordan]   Jordan, P., Zur Quantenmechanik der Gasentartung, Z. Phys., 44, 1927, 473.

[JW]   Jordan, P. and E. Wigner, Über das Paulische Äquivalenzverbot, Z. Phys., 47, 1928, 631.

[Jost]   Jost, J., *Riemannian Geometry and Geometric Analysis*, Third edition, Springer, New York, NY, 2002.

[KacM]   Kac, M., *Integration in Function Spaces and Some of its Applications*, Lezioni Fermiane, Accademia Nazionale dei Lincei, Pisa, 1980.

[KacV]   Kac, V. G., A sketch of lie superalgebra theory, Commun. Math. Phys., 53, 1977, 31–64.

[Kal]   Kallianpur, G., *Stochastic Filtering Theory*, Springer, New York, NY, 1980.

[Kant]   Kantorovitz, S., *Topics in Operator Semigroups*, Birkhäuser Boston, Springer, New York, NY, 2010.

[Kato]   Kato, T., Fundamental properties of Hamiltonian operators of Schrödinger type, Trans. Am. Math. Soc., 70, 1951, 195–211.

[KM]   Keller, J. B. and D. W. McLaughlin, The Feynman integral, Am. Math. Mon., 82, 1975, 451–465.

[Kenn]   Kennard, E. H., Zür Quantenmechanik einfacher Bewegungstypen, Z. Phys., 44, 1927, 326–352.

[Kiri]   Kirillov, A. A., *Lectures on the Orbit Method*, Amer. Math. Soc., Providence, RI, 2004.

[Kirwan]   Kirwan, F., Morse functions for which the stationary phase approximation is exact, Topology, 26, 1987, 37–40.

[Klau]   Klauder, J. R., The Feynman path integral: an historical slice. (Accessed May 9, 2021, at http://arxiv.org/abs/quant-ph/0303034.).

[Knapp]   Knapp, A., *Lie Groups: Beyond an Introduction*, Second edition, Birkhauser, Boston, MA, 2002.

[KN]   Kobayashi, S. and K. Nomizu, *Foundations of Differential Geometry*, Volumes I and II, Interscience Publishers, New York, NY, 1963.

[Koop]   Koopman, B. O., Hamiltonian systems and transformations in Hilbert spaces, Proc. Natl. Acad. Sci. USA, 17, 1931, 315–318.

[Lands]   Landsman, N. P., Between classical and quantum. (Accessed May 9, 2021, at http://arxiv.org/abs/quant-ph/0506082.).

[LaLi]   Landau, L. D. and E. M. Lifshitz, *The Classical Theory of Fields*, Third revised English edition, Pergamon Press, Oxford, England, 1971.

[Lang1]   Lang, S., *Linear Algebra*, Second edition, Addison-Wesley Publishing Company, Reading, MA, 1971.

[Lang2]   Lang, S., $SL_2(R)$, Springer, New York, NY, 1985.

[Lang3]   Lang, S., *Introduction to Differentiable Manifolds*, Second edition, Springer, New York, NY, 2002.

[Langm]   Langmann, E., Quantum theory of fermion systems: topics between physics and mathematics, in *Proceedings of the Summer School on Geometric Methods for Quantum Field Theory*, edited by H. O. Campo, A. Reyes and S. Paycha, World Scientific, Singapore, 2001.

[LM]   Lawson, H. B. and M. Michelsohn, *Spin Geometry*, Princeton University Press, Princeton, NJ, 1989.

[Lax]   Lax, P. D., *Functional Analysis*, Wiley-Interscience, New York, NY, 2002.

[Lee1]   Lee, J. M., *Riemannian Manifolds: An Introduction to Curvature*, Springer, New York, NY, 1997.

[Lee2]   Lee, J. M., *Introduction to Smooth Manifolds*, Springer, New York, NY, 2003.

[L-L]   Lévi-Leblond, J.-M., Nonrelativistic particles and wave equations, Commun. Math. Phys. 6, 1967, 286–311.

[LGM]   Liang, J.-Q., B.-H. Guo and G. Morandi, Extended Feynman formula for harmonic oscillator and its applications, Sci. China Ser. A, 34, 1991, 1346–1353.

[Lieb]   Lieb, E. H., The stability of matter: from atoms to stars, Bull. Am. Math. Soc., 22 1990, 1–49.

[LL]   Lieb, E. H. and M. Loss, *Analysis*, Amer. Math. Soc., Providence, RI, 1997.

[LR]   Lombardi, G. and R. Rebaudo, Eigenvalues and eigenvectors of a special class of band matrices. (Accessed May 9, 2021, at http://rendiconti.dmi.units.it/volumi/20/09.pdf.).

[Lop]   Lopuszánski, J., *An Introduction to Symmetry and Supersymmetry in Quantum Field Theory*, World Scientific, Singapore, 1991.

[Lucr]   Lucritius, *The Nature of Things*, Translated with Notes by A. E. Stallings, Penguin Classics, New York, NY, 2007.

[LS]   Lykken, J. and M. Spiropulu, Supersymmetry and the crisis in physics, Sci. Am., 310, 2014, 34–39.

[Mack1]   Mackey, G. W., Quantum mechanics and Hilbert space, Am. Math. Mon., 64, 1957, 45–57.

[Mack2]   Mackey, G. W., *Mathematical Foundations of Quantum Mechanics*, Dover Publications, Mineola, NY, 2004.

[MacL]   Mac Lane, S., Hamiltonian mechanics and geometry, Am. Math. Mon., 77, 1970, 570–586.

[Mar1]   Marsden, J., Darboux's theorem fails for weak symplectic forms, Proc. Am. Math. Soc., 32, 1972, 590–592.

[Mar2]   Marsden, J., *Applications of Global Analysis in Mathematical Physics*, Publish or Perish, Inc., Houston, TX, 1974.

[Mar3]   Marsden, J., *Lectures on Geometrical Methods in Mathematical Physics*, SIAM, Philadelphia, PA, 1981.

[Martin]   Martin, J. L., Generalized classical dynamics and the 'classical analogue' of a Fermi oscillator, Proc. R. Soc. Lond. A, 251, 1959, 536–542.

[Mazz1]   Mazzucchi, S., Feynman path integrals, in *Encyclopedia of Mathematical Physics*, Vol. 2, 307–313, edited by J.-P. Françoise, G. L. Naber and S. T. Tsou, Academic Press (Elsevier), Amsterdam, 2006.

[Mazz2]   Mazzucchi, S., *Mathematical Feynman Path Integrals and Their Applications*, World Scientific, Singapore, 2009.

[MS]   McDuff, D. and D. Salamon, *Introduction to Symplectic Topology*, Oxford University Press, Oxford, England, 1998.

[MR]   Mehra, J. and H. Rechenberg, *The Historical Development of Quantum Theory, Volume 2*, Springer, New York, NY, 1982.

[Mess1]   Messiah, A., *Quantum Mechanics, Volume I*, North Holland, Amsterdam, 1961.

[Mess2]   Messiah, A., *Quantum Mechanics, Volume II*, North Holland, Amsterdam, 1962.

[Milnor]   Milnor, J., *Morse Theory*, Princeton University Press, Princeton, NJ, 1963.

[MTW]   Misner, C. W., K. S. Thorne and J. A. Wheeler, *Gravitation*, W. H. Freeman and Company, San Francisco, CA, 1973.

[MP]   Mörters, P. and Y. Peres, *Brownian Motion*, Cambridge University Press, Cambridge, England, 2010.

[Moser]	Moser, J., On the volume elements on a manifold, Trans. Am. Math. Soc., 120, 1965, 286–294.
[Nab1]	Naber, G., *Spacetime and Singularities: An Introduction*, Cambridge University Press, Cambridge, England, 1988.
[Nab2]	Naber, G., *Topological Methods in Euclidean Spaces*, Dover Publications, Inc., Mineola, NY, 2000.
[Nab3]	Naber, G., *Topology, Geometry and Gauge Fields: Foundations*, Second edition, Springer, New York, NY, 2011.
[Nab4]	Naber, G., *Topology, Geometry and Gauge Fields: Interactions*, Second edition, Springer, New York, NY, 2011.
[Nab5]	Naber, G., *The Geometry of Minkowski Spacetime: An Introduction to the Mathematics of the Special Theory of Relativity*, Second edition, Springer, New York, NY, 2012.
[Nash]	Nash, C., *Differential Topology and Quantum Field Theory*, Academic Press (Elsevier), San Diego, CA, 2003.
[Nel1]	Nelson, E., Analytic vectors, Ann. Math., 70, 1959, 572–615.
[Nel2]	Nelson, E., A proof of Liouville's theorem, Proc. Am. Math. Soc., 12, 1961, 995.
[Nel3]	Nelson, E., Feynman integrals and the Schroedinger equation, J. Math. Phys., 5, 1964, 332–343.
[Nel4]	Nelson, E., *Dynamical Theories of Brownian Motion*, Second edition, Princeton University Press, Princeton, NJ, 2001. (Accessed May 9, 2021, at https://web.math.princeton.edu/ nelson/books/bmotion.pdf.).
[Nicol]	Nicolaescu, L. I., *Lectures on the Geometry of Manifolds*, Second edition, World Scientific, Singapore, 2007.
[OHUG]	Odom, B., D. Hanneke, B. D'Urso and G. Gabrielse, New measurement of the electron magnetic moment using a one-electron quantum cyclotron, Phys. Rev. Lett., 97, 2006, 030801-1-4.
[Olver]	Olver, P. J., *Applications of Lie Groups to Differential Equations*, Second edition, Springer, New York, NY, 2000.
[O'N]	O'Neill, B., *Semi-Riemannian Geometry with Applications to Relativity*, Academic Press, San Diego, CA, 1983.
[Osborn]	Osborn, H., *Vector Bundles, Volume 1, Foundations and Stiefel-Whitney Classes*, Academic Press, San Diego, CA, 1982.
[Ozawa]	Ozawa, M., Universally valid reformulation of the Heisenberg uncertainty principle on noise and disturbance in measurement, Phys. Rev. A, 67, 2003, 042105, 1–6. (Accessed May 9, 2021, at http://arxiv.org/pdf/quant-ph/0207121v1.pdf.).
[Pais]	Pais, A., *'Subtle is the Lord ...' The Science and the Life of Albert Einstein*, Oxford University Press, Oxford, England, 1982.
[Palais1]	Palais, R. S., *Seminar on the Atiyah-Singer Index Theorem*, Princeton University Press, Princeton, NJ, 1965. (Accessed May 9, 2021, at http://vmm.math.uci.edu/PalaisPapers/.).
[Palais2]	Palais, R. S., The Morse lemma for Banach spaces, Bull. Am. Math. Soc., 75, 1969, 968–971. (Accessed May 9, 2021, at http://vmm.math.uci.edu/PalaisPapers/.).
[Pankov]	Pankov, A., Introduction to spectral theory of Schrödinger operators. (Accessed May 9, 2021, at http://www.math.nsysu.edu.tw/ amen/posters/pankov.pdf.).
[PM]	Park, J. L. and H. Margenau, Simultaneous measurability in quantum theory, Int. J. Theor. Phys., 1, 1968, 211–283.
[Pauli1]	Pauli, W., Zur Quantenmechanik des Magnetischen Elektrons, Z. Phys., 43, 1927, 601–623.
[Pauli2]	Pauli, W., The connection between spin and statistics, Phys. Rev., 58, 1940, 716–722.

[Pauli3]   Pauli, W., Exclusion principle and quantum mechanics, in *Nobel Lectures: Physics*, 1942–1962, Elsevier, Amsterdam, 1964.

[Pazy]   Pazy, A., *Semigroups of Linear Operators and Applications to Partial Differential Equations*, Springer, New York, NY, 1983.

[Per]   Perrin, J., *Les Atomes*, New edition, Flammarion, Paris, 1991.

[Prug]   Prugovečki, E., *Quantum Mechanics in Hilbert Space*, Academic Press, Inc., Orlando, FL, 1971.

[Planck]   Planck, M., Zur Theorie des Gesetzes der Energiewerkeilung im Normalstetrum, Verh. Dtsch. Phys. Ges., 17, 1900, 237–245. ((English translation accessed July 10, 2021, at http://web.ihep.su/dbserv/compas/src/planck00b/eng.pdf.).

[Ratc]   Ratcliffe, J. G., *Foundations of Hyperbolic Manifolds*, Second edition, Springer, New York, NY, 2006.

[RS1]   Reed, M. and B. Simon, *Methods of Modern Mathematical Physics I: Functional Analysis*, Academic Press, Inc., Orlando, FL, 1980.

[RS2]   Reed, M. and B. Simon, *Methods of Modern Mathematical Physics II: Fourier Analysis, Self-Adjointness*, Academic Press, Inc., Orlando, FL, 1975.

[RiSz.N]   Riesz, F. and B. Sz.-Nagy, *Functional Analysis*, Dover Publications, Inc., Mineola, NY, 1990.

[Rob]   Robertson, H. P., A general formulation of the uncertainty principle and its classical interpretation, Phys. Rev., 34, 1929, 163–164.

[Rogers]   Rogers, A., *Supermanifolds: Theory and Applications*, World Scientific, Singapore, 2007.

[Ros]   Rosay, J.-P., A very elementary proof of the Malgrange–Ehrenpreis theorem, Am. Math. Mon., 98, 1991, 518–523.

[RosenS]   Rosenberg, S., *The Laplacian on a Riemannian Manifold*, Cambridge University Press, Cambridge, England, 1997.

[RosenJ]   Rosenberg, J., A selective history of the Stone–von Neumann theorem, in *Contemporary Mathematics*, 365, 331–353, Amer. Math. Soc., Providence, RI, 2004.

[Roy]   Royden, H. L., *Real Analysis*, Second edition, Macmillan Co., New York, NY, 1968.

[Roz]   Rozema, L. A., A. Darabi, D. H. Mahler, A. Hayat, Y. Soudagar and A. M. Steinberg, Violation of Heisenberg's measurement-disturbance relationship by weak measurements, Phys. Rev. Lett., 109, 2012, 100404, 1–5.

[Rud1]   Rudin, W., *Functional Analysis*, McGraw-Hill, New York, NY, 1973.

[Rud2]   Rudin, W., *Real and Complex Analysis*, Third edition, McGraw-Hill, New York, NY, 1987.

[Ryder]   Ryder, L. H., *Quantum Field Theory*, Second edition, Cambridge University Press, Cambridge, England, 2005.

[Saks]   Saks, S., *Theory of the Integral*, Second revised edition, Dover Publications, Mineola, NY, 2005.

[SN]   Sakuri, J. J. and J. Napolltano, *Modern Quantum Mechanics*, Second edition, Addison-Wesley, Boston, MA, 2011.

[Schl]   Schlosshauer, M., Decoherence, the measurement problem, and interpretations of quantum mechanics, Rev. Mod. Phys., 76, 4, 2005, 1267–1305.

[Schm1]   Schmudgen, K., On the Heisenberg commutation relation I, J. Funct. Anal., 50, 1983, 8–49.

[Schm2]   Schmudgen, K., On the Heisenberg commutation relation II, Publ. RIMS, Kyoto Univ., 19, 1983, 601–671.

[Schul]   Schulman, L. S., *Techniques and Applications of Path Integration*, Dover Publications, Inc., Mineola, NY, 2005.

[Segal]   Segal, I. E., Postulates for general quantum mechanics, Ann. Math., 48, 1947, 930–948.

[SDBS]   Sen, D., S. K. Das, A. N. Basu and S. Sengupta, Significance of Ehrenfest's theorem in quantum-classical relationship, Curr. Sci., 80, 2001, 536–541.

[Simm1]   Simmons, G. F., *Topology and Modern Analysis*, McGraw-Hill, New York, NY, 1963.

[Simm2]   Simmons, G. F., *Differential Equations with Applications and Historical Notes*, McGraw-Hill, New York, NY, 1972.

[Simon1]   Simon, B., The theory of semi-analytic vectors: a new proof of a theorem of Masson and McClary, Indiana Univ. Math. J., 20, 1971, 1145–1151.

[Simon2]   Simon, B., *Functional Integration and Quantum Physics*, Academic Press, New York, NY, 1979.

[Smir]   Smirnov, V. A., *Feynman Integral Calculus*, Springer, New York, NY, 2006.

[Socol]   Socolovsky, M., On the geometry of Spin1/2, Adv. Appl. Clifford Algebras, 11, 2001, 109–127.

[Sohn]   Sohnius, M. F., Introducing supersymmetry, Phys. Rep., 128, 1985, 39–204.

[Sp1]   Spivak, M., *Calculus on Manifolds*, W. A. Benjamin, New York, NY, 1965.

[Sp2]   Spivak, M., *A Comprehensive Introduction to Differential Geometry, Vol. I–V*, Third edition, Publish or Perish, Houston, TX, 1999.

[Sp3]   Spivak, M., *Physics for Mathematicians, Mechanics I*, Publish or Perish, Houston, TX, 2010.

[Srin]   Srinivas, M. D., Collapse postulate for observables with continuous spectra, Commun. Math. Phys. 71, 1980, 131–158.

[Str]   Strange, P., *Relativistic Quantum Mechanics with Applications in Condensed Matter Physics and Atomic Physics*, Cambridge University Press, Cambridge, England, 1998.

[SW]   Streater, R. F. and A. S. Wightman, *PCT, Spin and Statistics, and All That*, W. A. Benjamin, New York, NY, 1964.

[Szabo]   Szabo, R. J., *Equivariant Cohomology and Localization of Path Integrals*, Springer, New York, NY, 2000. (Accessed May 9, 2021, at http://arxiv.org/abs/hep-th/9608068.).

[Szegö]   Szegö, G., *Orthogonal Polynomials*, Amer. Math. Soc., Providence, RI, 1939.

[Takh]   Takhtajan, L. A., *Quantum Mechanics for Mathematicians*, Amer. Math. Soc., Providence, RI, 2008.

[TaylA]   Taylor, A. E., *Introduction to Functional Analysis*, John Wiley and Sons, Inc., London, England, 1967.

[TaylM]   Taylor, M. E., *Partial Differential Equations I*, Second edition, Springer, New York, NY, 2011.

[ter H]   ter Haar, D., *The Old Quantum Theory*, Pergamon Press, Oxford, England, 1967.

['t Ho1]   't Hooft, G., Gauge theories of the forces between elementary particles, Sci. Am., 242, 1980, 90–116.

['t Ho2]   't Hooft, G., How a wave function can collapse without violating Schrödinger's equation, and how to understand Born's rule. (Accessed May 9, 2021, at http://arxiv.org/abs/1112.1811.).

[Traut]   Trautman, A., Noether equations and conservation laws, Commun. Math. Phys. 6, 1967, 248–261.

[Tuyn]   Tuynman, G. M., *Supermanifolds and Supergroups: Basic Theory*, Kluwer Academic Publishers, New York, NY, 2005.

[TAE]   Twareque Ali, S. and M. Engliš, Quantization methods: a guide for physicists and analysts. (Accessed May 9, 2021, at http://arxiv.org/abs/math-ph/0405065.).

[Tych]   Tychonoff, A., Théorèmes d'Unicité pour l'Equation de la Chaleur, Mat. Sb., 42, 1935, 199–216.

[VdW]   van der Waerden, B. L., Editor, *Sources of Quantum Mechanics*, Dover Publications, Mineola, NY, 2007.

[Vara]   Varadarajan, V. S., *Supersymmetry for Mathematicians: An Introduction*, Amer. Math. Soc., Providence, RI.

[VH]   Van Hove, L., Sur Certaines Représentations Unitaires d'un Groupe Infini
            de Transformations, Proc. R. Acad. Sci. Belgium, 26, 1951, 1–102.
[Vog]   Vogan, D. A., Review of "Lectures on the orbit method, " by A. A. Kirillov, Bull. Am. Math.
            Soc., 42, 1997, 535–544.
[vonNeu]   von Neumann, J., *Mathematical Foundations of Quantum Mechanics*, Princeton
            University Press, Princeton, NJ, 1983.
[Wald]   Wald, R. M., *General Relativity*, University of Chicago Press, Chicago, IL, 1984.
[Warner]   Warner, F., *Foundations of Differentiable Manifolds and Lie Groups*, Springer, New York,
            NY, 1983.
[Watson]   Watson, G., Notes on generating functions of polynomials, Hermite polynomials, J. Lond.
            Math. Soc., 8, 1933, 194–199.
[Weinb]   Weinberg, S., *Dreams of a Final Theory: The Scientist's Search for the Ultimate Laws of
            Nature*, Vintage Books, New York, NY, 1993.
[Weins]   Weinstein, A., Symplectic structures on Banach manifolds, Bull. Am. Math. Soc., 75,
            1969, 1040–1041.
[Wiener]   Wiener, N., Differential space, J. Math. Phys., 2, 1923, 131–174.
[Wight]   Wightman, A. S., The spin-statistics connection: some pedagogical remarks in response
            to Neuenschwander's question, Mathematical Physics and Quantum Field Theory,
            Electronic Journal of Differential Equations, Conf. 04, 2000, 207–213.
[Wig1]   Wigner, E. P., On unitary representations of the inhomogeneous Lorentz group, Ann.
            Math., 40, 1, 1939, 149–204.
[Wig2]   Wigner, E. P., *Group Theory and its Application to the Quantum Mechanics of Atomic
            Spectra*, Academic Press, New York, NY, 1959.
[Wig3]   Wigner, E. P., Phenomenological distinction between unitary and antiunitary symmetry
            operators, J. Math. Phys., 1, 1960, 414–416.
[WW]   Wittaker, E. T. and G. N. Watson, *A Course of Modern Analysis*, Fourth edition, Cambridge
            University Press, Cambridge, England, 1965.
[Witt1]   Witten, E., Dynamical breaking of supersymmetry, Nucl. Phys. B, 185, 1981, 513–554.
[Witt2]   Witten, E., Supersymmetry and Morse theory, J. Differ. Geom., 17, 1982, 661–692.
[Yosida]   Yosida, K., *Functional Analysis*, Springer, New York, NY, 1980.

Index

www.ingramcontent.com/pod-product-compliance
Lightning Source LLC
Chambersburg PA
CBHW060946210326
41598CB00031B/4738